STUDY GUIDE

Kristi Mock

University of Toledo

CHEMISTRY

Eighth Edition

Jill K. Robinson

John E. McMurry

Robert C. Fay

Pearson

Courseware Portfolio Manager: Terry Haugen
Director of Portfolio Management: Jeanne Zalesky
Content Producer: Shercian Kinosian
Managing Producer: Kristen Flathman
Courseware Director, Content Development: Barbara Yien
Courseware Analysts: Cathy Murphy, Coleen Morrison, Jay McElroy
Courseware Editorial Assistant: Harry Misthos
Full-Service Vendor: SPi Global
Main Text Cover Designer: Gary Hespeneide
Manufacturing Buyer: Stacey Weinberger
Director of Product Marketing: Allison Rona
Senior Product Marketing Manager: Elizabeth Bell
Cover Photo Credit: Beauty of Science/Science Source

ScoutAutomatedPrintCode

Pearson www.pearson.com

ISBN 10: **0-135-45248-1**
ISBN 13: **978-0-13-545248-6**

ScoutAutomatedPrintCode

Table of Contents

To the Student

This study guide is intended to provide a bare-bones version of the material presented in *Chemistry*, Eighth Edition, by Robinson/McMurry/Fay. The reasoning behind this book is to get you working and thinking about the material. It is easy to read what someone else has written, but to write down concepts and equations will help you **begin** to internalize the material, making lectures more engaging, and exams easier to study for.

Each chapter of this study guide has been revised and rewritten to correspond directly, section-by-section, with the textbook. Most chapters contain the following sections:

- **Learning Objectives**—These were taken from the end of each chapter of the text. The Learning Objectives are the goals that the authors of your textbook (and your professor!) want you to accomplish as you study that chapter. This set of Learning Objectives should be used after you have gone through the chapter, worked out the problems in the Self-Test section at the end, and gone through the Practice Test in the textbook. It is a checklist you can use to keep track of the objectives you are comfortable with and those needing more work. It is also a good place to start before an exam. By reading through the objectives and honestly checking off the concepts you can explain, it gives you a starting point on what to study.
- **Chapter Summary**—This gives you a very brief overview of the entire chapter.
- **The Chapter in Detail**—Each section corresponds section-by-section to your textbook. It is separated each into a few sections:
    - Pre-lecture: This segment has a few activities that will help you familiarize yourself with the material and help you prepare for class. Students often say they want to hear from the professor first, but coming in with an idea of what will be discussed will help you engage more with the material, making time outside the classroom more productive so you can retain the information better and spend less time studying.
        - Learning Objectives: Here you will find the few specific learning objectives that pertain to that section and get an idea of what you are expected to learn.
        - Definitions: It is worthwhile to look these up in your book and write them down before class. Knowing the vocabulary your instructor will use can only help you get the most out of lecture.
        - Review: This is not included for every section, but when it is, it lists topics from earlier chapters that you will pull from to understand the new material.
        - Summary: Here you will find a condensed version of the material from the textbook including blanks that encourage you to think about what you learned from reading the chapter and questions to practice solving problems.
    - Post-lecture: The segment lists the problems in your textbook relating to the specific learning objectives of this section. Use these to test yourself on these concepts so you can check off the learning objectives in each section as you go to help keep on track and make sure you aren't missing any big ideas.

- **Self-Test**—A sample test covering that chapter's material is at the end of each chapter. Each Self-Test contains multiple choice, true–false, fill-in-the-blank, matching, and word problems to make sure that you're on top of your game. As an instructor who teaches from this book, I believe these are pretty fair questions. They are the kind of questions that I would set in an hour-long exam based on the book.

Finally, a Solutions Key is provided at the end of the study guide that provides worked-out solutions for all the problems in the book.

I hope that this will help you to engage with and enjoy chemistry while also strengthening the study skills you will need to take you through college.

Best of luck in this and all your future undertakings!

Acknowledgments
Behind this text is a whole team of people that made it possible. I would like to thank Terry Haugen for this opportunity; along with Shercian Kinosian, Saritha Radhasivan, and Kailash Jadli for trying to keep me on track. I am also indebted to the authors of the previous editions of this study guide: DonnaJean Freeden, Julie Klare, and James Zubricky. I've more or less modeled this edition of the study guide based on their work. Finally, a big thank you to Edith Kippenhan and Christine Hermann for proofreading and offering excellent suggestions to improve the text.

Finally, I want to thank my family. I am indebted to my sister, Sara Burditt, for her decisiveness; my dad, John Cook, for his never-ending encouragement and support; my husband, Dan Mock, for everything; and our children, Indiana and Aurelia Mock, my favorite syntheses.

Kristi Mock
The University of Toledo

CHAPTER ONE

Chemical Tools: Experimentation and Measurement

Learning Objectives

As a result of reading and studying this chapter, you should be able to:

- ☐ **LO 1.1** Identify the steps in the scientific method.
- ☐ **LO 1.2** Differentiate between a qualitative and quantitative measurement.
- ☐ **LO 1.3** Write numbers in scientific notation and use prefixes for multiples of SI units.
- ☐ **LO 1.4** Convert between different prefixes used in mass measurements.
- ☐ **LO 1.5** Convert between different prefixes used in length measurements.
- ☐ **LO 1.6** Convert between common units of temperature measurements.
- ☐ **LO 1.7** Convert between SI and metric units of volume.
- ☐ **LO 1.8** Convert between different prefixes used in volume measurements.
- ☐ **LO 1.9** Calculate mass, volume, or density using the formula for density.
- ☐ **LO 1.10** Predict whether a substance will float or sink in another substance based on density.
- ☐ **LO 1.11** Calculate kinetic energy of a moving object.
- ☐ **LO 1.12** Convert between common energy units.
- ☐ **LO 1.13** Specify the number of significant figures in a measurement.
- ☐ **LO 1.14** Evaluate the level of accuracy and precision in a data set.
- ☐ **LO 1.15** Report a measurement to the appropriate number of significant figures.
- ☐ **LO 1.16** Report the result of a mathematical calculation to the correct number of significant figures.
- ☐ **LO 1.17** Convert from one unit to another using conversion factors.

Chapter Summary

Chemistry is the study of the composition, properties, and transformations of matter. The basis for all experiments performed in science is the scientific method. Chemistry knowledge is acquired through experimentation, and experimentation requires both measurement and data analysis. The seven base units of the International System of Units (abbreviated as SI) are introduced in this chapter, and the ones that are most immediately useful are discussed in some detail. Units derived from these basic units are also introduced as well as prefixes used to enhance the usefulness of these basic units for very small and very large measurements. Scientific notation and the rules for significant figures are explained, and the process of dimensional analysis for the conversion of one unit to another is demonstrated.

The Chapter in Detail

Section 1.1 The Scientific Method: Nanoparticle Catalysts for Fuel Cells

PRE-LECTURE:

Learning Objectives: *These are the skills you should be proficient in after reading this section and practicing the associated problems. Check off the boxes below once you have gone through this section, answered the associated questions in your textbook, and feel confident you have mastered the skill.*

- ☐ **LO 1.1** Identify the steps in the scientific method.
- ☐ **LO 1.2** Differentiate between a qualitative and quantitative measurement.

Definitions: *To learn, you must do. This is your workbook. Fill in the definitions from the glossary of the textbook as a first step to understanding the material.*

— **chemistry**— ____________________

____

— **nanoscience**— ____________________

____

— **alloy**— ____________________

____

— **scientific method**— ____________________

____

— **observations**— ____________________

____

— **qualitative**— ____________________

____

— **quantitative**— ____________________

____

— **hypothesis**— ____________________

____

— **experiment**— ____________________

____

— **theory**— ____________________

____

Summary: *Read through this section and fill in the missing words and equations. Reference the textbook when you are not confident in your answers.*

The **scientific method** is an iterative approach that is used to try to understand our world. There are a few general elements that are common to the scientific method, but they are not in a rigid order and are often used more than once when scientists are exploring the world around us.

There are four main components to the scientific method:

- Making ___________.
- Formation of a ___________.
- Testing of the hypothesis through ___________.
- Development of a ___________ from the hypothesis that is consistent with the experimental data.

One of the beauties of science is that a theory can never be absolutely proven. There is always the possibility that new data will come to light and necessitate modification of the theory.

POST-LECTURE:

Work Problems: *One of the best ways to learn chemistry is through solving problems. Go to the textbook and work through the Worked Examples for this section, then go to the end of chapter and work the problems there. When you feel confident that you can correctly answer problems like the ones listed below for each objective, check off each of the Learning Objectives for this section above.*

LO 1.1: Problems 1.32, 1.34
LO 1.2: Problems 1.34, 1.35

Section 1.2 Measurements: SI Units and Scientific Notation

PRE-LECTURE:

Learning Objectives: *These are the skills you should be proficient in after reading this section and practicing the associated problems. Check off the boxes below once you have gone through this section, answered the associated questions in your textbook, and feel confident you have mastered the skill.*

☐ **LO 1.3** Write numbers in scientific notation and use prefixes for multiples of SI units.

Definitions: *To learn, you must do. This is your workbook. Fill in the definitions from the glossary of the textbook as a first step to understanding the material.*

— **SI unit**— ____

____

— **scientific notation**— ____

____

Summary: *Read through this section and fill in the missing words and equations. Reference the textbook when you are not confident in your answers.*

We use the fundamental units of the SI system so all scientists speak the same language. Match the name of the unit and abbreviation to the physical quantity.

Physical Quantity	Name of Unit	Abbreviation
mass	ampere	s
length	candela	mol
temperature	kelvin	m
amount of substance	kilogram	kg
time	meter	K
electric current	mole	cd
luminous intensity	second	A

Prefixes are used to indicate a very large or very small quantity. Fill in the symbol and factor for the prefixes below.

Prefix	Symbol	Factor
giga	G	10^9
mega		
kilo		
centi		
milli		
micro		
nano		
pico		

It is absolutely essential that you memorize and be able to correctly use these prefixes!

Another way scientists deal with very large and very small numbers is to write them in scientific notation. To do this:

- shift the decimal point to the right or left *n* places until you obtain a number between 1 and 10
  - if the decimal is shifted to the right, *n* is ___________,
  - if the decimal is shifted to the left, *n* is ___________,
- multiply the result by 10^n.

Workbook Example 1.1

PROBLEM:
Express the following measurements in scientific notation: a) 3652 m and b) 0.000 054 95 m.

SOLUTION:
We will use the rules given above.

For a), move the decimal point three places to the left to get the number 3.652 (which is between 1 and 10). Because the decimal point was moved to the left three (3) places, use the number 3 as the exponent. That gives the answer 3.652×10^3 m. *Let's take this one step further*: looking at the list of possible prefixes, 10^3 corresponds with *kilo*. Therefore, this number could also be expressed as 3.652 km.

Part b) has the number 0.000 054 95. In order to write this in scientific notation, move the decimal point five places to the right to obtain the number 5.495 (which is between 1 and 10). Because the decimal point was moved to the right, the exponent will be −5. That gives the answer 5.495×10^{-5} m.

If a number represents a measurement, it must have a unit. The difference between 20 feet, 20 meters, and 20 miles is important! Without the unit, a measured number is meaningless.

POST-LECTURE:

Work Problems: *One of the best ways to learn chemistry is through solving problems. Go to the textbook and work through the Worked Examples for this section, then go to the end of chapter and work the problems there. When you feel confident that you can correctly answer problems like the ones listed below for each objective, check off each of the Learning Objectives for this section above.*

LO 1.3: Worked Example 1.1; Problems 1.40, 1.44, 1.46, 1.48

Section 1.3 Mass and Its Measurement

PRE-LECTURE:

Learning Objectives: *These are the skills you should be proficient in after reading this section and practicing the associated problems. Check off the boxes below once you have gone through this section, answered the associated questions in your textbook, and feel confident you have mastered the skill.*

☐ **LO 1.4** Convert between different prefixes used in mass measurements.

Definitions: *To learn, you must do. This is your workbook. Fill in the definitions from the glossary of the textbook as a first step to understanding the material.*

— **mass**— ____

____

— **matter**—____

____

— **kilogram (kg)**—____

____

— **gram (g)**—____

____

— **milligram (mg)**—____

____

— **microgram (μg)**—____

____

Summary: *Read through this section and fill in the missing words and equations. Reference the textbook when you are not confident in your answers.*

Mass is the amount of ___________ in an object. Matter, in its turn, is everything that has ___________. The SI unit of mass is the kilogram. It is much too big for most uses in experimental chemistry so grams (g), milligrams (mg), and micrograms (μg) are commonly used.

Mass is frequently confused with weight. Weight has to do with gravitational pull, and it will vary depending on the strength of gravity; mass is a measurement of the amount of matter in an object. An object will have the same *mass* on Earth and on the moon, whereas it will have much less *weight* on the moon.

POST-LECTURE:

Work Problems: *One of the best ways to learn chemistry is through solving problems. Go to the textbook and work through the Worked Examples for this section, then go to the end of chapter and work the problems there. When you feel confident that you can correctly answer problems like the ones listed below for each objective, check off each of the Learning Objectives for this section above.*

LO 1.4: Problem 1.50

Section 1.4 Length and Its Measurement

PRE-LECTURE:

Learning Objectives: *These are the skills you should be proficient in after reading this section and practicing the associated problems. Check off the boxes below once you have gone through this section, answered the associated questions in your textbook, and feel confident you have mastered the skill.*

☐ **LO 1.5** Convert between different prefixes used in length measurements.

Definitions: *To learn, you must do. This is your workbook. Fill in the definitions from the glossary of the textbook as a first step to understanding the material.*

— **meter (m)**— ______________________________

____

— **centimeter (cm)**— ______________________________

____

— **millimeter (mm)**— ______________________________

____

— **micrometer (μm)**— ______________________________

____

— **nanometer (nm)**— ______________________________

____

— **picometer (pm)**— ______________________________

____

Summary: *Read through this section and fill in the missing words and equations. Reference the textbook when you are not confident in your answers.*

The _________ is the SI unit for length. The meter is equal to 39.37 inches and is too large for most measurements in chemistry. Thus, chemists more commonly use centimeters (1 cm = 2.54 cm), millimeters, micrometers, nanometers, and picometers to deal with most measurements.

POST-LECTURE:

Work Problems: *One of the best ways to learn chemistry is through solving problems. Go to the textbook and work through the Worked Examples for this section, then go to the end of chapter and work the problems there. When you feel confident that you can correctly answer problems like the ones listed below for each objective, check off each of the Learning Objectives for this section above.*

LO 1.5: Problems 1.49, 1.51

Section 1.5 Temperature and Its Measurement

PRE-LECTURE:

Learning Objectives: *These are the skills you should be proficient in after reading this section and practicing the associated problems. Check off the boxes below once you have gone through this section, answered the associated questions in your textbook, and feel confident you have mastered the skill.*

☐ **LO 1.6** Convert between common units of temperature measurements.

Definitions: *To learn, you must do. This is your workbook. Fill in the definitions from the glossary of the textbook as a first step to understanding the material.*

— **Celsius (°C)**— ____

____

— **Fahrenheit (°F)**— ____

____

— **kelvin (K)**— ____

____

Summary: *Read through this section and fill in the missing words and equations. Reference the textbook when you are not confident in your answers.*

Although the official SI unit for the measurement of temperature is the kelvin, both ___________ and ___________ are used in chemistry. The two scales are easily interconverted because both have _____ degrees between the freezing point of water and the boiling point of water. The difference in the scales is the temperature assigned the value zero: on the Celsius scale, 0 °C is the ___________ of water, on the kelvin scale, 0 K is known as ___________, the coldest temperature possible.

Write the equation used to convert a temperature in degrees Celsius to kelvin:

Write the equation used to convert from kelvins to degrees Celsius:

The Fahrenheit scale is still used predominantly in the United States. The Fahrenheit scale sets _____ °F as the freezing point of water and _____ °F as the boiling point of water. Whereas the Celsius and kelvin scales have 100 degrees between the freezing and boiling points of water, the Fahrenheit scale has 180° between them. An additional complication is the change of the freezing point of water from 0 °C to 32 °F.

To convert between degrees Fahrenheit and degrees Celsius requires two adjustments: one to adjust for the difference in degree size and the other to adjust for the difference in zero point. Write the equation used to convert from Celsius to Fahrenheit:

Write the equation used to convert from Fahrenheit to Celsius:

Workbook Example 1.2

PROBLEM:
Gallium, a metal whose salts are used to produce Blu-ray lasers, melts at 302.90 K. What is this temperature in °F?

SOLUTION:
This is a good time to use the "thinking approach." Not having been given a conversion from kelvins to degrees Fahrenheit, the easiest approach is to do the conversion from kelvins to degrees Celsius, then do the conversion from Celsius to Fahrenheit.

Zero on the Celsius scale is much higher than zero in the kelvin scale, so to convert from kelvins to degrees Celsius, subtract 273.15:

$$302.90\ \text{K} - 273.15 = 29.75\ °\text{C}$$

Now convert to Fahrenheit.

A Fahrenheit degree is smaller than a Celsius degree. Just think about the freezing point of water. Water freezes at 0° on the Celsius scale and at 32° on the Fahrenheit scale. Therefore, when converting from Celsius to Fahrenheit, the temperature should be higher.

Using the temperature conversion equation, we get

$$°\text{F} = \left(\frac{9}{5} \times 29.75\right) + 32 = 85.55\ °\text{F}$$

Gallium is actually really awesome; the melting point is lower than the temperature of the human body (98.6 °F), so gallium metal will actually melt in your hand!

Workbook Problem 1.1

You wish to impress upon a friend in Russia, which uses the Celsius scale, the misery of the heat wave you are enduring in Detroit. What is the Celsius equivalent of a balmy 102 °F? If you really want to feel hot, figure out the temperature in kelvins!

Strategy: Use the formula for the conversion of Fahrenheit to Celsius and then Celsius to kelvin.

Step 1: Use the formula for the conversion of °F to °C.

Step 2:	Use the formula for the conversion of °C to K.

POST-LECTURE:

Work Problems: *One of the best ways to learn chemistry is through solving problems. Go to the textbook and work through the Worked Examples for this section, then go to the end of chapter and work the problems there. When you feel confident that you can correctly answer problems like the ones listed below for each objective, check off each of the Learning Objectives for this section above.*

LO 1.6: Worked Example 1.2; Problems 1.54, 1.56, 1.60

Section 1.6 Derived Units: Volume and Its Measurement

PRE-LECTURE:

Learning Objectives: *These are the skills you should be proficient in after reading this section and practicing the associated problems. Check off the boxes below once you have gone through this section, answered the associated questions in your textbook, and feel confident you have mastered the skill.*

☐ **LO 1.7** Convert between SI and metric units of volume.
☐ **LO 1.8** Convert between different prefixes used in volume measurements.

Definitions: *To learn, you must do. This is your workbook. Fill in the definitions from the glossary of the textbook as a first step to understanding the material.*

— **cubic meter (m^3)**— ____

____

— **cubic decimeter (dm^3)**— ____

____

— **liter (L)**— ____

____

— **cubic centimeter (cm^3)**— ____

____

— **milliliter (mL)**— ____

____

Summary: *Read through this section and fill in the missing words and equations. Reference the textbook when you are not confident in your answers.*

Derived SI units are those made by combining the seven basic units. A simple example is speed, which would be expressed in SI units as meters per second, or m/s. Fill in the quantity and derived unit for the definitions below.

Quantity	Definition	Derived Unit (Name)
area	length times length	m^2
	change in speed per unit time	
	mass per unit volume	
	distance per unit time	
	mass times acceleration	
	force times distance	
	force per unit area	
	area times length	

Volume is also a derived unit; it comes from the unit of length: remember from geometry that the volume of a cube is equal to length × width × height. If this is done with the length in centimeters, the result is cm^3. This is exactly equivalent to a milliliter (1 mL and 1×10^{-3} L).

In the chemistry lab, volume is measured by using graduated cylinders, syringes, volumetric flasks, burets, and other tools.

POST-LECTURE:

Work Problems: *One of the best ways to learn chemistry is through solving problems. Go to the textbook and work through the Worked Examples for this section, then go to the end of chapter and work the problems there. When you feel confident that you can correctly answer problems like the ones listed below for each objective, check off each of the Learning Objectives for this section above.*

LO 1.7: Problems 1.64, 1.65
LO 1.8: Problem 1.63

Section 1.7 Derived Units: Density and Its Measurement

PRE-LECTURE:

Learning Objectives: *These are the skills you should be proficient in after reading this section and practicing the associated problems. Check off the boxes below once you have gone through this section, answered the associated questions in your textbook, and feel confident you have mastered the skill.*

☐ **LO 1.9** Calculate mass, volume, or density using the formula for density.
☐ **LO 1.10** Predict whether a substance will float or sink in another substance based on density.

Definitions: *To learn, you must do. This is your workbook. Fill in the definitions from the glossary of the textbook as a first step to understanding the material.*

— **density**— ______________________________

____

Summary: *Read through this section and fill in the missing words and equations. Reference the textbook when you are not confident in your answers.*

Density is calculated by dividing the mass of a substance by its volume. Write the formula for density:

For liquids and solids, density is expressed in g/mL or g/cm^3. (Recall that these units are identical: 1 mL = 1 cm^3.) Density is an intrinsic physical property that can be used to identify a substance. It can also be used to convert back and forth between mass and volume.

Density is temperature-dependent. For most substances, the volume increases as they are heated and the volume decreases as they are cooled, so when reporting a density, the temperature must also be noted.

The properties of water make it unique: down to a temperature of 4 °C water shrinks, then it begins to expand. Solid water (ice) is about 10% larger than the **same mass** of liquid water. This is why ice floats. More generally, substances with a lower density will float on top of substances with a higher density.

Workbook Example 1.3

PROBLEM:
A student determined that a metal cylinder has a mass of 25.478 g and a volume of 9.597 mL. What is the density of this metal cylinder?

SOLUTION:
Since we know both the mass and volume of the metal cylinder, we can determine the density by dividing the mass by the volume. Make sure to keep the units with the numbers, as this will prevent many careless errors.

$$Density = \frac{25.478\ g}{9.597\ mL} = 2.655 \frac{g}{mL}$$

If you had a chart of metal densities, you would be able to identify the metal directly from its density, because density is an intrinsic property. The density of aluminum (Al) is about 2.7 g/cm^3 at room temperature, making it likely that the cylinder is made of aluminum.

Workbook Example 1.4

PROBLEM:
A student is given containers of isopropyl alcohol (approximate density is 0.79 g/mL) and vegetable oil (approximate density is 0.93 g/mL). Vegetable oil will not mix with isopropyl alcohol (rubbing alcohol). If both liquids are poured into the same graduated cylinder, which will sink? Which will float? If an ice cube (approximate density is 0.91 g/mL) is dropped into the cylinder, where will it settle?

SOLUTION:
It is helpful to understand how density works: a substance that is more dense (has a higher density, or a higher number of grams per milliliter) will sink in something that is less dense. So, when the liquids are put into the same container, the more dense liquid (vegetable oil) will sink, and the less dense liquid (isopropyl alcohol) will float, because the substances do not mix. The ice cube has an intermediate density: it will sink in the alcohol, but float on the oil.

Workbook Problem 1.2

The density of lead at 25 °C is 11.34 g/mL. How many grams of lead are found in a volume of 53.43 mL?

Strategy: Using the information provided about density, set up a mathematical equation that allows you to solve for the mass of lead.

Step 1: Substitute the information given in the problem into the mathematical equation you set up.

Step 2: Solve for the mass of lead.

Step 3: Does your answer make physical sense?

POST-LECTURE:

Work Problems: *One of the best ways to learn chemistry is through solving problems. Go to the textbook and work through the Worked Examples for this section, then go to the end of chapter and work the problems there. When you feel confident that you can correctly answer problems like the ones listed below for each objective, check off each of the Learning Objectives for this section above.*

LO 1.9: Worked Example 1.3; Problems 1.66, 1.68, 1.74, 1.76
LO 1.10: Problem 1.31

Section 1.8 Derived Units: Energy and Its Measurement

PRE-LECTURE:

Learning Objectives: *These are the skills you should be proficient in after reading this section and practicing the associated problems. Check off the boxes below once you have gone through this section, answered the associated questions in your textbook, and feel confident you have mastered the skill.*

- ☐ **LO 1.11** Calculate kinetic energy of a moving object.
- ☐ **LO 1.12** Convert between common energy units.

Definitions: *To learn, you must do. This is your workbook. Fill in the definitions from the glossary of the textbook as a first step to understanding the material.*

— **energy**—____

____

— **kinetic energy (E_K)**— ____

____

— **potential energy (E_P)**—____

____

— **joule (J)**— ____

____

Summary: *Read through this section and fill in the missing words and equations. Reference the textbook when you are not confident in your answers.*

Energy is defined as the ability to do work or to supply heat. There are several different forms of energy (which we'll talk about further in Chapter 9), but the two major forms of energy that we come across in general chemistry are kinetic energy and potential energy.

Kinetic energy (E_K) is the energy of _______, and it can be calculated by the following equation:

where m is the ___________ of an object (measured in kg) and v is the ___________ of the object (measured in m/s). The final answer for energy is expressed with the units $kg\left(\frac{m}{s}\right)^2$, where $1\,kg\left(\frac{m}{s}\right)^2$, is equal to 1 joule (J), which is the SI unit for energy.

Potential energy (E_P) is ___________ energy and is the type of energy found in chemical bonds.

While most modern chemists use joules to describe energy, some chemists and biochemists still use the unit calories (cal) to describe energy. Write an equation that shows the relationship between joules and calories:

When we talk about calories in terms of the energy we get from food, we have another unit to deal with: Calories (where the capital C refers to food calories). Write an equation for the relationship between Calories and calories:

Workbook Example 1.5

PROBLEM:
I went to a Chicago Cubs game and caught a fly ball! Using a radar speed gun, I was able to clock the ball coming toward me at 45.15 m/s. If the mass of the baseball was 0.143 kilograms, how much kinetic energy did that baseball have?

SOLUTION:
Knowing the mass of the baseball (0.143 kg) and the speed of the baseball (45.15 m/s) will make solving this problem much easier: we can put our values into the kinetic energy equation and solve for E_K:

$$E_K = \frac{1}{2}mv^2$$

$$E_K = \frac{1}{2}(0.413\ kg)(45.15\ m/s)^2 = 145.75\ kg\ m^2/s^2$$

$$E_K = 146\ \text{kg}\ m^2/s^2$$

$$E_K = 146\ J$$

POST-LECTURE:

Work Problems: *One of the best ways to learn chemistry is through solving problems. Go to the textbook and work through the Worked Examples for this section, then go to the end of chapter and work the problems there. When you feel confident that you can correctly answer problems like the ones listed below for each objective, check off each of the Learning Objectives for this section above.*

LO 1.11: Worked Example 1.4; Problems 1.78, 1.79
LO 1.12: Problems 1.82, 1.83

Section 1.9 Accuracy, Precision, and Significant Figures in Measurement

PRE-LECTURE:

Learning Objectives: *These are the skills you should be proficient in after reading this section and practicing the associated problems. Check off the boxes below once you have gone through*

this section, answered the associated questions in your textbook, and feel confident you have mastered the skill.

- ☐ **LO 1.13** Specify the number of significant figures in a measurement.
- ☐ **LO 1.14** Evaluate the level of accuracy and precision in a data set.
- ☐ **LO 1.15** Report a measurement to the appropriate number of significant figures.

Definitions: *To learn, you must do. This is your workbook. Fill in the definitions from the glossary of the textbook as a first step to understanding the material.*

— **accuracy**— ____

____

— **precision**— ____

____

— **significant figures**— ____

____

Summary: *Read through this section and fill in the missing words and equations. Reference the textbook when you are not confident in your answers.*

Accuracy and **precision** are concepts that are related to the reliability of a measurement. *Accuracy* refers to the closeness of a number to its accepted value, while *precision* refers to the closeness of a set of measurements to one another. If a piece of metal is agreed to weigh 4.500 grams, and you weigh it three times, getting answers of 4.787 g, 4.784 g, and 4.791 g, you have measurements that are close to one another—they are ___________—but they do not agree with the accepted value: they are not ___________.

Significant figures are the number of digits recorded for a measurement. For the example masses in the previous paragraph, four significant figures are given. When recording measurements, be sure to include all the digits you are sure of plus _____ additional digit (on digital instruments the last digit that is given is an estimate).

When writing in scientific notation, only significant figures should be written. However, scientific notation is not always used, so you need to be able to determine if a number is significant or not. To do this, remember that all nonzero numbers are significant, and use the following rules to help determine which zeros are significant:

1. Zeros in the middle of a number are like any other digit; they are ___________ significant.
2. Zeros at the beginning of a number are ___________ significant.
3. Zeros at the end of a number and after the decimal are ___________ significant.
4. Zeros at the end of a number and before the decimal ___________ significant.

A special case is *exact numbers*. These are counting numbers that cannot have a fractional part: the number of students in a class, for example, is an exact number as there will never be a fraction of a student present. These numbers are said to have an infinite number of significant figures.

Workbook Example 1.6

PROBLEM:
How many significant figures do the following measurements have?

10.002 g, 0.00672 mL, 5.3000 m, 83,500 s

SOLUTION:
10.002 g—five significant figures (Rule 1)

0.006 72 mL—three significant figures (Rule 2); if you write the number as 6.72×10^{-3} mL, it is easier to see the number of significant figures. This could also be written as 6.72 μL.

5.3000 m—five significant figures (Rule 3)

83,500 s—uncertain (Rule 4); if the number is 8.3500×10^4, there are five significant figures. If the number is 8.350×10^4, there are four significant figures. If the number is 8.35×10^4, there are three significant figures.

POST-LECTURE:

Work Problems: *One of the best ways to learn chemistry is through solving problems. Go to the textbook and work through the Worked Examples for this section, then go to the end of chapter and work the problems there. When you feel confident that you can correctly answer problems like the ones listed below for each objective, check off each of the Learning Objectives for this section above.*

LO 1.13: Worked Example 1.5; Problems 1.84, 1.86
LO 1.14: Worked Example 1.6; Problem 1.12
LO 1.15: Problems 1.28, 1.29

Section 1.10 Significant Figures in Calculations

PRE-LECTURE:

Learning Objectives: *These are the skills you should be proficient in after reading this section and practicing the associated problems. Check off the boxes below once you have gone through this section, answered the associated questions in your textbook, and feel confident you have mastered the skill.*

☐ **LO 1.16** Report the result of a mathematical calculation to the correct number of significant figures.

Definitions: *To learn, you must do. This is your workbook. Fill in the definitions from the glossary of the textbook as a first step to understanding the material.*

— **rounded off**— ____

____

Summary: *Read through this section and fill in the missing words and equations. Reference the textbook when you are not confident in your answers.*

When working with numbers, you will often get an answer on a calculator that has far more digits displayed than are reasonably useful. For example, if you are calculating a density from measurements that are 1 g and 3 cm^3, a calculator will give you an answer of 0.333 333 333 333, but you should only report your answer as 0.3 g/cm^3. The following rules will help you figure out how many digits to keep:

1. When multiplying or dividing, ____

   ____

2. When adding or subtracting, ____

   ____

Once you have figured out how many digits to retain, you can round the number by using the rules below:

1. If the first digit you remove is less than 5, _________________________________

   ____

2. If the first digit you remove is 5 or greater, _________________________________

   ____

Be sure not to round or truncate any numbers until the calculation is complete.

Workbook Example 1.7

PROBLEM:
If a high-speed train in Japan covers 532 miles in 3.7 hours, what is the average speed of the train?

SOLUTION:

three significant figures

$$\frac{532\ miles}{3.7\ hours} = 140\frac{miles}{hour}$$

<u>*answer needs two significant figures*</u>

two significant figures

If this problem is plugged into a calculator, the answer provided will have more digits than can possibly be justified: 143.783 783 8 miles/hour suggests far more precision than the numbers permit.

Workbook Example 1.8

PROBLEM:
Using the correct number of significant figures, subtract 1.52 g from 103.7835.

SOLUTION:

103.7835 g	*ends four places past decimal point*
– 1.52 g	*ends two places past decimal point*
102.26 g	*ends two places past decimal point*

The final answer cannot have more digits to the right of the decimal than the smallest original number. However, in doing calculations, use all digits until the end of the problem. Only round off the final answer.

Workbook Example 1.9

PROBLEM:
The mass of a sample of granite is found to be 15.896 g. If the volume is determined to be 5.8 mL, what should you report for the density of the granite sample?

SOLUTION:

$$\text{Density} = \frac{15.896\ g}{5.8\ mL} = 2.740\,689\,65 \frac{g}{mL}$$

Decide how many significant figures should be in your answer. The denominator has only two significant figures; therefore, the answer must also have only two significant figures.

Round off your answer. The first digit to be dropped is less than 5. 2.740 689 65 g/mL becomes 2.7 g/mL. This situation is not unusual: the determination of the mass of a granite sample can be done far more precisely than the determination of the volume of an irregular sample.

Workbook Example 1.10

PROBLEM:
Round off the numbers to three significant figures:
(a) 83.567 g (b) 0.000 358 500 g (c) 28,572 g

SOLUTION:
(a) Because the digit to be dropped is greater than 5, the mass rounds to 83.6 g.
(b) Because the digit to be dropped is a 5, the number is truncated to 0.000 359 g.
(c) The rounding of this number to 28,600 g introduces ambiguity about the number of significant figures. To prevent this, the number should be reported as 2.86×10^4 g. It can also be reported as 28.6 kg.

Workbook Problem 1.3

A quantity of carbon dioxide gas, which has a density of 1.976 g/L, is found to have a mass of 4.843 65 g. Determine the volume of the gas, and report this volume with the correct number of significant figures.

Strategy: Set up a mathematical equation to determine the volume of the gas. Use the rules found above to determine the correct number of significant figures.

Step 1: Use the equation for density to find the volume of the gas.

Step 2:	Apply the rules for multiplication and division to determine the number of significant figures in your answer.
Step 3:	If necessary, use the rules round off numbers.

POST-LECTURE:

Work Problems: *One of the best ways to learn chemistry is through solving problems. Go to the textbook and work through the Worked Examples for this section, then go to the end of chapter and work the problems there. When you feel confident that you can correctly answer problems like the ones listed below for each objective, check off each of the Learning Objectives for this section above.*

LO 1.16: Worked Example 1.7; Problems 1.92, 1.93

Section 1.11 Converting from One Unit to Another

PRE-LECTURE:

Learning Objectives: *These are the skills you should be proficient in after reading this section and practicing the associated problems. Check off the boxes below once you have gone through this section, answered the associated questions in your textbook, and feel confident you have mastered the skill.*

☐ **LO 1.17** Convert from one unit to another using conversion factors.

Definitions: *To learn, you must do. This is your workbook. Fill in the definitions from the glossary of the textbook as a first step to understanding the material.*

— **dimensional-analysis method**—____

____

— **conversion factor**—____

____

Summary: *Read through this section and fill in the missing words and equations. Reference the textbook when you are not confident in your answers.*

A convenient way to carry out calculations that involve different units is the **dimensional-analysis method.** To do this, ___________ factors are used to convert one unit into another. Conversion factors are written from equivalent quantities as fractions that can be interchanged. An example is given below:

60 minutes = 1 hour

This tells you that there are 60 minutes per hour ("per" always implies division) OR 1 hour per 60 minutes.

$$\text{Conversion factors:} \frac{60\ minutes}{1\ hour} \text{ or } \frac{1\ hour}{60\ minutes}$$

The quantity in the numerator is equal to the quantity in the denominator, making this equivalent to multiplying by 1.

Choose the form of the conversion factor that will eliminate unwanted units and leave the desired units. If the units are correct, the numbers will necessarily follow. Make sure the units are multiplied and divided as the problem is worked: bizarre final units (g^2/cm^3, for example) are an indication that an error was made.

Workbook Example 1.11

PROBLEM:
On her first trip to the United States, a student rents the least expensive car available. When she gets on the highway, she discovers that the car's top speed is 43 miles per hour. When she calls home, what speed in kilometers per hour does she report to her family?

SOLUTION:
Using the conversion table on the inside back cover of the textbook, we learn that 1 mi = 1.6093 km. Set up the equation beginning with the information given (43 mi/h), and write the conversion factor so that the unit *mi* is cancelled.

$$\frac{43\ mi}{1\ h} \times \frac{1.6093\ km}{1\ mi} = 69\frac{km}{h}$$

Ballpark Check: There are more kilometers than miles in the same length, but less than double, so the answer in kilometers will be greater than the number of miles, but less than twice as much.

Workbook Example 1.12

PROBLEM:
In the United Kingdom and Ireland, the unit of "stone," though forbidden for use in trade, is still commonly used to report the weights of people. One stone is equal to 14 pounds. If a friend in Ireland is complaining about having put on the "freshman stone," and tells you that he now weighs 11 stone, how much does he weigh in kilograms?

SOLUTION:
Given that 14 lbs = 1 stone and 2.205 lbs = 1 kg, set up the equation knowing that you need to convert 11 stone to pounds, then pounds to kg. The units *stone* and *lbs* should be cancelled.

$$11\ stone \times \frac{14\ lbs}{1\ stone} \times \frac{1\ kg}{2.205\ lbs} = 70\ kg$$

Ballpark Check: A stone is 14 pounds, and there are about 2 pounds per kilogram, so the original number of stone will be multiplied by about 7 to get the equivalent quantity in kilograms. The answer will be close to the original number of stone multiplied by about 7. In this case, the number of significant figures in the original number determines the number of significant figures in the answer.

WARNING: It's easy to get the correct answer when using dimensional analysis without really understanding what you're doing.

Workbook Problem 1.4

If a runner completes a 5-mile race in 27 minutes 46 seconds, how long can she expect it will take her to run a 10-kilometer race?

Strategy: Using the conversion table on the inside back cover of your textbook, determine the conversion factor for meters to miles, and convert time to one unit. Use the time and distance for the 5-mile race to determine the runner's speed, convert that speed to kilometers per minute, and determine the time expected for the distance in kilometers.

Step 1: Calculate the time it takes her to run 5 miles.

Step 2: Use conversions to calculate the amount of time it will take for the athlete to run 10 km.

POST-LECTURE:

Work Problems: *One of the best ways to learn chemistry is through solving problems. Go to the textbook and work through the Worked Examples for this section, then go to the end of chapter and work the problems there. When you feel confident that you can correctly answer problems like the ones listed below for each objective, check off each of the Learning Objectives for this section above.*

LO 1.17: Worked Examples 1.8, 1.9; Problems 1.94, 1.98, 1.102, 1.106

Self-Test

This section is intended to test your knowledge of the material covered in this chapter. Think through these problems and make certain you understand what the problem is about—ask yourself if the answer makes sense. Many of these questions are linked to the chapter learning goals. Therefore, successful completion of these problems indicates that you have achieved the learning goals for this chapter. You will receive the greatest benefit from this section if you use it as a mock exam. Do not use your notes, book, Internet, or friends when taking the test. There is no grade and you only shortchange yourself by getting help with this activity. As you take the test, mark the problems with your level of confidence with the answer.

★ *Use a star for problems you know you got right.*
? *Use a question mark for those you think you got right.*
x *Use an x for those you are guessing on.*

If you get a question right that you marked with a star, you can be confident that you know the associated LO.

If you get a question right that you marked with a question mark, go back and work a few more problems and test yourself again tomorrow.

If you get a question wrong <u>or</u> mark it with an x, go back and read the section again, work a few problems today, a few tomorrow, and a few the next day before testing yourself again.

This will help you to discover which topics you have learned thoroughly and which topics you need to study in more detail.

<u>*True–False*</u>

1. There is not much difference between a theory and hypothesis.
2. The number 0.002 05 contains three significant figures.
3. *Weight* and *mass* are the same thing.
4. The unit most commonly used for laboratory volume measurements is the m^3.
5. 1×10^6 g is equal to 1 μg.
6. The number 4000 can be expressed in scientific notation as 4×10^3.
7. Kinetic energy is the energy of movement.
8. The SI unit for temperature is the degree Fahrenheit.
9. The temperature "absolute zero" describes 0 K.
10. Density describes an object's mass and its temperature.

<u>*Multiple Choice*</u>

11. Volume is
    a. anything physically real
    b. the pull of gravity on an object by the Earth or other celestial body
    c. how much space an object takes up
    d. the amount of matter in an object

12. The following measurements were made by a student during a laboratory exercise: 23.05 cm, 22.93 cm, 35.02 cm. The reference value in this experiment is 30.00 cm. These numbers represent data that are
    a. precise and accurate b. precise but not accurate
    c. accurate but not precise d. neither precise nor accurate

13. If 3.73 was multiplied by 153.2, the answer would contain
    a. three significant figures b. four significant figures
    c. five significant figures d. six significant figures

14. The SI unit for mass is
    a. oz b. lb c. g d. kg

15. The SI prefix *nano* corresponds to the multiplier
    a. 10^{-3} b. 10^{-6} c. 10^{-9} d. 10^{-12}

16. A Celsius degree is
    a. larger than a Fahrenheit degree, but the same as a kelvin.
    b. smaller than a kelvin and the same as a Fahrenheit degree.
    c. larger than a kelvin.
    d. smaller than a Fahrenheit degree.

<u>Matching</u>

17. scientific method	a. the amount of matter contained in an object
18. theory	b. how close a measurement is to an accepted value
19. mass	c. how well a number of independent measurements agree with one another
20. precision	d. an experiment-based method for increasing one's understanding of the natural world
21. accuracy	e. a way to carry out calculations that involve different units
22. dimensional analysis	f. a consistent explanation of known observations based on many independent experiments; useful, but cannot be proven

Fill in the Blanks

23. A _______________ is always open to modification by new experimental data.

24. The SI units for mass, temperature, and length are the _______________, the _______________, and the _______________.

25. There are three commonly used temperature scales. _______________ and _______________ have degrees of the same size, while _______________ has degrees that are smaller.

26. _______________ is the intrinsic property of a substance that can be determined by dividing a substance's mass by its volume.

27. If an instrument gives you measurements that average-out to the accepted reference value but are not close to each other, you would say that your data are _______________, but not _______________.

28. To write numbers that are very small or very large, it is helpful to use ________________.

29. When finished working a problem, it is often necessary to ________________ the answer to the correct number of ________________.

30. 1 inch = 2.54 cm is an example of a ________________ ________________.

31. When adding numbers together, the answer cannot have more significant figures on the right of the decimal point than ____.

32. The most appropriate prefix to use for 360,000,000 watts is ________________, abbreviated ________________.

33. If you know the density of a liquid and its volume, you can calculate its ________________.

34. A graduated cylinder and a syringe can both be used to measure ________________.

Problems

35. 246 cm = ____________ mm = ______________ m = ______________ km

36. Perform the following calculations and report your answer with the correct number of significant figures:

a. $\frac{8.3}{2.793} =$

b. $23.94562 \times 8.3421 =$

c. $3.2 + 5.3875 =$

d. $9434.21 - 4.199 =$

37. The freezing point of paraffin wax, commonly used in making preserves, is 56 °C. What is this in degrees Fahrenheit?

38. The land area of Zambia is 290,587 mi^2. Convert this to km^2.

39. How many nanometers are there in 6.73×10^{-7} m?

40. The boiling point of ethanol is 78.5 °C. Convert this temperature to °F and K.

41. In 2009, the world record for the 100-meter dash was 9.58 seconds, and the world record for the 100-meter freestyle swim was 47.85 seconds. How fast were the record holders moving in miles per hour?

42. The mass of an amorphous chunk of a yellow element is found to be 9.634 g. When it is placed in a graduated cylinder that contains 11.5 mL of water, the water level rises to 16.2 mL. What is the density of the substance in g/cm^3?

43. As of July 8 2014, the *Voyager 2* Spacecraft had completed 13,741 days of continuous operation and reached a distance of 19.1 billion kilometers from the Earth. If communications travel at the speed of light (3.00×10^8 m/s), how many hours does it take for data to get from the spacecraft to ground control?

44. A pycnometer is a device used to make precise density determinations. A pycnometer weighs 23.75 g when empty and 37.82 g when filled with water at 20 °C. The density of water at 20 °C is 0.9982 g/mL. When 6.345 g of beryllium are placed in an empty pycnometer and water at 20 °C is added it has a total mass of 40.72 g. What is the density of the beryllium?

45. If the density of titanium is 4.55 g/mL, and a titanium metal plate that is used to repair a broken skull weighs 10.67 g, what is the volume of the plate?

46. A rock weighing 7.13 g is dropped into a graduated cylinder with 20.0 mL of room temperature water in it. The water level rises to 22.7 mL. What is the density of the rock?

47. What is the physical basis of 0 °C? How is this different from 0 K?

48. When working through calculations, at what step are the numbers rounded or truncated to the correct number of significant figures? Why is it done this way?

Challenge Problem

49. Gallium has a melting point of 29.78 °C and a boiling point of 2403 °C. This very large liquid range coupled with the fact that gallium is a liquid near room temperature makes this metal ideal for the construction of a high-temperature thermometer.

    You need to measure the melting points of some metals and have decided to construct a thermometer that uses gallium as the liquid. However, you don't want the thermometer to be too awkward to handle, so you have decided to define the melting point of gallium as 0 °Ga and the boiling point as 1000 °Ga. To ensure the accuracy of your thermometer, you measure the melting point of copper. The melting point of copper on your thermometer is 447 °Ga. The reported melting point for copper is 1085 °C. Is your thermometer accurate?

    HINT: Think about how to convert from Fahrenheit to Celsius. How would you convert from Celsius to your new temperature system? You can then write the equation for conversion between °Ga and °C.

Test Yourself: *Take the Practice Test on pages 26 and 27 of your textbook. Again, use a testing environment using only those items that your instructor will allow you to have access to during the test and mark each question with a star, question mark, or x to denote your confidence level. Once you have finished you can check off the LO at the beginning of this chapter based on which questions you starred and got correct. Go back and spend more time with the topics from the other learning objectives.*

CHAPTER TWO

Atoms, Molecules, and Ions

Learning Objectives

As a result of reading and studying this chapter, you should be able to:

- ☐ **LO 2.1** Write symbols to represent element names.
- ☐ **LO 2.2** Identify the location of metals, nonmetals, and semimetals on the periodic table.
- ☐ **LO 2.3** Indicate the atomic number, group number, and period number for an element whose position in the periodic table is given.
- ☐ **LO 2.4** Identify groups as main group, transition metal group, or inner transition metal group.
- ☐ **LO 2.5** Specify the location and give examples of elements in the alkali metal, alkaline earth metal, halogen, noble gas groups.
- ☐ **LO 2.6** Classify an element as a metal, nonmetal, or semimetal using its properties.
- ☐ **LO 2.7** Determine the mass of the products in a reaction using the law of mass conservation.
- ☐ **LO 2.8** Demonstrate the law of multiple proportions using mass composition of two compounds of the same elements.
- ☐ **LO 2.9** Determine the formula of a compound given mass composition data for two compounds and the formula of one compound.
- ☐ **LO 2.10** Describe Thomson's cathode ray tube experiment and what it contributed to the current model of atomic structure.
- ☐ **LO 2.11** Describe Millikan's oil drop experiment and what it contributed to the current model of atomic structure.
- ☐ **LO 2.12** Describe Rutherford's gold foil experiment and what it contributed to the current model of atomic structure.
- ☐ **LO 2.13** Calculate the number of atoms in a sample given the size of the atom.
- ☐ **LO 2.14** Determine the mass number, atomic number, and number of protons neutrons and electrons from an isotope symbol.
- ☐ **LO 2.15** Calculate atomic weight given the fractional abundance and mass of each isotope.
- ☐ **LO 2.16** Convert between grams and numbers of moles or atoms using molar mass and Avogadro's number.
- ☐ **LO 2.17** Use data from a mass spectrum to calculate the atomic weight of an element.
- ☐ **LO 2.18** Classify matter as a mixture, pure substance, element, or compound.
- ☐ **LO 2.19** Convert between structural formulas, ball-and-stick models, and chemical formulas.
- ☐ **LO 2.20** Classify bonds as ionic or covalent.
- ☐ **LO 2.21** Determine the number of electrons and protons from chemical symbol and charge.
- ☐ **LO 2.22** Match the molecular representation of an ionic compound with its chemical formula.
- ☐ **LO 2.23** Convert between name and formula for ionic compounds.
- ☐ **LO 2.24** Convert between name and formula for binary molecular compounds.

Chapter Summary

This chapter begins with a brief description of the ways in which matter is categorized and how various properties, both chemical and physical, can be used in its identification. This discussion is followed by an introduction to the periodic table and how the elements are organized within it. Next, we begin to understand chemistry by examining Dalton's atomic theory and how his theory has developed and expanded since the early nineteenth century. Atoms will be described, and the three basic subatomic particles—protons, neutrons, and electrons—are introduced, along with the ways in which they vary to form ions and isotopes. After a quick introduction to the mole and how chemists measure the mass of atoms, we then turn to looking at compounds, mixtures, chemical-bond types, and chemical nomenclature.

The Chapter in Detail

Section 2.1 Chemistry and the Elements

PRE-LECTURE:

Learning Objectives: *These are the skills you should be proficient in after reading this section and practicing the associated problems. Check off the boxes below once you have gone through this section, answered the associated questions in your textbook, and feel confident you have mastered the skill.*

☐ **LO 2.1** Write symbols to represent element names.

Definitions: *To learn, you must do. This is your workbook. Fill in the definitions from the glossary of the textbook as a first step to understanding the material.*

— **element—**____

____

— **periodic table—**____________________________________

____

Summary: *Read through this section and fill in the missing words and equations. Reference the textbook when you are not confident in your answers.*

An element is a fundamental substance that can't be chemically changed or broken down. There are 118 known elements, 90 of which occur naturally; the rest have been produced through nuclear reactions.

Elements are given one- or two-letter symbols; the first letter is ___________, and if there is a second letter, it is written in ___________.

Workbook Example 2.1

PROBLEM:
Write out the full names for each of the following elements: F, Mg, I, Te, P.

SOLUTION:
Using the periodic table inside the front cover of the textbook, we find that the full names of the symbols are

F—fluorine	Mg—magnesium	I—iodine
Te—tellurium	P—phosphorus	

Workbook Example 2.2

PROBLEM:
Find the symbols for the following elements: iron, oxygen, potassium, chlorine, sodium.

SOLUTION:
Using the periodic table inside the front cover of the textbook, we find that the symbols for these elements are

iron—Fe	oxygen—O	potassium—K
chlorine—Cl	sodium—Na	

You may find it helpful to make flashcards, or find an app, to practice learning the names and symbols of the elements. You will generally have access to a periodic table, but it may have only the symbols of the elements, and not their names. If you are not given specific instructions on which elements to memorize, start with those in the top 4 periods (rows across the periodic table).

POST-LECTURE:

Work Problems: *One of the best ways to learn chemistry is through solving problems. Go to the textbook and work through the Worked Examples for this section, then go to the end of chapter and work the problems there. When you feel confident that you can correctly answer problems like the ones listed below for each objective, check off each of the Learning Objectives for this section above.*

LO 2.1: Problems 2.48, 2.50, 2.52

Section 2.2 Elements and the Periodic Table

PRE-LECTURE:

Learning Objectives: *These are the skills you should be proficient in after reading this section and practicing the associated problems. Check off the boxes below once you have gone through this section, answered the associated questions in your textbook, and feel confident you have mastered the skill.*

- ☐ **LO 2.2** Identify the location of metals, nonmetals, and semimetals on the periodic table.
- ☐ **LO 2.3** Indicate the atomic number, group number, and period number for an element whose position in the periodic table is given.
- ☐ **LO 2.4** Identify groups as main group, transition metal group, or inner transition metal group.

Definitions: *To learn, you must do. This is your workbook. Fill in the definitions from the glossary of the textbook as a first step to understanding the material.*

— **group—** ____

____

— **period—** ____

____

— **main group—** ____

____

— **transition metal group—** ____

____

— **inner transition metal group—** ____

____

Summary: *Read through this section and fill in the missing words and equations. Reference the textbook when you are not confident in your answers.*

Ten elements have been known since antiquity, but as the pace of chemical discovery increased in the 1700s and 1800s, efforts were made to categorize the elements by their properties. The most successful of these efforts was the **periodic table,** published in 1869 by Dmitri Mendeleev, a Russian chemist. If you are not familiar with it, there is one in the inside front cover of this book and your textbook.

The periodic table divides the elements into 18 vertical columns called **groups** and 7 horizontal rows called **periods.** The elements in each group have similar ________________. The number of elements in each period—two in period one, eight in periods two and three, eighteen in periods four and five, and so on—is an indication of the structure of atoms (we will see this in Chapter 5).

Groups 1A and 2A on the left, and 3A through 8A on the right side of the periodic table are the _______________. Their chemical properties are easy to predict based on their location on the periodic table. Groups 1B through 8B are known as the __________________. The 14 unnumbered groups shown separately on the bottom of the periodic table are the _________________.

The periodic table of the elements is the most important organizing principle of chemistry. It is worth printing out a black and white periodic table (or copying the one in the front of this study guide) and coloring in these groups to get a solid understanding of where the different types of elements lie on the table: use different colors for main-group elements, transition metals, and inner transition metals.

POST-LECTURE:

Work Problems: *One of the best ways to learn chemistry is through solving problems. Go to the textbook and work through the Worked Examples for this section, then go to the end of chapter and work the problems there. When you feel confident that you can correctly answer problems like the ones listed below for each objective, check off each of the Learning Objectives for this section above.*

LO 2.2: Problems 2.37, 2.62, 2.63
LO 2.3: Problem 2.36
LO 2.4: Problems 2.59, 2.61

Section 2.3 Some Common Groups of Elements and Their Properties

PRE-LECTURE:

Learning Objectives: *These are the skills you should be proficient in after reading this section and practicing the associated problems. Check off the boxes below once you have gone through this section, answered the associated questions in your textbook, and feel confident you have mastered the skill.*

- ☐ **LO 2.5** Specify the location and give examples of elements in the alkali metal, alkaline earth metal, transition metal, halogen, noble gas groups.
- ☐ **LO 2.6** Classify an element as a metal, nonmetal, or semimetal using its properties.

Definitions: *To learn, you must do. This is your workbook. Fill in the definitions from the glossary of the textbook as a first step to understanding the material.*

— **property—** ____

____

— **intensive property—** ____

____

— **extensive property—** ____

____

— **physical property—** ____

____

— **chemical property—** ____

____

Summary: *Read through this section and fill in the missing words and equations. Reference the textbook when you are not confident in your answers.*

Elements can be described in terms of their **properties.** Properties are themselves classified in a number of ways:

- __________ properties do not depend on the size of the sample. An example is melting point: ice melts at 0 °C, whether it is a snowflake or an iceberg.
- __________ properties do depend on the amount of the sample. To continue the previous example, the amount of energy needed to melt a snowflake or an iceberg is very different and depends on the sample size.
- __________ properties can be observed without changing the chemical makeup of the sample: melting is a physical property, as is boiling. Whether ice, liquid, or steam, the sample is still water. Physical changes tend to be readily reversible.

- __________ properties cannot be observed without a change in the composition of the sample. If hydrogen and oxygen are brought together and a spark is applied, water will result. Something has been learned about hydrogen (that it combines with oxygen to make water), but you no longer have hydrogen alone, because the sample has changed.

Workbook Example 2.3

PROBLEM:
There is overlap in the kinds of properties that are observed. Classify these as intensive or extensive, and also as physical or chemical.

Color Combustibility Smell Length Rusting

SOLUTION:
Using the descriptions of these properties above:
—Color is an *intensive physical* property.
—Combustibility is an *intensive chemical* property.
—Smell is an *intensive physical* property.
—Length is an *extensive physical* property.
—Rust is an *intensive chemical* property.

Four groups of the periodic table are listed below with examples of their similar properties.

- **Group 1A:** ____________. These are silvery, soft metals. All react strongly with water to form very alkaline solutions, which gives the group its name and also makes it impossible to find these metals in pure form naturally because of the ubiquitous presence of water.
- **Group 2A:** ____________. These are also lustrous, silvery metals that are not found free in nature. They also react with water, though not as strongly as the alkali metals.
- **Group 7A:** ____________. These are colorful, corrosive nonmetals. They are found in nature only in combination with other elements, including the alkali metals and the alkaline earth metals. These compounds are known as salts: the world "halogen" is from the ancient Greek word for salt.
- **Group 8A:** ____________. As the name suggests, these elements are almost completely unreactive.

It is a good idea to print out a new black and white periodic table and label these categories to help you remember them. We will continue asking you to add to this table throughout this study guide to help you understand the organizing power of the periodic table.

The elements are also divided into three major categories:

- __________ These are found on the left side of the periodic table, bounded by a zigzag line running from boron (B) at the top to astatine (At) at the bottom. They are generally shiny, malleable, ductile (able to be drawn into wire), and they conduct electricity and heat. All are solid except mercury.
- __________ These are found to the right of the zigzag line. They are typically dull in appearance (when solid), brittle, and poor conductors of heat and electricity. They have a variety of colors and are divided between solids, liquids (bromine, Br), and gases.

- __________ These nine elements are found sitting above the zigzag line, and also include germanium (Ge) and antimony (Sb) underneath the line.

Keep adding to the periodic table on which you labeled the categories above. This time use different shading to visually separate the metals, nonmetals, and semimetals. An example of this would be to leave the metals as is, use diagonal lines on the semimetals, and vertical lines on the nonmetals.

Workbook Example 2.4

PROBLEM:
Determine the type of the following elements (metal, nonmetal, or semimetal):

iridium   antimony   iodine

SOLUTION:
By determining the position of the element in the periodic table, you can determine the class of the element. (It is also helpful to know the symbol for each of the elements.)

Iridium—Ir; iridium is to the left of the zigzag line in the periodic table and is a metal.

Antimony—Sb; antimony is adjacent to the zigzag line in the periodic table and is a semimetal.

Iodine—I; iodine is found to the right of the zigzag line in the periodic table and is a nonmetal.

Workbook Example 2.5

PROBLEM:
Determine which class of elements each of these descriptions matches:
1. Corrosive yellow gas;
2. Shiny, malleable, conducts electricity

SOLUTIONS:
1. Gases are all nonmetals.
2. If something is malleable, that is, easily shaped without breaking, it is solid and a metal.

POST-LECTURE:

Work Problems: *One of the best ways to learn chemistry is through solving problems. Go to the textbook and work through the Worked Examples for this section, then go to the end of chapter and work the problems there. When you feel confident that you can correctly answer problems like the ones listed below for each objective, check off each of the Learning Objectives for this section above.*

LO 2.5: Problems 2.35, 2.64–2.67
LO 2.6: Worked Example 2.1; Problems 2.1–2.2, 2.68–2.71

Section 2.4 Observations Supporting Atomic Theory: The Conservation of Mass and the Law of Definite Proportions

PRE-LECTURE:

Learning Objectives: *These are the skills you should be proficient in after reading this section and practicing the associated problems. Check off the boxes below once you have gone through this section, answered the associated questions in your textbook, and feel confident you have mastered the skill.*

☐ **LO 2.7** Determine the mass of the products in a reaction using the law of mass conservation.

Definitions: *To learn, you must do. This is your workbook. Fill in the definitions from the glossary of the textbook as a first step to understanding the material.*

— **chemical compound—** ____________________

____

— **chemical formula—** ____________________

____

— **chemical equation—** ____________________

____

— **law of mass conservation—** ____________________

____

— **law of definite proportions—** ____________________

____

Summary: *Read through this section and fill in the missing words and equations. Reference the textbook when you are not confident in your answers.*

By carefully examining the behavior of gases, Englishman Robert Boyle (1627–1691) was able to propose an atomic theory as well as postulate the existence of a number of elements. Two other fundamental laws of chemistry came during the following century :

- **Law of mass conservation—** ____________________

  ____

  This law is a cornerstone of chemical science and makes it possible to perform reliable calculations about chemical reactions.

- **Law of definite proportions—** ____________________

  ____

  This means that specific mass ratios of elements will form specific compounds. For example, if hydrogen and water are combined to make water, 8 grams of oxygen will always combine with 1 gram of hydrogen. Where this gets tricky is when compounds can form more than one compound with each other. Carbon monoxide and carbon dioxide, for example, both only contain carbon and oxygen, but they will have different mass ratios.

Workbook Example 2.6

PROBLEM:
5.63 grams of methane are burned in oxygen to form carbon dioxide and water. If the reaction produces 15.5 grams of carbon dioxide and 12.7 grams of water, how much oxygen is required for the reaction to be completed?

SOLUTION:
According to the law of mass conservation, mass is neither created nor destroyed in chemical reactions. Therefore, the total mass of the reactants (the starting material) must equal the total mass of the products (the substances produced after the reaction has occurred). In this case:

$$\text{g methane} + \text{g oxygen} = \text{g carbon dioxide} + \text{g water}$$
$$(5.63\text{ g methane}) + (\text{x g oxygen}) = (15.5\text{ g carbon dioxide}) + (12.7\text{ g water})$$
$$\text{x g oxygen} = (15.5\text{ g} + 12.7\text{ g}) - 5.63\text{ g}$$
$$\text{x g oxygen} = 22.6\text{ g}$$

Remember that when adding and subtracting the answer can't have more digits to the right of the decimal than any of the original numbers.

Workbook Example 2.7

PROBLEM:
If 23 grams of sodium combine with 37.5 grams of chlorine to form sodium chloride, what is the mass ratio of the two elements in the compound? How much chlorine will be required to react with 8.5 grams of sodium?

SOLUTION:
The mass ratio will be the mass of either element divided by the mass of the other.

$$\frac{23\ g\ sodium}{37.5\ g\ chlorine} = \frac{0.6133\ g\ sodium}{1\ g\ chlorine} \quad or \quad \frac{37.5\ g\ chlorine}{23\ g\ sodium} = \frac{1.630\ g\ chlorine}{1\ g\ sodium}$$

To figure out how much chlorine will be needed to react with all of the sodium, either of the ratios can be used, but care must be taken to set up the ratio properly so the units cancel:

$$8.5\ g\ sodium \times \frac{1.630\ g\ chlorine}{1\ g\ sodium} = 13.86\ g\ chlorine \approx 14\ g\ chlorine$$

This answer on your calculator has more digits than it should: significant figure rules reduce the answer to 14 grams of chlorine.

POST-LECTURE:

Work Problems: *One of the best ways to learn chemistry is through solving problems. Go to the textbook and work through the Worked Examples for this section, then go to the end of chapter and work the problems there. When you feel confident that you can correctly answer problems like the ones listed below for each objective, check off each of the Learning Objectives for this section above.*

LO 2.7: Problems 2.78–2.79

Section 2.5 The Law of Multiple Proportions and Dalton's Atomic Theory

PRE-LECTURE:

Learning Objectives: *These are the skills you should be proficient in after reading this section and practicing the associated problems. Check off the boxes below once you have gone through this section, answered the associated questions in your textbook, and feel confident you have mastered the skill.*

☐ **LO 2.8** Demonstrate the law of multiple proportions using mass composition of two compounds of the same elements.

☐ **LO 2.9** Determine the formula of a compound given mass composition data for two compounds and the formula of one compound.

Definitions: *To learn, you must do. This is your workbook. Fill in the definitions from the glossary of the textbook as a first step to understanding the material.*

— **law of multiple proportions—** ____

____

— **atom—** ____

____

Summary: *Read through this section and fill in the missing words and equations. Reference the textbook when you are not confident in your answers.*

Around this time, John Dalton (1766–1844) proposed **the law of multiple proportions,** which said that two elements can combine in different ways, and when they do, the mass ratios are small, whole-number multiples of each other. This **law of multiple proportions** along with **the law of mass conservation** and **law of definite proportions** lead Dalton to develop a new atomic theory in 1808, which contains a number of precepts that have survived essentially unchanged to this day:

- Elements are made up of tiny particles called ___________.
- Each element is characterized by the mass of its atoms. Atoms of the same element have the ___________ masses, but atoms of different elements have ___________ masses.

- The chemical combination of elements to make different chemical compounds occurs when whole numbers of atoms join in fixed proportions.
- Chemical reactions only rearrange how atoms are ___________ in chemical compounds; the atoms themselves ___________ change.

Workbook Example 2.8

PROBLEM:
Nitrogen and oxygen form a large number of different compounds with each other. In one such compound, 2.8 grams of nitrogen combines with 3.2 grams of oxygen. In a different compound, 2.3 grams of nitrogen combines with 5.3 grams of oxygen. Do these compounds obey the law of multiple proportions?

SOLUTION:
Find the O:N mass ratio in each compound.

First compound: O:N mass ratio = $\dfrac{3.2\ g\ O}{2.8\ g\ N} = 1.14$

Second compound: O:N mass ratio = $\dfrac{5.3\ g\ O}{2.3\ g\ N} = 2.3$

Determine if the two O:N mass ratios are small, whole-number multiples of each other.

$$\frac{O:N\ mass\ ratio\ in\ second\ compound}{O:N\ mass\ ratio\ in\ first\ compound} = \frac{2.30}{1.14} = 2.00$$

Note: The ratios can be done in any way and give the same information. The ratio of the oxygen to nitrogen in the second compound is twice as high as in the first compound. If this was done the other way around, the answer obtained would state that the ratio in the first compound is half that of the second compound. This is the same answer. With a problem of this type, the important thing is to show that the ratios are small: 1:2 and 2:1 are both small, whole-number ratios.

Workbook Problem 2.1

Carbon and hydrogen form a large number of compounds. In one such compound, 3.43 g of carbon combines with 0.857 g of hydrogen. In another, 4.80 g of carbon combines with 0.400 g of hydrogen. Show that the law of multiple proportions is being obeyed.

Strategy: Determine the carbon-to-hydrogen mass ratios for both compounds and compare these ratios to each other.

Step 1: Determine the carbon-to-hydrogen ratio for the first compound.

Step 2: Determine the carbon-to-hydrogen ratio for the second compound.

Step 3:	Divide the ratio found for the first compound by the ratio for the second compound.
Step 4:	Can your answer be converted to a small, whole-number ratio?

POST-LECTURE:

Work Problems: *One of the best ways to learn chemistry is through solving problems. Go to the textbook and work through the Worked Examples for this section, then go to the end of chapter and work the problems there. When you feel confident that you can correctly answer problems like the ones listed below for each objective, check off each of the Learning Objectives for this section above.*

LO 2.8: Worked Example 2.2; Problems 2.80, 2.82
LO 2.9: Problems 2.83–2.84

Section 2.6 Atomic Structure: Electrons

PRE-LECTURE:

Learning Objectives: *These are the skills you should be proficient in after reading this section and practicing the associated problems. Check off the boxes below once you have gone through this section, answered the associated questions in your textbook, and feel confident you have mastered the skill.*

- ☐ **LO 2.10** Describe Thomson's cathode ray tube experiment and what it contributed to the current model of atomic structure.
- ☐ **LO 2.11** Describe Millikan's oil drop experiment and what it contributed to the current model of atomic structure.

Definitions: *To learn, you must do. This is your workbook. Fill in the definitions from the glossary of the textbook as a first step to understanding the material.*

— **electron—**____

____

Summary: *Read through this section and fill in the missing words and equations. Reference the textbook when you are not confident in your answers.*

In the late 1800s a number of physicists experimented with **cathode ray tubes.** These tubes produce a visible glow when high voltage is applied across the tube with a small amount of gas left in it.

J. J. Thomson's (1856–1940) experimentation with **cathode ray tubes** helped him to unlock the mystery of the ___________. He deduced:

- the glow was comprised of tiny negatively charged particles (what we now call electrons) because it flowed from the anode (negative electrode) to the cathode (positive electrode).
- that electrons are a fundamental part of matter because different metals were used to make electrodes and they all emitted electrons,
- that the electron's mass-to-charge ratio, e/m, was $1.758\ 820 \times 10^{8}$ C/g (C = coulombs, a measure of electrical current; we will talk about this in Chapter 19, Electrochemistry).

Robert Millikan (1868–1953) built on Thomson's work by constructing an apparatus that would enable him to coat tiny drops of oil with electrons and then determine the electrical charge on the drop. By the use of repeated measurements, he was finally able to determine that the charges on the drops were always multiples of $1.602\ 177 \times 10^{-19}$ C.

With the values for the charge of an electron, and the charge-to-mass ratio, it was then possible to calculate the mass of an electron. These infinitesimally tiny particles weigh a mere $9.109\ 390 \times 10^{-28}$ g. This calculation is shown in your textbook: note that it was determined using nothing more than simple unit analysis.

POST-LECTURE:

Work Problems: *One of the best ways to learn chemistry is through solving problems. Go to the textbook and work through the Worked Examples for this section, then go to the end of chapter and work the problems there. When you feel confident that you can correctly answer problems like the ones listed below for each objective, check off each of the Learning Objectives for this section above.*

LO 2.10: Problems 2.86–2.88
LO 2.11: Problems 2.89–2.90

Section 2.7 Atomic Structure: Protons and Neutrons

PRE-LECTURE:

Learning Objectives: *These are the skills you should be proficient in after reading this section and practicing the associated problems. Check off the boxes below once you have gone through this section, answered the associated questions in your textbook, and feel confident you have mastered the skill.*

☐ **LO 2.12** Describe Rutherford's gold foil experiment and what it contributed to the current model of atomic structure.
☐ **LO 2.13** Calculate the number of atoms in sample given the size of the atom.

Definitions: *To learn, you must do. This is your workbook. Fill in the definitions from the glossary of the textbook as a first step to understanding the material.*

— **nucleus—** ____

____

— **proton—** ____

____

— **neutron—** ____

____

Summary: *Read through this section and fill in the missing words and equations. Reference the textbook when you are not confident in your answers.*

Matter is electronically neutral, so if electrons are negatively charged, there must be other particles that provide positive charge to maintain electrical neutrality. Ernest Rutherford (1871–1937) successfully investigated this with the use of alpha (α) particles. These have the following characteristics:

- They are 7000 times more massive than electrons.
- They have a positive charge that is twice the magnitude of, but opposite in sign to, the charge on an electron (electrons have a charge of −1, alpha particles have a charge of +2).
- They are naturally emitted by some radioactive elements.

In his experiment, Rutherford directed a beam of alpha particles at a sheet of gold foil that was only a few atoms thick. Most passed through the foil, but a small number (0.005%) were deflected, and a tiny number bounced back at the source.

Rutherford explained this by proposing that atoms consist mostly of empty space, have a positive charge, and most of their mass is concentrated in a central core, which Rutherford called the **nucleus.** Most of the alpha particles passed through the empty space of the atoms, but those that were deflected were affected by both the mass and the charge of the particles within the atoms.

Further experiments showed that the nucleus itself contains not only positively charged protons but also uncharged neutrons.

To outline the nuclear model:

- Atoms contain three subatomic particles: __________, __________, and __________.
- Electrons and protons have equal but opposite charges, with electrons carrying ________ charges and protons carrying __________ charges.
- Protons and neutrons have almost the same mass, which is much __________ than the mass of an electron.
- The nucleus of the atom (protons and neutrons) provides most of the mass of the atom but very little of the volume.
- The electrons provide almost none of the mass of an atom but almost all of the volume.

POST-LECTURE:

Work Problems: *One of the best ways to learn chemistry is through solving problems. Go to the textbook and work through the Worked Examples for this section, then go to the end of chapter and work the problems there. When you feel confident that you can correctly answer problems like the ones listed below for each objective, check off each of the Learning Objectives for this section above.*

LO 2.12: Problems 2.91, 2.92
LO 2.13: Problems 2.93, 2.94

Section 2.8 Atomic Numbers

PRE-LECTURE:

Learning Objectives: *These are the skills you should be proficient in after reading this section and practicing the associated problems. Check off the boxes below once you have gone through this section, answered the associated questions in your textbook, and feel confident you have mastered the skill.*

☐ **LO 2.14** Determine the mass number, atomic number, and number of protons, neutrons, and electrons from an isotope symbol.

Definitions: *To learn, you must do. This is your workbook. Fill in the definitions from the glossary of the textbook as a first step to understanding the material.*

— **atomic number (Z)—** ____

____

— **mass number (A)—**____

____

— **isotope—** ____

____

Summary: *Read through this section and fill in the missing words and equations. Reference the textbook when you are not confident in your answers.*

Elements differ from each other according to the number of protons their nuclei contain. This is also known as their **atomic number,** or "*Z*."

In a neutral (uncharged) atom, the number of protons will be the ___________ as the number of electrons.

Most atomic nuclei also contain neutrons. The **mass number** of an atom, or "*A*," is found by adding the atomic number, *Z*, to the number of neutrons, *N*, as shown in the following equation. The mass of electrons is so small compared to the mass of the other particles that it can be ignored.

$$A = Z + N$$

As we said above, *elements* are defined by the amount of protons: any variation in the number of protons and you will have a different element. However, the number of neutrons in the nucleus can change without changing the identity of the element. Atoms with different numbers of neutrons are called **isotopes.** The number of neutrons has very little effect on the chemical properties of an atom. Many of the chemical properties of an atom are related to the number of electrons.

In other words (from above):

- The element is determined by the number of ____________________.
- The isotope is determined by the number of ____________________.
- The reactivity of an atom is determined by the number of ____________________.

Isotope symbols are written using the following convention, where "X" indicates the elemental symbol:

$${}^{A}_{Z}X$$

From this, it is possible to calculate the number of neutrons ($A - Z$). It is not difficult to remember which of these numbers is which: the smaller will always be the atomic number, the larger will always be the mass number. Sometimes the atomic number is omitted, because it does not vary and can be found in the periodic table.

Workbook Example 2.9

PROBLEM:
Determine the number of protons, neutrons, and electrons in the following isotopes:

$${}^{71}_{31}Ga \qquad {}^{191}_{77}Ir \qquad {}^{88}_{39}Y$$

SOLUTION:
The subscript is the atomic number (Z), and the superscript is the mass number (A). Remember that Z = the number of protons = the number of electrons, and $A - Z$ = the number of neutrons.

${}^{71}_{31}Ga$: $Z = 31$; number of protons = number of electrons = 31
$A - Z = 71 - 31 = 40$; number of neutrons = 40

${}^{191}_{77}Ir$: $Z = 77$; number of protons = number of electrons = 77
$A - Z = 191 - 77 = 114$; number of neutrons = 114

${}^{88}_{39}Y$: $Z = 39$; number of protons = number of electrons = 39
$A - Z = 88 - 39 = 49$; number of neutrons = 49

Workbook Problem 2.2

Oxygen is the most abundant element by mass in the biosphere. All classes of biomolecules, proteins, sugars, fats, and nucleic acids, contain oxygen, and oxygen makes up almost 90% of the mass of the world's oceans. Oxygen is made up of three stable isotopes, with the mass numbers 16, 17, and 18. Write the isotope symbols for each of these isotopes.

Strategy: Use the periodic table to determine the atomic number and the number of neutrons in each isotope. Give the standard symbol for each.

Step 1: Find the atomic number for oxygen.

Step 2: Calculate the mass number for each isotope.

Step 3: Write the isotope symbol with the mass number as a superscript and the atomic number as a subscript, both to the left of the symbol.

Workbook Problem 2.3

The metalloid X has been used as a cosmetic since at least 3000 B.C. When bound to sulfur atoms in a 2:3 ratio, it is known as "kohl." An atom of the metalloid contains 51 protons, and its most stable isotope contains 70 neutrons. Identify the metal and write the isotope symbol.

Strategy: Use the periodic table to identify the metal. Write the isotope symbol for the metal.

Step 1: Determine the atomic number from the information given.

Step 2: Knowing the atomic number, identify the metal by using the periodic table.

Step 3: Write the isotope symbol with the mass number as a superscript and the atomic number as a subscript, both to the left of the symbol.

POST-LECTURE:

Work Problems: *One of the best ways to learn chemistry is through solving problems. Go to the textbook and work through the Worked Examples for this section, then go to the end of chapter and work the problems there. When you feel confident that you can correctly answer problems like the ones listed below for each objective, check off each of the Learning Objectives for this section above.*

LO 2.14: Worked Example 2.4; Problems 2.100–2.102, 2.104, 2.106

Section 2.9 Atomic Weights and the Mole

PRE-LECTURE:

Learning Objectives: *These are the skills you should be proficient in after reading this section and practicing the associated problems. Check off the boxes below once you have gone through this section, answered the associated questions in your textbook, and feel confident you have mastered the skill.*

- ☐ **LO 2.15** Calculate atomic weight given the fractional abundance and mass of each isotope.
- ☐ **LO 2.16** Convert between grams and numbers of moles or atoms using molar mass and Avogadro's number.

Definitions: *To learn, you must do. This is your workbook. Fill in the definitions from the glossary of the textbook as a first step to understanding the material.*

— **unified atomic mass unit (u)—**______________________________________

____

— **atomic mass—** ____

____

— **atomic weight—**____

____

— **mole—** ____

____

— **molar mass—**____

____

— **Avogadro's number—** ____

____

Summary: *Read through this section and fill in the missing words and equations. Reference the textbook when you are not confident in your answers.*

Because the masses of individual atoms and their components are so tiny, grams are an inconvenient unit. As a result, **unified atomic mass units** are used instead. These are abbreviated ___ in chemistry and are known as daltons (Da) in biochemistry. A *unified atomic mass unit* is defined as exactly 1/12 the mass of an atom of ${}^{12}_{6}C$; $1\ u = 1.660\ 539 \times 10^{-24}$ g.

Most elements occur as a mixture of isotopes, so the atomic weight reported on the periodic table for an element is a weighted average of its naturally occurring isotopes.

The atomic weight of an element = Σ(___________ of each isotope × the ___________ of the isotope). The Greek letter Σ (sigma) means "the sum of."

Workbook Example 2.10

PROBLEM:
Calculate the mass in grams of a single atom of cadmium.

SOLUTION:
From the mass given in the periodic table, we know that 1 atom of Cd has a mass of 112.411 u. We also know that $1\ u = 1.6605 \times 10^{-24}$ g. The final unit we want to end up with is g/atom Cd. Therefore, we should set up our problem in the following manner:

$$\frac{112.411\,u}{1\,atom\,Cd} \times \frac{1.6605 \times 10^{-24}\ g}{1\,u} = 1.8666 \times 10^{-22}\ g$$

Workbook Example 2.11

PROBLEM:
Boron has two stable isotopes: B-10 and B-11. Boron-10 is 19.9% of a sample of boron and has an atomic mass of 10.0129 u. The remaining 80.1% is boron-11, with an atomic mass of 11.0093 u. Calculate the average atomic mass of boron.

SOLUTION:
To work out the average atomic mass of boron, multiply the individual masses by their abundance, and then add them together:

$$(0.199\times 10.0129\ u)+(0.801\times 11.0093\ u)=10.81\ u$$

This agrees with the value on the periodic table, which is always a good sign.

Workbook Problem 2.4

Lead has four naturally occurring isotopes. ^{208}Pb has an atomic mass of 207.977 u and an abundance 52.4% in an average lead sample. ^{207}Pb has an atomic mass of 206.976 and an abundance of 22.1%. ^{206}Pb has an atomic mass of 205.974 and an abundance of 24.1%. The remaining lead will be ^{204}Pb with an atomic mass of 203.973 u. Calculate the average atomic mass of Pb.

Strategy: Remember, the atomic mass of an element = Σ(mass of each isotope × the abundance of the isotope), and all abundances must add up to 100%.

Step 1: Use the information given to determine the abundance of ^{204}Pb.

Step 2: Use the information given to determine the average atomic mass.

Because the mass of individual atoms is so small, a group of them big enough that we can manipulate them in the macroscopic world has an enormous number of atoms. To go between mass and amount, chemists use the unit *mole*. There are 6.022 141 × 10^{23}, Avogadro's number, of items in a mole. For any element 1 mole of atoms = atomic weight in grams, also known as its molar mass.

Now we have two equalities, depending on whether we are working in the atomic scale or macroscale, which can be created from the number below the elemental symbol on the periodic table.

1 ____________ = atomic weight (u)
1 ____________ = atomic weight (g) = molar mass (g/mol)

Workbook Example 2.12

PROBLEM:
A 1.0-carat diamond weighs 0.20 grams. Assuming the diamond is pure carbon, how many moles and how many atoms are in the diamond?

SOLUTION:
The molar mass of carbon can be used to convert the amount of carbon in the diamond in grams to moles. From there Avogadro's number can be used to convert to atoms. Therefore, we should set up our problem in the following manner:

$$0.20\ g \times \frac{1\ mol\ C}{12.0107\ g} = 0.01665182\ mol\ C = 0.017\ mol\ C$$

$$0.01665182\ mol\ C \times \frac{6.022 \times 10^{23}\ atoms}{1\ mol} = 1.0 \times 10^{22}\ atoms\ of$$

Remember not to round in the middle of a calculation. That is why the second step starts with 0.01665182 and not the answer to the first part of the problem, 0.017 mol C.

Workbook Problem 2.5

An empty aluminum can weighs 14.9 grams. Assuming it is pure aluminum, how many moles and how many atoms of aluminum are in the can?

Strategy: The molar mass of aluminum can be used to convert the amount of aluminum in the can to moles. From there Avogadro's number can be used to convert to atoms.

Step 1: Use the molar mass of aluminum to convert the grams to moles.

Step 2: Convert moles to atoms using Avogadro's number.

POST-LECTURE:

Work Problems: *One of the best ways to learn chemistry is through solving problems. Go to the textbook and work through the Worked Examples for this section, then go to the end of chapter and work the problems there. When you feel confident that you can correctly answer problems*

like the ones listed below for each objective, check off each of the Learning Objectives for this section above.

LO 2.15: Worked Example 2.5; Problems 2.116, 2.118, 2.120
LO 2.16: Worked Example 2.6; Problems 2.124, 2.125

Section 2.10 Measuring Atomic Weight: Mass Spectrometry

PRE-LECTURE:

Learning Objectives: *These are the skills you should be proficient in after reading this section and practicing the associated problems. Check off the boxes below once you have gone through this section, answered the associated questions in your textbook, and feel confident you have mastered the skill.*

☐ **LO 2.17** Use data from a mass spectrum to calculate the atomic weight of an element.

Definitions: *To learn, you must do. This is your workbook. Fill in the definitions from the glossary of the textbook as a first step to understanding the material.*

— **mass spectrometer—**____

____

— **mass spectrum—**____

____

Summary: *Read through this section and fill in the missing words and equations. Reference the textbook when you are not confident in your answers.*

The instrument used most often to determine the mass of an atom is a mass spectrometer. To do this, a sample is vaporized and then bombarded with a beam of high-energy electrons. This beam knocks electrons out of the atoms in the sample, which creates an imbalance of charges and creates a ___________ charged particle, or ion. This ion is accelerated by an electric field and then a magnet is used to deflect the ion. Lighter ions are deflected ___________ than heavier ones. When the strength of the magnetic field is varied, it allows for the ions of different mass-to-charge ratios to pass through a slit in the detector. A mass spectrum is created by plotting the ion intensity versus the mass-to-charge ratio, m/z. The intensity (height) of the peaks is related to the individual isotope's abundance. Peak intensity is not the same as percent abundance.

Workbook Example 2.13

PROBLEM:
Calculate the atomic weight of the element that has two peaks; one with a m/z of 68.925580 and a relative intensity of 100 and the other with a m/z of 70.9247005 and an intensity of 66.37. What element is it?

SOLUTION:
First you must find the fractional abundance of each isotope from the relative intensities. To do this, you divide the relative intensity of an isotope by the sum of all the intensities. For the first isotope you would have 100/166.37 = 0.6010699 and for the second 66.37/166.37 = 0.3989301. Next you use the formula for finding atomic weight by using the isotopic masses and fractional abundances.

$$\text{Atomic weight} = (0.6010699)(68.925580) + (0.3989301)(70.9247005) = 69.72\ \text{u}$$

Gallium, Ga, has an atomic weight of 69.723 listed on the periodic table.

POST-LECTURE:

Work Problems: *One of the best ways to learn chemistry is through solving problems. Go to the textbook and work through the Worked Examples for this section, then go to the end of chapter and work the problems there. When you feel confident that you can correctly answer problems like the ones listed below for each objective, check off each of the Learning Objectives for this section above.*

LO 2.17: Worked Example 2.7; Problems 2.13, 2.14, 2.132, 2.133

Section 2.11 Mixtures and Chemical Compounds; Molecules and Covalent Bonds

PRE-LECTURE:

Learning Objectives: *These are the skills you should be proficient in after reading this section and practicing the associated problems. Check off the boxes below once you have gone through this section, answered the associated questions in your textbook, and feel confident you have mastered the skill.*

- ☐ **LO 2.18** Classify matter as a mixture, pure substance, element, or compound.
- ☐ **LO 2.19** Convert between structural formulas, ball-and-stick models, and chemical formulas.

Definitions: *To learn, you must do. This is your workbook. Fill in the definitions from the glossary of the textbook as a first step to understanding the material.*

— **mixture**—____

____

— **chemical bond**—____

____

— **covalent bond**—____

____

— **molecule**—____

____

Summary: *Read through this section and fill in the missing words and equations. Reference the textbook when you are not confident in your answers.*

There are many more types of substances found on Earth than there are elements on the periodic table. These substances can all be classified as either **pure substances** or **mixtures.**

Pure substances are either **elements** or **chemical compounds.** In general, elements cannot be changed into another substance, but we will see that in nuclear reactions that that is not always the case (Chapter 20). Chemical compounds are made up of atoms that are chemically bonded together. They are formed by chemical reactions and cannot be separated except by chemical means. They combine in a *specific ratio* every time, and also have properties that are very different from their constituent elements.

Mixtures, in contrast, are formed when two or more substances are mixed together without any chemical interaction. The individual substances maintain their properties and can be separated relatively easily.

Heterogeneous mixtures are not uniformly mixed: dirt in water will not be evenly distributed and will tend to settle out. If it is mixed, the liquid will be cloudy.

Homogeneous mixtures are uniform throughout, and they can be solids (metal alloys), liquids (solutions), or gases (air). As an example, sugar dissolved in water will have the same amount of sugar at the top of the water as at the bottom. The liquid will also be clear, because the sugar is perfectly dispersed in the liquid.

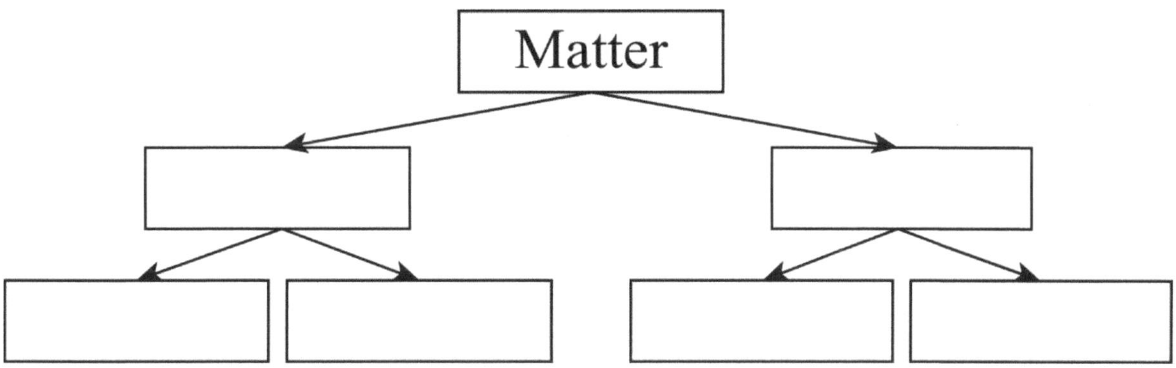

Workbook Example 2.14

PROBLEM:
Identify the following household items as elements, compounds, homogeneous mixtures, or heterogeneous mixtures:

salad   tea   metal utensils   water   soda can   air

SOLUTION:
Think about the composition of these things that you see in everyday life. Salad is a heterogeneous mixture, because it has areas of variable composition: two separate spoonfuls will have different components. Tea is a clear liquid and does not have constituents that settle: it is a homogeneous mixture. Metal utensils can be made of any number of things from stainless steel to sterling silver, but all are mixtures of metals in unvarying proportions, so these are homogeneous mixtures as well. Water contains elements bound together to form something with very different

properties: it cannot easily be separated back into its constituent parts, so it is a compound. A soda can, except for the paint on the outside, is made of aluminum, which is an element. Air is a mixture of elements not bonded to one another. Air contains nitrogen, oxygen, hydrogen, and many other components: it is a homogeneous mixture.

To form chemical compounds atoms join together by forming **chemical bonds.** One type of chemical bond is called a **covalent bond;** a covalent bond happens when two atoms share two electrons. When two or more atoms are joined together through covalent bonding they form a **molecule.** Covalent bonds typically form between nonmetals.

Covalent bonds:
- happen when atoms ____________ electrons.
- occur between __________________ elements.

There are a number of ways chemists represent molecules (see Figure 2.14 in the textbook), including *ball-and-stick models*, which show the bonds as "sticks" connecting the atoms, and *space-filling models* that show the atoms stuck together without explicitly showing the bonds.

Covalent bonds are commonly shown using **structural formulas:** these formulas use lines between atoms to show how the atoms are connected. This is particularly important in organic chemistry, which is the study of carbon compounds. In these compounds, the structure determines behavior, even if the atoms are the same.

Workbook Example 2.15

PROBLEM:
1-butanol has the structural formula shown below. Write the chemical formula.

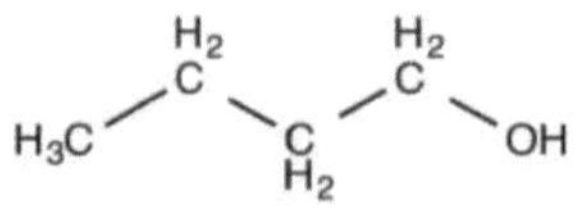

1-butanol

SOLUTION:
This structure has four carbon atoms, ten hydrogen atoms, and one oxygen atom. Its chemical structure is $C_4H_{10}O$.

On the black and white periodic table you are marking up, circle all the elements that exist as diatomic elements in nature. Memorize these.

POST-LECTURE:

Work Problems: *One of the best ways to learn chemistry is through solving problems. Go to the textbook and work through the Worked Examples for this section, then go to the end of chapter and work the problems there. When you feel confident that you can correctly answer problems like the ones listed below for each objective, check off each of the Learning Objectives for this section above.*

LO 2.18: Worked Example 2.8; Problems 2.15, 2.16, 2.42
LO 2.19: Worked Example 2.9; Problems 2.43, 2.44

Section 2.12 Ions and Ionic Bonds

PRE-LECTURE:

Learning Objectives: *These are the skills you should be proficient in after reading this section and practicing the associated problems. Check off the boxes below once you have gone through this section, answered the associated questions in your textbook, and feel confident you have mastered the skill.*

- ☐ **LO 2.20** Classify bonds as ionic or covalent.
- ☐ **LO 2.21** Determine the number of electrons and protons from chemical symbol and charge.
- ☐ **LO 2.22** Match the molecular representation of an ionic compound with its chemical formula.

Definitions: *To learn, you must do. This is your workbook. Fill in the definitions from the glossary of the textbook as a first step to understanding the material.*

— **ionic bond—**____

____

— **ions—** ____

____

— **cation—** ____

____

— **anion—**____

____

— **ionic solid—** ____

____

— **polyatomic ion—**____

____

— **structural formula—**____

____

Summary: *Read through this section and fill in the missing words and equations. Reference the textbook when you are not confident in your answers.*

Not all atoms share electrons. Ionic bonds are the result of electrons being *transferred* from one atom to another, which results in atoms with opposite charges. The attractions between opposite electrical charges hold the atoms together. **Ionic bonds** occur between metals and nonmetals.

Ionic bonds:
• happen when atoms ____________electrons.
• occur between __________________ elements.

Metals give up electrons easily to form ___________—positively charged particles. It is easy to remember that metals give up electrons if you recall that metals conduct electricity because they are not holding on to their electrons very tightly. Nonmetals readily gain electrons to form ___________— negatively charged particles.

When anions and cations interact, they do so generally on the basis of opposite charges. They do not form specific bonds, but they do all they can to neutralize their opposing charges. A cation will surround itself with anions, and *vice versa*. As a result, ionic bonds result in crystals (like those of common table salt) called **ionic solids** rather than forming molecules. When a formula is written for an ionic solid, it is written with the lowest whole-number ratio of atoms so as to show the neutralized charges, but it does not indicate that a specific anion is bonded exclusively to a specific cation.

Polyatomic ions also occur. These are charged (usually negatively, because they consist of nonmetals), covalently bonded groups of atoms that carry a charge. There are covalent bonds holding the atoms together within polyatomic ions. However, as units, the polyatomic ions can form ionic bonds.

Workbook Example 2.16

PROBLEM:
Identify the following as covalent molecules or ionic compounds:

a) CH_4 b) MgO c) KBr d) NO_2

SOLUTION:
The easiest way to differentiate between molecules and ionic compounds is to determine whether the formula contains both metals and nonmetals or only nonmetals. If only nonmetals are present, a covalent molecule has formed. If metals and nonmetals are both present, ionic bonding is occurring. CH_4 and NO_2 contain only nonmetals, so they are molecules. MgO and KBr are both combinations of metals and nonmetals. These are ionic compounds.

POST-LECTURE:

Work Problems: *One of the best ways to learn chemistry is through solving problems. Go to the textbook and work through the Worked Examples for this section, then go to the end of chapter and work the problems there. When you feel confident that you can correctly answer problems like the ones listed below for each objective, check off each of the Learning Objectives for this section above.*

LO 2.20: Worked Example 2.10; Problems 2.134, 2.135
LO 2.21: Problems 2.41, 2.138, 2.139
LO 2.22: Problem 2.45

Section 2.13 Naming Chemical Compounds

PRE-LECTURE:

Learning Objectives: *These are the skills you should be proficient in after reading this section and practicing the associated problems. Check off the boxes below once you have gone through this section, answered the associated questions in your textbook, and feel confident you have mastered the skill.*

☐ **LO 2.23** Convert between name and formula for ionic compounds.
☐ **LO 2.24** Convert between name and formula for binary molecular compounds.

Definitions: *To learn, you must do. This is your workbook. Fill in the definitions from the glossary of the textbook as a first step to understanding the material.*

— **oxoanion—**____

____

Summary: *Read through this section and fill in the missing words and equations. Reference the textbook when you are not confident in your answers.*

Special rules have been developed to name chemical compounds. The simplest compounds are **binary ionic compounds:** as the name suggests, these have only two elements and are ionic. They are named by writing the cation first, then the anion with the suffix _____.

The formula of a binary ionic compound must be written to give the smallest ratio of anion to cation that provides for electrical neutrality. What this means in plain English is that if you are balancing charges, make sure you don't include more than you need to: for example, Mg_3N_2 is correct, not Mg_6N_4. If all your subscripts have a common factor, you'll need to fix it.

To name chemical compounds, you will need to know the charges on the ions. For main group metals and nonmetals, this information can be taken straight from the periodic table.

- Main-group metals carry a positive charge equal to their group number. Alkali metals, for example, are found in group 1A, so they all carry charges of +1.
- Nonmetals carry negative charges equal to their group number minus 8. So, oxygen, which falls in group 6A, will carry a charge of $6 - 8 = -2$.

Write the charges that commonly occur for the elements in groups 1A, 2A, 3A, 5A, 6A, and 7A (atoms in groups 4A and 8A do not commonly form ions) on the black and white periodic table you are creating.

When transition metals are involved as cations, it is not always possible to take the charges from the periodic table. Because of this, the charge must be given when the name is written: for example, iron(II) chloride indicates that the iron atoms in the compound carry a charge of +2. When the formula is written, the charges on the iron atom must be balanced by negative charges. Because chlorine carries a charge of −1 when it is an ion (group 7A − 8 = −1), there will need to be two chlorines. The formula will be $FeCl_2$.

Add the charges of the transition metal ions from Figure 2.18 to the black and white periodic table you are creating.

Workbook Example 2.17

PROBLEM:
Name the following compounds: MgO $CuCl_2$ $RbCl$ SnO_2

SOLUTION:

MgO:	magnesium oxide	No Roman numeral is needed: magnesium is a group 2A metal and forms only Mg^{2+}.
$CuCl_2$:	copper (II) chloride	The charge on Cl is −1, and there are two of them. To have electrical neutrality, the charge on copper, which is a transition metal, must be +2. The Roman numeral (II) is used to indicate this charge.
$RbCl$:	rubidium chloride	Rubidium is a group 1A metal and forms only Rb^+.
SnO_2:	tin(IV) oxide	There are 2 O^{2-} ions with a total charge of −4. To have electrical neutrality, the charge on tin must be +4. Tin can form more than one kind of ion; the charge must be indicated by a Roman numeral.

Workbook Example 2.18

PROBLEM:
Write formulas for the following compounds:
a) cesium oxide b) magnesium nitride c) vanadium(V) bromide

SOLUTION:

Cs_2O Cesium is a group 1A metal and forms only Cs^+. Oxygen forms O^{2-} oxide ions, so two cesium ions are needed for each oxygen.

Mg_3N_2 Magnesium is a main group, group 2A metal that has a +2 charge, while nitride ions have a −3 charge (group 5 − 8 = −3). To have electrical neutrality, you need two Mg^{2+} and two N^{3-} ions [(3)(+2) + (2)(−3)] = 0. There will need to be six positive charges and six negative charges to achieve electrical neutrality.

VBr_5 Vanadium(V) has a +5 charge, while bromine has a −1 charge. To have electrical neutrality, you need 1 V^{5+} and 5 Br^-.

Workbook Problem 2.6

Name or write formulas for the following:

MgF_2	sodium chloride	chromium(III) nitride	CaO
vanadium(V) oxide	aluminum sulfide	FeF_3	$CuSe$

Strategy: Follow the rules outlined here and in your textbook.

Step 1: When naming compounds, determine if the metal is a main-group metal or a transition metal. If the metal is a transition metal, you must indicate the charge on the metal when naming the compound. The charge on the metal is determined from the number and charge on the anion. (Remember, the charges must all add up to zero.)

Step 2: When writing formulas, use the periodic table or the name of the compound to determine the charge on the metal. Use the periodic table to determine the charge on the anion. (Remember, the charges must all add up to zero.) Make sure the formula contains the smallest whole-number ratio of cation to anion.

Ionic compounds that contain polyatomic ions are named by following the rules for naming binary ionic compounds: cation, then anion. A number of polyatomic ions are listed in a table in your book: these should be memorized. Flashcards or an app are both great ways to do this!

Oxoanions have special naming conventions that may make their names easier to learn: first learn the name and formula of the ion whose name ends with *-ate* (including the charge on the anion). Other similar ions will be named using the following conventions:

- Add one O; add prefix _____.
- Remove one O; change ending to _____.
- Remove two Os; add prefix _____ and change *-ate* to ______.
- If a hydrogen is present in one of the oxoanions, add _________ in front of the ion name.

Workbook Example 2.19

PROBLEM:
Name the following compounds: a) $NaClO_4$ b) MgS_2O_3 c) $Fe_2(CO_3)_3$ d) $Mn(NO_3)_2$

SOLUTION:

$NaClO_4$: sodium perchlorate

MgS_2O_3: magnesium thiosulfate

$Fe_2(CO_3)_3$: iron(III) carbonate. The carbonate anion has a −2 charge, and there are three of them, for a total charge of −6. Therefore, there must be a +6 charge spread over the two iron cations, so the charge on the iron cation must be +3. Each iron cation has a charge of +3 (+6/2). The charge on the iron is indicated with Roman numerals because iron is a transition metal.

$Mn(NO_3)_2$: manganese(II) nitrate.

Workbook Example 2.20

PROBLEM:
Write formulas for the following compounds:
iron(III) sulfate cesium nitrite sodium acetate

SOLUTION:

iron(III) sulfate: $Fe_2(SO_4)_3$ The sulfate anion has a −2 charge. Iron(III) by definition has a charge of +3, so the charges must be balanced. Three sulfate ions are needed

		to give a charge of −6, which will balance two Fe^{3+} ions, with a charge of +6.
cesium nitrite:	$CsNO_2$	Cesium is a group 1A metal, so it carries a charge of +1, as does the nitrite ion.
sodium acetate:	$NaCH_3CO_2$	The acetate anion has a −1 charge, and is thus perfectly balanced by the +1 charge of the group 1A sodium ion.

Workbook Problem 2.7

Name or write formulas for the following:

$Fe(NO_3)_3$ potassium chromate magnesium phosphate $Al_2(SO_4)_3$ ammonium nitrate

Strategy: Follow the rules listed above and in your textbook.

Step 1: When naming a compound, identify the cation and the anion. Write the cation first.

Step 2: Determine the charge on the cation, and write the formula to make sure all charges are neutralized.

Binary molecular compounds contain only two nonmetals. They are named by first naming the element closest to the metals on the periodic table, followed by the element farthest from the metals with an *–ide*.

Because these are electron-sharing interactions, they can occur in many different ways (remember the law of multiple proportions?). Numerical prefixes must be used to identify the number of each element in the molecule. *Mono-* is not generally used for the first element. When a prefix ends in *a* or *o* (not i) and the anion begins with a vowel, the *a* or *o* of the prefix is dropped. These prefixes should already be familiar to you, and you should memorize them.

Fill in the prefixes below from Table 2.6 in the textbook.

Prefix	Meaning
	1
	2
	3
	4
	5
	6
	7
	8
	9
	10

Careful! If a binary molecular compound contains hydrogen, it will be named using these conventions if it is a gas, but it will be named as an acid if it is dissolved in water (aqueous). This will be indicated in parentheses after the formula: HCl (*g*) is hydrogen chloride, but HCl (*aq*) is hydrochloric acid!

Workbook Example 2.21

PROBLEM:
Name the following compounds: SF_6 N_2O_5 P_4S_{10} NI_3

SOLUTION:

SF_6: sulfur hexafluoride
N_2O_5: dinitrogen pentoxide
P_4S_{10}: tetraphosphorus decasulfide
NI_3: nitrogen triiodide

Workbook Example 2.22

PROBLEM:
Write formulas for the following compounds:
hydrogen chloride (gas) xenon hexafluoride diphosphorus trioxide

SOLUTION:
hydrogen chloride (gas): HCl(*g*) gaseous, so it is named as a binary molecule

xenon hexafluoride: XeF_6

diphosphorus trioxide: P_2O_3

Workbook Problem 2.8

Name or write formulas for the following:
a) PF_6 b) dinitrogen tetroxide c) selenium dioxide d) NBr_3 e) HF(*g*)

Strategy: Follow the rules given here and in your textbook.

Step 1: When naming the molecules, remember that the element that is more anion-like uses the *-ide* suffix. Also, remember to use numerical prefixes to indicate the number of each atom.

Step 2: Add numerical prefixes.

POST-LECTURE:

Work Problems: *One of the best ways to learn chemistry is through solving problems. Go to the textbook and work through the Worked Examples for this section, then go to the end of chapter and work the problems there. When you feel confident that you can correctly answer problems like the ones listed below for each objective, check off each of the Learning Objectives for this section above.*

LO 2.23: Worked Examples 2.11, 2.12; Problems 2.146, 2.148, 2.150, 2.152
LO 2.24: Worked Example 2.13; Problems 2.161, 2.162

Self-Test

This section is intended to test your knowledge of the material covered in this chapter. Think through these problems and make certain you understand what the problem is about—ask yourself if the answer makes sense. Many of these questions are linked to the chapter learning goals. Therefore, successful completion of these problems indicates that you have achieved the learning goals for this chapter. You will receive the greatest benefit from this section if you use it as a mock exam. Do not use your notes, book, Internet, or friends when taking the test. There is no grade and you only shortchange yourself by getting help with this activity. As you take the test, mark the problems with your level of confidence with the answer.

★ *Use a star for problems you know you got right.*
? *Use a question mark for those you think you got right.*
x *Use an x for those you are guessing on.*

If you get a question right that you marked with a star, you can be confident that you know the associated LO.

If you get a question right that you marked with a question mark, go back and work a few more problems and test yourself again tomorrow.

If you get a question wrong or mark it with an x, go back and read the section again, work a few problems today, a few tomorrow, and a few the next day before testing yourself again.

This will help you to discover which topics you have learned thoroughly and which topics you need to study in more detail.

True–False

1. Isotopes are atoms of the same element that carry a different charge.
2. The mass of the products of a chemical reaction will always be the same as the mass of the reactants.
3. The smallest unit of matter that retains its identity as an element is an atom.
4. Two nonmetals will only ever form one type of compound.
5. Electrons and neutrons are found in the nucleus of an atom, protons in a diffuse cloud outside the nucleus.

6. The atomic number—the number of protons in an element—is what gives an element its identity.
7. All atoms of the same element have the same infinitesimally tiny mass.
8. Ionic compounds are named using prefixes to denote the number of each element.
9. Main-group metals such as sodium can carry a variety of different charges.

Multiple Choice

10. The isotope ${}^{70}_{31}Ga$ contains:
    a. 70 protons, 31 neutrons, and 70 electrons
    b. 31 protons, 70 neutrons, and 31 electrons
    c. 31 protons, 39 neutrons, and 31 electrons
    d. 31 protons, 39 neutrons, and 39 electrons

11. The compound copper(I) selenide is
    a. covalent, with formula CuS
    b. covalent, with formula CuSe
    c. ionic, with formula CuSe
    d. ionic, with formula Cu_2Se

12. The proper configuration of an atom is
    a. neutrons and protons in the nucleus, electrons outside the nucleus.
    b. neutrons and electrons in the nucleus, protons outside the nucleus.
    c. protons and electrons in the nucleus, neutrons outside the nucleus.
    d. none of the above

13. The compound Na_2SO_4 is
    a. covalent, called sodiosulfuric acid
    b. ionic, called sodium sulfate
    c. covalent, called sodium sulfide
    d. ionic, called sodium sulfoxide

14. Covalent bonding involves:
    a. interactions between electrons
    b. formation of molecules
    c. nonmetals bonding to nonmetals
    d. all of the above

15. Ionic bonding requires:
    a. similarly sized atoms
    b. metal and nonmetal atoms
    c. oppositely charged particles
    d. isotopes

16. In a homogenous mixture:
    a. the properties of the mixture are completely different from the properties of the components.
    b. the composition is uniform throughout, and the components retain their separate identities.
    c. the composition is variable, with different proportions in different locations.
    d. only liquids can be involved.

17. Which of the following correctly identifies the subatomic particles?
    a. proton, positive charge, mass of 1 u
    b. neutron, negative charge, mass of 1 u
    c. neutron, no charge, negligible mass
    d. electron, negative charge, mass of 1 u

18. The correct name of N_2O_5 is
    a. nitrogen oxide
    b. oxygen nitrate
    c. dinitrogen pentoxide
    d. nitropentoxygen

19. The correct name of $NaC_2H_3O_2$ is
   a. sodiumcarbohydrooxide
   b. sodium acetate
   c. monosodium monoacetate
   d. none of the above

Fill in the Blanks

20. Elements are made up of _____________, which contain three fundamental particles: ____.

21. The *identity of an element* is defined by the number of _______________ it contains, also called the ______________________; the other particles can vary.

22. When the number of ______________changes, the mass of the element changes. The result is called a(n) _____________. When the number of ______________ changes, the charge changes: the result is known as a(n) ________________.

23. Very few pure elements exist in the world around us; mostly we are surrounded by _______________, which are also considered pure substances, and _______________.

24. When nonmetals combine, the result is a(n)_____________ bond. In this type of bond, electrons are ______________.

25. When a(n) _________________ and a(n)__________________ combine, the result is a(n) _______________ bond. In this type of bond, atoms are held together by opposite ___________.

26. When a covalently bonded group of atoms also carries a charge, it is known as a(n) _________________. Some of the most common of these are the ______________ ion, NO_3^-, the _______________ ion, NH_4^+, and the ______________ ion, SO_4^{2-}.

27. In molecules, the number of each element is denoted using a series of prefixes. One of an element is denoted mono, two is ____________, three is ___________, four is ____________, and so on.

28. When chemical reactions occur, the mass of the ______________ will always be the same as the mass of the products. This is known as the ________________________.

Matching

29. ionic compound	a. compounds containing two elements
30. molecules	b. number of protons in an element
31. atomic number	c. negatively charged subatomic particle
32. polyatomic ions	d. discoverer of atomic nucleus
33. binary compounds	e. bond formed by sharing electrons
34. heterogeneous mixture	f. compound formed by opposite charges
35. structural formula	g. 1/12 the mass of a C-12 atom
36. electrons	h. positively charged ion
37. proton	i. covalently bonded units
38. anion	j. provide information about molecular connectivity
39. atoms	k. uncharged subatomic particle
40. neutron	l. positively charged subatomic particle
41. Ernest Rutherford	m. covalently bonded units that carry a charge
42. isotopes	n. discoverer of the electron
43. J. J. Thomson	o. negatively charged ion
44. covalent bond	p. same element, varying masses
45. cation	q. proportions of components vary within mixture.
46. atomic mass unit	r. smallest particle of any element

Problems

47. Hydrogen gas reacts explosively with oxygen to form water vapor. If an unknown quantity of hydrogen gas is reacted with 3.2 L of oxygen (density, 1.43 g/L), and 5.15 g of water are formed, what volume of hydrogen gas (density, 0.0893 g/L) was in the initial mixture?

48. Phosphorus and oxygen form two compounds, one of which is a white crystalline solid, the other of which is a waxy substance that smells like garlic. In the first compound, 4.3 g phosphorus is combined with 11.4 g oxygen. In the second, 3.34 g phosphorus is combined with 5.33 g oxygen. Demonstrate that these compounds obey the law of multiple proportions.

49. How many protons, neutrons, and electrons are present in the following elements?

a. ${}^{97}_{42}Mo$ b. ${}^{235}_{92}U^{6+}$ c. ${}^{235}_{93}Np$ d. ${}^{202}_{80}Hg^{4+}$

50. Neon has three stable isotopes and an average atomic mass of 20.18 u. If neon-20 has an atomic mass of 19.992 and an abundance of 90.48%, and neon-21 has an atomic mass of 20.993 u and an abundance of 0.27%, what is the abundance and atomic mass of neon-22?

51. Magnesium, Mg, has three naturally occurring isotopes: ^{24}Mg has an abundance of 78.99% and an atomic mass of 23.985 u, ^{25}Mg with an abundance of 10.00% and an atomic mass of 24.986 u, and ^{26}Mg with an abundance of 11.01% and an atomic mass of 25.983 u. Calculate the average atomic mass of magnesium.

52. Naphthalene is the compound that gives mothballs their distinctive smell. Given its structural formula below, what is its chemical formula?

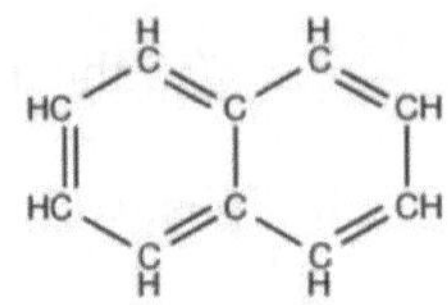

naphthalene

53. Write the formula and name for the compound that forms between each of the following pairs of elements:
a. Li, Cl b. Na, S c. Ca, O d. Mg, N

54. Name the following compounds, and identify them as ionic or covalent:
a. CO b. $SrCl_2$ c. CCl_4 d. PtO_2

55. What is the mass (in grams) of 9500 atoms of gold?

56. Formulas for chemical compounds can be worked out using the percent masses of each component. For the organic compound dichloromethane (CCl_2H_2), what percentage of the total mass is provided by each component?

57. Write a formula for the compound that would form between calcium and the following ions:
a. fluoride b. phosphate c. nitrate d. carbonate

58. Name the following compounds:
a. $SnCl_4$ b. Na_3PO_4 c. $Mg(C_2H_3O_2)_2$ d. HCl
e. Rb_2S f. CO_2 g. $CrPO_4$ h. F_2O
i. P_2O_5 j. $NiCl_2$ k. $Fe(OH)_3$ l. HCl
m. PbI_2 n. $NaNO_3$ o. $BaBr_2$ p. $Al(NO_3)_3$

59. Write formulas for the following compounds:
a. calcium nitrite b. sodium chloride c. dichlorine monoxide
d. dihydrogen monoxide e. nickel(II) hydride f. boron tribromide
g. copper(II) chloride h. xenon tetrafluoride i. nitrogen dioxide
j. barium nitride k. carbon tetrabromide l. phosphorous trichloride
m. lithium permanganate n. sodium phosphite o. sodium hydrogen phosphate

<u>*Challenge Problem*</u>

60. The smallest object that can be resolved by the human eye is about 0.2 mm across. How many atoms does a sphere of osmium metal (density = 22.61g/cm^3) with a diameter of 0.2 mm contain? How many atoms would a sphere of aluminum (density 2.7 g/cm^3) contain? If these atoms were cubes with an of edge length 0.1 mm, what would be the volume of these atoms?

Test Yourself: *Take the Practice Test on pages 73 and 74 of your textbook. Again, use a testing environment using only those items that your instructor will allow you to have access to during the test and mark each question with a star, question mark, or x to denote your confidence level. Once you have finished you can check off the LO at the beginning of this chapter based on which questions you starred and got correct. Go back and spend more time with the topics from the other learning objectives.*

CHAPTER THREE

Mass Relationships in Chemical Reactions

Learning Objectives

As a result of reading and studying this chapter, you should be able to:

- ☐ **LO 3.1** Visualize bonds broken and formed in a chemical reaction and relate numbers of molecules or atoms to the balanced reaction.
- ☐ **LO 3.2** Balance a chemical reaction given the formulas of reactants and products.
- ☐ **LO 3.3** Calculate formula weight, molecular weight, and molar mass given a chemical formula or structure.
- ☐ **LO 3.4** Convert between mass, moles, and molecules or atoms of a substance.
- ☐ **LO 3.5** Relate the amount (moles or mass) of reactants and products in a balanced equation using stoichiometry.
- ☐ **LO 3.6** Calculate the percent yield of a reaction.
- ☐ **LO 3.7** Determine the relative amounts of atoms or molecules in the reactants and products of a balanced reaction given a molecular representation.
- ☐ **LO 3.8** Determine which reactant is limiting and calculate the theoretical yield of the product and the amount of excess reactant.
- ☐ **LO 3.9** Calculate percent yield when one reactant is limiting.
- ☐ **LO 3.10** Calculate the percent composition, given a chemical formula or structure.
- ☐ **LO 3.11** Determine the empirical and molecular formula, given the mass percent composition and molecular weight of a compound.
- ☐ **LO 3.12** Determine the empirical and molecular formula, given combustion analysis data and molecular weight.
- ☐ **LO 3.13** Determine empirical and molecular formulas using data from a mass spectrum and combustion analysis.

Chapter Summary

The central concern of chemistry is the change of one substance into another—chemical reactions are at the heart of the science. In this chapter, you begin your study of chemical reactions by learning how to balance chemical equations. Using balanced chemical equations, along with the concepts of the mole and stoichiometry, you will learn how to calculate the amounts of reactants needed for a reaction and the theoretical amount of products produced in a reaction. These calculations are carried out by using molar masses for the conversion between grams and moles. You will also learn how to determine an empirical formula from the percent composition of a compound and vice versa. Finally, you will see how combustion analysis, a type of elemental analysis, along with mass spectrometry, can be used to determine both the empirical and molecular formula of a compound.

The Chapter in Detail

Section 3.1 Representing Chemistry on Different Levels

PRE-LECTURE:

Summary: *Read through this section and fill in the missing words and equations. Reference the textbook when you are not confident in your answers.*

Chemists try to understand a world that is much too small to see or manipulate. We work with things on a macroscopic level in order to understand what is going on at the atomic scale. Because of this duality, our symbols often represent what is happening on both levels. The equation $2\,H_2 + O_2 \rightarrow 2\,H_2O$ can be read as two hydrogen molecules react with one oxygen molecule to form two molecules of water when referring to the ___________ level OR as two moles of hydrogen react with one mole of oxygen to form two moles of water when referring to the ___________ level.

Section 3.2 Balancing Chemical Equations

PRE-LECTURE:

Learning Objectives: *These are the skills you should be proficient in after reading this section and practicing the associated problems. Check off the boxes below once you have gone through this section, answered the associated questions in your textbook, and feel confident you have mastered the skill.*

☐ **LO 3.1** Visualize bonds broken and formed in a chemical reaction, and relate numbers of molecules or atoms to the balanced reaction.

☐ **LO 3.2** Balance a chemical reaction given the formulas of reactants and products.

Definitions: *To learn, you must do. This is your workbook. Fill in the definitions from the glossary of the textbook as a first step to understanding the material.*

— **balanced equation—** ____

____

— **formula unit—** ____

____

— **coefficient—** ____

____

Review: *Before going through the section make sure you are familiar with these concepts:*

- Ch. 2.4 law of conservation of mass
- Ch. 2.5 Dalton's atomic theory

Summary: *Read through this section and fill in the missing words and equations. Reference the textbook when you are not confident in your answers.*

The law of mass conservation states that the total mass of products and reactants in a chemical reaction must be equal: matter is neither created nor destroyed. Dalton's atomic theory states that, in chemical reactions, atoms are simply rearranged, not altered. According to these rules, the number and kind of each atom must be the same on both sides of a chemical reaction.

To balance an equation:

1. Write an unbalanced equation using the correct chemical ___________ ___________ for each reactant and product.
2. Find suitable ___________—the numbers placed before formulas to indicate how many formula units of each substance are required to balance the equation.
3. Report coefficients to their ___________ whole-number values.
4. ___________ your answer.

Workbook Example 3.1

PROBLEM:
Methane (CH_4) burns in oxygen to produce carbon dioxide and water. Write and balance the equation.

SOLUTION:
Using the rules we established above:

Step 1: Write the unbalanced equation.

$$CH_4 + O_2 \rightarrow CO_2 + H_2O$$

Step 2: Use coefficients to balance the equation. Think about one element at a time. (HINT: It usually helps to save oxygen for last. Balance any element that shows up in more than one reactant or product last.)

Notice that there is one carbon (C) atom in CH_4 and one in CO_2, so carbon is already balanced. There are four hydrogen atoms on the reactant side, but only two hydrogen atoms on the product side (H_2O). Place a 2 in front of H_2O.

$$CH_4 + O_2 \rightarrow CO_2 + 2\ H_2O$$

You now have four hydrogen atoms on each side and one carbon atom on each side; oxygen is still unbalanced: there are two oxygen atoms on the reactant side and four on the product side. Place a 2 in front of O_2.

$$CH_4 + 2\ O_2 \rightarrow CO_2 + 2\ H_2O$$

Step 3: The coefficients are already reduced to their smallest whole-number ratio.

Step 4: Check your answer.

Reactant side	Product side
1 C	1 C
4 O	4 O
4 H	4 H

Workbook Example 3.2

PROBLEM:
Write a balanced equation for the combustion of 1-pentanol ($C_5H_{11}OH$).

SOLUTION:
The term *combustion* means reaction with oxygen. When hydrocarbons (compounds that contain primarily C and H), such as methane and 1-pentanol, undergo combustion, carbon dioxide and water are produced.

Step 1: Write the unbalanced equation.

$$C_5H_{11}OH + O_2 \rightarrow CO_2 + H_2O$$

Step 2: Use coefficients to balance the equation. Begin with carbon. There are five carbon atoms on the reactant side, but only one carbon atom on the product side. Begin by placing a 5 in front of CO_2.

$$C_5H_{11}OH + O_2 \rightarrow 5\ CO_2 + H_2O$$

There are 12 hydrogen atoms on the reactant side—don't be confused by the fact that they are broken up—but only two hydrogen atoms on the product side. Place a 6 in front of H_2O.

$$C_5H_{11}OH + O_2 \rightarrow 5\ CO_2 + 6\ H_2O$$

Now balance oxygen. There are 3 oxygen atoms on the reactant side (don't forget the oxygen atom in methanol) and 16 oxygen atoms on the product side. To get an even number of oxygen atoms will require that the number of pentanol molecules is doubled, and all of the coefficients that were based on it.

$$2\ C_5H_{11}OH + O_2 \rightarrow 10\ CO_2 + 12\ H_2O$$

There are now 4 oxygen atoms on the reactant side and a whopping 32 oxygen atoms on the product side. To balance this number of oxygens on the product side, subtract the 2 oxygen atoms that are associated with the methanol molecules, which leaves 30 oxygen atoms that need balancing. Put a 15 in front of the molecular oxygen (O_2).

$$2\ C_5H_{11}OH + 15\ O_2 \rightarrow 10\ CO_2 + 12\ H_2O$$

Step 3: Reduce the coefficients to their smallest whole-number ratio.

The ratio is 2:15:10:12, which is the smallest whole-number ratio.

Step 4: Check your answer.

Reactant side	Product side
10 C	10 C
24 H	24 H
32 O	32 O

Workbook Problem 3.1

Write a balanced chemical equation for the combustion of hexane (C_6H_{14}).

Strategy: Remember, the term *combustion* is used to indicate reaction with oxygen. When hydrocarbons (compounds that contain primarily C and H) undergo a combustion reaction, carbon dioxide and water are the products.

Step 1: Write the unbalanced chemical equation.

Step 2: Use coefficients to balance the equation. (Remember, it helps to save oxygen for last.)

Step 3: Reduce the coefficients to their smallest whole-number ratio.

Step 4: Check your answer.

POST-LECTURE:

Work Problems: *One of the best ways to learn chemistry is through solving problems. Go to the textbook and work through the Worked Examples for this section, then go to the end of chapter and work the problems there. When you feel confident that you can correctly answer problems like the ones listed below for each objective, check off each of the Learning Objectives for this section above.*

LO 3.1: Worked Example 3.1; Problems 3.29, 3.30
LO 3.2: Worked Example 3.2; Problems 3.28, 3.40

Section 3.3 Molecular Weight and Molar Mass

PRE-LECTURE:

Learning Objectives: *These are the skills you should be proficient in after reading this section and practicing the associated problems. Check off the boxes below once you have gone through this section, answered the associated questions in your textbook, and feel confident you have mastered the skill.*

☐ **LO 3.3** Calculate formula weight, molecular weight, and molar mass given a chemical formula or structure.

☐ **LO 3.4** Convert between mass, moles, and molecules or atoms of a substance.

Definitions: *To learn, you must do. This is your workbook. Fill in the definitions from the glossary of the textbook as a first step to understanding the material.*

— **formula weight—** ____

____

— **molecular weight—**____

____

Review: *Before going through the section make sure you are familiar with these concepts:*

- Ch. 2.9 atomic mass and molar mass

Summary: *Read through this section and fill in the missing words and equations. Reference the textbook when you are not confident in your answers.*

Atoms and molecules are too small to count individually, so in the lab we use mass to easily find the amount of substance we are working with. However, chemical reactions do not occur based on the mass of a substance, but on the ratio of atoms and molecules as represented in the balanced equation. We need a way to go back and forth between the mass of a substance and the moles of a substance represented in the balanced equation. To do this we can use __________ __________ for molecular compounds (with covalent bonds) or the more generalized term __________ __________ for any substance (with ionic or covalent bonds).

To find either the molecular weight or formula weight, multiply the number of each type of atom in one unit by each of the atom's atomic weights. For water, H_2O, there are two atoms of hydrogen so you multiply the atomic weight of hydrogen by two, [2(1.008)], and add that to the number of oxygen atoms times the atomic weight of oxygen [1(15.9994)];

$$2(1.008) + 1(15.9994) = 18.0154 \text{ u/molecule } H_2O$$

Just as there is a relationship between the atomic weight in unified atomic mass units and atomic weight in grams, there is a relationship between the molecular weight or formula weight and molar mass.

1 ____________ = molecular weight (u)
1 ____________ = molecular weight (g)

Using the relationship above where 1 mole = molecular weight (g), we can write the conversion factor molar mass in units of g/mol.

Workbook Example 3.3

PROBLEM:
What is the mass of one mole of hydrogen?

SOLUTION:
The definition of the *mass of one mole of a substance* is the amount of that substance contained in the molecular mass (from the periodic table) in grams. But is that the actual mass?

Hydrogen exists as a diatomic molecule: H_2. The molecular mass of hydrogen from the periodic table is 1.008 u, so the mass of a molecule of hydrogen would be (1.008×2), or 2.016 u. By definition, this means that one mole of hydrogen will have a mass of 2.016 grams.

Just this once, let's do the conversion to see if it comes out the same:

$$\frac{2.016\,u}{1\,molecule} \times \frac{6.022 \times 10^{23}\ molecules}{1\,mole} \times \frac{1.66054 \times 10^{-24}\ g}{1\,u} = \frac{2.016\ g}{1\,mol}$$

Good news: the lengthy conversion from u to moles, with a conversion factor in it that no one wants to memorize, is completely unnecessary. When you need to know the mass of one mole of a substance, you can pull it straight off the periodic table, no conversions necessary.

Moles make it possible to use macroscopic quantities of substances for microscopic purposes: one mole of hydrogen molecules will have exactly the same number of particles as a mole of water, or oxygen, or table salt. If you need to react substances together, you will have the same number of particles if you have the same number of moles, even if the masses are wildly different.

By using the new conversion factor of molar mass along with Avogadro's number, we can convert between grams, moles, and number of atoms or molecules. To go from grams to moles, use molar mass; from moles to atoms or molecules (particles); use Avogadro's number.

grams	Use ____________ as a conversion factor	moles	Use ____________ as a conversion factor	particles

POST-LECTURE:

Work Problems: *One of the best ways to learn chemistry is through solving problems. Go to the textbook and work through the Worked Examples for this section, then go to the end of chapter and work the problems there. When you feel confident that you can correctly answer problems like the ones listed below for each objective, check off each of the Learning Objectives for this section above.*

LO 3.3: Worked Example 3.3; Problems 3.31, 3.44, 3.46
LO 3.4: Worked Example 3.4; Problems 3.48, 3.50, 3.56

Section 3.4 Stoichiometry: Relating Amounts of Reactants and Products

PRE-LECTURE:

Learning Objectives: *These are the skills you should be proficient in after reading this section and practicing the associated problems. Check off the boxes below once you have gone through this section, answered the associated questions in your textbook, and feel confident you have mastered the skill.*

☐ **LO 3.5** Relate the amount (moles or mass) of reactants and products in a balanced equation using stoichiometry.

Definitions: *To learn, you must do. This is your workbook. Fill in the definitions from the glossary of the textbook as a first step to understanding the material.*

— **stoichiometry—**____

____

Summary: *Read through this section and fill in the missing words and equations. Reference the textbook when you are not confident in your answers.*

Stoichiometry is chemical arithmetic. As moles cannot be measured directly in the lab, moles must be converted back to grams for practical use. Thus, it helps chemists to get from the mass of a substance that is weighed out in lab to the expected mass of another substance through the ratios from the balanced chemical equation. Remember that you must always be in units of ___________ to move from one substance to another.

Workbook Example 3.4

PROBLEM:
How many grams of each substance will be required/produced if one mole of hydrogen reacts with one mole of chlorine to produce hydrogen chloride gas by the following unbalanced equation?

$$H_2(g) + Cl_2(g) \rightarrow HCl(g)$$

SOLUTION:
First, the equation must be balanced. This is done by putting a 2 in front of the hydrogen chloride (HCl) gas produced.

$$H_2(g) + Cl_2(g) \rightarrow 2\ HCl(g)$$

One mole of any substance has a mass equal to its molecular mass in grams, so one mole of hydrogen will have a mass of 2.016 g. One mole of chlorine gas will be equal to 2 × 35.45 g/mol, or 70.9 g. Despite this huge mass difference, the number of particles of each substance—hydrogen and chlorine—is the same. The chlorine molecules are just *much* heavier.

To determine how much hydrogen chloride gas is produced, there is a coefficient to be considered. The coefficient tells you that for every mole of hydrogen or chlorine gas, two moles of hydrogen chloride gas are produced. This makes sense, because when you break apart one molecule of hydrogen and one molecule of chlorine, you have four atoms. Redistribute them, and you have made two hydrogen chloride molecules.

So, when we added together one mole of hydrogen and one mole of chlorine, we made two moles of hydrogen chloride. This has a molecular mass of 1.008 + 35.45, or 36.46 g/mol. If there are two moles, there will be 72.9 grams of hydrogen chloride gas.

Take note that once again, the law of mass conservation is at work. On the reactants side, 2.016 g + 70.9 g = 72.9 grams of reactants. On the product side, 72.9 grams of hydrogen chloride gas is produced.

Workbook Example 3.5

PROBLEM:
When concentrated sulfuric acid is added to table sugar, a dehydration reaction occurs according to the following unbalanced equation:

$$C_{12}H_{22}O_{11} + H_2SO_4 \rightarrow C + H_2O + H_2SO_4$$

If 0.50 moles of sucrose are to be reacted with an excess of sulfuric acid, how many grams of sucrose are needed, and how many grams of carbon will result?

SOLUTION:
First, balance the equation:

$$C_{12}H_{22}O_{11} + H_2SO_4 \rightarrow 12\ C + 11\ H_2O + H_2SO_4$$

Molar mass of sucrose:

12 moles carbon	×	12.01 g =	144.12 g/mol
22 moles hydrogen	×	1.008 g =	22.176 g/mol
11 moles oxygen	×	16.00 g =	176.00 g/mol
		Total =	342.30 g/mol

This molar mass also represents a conversion factor of $\frac{342.3\ g}{1\ mol}$ or $\frac{1\ mol}{342.3\ g}$.

To determine the grams of sucrose in 0.50 moles:

$$0.50\ mol \times \frac{342.3\ g}{1\ mol} = 171.15\frac{g}{mol} = 170\frac{g}{mol}$$

It will take a couple of steps to determine the number of grams of carbon that will be produced. First, determine how many moles of carbon will be produced using the ratios from the balanced equation:

$$0.50\ mol\ sucrose \times \frac{12\ mol\ C}{1\ mol\ sucrose} = 6.0\ mol\ C$$

Now convert that number of moles to grams:

$$6.0\ mol\ C \times \frac{12.01\ g}{1\ mol} = 72\ g\ C$$

72 grams of carbon will be produced from the reaction of 0.5 moles of sucrose with sulfuric acid. With practice, it will become possible for you to do this without breaking it up into separate steps. Once you have a balanced equation, you can set up a problem like this to go straight from moles of sucrose to grams of carbon:

$$0.60\ moles\ sucrose \times \frac{12\ mol\ C}{1\ mol\ sucrose} \times \frac{12.01\ g\ C}{1\ mol\ C} = 72\ g\ C$$

Start out by breaking things up into as many steps as you need to when you are starting out, but with practice, you will get faster and be able to do it in one step!

The basic steps needed to complete stoichiometry problems can be summarized like this:

- _______________________: an unbalanced or incorrectly balanced equation will make it impossible to solve a problem correctly.
- _______________________: if you are given grams of one substance and need grams of another substance, you will need the formula weights of both.
- Set up a _______________________ problem. Keep both units and identifiers with all numbers. For example, don't just write 12.01 g/mol, write 12.01 g carbon/mol. You will have numerous conversion factors of this type, so you want to make sure that not only the units, but also the substances cancel out.
- ___________, making sure that your units are correct and that your answer makes sense.

Workbook Problem 3.2

When elemental iron rusts, it combines with oxygen to form iron(III) oxide. How many grams of rust will be produced if 14 grams of metal nails are left out on a construction site all winter? How many grams of oxygen will have been consumed in this reaction?

Strategy: You can calculate the grams of iron(III) oxide produced by using the molar mass of iron(III) oxide and the mole ratio of iron-to-iron(III) oxide from the balanced chemical equation as conversion factors.

g iron → moles iron → moles iron(III) oxide → g iron(III) oxide

Step 1:	Write an unbalanced chemical equation from the information given above.
Step 2:	Balance the equation.
Step 3:	From the balanced chemical equation, determine the mole to mole ratio of iron-to-iron(III) oxide.
Step 4:	Find the atomic mass of iron.
Step 5:	Determine the molecular mass for iron(III) oxide.
Step 6:	From the information in Steps 3, 4, and 5, create conversion factors. Use these conversion factors, and set up a dimensional-analysis problem to convert from grams of iron, to moles of iron, to moles of iron oxide, to grams of iron oxide.
Step 7:	Using the law of mass conservation, how many grams of oxygen were added to the iron?

POST-LECTURE:

Work Problems: *One of the best ways to learn chemistry is through solving problems. Go to the textbook and work through the Worked Examples for this section, then go to the end of chapter and work the problems there. When you feel confident that you can correctly answer problems like the ones listed below for each objective, check off each of the Learning Objectives for this section above.*

LO 3.5: Worked Example 3.5; Problems 3.62, 3.64

Section 3.5 Yields of Chemical Reactions

PRE-LECTURE:

Learning Objectives: *These are the skills you should be proficient in after reading this section and practicing the associated problems. Check off the boxes below once you have gone through this section, answered the associated questions in your textbook, and feel confident you have mastered the skill.*

☐ **LO 3.6** Calculate the percent yield of a reaction.

Definitions: *To learn, you must do. This is your workbook. Fill in the definitions from the glossary of the textbook as a first step to understanding the material.*

— **yield—** ____

____

— **percent yield—** ____

____

Summary: *Read through this section and fill in the missing words and equations. Reference the textbook when you are not confident in your answers.*

When stoichiometric calculations are done, we get an ideal result, known as the ___________ **yield.** Running an experiment will provide the ___________ **yield.** The actual yield will always be less than the theoretical yield. In the workbook problem above, for instance, there will probably be little pockets of unrusted iron within the nails, which would make the final mass lower than you calculated by assuming that every atom of elemental iron became iron oxide.

A report of an experiment is often given in terms of reaction efficiency or **percent yield.** This is calculated very simply:

Percent yield =

Workbook Example 3.6

PROBLEM:
When 15.0 grams of hydrogen react with oxygen, 120 grams of water result. What is the percent yield of this reaction?

SOLUTION:
Write and balance the equation.

$$2\ H_2 + O_2 \rightarrow 2\ H_2O$$

Figure out the atomic or molecular weights of all the relevant components:

H_2	=	2.02 g/mol
H_2O	=	18.02 g/mol

Because you are not given any information of oxygen in the problem, it's assumed that the oxygen is the excess reagent. In that case, you don't need the molecular weight for oxygen.

Set up a dimensional-analysis equation to determine the theoretical yield of the reaction:

$$15.0\ g\ H_2 \times \frac{1\ mol\ H_2}{2.02\ g} \times \frac{2\ mol\ H_2O}{2\ mol\ H_2} \times \frac{18.02\ g\ H_2O}{1\ mol\ H_2O} = 134\ g\ H_2O$$

Now solve for the percent yield:

$$percent\ yield = \frac{actual\ yield}{theoretical\ yield} \times 100\% = \frac{120\ g}{134\ g} \times 100\% = 89.6\%$$

Workbook Example 3.7

PROBLEM:
If burning hydrogen in oxygen to form water generally has a 95% yield, how many grams of water can be expected from the combustion of 23 grams of hydrogen?

SOLUTION:
Use the balanced equation from above, $2\ H_2 + O_2 \rightarrow 2\ H_2O$, and use it to determine the theoretical yield of the equation if 23 grams of hydrogen are used:

$$23\ g\ H_2 \times \frac{1\ mol\ H_2}{2.02\ g\ H_2} \times \frac{2\ mol\ H_2O}{2\ mol\ H_2} \times \frac{18.02\ g\ H_2O}{1\ mol\ H_2O} = 205\ g$$

Now the actual yield can be estimated: the actual yield will be 95% of the theoretical yield:

$$205\ g \times 0.95 = 195\ g$$

Workbook Problem 3.3

The last step in ammonia production is a high-pressure reaction that proceeds according to the following unbalanced reaction:

$$H_2 + N_2 \rightarrow NH_3$$

If 18.0 grams of nitrogen gas are reacted, and 14.6 grams of ammonia result, what is the percent yield?

Strategy: To determine the percent yield, first balance the equation, then determine the theoretical yield of ammonia in the reaction.

Step 1:	Balance the equation.
Step 2:	Calculate the theoretical yield using dimensional analysis.
Step 3:	Calculate the percent yield using the actual yield stated in the problem and the theoretical yield just calculated.

POST-LECTURE:

Work Problems: *One of the best ways to learn chemistry is through solving problems. Go to the textbook and work through the Worked Examples for this section, then go to the end of chapter and work the problems there. When you feel confident that you can correctly answer problems like the ones listed below for each objective, check off each of the Learning Objectives for this section above.*

LO 3.6: Worked Example 3.6; Problems 3.11–3.12, 3.72

Section 3.6 Reactions with Limiting Amounts of Reactants

PRE-LECTURE:

Learning Objectives: *These are the skills you should be proficient in after reading this section and practicing the associated problems. Check off the boxes below once you have gone through this section, answered the associated questions in your textbook, and feel confident you have mastered the skill.*

- ☐ **LO 3.7** Determine the relative amounts of atoms or molecules in the reactants and products of a balanced reaction given a molecular representation.
- ☐ **LO 3.8** Determine which reactant is limiting and calculate the theoretical yield of the product and the amount of excess reactant.
- ☐ **LO 3.9** Calculate percent yield when one reactant is limiting.

Definitions: *To learn, you must do. This is your workbook. Fill in the definitions from the glossary of the textbook as a first step to understanding the material.*

— **limiting reactant—** ____

____

Summary: *Read through this section and fill in the missing words and equations. Reference the textbook when you are not confident in your answers.*

To maximize yields, reactions are often carried out with much more of one reactant than necessary based on the balanced equation. In this situation, there is a **limiting reactant** and an ___________ reactant. The limiting reactant determines the theoretical yield of the reaction: once it is used up, the reaction stops.

Problems that involve a limiting reactant can be complex: proceed with caution, because no information can be obtained from the masses of the reactants: only their mole ratios contain useful information.

To work these problems, convert all reactant quantities to moles and run the reaction mathematically, once with each reactant. The reactant that gives you the fewest moles of product is your limiting reactant, and that amount is the theoretical yield of the reaction.

Workbook Example 3.8

PROBLEM:
15 grams of hydrogen gas and 37 grams of chlorine gas are reacted together to form hydrogen chloride gas. Which is the limiting reactant? How many grams of hydrogen chloride gas can be produced (theoretical yield)? How many grams of the excess reactant will be left over?

SOLUTION: First, write a balanced chemical equation.

$$H_2 + Cl_2 \rightarrow 2\ HCl$$

Determine the number of moles of each reactant present.

$$15\ g\ H_2 \times \frac{1\ mol\ H_2}{2.02\ g\ H_2} = 7.4\ mol\ H_2$$

$$37\ g\ Cl_2 \times \frac{1\ mol\ Cl_2}{71\ g\ Cl_2} = 0.52\ mol\ Cl_2$$

The reactants are required in a 1:1 mole ratio, so in this example it is possible to see that chlorine will be the limiting reactant and hydrogen is the excess reactant despite the fact that by mass there is a great deal more chlorine than hydrogen.

These ratios will not always be this straightforward: a foolproof way to find the limiting and excess reactants is to run the reaction stoichiometry with each reactant and look at the amount of product produced by each:

$$7.4\ mol\ H_2 \times \frac{2\ mol\ HCl}{1\ mol\ H_2} = 14.8\ mol\ HCl$$

$$0.52\ mol\ Cl_2 \times \frac{2\ mol\ HCl}{1\ mol\ Cl_2} = 1.04\ mol\ HCl$$

This method works equally well for simple and complex reactions: the number of moles of chlorine available can produce far less than the number of moles of hydrogen. Chlorine is your limiting reactant, and the theoretical yield of the reaction is 1.04 mol HCl gas.

$$1.04\ mol\ HCl \times \frac{36.5\ g\ HCl}{1\ mol\ HCl} = 38.0\ g\ HCl$$

How many grams of hydrogen will be left over? There are several ways of getting at this information. The most straightforward is to do a quick calculation using the law of mass conservation: we have obtained 38 grams of products from 52 g (15 g + 37 g) of reactants: 14 grams of reactants must be left over. As hydrogen is in the excess, there are 14 grams of hydrogen left over.

Another way to get at this is to use mole quantities: 0.52 moles of chlorine reacted, so you can set a stoichiometric equation to get at the number of moles of hydrogen that were used. This is extreme overkill for the simple ratio here, but is very useful when the ratio is less straightforward:

$$0.52\ mol\ Cl_2 \times \frac{1\ mol\ H_2}{1\ mol\ Cl_2} = 0.52\ mol\ H_2$$

Now that the number of moles of hydrogen used up is known, you can simply subtract from the original number of moles of hydrogen and convert to grams:

$$(7.4\ mol\ H_2 - 0.52\ mol\ H_2) \times \frac{2.02\ g\ H_2}{1\ mol\ H_2} = 14\ g\ H_2$$

Workbook Problem 3.4

Aluminum metal and hydrochloric acid react to make aluminum chloride and hydrogen gas. If 5.3 grams of aluminum are reacted with 9.2 grams of hydrochloric acid in solution, how many grams of aluminum chloride can be obtained? What is the limiting reactant? How much of the excess reactant (in grams) is left over? What is the percent yield of the reaction if 8.3 grams of aluminum chloride are produced?

Strategy: Generate a balanced equation and determine the number of moles available for each reactant. Next, run the reaction mathematically with each of the reactants and see which gives the smallest number of moles of product. This will determine the limiting reactant and the theoretical yield, and make it possible to calculate the percent yield and the amount of excess reactant left over. Next, compare the mole ratio of the actual amounts to the mole ratio required according to the *balanced chemical equation*. You can then determine which reactant is in excess, and which reactant is present in a limiting amount.

Step 1: Using the information given, write an unbalanced chemical equation.

Step 2:	Balance the chemical equation.
Step 3:	Determine the number of moles of each reactant present.
Step 4:	Determine which reactant gives the smallest number of moles of product. This gives the limiting reactant and the theoretical yield.
Step 5:	Determine how many moles of the excess reactant were used.
Step 6:	Determine how many moles of the excess reactant remain.
Step 7:	Determine how many grams of the excess reactant remain.
Step 8:	Determine the percent yield of the reaction.

POST-LECTURE:

Work Problems: *One of the best ways to learn chemistry is through solving problems. Go to the textbook and work through the Worked Examples for this section, then go to the end of chapter and work the problems there. When you feel confident that you can correctly answer problems*

like the ones listed below for each objective, check off each of the Learning Objectives for this section above.

LO 3.7: Worked Example 3.7; Problems 3.13, 3.14, 3.32
LO 3.8: Worked Example 3.8; Problems 3.74, 3.77, 3.84
LO 3.9: Problems 3.80, 3.82

Section 3.7 Percent Composition and Empirical Formulas

PRE-LECTURE:

Learning Objectives: *These are the skills you should be proficient in after reading this section and practicing the associated problems. Check off the boxes below once you have gone through this section, answered the associated questions in your textbook, and feel confident you have mastered the skill.*

- ☐ **LO 3.10** Calculate the percent composition, given a chemical formula or structure.
- ☐ **LO 3.11** Determine the empirical and molecular formula, given the mass percent composition and molecular weight of a compound.

Definitions: *To learn, you must do. This is your workbook. Fill in the definitions from the glossary of the textbook as a first step to understanding the material.*

— **percent composition—** ____

____

— **empirical formula—** ____

____

— **molecular formula—**____

____

Summary: *Read through this section and fill in the missing words and equations. Reference the textbook when you are not confident in your answers.*

Determining the composition of a new compound requires that the elements in the compound and their relative abundances be known. This is expressed as **percent composition.** This can also be calculated for a known compound, if desired.

Percent composition can be converted into mole ratios by figuring out how many moles of each element would be present in 100 grams of the unknown compound. This mole ratio can then be expressed as an ___________ **formula:** the simplest ratio of the elements to one another.

If the molecular mass is known, the empirical formula can be converted into a ___________ **formula:** the actual numbers of each element in a molecule of the compound. Sometimes, the two are the same, but other times the molecular formula will be a whole-number multiple of the empirical formula.

Workbook Example 3.9

PROBLEM:
A simple hydrocarbon is found to be 75% C by mass and 25% H by mass. What is the empirical formula of the compound? If its molecular mass is determined to be 16.04 g/mol, what is its molecular formula?

SOLUTION:
Assume that 100 grams of the compound are present. 75% of 100 g is 75 grams of carbon; 25% of 100 grams is 25 g hydrogen. Convert to moles:

$$75\ g\ C \times \frac{1\ mol\ C}{12\ g\ C} = 6.25\ mol\ C$$

$$25\ g\ H \times \frac{1\ mol\ H}{1\ g\ H} = 25\ mol\ H$$

This makes it possible to find the ratio of one element to the other. To determine the empirical formula, divide to find the smallest ratio. Be sure to keep element names or symbols with the numbers to avoid confusion:

$$\frac{25\ mol\ H}{6.25\ mol\ C} = \frac{4\ mol\ H}{1\ mol\ C}$$

This tells us that there are four hydrogens for every one carbon, which gives an empirical formula of CH_4.

If the unknown hydrocarbon has a molecular mass of 16.04 g/mol, we can also determine the molecular formula of the compound. The empirical formula has a molecular weight of 16.04 g/mol, so in this case, the empirical formula is the same as the molecular formula.

Workbook Example 3.10

PROBLEM:
A compound is found to be 36.8% N and 63.2% O. What is the empirical formula of the compound?

SOLUTION:
Assume that 100 grams of the compound are present. 36.8% of 100 g is 36.8 grams of nitrogen; 63.2% of 100 grams is 63.2 g oxygen. Convert to moles:

$$63.2\ g\ O \times \frac{1\ mol\ O}{16\ g\ O} = 3.95\ mol\ O$$

$$36.8\ g\ N \times \frac{1\ mol\ N}{14\ g\ N} = 2.63\ mol\ N$$

There is no obvious whole-number ratio here, but continue determining the ratio:

$$\frac{3.95\ mol\ O}{2.63\ mol\ N} = \frac{1.5\ mol\ O}{1\ mol\ N}$$

A formula can be written with this, but it will need a little extra work:

$NO_{1.5}$ is not a proper formula, as partial atoms are not allowed. Multiply through to get rid of the fraction (this sometimes takes a little trial and error) and make sure that the formula still is the smallest allowable: N_2O_3 is an allowable formula that maintains the ratio of one-and-a-half oxygen atoms for each nitrogen atom.

Workbook Example 3.11

PROBLEM:
A slightly more complex compound is determined to be 27.0% Na, 16.5% N, and 56.5% O. What is the empirical formula of the compound?

SOLUTION: Assume that 100 grams of the compound are present. This will be 27.0 g Na, 16.5 g N, and 56.5 g O. Convert to moles:

$$27\ g\ Na \times \frac{1\ mol\ Na}{23\ g\ Na} = 1.174\ mol\ Na$$

$$16.5\ g\ N \times \frac{1\ mol\ N}{14\ g\ N} = 1.179\ mol\ N$$

$$56.5\ g\ O \times \frac{1\ mol\ O}{16\ g\ O} = 3.531\ mol\ O$$

These are essentially the same number of moles of Na and N, so there will be the same number of each of these in the product. What remains to determine is the ratio of oxygen to the other elements:

$$\frac{3.531\ mol\ O}{1.179\ mol\ N} = \frac{3\ mol\ O}{1\ mol\ N}$$

So there are three oxygen atoms for every one nitrogen atom and one sodium atom for each nitrogen atom: $NaNO_3$. This is sodium nitrate. The most important thing with these problems is to not confuse the elements with one another.

Workbook Problem 3.5

Determine the empirical formula of a compound that is 18.9% Li, 16.2% C, and 64.9% O.

Strategy: Assume you have a 100-gram sample and convert the percentages of the elements into grams.

Step 1: Convert the grams of each element into moles of each element.

Step 2: Find the mole ratios.

Step 3: If necessary, convert the ratios to whole numbers, and write the empirical formula.

Workbook Example 3.12

PROBLEM:
What is the percent composition of sulfuric acid?

SOLUTION:
The formula for sulfuric acid is H_2SO_4. Convert this into a mass percent by determining the mass of each element and the total molecular mass:

H: 2 × 1.008 g	=	2.0016 g
S; 1 × 32.06 g	=	32.06 g
O: 4 × 16.00 g	=	64.00 g
Molecular mass	=	98.06 g/mol

Mass % of each element:

$$\%H = \frac{2.0016\ g}{98.06\ g} \times 100\% = 2.04\% H$$

$$\%S = \frac{32.06\ g}{98.06\ g} \times 100\% = 32.7\% S$$

$$\%O = \frac{64.00\ g}{98.06\ g} \times 100\% = 65.3\% O$$

Workbook Problem 3.6

Determine the percent composition of sodium hypochlorite.

Strategy: Write the formula for sodium hypochlorite and determine the mole ratio of the elements in the compound.

Step 1: Determine the formula of the compound.

Step 2: Determine the molecular mass and the mass of each component.

Step 3: Determine the percent composition of each component.

POST-LECTURE:

Work Problems: *One of the best ways to learn chemistry is through solving problems. Go to the textbook and work through the Worked Examples for this section, then go to the end of chapter and work the problems there. When you feel confident that you can correctly answer problems like the ones listed below for each objective, check off each of the Learning Objectives for this section above.*

LO 3.10: Problems 3.86, 3.87
LO 3.11: Worked Example 3.9; Problems 3.88, 3.90

Section 3.8 Determining Empirical Formulas: Elemental Analysis

PRE-LECTURE:

Learning Objectives: *These are the skills you should be proficient in after reading this section and practicing the associated problems. Check off the boxes below once you have gone through this section, answered the associated questions in your textbook, and feel confident you have mastered the skill.*

☐ **LO 3.12** Determine the empirical and molecular formula, given combustion analysis data and molecular weight.

Summary: *Read through this section and fill in the missing words and equations. Reference the textbook when you are not confident in your answers.*

As we mentioned before, when a hydrocarbon burns, it produces ___________ and ___________. This useful fact can be used to determine the ratios of the elements to one another: the moles of carbon dioxide and water produced by such a reaction can be used to determine the moles of hydrogen and carbon in the original compound, and thus the empirical formula of the compound.

Determine moles of CO_2—this will be the same as the number of moles of carbon.

Determine moles of H_2O—this will be one-half the number of moles of hydrogen.

If there is another element present, it can be determined by subtracting the masses of carbon and hydrogen (go from moles to grams to determine this) from the total mass of the unknown compound.

If the molecular mass is also known, it is possible to determine not only the empirical formula, but the ___________ formula as well.

Workbook Example 3.13

PROBLEM:
When a 7.03-gram sample of a hydrocarbon is burned, 21.32 grams of carbon dioxide and 10.82 grams of water are produced. If the compound contains only carbon and hydrogen, and its molecular mass is 58.0 g/mol, what is the molecular formula of the compound?

SOLUTION:
Convert the grams of carbon dioxide and water into moles of carbon and hydrogen, then use the ratio to generate an empirical formula. From the molecular mass, develop the molecular formula from the empirical formula if necessary.

$$21.32\ g\ CO_2 \times \frac{1\ mol\ CO_2}{44.01\ g\ CO_2} \times \frac{1\ mol\ C}{1\ mol\ CO_2} = 0.484\ mol\ C$$

$$10.82\ g\ H_2O \times \frac{1\ mol\ H_2O}{18.00\ g\ H_2O} \times \frac{2\ mol\ H}{1\ mol\ H_2O} = 1.202\ mol\ H$$

Calculate the mole ratio of hydrogen to carbon:

$$\frac{1.202\ mol\ H}{0.484\ mol\ C} = \frac{2.5\ H}{1\ C}$$

The initial molecular formula is $C_1H_{2.5}$, which, if we multiply everything by 2, becomes a proper empirical formula, C_2H_5.

In order to get to the actual molecular formula, we need to get a ratio of the molar masses. In the problem, we were told that the molar mass is 58 g/mol, and the molar mass of the empirical formula we just calculated is 29 g/mol. If we get a ratio of the two molar masses

$$\frac{58\ g/mol}{29\ g/mol} = \frac{2}{1}$$

this tells us that we need to multiply the empirical formula by 2 in order to get to the molecular formula. The molecular formula will be C_4H_{10}.

Workbook Example 3.14

PROBLEM:
A compound known as methyl butanoate is one of the compounds responsible for the distinctive smell of pineapple. It contains carbon, hydrogen, and oxygen. When combusted, a 3.73 g sample of this compound produced 8.03 g carbon dioxide and 3.27 g water. If the molecular weight of the compound is 102 g/mol, what is the molecular formula?

SOLUTION:
Convert the grams of carbon dioxide and water into moles of carbon and hydrogen. Convert these back into grams to determine the weight of oxygen in the compound, then determine the number of moles of oxygen. Use these three quantities to develop mole ratios and an empirical formula. Compare the empirical formula weight and the molecular formula weight to determine the molecular formula.

$$8.03\ g\ CO_2 \times \frac{1\ mol\ CO_2}{44.01\ g\ CO_2} \times \frac{1\ mol\ C}{1\ mol\ CO_2} \times \frac{12.01\ g\ C}{1\ mol\ C} = 2.19\ g\ C$$

$$3.27\ g\ H_2O \times \frac{1\ mol\ H_2O}{18.00\ g\ H_2O} \times \frac{2\ mol\ H}{1\ mol\ H_2O} \times \frac{1.008\ g\ H}{1\ mol\ H} = 0.366\ g\ H$$

Now determine the amount of O by subtracting the masses of carbon and hydrogen from the mass of the starting sample:

$$3.73\ g\ compound - 2.19\ g\ C - 0.37\ g\ H = 1.17\ g\ O$$

Convert the mass of O to moles of O.

$$1.17\ g\ O \times \frac{1\ mol\ O}{16.00\ g\ O} = 0.0731\ mol\ O$$

In order to calculate the mole ratios, we also need the moles of hydrogen and carbon:

$$0.366\ g\ H \times \frac{1\ mol\ H}{1.008\ g\ H} = 0.363\ mol\ H$$

$$2.19\ g\ C \times \frac{1\ mol\ C}{12.01\ g\ C} = 0.182\ mol\ C$$

We can now determine the mole ratios between carbon, hydrogen, and oxygen. Because oxygen has the least amount of moles, we're going to use that as our denominator, so that we always have one mole of oxygen:

$$\frac{0.363\ mol\ H}{0.0731\ mol\ O} = \frac{5\ mol\ H}{1\ mol\ O}$$

$$\frac{0.182\ mol\ C}{0.0731\ mol\ O} = \frac{2.5\ mol\ C}{1\ mol\ O}$$

This gives a preliminary formula of $C_{2.5}H_5O$ and a real empirical formula of $C_5H_{10}O_2$. The empirical formula has a molecular mass of 102 g/mol, so this is also the molecular formula.

Workbook Problem 3.7

2-butene-1-thiol is one of the compounds that gives skunk spray such a memorable odor. It contains only carbon, hydrogen, and sulfur. When 2.85 grams of this compound were burned (over a weekend, to make sure that all the custodial staff didn't quit), 5.70 grams of carbon dioxide and 2.31 grams of water were produced. If the compound has a molecular mass of 88 g/mol, determine the empirical and molecular formulas of this compound.

Strategy: Convert the grams of carbon dioxide and water into moles of carbon and hydrogen. Convert these back into grams to determine the weight of sulfur in the compound, then determine the number of moles of sulfur. Use these three quantities to develop mole ratios and an empirical formula. Compare the empirical formula weight and the molecular formula weight to determine the molecular formula.

Step 1: Find the molar amounts of C and H in CO_2 and H_2O.

Step 2: Carry out mole-to-gram conversions to find the number of grams of C and H in the original sample.

Step 3: Subtract the masses of C and H from the mass of the starting sample to determine the mass of S.

Step 4: Convert the mass of S to moles of S.

Step 5: Find the mole ratios.

Step 6: Use the ratios above to write the empirical formula of the compound.

Step 7: Compare the empirical formula mass to the molecular formula weight to see if a conversion is needed.

Step 8: Write the molecular formula.

POST-LECTURE:

Work Problems: *One of the best ways to learn chemistry is through solving problems. Go to the textbook and work through the Worked Examples for this section, then go to the end of chapter and work the problems there. When you feel confident that you can correctly answer problems like the ones listed below for each objective, check off each of the Learning Objectives for this section above.*

LO 3.12: Worked Example 3.10; Problems 3.92, 3.94, 3.100

Section 3.9 Determining Molecular Weights: Mass Spectrometry

PRE-LECTURE:

Learning Objectives: *These are the skills you should be proficient in after reading this section and practicing the associated problems. Check off the boxes below once you have gone through this section, answered the associated questions in your textbook, and feel confident you have mastered the skill.*

☐ **LO 3.13** Determine empirical and molecular formulas using data from a mass spectrum and combustion analysis.

Review: *Before going through the section make sure you are familiar with these concepts:*

- Ch. 2.10 mass spectrometry

Summary: *Read through this section and fill in the missing words and equations. Reference the textbook when you are not confident in your answers.*

Mass spectrometry works the same whether determining atomic weights or molecular weights. A sample is vaporized and ___________ before electric and ___________ forces are used to separate the resulting ions by their ___________ ratio, which allows them to be quantified. The resulting mass spectrum may be a little more complicated than that for a sample of monatomic ions, but it is usually the peak with the highest intensity that represents the ionized molecule.

Data from elemental analysis can be combined with the data from the mass spectrometer to determine the molecular formula of the compound.

POST-LECTURE:

Work Problems: *One of the best ways to learn chemistry is through solving problems. Go to the textbook and work through the Worked Examples for this section, then go to the end of chapter and work the problems there. When you feel confident that you can correctly answer problems like the ones listed below for each objective, check off each of the Learning Objectives for this section above.*

LO 3.13: Worked Example 3.11; Problems 3.106, 3.107

Self-Test

This section is intended to test your knowledge of the material covered in this chapter. Think through these problems and make certain you understand what the problem is about—ask yourself if the answer makes sense. Many of these questions are linked to the chapter learning goals. Therefore, successful completion of these problems indicates that you have achieved the learning goals for this chapter. You will receive the greatest benefit from this section if you use it as a mock exam. Do not use your notes, book, Internet, or friends when taking the test. There is no grade and you only shortchange yourself by getting help with this activity. As you take the test, mark the problems with your level of confidence with the answer.

★ *Use a star for problems you know you got right.*
? *Use a question mark for those you think you got right.*
x *Use an x for those you are guessing on.*

If you get a question right that you marked with a star, you can be confident that you know the associated LO.

If you get a question right that you marked with a question mark, go back and work a few more problems and test yourself again tomorrow.

If you get a question wrong <u>or</u> mark it with an x, go back and read the section again, work a few problems today, a few tomorrow, and a few the next day before testing yourself again.

This will help you to discover which topics you have learned thoroughly and which topics you need to study in more detail.

<u>True–False</u>

1. When balancing equations, it is essential that there are the same number of molecules on each side.

2. Chemical equations describe behavior on both the atomic (microscopic) level and the bulk (macroscopic) level.

3. A mole of two different substances contains the same number of particles.

4. The coefficients in a balanced chemical equation show the mass ratios of the reactants and products.

5. Theoretical yields of chemical reactions are determined in the experimental laboratory.

6. When reactants are not present in stoichiometric ratios, the limiting reactant will determine the maximum yield of the reaction.

7. Combustion analysis can be used to determine the molecular formula of an unknown compound.

8. The empirical formula and the molecular formula for a compound are sometimes the same.

<u>Matching</u>

9. balanced	a. sum of all the masses of the atoms in a molecule
10. yield	b. the smallest ratio of elements in a compound
11. mole	c. the amount of product formed in a reaction
12. coefficient	d. the number in front of a reactant or product that shows its mole ratio with the other components
13. stoichiometry	e. reaction with oxygen
14. molecular mass	f. indication of the number of each type of atom in a molecule

15. empirical formula

16. subscript

17. combustion

18. actual yield

g. having the same number of all atom types in both products and reactants

h. experimentally determined amount of product in a reaction.

i. the amount of a substance equal to the molecular or atomic mass in grams

j. the use of mole ratios and molecular weights to carry out chemical calculations

Fill in the Blanks

19. To be useful in making calculations, a chemical equation must be ______________.

20. The mole ratios of the components in a reaction are indicated by ______________________ in front of the components. These can be changed if necessary, but the ________________ within the components cannot be changed, because it will change the identity of the component.

21. To convert amounts of substances from grams to moles, the conversion factor is __________ ________.

22. If reactants are not present in stoichiometric ratios, the amount of product will be determined by the ________________ ________________.

23. The reactant that does not determine the yield of the reaction is known as the ___________ ______________.

24. The _____________________ __________ of a reaction can be calculated using a balanced equation, but it is rarely, if ever, achieved. In the lab, the ____________ __________ will be obtained. The two together can be used to calculate the ______________ __________ of the reaction.

25. It is possible to determine the _________________ __________________ of a compound from the mass percentages of the elements in it, but to determine the _________________ ______________ of the compound, the _________________ ________ must be known.

26. When a hydrocarbon is burned in oxygen, the primary products are ____________ _____________ and _________.

27. The number of particles in one mole of a substance is known as ________________ __________. The number is ______________.

Problems

28. Balance the following equations:
    a. $N_2 + O_2 \rightarrow N_2O_5$
    b. $C_3H_8 + O_2 \rightarrow CO_2 + H_2O$
    c. $H_3PO_4 + NaOH \rightarrow Na_3PO_4 + H_2O$
    d. $Ba(NO_3)_2 + NaOH \rightarrow Ba(OH)_2 + NaNO_3$

29. Determine the molecular or formula mass of the following compounds, and convert the given amounts to moles:
    a. 45 g calcium chloride
    b. 23 g sucrose ($C_{12}H_{22}O_{11}$)
    c. 67 g phosphorus trifluoride
    d. 5 g ammonium carbonate

30. Using the balanced equation for the production of dinitrogen pentoxide (the answer for problem 28a): if 4.0 moles of nitrogen and 10.0 moles of oxygen are combined, what is the theoretical yield of the reaction in moles and grams? If 376 grams of product are obtained, what is the percent yield?

31. When sodium hydrogen carbonate is heated, it decomposes according to the following unbalanced equation:

$$NaHCO_3 \rightarrow Na_2CO_3 + CO_2 + H_2O$$

    If 20.0 grams of sodium hydrogen carbonate are heated, what is the theoretical yield of water in grams? If the reaction has a percent yield of 87%, what is the maximum quantity of water that can be obtained?

32. When 6.7 grams of methane (CH_4) are burned in 20.0 grams of oxygen, what is the limiting reactant? What is the theoretical yield of water and carbon dioxide? How many grams of the excess reactant will remain when the reaction is complete?

33. Determine the percent composition of the following compounds:
    a. carbon dioxide
    b. sodium phosphate
    c. acetic acid, CH_3COOH
    d. adenosine triphosphate: $C_{10}H_{16}N_5O_{13}P_3$

34. A compound that contains 40.0% carbon, 6.7% hydrogen, and 53.3% oxygen has a molecular mass of 180 g/mol. What is the molecular formula of this compound?

35. Isopentyl acetate is a compound responsible for the distinctive smell of bananas. It contains only carbon, hydrogen, and oxygen. When 1.57 grams of this compound were subjected to combustion analysis, 3.71 grams of CO_2 and 1.51 grams of H_2O were measured. What is the empirical formula of this compound?

Test Yourself: *Take the Practice Test on pages 108 and 109 of your textbook. Again, use a testing environment using only those items that your instructor will allow you to have access to during the test and mark each question with a star, question mark, or x to denote your confidence level. Once you have finished you can check off the LO at the beginning of this chapter based on which questions you starred and got correct. Go back and spend more time with the topics from the other learning objectives.*

CHAPTER FOUR

Reactions in Aqueous Solution

Learning Objectives

As a result of reading and studying this chapter, you should be able to:

- ☐ **LO 4.1** Calculate the molarity of a solution given the mass of solute and total volume.
- ☐ **LO 4.2** Calculate the amount of solute in a given volume of solution with a known molarity.
- ☐ **LO 4.3** Describe the proper technique for preparing solutions of known molarity.
- ☐ **LO 4.4** Calculate the concentration of a solution that has been diluted.
- ☐ **LO 4.5** Describe the proper technique for diluting solutions.
- ☐ **LO 4.6** Classify a substance as a strong, weak, or nonelectrolyte. (Table 4.1)
- ☐ **LO 4.7** Calculate the concentration of ions in a strong electrolyte solution.
- ☐ **LO 4.8** Classify a reaction as a precipitation, acid–base neutralization, or oxidation–reduction (redox) reaction.
- ☐ **LO 4.9** Write a net ionic equation and identify spectator ions given the molecular equation.
- ☐ **LO 4.10** Use the solubility guidelines to predict the solubility of an ionic compound in water.
- ☐ **LO 4.11** Predict whether a precipitation reaction will occur and write the ionic and net ionic equations.
- ☐ **LO 4.12** Convert between name and formula for an acid.
- ☐ **LO 4.13** Classify acids as strong or weak based on the molecular picture of dissociation.
- ☐ **LO 4.14** Write the ionic equation and net ionic equation for an acid–base neutralization reaction.

Chapter Summary

Reactions often occur in solution. Water is ubiquitous in nature and so many reactions occur in aqueous solution, which means that a solid has been dissolved in water. In this chapter, we begin by describing how we measure the concentration of a solution by defining the unit *molarity* and then keep going by applying molarity to stoichiometric problems. We then move from the quantitative to the qualitative world to talk about three different types of chemical reactions in aqueous solution: precipitation reactions, acid–base reactions, and oxidation–reduction reactions. These reactions are often written as net ionic equations. Therefore, it is necessary to know the difference between strong and weak electrolytes and nonelectrolytes. You can predict the outcome of each of these kinds of reactions if you know the solubility rules, you can recognize acids and bases, and you know how to assign oxidation numbers to compounds. Finally, you will apply the concepts you learned in earlier chapters about stoichiometry to the reactions that are discussed in this chapter.

The Chapter in Detail

Section 4.1 Solution Concentration: Molarity

PRE-LECTURE:

Learning Objectives: *These are the skills you should be proficient in after reading this section and practicing the associated problems. Check off the boxes below once you have gone through this section, answered the associated questions in your textbook, and feel confident you have mastered the skill.*

- ☐ **LO 4.1** Calculate the molarity of a solution given the mass of solute and total volume.
- ☐ **LO 4.2** Calculate the amount of solute in a given volume of solution with a known molarity.
- ☐ **LO 4.3** Describe the proper technique for preparing solutions of known molarity.

Definitions: *To learn, you must do. This is your workbook. Fill in the definitions from the glossary of the textbook as a first step to understanding the material.*

— **molarity (M)**— ____

____

— **solute**— ____

____

Summary: *Read through this section and fill in the missing words and equations. Reference the textbook when you are not confident in your answers.*

The standard unit for reporting concentration in chemistry is molarity. *Molarity* is defined as the moles of solute per liter of solution. When written in equation form, it can also be rearranged, as needed, to be used as a conversion factor:

Molarity =

To make a solution of a specific molarity, the moles of solute needed must be converted to grams so that the proper amount can be weighed out on a balance. This amount of solute is then placed in a properly sized volumetric flask, and the flask is partially filled with water. Once the solute is dissolved completely, solvent is added up to a calibration line to reach the desired volume of solution. The reasons for this procedure will be demonstrated in the following examples.

Workbook Example 4.1

PROBLEM:
How many grams of sucrose (MW = 342.3 g/mol) are needed to make 0.500 L of a 0.50 M solution? How would you make this solution?

SOLUTION:
Molarity refers to moles, not grams, so it is necessary to first figure out how many moles of sucrose are needed, then convert that number to grams:

$$moles = 0.50\ M \times 0.500\ L = 0.25\ mol$$

Notice that the volume must be converted to liters. The reason for this is more obvious if molarity is expanded to moles/liter:

$$moles = \frac{0.50\ mol}{1\ L} \times 0.500\ L = 0.25\ mol$$

Always be *very careful* with units. They matter!

Now convert to grams:

$$0.25\ mol\ sucrose \times \frac{342.3\ g\ sucrose}{1\ mol\ sucrose} = 85.6\ g\ sucrose$$

To make up this solution, put 85.6 grams of sucrose in a 500 mL volumetric flask (remember that 0.5 L and 500 mL are the same), add enough water to dissolve the sucrose, and then fill the volumetric flask to 500 mL after the solute has been completely dissolved.

Note: 85.6 grams of sucrose is about a half-cup of sugar, and 500 mL is about two cups of water. If the solution were to be made by dropping the sugar into 500 mL of water, the volume of the final solution would be far more than 500 mL, and the concentration would be uncertain. This is why volumetric glassware is used to make these solutions.

Workbook Example 4.2

PROBLEM:
How many milliliters of 0.40 M salt solution contain 0.112 moles of salt?

SOLUTION:
The identity of the chemical compound here generically referred to as "salt" is not given, nor is it required. Don't let that throw you off!

Since you know the amount of moles and the molarity of the solution, we will use the definition of *molarity* as a conversion factor:

$$0.112\ mol\ salt \times \frac{1\ L}{0.40\ mol} = 0.28\ L$$

We write the molarity expression with L on top and mol on the bottom so the moles will cancel out.

To convert liters to milliliters, remember that there are 1000 milliliters in 1 liter of solution:

$$0.28\ L \times \frac{1000\ mL}{1\ L} = 280\ mL$$

Workbook Example 4.3

PROBLEM:
What is the molarity of a solution of barium chloride made by dissolving 15 grams of the salt into 250 mL of solution?

SOLUTION:
First, determine the number of moles of barium chloride in 15 grams:

$$15\ g\ BaCl_2 \times \frac{1\ mol\ BaCl_2}{208.3\ g\ BaCl_2} = 0.072\ mol\ BaCl_2$$

The molarity of the solution can now be calculated:

$$\frac{0.072\ mol\ BaCl_2}{0.250\ L} = 0.29\ M$$

Workbook Problem 4.1

How many grams of sodium chloride are needed to make 500 milliliters of a 0.30 M sodium chloride solution?

Strategy: Using the definition of *molarity* and the volume of solution, determine moles and grams of solute.

Step 1: First, determine the number of moles found in 500 mL of a 0.30 M solution.

Step 2: Determine the number of grams of sodium chloride from the number of moles.

POST-LECTURE:

Work Problems: *One of the best ways to learn chemistry is through solving problems. Go to the textbook and work through the Worked Examples for this section, then go to the end of chapter and work the problems there. When you feel confident that you can correctly answer problems like the ones listed below for each objective, check off each of the Learning Objectives for this section above.*

LO 4.1: Worked Example 4.1; Problems 4., 4.2, 4.50, 4.51
LO 4.2: Worked Example 4.2; Problems 4.3–4.4, 4.46, 4.53
LO 4.3: Problems 4.55, 4.56

Section 4.2 Diluting Concentrated Solutions

PRE-LECTURE:

Learning Objectives: *These are the skills you should be proficient in after reading this section and practicing the associated problems. Check off the boxes below once you have gone through this section, answered the associated questions in your textbook, and feel confident you have mastered the skill.*

- ☐ **LO 4.4** Calculate the concentration of a solution that has been diluted.
- ☐ **LO 4.5** Describe the proper technique for diluting solutions.

Summary: *Read through this section and fill in the missing words and equations. Reference the textbook when you are not confident in your answers.*

Some chemicals are sold as concentrated solutions (often called *stock solutions*) that must be diluted before use. Remember, when a solution is diluted, the amount of *solute* does not change, only the amount of *solvent* does. This leads to the equation:

$$M_iV_i =$$

Written out, this means the initial molarity times the initial volume equals the final molarity times the final volume.

A word of caution: when performing dilutions in the lab, you need to be cautious to always add strong acids to water, and not the other way around. When acids are diluted, they generate large quantities of heat: if water is added to acid, the heat can cause the water to boil and splatter the acid or cause the glass to crack. Be careful!

Workbook Example 4.4

PROBLEM:
A lab protocol asks for 100. mL of 5.0 M HCl. If the stock solution is 12 M, how much is needed for the dilution, and how would you perform it?

SOLUTION: Use the above equation, rearrange it to solve for the initial volume:

$$M_iV_i = M_fV_f$$

$$V_i = \frac{M_fV_f}{M_i}$$

$$V_i = \frac{5.0\ M \times 100\ mL}{12\ M} = 42\ mL$$

Note: No conversion was done of the volumes here. When this equation is used, molarity cancels out, leaving behind whatever volume unit is being used. If this makes you uncomfortable, by all means convert everything to liters, but it isn't strictly necessary.

To make this solution, 42 mL of acid would be carefully measured, then poured slowly into a volumetric flask containing about 40 mL of water. Once the acid has been diluted by being added to water, solvent is added to make up to volume.

Workbook Problem 4.2

What is the final concentration when 50 mL of 10 M magnesium chloride is diluted to 1.0 L?

Strategy: Remember that the number of moles of solute do not change when a solution is diluted.

Step 1: Rearrange the dilution formula to solve for the final concentration.

POST-LECTURE:

Work Problems: *One of the best ways to learn chemistry is through solving problems. Go to the textbook and work through the Worked Examples for this section, then go to the end of chapter and work the problems there. When you feel confident that you can correctly answer problems like the ones listed below for each objective, check off each of the Learning Objectives for this section above.*

LO 4.4: Worked Example 4.3; Problems 4.5, 4.6, 4.59, 4.60
LO 4.5: Problems 4.60, 4.61

Section 4.3 Electrolytes in Aqueous Solution

PRE-LECTURE:

Learning Objectives: *These are the skills you should be proficient in after reading this section and practicing the associated problems. Check off the boxes below once you have gone through this section, answered the associated questions in your textbook, and feel confident you have mastered the skill.*

☐ **LO 4.6** Classify a substance as a strong, weak, or nonelectrolyte. (Table 4.1)
☐ **LO 4.7** Calculate the concentration of ions in a strong electrolyte solution.

Definitions: *To learn, you must do. This is your workbook. Fill in the definitions from the glossary of the textbook as a first step to understanding the material.*

— **electrolyte**—____

____

— **nonelectrolyte**— ____________________________________

____

— **dissociate**— _______________________________________

____

— **strong electrolyte—**____

____

— **weak electrolyte—**____

____

Summary: *Read through this section and fill in the missing words and equations. Reference the textbook when you are not confident in your answers.*

Substances that dissolve in water to form ___________ are known as **electrolytes;** solutions that contain these compounds will conduct electricity. Electrolytes are soluble ionic compounds such as sodium chloride (table salt). Acids and bases are also electrolytes, because they dissociate to form ions—in particular, the ions that distinguish them as being acids (H^+) or bases (OH^-).

Nonelectrolytes may dissolve in water, but they will not conduct electricity when dissolved because they do not dissociate into ions. Covalent molecules such as sugars do not dissociate when they dissolve.

Electrolytes can be either **strong** or **weak:** this is defined simply by how much they dissociate. That is, a ___________ **electrolyte** dissolves almost completely and a ___________ electrolyte dissolves very little. By extension, a strong acid dissociates completely and a weak acid only partially.

Dissociation is a dynamic process: if a weak acid is dissolved in water, there will be both dissociated ions and whole molecules present. The weak acid will conduct electricity to a certain extent. In the weak acid solution, whole molecules dissociate and dissociated ions reform constantly in equilibrium between the forward and reverse reactions. The process is represented by a double arrow in the following equation for the dissociation of a weak acid:

$$CH_3COOH \rightleftharpoons CH_3COO^- + H^+$$

Nonelectrolytes dissolve in water, but will ___________ dissociate into ions.

Workbook Example 4.5

PROBLEM:
$Mg(OH)_2$ is a strong base. What is the concentration of OH^- ions in a 0.450 M solution of $Mg(OH)_2$?

SOLUTION:
A strong electrolyte dissociates completely, so $Mg(OH)_2$ will not be found as units, but only as dissociated ions:

$$Mg(OH)_2 \rightarrow Mg^{2+} + 2OH^-$$

This makes it possible to calculate the concentration of OH^- ions directly:

$$\frac{0.450\ mol\ Mg(OH)_2}{1\ L} \times \frac{2\ mol\ OH^-}{1\ mol\ Mg(OH)_2} = \frac{0.900\ mol\ OH^-}{1\ L} = 0.900\ M\ OH^-$$

Workbook Problem 4.3

Determine the total molar concentration of ions in a 0.75 M solution of aluminum chloride.

Strategy: $AlCl_3$ is a strong electrolyte and completely dissociates in water.

Step 1: Determine the total number of moles of ions formed when $AlCl_3$ completely dissociates in water.

Step 2: Create a conversion factor that relates the total number of moles of ions in solution to 1 mol of $AlCl_3$.

Step 3: Use the conversion factor to calculate the molar concentration of ions in solution.

POST-LECTURE:

Work Problems: *One of the best ways to learn chemistry is through solving problems. Go to the textbook and work through the Worked Examples for this section, then go to the end of chapter and work the problems there. When you feel confident that you can correctly answer problems like the ones listed below for each objective, check off each of the Learning Objectives for this section above.*

LO 4.6: Problems 4.62, 4.63
LO 4.7: Worked Example 4.4; Problems 4.7, 4.8, 4.68, 4.70

Section 4.4 Types of Chemical Reactions in Aqueous Solutions

PRE-LECTURE:

Learning Objectives: *These are the skills you should be proficient in after reading this section and practicing the associated problems. Check off the boxes below once you have gone through this section, answered the associated questions in your textbook, and feel confident you have mastered the skill.*

☐ **LO 4.8** Classify a reaction as a precipitation, acid–base neutralization, or oxidation–reduction (redox) reaction.

Definitions: *To learn, you must do. This is your workbook. Fill in the definitions from the glossary of the textbook as a first step to understanding the material.*

— **precipitation reaction—** ____

____

— **acid–base neutralization reaction—** ____________________________________

____

— **oxidation–reduction (redox) reaction—** _________________________________

____

Summary: *Read through this section and fill in the missing words and equations. Reference the textbook when you are not confident in your answers.*

There are three types of reactions we're concerned with in this chapter: precipitation reactions, acid–base neutralization reactions, and oxidation–reduction reactions.

In **precipitation reactions,** two soluble components interact to form an ___________ product, which then falls out of solution.

In **acid–base neutralization reactions,** an acid and a base react to form ___________ and a ___________ (*salt* is simply a generic term that refers to an ionic compound). In acid–base reactions, both the acidic and basic components of the mixture (H^+ and OH^-) are removed from the reaction solution when they combine to form a salt in neutral water.

In **oxidation–reduction reactions,** also called **redox** reactions, ___________ are transferred between components in a reaction mixture. You can recognize redox reactions by looking for elements on one side of the reaction and their ions, sometimes written as ionic compounds, on the other side of the reaction.

POST-LECTURE:

Work Problems: *One of the best ways to learn chemistry is through solving problems. Go to the textbook and work through the Worked Examples for this section, then go to the end of chapter and work the problems there. When you feel confident that you can correctly answer problems like the ones listed below for each objective, check off each of the Learning Objectives for this section above.*

LO 4.8: Problems 4.72, 4.73

Section 4.5 Aqueous Reactions and Net Ionic Equations

PRE-LECTURE:

Learning Objectives: *These are the skills you should be proficient in after reading this section and practicing the associated problems. Check off the boxes below once you have gone through this section, answered the associated questions in your textbook, and feel confident you have mastered the skill.*

☐ **LO 4.9** Write a net ionic equation and identify spectator ions given the molecular equation.

Definitions: *To learn, you must do. This is your workbook. Fill in the definitions from the glossary of the textbook as a first step to understanding the material.*

— **molecular equation—** ____

____

— **ionic equation—** ____

____

— **spectator ion—** ____

____

— **net ionic equation—** ____

____

Summary: *Read through this section and fill in the missing words and equations. Reference the textbook when you are not confident in your answers.*

In order to focus on the chemistry that occurs in a reaction, chemists generally break down the molecular equation into a net ionic equation.

__________ **equations** are written as if the components are not dissociated; ions are kept together as neutral compounds.

__________ **equations** take the molecular equations and break the soluble ionic compounds into ions. Solids, liquids, and gases are not broken down in solution and thus are not broken down.

__________ **equations** are written so that only the ions that participate in the reaction are shown. Generally, ionic reactions contain ions that do not participate, but are unchanged as the reaction proceeds. These are known as **spectator ions;** they are removed when a net ionic equation is written.

Workbook Example 4.6

PROBLEM:
When solutions of lead(II) nitrate and sodium chloride are mixed, solid lead chloride and aqueous sodium nitrate are produced. Write molecular, ionic, and net ionic equations for this reaction.

SOLUTION:
Molecular Equation:

$$Pb(NO_3)_2(aq) + NaCl(aq) \rightarrow PbCl_2(s) + NaNO_3(aq)$$

Balanced Molecular Equation:

$$Pb(NO_3)_2(aq) + 2\ NaCl(aq) \rightarrow PbCl_2(s) + 2\ NaNO_3(aq)$$

Ionic Equation:

$$Pb^{2+}(aq) + 2\,NO_3^-(aq) + 2\,Na^+(aq) + 2\,Cl^-(aq) \rightarrow PbCl_2(s) + 2\,Na^+(aq) + 2\,NO_3^-(aq)$$

Spectator Ions:

$$Na^+ \text{ and } NO_3^-$$

Net Ionic Equation:

$$Pb^{2+}(aq) + 2\,Cl^-(aq) \rightarrow PbCl_2(s)$$

Workbook Problem 4.4

When solutions of nitric acid and potassium hydroxide are mixed, a neutralization reaction occurs. Write the net ionic equation for this reaction.

Strategy: From the information given, write a molecular equation and determine if any of the reactants or products is a strong electrolyte.

Step 1: Write and balance the molecular equation.

Step 2: Write the ionic equation (from Step 1).

Step 3: List the spectator ions in the complete ionic equation.

Step 4: Write the net ionic equation.

POST-LECTURE:

Work Problems: *One of the best ways to learn chemistry is through solving problems. Go to the textbook and work through the Worked Examples for this section, then go to the end of chapter and work the problems there. When you feel confident that you can correctly answer problems like the ones listed below for each objective, check off each of the Learning Objectives for this section above.*

LO 4.9: Worked Example 4.5; Problems 4.9, 4.10, 4.74

Section 4.6 Precipitation Reactions and Solubility Guidelines

PRE-LECTURE:

Learning Objectives: *These are the skills you should be proficient in after reading this section and practicing the associated problems. Check off the boxes below once you have gone through this section, answered the associated questions in your textbook, and feel confident you have mastered the skill.*

- ☐ **LO 4.10** Use the solubility guidelines to predict the solubility of an ionic compound in water.
- ☐ **LO 4.11** Predict whether a precipitation reaction will occur and write the ionic and net ionic equations.

Definitions: *To learn, you must do. This is your workbook. Fill in the definitions from the glossary of the textbook as a first step to understanding the material.*

— **solubility—**____

____

Summary: *Read through this section and fill in the missing words and equations. Reference the textbook when you are not confident in your answers.*

Being able to predict the results of precipitation reactions requires an understanding of **solubility.** Ionic compounds that are highly soluble will not precipitate, or fall out of solution, while those with low solubility will fall out of solution.

Solubility guidelines are called that because they are only partially predictive; however, these guidelines will get you through most problems that have to do with precipitation reactions. Summarize the solubility rules below:

Soluble compounds without common exceptions:

-
-
-
-
-

Soluble compounds and their common exceptions (making them insoluble):

-
-

Insoluble compounds and their common exceptions (making them soluble):

-
-
-
-

Workbook Example 4.7

PROBLEM:
For which of the following combinations of solutions would precipitations occur?
Na_2CO_3 and $Pb(NO_3)_2$, NH_4Cl and $NaCH_3CO_2$, $AgNO_3$ and $MgCl_2$.

SOLUTION:
Exchange the cations and anions and examine the possible products.

Na_2CO_3 and $Pb(NO_3)_2$ can yield $NaNO_3$ and $PbCO_3$.

Anything attached to lead is a suspect for precipitation, and $NaNO_3$ contains two ions identified as always soluble. $PbCO_3$ will precipitate.

NH_4Cl and $NaCH_3CO_2$ can yield $NH_4CH_3CO_2$ and NaCl

Anything attached to an ammonium ion will be soluble as will anything attached to a sodium ion: there is no cation here that can generate a precipitate.

$AgNO_3$ and $MgCl_2$ can yield AgCl and $Mg(NO_3)_2$.

Although chlorides are usually soluble, silver is always suspicious: silver chloride will precipitate out. Although magnesium is not listed as "always soluble," it is attached to a nitrate, which is always soluble; therefore, magnesium nitrate will not precipitate in this reaction.

POST-LECTURE:

Work Problems: *One of the best ways to learn chemistry is through solving problems. Go to the textbook and work through the Worked Examples for this section, then go to the end of chapter and work the problems there. When you feel confident that you can correctly answer problems like the ones listed below for each objective, check off each of the Learning Objectives for this section above.*

LO 4.10: Problems 4.76, 4.77
LO 4.11: Worked Examples 4.6, 4.7; Problems 4.11–4.14, 4.78, 4.80

Section 4.7 Acids, Bases, and Neutralization Reactions

PRE-LECTURE:

Learning Objectives: *These are the skills you should be proficient in after reading this section and practicing the associated problems. Check off the boxes below once you have gone through this section, answered the associated questions in your textbook, and feel confident you have mastered the skill.*

- ☐ **LO 4.12** Convert between name and formula for an acid.
- ☐ **LO 4.13** Classify acids as strong or weak based on the molecular picture of dissociation.
- ☐ **LO 4.14** Write the ionic equation and net ionic equation for an acid–base neutralization reaction.

Definitions: *To learn, you must do. This is your workbook. Fill in the definitions from the glossary of the textbook as a first step to understanding the material.*

— **acid—** ______________________________

____

— **base—** ______________________________

____

— **hydronium ion, H_3O^+ —** ______________________________

____

— **strong acid—** ______________________________

____

— **weak acid—** ______________________________

____

— **monoprotic acid—** ______________________________

____

— **diprotic acid—** ______________________________

____

— **triprotic acid—** ______________________________

____

— **strong base—** ______________________________

____

— **weak base—** ______________________________

____

— **oxoacid—** ______________________________

____

— **salt—** ______________________________

____

Review: *Before going through the section make sure you are familiar with these concepts:*

- Ch. 2.13 naming oxoanions

Summary: *Read through this section and fill in the missing words and equations. Reference the textbook when you are not confident in your answers.*

Acids dissociate in water to form __________ ions and bases dissociate in water to form __________ ions.

H^+, as a bare proton, is an exceptionally strong ion. As a result, it does not truly exist free in solution, but binds itself to a water molecule to form a **hydronium ion** with formula H_3O^+.

Different acids will release different amounts of protons into a solution. Hydrochloric acid, HCl, is called a **monoprotic acid** because it has only _____ proton to release into solution, however, sulfuric acid, H_2SO_4, is a **diprotic acid,** and phosphoric acid, H_3PO_4, is a **triprotic acid.** In addition, protons differ in their strength depending on conditions. For example, it is more difficult for a proton to be removed from an already-charged unit: H_2SO_4 will dissociate completely to HSO_4^- + H^+, but HSO_4^- is a *weak* acid.

Acids and bases can be strong and weak, just as other electrolytes, and the definition means the same thing: it has only to do with the amount of dissociation and not the corrosiveness of a substance. A concentrated solution of a weak acid could cause a more serious burn than a dilute solution of a strong acid.

The following is a list of strong and weak acids and bases (there are many more than shown here):

- Strong acids: perchloric ($HClO_4$), sulfuric (H_2SO_4), hydrochloric (HCl), hydrobromic (HBr), hydroiodic (HI), and nitric (HNO_3). These acids dissociate fully.
- Weak acids: hydrofluoric (HF), phosphoric (H_3PO_4), and acetic (CH_3CO_2H).
- Strong bases: hydroxide metal salts are all strong bases.
- Weak bases: ammonia (NH_3), and any other compound that has the ability to remove hydrogen ions from solution.

Workbook Example 4.8

PROBLEM:
Determine if the following are acids, bases, or neutral salts: $Ca(OH)_2$, HCl(*aq*), NaCl, NaOH, $CaBr_2$

SOLUTION:
An acid will produce H^+ ions, a base will produce OH^- ions, and a neutral salt will produce neither.

$Ca(OH)_2$—produces OH^- in solution; base
HCl(*aq*)—produces H^+ in solution; acid
NaCl—produces Na^+ and Cl^- in solution; neutral salt
NaOH—produces OH^- in solution; base
$CaBr_2$—produces Ca^{2+} and Br^- in solution; neutral salt

Workbook Problem 4.5

Determine the reaction type, and write balanced molecular, ionic, and net ionic equations for the following reactions:

a. $LiOH$ and H_2SO_4
b. NH_4Cl and $AgNO_3$

Strategy: Determine the reaction type by examining the reactants—are they acids or bases or neutral?

Step 1: Write the balanced molecular equation for each reaction.

Step 2: Determine the presence of any strong electrolytes. Write the complete ionic equation, showing the strong electrolytes in terms of their free ions.

Step 3: Write the net ionic equation by removing spectator ions.

Many acids are oxoacids. Their formulas can be written by adding enough H^+ ions to neutralize the charge of oxoanion. To name them, the *-ate* or *-ite* ending of the oxoanion is converted to *-ic acid* or *-ous acid.* Use the following sentence to help you remember which suffixes go together. "Ick I ate a poisonous bite!" The *-ic* and *-ate* are found in the beginning of the sentence while the *-ous* and *-ite* are found at the end. Acids that do not contain oxygen have the prefix *hydro-* and the suffix *-ic acid.*

When an acid and base are mixed together they will react to form a salt and water. If mixed in the right proportions, all of the acid and base will be converted to the neutral substance water. The net ionic equation for the mixing of a strong acid and strong base highlights this reaction:

POST-LECTURE:

Work Problems: *One of the best ways to learn chemistry is through solving problems. Go to the textbook and work through the Worked Examples for this section, then go to the end of chapter and work the problems there. When you feel confident that you can correctly answer problems like the ones listed below for each objective, check off each of the Learning Objectives for this section above.*

LO 4.12: Worked Example 4.8; Problems 4.15, 4.16
LO 4.13: Problem 4.40
LO 4.14: Worked Example 4.9; Problems 4.17, 4.18, 4.96, 4.98

Section 4.8 Solution Stoichiometry

PRE-LECTURE:

Learning Objectives: *These are the skills you should be proficient in after reading this section and practicing the associated problems. Check off the boxes below once you have gone through this section, answered the associated questions in your textbook, and feel confident you have mastered the skill.*

☐ **LO 4.15** Convert between moles and volume using molarity in stoichiometry calculations.

Review: *Before going through the section make sure you are familiar with these concepts:*

- Ch. 3.4 stoichiometry

Summary: *Read through this section and fill in the missing words and equations. Reference the textbook when you are not confident in your answers.*

Molarity can be used as a conversion factor in stoichiometry calculations to determine volumes of solutions needed in reactions.
Instead of always using mass or amount in stoichiometry calculations, molarity can be used as a conversion factor when volume of solution is known.

volume	Use ____________ as a conversion factor	moles

Workbook Example 4.9

PROBLEM:
Silver nitrate and calcium chloride, both soluble, react to form silver chloride, an insoluble solid. How much 0.10 M calcium chloride is needed to completely react with 100 mL of 0.5 M silver nitrate?

SOLUTION:
First, write the balanced equation for the reaction.

$$CaCl_2 + 2\ AgNO_3 \rightarrow 2\ AgCl + Ca(NO_3)_2$$

Next, calculate the moles of silver nitrate present, remembering to first convert 100 mL to L:

$$0.100\ L \times \frac{0.50\ mol}{1\ L} = 0.050\ mol\ silver\ nitrate$$

Using the balanced equation, calculate the number of moles of $CaCl_2$ needed to react with 0.050 mol silver nitrate:

$$0.050\ mol\ AgNO_3 \times \frac{1\ mol\ CaCl_2}{2\ mol\ AgNO_3} = 0.025\ mol\ CaCl_2$$

The volume of the $CaCl_2$ solution can now be calculated:

$$0.025\ mol\ CaCl_2 \times \frac{1\ L}{0.10\ mol\ CaCl_2} = 0.250\ L = 250\ mL\ CaCl_2$$

Workbook Problem 4.6

Hydrochloric acid reacts with zinc metal to produce a solution of zinc chloride and hydrogen gas. What volume of 5.0 M hydrochloric acid is needed to react with 4.75 g of zinc? If the density of hydrogen gas is 0.0899 g/L, what volume of hydrogen gas will be produced?

Strategy: To solve any stoichiometry problem, we need to know the mole ratios of reactants and products as determined by the balanced chemical equation. Knowing the actual number of moles present of one reactant or product, we can then determine the number of moles needed of all other reactants and products.

Step 1: Write a balanced chemical equation.

Step 2: Determine the number of moles of zinc present.

Step 3: Determine the number of moles of HCl needed.

Step 4: Calculate the volume of 5.0 M HCl needed for this reaction.

Step 5: To calculate the volume of H_2 gas produced, we first need to calculate the moles, and then the grams of H_2 gas produced.

Step 6: Determine the volume of H_2 gas produced, using the grams of H_2 produced and the density of H_2 gas.

POST-LECTURE:

Work Problems: *One of the best ways to learn chemistry is through solving problems. Go to the textbook and work through the Worked Examples for this section, then go to the end of chapter and work the problems there. When you feel confident that you can correctly answer problems like the ones listed below for each objective, check off each of the Learning Objectives for this section above.*

LO 4.15: Worked Example 4.10; Problems 4.19, 4.20, 4.100, 4.101

Section 4.9 Measuring the Concentration of a Solution: Titration

PRE-LECTURE:

Learning Objectives: *These are the skills you should be proficient in after reading this section and practicing the associated problems. Check off the boxes below once you have gone through this section, answered the associated questions in your textbook, and feel confident you have mastered the skill.*

- ☐ **LO 4.16** Determine the concentration of a solution using titration data.
- ☐ **LO 4.17** Interpret molecular representations of substances in solution in a titration procedure.

Definitions: *To learn, you must do. This is your workbook. Fill in the definitions from the glossary of the textbook as a first step to understanding the material.*

— **titration—** ____

____

Review: *Before going through the section make sure you are familiar with these concepts:*

- Ch. 3.4 stoichiometry

Summary: *Read through this section and fill in the missing words and equations. Reference the textbook when you are not confident in your answers.*

To determine the concentration of an unknown solution, a procedure called **titration** can be used. A carefully measured volume of the unknown solution is slowly reacted with a solution of known concentration. An **indicator**—a molecule that changes color when the reaction is complete—is generally used to indicate the ___________ of the reaction.

Workbook Example 4.10

PROBLEM:
Apple juice contains malic acid, a weak acid with the formula $C_4H_6O_5$ that contains two acid groups. If 27 mL of 0.10 M sodium hydroxide is needed to neutralize 15 mL of a malic acid solution, what is the concentration of the solution?

SOLUTION:
First, write a balanced equation for the neutralization reaction.

$$C_4H_6O_5(aq) + 2\ NaOH(aq) \rightarrow Na_2C_4H_4O_5(aq) + 2\ H_2O(l)$$

From the information given, we can calculate the number of moles of sodium hydroxide used to reach the equivalence point of the titration.

$$0.027\ L\ NaOH \times \frac{0.10\ mol}{1\ L} = 0.0027\ mol\ NaOH$$

From the balanced equation, calculate the moles and molarity of malic acid.

$$0.0027\ mol\ NaOH \times \frac{1\ mol\ malic\ acid}{2\ mol\ NaOH} = 0.0014\ mol\ malic\ acid$$

$$\frac{0.0014\ mol\ malic\ acid}{0.015\ L} = 0.093\ M\ malic\ acid$$

Workbook Problem 4.7

A student determines that 16.37 mL of sulfuric acid is required to neutralize a 100.0 mL solution that contains 0.153 g of NaOH. Determine the molarity of the sulfuric acid solution.

Strategy: Balance the equation to find mole ratios, determine moles of known compounds, then calculate moles of the unknown and determine concentration.

Step 1: Write a balanced chemical equation for the reaction.

Step 2: Calculate the number of moles of NaOH present.

Step 3: Determine the number of moles of sulfuric acid needed to react with that quantity of NaOH.

Step 4: Calculate the molarity of the acid solution.

POST-LECTURE:

Work Problems: *One of the best ways to learn chemistry is through solving problems. Go to the textbook and work through the Worked Examples for this section, then go to the end of chapter and work the problems there. When you feel confident that you can correctly answer problems like the ones listed below for each objective, check off each of the Learning Objectives for this section above.*

LO 4.16: Worked Example 4.11; Problems 4.21, 4.102, 4.103
LO 4.17: Problems 4.22, 4.42, 4.43

Section 4.10 Oxidation–Reduction (Redox) Reactions

PRE-LECTURE:

Learning Objectives: *These are the skills you should be proficient in after reading this section and practicing the associated problems. Check off the boxes below once you have gone through this section, answered the associated questions in your textbook, and feel confident you have mastered the skill.*

☐ **LO 4.18** Assign oxidation numbers to atoms in a compound.

Definitions: *To learn, you must do. This is your workbook. Fill in the definitions from the glossary of the textbook as a first step to understanding the material.*

— **oxidation—** ____

____

— **reduction—** ____

____

— **oxidation number—** ____

____

Summary: *Read through this section and fill in the missing words and equations. Reference the textbook when you are not confident in your answers.*

Originally, **oxidation** was defined as the gaining of oxygen and **reduction** was defined as the removal of oxygen. Although these definitions have been significantly broadened, this is still a useful first step in recognizing redox reactions.

Oxidation is now defined as *the* ________ *of electrons.* When an element combines with oxygen, to use the historical definition, the metal loses electrons to become a cation, and the oxygen gains those same electrons to become an oxide ion:

$$2\ Mg + O_2 \rightarrow 2\ MgO$$

When electrons are lost from a chemical species, the result is a change in its charge *toward positive.* It is described this way because some elements will have more than one charge: a change from −2 to −1 is *oxidation*, despite the fact that no positively charged species is generated.

Reduction is now defined as *the* ________ *of electrons*. If the reaction above is reversed, the Mg^{2+} ions gain two electrons to become elemental magnesium.

$$2\ MgO \rightarrow 2\ Mg + O_2$$

In reactions of this kind, the number of electrons lost by one species must be the number gained somewhere else: electrons do not just fly away!

So how do we keep track of all this? We use a method of electron "bookkeeping" that uses oxidation numbers to show where electrons are going in a reaction. Sometimes, the oxidation number indicates a charge, but not always.

The rules for assigning oxidation numbers:

1.

2.

3.

   a.

   b.

   c.

4.

Workbook Example 4.11

PROBLEM:
Determine the oxidation number of each of the atoms in the following: HCl, NaOH, NH_3, H_3PO_4

SOLUTION:
For HCl, halogens are −1, hydrogen is +1. When these are added together, they give 0, correct for a neutral molecule.

H = +1
Cl = −1

For NaOH, Na carries its usual charge as its oxidation number, which is +1. Oxygen is −2, leaving hydrogen. Hydrogen's charge depends on whether it is bonded to a metal or a nonmetal, and here we have both as possibilities. However, this is a neutral compound, so the oxidation numbers must add up to 0, and hydrogen must carry an oxidation state of +1.

Na = +1
O = −2
H = +1

In NH_3, hydrogen is bound to a nonmetal and thus each carries an oxidation number of +1. This is a neutral compound, so N must carry an oxidation number of –3 to balance the three +1 charges on the hydrogen atoms.

$$N = -3$$
$$H = +1$$

In H_3PO_4, there are a number of things to consider. This compound is all nonmetals, so the hydrogen atoms will carry oxidation numbers of +1. Oxygen carries a charge of –2, and there are four of them. This leaves phosphorus to balance things out $(4 \times -2) + (3 \times +1) + P = 0$, so phosphorus carries an oxidation number of +5. This is clearly not a real charge; it is results of this kind that remind us that these are useful bookkeeping methods, but they do not necessarily reflect ionic reality.

Workbook Problem 4.8

Determine the oxidation number for each element in VO_4^{3-} and $NaVO_3$.

Strategy: Follow the rules above for determining oxidation states.

Step 1: Identify the elements that have a fixed oxidation state.

Step 2: Determine the oxidation number of the remaining elements, keeping in mind that the sum of all oxidation numbers must be equal to the ionic charge or to zero for a neutral molecule.

POST-LECTURE:

Work Problems: *One of the best ways to learn chemistry is through solving problems. Go to the textbook and work through the Worked Examples for this section, then go to the end of chapter and work the problems there. When you feel confident that you can correctly answer problems like the ones listed below for each objective, check off each of the Learning Objectives for this section above.*

LO 4.18: Worked Example 4.12; Problems 4.23, 4.24, 4.108, 4.110

Section 4.11 Identifying Redox Reactions

PRE-LECTURE:

Learning Objectives: *These are the skills you should be proficient in after reading this section and practicing the associated problems. Check off the boxes below once you have gone through this section, answered the associated questions in your textbook, and feel confident you have mastered the skill.*

☐ **LO 4.19** Identify redox reactions, oxidizing agents, and reducing agents.

Definitions: *To learn, you must do. This is your workbook. Fill in the definitions from the glossary of the textbook as a first step to understanding the material.*

— **half-reaction—** ____

____

— **oxidizing agent—**____

____

— **reducing agent—**____

____

Summary: *Read through this section and fill in the missing words and equations. Reference the textbook when you are not confident in your answers.*

When one species in a reaction loses electrons, another gains them. Oxidation and reduction must occur together. When a metal and a nonmetal react with one another, the metal will be the reducing agent, and the nonmetal will be the oxidizing agent (this is easy to remember as oxygen is a nonmetal). Metals can also oxidize and reduce one another, depending on the strength of their attraction for electrons.

A reducing agent is oxidized and gives up electrons. An oxidizing agent is reduced and gains electrons. To help keep this straight fill out the chart below from the bottom of pg. 139 in your textbook:

Reducing Agent	Oxidizing Agent

Workbook Example 4.12

PROBLEM:
Identify the species oxidized, the species reduced, the oxidizing agent, and the reducing agent in the following reaction:

$$Zn(s) + 2\,H^{+}(aq) + 2\,Cl^{-}(aq) \rightarrow Zn^{2+} + 2\,Cl^{-}(aq) + H_2(g)$$

SOLUTION:
To solve this problem, look for atoms that have had a change in oxidation number:

$$Zn \rightarrow Zn^{2+}$$

The oxidation number of zinc goes from 0 to +2. Zinc is oxidized and donates electrons to hydrogen, therefore it is the reducing agent.

$$H^{+} \rightarrow H_2$$

The oxidation number of hydrogen goes from +1 to 0. Hydrogen is reduced and accepts electrons from zinc, therefore it is the oxidizing agent.

Workbook Problem 4.9

Identify the species oxidized, the species reduced, the oxidizing agent, and the reducing agent in the following reaction:

$$Cu(s) + 2\ Ag^{+}(aq) + 2\ NO_3^{-}(aq) \rightarrow 2\ Ag(s) + Cu^{2+}(aq) + 2\ NO_3^{-}(aq)$$

Strategy: Determine the oxidation number of all species present.

Step 1: Identify the species that have a change in oxidation number.

Step 2: Based on the change in oxidation number, identify the species oxidized, the species reduced, and the oxidizing and reducing agents.

POST-LECTURE:

Work Problems: *One of the best ways to learn chemistry is through solving problems. Go to the textbook and work through the Worked Examples for this section, then go to the end of chapter and work the problems there. When you feel confident that you can correctly answer problems like the ones listed below for each objective, check off each of the Learning Objectives for this section above.*

LO 4.19: Worked Example 4.13; Problems 4.25–4.26, 4.118–4.119

Section 4.12 The Activity Series of the Elements

PRE-LECTURE:

Learning Objectives: *These are the skills you should be proficient in after reading this section and practicing the associated problems. Check off the boxes below once you have gone through this section, answered the associated questions in your textbook, and feel confident you have mastered the skill.*

☐ **LO 4.20** Use the location of elements in the periodic table and activity series to predict if a redox reaction will occur.
☐ **LO 4.21** Develop an activity series and predict if a redox reaction will occur based on experimental data provided.

Definitions: *To learn, you must do. This is your workbook. Fill in the definitions from the glossary of the textbook as a first step to understanding the material.*

— **activity series—**____

____

Summary: *Read through this section and fill in the missing words and equations. Reference the textbook when you are not confident in your answers.*

Predicting which metals will accept electrons and which will donate electrons when they are together in aqueous solution can be done by using the *activity series*. A substance that is listed above another on the activity series will oxidize, giving up electrons, and reducing any substance listed below it. So, if you are looking to see what will happen if you put magnesium ribbon into a copper sulfate solution, you will see that magnesium is ___________ copper; therefore you will wind up with magnesium ions and copper metal.

Also in the activity series is H^+, though it is not a metal. Metals above H^+ will react with acids to form metal ions and hydrogen gas; those below it will not. The metals at the bottom of the activity series are the least-reactive metals (i.e., the least likely to give up electrons) and those at the top are the most reactive. Those at the very top are so eager to ionize that they will react with water to form hydrogen gas and hydroxide ions.

Partial Activity Series for Oxidation Reactions

Strongly reducing	$Li \rightarrow Li^+ + e^-$	
	$K \rightarrow K^+ + e^-$	
	$Ba \rightarrow Ba^{2+} + 2\ e^-$	These elements react rapidly with aqueous acid (H^+) or H_2O to release H_2 gas.
	$Ca \rightarrow Ca^{2+} + 2\ e^-$	
	$Na \rightarrow Na^+ + e^-$	
	$Mg \rightarrow Mg^{2+} + 2\ e^-$	
	$Al \rightarrow Al^{3+} + 3\ e^-$	
	$Mn \rightarrow Mn^{2+} + 2\ e^-$	These elements react with aqueous H^+ ions or with steam to release H_2 gas.
	$Zn \rightarrow Zn^{2+} + 2\ e^-$	
	$Cr \rightarrow Cr^{3+} + 3\ e^-$	
	$Fe \rightarrow Fe^{2+} + 2\ e^-$	
	$Co \rightarrow Co^{2+} + 2\ e^-$	
	$Ni \rightarrow Ni^{2+} + 2\ e^-$	These elements react with aqueous H^+ ions to release H_2 gas.
	$Sn \rightarrow Sn^{2+} + 2\ e^-$	
	$\mathbf{H_2 \rightarrow 2\ H^+ + 2\ e^-}$	
	$Cu \rightarrow Cu^{2+} + 2\ e^-$	
	$Ag \rightarrow Ag^+ + e^-$	These elements do not react with aqueous H^+ ions to release H_2 gas.
	$Hg \rightarrow Hg^{2+} + 2\ e^-$	
Weakly reducing	$Pt \rightarrow Pt^{2+} + 2\ e^-$	
	$Au \rightarrow Au^{3+} + 3\ e^-$	

Workbook Example 4.13

PROBLEM:
Using the activity series, write balanced chemical equations for the following reactions:

$Cu(s) + Co^{2+}(aq) \rightarrow$ $Mn(s) + Zn^{2+}(aq) \rightarrow$ $Sn(s) + HCl(aq) \rightarrow$

$Pt(s) + Hg^{2+}(aq) \rightarrow$ $Al(s) + Zn^{2+}(aq) \rightarrow$ $Ni(s) + Mn^{2+}(aq) \rightarrow$

$Ni(s) + HCl(aq) \rightarrow$ $Pt(s) + HBr(aq) \rightarrow$

SOLUTION:
To predict the outcome of these reactions, remember that any element higher in the activity series will react with the ion of any element lower in the activity series. Also remember that metals above the H^+ ion in the activity series will displace the hydrogen ion from an acid to form H_2 gas.

$Cu(s) + Co^{2+}(aq) \rightarrow$ no reaction — Cu is below Co^{2+} in the activity series

$Mn(s) + Zn^{2+}(aq) \rightarrow Mn^{2+}(aq) + Zn(s)$

$Sn(s) + 2\ HCl(aq) \rightarrow SnCl_2(aq) + H_2(g)$

$Pt(s) + Hg^{2+}(aq) \rightarrow$ no reaction — Pt is below Hg^{2+} in the activity series

$2\ Al(s) + 3\ Zn^{2+}(aq) \rightarrow 2\ Al^{3+}(aq) + 3\ Zn(s)$

$Ni(s) + Mn^{2+}(aq) \rightarrow$ no reaction — Ni is below Mn^{2+} in the activity series

$Ni(s) + 2\ HCl(aq) \rightarrow NiCl_2(aq) + H_2(g)$

$Pt(s) + HBr(aq) \rightarrow$ no reaction — Pt is below H^+ in the activity series.

Workbook Problem 4.10

Using the activity series, write balanced chemical equations for the following reactions:

a. $Cu(s) + HCl(aq) \rightarrow$
b. $Mn(s) + ZnCl_2(aq) \rightarrow$
c. $Mg(s) + Al(NO_3)_3(aq) \rightarrow$

Strategy: Remember that any element higher in the activity series will react with the ion of any element lower in the activity series. Also remember that metals above the H^+ ion in the activity series will displace the hydrogen ion from an acid to form H_2 gas.

Step 1: Predict the outcome of these reactions.

Step 2: Balance the equations.

POST-LECTURE:

Work Problems: *One of the best ways to learn chemistry is through solving problems. Go to the textbook and work through the Worked Examples for this section, then go to the end of chapter and work the problems there. When you feel confident that you can correctly answer problems like the ones listed below for each objective, check off each of the Learning Objectives for this section above.*

LO 4.20: Worked Example 4.14; Problems 4.27, 4.28, 4.45, 4.120
LO 4.21: Problems 4.122, 4.123

Section 4.13 Redox Titrations

PRE-LECTURE:

Learning Objectives: *These are the skills you should be proficient in after reading this section and practicing the associated problems. Check off the boxes below once you have gone through this section, answered the associated questions in your textbook, and feel confident you have mastered the skill.*

☐ **LO 4.22** Use a redox titration to determine the concentration of an oxidizing or reducing agent in solution.

Review: *Before going through the section make sure you are familiar with these concepts:*

- Ch. 3.4 stoichiometry
- Ch. 4.9 titrations

Summary: *Read through this section and fill in the missing words and equations. Reference the textbook when you are not confident in your answers.*

Just as the concentrations of acids and bases in solution can be determined by titration using an indicator, the concentration of an oxidizing or reducing agent can be similarly determined if the following conditions are met:

- The reaction must have a _____% yield
- A color change must signal the endpoint (an indicator can be used if necessary)

Because many redox reagents are strongly colored, this can be a useful method. A balanced equation is required, of course, as is a known quantity of one of the reactants.

Workbook Problem 4.11

A 50 mL solution of iodine and starch was titrated with an acidic 0.15 M thiosulfate solution. After 15.7 mL of the thiosulfate solution had been added, the blue color of the iodine:starch complex completely disappeared, which indicates that the iodine had been completely reduced to iodide ions. What was the original molar concentration of the iodine in solution?

The unbalanced equation is

$$I_2 + S_2O_3^{2-} \rightarrow I^- + S_4O_6^{2-}$$

Strategy: Balance the equation, proceed as for an acid–base titration.

Step 1: Determine the number of moles of thiosulfate ion that were required to react with the iodine solution to completely reduce the iodine.

Step 2: Balance the equation.

Step 3: Use mole ratios to determine the amount of iodine present in the reaction mixture.

Step 4: Determine the original concentration of I_2 in solution.

POST-LECTURE:

Work Problems: *One of the best ways to learn chemistry is through solving problems. Go to the textbook and work through the Worked Examples for this section, then go to the end of chapter and work the problems there. When you feel confident that you can correctly answer problems like the ones listed below for each objective, check off each of the Learning Objectives for this section above.*

LO 4.22: Worked Example 4.15; Problems 4.29, 4.124, 4.126, 4.132

Section 4.14 Some Applications of Redox Reactions

PRE-LECTURE:

Learning Objectives: *There are no specific learning objectives for this section; the information is a continuation of the previous section.*

Summary: *Read through this section and fill in the missing words and equations. Reference the textbook when you are not confident in your answers.*

Reactions such as the ones discussed here may seem like they are rare and confined to the laboratory, but in reality, electrons move around in many different types of reactions that occur in daily life. All of the following have a redox component:

- ___________: the oxidation state of oxygen shifts from 0, when it is an element, to −2 after a reaction occurs. This causes an oxidation in whatever is being burned.
- ___________: strong oxidizing agents such as peroxides, hypochlorite, or elemental chlorine can destroy colored molecules.
- ___________: not surprisingly, the ability to generate a current, which is a flow of electrons, relies on this type of reaction.
- ___________: most metals are found in the Earth's crust as oxides, and thus must be reduced to get their elemental forms.
- ___________: a metal deteriorates by losing electrons.
- ___________ ___________: This is a combustion reaction that operates in an extremely controlled manner, but when reduced to its basics, glucose is combined to form carbon dioxide (which you then breathe out), water, and the heat that maintains your body temperature.

Self-Test

This section is intended to test your knowledge of the material covered in this chapter. Think through these problems and make certain you understand what the problem is about—ask yourself if the answer makes sense. Many of these questions are linked to the chapter learning goals. Therefore, successful completion of these problems indicates that you have achieved the learning goals for this chapter. You will receive the greatest benefit from this section if you use it as a mock exam. Do not use your notes, book, Internet, or friends when taking the test. There is no grade and you only shortchange yourself by getting help with this activity. As you take the test, mark the problems with your level of confidence with the answer.

★ *Use a star for problems you know you got right.*
? *Use a question mark for those you think you got right.*
x *Use an x for those you are guessing on.*

If you get a question right that you marked with a star, you can be confident that you know the associated LO.

If you get a question right that you marked with a question mark, go back and work a few more problems and test yourself again tomorrow.

If you get a question wrong or mark it with an x, go back and read the section again, work a few problems today, a few tomorrow, and a few the next day before testing yourself again.

This will help you to discover which topics you have learned thoroughly and which topics you need to study in more detail.

True–False

1. In acid–base reactions, an acid and a base react to form carbon dioxide and water.
2. Sodium nitrate will precipitate out of aqueous solution.
3. When electrons move from one species to another in chemical reactions, this is known as oxidation (if electrons are lost) and reduction (if electrons are gained).
4. Weak electrolytes do not fully dissociate.
5. Spectator ions do not participate in precipitation reactions.
6. Acids release nitronium ions into solution.
7. NaCl is a strong electrolyte.
8. Oxidations are always equal to ionic charge.
9. An activity series makes it possible to predict the outcome of precipitation reactions.
10. Redox reactions can be broken into half-reactions to make them easier to balance.

Matching

11. reduction	a. dissociates only slightly
12. net ionic equation	b. reactions involving transfer of electrons
13. oxidation number	c. H_3O^+
14. weak electrolyte	d. ranking of elements by reducing ability
15. redox reaction	e. substance that gives up electrons
16. dissociate	f. reaction description without spectators
17. spectator	g. substance that accepts electrons
18. hydroxide ion	h. gain of electrons
19. activity series	i. loss of electrons
20. electrolyte	j. ion that does not participate in a reaction
21. reducing agent	k. reaction in which two soluble salts form an insoluble product

22. precipitation reaction	l. measure of electron richness
23. oxidation	m. substance that will conduct electricity when dissolved
24. hydronium ion	n. OH^-
25. oxidizing agent	o. separate into ions in solution
26. strong electrolyte	p. substance that dissociates to produce hydroxide ions in solution
27. base	q. substance that dissociates to produce hydronium ions in solution
28. acid	r. completely dissociates

Fill in the Blanks

29. When ________________ reactants yield a(n) __________________ product, removing some of the dissolved ions, a(n) ___________________________ reaction has occurred.

30. In neutralization reactions, a(n) ________ and a(n) _________ react to form __________ and ________.

31. In __________ reactions, electrons are transferred between reactants.

32. Acids, bases, or electrolytes that dissociate completely in solution are known as ____________, while those that dissociate incompletely are known as ________.

33. A(n) _______________________________ shows all substances in their undissociated forms, a(n) _______________________________ shows all substances dissociated, and a(n) ____________________________ shows only those substances that have undergone a change.

34. In the Arrhenius definition, acids dissociate in aqueous solution to form an anion and a(n) __________ ion, while bases dissociate to form a cation and a(n) _______________ ion.

35. When a substance is oxidized, it __________ electrons; when it is reduced, it __________ electrons.

Problems

36. Classify each reaction as a precipitation, neutralization, or redox reaction, and justify your answer.
    a. $HNO_3(aq) + KOH(aq) \rightarrow KNO_3(aq) + H_2O(l)$
    b. $Cu(NO_3)_2(aq) + 2\ NaOH(aq) \rightarrow Cu(OH)_2(s) + 2\ NaNO_3(aq)$
    c. $Cu(NO_3)_2(aq) + Zn(s) \rightarrow Zn(NO_3)_2(aq) + Cu(s)$

37. If 15 g of Na_3PO_4, a strong electrolyte, are dissolved in 250 mL of solution, what is the total molar concentration of ions?

38. Write balanced molecular, ionic, and net ionic equations for the following reactions in aqueous solution:
    a. lead(II) nitrate + sodium chloride
    b. sulfuric acid + lithium hydroxide
    c. potassium chloride + silver acetate

39. Given 1.00 M solutions of $Ca(NO_3)_2$ and KOH, how much of each need to be combined to form 1.8 g of $Ca(OH)_2$?

40. Predict reactions for each of the following. If a reaction occurs, write the net ionic equation. If no reaction occurs, list the possible soluble products.
    a. $Cu(CH_3CO_2)_2 + Pb(NO_3)_2$
    b. $NaNO_3 + (NH_4)_2SO_4$
    c. $CaCl_2 + AgNO_3$

41. Assign oxidation numbers to all of the atoms in the following:
    a. Cl_2
    b. NaCl
    c. CO_3^{2-}
    d. Cr_2O_3
    e. ClO_3^-

42. For each unbalanced equation given below, identify the oxidizing agent and the reducing agent.
    a. $C_2H_6 + O_2 \rightarrow CO_2 + H_2O$
    b. $Na + Cl_2 \rightarrow NaCl$
    c. $Mg + Fe^{2+} \rightarrow Mg^{2+} + Fe$
    d. $Ca + H^+ \rightarrow Ca^{2+} + H_2$

43. Given the activity series provided here, predict the outcome of the following reactions. When a reaction occurs, balance the equation.
    a. chromium and copper(II) nitrate
    b. tin and silver nitrate
    c. copper(II) nitrate and cobalt
    d. silver and hydrochloric acid

Activity Series
$Li \rightarrow Li^+ + e^-$
$Ba \rightarrow Ba^{2+} + 2e^-$
$Na \rightarrow Na^+ + e^-$
$Cr \rightarrow Cr^{3+} + 3e^-$
$Co \rightarrow Co^{2+} + 2e^-$
$Sn \rightarrow Sn^{2+} + 2e^-$
$H_2 \rightarrow 2H^+ + 2e^-$
$Cu \rightarrow Cu^{2+} + 2e^-$
$Ag \rightarrow Ag^+ + e^-$

44. Calculate the molarity of a Na_2SO_3 solution if 15.0 mL of the solution reacts completely with 23.7 mL of 0.124 M $K_2Cr_2O_7$ in acid solution according to the following balanced equation:

$$3SO_3^{2-}(aq) + Cr_2O_7^{2-}(aq) + 8H^+(aq) \rightarrow 3SO_4^{2-}(aq) + 2Cr^{3+}(aq) + 4H_2O(l)$$

45. 5.3 g of potassium sulfate are combined with 4.9 g of lead(II) nitrate in aqueous solution.
    a. Write the molecular, complete ionic, and net ionic equations for this reaction.
    b. How many grams of solid will be produced?
    c. How much of the excess reactant will be left over?
    d. If the experimental yield is 4.37 g, what is the percent yield?
    e. What will be the molar concentration of ions remaining in the solution if the reaction mixture is made up to 200 mL?

<u>*Challenge Problem*</u>

46. When sodium or lithium is added to water, the following rather exciting reaction occurs according to the following unbalanced equation:

$$M(s) + H_2O(l) \rightarrow MOH(aq) + H_2(g)$$

Enough heat is typically generated by this reaction that the hydrogen gas burns as it is generated. If a 4 g mixture of lithium and sodium is added to water and the resulting solution is titrated, 59.7 mL of 7.5 M HCl is required to neutralize the solution. How many grams of lithium and sodium were contained in the original mixture?

Test Yourself: *Take the Practice Test on pages 151 and 152 of your textbook. Again, use a testing environment using only those items that your instructor will allow you to have access to during the test and mark each question with a star, question mark, or x to denote your confidence level. Once you have finished you can check off the LO at the beginning of this chapter based on which questions you starred and got correct. Go back and spend more time with the topics from the other learning objectives.*

CHAPTER FIVE

Periodicity and the Electronic Structure of Atoms

Learning Objectives

As a result of reading and studying this chapter, you should be able to:

- ☐ **LO 5.1** Label the wavelength, frequency, and amplitude in an electromagnetic wave.
- ☐ **LO 5.2** Convert between wavelength and frequency of electromagnetic radiation.
- ☐ **LO 5.3** Calculate the energy of electromagnetic radiation in units of J/photon or kJ/mol, when given the frequency or wavelength.
- ☐ **LO 5.4** Describe the photoelectric effect and explain how it supports the theory of particle-like properties of light.
- ☐ **LO 5.5** Calculate the frequency or wavelength of radiation needed to produce the photoelectric effect given the work function of a metal.
- ☐ **LO 5.6** Describe the difference between a continuous spectrum and a line spectrum.
- ☐ **LO 5.7** Compare the wavelength and frequency of different electron transitions in the Bohr model of the atom.
- ☐ **LO 5.8** Use the Balmer–Rydberg equation to calculate the wavelength and energies of electron transitions in the hydrogen atom.
- ☐ **LO 5.9** Calculate the wavelength of a moving object using the de Broglie equation.
- ☐ **LO 5.10** Explain why the wavelength of macroscopic objects cannot be observed.
- ☐ **LO 5.11** Calculate the uncertainty in the position of the moving object if the velocity is known.
- ☐ **LO 5.12** Identify and write sets of quantum numbers that describe electrons in different types of orbitals.
- ☐ **LO 5.13** Identify an orbital based on its shape and describe it using a set of quantum numbers.
- ☐ **LO 5.14** Locate the nodal planes in different types of orbitals and different shells.
- ☐ **LO 5.15** Assign a set of four quantum numbers for electrons in an atom.
- ☐ **LO 5.16** Explain how electron shielding gives the order of subshells from lowest to highest in energy.
- ☐ **LO 5.17** Predict the order of filling of subshells based upon energy.
- ☐ **LO 5.18** Assign electron configurations to atoms in their ground state.
- ☐ **LO 5.19** Draw orbital filling diagrams for the ground state of an atom and determine the number of unpaired electrons.
- ☐ **LO 5.20** Identify atoms from orbital filling diagrams or electron configurations.
- ☐ **LO 5.21** Explain the periodic trend in atomic radii.
- ☐ **LO 5.22** Predict the relative size of atoms based upon their position in the periodic table.

Chapter Summary

The birth of quantum mechanics at the turn of the last century greatly expanded the understanding of the atomic nature of matter. This discovery blurred the lines between matter and energy, demonstrating that at the smallest levels matter and energy share properties: matter is wave like, and light is particle-like. The observation of line spectra and the development of the Balmer–Rydberg equation, though it was some years before it was fully explained, was the first step in understanding the quantization of energy: it led to the Bohr model of the hydrogen atom, which was soon greatly expanded by Schrödinger's wave functions. The variety of orbital shapes, and the development of the system of quantum numbers to identify each electron in an element, helped to solidify the remarkable depth of information contained in the periodic table. These discoveries also resolved reactivity and periodic properties of the elements that seemed paradoxical, such as the decrease in atomic radius as the atomic number increases across the periodic table.

The Chapter in Detail

Section 5.1 Wave Properties of Radiant Energy and the Electromagnetic Spectrum

PRE-LECTURE:

Learning Objectives: *These are the skills you should be proficient in after reading this section and practicing the associated problems. Check off the boxes below once you have gone through this section, answered the associated questions in your textbook, and feel confident you have mastered the skill.*

☐ **LO 5.1** Label the wavelength, frequency, and amplitude in an electromagnetic wave.
☐ **LO 5.2** Convert between wavelength and frequency of electromagnetic radiation.

Definitions: *To learn, you must do. This is your workbook. Fill in the definitions from the glossary of the textbook as a first step to understanding the material.*

— **radiant energy—** ____________________

____

— **electromagnetic spectrum—** ____________________

____

— **frequency (ν)—** ____________________

____

— **hertz (Hz)—** ____________________

____

— **wavelength (λ)—** ____________________

____

— **amplitude—** ____________________

____

Summary: *Read through this section and fill in the missing words and equations. Reference the textbook when you are not confident in your answers.*

Although we experience them quite differently, visible light, infrared and ultraviolet radiation, and X-rays are all forms of **electromagnetic energy.** Collectively, they make up the **electromagnetic spectrum.**

Electromagnetic energy travels in waves. Electromagnetic waves can be characterized using four properties:

- *Frequency:* the number of waves that pass by a certain point in one second.
  - Units of measurement are per second, abbreviated s^{-1} or 1/s, or given the name hertz (Hz).
  - Indicated by the Greek lowercase letter *nu* (ν)
- *Wavelength:* the length of one complete wave, usually measured from peak to peak.
  - Units used are meters or nanometers (1×10^{-9} m), and
  - Indicated by the Greek letter lambda (λ)
- *Amplitude:* The height of a wave measured from the centerline between peak and trough and proportional to its intensity.
- *Speed:* when wavelength (m) and frequency (1/s) are multiplied together, the result is m/s. In a vacuum, all electromagnetic radiation moves at the same speed: 3.00×10^8 m/s, which is the speed of light, indicated by the letter "c."

Label the wave below:

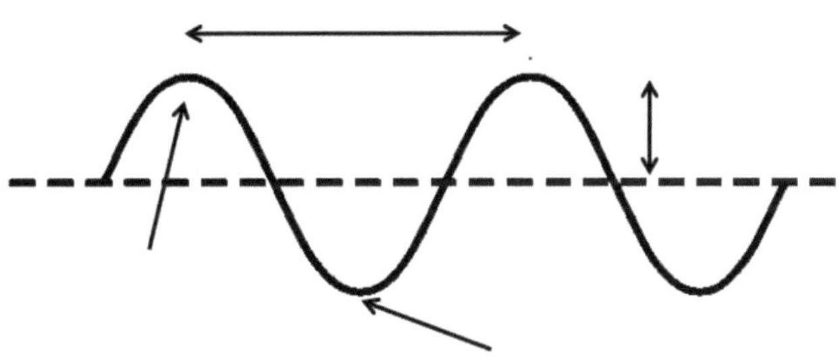

Because all electromagnetic radiation has the same speed, it is possible to use this speed of light as a conversion factor to convert between wavelength and frequency:

$$c = \lambda \nu$$

Rearrange it to solve for wavelength:

$\lambda =$

and again to solve for frequency:

$\nu =$

It is also helpful to remember the results of this relationship: a longer wavelength is lower in frequency, and a shorter wavelength is higher in frequency.

Workbook Example 5.1

PROBLEM:
Blu-ray lasers typically operate at 473 nm. What is their frequency?

SOLUTION:
Although it is not stated in the problem, you have a three-variable equation and two of the variables: wavelength and speed. Remember that c is a constant. Also remember that the speed of light is in units of m/s, which means that the 473 nm *must* be converted to m!

$$\nu = \frac{3.00\times10^{8}\ m/s}{4.73\times10^{-7}\ m} = 6.3\times10^{14}\ Hz$$

Note: You will quickly notice that a frequency in this range (10^{14} Hz) is typical of visible light.

Workbook Problem 5.1

Your college radio station transmits at 88.5 MHz. What is the wavelength associated with this frequency?

Strategy: Use the relationship between speed, wavelength, and frequency.

Step 1: Choose the appropriate form of the equation and solve.

POST-LECTURE:

Work Problems: *One of the best ways to learn chemistry is through solving problems. Go to the textbook and work through the Worked Examples for this section, then go to the end of chapter and work the problems there. When you feel confident that you can correctly answer problems like the ones listed below for each objective, check off each of the Learning Objectives for this section above.*

LO 5.1: Problems 5.2, 5.26
LO 5.2: Worked Example 5.1; Problems 5.38, 5.40

Section 5.2 Particlelike Properties of Radiant Energy: The Photoelectric Effect and Planck's Postulate

PRE-LECTURE:

Learning Objectives: *These are the skills you should be proficient in after reading this section and practicing the associated problems. Check off the boxes below once you have gone through*

this section, answered the associated questions in your textbook, and feel confident you have mastered the skill.

- ☐ **LO 5.3** Calculate the energy of electromagnetic radiation in units of J/photon or kJ/mol, when given the frequency or wavelength.
- ☐ **LO 5.4** Describe the photoelectric effect and explain how it supports the theory of particle-like properties of light.
- ☐ **LO 5.5** Calculate the frequency or wavelength of radiation needed to produce the photoelectric effect given the work function of a metal.

Definitions: *To learn, you must do. This is your workbook. Fill in the definitions from the glossary of the textbook as a first step to understanding the material.*

— **photoelectric effect—** ____

____

— **photon—** ____

____

— **Planck's postulate—** ____

____

— **quantized—** ____

____

— **quantum—** ____

____

Summary: *Read through this section and fill in the missing words and equations. Reference the textbook when you are not confident in your answers.*

The *photoelectric effect* refers to the fact that light of a specific frequency can eject electrons from the surface of a clean metal. In 1905, Albert Einstein proposed that this was due to light behaving like streams of small particles or **photons.** Below that specific threshold frequency, there is not enough energy per photon to eject an electron.

Write the equation that relates energy to frequency:

E =

or, since $c = \lambda\nu$, we can relate the energy of a photon to its frequency. Write that equation below:

E =

where "h" is Planck's constant, which has a value of 6.626×10^{-34} J·s. This gives the energy of one photon of a specific frequency or wavelength.

Higher energy waves have higher frequencies and shorter wavelengths than lower energy waves. This is borne out in the way humans experience the electromagnetic spectrum: no one ever got a radiowave burn, but UV radiation damages skin, and X-rays can do even deeper damage, because they have enough energy to pass right through the body and potentially damage living tissue.

Energy and intensity are not the same. A high-intensity low-energy beam cannot eject electrons from a metal while a low-intensity high-energy beam can (even if it is just a few).

This work led to the conclusion that the behavior of electromagnetic radiation (and thus light) is more complex than previously realized, and electromagnetic radiation can be described as waves and as small particles.

Workbook Example 5.2

PROBLEM:
If one mole of photons has an energy of 53.7 kJ, what is the wavelength of the light from which the photons came?

SOLUTION:
We need to rearrange the equation $E = h\nu$ to calculate the wavelength. However, the given energy must first be converted from kJ/mol to J/photon.

$$\frac{53.7\ kJ}{1\ mol} \times \frac{1\ mol}{6.022 \times 10^{23}\ photons} \times \frac{1000\ J}{1\ kJ} = 8.92 \times 10^{-20}\ J/photon$$

Now, rearrange the equation to solve for wavelength and put in the known values:

$$\lambda = \frac{hc}{E} = \frac{(6.626 \times 10^{-34}\ J \cdot s)(3.00 \times 10^{8}\ m/s)}{8.92 \times 10^{-20}\ J/photon} = 2.23 \times 10^{-6}\ m = 2230\ nm$$

Workbook Problem 5.2

The threshold frequency of sodium metal is 4.4×10^{14} Hz. What is the wavelength and color of light to which this corresponds and what is the energy of one mole of photons at this frequency?

Strategy: Use the equation for the energy of a photon and multiply by Avogadro's number, then convert frequency to wavelength.

Step 1: Determine the energy of one photon, then convert to one mole of photons.

Step 2: Convert the frequency to wavelength.

Step 3: Consult your textbook for the color of this wavelength of light.

POST-LECTURE:

Work Problems: *One of the best ways to learn chemistry is through solving problems. Go to the textbook and work through the Worked Examples for this section, then go to the end of chapter and work the problems there. When you feel confident that you can correctly answer problems like the ones listed below for each objective, check off each of the Learning Objectives for this section above.*

LO 5.3: Worked Example 5.2; Problems 5.42, 5.44, 5.46
LO 5.4: Problems 5.52, 5.53
LO 5.5: Worked Example 5.3; Problems 5.54, 5.55

Section 5.3

PRE-LECTURE:

Learning Objectives: *These are the skills you should be proficient in after reading this section and practicing the associated problems. Check off the boxes below once you have gone through this section, answered the associated questions in your textbook, and feel confident you have mastered the skill.*

- ☐ **LO 5.6** Describe the difference between a continuous spectrum and a line spectrum.
- ☐ **LO 5.7** Compare the wavelength and frequency of different electron transitions in the Bohr model of the atom.
- ☐ **LO 5.8** Use the Balmer–Rydberg equation to calculate the wavelength and energies of electron transitions in the hydrogen atom.

Definitions: *To learn, you must do. This is your workbook. Fill in the definitions from the glossary of the textbook as a first step to understanding the material.*

— **line spectrum—** ______________________________

____

— **Balmer–Rydberg equation—** ______________________________

____

Summary: *Read through this section and fill in the missing words and equations. Reference the textbook when you are not confident in your answers.*

Sunlight has a continuous spectrum; it contains all wavelengths of visible light, and it even extends outside of the visible range to ultraviolet and infrared. This is different from the light given off by "excited" atoms. Excitation of atoms can occur when an atom is subjected to heat or electricity. When this happens, they give off light only in the specific wavelengths that are unique to that element, visualized as a **line spectrum.**

These line spectra were not understood until decades after their discovery. Niels Bohr (1885–1962) put together Plank's postulate of quantized energy, Einstein's concept of photons, and Rutherford's nuclear model to propose an atomic model that explained the existence of the line spectra. In this model, he predicted that the center of the atom held small positively charged particles with electrons orbiting around it like planets orbit the sun. This leads to a quantization of energy as each orbit corresponded to a specific energy, and changes between these orbitals lead to the emission of a photon whose energy was equal to the change in energy between orbitals.

In 1885, Johannes Balmer was able to mathematically describe the pattern in the visible atomic line spectrum of the hydrogen atom. His equation was later extended by Johannes Rydberg to describe all the lines of the spectrum, including those outside the visible range.

Write the Balmer–Rydberg equation below:

$$\frac{1}{\lambda} =$$

where R is the *Rydberg constant,* $1.097 \times 10^{-2}\ nm^{-1}$, and n and m are integers such that $n > m$.

Workbook Example 5.3

PROBLEM:
Calculate the wavelength of light emitted when an electron falls from the $n = 8$ level to the $n = 3$ level in the hydrogen atom.

SOLUTION:
We solve this problem by using the Balmer–Rydberg equation. Assign the higher energy level to be "n," and the lower level to be "m."

$$\frac{1}{\lambda} = R\left[\frac{1}{m^2} - \frac{1}{n^2}\right]$$

$$\frac{1}{\lambda} = 1.097 \times 10^{-2}\ nm^{-1}\left[\frac{1}{3^2} - \frac{1}{8^2}\right] = 0.00105\ nm^{-1}$$

Don't forget to invert the answer— nm^{-1} is not the desired unit!

$$\frac{1}{\lambda} = \frac{1}{0.00105\ nm^{-1}} = 955\ nm$$

This is in the infrared range.

Workbook Problem 5.3

The Pfund series of spectral bands has $m = 5$ and $n > m$. What are the longest and shortest wavelengths in this series?

Strategy: Determine the values of n that will make λ the longest and shortest. Remember that the value of λ is greatest when the value of n is smallest, and the value λ is smallest when the value of n is greatest.

Step 1: Determine the value of n that will make λ the shortest and longest.

Step 2: Use the Balmer–Rydberg equation with $m = 5$ and solve for the shortest wavelength.

Step 3: Solve for λ using the Balmer–Rydberg equation and the values of m that make λ the longest.

POST-LECTURE:

Work Problems: *One of the best ways to learn chemistry is through solving problems. Go to the textbook and work through the Worked Examples for this section, then go to the end of chapter and work the problems there. When you feel confident that you can correctly answer problems like the ones listed below for each objective, check off each of the Learning Objectives for this section above.*

LO 5.6: Problems 5.56, 5.57
LO 5.7: Problem 5.27
LO 5.8: Worked Example 5.4; Problems 5.58, 5.60, 5.62, 5.64

Section 5.4 Wavelike Properties of Matter: de Broglie's Hypothesis

PRE-LECTURE:

Learning Objectives: *These are the skills you should be proficient in after reading this section and practicing the associated problems. Check off the boxes below once you have gone through this section, answered the associated questions in your textbook, and feel confident you have mastered the skill.*

☐ **LO 5.9** Calculate the wavelength of a moving object using the de Broglie equation.
☐ **LO 5.10** Explain why the wavelength of macroscopic objects cannot be observed.

Definitions: *To learn, you must do. This is your workbook. Fill in the definitions from the glossary of the textbook as a first step to understanding the material.*

— **de Broglie equation**—____

____

Summary: *Read through this section and fill in the missing words and equations. Reference the textbook when you are not confident in your answers.*

In 1925, Louis de Broglie suggested that if light can behave like a particle, it is only logical that particles can also behave like ___________, thus all matter will have wavelike properties. He supported his proposal with a mathematical derivation using Einstein's now-famous relationship between mass and energy: $E = mc^2$. Using the known relationship between wavelength and frequency, and substituting simple velocity (v) for the speed of light, de Broglie suggested that the following equation can be used to solve for the wavelength of any species of matter:

$$\lambda = \frac{h}{mv}$$

In this equation, the units for h are rearranged from joules to kg · m²/s or mass times velocity. For problems involving the de Broglie equation, use $h = 6.626 \times 10^{-34}$ kg · m²/s.

Although this discovery doesn't mean much on the macroscale where we interact with matter everyday, it is important at the subatomic level with particles like electrons. Electrons have since been shown to have wave like properties as well behave like particles.

Workbook Example 5.4

PROBLEM:
The muzzle velocity of a 50-gram sniper bullet is 860 m/s. The sniper must consider wind, movement of target, humidity, and altitude. What is the de Broglie wavelength of the ammunition at this speed? Could the de Broglie wavelength contribute to a missed shot?

SOLUTION:
Using the de Broglie equation, we can solve for wavelength:

$$\lambda = \frac{h}{mv} = \frac{6.626\times10^{-34}\ kg \cdot m^2/s}{(0.050\ kg)(860\ m/s)} = 1.5\times10^{-35}\ m$$

The de Broglie wavelength of the bullet is less than the diameter of an atom. But even if I might miss my shot, I'd still give it a try!

POST-LECTURE:

Work Problems: *One of the best ways to learn chemistry is through solving problems. Go to the textbook and work through the Worked Examples for this section, then go to the end of chapter and work the problems there. When you feel confident that you can correctly answer problems like the ones listed below for each objective, check off each of the Learning Objectives for this section above.*

LO 5.9: Worked Example 5.5; Problems 5.66, 5.67, 5.71
LO 5.10: Problem 5.68, 5.69

Section 5.5 The Quantum Mechanical Model of the Atom: Heisenberg's Uncertainty Principle

PRE-LECTURE:

Learning Objectives: *These are the skills you should be proficient in after reading this section and practicing the associated problems. Check off the boxes below once you have gone through this section, answered the associated questions in your textbook, and feel confident you have mastered the skill.*

☐ **LO 5.11** Calculate the uncertainty in the position of the moving object if the velocity is known.

Definitions: *To learn, you must do. This is your workbook. Fill in the definitions from the glossary of the textbook as a first step to understanding the material.*

— **quantum mechanical model—** ____

____

— **Heisenberg uncertainty principle—** ________________________________

____

Summary: *Read through this section and fill in the missing words and equations. Reference the textbook when you are not confident in your answers.*

The Bohr model was an essential step in the understanding of the structure of matter, but, like the Balmer–Rydberg equation, it only works for hydrogen. Its most important insight was the idea that the electrons in an atom are confined to specific energy levels.

Erwin Schrödinger proposed the **quantum mechanical model** of the atom in 1926. In this model, the idea of electrons as particles is abandoned in favor of their wavelike properties.

Just one year later, Werner Heisenberg provided support for this idea by demonstrating that it is impossible to know simultaneously both __________ an electron is and what __________. Any attempt to "see" an electron's position would require that it interact with photons of light. As these photons would transfer energy to the electron, the speed of the electron would be altered by the interaction.

This idea is called the Heisenberg uncertainty principle, and it can be expressed mathematically:

$$(\Delta x)(\Delta mv) \geq \frac{h}{4\pi}$$

The uncertainty is only evident on the atomic level, because on the macroscale the mass of an object is so large that the change in position and velocity are very small. Thus, we have no problem measuring both position and velocity for large objects.

Workbook Example 5.5

PROBLEM:
The mass of an electron is 9.11×10^{-31} kilograms, and its velocity, known to within 15%, is 2.50×10^6 m/s. What is the uncertainty in its position?

SOLUTION:
First, determine the uncertainty in the velocity:

$$0.15\left(2.5\times10^{6}\ {}^{m}\!/_{s}\right) = 3.75\times10^{5}\ {}^{m}\!/_{s}$$

The Heisenberg uncertainty principle equation can be rearranged to solve for Δx:

$$(\Delta x)(\Delta mv) \geq \frac{h}{4\pi}$$

$$\Delta x \geq \frac{h}{4\pi(\Delta mv)}$$

$$\Delta x \geq \frac{6.626\times10^{-34}\ {}^{kg\cdot m^2}\!/_{s}}{4\pi\left(9.11\times10^{-31}\ kg\right)\left(3.75\times10^{5}\ {}^{m}\!/_{s}\right)}$$

$$\Delta x \geq 1.54\times10^{-10}\ m\ or\ 154\ pm$$

Although these numbers are quite small, so are the objects in question. Like de Broglie wavelengths, these calculations lose their relevance in the macroscopic world.

POST-LECTURE:

Work Problems: *One of the best ways to learn chemistry is through solving problems. Go to the textbook and work through the Worked Examples for this section, then go to the end of chapter and work the problems there. When you feel confident that you can correctly answer problems like the ones listed below for each objective, check off each of the Learning Objectives for this section above.*

LO 5.11: Problems 5.72, 5.73

Section 5.6 The Quantum Mechanical Model of the Atom: Orbitals and Quantum Numbers

PRE-LECTURE:

Learning Objectives: *These are the skills you should be proficient in after reading this section and practicing the associated problems. Check off the boxes below once you have gone through this section, answered the associated questions in your textbook, and feel confident you have mastered the skill.*

☐ **LO 5.12** Identify and write sets of quantum numbers that describe electrons in different types of orbitals.

Definitions: *To learn, you must do. This is your workbook. Fill in the definitions from the glossary of the textbook as a first step to understanding the material.*

— **wave function—**____

____

— **orbital—**____

____

— **quantum number—** _____________________________________

____

— **shell—**____

____

— **principal quantum number (*n*)—** ____________________________

____

— **angular momentum quantum number (*l*)—**______________________

____

— **subshell—**____

____

— **magnetic quantum number (m_l)—** ___________________________

____

Summary: *Read through this section and fill in the missing words and equations. Reference the textbook when you are not confident in your answers.*

Although both the position and momentum of an electron cannot be precisely known, its probable location can be calculated using a differential equation known as Schrödinger's wave equation. The solutions to this equation are called *wave functions* or *orbitals*, and they are represented by the Greek letter ψ (psi). ψ^2 defines the probability of finding the electron in a specific area around the nucleus.

Three parameters are required for wave functions: these are called the **quantum numbers,** and they are defined using the following rules:

The **principal quantum number** is designated *n*. It defines the *energy level* of the electron. It also defines the *shell* of the electron. Electrons with $n = 5$, for example, are said to be in the fifth shell.

The **angular momentum quantum number** is designated *l*. It defines the *shape* of the orbital. These numbers define the *subshell* of the electron by the following system (enter the letter of the subshell below:

$l = 0$ ___ subshell $l = 1$ ___ subshell

$l = 2$ ___ subshell $l = 3$ ___ subshell

The **magnetic quantum number** is designated "m_l." It defines the *orientation* of an orbital.

The possible values for these numbers are:

- $n =$ ________________
- $l =$ ________________
- $m_l =$ ________________

The energy levels for hydrogen electrons are dependent only on the principal quantum number. In multielectron atoms, things are complicated by the fact that electrons can interact with one another as well as with the nucleus. In general, as energy is required to separate positive and negative charges, an electron with a larger value of *n* will be higher in energy.

Workbook Example 5.6

PROBLEM:
Give the values of the quantum numbers of a 2*p* subshell.

SOLUTION:
For a 2*p* subshell, $n = 2$. The fact that it is a *p* orbital says that *l* must be 1. Because m_l can be any number from $-l$ to l, it has possible values of −1, 0, and 1. There are three orbitals in a 2*p* subshell, because there are three possible orientations (as shown by the three magnetic quantum numbers).

Workbook Problem 5.4

Determine the subshell for an electron having $n = 3$, $l = 0$. How many orbitals are in this subshell?

Strategy: Determine the subshell associated with a value of $l = 0$.

Step 1: Identify the subshell with the value of *n* and the letter designation for $l = 0$.

Step 2:	Determine the number of orbitals in this subshell.

POST-LECTURE:

Work Problems: *One of the best ways to learn chemistry is through solving problems. Go to the textbook and work through the Worked Examples for this section, then go to the end of chapter and work the problems there. When you feel confident that you can correctly answer problems like the ones listed below for each objective, check off each of the Learning Objectives for this section above.*

LO 5.12: Worked Example 5.6; Problems 5.76, 5.78, 5.80, 5.86

Section 5.7 The Shapes of Orbitals

PRE-LECTURE:

Learning Objectives: *These are the skills you should be proficient in after reading this section and practicing the associated problems. Check off the boxes below once you have gone through this section, answered the associated questions in your textbook, and feel confident you have mastered the skill.*

☐ **LO 5.13** Identify an orbital based on its shape and describe it using a set of quantum numbers.
☐ **LO 5.14** Locate the nodal planes in different types of orbitals and different shells.

Definitions: *To learn, you must do. This is your workbook. Fill in the definitions from the glossary of the textbook as a first step to understanding the material.*

— **radical distribution plot—** ______________________________

____

— **node—** ______________________________________

____

— **phase—**______________________________________

____

Summary: *Read through this section and fill in the missing words and equations. Reference the textbook when you are not confident in your answers.*

The probability of finding an electron is greatest near the nucleus, but quickly decreases as distance from the nucleus increases. As a result, there is no definite boundary to the atom, and thus no definite size. The imaginary boundary is set so that the region includes a 90% chance of the electron being found in that volume.

The s orbitals have a ___________ shape. The size increases in successively higher shells, with nodes—areas of zero probability—occurring in shells with an $n > 1$. An area of zero probability corresponds to an area of zero amplitude in a wave. There is only one s orbital per shell ($l = 0$, so $m_l = 0$). Fill in the trend for orbital size and energy:

s orbital size and energy =

The p orbitals are commonly described as __________-shaped, with a node near the nucleus. As expected from the quantum number rules, there are three p orbitals per shell: $l = 1$, so $m_l = -1$, 0, or 1. These are defined as occurring along the x-axis, y-axis, and z-axis of a Cartesian coordinate system. The two lobes of a p orbital have different phases, which is important for covalent bonding (we will look at this in Chapter 8): only electrons that are in the same phase can interact. As the principal quantum number increases, the size of the p orbitals also ___________.

The d orbitals have a more complex shape. With an angular momentum quantum number of 2, there are five orbitals in each d subshell: $m_l = -2, -1, 0, 1, 2$. Four of the orbitals are shaped like paired p orbitals, 90 degrees rotated from one another; the fifth is shaped like a p orbital with a doughnut-shaped "skirt" in the xy-plane. As in the p orbitals, the alternating lobes have different phases.

The f orbitals have eight lobes and three nodal planes through the nucleus: $l = 3$, so $m_l = -3, -2, -1, 0, 1, 2, 3$. This additional complexity makes them difficult to describe, and elements that contain them will not be much discussed at this level of study.

POST-LECTURE:

Work Problems: *One of the best ways to learn chemistry is through solving problems. Go to the textbook and work through the Worked Examples for this section, then go to the end of chapter and work the problems there. When you feel confident that you can correctly answer problems like the ones listed below for each objective, check off each of the Learning Objectives for this section above.*

LO 5.13: Worked Example 5.7; Problems 5.13, 5.28, 5.93
LO 5.14: Problems 5.90–5.92

Section 5.8 Electron Spin and the Pauli Exclusion Principle

PRE-LECTURE:

Learning Objectives: *These are the skills you should be proficient in after reading this section and practicing the associated problems. Check off the boxes below once you have gone through this section, answered the associated questions in your textbook, and feel confident you have mastered the skill.*

☐ **LO 5.15** Assign a set of four quantum numbers for electrons in an atom.

Definitions: *To learn, you must do. This is your workbook. Fill in the definitions from the glossary of the textbook as a first step to understanding the material.*

— **spin quantum number (m_s)**—____

____

— **Pauli exclusion principle**—____

____

Summary: *Read through this section and fill in the missing words and equations. Reference the textbook when you are not confident in your answers.*

The three quantum numbers discussed earlier define the orbitals available in an atom, but they do not allow for a unique description of the electrons in an orbital.

The property that distinguishes the individual electrons in an orbital is their "spin." Electrons will spin either clockwise or counterclockwise, which causes them to have a tiny magnetic field. Each orbital can hold one each: an electron spinning clockwise and an electron spinning counterclockwise.

To describe this, a fourth quantum number, the *spin quantum number*—designated m_s—is used. It can have values of _____ or _____, referred to as "up" and "down" spins, and it is not constrained by the other quantum numbers.

Wolfgang Pauli proposed the **Pauli exclusion principle,** which states that no two electrons in the same atom can have the ___________ set of quantum numbers. Each orbital has the ability to hold two paired electrons: ↑↓.

Workbook Example 5.7

PROBLEM:
State the four quantum numbers for each of the electrons in a full 2*p* subshell.

SOLUTION:
A 2*p* orbital has a principal quantum number *n* of 2, an angular momentum quantum number *l* of 1, and magnetic quantum numbers m_l of −1, 0, and 1. For each of these orbitals, two electrons with +½ and −½ are possible. So the quantum numbers possible in this subshell are:

n	*l*	m_l	m_s
2	1	−1	+ ½
2	1	−1	− ½
2	1	0	+ ½
2	1	0	− ½
2	1	+1	+ ½
2	1	+1	− ½

There are six unique sets of quantum numbers for a *p* subshell, so it can hold six electrons.

POST-LECTURE:

Work Problems: *One of the best ways to learn chemistry is through solving problems. Go to the textbook and work through the Worked Examples for this section, then go to the end of chapter and work the problems there. When you feel confident that you can correctly answer problems like the ones listed below for each objective, check off each of the Learning Objectives for this section above.*

LO 5.15: Problems 5.80, 5.81, 5.86–5.89

Section 5.9 Orbital Energy Levels in Multielectron Atoms

PRE-LECTURE:

Learning Objectives: *These are the skills you should be proficient in after reading this section and practicing the associated problems. Check off the boxes below once you have gone through this section, answered the associated questions in your textbook, and feel confident you have mastered the skill.*

- ☐ **LO 5.16** Explain how electron shielding gives the order of subshells from lowest to highest in energy.
- ☐ **LO 5.17** Predict the order of filling of subshells based upon energy.

Definitions: *To learn, you must do. This is your workbook. Fill in the definitions from the glossary of the textbook as a first step to understanding the material.*

— **effective nuclear charge (Z_{eff})**—____

____

Summary: *Read through this section and fill in the missing words and equations. Reference the textbook when you are not confident in your answers.*

In hydrogen atoms, only the distance from the nucleus determines the energy level of an electron. In multielectron atoms, electron–electron repulsions also play a part in the energy level of an electron.

Outer-shell electrons are farther from the nucleus and held less tightly. The electrons that are closer to the nucleus also shield the outer electrons from the nuclear charge. These outer electrons are not subject to the full charge of the nucleus, only to an **effective nuclear charge, Z_{eff}.**

$$Z_{eff} = Z_{actual} - \text{Electron shielding}$$

We can see how this relates to the shapes of the orbitals: a lower angular momentum quantum number will correspond to less electron shielding. For example, a 2*s* electron and a 2*p* electron are in the same shell, but because of the shape of the orbital, a 2*s* electron will spend more time close to the nucleus than a 2*p* electron. Therefore, the 2*s* electron is lower in energy than the 2*p* electron. Draw arrows to show the direction in which effective nuclear charge and energy increase within a shell:

Attraction to the nucleus, Z_{eff}				
Orbital type =	ns	np	nd	nf
Energy				

POST-LECTURE:

Work Problems: *One of the best ways to learn chemistry is through solving problems. Go to the textbook and work through the Worked Examples for this section, then go to the end of chapter and work the problems there. When you feel confident that you can correctly answer problems like the ones listed below for each objective, check off each of the Learning Objectives for this section above.*

LO 5.16: Problems 5.94–5.97
LO 5.17: Problems 5.100–5.103

Section 5.10 Electron Configurations of Multielectron Atoms

PRE-LECTURE:

Learning Objectives: *These are the skills you should be proficient in after reading this section and practicing the associated problems. Check off the boxes below once you have gone through this section, answered the associated questions in your textbook, and feel confident you have mastered the skill.*

☐ **LO 5.18** Assign electron configurations to atoms in their ground state.

Definitions: *To learn, you must do. This is your workbook. Fill in the definitions from the glossary of the textbook as a first step to understanding the material.*

— **electron configuration—**____

— **aufbau principle—** ____

— **ground-state electron configuration—** ____________________________

— **Hund's rule—** ____

— **degenerate—** ____

Summary: *Read through this section and fill in the missing words and equations. Reference the textbook when you are not confident in your answers.*

Given the information above, we can now describe the **ground-state electron configuration,** or the lowest-energy configuration, of any atom. Orbitals with the same energy level, such as the three orbitals in a *p* subshell, are referred to as **degenerate.**

Three rules are used, collectively known as the **aufbau principle** (*aufbau* means "building up"):

- ____________________
- ____________________
- ____________________

There are two ways of representing these configurations. The first gives the principal quantum number, *n*, followed by the letter designation of the subshell, with the number of electrons in the subshell represented by a superscript:

H: $1s^1$ He: $1s^2$ Li: $1s^22s^1$ Be: $1s^22s^2$

One of the consequences of the rules about quantum numbers is that there is no $1p$ subshell: with a principal quantum number of 1, the only angular momentum quantum number allowed is 0, because the number must be less than *n*. Each shell has *n* subshells. For the $n = 1$ shell there is one subshell, the *s* subshell; for the $n = 2$ shell there are two subshells, the *s* subshell and the *p* subshell; for the $n = 3$ shell there are three subshells, the *s* subshell, the *p* subshell, and the *d* subshell; and for the $n = 4$ shell there are four subshells, the *s* subshell, the *p* subshell, the *d* subshell and the *f* subshell.

The *s* subshells hold 2 electrons, *p* subshells hold 6 electrons, *d* subshells hold 10 electrons, and *f* subshells hold 14. These numbers can all be pulled straight from the periodic table, as will be explained in a later section. After argon, the energy levels begin to cross one another, but these can still be taken from the periodic table, as will also be explained later.

The second way of representing these configurations uses *orbital filling diagrams* with the electrons represented by arrows. This method better shows the functioning of Hund's Rule:

	$1s$	$2s$	$2p$		
C:	↑↓	↑↓	↑ ↑ _	or	$1s^2\,2s^2\,2p^2$
N:	↑↓	↑↓	↑ ↑ ↑	or	$1s^2\,2s^2\,2p^3$
O:	↑↓	↑↓	↑↓ ↑ ↑	or	$1s^2\,2s^2\,2p^4$

Electron configurations can get fairly long and can be quite tedious to write. A shorthand version that gives the symbol of the noble gas from the previous period in square brackets and then lists all of the remaining electrons can be used.

Cl: ____________________ or [Ne] $3s^2\,3p^5$

Workbook Example 5.8

PROBLEM:
Give the electron configuration for neon.

SOLUTION:
The electron configuration for neon, with 10 electrons, is $1s^2\,2s^2\,2p^6$.

Noble gases, which are known to be unreactive, all have completely full energy levels: there are no more subshells available in the $n = 2$ energy level. This pattern continues with other noble gases as well.

Workbook Example 5.9

PROBLEM:
Give the orbital filling diagram for magnesium.

SOLUTION:
The orbital filling diagram for magnesium is

Mg:	↑↓	↑↓	↑↓ ↑↓ ↑↓	↑↓		or	$1s^2\,2s^2\,2p^6\,3s^2$.
	$1s$	$2s$	$2p$	$3s$			

For elements that have more electrons than a noble gas, the configuration of the noble gas can be inserted in brackets. So, the electron configuration of magnesium can also be written as

Mg: [Ne] $3s^2$

Workbook Problem 5.5

Give the ground-state electronic configuration, both complete and shorthand, for calcium. Also draw the orbital filling diagram.

Strategy: Using the periodic table, determine the number of electrons in calcium.

Step 1: Determine the number of electrons in calcium.

Step 2: Use the aufbau principle to determine the ground-state electronic configuration.

Step 3:	Determine the noble gas in the previous row and remove the electrons associated with it to a shorthand notation. Specifically name the electrons in any unfilled subshells.
Step 4:	Draw the orbital filling diagram.

POST-LECTURE:

Work Problems: *One of the best ways to learn chemistry is through solving problems. Go to the textbook and work through the Worked Examples for this section, then go to the end of chapter and work the problems there. When you feel confident that you can correctly answer problems like the ones listed below for each objective, check off each of the Learning Objectives for this section above.*

LO 5.18: Worked Example 5.8; Problems 5.104, 5.105, 5.109, 5.110

Section 5.11 Anomalous Electron Configurations

PRE-LECTURE:

Learning Objectives: *These are the skills you should be proficient in after reading this section and practicing the associated problems. Check off the boxes below once you have gone through this section, answered the associated questions in your textbook, and feel confident you have mastered the skill.*

☐ **LO 5.18** Assign electron configurations to atoms in their ground state.

Summary: *Read through this section and fill in the missing words and equations. Reference the textbook when you are not confident in your answers.*

Completely filled and exactly half-filled subshells confer stability. As a result, when energy levels are very close together, electrons will move from lower-energy shells into higher-energy shells to gain these configurations and become more stable. The larger the atom, the closer together the energy levels are found. Thus, for some of the transition metals, electron configurations occur that are not accounted for by the previous rules. The two elements from period 4 that have anomalous electron configurations are ___________ and ___________.

Workbook Example 5.10

PROBLEM:
Give the orbital filling diagram of chromium.

SOLUTION:
Chromium is the smallest element that will transfer electrons to gain stability. Its electron configuration is [Ar]$4s^1 3d^5$. The reason for the increased stability can be seen in the orbital filling diagram:

Cr:	↑↓	↑↓	↑↓ ↑↓ ↑↓	↑↓	↑↓ ↑↓ ↑↓	↑	↑ ↑ ↑ ↑ ↑
	1*s*	2*s*	2*p*	3*s*	3*p*	4*s*	3*d*

POST-LECTURE:

Work Problems: *One of the best ways to learn chemistry is through solving problems. Go to the textbook and work through the Worked Examples for this section, then go to the end of chapter and work the problems there. When you feel confident that you can correctly answer problems like the ones listed below for each objective, check off each of the Learning Objectives for this section above.*

LO 5.18: Worked Example 5.8; Problems 5.104, 5.105, 5.109, 5.110

Section 5.12 Electron Configurations and the Periodic Table

PRE-LECTURE:

Learning Objectives: *These are the skills you should be proficient in after reading this section and practicing the associated problems. Check off the boxes below once you have gone through this section, answered the associated questions in your textbook, and feel confident you have mastered the skill.*

- ☐ **LO 5.19** Draw orbital filling diagrams for the ground state of an atom and determine the number of unpaired electrons.
- ☐ **LO 5.20** Identify atoms from orbital filling diagrams or electron configurations.

Definitions: *To learn, you must do. This is your workbook. Fill in the definitions from the glossary of the textbook as a first step to understanding the material.*

— **valence shell—** ______________________________

____

— ***s*-block element—** ______________________________

____

— ***p*-block element—** ______________________________

____

— ***d*-block element—** ______________________________

____

— ***f*-block element—** ______________________________

____

Summary: *Read through this section and fill in the missing words and equations. Reference the textbook when you are not confident in your answers.*

The electrons that cannot be listed as the noble gas in brackets in the shorthand electron configuration are **valence shell** electrons. They are the most loosely held electrons, and they determine the __________ behavior of an element. All of the elements in a given group will have the same valence shell configuration: only the principal quantum number will change. It is the periodic trend in valence electrons that gives the elements in the same group similar properties.

You can use the periodic table to help remember the order of orbital filling by starting at hydrogen and adding an electron for each subsequent element you pass on your way to the element whose electron configuration you are writing. To help with this it is important to understand that the periodic table is divided into "blocks" based on what type of subshell the outermost valence electrons are in:

- *s*-block elements: Groups 1A and 2A (filling of an *s* orbital); n = row number.
- *p*-block elements: Groups 3A through 8A (filling of *p* orbitals; *ns* orbitals are filled); n = row number.
- *d*-block elements: transition metals; n = row number −1.
- *f*-block elements: lanthanide and actinide elements; n = row number −2.

Go to the periodic table at the front of this guide and color in the "blocks" on your periodic table Don't forget to label your *d*- and *f*-blocks with a note to subtract 1 or 2 from the row number.

Workbook Example 5.11

PROBLEM:
Using only the periodic table, write the shorthand electron configuration for technetium (Tc).

SOLUTION:
Technetium is in the *d*-block, in the fifth row of the periodic table. The noble gas that is before it in electron configuration is krypton (the end of row 4). The electron configuration will begin with krypton. Because technetium is in the fifth row of the periodic table, its *s* electrons that will immediately follow will be 5*s*, and this subshell is filled. The *d* electrons will be in the $n - 1$ energy level, so they will be 4*d* electrons. There will be five of them.

$$\text{Tc} = [\text{Kr}]\ 5s^2\ 4d^5$$

POST-LECTURE:

Work Problems: *One of the best ways to learn chemistry is through solving problems. Go to the textbook and work through the Worked Examples for this section, then go to the end of chapter and work the problems there. When you feel confident that you can correctly answer problems like the ones listed below for each objective, check off each of the Learning Objectives for this section above.*

LO 5.19: Worked Example 5.8; Problems 5.15, 5.16, 5.106–5.108
LO 5.20: Problems 5.16, 5.31, 5.32

Section 5.13 Electron Configurations and Periodic Properties: Atomic Radii

PRE-LECTURE:

Learning Objectives: *These are the skills you should be proficient in after reading this section and practicing the associated problems. Check off the boxes below once you have gone through this section, answered the associated questions in your textbook, and feel confident you have mastered the skill.*

- ☐ **LO 5.21** Explain the periodic trend in atomic radii.
- ☐ **LO 5.22** Predict the relative size of atoms based upon their position in the periodic table.

Summary: *Read through this section and fill in the missing words and equations. Reference the textbook when you are not confident in your answers.*

Although atoms cannot be said to have a definite size, atomic radii can be assigned to atoms by measuring covalent bonds. When two identical atoms are covalently bonded together, half of the bond length is known as the atomic radius.

As *n* increases down through a group, atomic radii ___________ as larger valence shell orbitals are being occupied.

Not as obvious, is that when going across the periodic table from left to right, atomic radius ___________. This occurs because the effective nuclear charge felt by each added electron increases due to the increasing positive charge in the nucleus as the atomic number increases. Although the number of electrons increases as the number of protons increases, electrons within the same subshell only shield each other weakly.

Add arrows to your periodic table to help remind you of the trend of atomic radii.

Workbook Example 5.12

PROBLEM:
Choose the smaller of the following pairs:

K and Fe	N and Be	S and Al	Br and Ca

SOLUTION:
These element pairs are all in the same row as one another, so in each case the element that is rightmost in the periodic table will be smaller.

K > Fe	N < Be	S < Al	Br < Ca

POST-LECTURE:

Work Problems: *One of the best ways to learn chemistry is through solving problems. Go to the textbook and work through the Worked Examples for this section, then go to the end of chapter and work the problems there. When you feel confident that you can correctly answer problems like the ones listed below for each objective, check off each of the Learning Objectives for this section above.*

LO 5.21: Problems 5.118, 5.119
LO 5.22: Worked Example 5.9; Problems 5.120–5.123

Self-Test

This section is intended to test your knowledge of the material covered in this chapter. Think through these problems and make certain you understand what the problem is about—ask yourself if the answer makes sense. Many of these questions are linked to the chapter learning goals. Therefore, successful completion of these problems indicates that you have achieved the learning goals for this chapter. You will receive the greatest benefit from this section if you use it as a mock exam. Do not use your notes, book, Internet, or friends when taking the test. There is no grade and you only shortchange yourself by getting help with this activity. As you take the test, mark the problems with your level of confidence with the answer.

- ★ *Use a star for problems you know you got right.*
- ? *Use a question mark for those you think you got right.*
- x *Use an x for those you are guessing on.*

If you get a question right that you marked with a star, you can be confident that you know the associated LO.

If you get a question right that you marked with a question mark, go back and work a few more problems and test yourself again tomorrow.

If you get a question wrong or mark it with an x, go back and read the section again, work a few problems today, a few tomorrow, and a few the next day before testing yourself again.

This will help you to discover which topics you have learned thoroughly and which topics you need to study in more detail.

True–False

1. The amplitude of a wave is the number of crests that pass a certain point per second.
2. Each element has a unique spectral "signature."
3. The photoelectric effect provides evidence for the particle-like nature of light.
4. Even at the size of everyday objects, de Broglie wavelengths are relevant.
5. The Bohr model of the atom holds true for all elements.
6. The energy of a photon can be calculated from its frequency.
7. The set of numbers {4, 4, 3, +½} is an allowed set of quantum numbers.
8. Electrons reside in *f* orbitals in common nonmetals.
9. Electrons in the same subshell do not effectively shield one another from nuclear charge.
10. Electrons that share an orbital must have parallel spins.

Multiple Choice

11. Electromagnetic energy is characterized by
    a. wavelength
    b. amplitude
    c. frequency
    d. all of the above

12. The Balmer–Rydberg equation accounts for the spectral lines of
    a. helium
    b. hydrogen
    c. halfnium
    d. any element

13. For photons to knock electrons out of a metal, they must
    a. have sufficient intensity
    b. have values close to Planck's constant
    c. be above the threshold frequency
    d. occur in large numbers

14. de Broglie determined that at the atomic level
    a. matter has a wavelike nature
    b. electrons are very slow moving
    c. elements are sometimes photons
    d. noble gases are mostly energy

15. Which of the following are permitted sets of quantum numbers?
    a. 3, 4, −2, +½
    b. 4, 3, −2, −½
    c. 4, 2, 2, 1
    d. 4, 2, 2, +⅔

16. Frequency, wavelength, and energy are related such that:
    a. low-frequency electromagnetic energy has a short wavelength and high-energy photons
    b. high-frequency electromagnetic energy has a long wavelength and high-energy photons
    c. low-frequency electromagnetic energy has a long wavelength and low-energy photons
    d. high-frequency electromagnetic energy has a short wavelength and low-energy photons

17. The Heisenberg uncertainty principle states that
    a. matter and energy are the same thing
    b. the position of an electron or its velocity can be known, but not both at the same time
    c. electrons need to fill degenerate orbitals with perpendicular spins first
    d. everything with momentum has a wavelength

18. The alkali metals are
    a. *s*-block elements
    b. *p*-block elements
    c. *d*-block elements
    d. *f*-block elements

19. The electron configuration for F is
    a. $1s^2\,1p^6\,1d^7$
    b. $1s^2\,2s^2\,2p^6\,3s^2\,3p^5$
    c. $[Ar]\,3s^2\,3p^5$
    d. $1s^2\,2s^2\,2p^5$

20. The size of elements *increases*
    a. down the groups and left to right along the periods
    b. up the groups and left to right along the periods
    c. down the groups and right to left along the periods
    d. up the groups and right to left along the periods

Matching

21. degenerate	a. can hold two electrons with opposite spins
22. wavelength	b. equal in energy
23. frequency	c. unit of s^{-1} or 1/s
24. amplitude	d. height of a wave from the midline
25. hertz	e. number of crests that pass a certain point per second
26. photon	f. set of rules that guide the filling of atomic orbitals
27. orbital	g. distance from peak to peak of a wave
28. aufbau principle	h. ejection of electrons from a clean metal surface by certain wavelengths of light
29. photoelectric effect	i. packet of light energy

Fill in the Blanks

30. When discussing electromagnetic energy, the Greek letter λ indicates ________________, while the Greek letter ν indicates __________________.

31. When excited with electricity or heat, elements give off a characteristic __________________.

32. The photoelectric effect was explained by Einstein and Planck as demonstrating that light contained ______________, or particles of light.

33. In the Bohr model of the atom, a transition from a higher energy level to a lower energy level corresponds with the ______________ of a photon with specific energy.

34. An electron can be specifically described using four ________________ ______________.

35. __________________ are spherical. Their energy depends only on their distance from the nucleus.

36. __________________ subshells are dumbbell shaped, with nodes at the nucleus. Each of the orbitals can hold _____ electrons, for a total of ______ electrons in the subshell.

37. Hund's rule states that when __________________ orbitals are being filled, they must first fill with electrons having ____________________________________.

38. The outermost electrons in an atom are called the ____________________. These determine the chemical reactivity of the element.

39. One-half of the bond length when two identical atoms are covalently bonded is known as the ___________ _____________________.

Problems

40. Cellular telephones commonly transmit at 835.6 MHz. What is the wavelength of this energy? How much energy is contained in one mole of cell phone photons?

41. Gamma rays have a wavelength of 2.15×10^{-5} nm. What is the frequency of this energy? How much energy is contained in one mole of gamma ray photons?

42. A 432-nm laser emits a quick pulse that contains 2.5 mJ of energy. How many photons were in the pulse?

43. The ionization energy of sodium metal is 495.8 kJ/mol. If this much energy is required to eject 1 mol of electrons from the surface, what frequency and wavelength of light corresponds to the energy of photons required? Will light with wavelength 540 nm dislodge electrons from sodium? Will light with wavelength 195 nm?

44. An electron has an uncertainty in its position of 327 pm. What is the uncertainty in its velocity?

45. If a hydrogen electron in the $n = 6$ energy level returns to ground state, what is the wavelength of the light emitted? What is the energy of the photon emitted?

46. Without using shorthand notation, write the electron structure for Yttrium.

47. Write the four quantum numbers for every electron in the ground state of neon.

48. Write the first two quantum numbers for the highest-energy electron in bromine.

49. When electrons are unpaired, the weak magnetic field they generate is reinforced by any other unpaired electrons in the same atom. This leads to a phenomenon called *paramagnetism*: the more unpaired electrons, the more the atom is attracted to a magnetic field. Draw the orbital filling diagrams for Si, P, and Cl and rank them from most paramagnetic to least paramagnetic.

50. If an excited hydrogen atom falls to $n = 2$ and emits a photon of light with 4.58×10^{-19} J, at what energy level did it begin?

51. If an electron is accelerated to 2.57×10^{6} m/s, what is its de Broglie wavelength? The mass of an electron is 9.109×10^{-31} kg.

52. Identify the atoms with the following electron configurations:
    a. $[Ar]4s^{1}3d^{10}$
    b. $[Xe]6s^{2}4f^{14}5d^{10}6p^{5}$
    c. $[Kr]\,4d^{10}$
    d. $[Ar]4s^{1}3d^{5}$

53. Group the following atoms in order of increasing atomic radius: Os, Hs, Fe, Ru

54. Group the following atoms in order of increasing atomic radius: S, Mg, Cl, Na

Challenge Problems

55. *Watts* are units of power defined as J/s. If an incandescent bulb consumes 40 watts at 550 nm for 4 hours, how many photons were emitted if it were 100% efficient? If only 5% of the electricity consumed by the bulb was converted to light, and the rest was converted to heat, how long would the bulb need to be lit to heat 200 g of water from 20 °C to 95 °C? The specific heat of water is 4.184 J/g · °C. Use the units of specific heat to figure out how to determine the temperature change.

56. If a laser consumes 1500 W of power and produces 2.73×10^{20} photons per second, what would the wavelength be if the efficiency of the laser were 100%? 80%? 20%? If the wavelength at which the laser operates is 1542 nm, how efficient is it?

Test Yourself: *Take the Practice Test on pages* 199–201 *of your textbook. Again, use a testing environment using only those items that your instructor will allow you to have access to during the test and mark each question with a star, question mark, or x to denote your confidence level. Once you have finished you can check off the LO at the beginning of this chapter based on which questions you starred and got correct. Go back and spend more time with the topics from the other learning objectives.*

CHAPTER SIX

Ionic Compounds: Periodic Trends and Bonding Theory

Learning Objectives

As a result of reading and studying this chapter, you should be able to:

- ☐ **LO 6.1** Write ground-state electron configurations for main group and transition metal ions.
- ☐ **LO 6.2** Determine the number of unpaired electrons in main group and transition metal ions.
- ☐ **LO 6.3** Predict the relative size of anions, cations, and neutral atoms.
- ☐ **LO 6.4** Predict the relative size of isoelectronic ions.
- ☐ **LO 6.5** Order elements from lowest to highest ionization energy.
- ☐ **LO 6.6** Explain the periodic trend in ionization energy.
- ☐ **LO 6.7** Compare successive ionization energies for different elements.
- ☐ **LO 6.8** Identify elements based on values of successive ionization energies.
- ☐ **LO 6.9** Compare the value of electron affinity for different elements.
- ☐ **LO 6.10** Explain the periodic trend in electron affinity.
- ☐ **LO 6.11** Predict whether a precipitation reaction will occur and write the ionic and net ionic equations.

Chapter Summary

Chemical bonds are the forces that hold atoms together. There are two types: ionic bonds and covalent bonds. This chapter concentrates on the reasons ionic bonds are formed and the energies involved in the formation of these bonds. You will learn that the underlying reason for the formation of ions and ionic compounds is the valence shell configuration of the elements, which determines the amount of energy needed for losing or gaining electrons. From the periodic trends in the energies for ion formation, you will learn how to predict which elements form cations or anions, and if the formation of ionic compounds is energetically favorable. You will also learn about the Born–Haber cycle, a series of hypothetical steps used to calculate the overall energy involved in the formation of ionic compounds.

The Chapter in Detail

Section 6.1 Electron Configurations of Ions

PRE-LECTURE:

Learning Objectives: *These are the skills you should be proficient in after reading this section and practicing the associated problems. Check off the boxes below once you have gone through this section, answered the associated questions in your textbook, and feel confident you have mastered the skill.*

- ☐ **LO 6.1** Write ground-state electron configurations for main group and transition metal ions.

☐ **LO 6.2** Determine the number of unpaired electrons in main group and transition metal ions.

Definitions: *To learn, you must do. This is your workbook. Fill in the definitions from the glossary of the textbook as a first step to understanding the material.*

— **isoelectronic—** ____

____

Review: *Before going through the section make sure you are familiar with these concepts:*

- Ch. 5.10 electron configurations

Summary: *Read through this section and fill in the missing words and equations. Reference the textbook when you are not confident in your answers.*

Metal elements reside on the __________-hand side of the periodic table and tend to give up electrons to form __________. When they do so, they adopt the electron configurations of noble gases: alkali metals lose one electron to ionize and adopt the electron configurations of the noble gases previous to them on the periodic table, alkaline earth metals lose two electrons and adopt the electron configuration of the previous noble gas. In groups IA and 2A, the topmost energy level is emptied to achieve noble gas configuration.

Nonmetals, on the __________ side of the periodic table, __________ electrons and move to the electron configurations of the noble gases after them. Halogens all gain one electron when they are ionized; group 6A elements gain two electrons and adopt the same electron configurations as the noble gas elements in family 8A. In these elements, the topmost energy level is filled to gain noble gas configuration.

Transition metals ionize so that they lose their _____ electrons first, then they begin losing *d* electrons. This sometimes gives multiple stable ions, which accounts for the multiple ionization states of some transition metals.

Workbook Example 6.1

PROBLEM:
Write shorthand electron configurations for the following elements and their ions:

Mg^{2+} F^- S^{2-} Fe^{3+}

SOLUTION:

Mg =	$[Ne]3s^2$	Mg^{2+} – [Ne]	Noble gas configuration
F =	$[He]2s^22p^5$	F^- – [Ne]	Noble gas configuration
S =	$[Ne]3s^23p^4$	S^{2-} – [Ar]	Noble gas configuration
Fe =	$[Ar]4s^23d^6$	Fe^{3+} – $[Ar]3d^5$	Half-filled *d* orbital provides stability

POST-LECTURE:

Work Problems: *One of the best ways to learn chemistry is through solving problems. Go to the textbook and work through the Worked Examples for this section, then go to the end of chapter and work the problems there. When you feel confident that you can correctly answer problems like the ones listed below for each objective, check off each of the Learning Objectives for this section above.*

LO 6.1: Worked Example 6.1; Problems 6.21, 6.40, 6.42, 6.44
LO 6.2: Problems 6.46, 6.47

Section 6.2 Ionic Radii

PRE-LECTURE:

Learning Objectives: *These are the skills you should be proficient in after reading this section and practicing the associated problems. Check off the boxes below once you have gone through this section, answered the associated questions in your textbook, and feel confident you have mastered the skill.*

- ☐ **LO 6.3** Predict the relative size of anions, cations, and neutral atoms.
- ☐ **LO 6.4** Predict the relative size of isoelectronic ions.

Summary: *Read through this section and fill in the missing words and equations. Reference the textbook when you are not confident in your answers.*

Cations are smaller than neutral atoms. When cations are formed, electrons are removed from valence shell orbitals. In the case of groups 1A and 2A, the outermost energy levels electrons are removed entirely. In addition, the effective nuclear charge (Z_{eff}) on all the remaining electrons increases, and those electrons are held more closely. Add < or > to the box below to show the change in radius between a neutral atom and its cation:

radius of neutral atom		radius of cation

Anions are larger than neutral atoms. When anions are formed, valence shell orbitals are completed. There are more electrons than protons, so the effective nuclear charge felt by the electrons will decrease. The additional electrons will also add to the electron–electron repulsions felt by the outer electrons. The changes can be dramatic: a chlorine ion is almost twice the size of a neutral chlorine atom. Add < or > to the box below to show the change in radius between a neutral atom and its anion:

radius of neutral atom		radius of anion

Add the trend of ionic radii to your periodic table.

Workbook Example 6.2

PROBLEM:
Which atom or ion in the following pairs has the larger radius:
a. Cr^{2+} or Cr^{4+} b. P or P^{3-} c. S or S^{2-} d. Na or Na^{+}

SOLUTION:
a. Cr^{2+}: The radius of Cr^{4+} is smaller, because both of the *s* and *d* orbitals are empty, so the pull on the remaining electrons is correspondingly stronger.
b. P^{3-}: The addition of three electrons to a neutral phosphorus atom will greatly decrease the Z_{eff} felt by the electrons and increase electron–electron repulsions.
c. S^{2-}: The addition of electrons causes a decrease in Z_{eff} so that the outer electrons are more loosely held, and the radius is greater.
d. Na: The removal of electrons causes an increase in Z_{eff}, which causes a decrease in the radius.

POST-LECTURE:

Work Problems: *One of the best ways to learn chemistry is through solving problems. Go to the textbook and work through the Worked Examples for this section, then go to the end of chapter and work the problems there. When you feel confident that you can correctly answer problems like the ones listed below for each objective, check off each of the Learning Objectives for this section above.*

LO 6.3: Worked Example 6.2; Problems 6.22, 6.48, 6.49, 6.51
LO 6.4: Problems 6.52, 6.54

Section 6.3 Ionization Energy

PRE-LECTURE:

Learning Objectives: *These are the skills you should be proficient in after reading this section and practicing the associated problems. Check off the boxes below once you have gone through this section, answered the associated questions in your textbook, and feel confident you have mastered the skill.*

☐ **LO 6.5** Order elements from lowest to highest ionization energy.
☐ **LO 6.6** Explain the periodic trend in ionization energy.

Definitions: *To learn, you must do. This is your workbook. Fill in the definitions from the glossary of the textbook as a first step to understanding the material.*

— **ionization energy (E_i)**— ______________________________

____

— **core electron**— ______________________________

____

Summary: *Read through this section and fill in the missing words and equations. Reference the textbook when you are not confident in your answers.*

Ionization energy (E_i) is the amount of energy required to create a +1 cation from a gas-phase atom. It is always positive: energy must always be added to an atom to remove an electron, no matter how stable the resulting ion will be.

Ionization energy shows periodic trends:

- The lowest ionization energies are those for the alkali metals, because the loss of one electron empties the highest energy level and noble gas configuration is achieved.
- The highest ionization energies are those of the noble gases, because the loss of an electron at this level means that you would have to break into a filled energy level.
- Ionization energy ___________ down a group: it takes less energy to remove the outermost electron from a large atom than from a small atom, because it is farther from the nucleus.

The periodic trends are not perfectly smooth:

- Group 3A ionization energies are ___________ than those of group 2A. The *s* subshell electron removed to ionize an element from group 2A is closer to the nucleus and therefore harder to remove than the *p* subshell electron removed to ionize a group 3A element.
- Group 6A ionization energies are ___________than those of group 5A. The loss of one electron from an element in group 6A creates a more-stable, half-filled *p* subshell.

Add the general trend of ionization energy to your periodic table.

Workbook Example 6.3

PROBLEM:
Draw the electron-filling diagrams for Be, B, N, and O. Explain how these demonstrate the reasons behind the lower E_i for group 3A compared to group 2A, and group 6A compared to group 5A.

SOLUTION:

Be:	↑↓ (1*s*) ↑↓ (2*s*)		Ionization requires disruption of filled subshell
B:	↑↓ (1*s*) ↑↓ (2*s*)	↑ __ __ (2*p*)	Ionization removes single, well-shielded electron
N:	↑↓ (1*s*) ↑↓ (2*s*)	↑ ↑ ↑ (2*p*)	Ionization requires disruption of half-filled subshell
O:	↑↓ (1*s*) ↑↓ (2*s*)	↑↓ ↑ ↑ (2*p*)	Ionization creates stable half-filled subshell

POST-LECTURE:

Work Problems: *One of the best ways to learn chemistry is through solving problems. Go to the textbook and work through the Worked Examples for this section, then go to the end of chapter and work the problems there. When you feel confident that you can correctly answer problems like the ones listed below for each objective, check off each of the Learning Objectives for this section above.*

LO 6.5: Worked Example 6.3; Problems 6.6, 6.58, 6.59
LO 6.6: Problem 6.56

Section 6.4 Higher Ionization Energies

PRE-LECTURE:

Learning Objectives: *These are the skills you should be proficient in after reading this section and practicing the associated problems. Check off the boxes below once you have gone through this section, answered the associated questions in your textbook, and feel confident you have mastered the skill.*

☐ **LO 6.7** Compare successive ionization energies for different elements.
☐ **LO 6.8** Identify elements based on values of successive ionization energies.

Summary: *Read through this section and fill in the missing words and equations. Reference the textbook when you are not confident in your answers.*

It is possible to remove sequential electrons from an atom, and the energy required ___________ for each successive electron that is removed. When all of the valence electrons have been removed, there will be a relatively large jump in ionization energy.

Workbook Example 6.4

PROBLEM:
Identify the following fourth-period element from these ionization energies:

E_i = 590 kJ/mol	E_{i2} = 1145 kJ/mol	E_{i3} = 4912 kJ/mol
E_{i4} = 6491 kJ/mol	E_{i5} = 8153 kJ/mol	

SOLUTION:
To determine which group this element is in, it is necessary to look for a large jump in the ionization energies. We find this between E_{i2} and E_{i3}, which indicates that two electrons can be lost fairly easily, but once two electrons are lost, the ion has noble gas configuration and is difficult to ionize further. This describes the alkaline earth metals; the period four alkaline earth metal is calcium.

Workbook Problem 6.1

From the following list of ionization energies, identify the group of the following element:

E_{i1} = 1012 kJ/mol	E_{i2} = 1903 kJ/mol	E_{i3} = 2912 kJ/mol
E_{i4} = 4956 kJ/mol	E_{i5} = 6273 kJ/mol	E_{i6} = 22 233 kJ/mol

Strategy:	Look for a relatively large jump between ionization energies: this will show the number of valence electrons and thus the distance from the nearest noble gas.
Step 1:	Find the large gap in ionization energies and identify the group number.

POST-LECTURE:

Work Problems: *One of the best ways to learn chemistry is through solving problems. Go to the textbook and work through the Worked Examples for this section, then go to the end of chapter and work the problems there. When you feel confident that you can correctly answer problems like the ones listed below for each objective, check off each of the Learning Objectives for this section above.*

LO 6.7: Worked Example 6.4; Problems 6.24, 6.60, 6.62
LO 6.8: Problems 6.8, 6.25, 6.64, 6.65

Section 6.5 Electron Affinity

PRE-LECTURE:

Learning Objectives: *These are the skills you should be proficient in after reading this section and practicing the associated problems. Check off the boxes below once you have gone through this section, answered the associated questions in your textbook, and feel confident you have mastered the skill.*

- ☐ **LO 6.9** Compare the value of electron affinity for different elements.
- ☐ **LO 6.10** Explain the periodic trend in electron affinity.

Definitions: *To learn, you must do. This is your workbook. Fill in the definitions from the glossary of the textbook as a first step to understanding the material.*

— **electron affinity (E_{ea})**—____

____

Summary: *Read through this section and fill in the missing words and equations. Reference the textbook when you are not confident in your answers.*

Electron affinity (E_{ea}) is a measure of the energy change when an electron is added to an isolated gas-phase atom. Electron affinities are often negative; a negative energy value is associated with the ___________ of energy. Energy is often released when an electron is added because the attraction between the negative electron and the positive nucleus outweighs the repulsion between electrons. The more stable the anion, the more energy is released. Electron affinity gets more negative as you go from left to right across the periodic table, but contrary to other trends, there is not much change as you go down the group.

Workbook Example 6.5

PROBLEM:
Using electron filling diagrams, rank the E_{ea} of Si, P, S, and Cl.

SOLUTION:

	1s	2s	2p	3s	3p	
Si:	↑↓	↑↓	↑↓ ↑↓ ↑↓	↑↓	↑ ↑ __	Ionization adds to stability by half-filling the 3*p* subshell.
P:	↑↓	↑↓	↑↓ ↑↓ ↑↓	↑↓	↑ ↑ ↑	Ionization destabilizes the half-filled subshell.
S:	↑↓	↑↓	↑↓ ↑↓ ↑↓	↑↓	↑↓ ↑ ↑	Ionization moves toward the noble gas configuration.
Cl:	↑↓	↑↓	↑↓ ↑↓ ↑↓	↑↓	↑↓ ↑↓ ↑	Ionization completes noble gas configuration.

Based on the electron-filling diagrams, and the general trend that the electron affinities increase in moving toward the halogens, one would expect Cl to have the largest negative E_{ea}, followed by S, then Si, with P having the smallest negative E_{ea} due to the destabilizing effect of adding an electron to a half-filled *p* subshell. This is in fact the case.

POST-LECTURE:

Work Problems: *One of the best ways to learn chemistry is through solving problems. Go to the textbook and work through the Worked Examples for this section, then go to the end of chapter and work the problems there. When you feel confident that you can correctly answer problems like the ones listed below for each objective, check off each of the Learning Objectives for this section above.*

LO 6.9: Worked Example 6.5; Problems 6.68, 6.69
LO 6.10: Problems 6.70–6.73

Section 6.6 The Octet Rule

PRE-LECTURE:

Learning Objectives: *These are the skills you should be proficient in after reading this section and practicing the associated problems. Check off the boxes below once you have gone through this section, answered the associated questions in your textbook, and feel confident you have mastered the skill.*

☐ **LO 6.11** Use the octet rule to predict charges on main-group ions, electron configurations of main-group ions, and formulas for ionic compounds.

Definitions: *To learn, you must do. This is your workbook. Fill in the definitions from the glossary of the textbook as a first step to understanding the material.*

— **octet rule—** ______________________________

____

Summary: *Read through this section and fill in the missing words and equations. Reference the textbook when you are not confident in your answers.*

From the past few sections it can be seen that:

- alkali metals (group 1A) have low first ionization energies, and so tend to lose their ns^1 electrons in reactions to gain a noble gas configuration.
- alkaline earth metals (group 2A) have low first and second ionization energies, and so tend to lose all of their *ns* electrons in reactions so as to gain noble gas configuration.
- halogens (group 7A) have a large negative electron affinity, and so tend to ionize readily to obtain noble gas configuration.
- noble gases (group 8A) do not tend to gain or lose electrons. They are largely inert.

All of these observations can be summarized in the **octet rule:** main-group elements tend to undergo reactions that leave them with ___________ outer-shell electrons.

Why does it work this way? The Group 1A and 2A metals ___________ valence electrons easily, because the electrons they lose have a low Z_{eff}: they are the only electrons in the outermost *s* orbital. The group 7A elements ___________ electrons easily because the electrons have a high Z_{eff} and can hold an additional electron. In both cases, the atoms have filled their outer *s* and *p* subshell and thus have eight electrons in their outermost shell.

Workbook Example 6.6

PROBLEM:
Which noble gas configurations are adopted by the atoms in the following ionic compounds:

Li_2O AlN $CaBr_2$

SOLUTION:
In lithium oxide, the two lithium atoms each lose an electron and adopt the electron configuration of helium. The two lost electrons are gained by oxygen, which then has the electron configuration of neon.

In aluminum nitride, the aluminum metal gives up three valence electrons to adopt the electron configuration of neon, while the nitrogen takes on the three electrons and also has the electron structure of neon.

When calcium loses two electrons, it gains the electron structure of argon. The bromine atoms each take one of the lost electrons to take on the electron structure of krypton.

Workbook Problem 6.2

Will the following elements be more likely to gain or lose electrons when they form ions? How many electrons will they gain or lose?

Rb Sc Se As

Strategy: Find these elements on the periodic table and locate the nearest noble gas. Remembering the general rule that metals lose electrons and nonmetals gain them, and that main-group elements ionize in predictable ways, determine the most likely ionized form of the element.

Step 1:	Identify the group number of the elements and identify them as metals or nonmetals.
Step 2:	Find the nearest noble gas.
Step 3:	Identify the number of electrons that must be gained/lost to achieve the electron configuration of that noble gas.
Step 4:	Write the ionized form of the element.

POST-LECTURE:

Work Problems: *One of the best ways to learn chemistry is through solving problems. Go to the textbook and work through the Worked Examples for this section, then go to the end of chapter and work the problems there. When you feel confident that you can correctly answer problems like the ones listed below for each objective, check off each of the Learning Objectives for this section above.*

LO 6.11: Worked Example 6.6; Problems 6.31, 6.74, 6.76, 6.78

Section 6.7 Ionic Bonds and the Formation of Ionic Solids

PRE-LECTURE:

Learning Objectives: *These are the skills you should be proficient in after reading this section and practicing the associated problems. Check off the boxes below once you have gone through this section, answered the associated questions in your textbook, and feel confident you have mastered the skill.*

- ☐ **LO 6.12** Match the formula of an ionic compound with its molecular image.
- ☐ **LO 6.13** Draw a Born–Haber cycle and calculate the energy change that occurs when an ionic compound is formed from its elements.

Definitions: *To learn, you must do. This is your workbook. Fill in the definitions from the glossary of the textbook as a first step to understanding the material.*

— **ionic bond**—________________________________

____

— **Born-Haber cycle**— ________________________________

____

Summary: *Read through this section and fill in the missing words and equations. Reference the textbook when you are not confident in your answers.*

An ionic bond occurs when an electron is transferred from one atom to another; the resulting electrical charges on the ions that are produced attract one another.

Reactions tend to occur when energy is ___________, because the newly formed substance is lower in energy than the species from which it was created. However, when you look at the energy involved to transfer an electron, the result is often a positive energy value, and thus energetically unfavorable. Nevertheless, these reactions occur! They occur because electron transfer is only part of the story. When an ionic bond is formed, the substance becomes more stable due to the attraction between the electrical charges. This stability releases energy and thereby decreases the energy of the products.

These reactions proceed rapidly and cannot be observed to occur in stages. However, when dealing with reaction energetics, it is possible to break the reaction into steps mathematically. In the case of ionic solid formation, this is done by what is known as the **Born–Haber cycle.** There are five common steps in this cycle:

Sublimation of the metal atoms: conversion of the metal to gas	Unfavorable
Nonmetal molecules broken into individual atoms	Unfavorable
Ionization of the gaseous metal atoms	Unfavorable
Ionization of the gaseous nonmetal atoms	Favorable
Formation of ionic bond	Very favorable

Each of these steps can be assigned numerical values. The energetically favorable steps will have ___________values, because they are releasing energy, and the unfavorable steps will have ___________ values, because they are consuming energy.

Workbook Example 6.7

PROBLEM:
Write the Born–Haber cycle for the formation of lithium fluoride from lithium metal and gaseous fluorine.

SOLUTION:

Sublimation of lithium	$Li(s) \rightarrow Li(g)$	+159.4	kJ/mol
Splitting of fluorine molecules	$\frac{1}{2} F_2(g) \rightarrow F(g)$	+158	kJ/mol
Ionization of lithium	$Li(g) \rightarrow Li^+(g) + e^-$	+520	kJ/mol
Ionization of fluorine	$F(g) + e^- \rightarrow F^-(g)$	–328	kJ/mol
Formation of ionic solid	$Li^+(g) + F^-(g) \rightarrow LiF(s)$	–1036	kJ/mol
	Total:	–527	kJ/mol

Workbook Problem 6.3

Write the Born–Haber cycle for the formation of sodium bromide using the following information:

Sublimation of Na:	+107.3 kJ/mol	Vaporization of $Br_2(l)$:	+ 30.9 kJ/mol
Breaking of Br_2 bond:	+224 kJ/mol	Ionization of Na:	+495.8 kJ/mol
Ionization of Br:	–325 kJ/mol	Formation of NaBr:	–747 kJ/mol

Strategy:	Write and sum the energies of each of the steps in the reaction to calculate the overall energy change of the formation of sodium bromide from its elements.
Step 1:	Write each step of the reaction with its energy required or produced per mole.
Step 2:	Sum the energy terms to find the overall energy change of the reaction.

POST-LECTURE:

Work Problems: *One of the best ways to learn chemistry is through solving problems. Go to the textbook and work through the Worked Examples for this section, then go to the end of chapter and work the problems there. When you feel confident that you can correctly answer problems like the ones listed below for each objective, check off each of the Learning Objectives for this section above.*

LO 6.12: Problems 6.26–6.28
LO 6.13: Worked Example 6.7; Problems 6.32, 6.33, 6.82, 6.84, 6.86

Section 6.8 Lattice Energies in Ionic Solids

PRE-LECTURE:

Learning Objectives: *These are the skills you should be proficient in after reading this section and practicing the associated problems. Check off the boxes below once you have gone through this section, answered the associated questions in your textbook, and feel confident you have mastered the skill.*

☐ **LO 6.14** Predict the relative magnitude of lattice energy given the formula or molecular representation of an ionic compound.

Definitions: *To learn, you must do. This is your workbook. Fill in the definitions from the glossary of the textbook as a first step to understanding the material.*

— **lattice energy (U)**—____

____

— **Coulomb's law**—____

____

Summary: *Read through this section and fill in the missing words and equations. Reference the textbook when you are not confident in your answers.*

Ionic solids have a structure that is referred to as a *crystal lattice.* The energy required to break a crystal lattice into gaseous ions is called **lattice energy,** abbreviated *U.* It has the same magnitude, but the opposite sign of the final step of the Born–Haber cycle: just as that last step—the formation of the ionic solid from gaseous ions—is always negative, the reverse will always be positive.

$Li^+(g) + F^-(g) \rightarrow LiF(s)$	$-U = -1007$ kJ/mol
$LiF(s) \rightarrow Li^+(g) + F^-(g)$	$U = +1007$ kJ/mol

The strength of ionic interactions is described by **Coulomb's law,** which has the mathematical form:

$$F = k\frac{z_1 z_2}{d^2}$$

where k is a mathematical constant that is different for each substance, z_1 and z_2 are the charges on the ions, and d is the distance between the centers of the ions.

The practical result of this solid structure is that the lattice energy is largest—meaning the ions are most difficult to separate—when the charges are large and the distance is small. Small, highly charged ions will have the strongest crystal lattice.

If one ion is changed and the other is held constant:

- the lattice energy will ___________ as the ion grows (assuming no change in the charge of the ion).
- the lattice energy will ___________ as charge increases (assuming little change in the distance between ions).

Workbook Example 6.8

PROBLEM:
Rank the following compounds from strongest to weakest lattice energy:

LiCl   CsCl   KCl   NaCl   RbCl

SOLUTION:
The anion in each of these cases is chlorine, so only the size of the cation varies. According to Coulomb's law, the smaller the distance between the ions, the stronger the lattice energy; so the strongest lattice energy will occur with the smallest cation, and the weakest lattice energy with the largest cation.

LiCl > NaCl > KCl > RbCl > CsCl

Workbook Example 6.9

PROBLEM:
Rank the following compounds from strongest to weakest lattice energy:

$MgCl_2$ $TiCl_4$ $CaCl_2$ KCl $AlCl_3$

SOLUTION:
This is a slightly more complex situation than the previous example, but changes in charge have a huge effect on the lattice energy. Ti, with a charge of +4, will have a stronger lattice energy than Al, with a charge of +3. Only when the charges are the same does the size of ions play a role. Mg is smaller than Ca, so it will form a stronger lattice.

$TiCl_4 > AlCl_3 > MgCl_2 > CaCl_2 > KCl$

Workbook Problem 6.4

Which compound in each of the following pairs has the greatest lattice energy?

$NaCl$ or KBr AlF_3 or $MgCl_2$ $CuBr_2$ or $CuCl$

Strategy: Determine the relative sizes and charges of the ions and assign relative strengths based on charge and size.

Step 1: Identify the relative charges in the two sets of ions and how they change from one set to the other.

Step 2: If necessary, identify the relative ion sizes of the two sets of ions and show how they change from one to the other.

Step 3: Determine the relative lattice energies of the ion pairs.

POST-LECTURE:

Work Problems: *One of the best ways to learn chemistry is through solving problems. Go to the textbook and work through the Worked Examples for this section, then go to the end of chapter and work the problems there. When you feel confident that you can correctly answer problems like the ones listed below for each objective, check off each of the Learning Objectives for this section above.*

LO 6.14: Worked Example 6.8; Problems 6.29, 6.30, 6.98, 6.99

Self-Test

This section is intended to test your knowledge of the material covered in this chapter. Think through these problems and make certain you understand what the problem is about—ask yourself if the answer makes sense. Many of these questions are linked to the chapter learning goals. Therefore, successful completion of these problems indicates that you have achieved the learning goals for this chapter. You will receive the greatest benefit from this section if you use it as a mock exam. Do not use your notes, book, Internet, or friends when taking the test. There is no grade and you only shortchange yourself by getting help with this activity. As you take the test, mark the problems with your level of confidence with the answer.

★ *Use a star for problems you know you got right.*
? *Use a question mark for those you think you got right.*
x *Use an x for those you are guessing on.*

If you get a question right that you marked with a star, you can be confident that you know the associated LO.

If you get a question right that you marked with a question mark, go back and work a few more problems and test yourself again tomorrow.

If you get a question wrong or mark it with an x, go back and read the section again, work a few problems today, a few tomorrow, and a few the next day before testing yourself again.

This will help you to discover which topics you have learned thoroughly and which topics you need to study in more detail.

True–False

1. Main-group metals tend to form cations that give them a noble gas electron configuration.
2. Nonmetals lose electrons to adopt lower energy levels.
3. Transition metals always ionize the same way.
4. When ions form, their radii increase when they are forming anions, and decrease when they are forming cations.
5. Ionization energy is always positive.
6. Ionization energies show no periodic properties.
7. Electron affinity applies primarily to metals, and it is difficult to determine for nonmetals.
8. The octet rule can be used to predict which ions will be formed in reactions.
9. The Born–Haber cycle provides information about the strength of covalent bonds.
10. Lattice energy is related only to ionic size.

Multiple Choice

11. Main-group elements form ions by
    a. losing electrons.
    b. gaining electrons.
    c. working toward a half-filled *d* orbital.
    d. losing or gaining electrons to obtain a noble gas configuration.

12. Which +3 ion has the electron configuration $1s^2 2s^2 2p^6$?
    a. B^{3+} b. Al^{3+} c. P^{3+} d. Fe^{3+}

13. Rank the following elements in order of greatest to least first ionization energy (IE_1):
    Rb Li Cs K Na

    a. Na > Li > K > Cs > Rb
    b. Cs > Rb > K > Na > Li
    c. Li > Na > K > Rb > Cs
    d. Li > K > Na > Cs > Rb

14. Given the following ionization energies (shown in kJ/mol), which element is this most likely to be?
    IE_1: 738 IE_2: 1451 IE_3: 7733 IE_4: 10 540

    a. Na b. Mg c. Al d. Si

15. Ionization energies are always positive, but electron affinities are generally negative because
    a. energy is released when an atom achieves a more stable configuration.
    b. an electron can only be added to a cation.
    c. energy is required to hold an extra electron onto an atom.
    d. atoms will not accept extra electrons.

16. Born–Haber cycles do NOT require which of the following?
    a. ionization energy of the metal
    b. formation of the ionic solid
    c. liquefying the nonmetal
    d. sublimation of the metal

17. Lattice energies vary with
    a. ion charge. b. electron affinity. c. ionic radius. d. a and c

18. Which of the following will have the highest lattice energy?
    a. AlN b. $MgCl_2$ c. NCl_3 d. LiCl

Matching

19. ionization energy
20. Coulomb's law
21. electron affinity
22. lattice energy

a. the energy change that occurs when an electron is added to an isolated atom in the gaseous state.
b. sum of interaction energies of ions in a crystal.
c. main-group elements tend to undergo reactions that leave them with eight valence electrons.
d. mathematical relationship between charge interaction and energy.

23. Born–Haber cycle — e. the amount of energy required to remove the outermost electron from an isolated atom in the gaseous state.

24. core electron — f. a bond maintained by electrostatic interactions.

25. ionic bond — g. electrons that form the inner shell of an element and do not participate in reactions.

26. octet rule — h. a series of hypothetical steps, each of which contributes to the overall energy change during the formation of an ionic compound.

Fill in the Blanks

27. Chlorine and potassium will both ionize to have the electron configuration of __________.

28. When metals ionize to form cations, the atomic radius ________________.

29. Valence electrons are strongly shielded by inner __________ electrons, making them relatively ____________ to remove.

30. Multiple ionization energies are possible with multielectron atoms. A very large _____________ in ionization energies indicates that all of the _______________ electrons have been removed.

31. Electron affinities are measured when an electron is ____________ to an isolated atom in the ___________ state.

32. The higher the ionic ________, the stronger the lattice energy, is a use of _______________ law.

33. Main-group elements tend to undergo reactions that leave them with ____________ outer-shell electrons. This is the __________________.

Problems

34. Write the electron configurations of the following elements first as neutral atoms, and then as their most stable ion.

P Ti Mn K Se

35. Rank these ions and atoms in order from smallest to largest.

Cl^- Al Li^+ Cl Al^{3+} Li

36. Identify the following third-period element from these ionization energies:

E_{i1} = 578 kJ/mol E_{i2} = 1817 kJ/mol E_{i3} = 2745 kJ/mol
E_{i4} = 11 575 kJ/mol E_{i5} = 14 830 kJ/mol

37. Draw orbital diagrams for the following elements and use them to rank these elements by lowest to highest electron affinity. Justify your answer.

Ga Br Ge

38. Calculate the heat of sublimation for aluminum, given that the net energy change for the formation of $AlCl_3(s)$ from the elements, $Al(s) + 3/2\ Cl_2(g) \rightarrow AlCl_3(s)$, is −704.2 kJ/mol. Use the following information:

    Al: E_{i1} = 578 kJ/mol, E_{i2} = 1817 kJ/mol, E_{i3} = 2745 kJ/mol
    Cl: E_{ea} = −348.6 kJ/mol
    Bond-dissociation energy for Cl_2 = 244 kJ/mol
    Lattice energy for $AlCl_3(s)$ = 5492 kJ/mol

39. Using the octet rule, write and balance equations for the reactions of Mg, Li, and Al each with Cl, O, and P.

40. Write the Born–Haber cycle for the formation of KF from its elements given the following information:

Sublimation of K:	+89.2 kJ/mol	$F_2 \rightarrow 2\ F(g)$:	+158 kJ/mol
Ionization of K:	+418.8 kJ/mol	Ionization of F:	−328 kJ/mol
Formation of KF:	−821 kJ/mol		

41. Rank the following from strongest to weakest lattice energy and justify your answers:

    LiCl $MgCl_2$ KCl $AlCl_3$ CsCl

<u>*Challenge Problem*</u>

42. A 1.53-g sample of an alkaline earth metal was reacted with 4.75 L of chlorine gas (3.17 g/L). The resulting metal chloride was analyzed by dissolving a 5.25-g sample in water and adding an excess of silver nitrate. The analysis yielded 18.8 g of silver chloride.
    a. What is the percent chlorine in the alkaline earth chloride?
    b. What is the identity of the alkaline earth metal?
    c. Write the balanced chemical equations for all reactions.
    d. What was the limiting reactant in the reaction between the alkaline earth metal and chlorine? If the metal is in excess, how many grams did not react? If chlorine is in excess, how many liters remained?

Test Yourself: *Take the Practice Test on pages 231 and 232 of your textbook. Again, use a testing environment using only those items that your instructor will allow you to have access to during the test and mark each question with a star, question mark, or x to denote your confidence level. Once you have finished you can check off the LO at the beginning of this chapter based on which questions you starred and got correct. Go back and spend more time with the topics from the other learning objectives.*

CHAPTER SEVEN

Covalent Bonding and Electron-Dot Structures

Learning Objectives

As a result of reading and studying this chapter, you should be able to:

- ☐ **LO 7.1** Describe the difference between an ionic and covalent bond.
- ☐ **LO 7.2** Describe the changes in energy that occur as two nuclei approach to form a covalent bond.
- ☐ **LO 7.3** Predict trends in bond length and bond dissociation energy based on bond order and atomic size.
- ☐ **LO 7.4** Rank elements by increasing value of electronegativity.
- ☐ **LO 7.5** Classify bonds as nonpolar covalent, polar covalent, or ionic.
- ☐ **LO 7.6** Interpret electrostatic potential maps to determine regions of high and low electron density.
- ☐ **LO 7.7** Calculate the percent ionic character of a bond.
- ☐ **LO 7.8** Explain the different physical properties of ionic and covalent compounds.
- ☐ **LO 7.9** Draw an electron-dot structure that satisfies the octet rule.
- ☐ **LO 7.10** Use the five-step procedure to draw electron-dot structures for molecules, including those with expanded octets and those containing multiple bonds.
- ☐ **LO 7.11** Draw electron-dot structures for radicals.
- ☐ **LO 7.12** Draw electron-dot structures for molecules with more than one central atom.
- ☐ **LO 7.13** Draw resonance structures and use curved arrows to depict how one structure can be converted to another.
- ☐ **LO 7.14** Describe the bonding in a molecule using the resonance hybrid model.
- ☐ **LO 7.15** Calculate the formal charge on atoms in an electron-dot structure.
- ☐ **LO 7.16** Use the formal charge to evaluate the contribution of different resonance structures to the resonance hybrid.

Chapter Summary

Covalent bond formation is the result of electron sharing between atoms. Covalent bonds have specific lengths and energies; energy is released when a bond is formed and absorbed when a bond is broken. Covalent compounds generally obey the octet rule, but atoms in the third period and below can form expanded octets as well. These structures can be represented by electron-dot structures, which can also be used to predict the existence of resonance hybrids. When dissimilar atoms participate in covalent bonding, differences in electronegativity can lead to a polar bond. These polarity differences can be predicted with the periodic table.

The Chapter in Detail

Section 7.1 Covalent Bonding in Molecules

PRE-LECTURE:

Learning Objectives: *These are the skills you should be proficient in after reading this section and practicing the associated problems. Check off the boxes below once you have gone through this section, answered the associated questions in your textbook, and feel confident you have mastered the skill.*

- ☐ **LO 7.1** Describe the difference between an ionic and covalent bond.
- ☐ **LO 7.2** Describe the changes in energy that occur as two nuclei approach to form a covalent bond.

Definitions: *To learn, you must do. This is your workbook. Fill in the definitions from the glossary of the textbook as a first step to understanding the material.*

— **bond length—** ____

____

— **lines between atoms—** ______________________________________

____

Summary: *Read through this section and fill in the missing words and equations. Reference the textbook when you are not confident in your answers.*

Hydrogen forms the simplest covalent molecule and provides a good model for covalent bonding: as the hydrogen atoms approach each other, the positive nuclei __________ each other and the electrons __________ each other, but the nuclei pull on the electrons of both atoms. The attractive forces are __________ than the repulsive forces, and a covalent bond is formed with the electrons shared between the nuclei.

There is an optimum distance for atoms to form a bond: too far away and no bond can form; too close and the nuclei repel. Every covalent bond has a characteristic bond length that leads to maximum stability. These can be roughly predicted from atomic radii.

POST-LECTURE:

Work Problems: *One of the best ways to learn chemistry is through solving problems. Go to the textbook and work through the Worked Examples for this section, then go to the end of chapter and work the problems there. When you feel confident that you can correctly answer problems like the ones listed below for each objective, check off each of the Learning Objectives for this section above.*

LO 7.1: Problem 7.40
LO 7.2: Problems 7.33, 7.34, 7.47

Section 7.2 Strengths of Covalent Bonds

PRE-LECTURE:

Learning Objectives: *These are the skills you should be proficient in after reading this section and practicing the associated problems. Check off the boxes below once you have gone through this section, answered the associated questions in your textbook, and feel confident you have mastered the skill.*

☐ **LO 7.3** Predict trends in bond length and bond dissociation energy based on bond order and atomic size.

Definitions: *To learn, you must do. This is your workbook. Fill in the definitions from the glossary of the textbook as a first step to understanding the material.*

— **Bond dissociation energy (D)—** ____

____

— **bond order—** ____

____

Summary: *Read through this section and fill in the missing words and equations. Reference the textbook when you are not confident in your answers.*

The energy of a covalently bonded molecule is lower than that of the individual molecules: energy is ___________ when they come together. This same amount of energy has to be put into the bond to break it. This is the **bond dissociation energy (D).**

Bond dissociation energies are defined as the amount of energy needed to break a bond in an isolated, gaseous molecule. They are always positive and ___________ in magnitude, but ___________ in sign to the energy released when the bond forms. On the image below, label the arrows to show the direction in which energy is released and absorbed.

Although atoms in different molecules are in different environments, bond dissociation energies tend to be the same: a carbon to carbon single bond (C–C) always requires about the same energy to break no matter what molecule it is in.

The shorter the bond, the ___________ it tends to be. It is possible for atoms to share more than one pair of electrons, and when this happens, multiple bonds are formed. These are shorter and stronger than single bonds. Double bonds are formed when _____ pairs of electrons are shared, and triple bonds are formed when _____ pairs of electrons are shared. The **bond order** is the number of electron pairs: a double bond has a bond order of ___, a triple bond has a bond order of ___.

POST-LECTURE:

Work Problems: *One of the best ways to learn chemistry is through solving problems. Go to the textbook and work through the Worked Examples for this section, then go to the end of chapter and work the problems there. When you feel confident that you can correctly answer problems like the ones listed below for each objective, check off each of the Learning Objectives for this section above.*

LO 7.3: Problems 7.42, 7.44

Section 7.3 Polar Covalent Bonds: Electronegativity

PRE-LECTURE:

Learning Objectives: *These are the skills you should be proficient in after reading this section and practicing the associated problems. Check off the boxes below once you have gone through this section, answered the associated questions in your textbook, and feel confident you have mastered the skill.*

- ☐ **LO 7.4** Rank elements by increasing value of electronegativity.
- ☐ **LO 7.5** Classify bonds as nonpolar covalent, polar covalent, or ionic.
- ☐ **LO 7.6** Interpret electrostatic potential maps to determine regions of high and low electron density.
- ☐ **LO 7.7** Calculate the percent ionic character of a bond.

Definitions: *To learn, you must do. This is your workbook. Fill in the definitions from the glossary of the textbook as a first step to understanding the material.*

— **polar covalent bond—** ____

____

— **electronegativity (EN)—** ____

____

— **dipole moment—** ____

____

— **percent ionic character—** ____

____

Summary: *Read through this section and fill in the missing words and equations. Reference the textbook when you are not confident in your answers.*

Chemical bonding exists on a continuum from nonpolar covalent bonds, such as those found in diatomic molecules, to ionic molecules. In between are molecules in which electrons are __________ shared: these are known as **polar covalent** molecules.

In polar covalent molecules, one atom pulls harder on the electrons, and will therefore carry a slight __________ charge. The other atom, deprived of its share of the electrons, will have a slight ___________ charge. These partial charges are represented with the lowercase Greek delta: δ– for the electron-rich atom, δ+ for the electron-poor atom.

Bond polarity is the result of differences in **electronegativity**—the ability of an atom in a molecule to attract the shared electrons in a covalent bond. ___________ is the most electronegative element, and it is given a unitless value of 4. Electronegativity ___________ down a group and left across the periodic table.

Add the trend of electronegativity to your periodic table on the inside front cover of this study guide.

Bond polarity is generally predicted using the following:

- Atoms with similar electronegativities form ___________________ bonds.
- Atoms whose electronegativities differ by more than 2.0 tend to form ______________ bonds.
- Atoms whose electronegativities differ by less than 2.0 tend to form ______________ bonds.

Workbook Example 7.1

PROBLEM:
Using the electronegativity values in Figure 7.4 on pg. 243 in the textbook, predict whether the following bonds are polar, nonpolar, or ionic:

Cl–Cl   O–F   Mg–Se   Li–F

SOLUTION:

Cl–Cl:	nonpolar	3.0–3.0 = 0
O–F:	polar	4.0–3.5 = 0.5
Mg–Se:	polar	2.4–1.2 = 1.2
Li–F:	ionic	4.0–1.0 = 3

Workbook Problem 7.1

Predict whether the following bonds are polar, nonpolar, or ionic:

H–F   F–Cl   C–O   K–Cl   Ge–Br

Strategy: First, make your predictions based on periodic properties; then confirm your answer with the numbers in Figure 7.4 in the textbook.

Step 1: Predict the polarity of the bonds based on their locations, then check with electronegativity values from Figure 7.4 in the textbook:

Step 2:	Find the difference between the numbers to check.

POST-LECTURE:

Work Problems: *One of the best ways to learn chemistry is through solving problems. Go to the textbook and work through the Worked Examples for this section, then go to the end of chapter and work the problems there. When you feel confident that you can correctly answer problems like the ones listed below for each objective, check off each of the Learning Objectives for this section above.*

LO 7.4: Problem 7.48
LO 7.5: Worked Example 7.1; Problems 7.50, 7.54
LO 7.6: Problems 7.33, 7.34
LO 7.7: Worked Example 7.2; Problems 7.60, 7.61

Section 7.4 A Comparison of Ionic and Covalent Compounds

PRE-LECTURE:

Learning Objectives: *These are the skills you should be proficient in after reading this section and practicing the associated problems. Check off the boxes below once you have gone through this section, answered the associated questions in your textbook, and feel confident you have mastered the skill.*

☐ **LO 7.8** Explain the different physical properties of ionic and covalent compounds.

Summary: *Read through this section and fill in the missing words and equations. Reference the textbook when you are not confident in your answers.*

Ionic compounds are solids with very high melting points because the entire crystal lattice must be disrupted for the compound to melt.

Covalent compounds are low-melting-point solids, liquids, or gases. The atomic interactions within the molecules are __________, but the interactions between molecules are fairly __________.

POST-LECTURE:

Work Problems: *One of the best ways to learn chemistry is through solving problems. Go to the textbook and work through the Worked Examples for this section, then go to the end of chapter and work the problems there. When you feel confident that you can correctly answer problems like the ones listed below for each objective, check off each of the Learning Objectives for this section above.*

LO 7.8: Problems 7.35, 7.62, 7.63

Section 7.5 Electron-Dot Structures: The Octet Rule

PRE-LECTURE:

Learning Objectives: *These are the skills you should be proficient in after reading this section and practicing the associated problems. Check off the boxes below once you have gone through this section, answered the associated questions in your textbook, and feel confident you have mastered the skill.*

☐ **LO 7.9** Draw an electron-dot structure that satisfies the octet rule.

Definitions: *To learn, you must do. This is your workbook. Fill in the definitions from the glossary of the textbook as a first step to understanding the material.*

— **electron-dot structure—** ____

____

— **lone pair—** ____

____

— **bonding pair—** ____

____

— **single bond—** ____

____

— **double bond—** ____

____

— **triple bond—** ____

____

— **coordinate covalent bond—** ____

____

Review: *Before going through the section make sure you are familiar with these concepts:*

- Ch. 5.12 valence electrons
- Ch. 6.6 the octet rule

Summary: *Read through this section and fill in the missing words and equations. Reference the textbook when you are not confident in your answers.*

Lewis structures, **or electron-dot structures,** offer a simplified representation of the bonding in a molecule. Dots are used to depict ___________ electrons and show how they combine to complete octets.

The octet rule is the guiding principle for these structures, just as it was for ionic structures. Atoms ___________ one or more electrons to fill their valence shell and obey the octet rule. Dots between atoms represent covalent bonds (often drawn as single line denoting the two shared electrons) and are called bonding pairs. Electrons not shared between atoms are nonbonding pairs or lone pairs.

Sometimes an atom donates a lone pair of electrons to a bond, so that both of the electrons in that bond come from one atom. These are called __________ **covalent bonds:** they are frequently formed by N, O, P, and S, which have lone pairs to donate.

POST-LECTURE:

Work Problems: *One of the best ways to learn chemistry is through solving problems. Go to the textbook and work through the Worked Examples for this section, then go to the end of chapter and work the problems there. When you feel confident that you can correctly answer problems like the ones listed below for each objective, check off each of the Learning Objectives for this section above.*

LO 7.9: Worked Example 7.3; Problem 7.3

Section 7.6 Procedure for Drawing Electron-Dot Structures

PRE-LECTURE:

Learning Objectives: *These are the skills you should be proficient in after reading this section and practicing the associated problems. Check off the boxes below once you have gone through this section, answered the associated questions in your textbook, and feel confident you have mastered the skill.*

☐ **LO 7.10** Use the five-step procedure to draw electron-dot structures for molecules, including those with expanded octets and those containing multiple bonds.

Definitions: *To learn, you must do. This is your workbook. Fill in the definitions from the glossary of the textbook as a first step to understanding the material.*

— **electron deficient—**____

____

— **expanded octet—**____

____

Summary: *Read through this section and fill in the missing words and equations. Reference the textbook when you are not confident in your answers.*

Many compounds contain only hydrogen and second-row elements. Structures for these are straightforward to draw because the octet rule almost always applies and the number of bonds formed by each element is easy to predict. Atoms of second-row elements are bonded to one another to form the core of the molecule, and the hydrogens are found around the periphery.

The general method that chemists use to draw Lewis structures of molecules are:

Step 1. ____

- add 1 for each _______________ charge in an anion.
- subtract 1 for each _______________ charge in a cation.

Step 2. ____

____

- ____
- ____

  ____
- elements in group 3A are often _______________________
- elements in the third row and lower can _____________________

Step 3. ____

____

Step 4. If unassigned electrons remain after Step 3, ______________________

Step 5. Form _______________ to fill an incomplete octet on the central atom.

Workbook Example 7.2

PROBLEM:
Draw the electron-dot structure for CH_4.

SOLUTION:

1. Calculate the total number of valence electrons.

Carbon has four valence electrons and each hydrogen has one.

$$\begin{array}{l} \text{C} - 1 \text{ atom} \times 4\ e^- = 4\ e^- \\ \underline{\text{H} - 4 \text{ atoms} \times 1\ e^- = 4\ e^-} \\ \qquad\qquad 8 \text{ valence } e^- \text{ total} \end{array}$$

2. Determine how the atoms are connected.

Since each hydrogen can only form one bond, the carbon goes in the center.

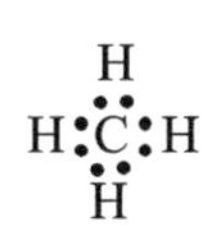

The paired electrons can also be represented as bonds:

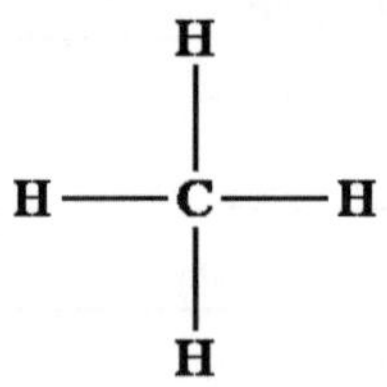

3&4. Subtract the number of valence electrons used for bonding from the total number calculated in Step 1 and assign the rest to terminal atoms.

4 bonds × 2 electrons each = 8 total e^- in the structure

Both the structure and the value calculated in Step 1 are the same; no additional electrons need to be added to the terminal atoms. This makes sense because each hydrogen can only hold a maximum of two electrons.

5. Form multiple bonds as needed to complete the octet of the central atom.

All the electrons are assigned and all the octets are full (hydrogen can only holds 2 e^-); no multiple bonds are necessary for this molecule.

Workbook Problem 7.2

Draw electron-dot structures for the following compounds.

HF CH_2Br_2 Cl_2

Strategy: Share valence electrons to complete the octect of each atom.

Step 1: Calculate the total number of valence electrons.

Step 2: Determine how the atoms are connected.

Steps 3 and 4: Subtract the number of valence electrons used for bonding from the total number calculated in Step 1 and assign the rest to terminal atoms.

Step 5: Form multiple bonds as needed to complete the octet of the central atom.

Workbook Example 7.3

PROBLEM:
Draw the electron-dot structure for SO_4^{2-}.

SOLUTION:
To draw the electron-dot structure of SO_4^{2-}:

1. Calculate the total number of valence electrons.

S – 1 atom	$\times 6\ e^- =$	$6\ e^-$
O – 4 atoms	$\times 6\ e^- =$	$24\ e^-$
−2 charge –	$2 \times 1\ e^- =$	$2\ e^-$
		32 valence e^- total

2. Determine how the atoms are connected.

The sulfur atom can form an expanded octet; placing it in the center gives the most symmetrical arrangement.

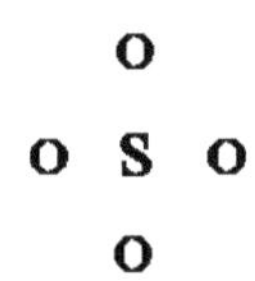

3&4. Subtract the number of valence electrons used for bonding from the total number calculated in Step 1 and assign the rest to terminal atoms.

This leaves 24 electrons to distribute around the oxygens.

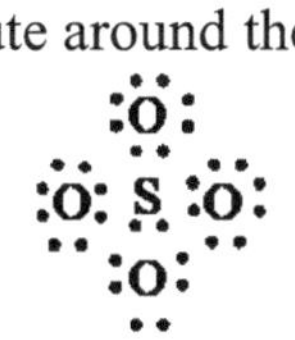

5. Form multiple bonds as needed to complete the octet of the central atom.

This accounts for all of the electrons, but oxygens typically form two bonds, not one. Two of the oxygens are ionized, making them stable with only one bond, but the other two will be double-bonded to the sulfur, giving it an expanded octet of 10 electrons. This will make more sense after looking at sections 7.9 and 7.10.

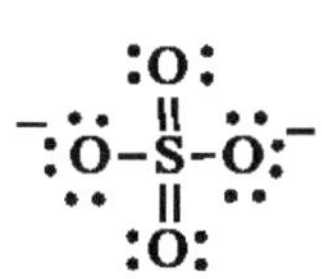

Workbook Problem 7.3

Draw the electron-dot structure for N_2.

Strategy: Share valence electrons to complete the octect of each atom.

Step 1: Calculate the total number of valence electrons.

Step 2: Determine how the atoms are connected

Steps 3 and 4: Subtract the number of valence electrons used for bonding from the total number calculated in Step 1 and assign the rest to terminal atoms.

Step 5: Form multiple bonds as needed to complete the octet of the central atom.

POST-LECTURE:

Work Problems: *One of the best ways to learn chemistry is through solving problems. Go to the textbook and work through the Worked Examples for this section, then go to the end of chapter and work the problems there. When you feel confident that you can correctly answer problems like the ones listed below for each objective, check off each of the Learning Objectives for this section above.*

LO 7.10: Worked Examples 7.4, 7.5; Problems 7.66, 7.68, 7.70, 7.74

Section 7.7 Drawing Electron-Dot Structures for Radicals

PRE-LECTURE:

Learning Objectives: *These are the skills you should be proficient in after reading this section and practicing the associated problems. Check off the boxes below once you have gone through this section, answered the associated questions in your textbook, and feel confident you have mastered the skill.*

☐ **LO 7.11** Draw electron-dot structures for radicals.

Definitions: *To learn, you must do. This is your workbook. Fill in the definitions from the glossary of the textbook as a first step to understanding the material.*

— **radicals—**____

____

Summary: *Read through this section and fill in the missing words and equations. Reference the textbook when you are not confident in your answers.*

Most molecules or ions that we deal with in chemistry have an even number of electrons that are paired up in orbitals. **Free radicals** are ions or molecules that have an __________ number of electrons, which means that one orbital will be half-filled. Because of this, free radicals are very __________ and will seek to undergo some sort of reaction in order to form a more stable product.

The rules that we use to draw Lewis structures apply to free radicals, but you will also want to remember that the __________ electronegative atom will have the odd number of electrons.

POST-LECTURE:

Work Problems: *One of the best ways to learn chemistry is through solving problems. Go to the textbook and work through the Worked Examples for this section, then go to the end of chapter and work the problems there. When you feel confident that you can correctly answer problems like the ones listed below for each objective, check off each of the Learning Objectives for this section above.*

LO 7.11: Worked Example 7.6; Problems 7.9, 7.10

Section 7.8 Electron-Dot Structures of Compounds Containing Only Hydrogen and Second-Row Elements

PRE-LECTURE:

Learning Objectives: *These are the skills you should be proficient in after reading this section and practicing the associated problems. Check off the boxes below once you have gone through this section, answered the associated questions in your textbook, and feel confident you have mastered the skill.*

☐ **LO 7.12** Draw electron-dot structures for molecules with more than one central atom.

Definitions: *To learn, you must do. This is your workbook. Fill in the definitions from the glossary of the textbook as a first step to understanding the material.*

— **organic compounds—**____

____

Summary: *Read through this section and fill in the missing words and equations. Reference the textbook when you are not confident in your answers.*

We use Lewis structures to help us visualize **organic molecules,** or compounds that contain __________. Organic compounds typically contain carbon, hydrogen, nitrogen, and oxygen and are essential to the study of the chemistry of biomolecules such as carbohydrates, lipids, nucleic acids, proteins, and others. The rules that we use to draw the Lewis structures for organic molecules are the same as those stated above. Often information about the connectivity between atoms is given in the formula.

Workbook Example 7.4

PROBLEM:
Draw the electron-dot structure for CH_3NH_2.

SOLUTION:
To draw the electron-dot structure of CH_3NH_2:

1. Calculate the total number of valence electrons.

$$C - 1 \times 4\ e^- = 4\ e^-$$
$$N - 1 \times 5\ e^- = 5\ e^-$$
$$\underline{H - 5 \times 1\ e^- = 5\ e^-}$$
$$14 \text{ valence } e^- \text{ total}$$

2. Determine how the atoms are connected.

Often structural information is contained in the chemical formula. Since there are three hydrogens listed after the carbon it is a clue that three hydrogens are attached to the carbon. There are two hydrogens listed after the nitrogen in the chemical formula and therefore it has two hydrogens attached to it.

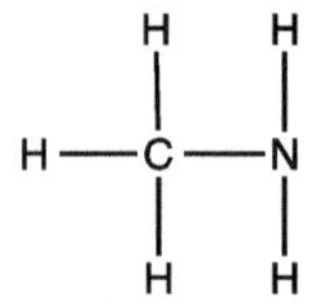

3&4. Subtract the number of valence electrons used for bonding from the total number calculated in Step 1 and assign the rest to terminal atoms.

$$6 \text{ bonds} \times 2 \text{ electrons each} = 12 \text{ total } e^- \text{ in the structure}$$

There are 14 electrons in the molecule, but only 12 assigned so far. The other two electrons can be used to complete the octet on nitrogen.

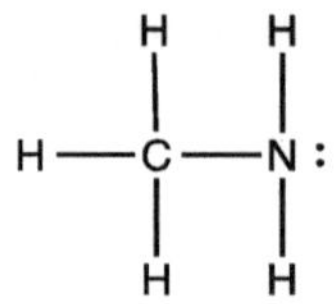

5. Form multiple bonds as needed to complete the octet of the central atom.

All the electrons are assigned and all the octets are full; no multiple bonds are necessary for this molecule.

Workbook Problem 7.4

Draw electron-dot structures for the following compounds.

$CH_3CH_2CH_3$ CH_3OH C_2H_4

Strategy: Share valence electrons to complete the octect of each atom.

Step 1: Calculate the total number of valence electrons.

Step 2: Determine how the atoms are connected.

Steps 3 and 4: Subtract the number of valence electrons used for bonding from the total number calculated in Step 1 and assign the rest to terminal atoms.

Step 5: Form multiple bonds as needed to complete the octet of the central atom.

POST-LECTURE:

Work Problems: *One of the best ways to learn chemistry is through solving problems. Go to the textbook and work through the Worked Examples for this section, then go to the end of chapter and work the problems there. When you feel confident that you can correctly answer problems like the ones listed below for each objective, check off each of the Learning Objectives for this section above.*

LO 7.12: Worked Examples 7.7, 7.8; Problems 7.38, 7.80, 7.82

Section 7.9 Electron-Dot Structures and Resonance

PRE-LECTURE:

Learning Objectives: *These are the skills you should be proficient in after reading this section and practicing the associated problems. Check off the boxes below once you have gone through this section, answered the associated questions in your textbook, and feel confident you have mastered the skill.*

- ☐ **LO 7.13** Draw resonance structures and use curved arrows to depict how one structure can be converted to another.
- ☐ **LO 7.14** Describe the bonding in a molecule using the resonance hybrid model.

Definitions: *To learn, you must do. This is your workbook. Fill in the definitions from the glossary of the textbook as a first step to understanding the material.*

— **delocalized—** ____

____

— **resonance theory—** ____

____

— **resonance hybrid—** ____

____

Summary: *Read through this section and fill in the missing words and equations. Reference the textbook when you are not confident in your answers.*

It is often the case, as in the sulfate ion (from Example 7.3), that there is more than one possibility for an electron structure. When deciding which of the sulfur atoms was double-bonded to oxygen, there was no reason that vertical double bonds were any more or less valid than horizontal double bonds. When this occurs, the "real" structure is an average of all possible structures. This going back and forth between equivalent structures is known as ____________. It is indicated by a single, straight, double-headed arrow, ↔. Resonance structures differ *only* in placement of valence ____________, and not the movement of atoms!

Workbook Example 7.5

PROBLEM:
Draw the resonance structures for NO_3^-.

SOLUTION:

1. Calculate the total number of valence electrons.

$$\begin{array}{r} N - 1 \times 5\ e^- = \ \ 5\ e^- \\ O - 3 \times 6\ e^- = 18\ e^- \\ \underline{-1\ \text{charge} - 1 \times 1\ e^- = \ \ 1\ e^-} \\ 24\ \text{valence } e^-\ \text{total} \end{array}$$

2. Determine how the atoms are connected.

Nitrogen is less electronegative than oxygen, and by putting the nitrogen atom in the center, it allows for symmetry of the molecule, so N is the central atom.

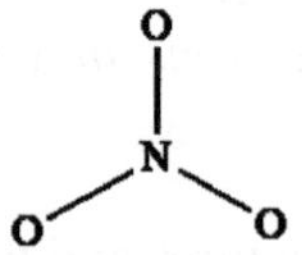

3&4. Subtract the number of valence electrons used for bonding from the total number calculated in Step 1 and assign the rest to terminal atoms.

The molecule drawn has 3 bonds and therefore 6 electrons. It was calculated that the molecule will have 24 electrons, so 18 more need to be placed on the molecule.

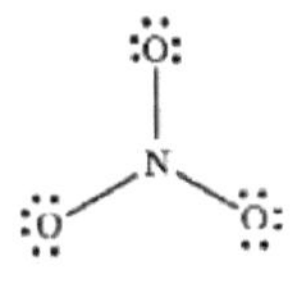

5. Form multiple bonds as needed to complete the octet of the central atom.

There are no more electrons to distribute, however, nitrogen does not have a completed octet. To form a completed octet, we can borrow a pair of lone electrons from one of the oxygen atoms. However, since all three N–O bonds are equal, we must draw three different resonance structures, each of which has a N=O bond between nitrogen and each of the different oxygen atoms.

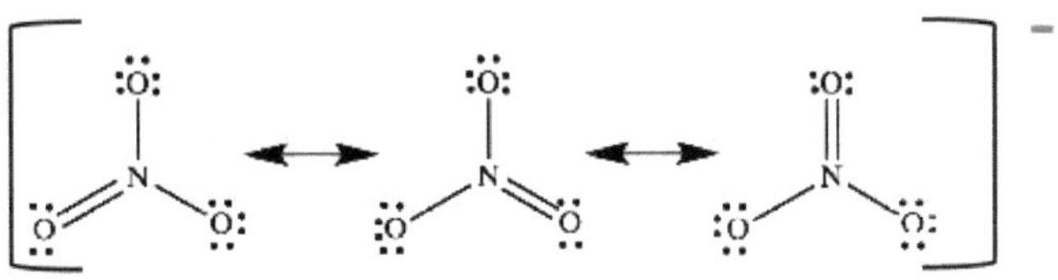

Workbook Problem 7.5

Draw as many resonance structures as possible for the carbonate (CO_3^{2-}) ion.

Strategy: Share valence electrons to complete the octect of each atom, use multiple bonds as needed.

Step 1: Calculate the total number of valence electrons.

Step 2: Determine how the atoms are connected

Steps 3 and 4: Subtract the number of valence electrons used for bonding from the total number calculated in Step 1 and assign the rest to terminal atoms.

Step 5: Form multiple bonds as needed to complete the octet of the central atom.

POST-LECTURE:

Work Problems: *One of the best ways to learn chemistry is through solving problems. Go to the textbook and work through the Worked Examples for this section, then go to the end of chapter and work the problems there. When you feel confident that you can correctly answer problems*

like the ones listed below for each objective, check off each of the Learning Objectives for this section above.

LO 7.13: Worked Examples 7.9, 7.10; Problems 7.84, 7.86, 7.88, 7.91
LO 7.14: Problems 7.90, 7.93

Section 7.10 Formal Charges

PRE-LECTURE:

Learning Objectives: *These are the skills you should be proficient in after reading this section and practicing the associated problems. Check off the boxes below once you have gone through this section, answered the associated questions in your textbook, and feel confident you have mastered the skill.*

- ☐ **LO 7.15** Calculate the formal charge on atoms in an electron-dot structure.
- ☐ **LO 7.16** Use the formal charge to evaluate the contribution of different resonance structures to the resonance hybrid.

Definitions: *To learn, you must do. This is your workbook. Fill in the definitions from the glossary of the textbook as a first step to understanding the material.*

— **formal charge—** ____

____

Summary: *Read through this section and fill in the missing words and equations. Reference the textbook when you are not confident in your answers.*

Sometimes it is difficult to distinguish which of a number of resonance possibilities is the "best," meaning it is the most stable configuration. Formal charges make it possible to distinguish between structures and find the one that is the most stable.

To determine the formal charge on an atom, take the number of valence electrons in a neutral, unbonded atom and subtract from it one-half the number of bonded electrons and all of the nonbonded electrons:

Formal charge =

For an ion, the overall formal charge of the molecule should be equal to the charge on the ion.

The most stable structures are those with:

- ____
  ____
- ____

Workbook Example 7.6

PROBLEM:
Distinguish the best of these two structures for carbon dioxide by checking their formal charges.

SOLUTION:
Symmetry makes the structure on the right appear more favorable, but both structures satisfy the octets of all the atoms. Their formal charges will make it possible to distinguish between them.

For the structure on the left:

Left oxygen	$= 6 - \frac{1}{2}(2) - 6 = -1$
Carbon	$= 4 - \frac{1}{2}(8) - 0 = 0$
Right oxygen	$= 6 - \frac{1}{2}(6) - 2 = +1$

For the structure on the right:

Oxygen (both)	$= 6 - \frac{1}{2}(4) - 4 = 0$
Carbon	$= 4 - \frac{1}{2}(8) - 0 = 0$

The formal charges confirm that the intuitive choice is correct: although both structures have an overall charge of 0, the asymmetrical structure has two of the same atoms carrying opposite charges, which is an unstable situation.

Workbook Problem 7.6
Draw three resonance structures for the cyanate ion (CNO^-) and use formal charges to determine which is the most stable. Carbon is the central atom.

Strategy: Draw electron-dot structures, then check the formal charges on the atoms to determine which is the most stable.

Step 1: Determine the number of valence electrons in the ion.

Step 2: Determine three reasonable electron-dot structures with carbon as the central atom and the number of valence electrons from Step 1.

Step 3: Calculate the formal charge on each of the atoms in each of the structures.

Step 4:	Choose the most stable structure based on your comparison of the formal charges.

POST-LECTURE:

Work Problems: *One of the best ways to learn chemistry is through solving problems. Go to the textbook and work through the Worked Examples for this section, then go to the end of chapter and work the problems there. When you feel confident that you can correctly answer problems like the ones listed below for each objective, check off each of the Learning Objectives for this section above.*

LO 7.15: Worked Example 7.11; Problems 7.94, 7.96, 7.100
LO 7.16: Worked Example 7.12; Problems 7.98, 7.102, 7.105

Self-Test

This section is intended to test your knowledge of the material covered in this chapter. Think through these problems and make certain you understand what the problem is about—ask yourself if the answer makes sense. Many of these questions are linked to the chapter learning goals. Therefore, successful completion of these problems indicates that you have achieved the learning goals for this chapter. You will receive the greatest benefit from this section if you use it as a mock exam. Do not use your notes, book, Internet, or friends when taking the test. There is no grade and you only shortchange yourself by getting help with this activity. As you take the test, mark the problems with your level of confidence with the answer.

★ *Use a star for problems you know you got right.*
? *Use a question mark for those you think you got right.*
x *Use an x for those you are guessing on.*

If you get a question right that you marked with a star, you can be confident that you know the associated LO.

If you get a question right that you marked with a question mark, go back and work a few more problems and test yourself again tomorrow.

If you get a question wrong or mark it with an x, go back and read the section again, work a few problems today, a few tomorrow, and a few the next day before testing yourself again.

This will help you to discover which topics you have learned thoroughly and which topics you need to study in more detail.

True–False

1. Covalent bonds involve transfer of electrons from one atom to another.

2. Bond dissociation energies are always positive.

3. Covalent compounds have lower melting points and weaker intermolecular interactions than ionic compounds.

4. Bond polarity is usually the result of differences in electronegativity.

5. Electron-dot structures give a simplified, but useful picture of bonding.

6. Valences of main-group elements cannot be predicted based on their position on the periodic table alone.

7. Only the structures of binary compounds can be predicted with electron-dot modeling.

8. It is not possible to model charged compounds using electron-dot structures.

9. The more symmetrical an atomic structure is, the more likely it is to be stable.

10. Free radicals contain an unpaired electron in an orbital.

Fill in the Blanks

11. Covalent bonds have characteristic _________ _________. If the atoms are too far away, no ________ can form. Too close, and the _______________ repel each other.

12. The energy required to break a bond in an isolated gaseous molecule is the ________ _____________ ___________. It is always ________________, and equal in magnitude to the energy ________________ when the bond forms.

13. Chemical bonds exist on a continuum from pure covalent to __________ ________________, to pure ____________.

14. In electron-dot structures, dots shared between atoms represent ______________ electrons, and dots not shared between atoms represent _____________ electrons.

15. Second-row elements are sometimes exceptions to the ______________ ________, because they can form _________________ _____________.

16. Sometimes multiple structures can be drawn for a compound. In this situation, ____________ ____________ can be used to distinguish the structure that is the most stable.

Matching

17. bond length	a. the energy required to break a chemical bond in an isolated molecule in the gaseous state

18. bond dissociation energy	b. a bond in which the electrons are more strongly attracted to one atom
19. coordinate covalent bond	c. an average of several valid electron-dot structures for a molecule
20. polar covalent bond	d. a pair of electrons not participating in a chemical bond.
21. lone pair	e. "bookkeeping" charge assigned to an atom
22. electronegativity	f. pair of electrons participating in a chemical bond
23. bonding pair	g. the strength with which an atom pulls on the electrons associated with it
24. formal charge	h. the distance at which attractive forces and nuclear repulsions are in balance
25. resonance hybrid	i. a bond in which one atom donates a pair of electrons to an atom with a vacant orbital

Problems

26. Identify each of the following compounds as primarily ionic or covalent:

LiBr HCl MgO NF_3

27. Draw electron-dot structures for H_2O_2, NH_4^+, Cl_2CO, SF_4, and XeF_4. Use resonance structures, if necessary.

28. Two white crystalline compounds are tested. The first has a melting point of 800 °C, a solubility of 36 g/100 mL of water, and conducts electricity in solution. The second has a melting point of 180 °C, a solubility of 200 g/100 mL of water, and does not conduct electricity in solution. What conclusions can you make about the chemical bonds in these compounds?

29. Draw all resonance structures for CO_3^{2-}, and assign formal charges to each atom.

30. Rank the following bonds from most polar to least polar, using only the periodic table.

C—H Al—F Li—Cl Cl—Cl H—Cl

Justify your answers.

31. Rank the following atoms in order of increasing electronegativity.

Cs F Fe O

32. Given the following three structures for nitrous oxide, which is likely to contribute most to the structure? Use formal charges to justify your answer.

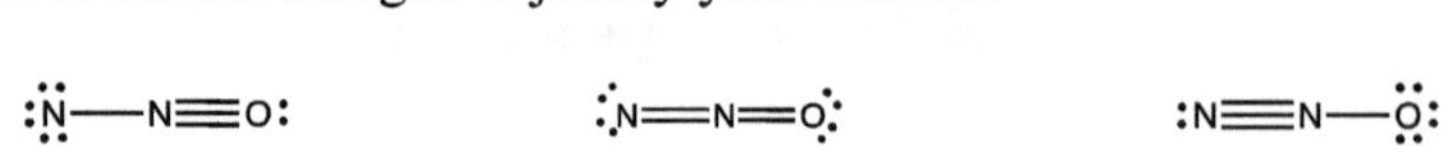

Test Yourself: *Take the Practice Test on pages* 268–270 *of your textbook. Again, use a testing environment using only those items that your instructor will allow you to have access to during the test and mark each question with a star, question mark, or x to denote your confidence level. Once you have finished you can check off the LO at the beginning of this chapter based on which questions you starred and got correct. Go back and spend more time with the topics from the other learning objectives.*

CHAPTER EIGHT

Covalent Compounds: Bonding Theories and Molecular Structure

Learning Objectives

As a result of reading and studying this chapter, you should be able to:

- ☐ **LO 8.1** Use the VSEPR model to predict shape and bond angles in a molecule or ion with one central atom.
- ☐ **LO 8.2** Use the VSEPR model to predict bond angles in a molecule with more than one central atom.
- ☐ **LO 8.3** Describe the difference between a sigma and a pi bond in valence bond theory.
- ☐ **LO 8.4** Write an electron-dot structure for a molecule and determine hybridization and bond angles.
- ☐ **LO 8.5** Identify which orbitals overlap to form σ and π bonds in molecules.
- ☐ **LO 8.6** Predict whether a given molecule has a dipole moment and draw its direction.
- ☐ **LO 8.7** Interpret electrostatic potential maps of molecules.
- ☐ **LO 8.8** Identify the types of intermolecular forces experienced by a molecule.
- ☐ **LO 8.9** Relate the strength of intermolecular forces to physical properties such as melting point and boiling point.
- ☐ **LO 8.10** Sketch the hydrogen bonding that occurs between two molecules.
- ☐ **LO 8.11** Interpret the molecular orbital diagram for a first-row diatomic molecule or ion.
- ☐ **LO 8.12** Interpret the molecular orbital diagram for a second-row diatomic molecule or ion. Calculate the bond order and predict magnetic properties.
- ☐ **LO 8.13** Draw orbital overlap diagrams for molecules and describe the use of both valence bond theory and molecular orbital theory.

Chapter Summary

In Chapter 7, you learned about formation of a covalent bond and how to draw electron-dot structures to represent molecules; Chapter 8 continues this by expanding upon the two-dimensional Lewis structures that you drew in the last chapter. You will also learn in this chapter how to transfer this information to a three-dimensional image by using the valence shell electron-pair repulsion model, or VSEPR, which takes both atoms and unbonded electrons into account. You will then be able to enhance your picture of the three-dimensional model by using valence bond theory, which describes how electrons are spread throughout hybrid orbitals that permit the electrons in pairs and bonds to maximize their distances from each other. We end our overview of molecular modeling by looking at molecular orbital theory, which provides mathematical precision to the visualization of molecular structure and solves some problems that the previous models cannot. In addition to the above, we will talk about how three-dimensional molecules interact through intermolecular forces.

The Chapter in Detail

Section 8.1 Molecular Shapes: The VSEPR Model

PRE-LECTURE:

Learning Objectives: *These are the skills you should be proficient in after reading this section and practicing the associated problems. Check off the boxes below once you have gone through this section, answered the associated questions in your textbook, and feel confident you have mastered the skill.*

- ☐ **LO 8.1** Use the VSEPR model to predict the shape and bond angles in a molecule or ion with one central atom.
- ☐ **LO 8.2** Use the VSEPR model to predict bond angles in a molecule with more than one central atom.

Definitions: *To learn, you must do. This is your workbook. Fill in the definitions from the glossary of the textbook as a first step to understanding the material.*

— **valence shell electron pair repulsion (VSEPR) model—** ____________________

____

— **bond angle—** ____

____

Summary: *Read through this section and fill in the missing words and equations. Reference the textbook when you are not confident in your answers.*

Electron-dot structures are good for representing the connectivity between atoms; however, it is important to remember that molecules have a three-dimensional shape. Their three-dimensional shapes are very important for their biological functions, and these shapes can often be predicted by using the **valence shell electron-pair repulsion (VSEPR) model**.

Electrons both in lone pairs and in bonds try to repel each other because of their negative charges. The result of these repulsions between electrons is molecular shape, which can be predicted by counting the number of electron clouds. An electron cloud can be a:

- ____________ bond
- ____________ bond
- ____________ bond
- ____________
- ____________

*Note: double or triple bonds count as <u>one</u> electron cloud

Different molecular shapes are formed based on the number of clouds around a central atom.

- Two charge clouds:
  - Linear arrangement

- Bond angles = ___________
- Molecular shape = ____________________

- Three charge clouds:
  - Trigonal planar arrangement
  - Bond angles = ___________
  - Molecular shapes:
    - 3 bonds, 0 lone pairs = ____________________
    - 2 bonds, 1 lone pair = ____________________
- Four charge clouds:
  - Tetrahedral arrangement
  - Bond angles = ___________
  - Molecular shapes:
    - 4 bonds, 0 lone pairs = ____________________
    - 3 bonds, 1 lone pair = ____________________
    - 2 bonds, 2 lone pairs = ____________________
- Five charge clouds:
  - Trigonal bipyramidal arrangement
  - Bond angles = ___________
    - Atoms that are 180° from one another = ____________________
    - Atoms that are 90° from the others = ____________________
  - Molecular shapes:
    - 5 bonds, 0 lone pairs = ____________________
    - 4 bonds, 1 lone pair = ____________________
    - 3 bonds, 2 lone pairs = ____________________
    - 2 bonds, 3 lone pairs = ____________________
- Six charge clouds:
  - Octahedral arrangement
  - Bond angles = ___________
  - Molecular shapes:
    - 6 bonds, 0 lone pairs = ____________________
    - 5 bonds, 1 lone pair = ____________________
    - 4 bonds, 2 lone pairs = ____________________

For molecules with more than one central atom, use the VSEPR model to predict the shape of each atom as if it were a single central atom, and combine them to visualize the entire molecule.

Workbook Example 8.1

PROBLEM:
Predict the shape of XeF_2.

SOLUTION:
First, draw the electron-dot structure of the molecule. There are 8 valence electrons from xenon and 14 from fluorine, for a total of 22.

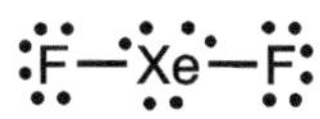

There are five charge clouds around xenon, two of which are attached to atoms. Referring back to the list above, molecules with five electron groups, two in bonds and three lone pairs, are linear.

Workbook Example 8.2

PROBLEM:
Ethanol has the formula CH_3CH_2OH. It is the component of alcoholic beverages that gives them their vivifying effects. The structural formula for this compound is

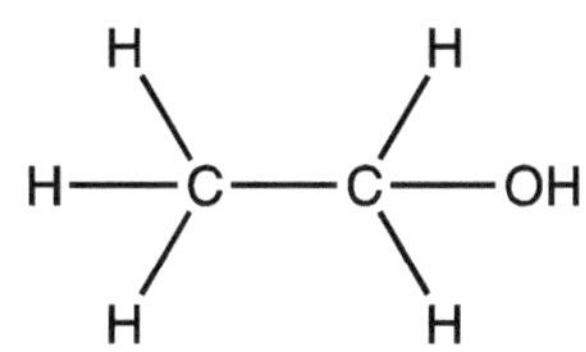

Describe the geometry around the carbon and oxygen atoms.

SOLUTION:
The two carbons each have four bonds and no nonbonding pairs: the geometry around these atoms is tetrahedral. The oxygen atom is more difficult to see because for these larger structures we often do not include the lone pairs in the drawing; they are understood to be there (you will get there with practice). Thus, the oxygen atom has two bonds and two nonbonding pairs: the geometry around this atom is bent.

Workbook Problem 8.1

Predict the molecular shape of CS_2 and SF_4.

Strategy: Draw the electron-dot structure for each of the molecules and use the list above to determine the molecular shape.

Step 1: Draw the electron-dot structure for the molecules, using the number of valence electrons and the most symmetrical structure.

Step 2:	Determine the number of charge clouds around the central atom. How many are used for bonding and how many are nonbonding?
Step 3:	Based on the chart, determine the shape.

POST-LECTURE:

Work Problems: *One of the best ways to learn chemistry is through solving problems. Go to the textbook and work through the Worked Examples for this section, then go to the end of chapter and work the problems there. When you feel confident that you can correctly answer problems like the ones listed below for each objective, check off each of the Learning Objectives for this section above.*

LO 8.1: Worked Example 8.1; Problems 8.42–8.49
LO 8.2: Worked Example 8.2; Problems 8.50–8.55

Section 8.2 Valence Bond Theory

PRE-LECTURE:

Learning Objectives: *These are the skills you should be proficient in after reading this section and practicing the associated problems. Check off the boxes below once you have gone through this section, answered the associated questions in your textbook, and feel confident you have mastered the skill.*

☐ **LO 8.3** Describe the difference between a sigma and a pi bond in valence bond theory.

Definitions: *To learn, you must do. This is your workbook. Fill in the definitions from the glossary of the textbook as a first step to understanding the material.*

— **valence bond theory—** ____

____

— **sigma (σ) bond—**____

____

Summary: *Read through this section and fill in the missing words and equations. Reference the textbook when you are not confident in your answers.*

VSEPR theory provides information about molecular shape, but says nothing about the nature of covalent bonds. Valence bond theory helps us to understand covalent bonding.

Valence bond theory states that when atoms approach one another, singly occupied electronic orbitals will overlap, allowing the now-paired electrons to be influenced by both nuclei. The orbitals must be in the same phase, and they will contain electrons of opposite spin.

If anything other than *s* orbitals overlaps, there will be a directionality to the bond.

Bonds resulting from head-on overlap of orbitals are called ___________**bonds.**

POST-LECTURE:

Work Problems: *One of the best ways to learn chemistry is through solving problems. Go to the textbook and work through the Worked Examples for this section, then go to the end of chapter and work the problems there. When you feel confident that you can correctly answer problems like the ones listed below for each objective, check off each of the Learning Objectives for this section above.*

LO 8.3: Problem 8.58

Section 8.3 Hybridization and sp^3 Hybrid Orbitals

PRE-LECTURE:

Learning Objectives: *These are the skills you should be proficient in after reading this section and practicing the associated problems. Check off the boxes below once you have gone through this section, answered the associated questions in your textbook, and feel confident you have mastered the skill. The following LOs are addressed in this, and the next few sections.*

- ☐ **LO 8.4** Write an electron-dot structure for a molecule and determine its hybridization and bond angles.
- ☐ **LO 8.5** Identify which orbitals overlap to form σ and π bonds in molecules.

Definitions: *To learn, you must do. This is your workbook. Fill in the definitions from the glossary of the textbook as a first step to understanding the material.*

— **hybrid atomic orbital—** ____

____

— **sp^3 hybrid orbital—** ____

____

Summary: *Read through this section and fill in the missing words and equations. Reference the textbook when you are not confident in your answers.*

For carbon in particular, the electron structure of the atom does not mesh easily with the VSEPR structure of molecules. Carbon has two unpaired *p* electrons and no unpaired *s* electrons, but it forms four bonds.

Linus Pauling provided a mathematical solution, showing that the Schrödinger wave equations for *s* and *p* orbitals can be combined to form **hybrid orbitals.** In the case of carbon (and others), one *s* combines with three *p* to form four sp^3 orbitals.

Each sp^3 orbital has two lobes, one large and one small lobes. The large lobe points toward the corners of a ___________. Bonds formed by overlapping sp^3 orbitals are very strong: a tetrahedral arrangement of charge clouds always implies _____hybridization.

Section 8.4 Other Kinds of Hybrid Orbitals

PRE-LECTURE:

Learning Objectives: *These are the skills you should be proficient in after reading this section and practicing the associated problems. Check off the boxes below once you have gone through this section, answered the associated questions in your textbook, and feel confident you have mastered the skill.*

- ☐ **LO 8.4** Write an electron-dot structure for a molecule and determine its hybridization and bond angles.
- ☐ **LO 8.5** Identify which orbitals overlap to form σ and π bonds in molecules.

Definitions: *To learn, you must do. This is your workbook. Fill in the definitions from the glossary of the textbook as a first step to understanding the material.*

— **sp² hybrid orbital—** ____

____

— **pi (π) bond—** ____

____

— **sp hybrid orbital—** ____

____

Summary: *Read through this section and fill in the missing words and equations. Reference the textbook when you are not confident in your answers.*

Atoms with three electron clouds are formed by ___ hybridization: the combination of one *s* orbital and two *p* orbitals. The hybridized orbitals interact head-on to form σ bonds between the nuclei, but this leaves an empty *p* orbital with its lobes above and below the plane of the hybrid orbitals. The unhybridized *p* orbitals can interact sideways, thus forming a _____ bond, which shares electrons not between the nuclei, but above and below a line that connects the nuclei. The combination of the σ and π bonds results in four electrons being shared: a double bond.

Atoms with two electron clouds are formed by _____ hybridization: the combination of one *s* orbital and one *p* orbital. The two unhybridized *p* orbitals on an *sp* hybridized atom can interact sideways with other unhybridized *p* orbitals, forming two π bonds. As mentioned above for the π bond that forms in sp^2 hybridized atoms, the electrons are shared in a plane above and below a line that connects the nuclei. Since two π bonds can exist with *sp* hybridization, one occurs above and below the line between the nucleus, and another exists in front of and behind that line. The combination of one σ and two π bonds results in six electrons being shared: a triple bond.

Fill in the chart below.

Number of Charge Clouds	Arrangement of Charge Clouds	Hybridization
2		
3		
4		

Atoms with five and six charge clouds have recently been found to be more complex than can be explained by valence bond theory.

Workbook Example 8.3

PROBLEM:
Determine the hybridization for the central atom in H_2CO (formaldehyde) and its bond types.

SOLUTION:
This molecule has the following electron-dot structure.

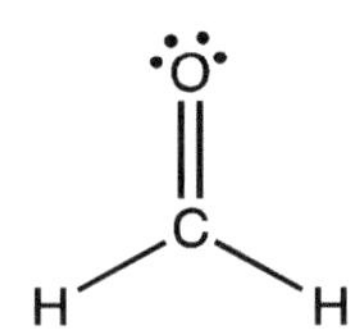

With three charge clouds, the central carbon atom is sp^2 hybridized. It has one σ bond with each hydrogen atom, and one σ and one π bond with the oxygen atom.

Workbook Problem 8.2

Describe the hybridization and bonding of each of the carbon atoms in the structure below.

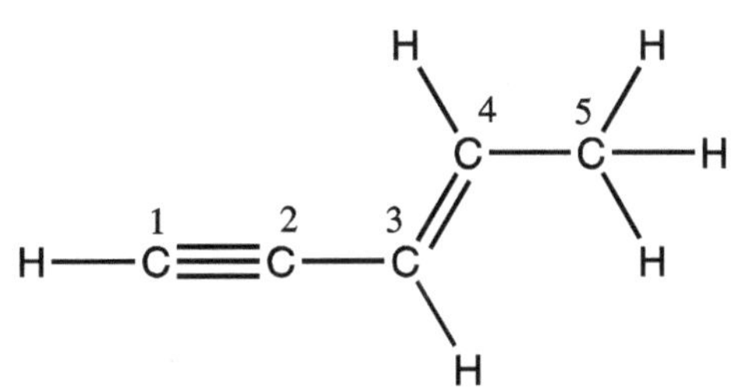

Strategy: Determine the number of electron clouds around each carbon to determine its hybridization, then identify the number of σ and π bonds.

Step 1: Determine the number of electron clouds around each carbon and its hybridization.

Step 2: Determine the bond types present.

POST-LECTURE:

Work Problems: *One of the best ways to learn chemistry is through solving problems. Go to the textbook and work through the Worked Examples for this section, then go to the end of chapter and work the problems there. When you feel confident that you can correctly answer problems like the ones listed below for each objective, check off each of the Learning Objectives for this section above.*

LO 8.4: Problems 8.32, 8.33, 8.62–8.65
LO 8.5: Worked Examples 8.3–8.5; Problems 8.66–8.72

Section 8.5 Polar Covalent Bonds and Dipole Moments

PRE-LECTURE:

Learning Objectives: *These are the skills you should be proficient in after reading this section and practicing the associated problems. Check off the boxes below once you have gone through this section, answered the associated questions in your textbook, and feel confident you have mastered the skill.*

☐ **LO 8.6** Predict whether a given molecule has a dipole moment, and draw its direction.
☐ **LO 8.7** Interpret electrostatic potential maps of molecules.

Definitions: *To learn, you must do. This is your workbook. Fill in the definitions from the glossary of the textbook as a first step to understanding the material.*

— **dipole**— ____

____

Review: *Before going through the section make sure you are familiar with these concepts:*

- LO 7.5 Classify bonds as nonpolar covalent, polar covalent, or ionic.
- LO 7.6 Interpret electrostatic potential maps to determine regions of high and low electron density.

Summary: *Read through this section and fill in the missing words and equations. Reference the textbook when you are not confident in your answers.*

As seen in Chapter 7, polar covalent bonds result from the difference in electronegativity between bonded atoms: these differences lead to electrons being displaced toward one atom in the bond, making it electron-rich and giving it a slight negative charge, while the other end of the bond becomes electron-poor with a slight positive charge. The dipole is indicated by a crossed arrow (⟼) with the crossed end at the positive end of the bond and the arrow pointing in the direction in which the electrons are concentrated, the negative end of the molecule.

The net sum of the individual bond dipoles defines the overall polarity of a molecule, or the molecular dipole.

- Nonpolar molecules
  - are made of ____________bonds or
  - have a ____________ arrangement of bonds with bond dipoles of equal strength

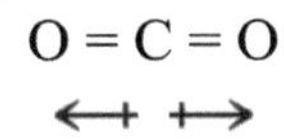

- Polar molecules
  - have an ____________ arrangement of bonds and
  - lone pairs or ____________ bond strengths.

Fill in the figure below with a strategy for predicting molecular polarity.

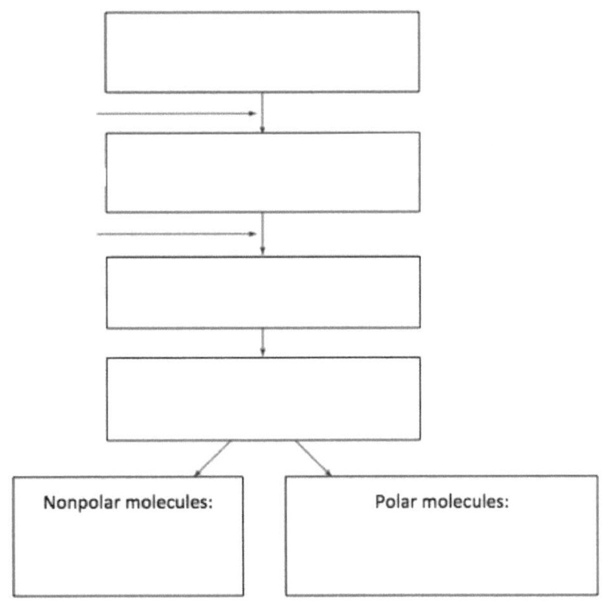

Workbook Example 8.4

PROBLEM:
Which of the following would be expected to be polar, and which nonpolar:

Cl_2 CH_4 CH_3Cl HCl

SOLUTION:
CH_4 and Cl_2 are both symmetrical molecules, which makes them nonpolar. CH_3Cl and HCl are asymmetrical, and contain polar bonds. Drawing the electron-dot structures can help to confirm this initial determination:

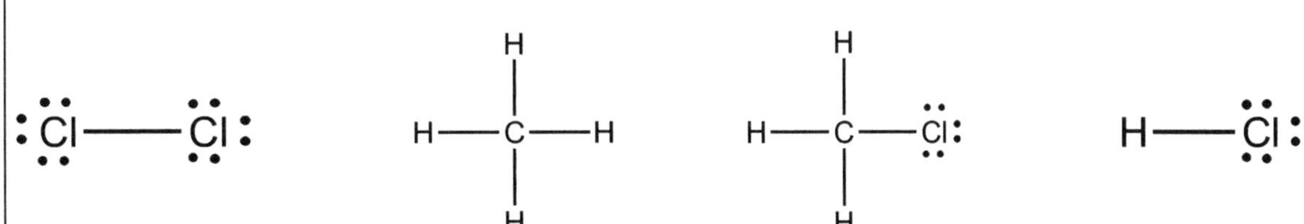

Workbook Problem 8.3

Indicate which of the following are likely to have dipole moments, and, if a dipole is present, indicate its direction:

H_2CO—formaldehyde $N(CH_3)_3$—trimethylamine HBr—hydrogen bromide

Strategy: Look for asymmetrical molecules with electronegativity differences between bonded atoms.

Step 1: Draw the electron structure for each molecule.

Step 2: Draw arrows that represent the flow of electrons in each bond.

Step 3: Consider the net sum of these bond dipoles to find the molecular dipole.

POST-LECTURE:

Work Problems: *One of the best ways to learn chemistry is through solving problems. Go to the textbook and work through the Worked Examples for this section, then go to the end of chapter and work the problems there. When you feel confident that you can correctly answer problems like the ones listed below for each objective, check off each of the Learning Objectives for this section above.*

LO 8.6: Worked Example 8.6; Problems 8.35, 8.78, 8.80
LO 8.7: Problems 8.34, 8.36, 8.37

Section 8.6 Intermolecular Forces

PRE-LECTURE:

Learning Objectives: *These are the skills you should be proficient in after reading this section and practicing the associated problems. Check off the boxes below once you have gone through this section, answered the associated questions in your textbook, and feel confident you have mastered the skill.*

- ☐ **LO 8.8** Identify the types of intermolecular forces experienced by a molecule.
- ☐ **LO 8.9** Relate the strength of intermolecular forces to physical properties such as melting point and boiling point.
- ☐ **LO 8.10** Sketch the hydrogen bonding that occurs between two molecules.

Definitions: *To learn, you must do. This is your workbook. Fill in the definitions from the glossary of the textbook as a first step to understanding the material.*

— **intermolecular force—** ____

____

— **van der Waals forces—** ____

____

— **London dispersion forces—** ____

____

— **polarizability—** ____

____

— **dipole–dipole force—** ____

____

— **deoxyribonucleic acid (DNA)—** ____

____

— **hydrogen bond—** ____

____

— **ion–dipole force—** ____

____

Summary: *Read through this section and fill in the missing words and equations. Reference the textbook when you are not confident in your answers.*

Covalent and ionic bonds are **intramolecular forces:** they occur *within* molecules. **Intermolecular** forces occur *between* molecules.

Intermolecular forces are important in many macroscopic properties, and they are responsible for the phase in which matter is found. Increasing the temperature ___________ the amount of kinetic energy in a sample. Solids are made up of particles that are held together rather tightly. This is because the particles have strong interactions that hold them together, and they do not have enough energy to break apart or even slide around one another. At that same temperature (with the same amount of energy), the particles that make up liquids are able to gain a little freedom to move around one another, but they are unable to completely break apart and are still held close together. Again, holding the temperature constant (holding energy constant), if you look at gaseous particles, they have enough energy to break free because the interactions between the particles are so weak. Label the particulate view below with the type of matter it represents.

There are a number of intermolecular forces. The four general categories are described below:

- **Ion–dipole forces:** these occur between __________ and __________ __________. The charge on the ion will attract the portion of a polar molecule that has an opposite partial charge. These are particularly important in aqueous solutions of ionic substances. Ion–dipole forces have moderate strength: 10–50 kJ/mol. If the particles are both ions, the interactions are very strong; we call these ionic bonds.
- **Dipole–dipole forces:** are found in neutral, __________ molecules; the stronger the dipole, the stronger the interaction. Increasing the strength of dipole moments correlates roughly with an increase in boiling points because the stronger dipole–dipole interactions require more kinetic energy to disrupt them. Dipole–dipole forces are weak: 3–4 kJ/mol.
- **London dispersion forces:** these forces are found in all types of molecules, but are the only ones that occur between ______________ molecules. They are the result of varying electron densities; average electron densities are uniform, but a small variation in electron density can create a small dipole that can then induce an opposite dipole in another, nearby molecule. Such *polarizability* increases with molecular size and varies with molecular shape; an extended molecule has more opportunity to create dipoles than one that is compact. Polarizability is related to boiling point: more dispersion forces mean a higher boiling point. London dispersion forces are weak: 1–10 kJ/mol.
- **Hydrogen bonds:** are unusually strong dipole–dipole forces between a hydrogen atom that is bonded to a ________, ________, or ________ atom and another electronegative atom that has an unshared electron pair. Molecules that can form hydrogen bonds have boiling points that are unusually high for their small size because these strong interactions must be broken for the molecules to enter the gas phase. Hydrogen bonds have moderate strength as compared to ionic and covalent bonds: 10–40 kJ/mol.

Workbook Example 8.5

PROBLEM:
Identify the intermolecular forces experienced by the following molecules, and rank them in order of increasing intermolecular forces:

NH_3 CH_4 HBr $CH_3CH_2CH_2CH_3$

SOLUTION:
NH_3 can form hydrogen bonds. It is polar, and it will have a dipole. It will also experience dipole–dipole and dispersion forces.

CH_4 cannot form hydrogen bonds, and it does not have a dipole. It will experience only dispersion forces.

HBr cannot form hydrogen bonds, but it is polar, and it will have a dipole. This molecule will have both dipole–dipole and dispersion forces.

$CH_3CH_2CH_2CH_3$ cannot form hydrogen bonds, and it does not have a dipole. It will experience only dispersion forces.

CH_4 will experience the weakest intermolecular forces. $CH_3CH_2CH_2CH_3$ will have stronger dispersion forces because it is larger. HBr will have the next strongest forces because it has both dipole–dipole and dispersion forces. NH_3, with hydrogen bonds, will have the strongest intermolecular interactions. Below is the ranking in order of increasing boiling point:

$$CH_4 < CH_3CH_2CH_2CH_3 < HBr < NH_3$$

Workbook Problem 8.4

Determine the intermolecular forces that are present in the following molecules and order them by increasing strength.

CH_3OH Xe CH_3Cl

Strategy: Examine the structures of the molecules and determine which interactions are present, then rank them in order of strength.

Step 1: List the interactions experienced by each molecule.

Step 2: Rank the molecules in order of increasing strength of interactions.

POST-LECTURE:

Work Problems: *One of the best ways to learn chemistry is through solving problems. Go to the textbook and work through the Worked Examples for this section, then go to the end of chapter and work the problems there. When you feel confident that you can correctly answer problems like the ones listed below for each objective, check off each of the Learning Objectives for this section above.*

LO 8.8: Worked Example 8.8; Problems 8.15, 8.88, 8.89
LO 8.9: Problems 8.16, 8.90, 8.91
LO 8.10: Worked Example 8.7; Problems 8.14, 8.94, 8.95

Section 8.7 Molecular Orbital Theory: The Hydrogen Molecule

PRE-LECTURE:

Learning Objectives: *There are no specific learning objectives for this section; the information in this section is used as a basis for understanding subsequent sections.*

Definitions: *To learn, you must do. This is your workbook. Fill in the definitions from the glossary of the textbook as a first step to understanding the material.*

— **molecular orbital (MO) theory—** ______________________________

____

— **molecular orbital—** ______________________________

____

— **bonding molecular orbital—** ______________________________

____

— **antibonding molecular orbital—** ______________________________

____

Summary: *Read through this section and fill in the missing words and equations. Reference the textbook when you are not confident in your answers.*

The valence bond model does not always agree with experimental observations. A more complex model, known as **molecular orbital theory,** provides results that come closer to observed molecular behavior. In contrast to the previous models described, which focused on atomic orbitals, molecular orbital theory looks at the molecule as a whole.

When two hydrogen atoms interact, the wave portions of the electrons will interfere with each other. They can interfere constructively and add together, or they can interfere destructively and subtract from one another.

If the 1*s* orbitals of hydrogen atoms add together, they form a single, egg-shaped *bonding* orbital. If they subtract from each other, they form two small orbitals with a node between the nuclei, which is an ____________ orbital, denoted σ^*. The bonding orbital is stable and low in energy, and the antibonding orbital is high energy and unstable. If more electrons are added to an H_2 molecule, they have to be added to the antibonding orbital, thus destabilizing the molecule.

Bond order is defined as:

Bond order =

Workbook Example 8.6

PROBLEM:
What is the bond order of an He_2^{2+} ion? Is it stable or unstable?

SOLUTION:
This molecule has two electrons, just as hydrogen does. The electrons will first go into the bonding orbital; no electrons are left to go into the antibonding orbital. The bond order will be

$$\text{Bond order} = \frac{2 \text{ bonding electrons} - 0 \text{ antibonding electrons}}{2} = 1$$

Contrary to expectations, this will be a stable ion because it will form one bond.

Key ideas in Molecular Orbital Theory:

- A molecular orbital describes the regions in space where electrons are most likely to be found.
- The number of atomic orbitals that are combined ___________ the number of molecular orbitals produced.
- Molecular orbitals with energy ___________ than the starting atomic orbital are bonding orbitals. Molecular orbitals with energy ___________ than the starting atomic orbital are antibonding orbitals.
- Electrons occupy orbitals in order, starting with those of ___________ energy.
- Bond order is calculated by finding the difference in the number of bonding and antibonding orbitals and dividing by ___________.

Section 8.8 Molecular Orbital Theory: Other Diatomic Molecules

PRE-LECTURE:

Learning Objectives: *These are the skills you should be proficient in after reading this section and practicing the associated problems. Check off the boxes below once you have gone through this section, answered the associated questions in your textbook, and feel confident you have mastered the skill.*

☐ **LO 8.11** Interpret the molecular orbital diagram for a first-row diatomic molecule or ion.

☐ **LO 8.12** Interpret the molecular orbital diagram for a second-row diatomic molecule or ion. Calculate the bond order and predict magnetic properties.

Definitions: *To learn, you must do. This is your workbook. Fill in the definitions from the glossary of the textbook as a first step to understanding the material.*

— **paramagnetic—**____

____

— **diamagnetic—** ____

____

Summary: *Read through this section and fill in the missing words and equations. Reference the textbook when you are not confident in your answers.*

Electron-dot structure and valence bond theory fail to explain why experiments show that the oxygen molecule is paramagnetic. Molecular orbital theory can be used to understand why this is so; when a molecular orbital diagram is drawn, we see two electrons in __________ _____ *2p* orbitals.

POST-LECTURE:

Work Problems: *One of the best ways to learn chemistry is through solving problems. Go to the textbook and work through the Worked Examples for this section, then go to the end of chapter and work the problems there. When you feel confident that you can correctly answer problems like the ones listed below for each objective, check off each of the Learning Objectives for this section above.*

LO 8.11: Worked Example 8.9; Problems 8.17, 8.18
LO 8.12: Worked Example 8.10; Problems 8.98, 8.100, 8.102

Section 8.9 Combining Valence Bond Theory and Molecular Orbital Theory

PRE-LECTURE:

Learning Objectives: *These are the skills you should be proficient in after reading this section and practicing the associated problems. Check off the boxes below once you have gone through this section, answered the associated questions in your textbook, and feel confident you have mastered the skill.*

☐ **LO 8.13** Draw orbital overlap diagrams for molecules and describe the use of both valence bond theory and molecular orbital theory.

Summary: *Read through this section and fill in the missing words and equations. Reference the textbook when you are not confident in your answers.*

Both of these theories are useful to chemists. The simplicity of valence bond theory makes it a desirable tool whenever it can be useful; but where it fails to provide answers, the mathematical precision of molecular orbital theory can save the day.

For example, sigma (σ) bonds are "local" bonds—they occur __________ the nuclei of two atoms and go no further. Valence bond theory describes these bonds well. On the other hand, pi (π) bonds are not local; they occur outside of the internuclear space, and as a result, they can spread over more than two atoms in resonance structures. When resonance structures occur, electrons are said to be "delocalized." Valence bond theory deals with this awkwardly at best; molecular orbital theory gives a better picture.

POST-LECTURE:

Work Problems: *One of the best ways to learn chemistry is through solving problems. Go to the textbook and work through the Worked Examples for this section, then go to the end of chapter and work the problems there. When you feel confident that you can correctly answer problems like the ones listed below for each objective, check off each of the Learning Objectives for this section above.*

LO 8.13: Worked Example 8.11; Problems 8.21, 8.22, 8.108, 8.109

Self-Test

This section is intended to test your knowledge of the material covered in this chapter. Think through these problems and make certain you understand what the problem is about—ask yourself if the answer makes sense. Many of these questions are linked to the chapter learning goals. Therefore, successful completion of these problems indicates that you have achieved the learning goals for this chapter. You will receive the greatest benefit from this section if you use it as a mock exam. Do not use your notes, book, Internet, or friends when taking the test. There is no grade and you only shortchange yourself by getting help with this activity. As you take the test, mark the problems with your level of confidence with the answer.

★ *Use a star for problems you know you got right.*
? *Use a question mark for those you think you got right.*
x *Use an x for those you are guessing on.*

If you get a question right that you marked with a star, you can be confident that you know the associated LO.

If you get a question right that you marked with a question mark, go back and work a few more problems and test yourself again tomorrow.

If you get a question wrong or mark it with an x, go back and read the section again, work a few problems today, a few tomorrow, and a few the next day before testing yourself again.

This will help you to discover which topics you have learned thoroughly and which topics you need to study in more detail.

True–False

1. Bond polarity is usually the result of differences in electronegativity.

2. Intermolecular forces occur within a molecule.

3. The number of electron clouds in a molecule can be used to determine its three-dimensional shape.

4. Methane (CH_4) molecules have a pyramidal shape.

5. Molecular shape can be used to determine orbital hybridization.

6. When orbitals overlap head-on, π bonds are formed.

7. Molecular orbital theory can be used to explain molecular properties that cannot be explained by valence bond theory.

8. Multiple bonds can be explained by electron-dot structures and valence bond theories.

9. If a molecule has polar bonds, it will always have a dipole moment.

10. Intermolecular forces all are equal in strength.

11. Intermolecular forces can be related to melting and boiling points.

Fill in the Blanks

12. A molecule with four electron clouds will have a ____________________ electronic shape, but the molecular shape may also be ____________ or ______________ ________________.

13. A trigonal planar arrangement of charge clouds always implies _________________ hybridization.

14. _______________ bonds are always local, but _______________ bonds can be delocalized over more than two atoms.

15. ________________ is a measure of the strength with which an atom pulls on its electrons.

16. There are several types of intermolecular forces. ____________________ result from the interactions of two polar molecules. __________________ occur in all molecules and are the result of temporary charge imbalances. ________________ are the strongest intermolecular forces, and they result from interactions between a H atom bonded to an N, O, or F, and a nonbonding pair on another atom.

Matching

17. VSEPR model	a.	intermolecular force characterized by the attraction between a hydrogen atom with a partial positive charge and an unbonded electron pair
18. hybrid orbitals	b.	attractive force caused by temporary dipoles
19. σ bond	c.	substance that is weakly repelled by a magnetic field because all of its electrons are spin-paired
20. π bond	d.	attracted to a magnetic field as a result of unpaired electrons
21. molecular orbital	e.	an area in which the electrons in an atom or bond are most likely to be found
22. paramagnetic	f.	a bond formed by sideways interactions with parallel orbitals in which the electrons do not reside between the nuclei
23. diamagnetic	g.	a bond formed by head-on overlap of orbitals in which the electrons reside primarily between the two nuclei
24. London dispersion forces	h.	a combined solution for the Schrödinger wave equations that gives a new set of values

25. hydrogen bonding

i. a method for determining the three-dimensional structures of molecules using electron repulsions

Problems

26. Determine the electron geometry and molecular geometry for the following molecules:

NF_3 H_2S IF_4^- PF_5

27. Predict the geometry and hybridization of the carbon atoms in acetaldehyde:

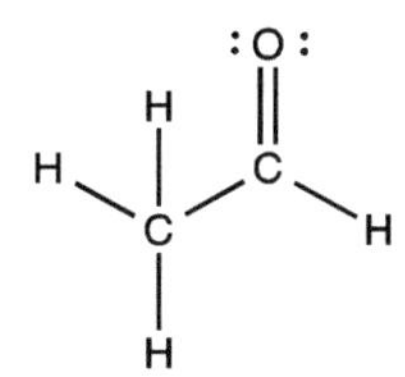

What types of bonds are found in this molecule? Which atoms are linked by these bond types?

28. Sketch bonding and antibonding orbitals that come from the combination of two *s* orbitals. Indicate *where* they are interfering and *how* they are interfering (constructively or destructively).

Challenge Problem

29. 360 mg of an unknown monoprotic acid requires 78.24 mL of 0.10 M NaOH to be titrated to neutrality. Elemental analysis shows that the acid is 71.1% oxygen, 26.7% carbon, and 2.22% hydrogen. Determine the formula of the acid, and draw its electron-dot structure. Using VSEPR theory, provide the molecular shape around each of the central atoms. Using valence bond theory, explain how the oxygen and carbon atoms are hybridized, and in which bond types (σ and π) each are participating.

Test Yourself: *Take the Practice Test on pages 317–319 of your textbook. Again, use a testing environment using only those items that your instructor will allow you to have access to during the test and mark each question with a star, question mark, or x to denote your confidence level. Once you have finished you can check off the LO at the beginning of this chapter based on which questions you starred and got correct. Go back and spend more time with the topics from the other learning objectives.*

CHAPTER NINE

Thermochemistry: Chemical Energy

Learning Objectives

As a result of reading and studying this chapter, you should be able to:

- ☐ **LO 9.1** Calculate the kinetic energy of an object in motion.
- ☐ **LO 9.2** Convert between different units of energy.
- ☐ **LO 9.3** Identify state functions.
- ☐ **LO 9.4** Identify the sign of heat and work.
- ☐ **LO 9.5** Calculate PV work.
- ☐ **LO 9.6** Calculate the internal energy change (ΔE) for a reaction.
- ☐ **LO 9.7** Calculate the amount of heat transferred given a thermochemical equation and the amount of reactant or product.
- ☐ **LO 9.8** Classify reaction as endothermic or exothermic, and give the sign of ΔH.
- ☐ **LO 9.9** Calculate heat capacities, temperature changes, or heat transfer using the equation for heat capacity (C), specific heat (c), or molar heat capacity (C_m).
- ☐ **LO 9.10** Calculate enthalpy changes in a calorimetry experiment.
- ☐ **LO 9.11** Use Hess's Law to find ΔH for an overall reaction, given reaction steps and their ΔH values.
- ☐ **LO 9.12** Identify standard states of elements.
- ☐ **LO 9.13** Write standard enthalpy of formation reactions (ΔH°_f) for compounds from their elements.
- ☐ **LO 9.14** Use values of (ΔH°_f) for elements and compounds to calculate ΔH° for a reaction.
- ☐ **LO 9.15** Use bond dissociation energies to estimate ΔH° for a reaction.
- ☐ **LO 9.16** Predict the sign of the entropy change (ΔS) given the chemical equation or a molecular diagram.
- ☐ **LO 9.17** Predict the sign of ΔG, ΔH, and ΔS, using the equation for Gibbs free-energy change (ΔG).
- ☐ **LO 9.18** Use the Gibbs free-energy equation to calculate the temperature at which a reaction crosses over from spontaneous to nonspontaneous.

Chapter Summary

In this chapter, we introduce you to the concept of thermochemistry: the heat changes that take place during reactions. You'll begin the study of this topic by learning the difference between heat and energy, and the types of energy changes that can take place. You are then introduced to the law of conservation of energy, which is the first law of thermodynamics, and the concept of state functions. With this background knowledge, you will learn how to calculate the internal energy of a system using *PV* work and also how the internal energy of a system is related to the enthalpy (ΔH) of the system. You will spend much of the rest of the chapter exploring how to use specific heat calculations in the laboratory and how to use Hess's law, standard heats of formation, and bond dissociation energies to calculate heats of reaction. Finally, you are introduced to the topics of entropy and free energy, topics that will be explored in more detail in the later chapters.

The Chapter in Detail

Section 9.1 Energy and Its Conservation

PRE-LECTURE:

Learning Objectives: *These are the skills you should be proficient in after reading this section and practicing the associated problems. Check off the boxes below once you have gone through this section, answered the associated questions in your textbook, and feel confident you have mastered the skill.*

- ☐ **LO 9.1** Calculate the kinetic energy of an object in motion.
- ☐ **LO 9.2** Convert between different units of energy.

Definitions: *To learn, you must do. This is your workbook. Fill in the definitions from the glossary of the textbook as a first step to understanding the material.*

— **thermochemistry—** ____

____

— **energy—** ____

____

— **kinetic energy (E_K)—** ____

____

— **joule (J)—** ____

____

— **potential energy (E_P)—** ____

____

— **conservation of energy law—** ____

____

— **temperature—** ____

____

— **heat—** ____

____

— **first law of thermodynamics—** ____

____

Summary: *Read through this section and fill in the missing words and equations. Reference the textbook when you are not confident in your answers.*

Energy is the capacity to do __________ or supply __________. Energy can be changed from one form to another, but it cannot be created or destroyed. This is the conservation of energy law, also known as the first law of thermodynamics.

Kinetic energy (E_K) is the energy of motion, and is described by the following formula:

$E_K =$

where $m =$ ________ (measured in kg) and $v =$ ________ (measured in m/s). Thermal energy, the energy of molecular motion, is a type of kinetic energy. It is measured by finding an object's temperature. Fast-moving molecules have a ________ temperature than slow-moving molecules. When thermal energy is transferred from one object to another, it is transferred as heat.

Potential energy (E_P) is ________ energy. This can be the energy of position, as in a marble sitting at the top of a hill, or chemical potential energy, which is the capacity to react and thereby release energy. A bucket of gasoline has a great deal of chemical potential energy: a small spark will release a tremendous amount of energy as the hydrocarbons react with the oxygen in the air.

Workbook Example 9.1

PROBLEM:
Describe the energy changes as a rubber band is shot.

SOLUTION:
As the rubber band is stretched, potential energy is added to it. When the rubber band is released, the potential energy is converted to kinetic energy as it flies across the room.

POST-LECTURE:

Work Problems: *One of the best ways to learn chemistry is through solving problems. Go to the textbook and work through the Worked Examples for this section, then go to the end of chapter and work the problems there. When you feel confident that you can correctly answer problems like the ones listed below for each objective, check off each of the Learning Objectives for this section above.*

LO 9.1: Problems 9.46, 9.47
LO 9.2: Problems 9.52, 9.53

Section 9.2 Internal Energy and State Functions

PRE-LECTURE:

Learning Objectives: *These are the skills you should be proficient in after reading this section and practicing the associated problems. Check off the boxes below once you have gone through this section, answered the associated questions in your textbook, and feel confident you have mastered the skill.*

- ☐ **LO 9.3** Identify state functions.
- ☐ **LO 9.4** Identify the sign of heat and work.

Definitions: *To learn, you must do. This is your workbook. Fill in the definitions from the glossary of the textbook as a first step to understanding the material.*

— **system—** ____

____

— **surroundings—** ____

____

— **internal energy (E)—** ____

____

— **state function—** ____

____

Summary: *Read through this section and fill in the missing words and equations. Reference the textbook when you are not confident in your answers.*

When accounting for the energy changes in a reaction, the reaction is thought of as being isolated. The reactants and products are referred to as the ________, while everything else is the ________.

In a completely isolated system—a theoretical situation—no energy is transferred to the surroundings, so the internal energy of the system is constant. (This is the first law of thermodynamics stated in a different way.)

It is not really possible to completely isolate a system, so energy will flow into and out of any system. When this is measured, it is expressed as $\Delta E = E_{\text{final}} - E_{\text{initial}}$.

Changes in energy are expressed in terms of the system. So, if energy flows out to the surroundings, the energy change is negative: the system has lost energy. If energy flows into the system, the energy change is positive: the system has gained energy. What makes this difficult to get a handle on is that in general, *you* are a part of the surroundings, but we measure change from the view of the system. When you hold a glass of cold water in your hand, the system of the water is gaining energy, while you are losing it. So, a process that generates heat has a negative ΔE, while a process that requires heat has a positive ΔE.

- $-\Delta E$ = energy flowing ________________
- $+\Delta E$ = energy flowing ________________

Internal energy is a state function: it depends only on current conditions, and not at all on how the system came to be in that condition. State functions are reversible: if you start at sea level and climb 2000 feet to the top of a mountain, your current altitude will change, but if you then return to sea level, your overall change in altitude is zero; altitude is a state function.

Workbook Example 9.2

PROBLEM:
Which of the following are state functions:

Temperature Volume Mass Work

SOLUTION:
Temperature is a state function: if you warm a beaker in your hands, then put it back down on the lab bench, it will return to its previous temperature. The process is reversible, and you have a $\Delta T = 0$.

A balloon carried outside on a cold day will shrink and look miserable, but when brought back indoors, it will resume its previous cheerful volume. It is a reversible process, $\Delta V = 0$, so volume is a state function.

Mass is a state function: how an object got its mass is irrelevant.

Work is *not* a state function. If an object has been moved from one side of a room to the other and back again, its overall position has not changed, but work was still expended moving it from one place to the other. The sofa is in the same place it was, but you are tired!

POST-LECTURE:

Work Problems: *One of the best ways to learn chemistry is through solving problems. Go to the textbook and work through the Worked Examples for this section, then go to the end of chapter and work the problems there. When you feel confident that you can correctly answer problems like the ones listed below for each objective, check off each of the Learning Objectives for this section above.*

LO 9.3: Problems 9.48, 9.49
LO 9.4: Problems 9.34–9.36

Section 9.3 Expansion Work

PRE-LECTURE:

Learning Objectives: *These are the skills you should be proficient in after reading this section and practicing the associated problems. Check off the boxes below once you have gone through this section, answered the associated questions in your textbook, and feel confident you have mastered the skill.*

☐ **LO 9.5** Calculate PV work.

Definitions: *To learn, you must do. This is your workbook. Fill in the definitions from the glossary of the textbook as a first step to understanding the material.*

— **work (*w*)**—____

____

Summary: *Read through this section and fill in the missing words and equations. Reference the textbook when you are not confident in your answers.*

In physics (and by extension, chemistry), *work* is defined as force × distance ($w = F \times d$). In a system in which a cylinder is fitted with a piston, changes in the pressure of the gas can force the piston upward, doing work. In this case, the equation used is:

$w =$

This is referred to as "PV work" and has the units of $L \cdot atm$ (we use J as the units of energy in chemistry; to convert to J, 1 L · atm = 101 J). An expanding system is doing work, so it is losing energy to the surroundings. This will have a negative value. A contracting system is having work done on it by the ________ and so is gaining energy and will have a ________ value. If the volume of the system does not change, $\Delta V = 0$, and no PV work was done.

Workbook Example 9.3

PROBLEM:
Calculate the work done when the volume of a reaction changes from 3.0 L to 12 L against a pressure of 15 atm.

SOLUTION:
Using the equation $w = -P\Delta V$:

$$w = -(15\ atm)(12\ L - 3.0\ L) = -135\ L \cdot atm$$

$$-135\ L \cdot atm \times \frac{101\ J}{1\ L \cdot atm} = -13{,}635\ J \text{ or } -14\ kJ$$

Workbook Problem 9.1

If 7.4 L of hydrogen gas is reacted with 3.7 L of oxygen gas at 13.0 atm, 7.4 L of water vapor is produced. How many kilojoules of work have been done? Have they been done *by* the system or *to* the system?

Strategy: Use the equation $w = -P\Delta V$, then convert to kilojoules.

Step 1: Use $w = -P\Delta V$:

Step 2: Convert from $L \cdot atm$ to joules to kilojoules.

Step 3: What does the sign tell you? Was work done by or to the system?

POST-LECTURE:

Work Problems: *One of the best ways to learn chemistry is through solving problems. Go to the textbook and work through the Worked Examples for this section, then go to the end of chapter and work the problems there. When you feel confident that you can correctly answer problems like the ones listed below for each objective, check off each of the Learning Objectives for this section above.*

LO 9.5: Worked Example 9.1; Problems 9.37, 9.54, 9.56, 9.60

Section 9.4 Energy and Enthalpy

PRE-LECTURE:

Learning Objectives: *These are the skills you should be proficient in after reading this section and practicing the associated problems. Check off the boxes below once you have gone through this section, answered the associated questions in your textbook, and feel confident you have mastered the skill.*

☐ **LO 9.6** Calculate the internal energy change (ΔE) for a reaction.

Definitions: *To learn, you must do. This is your workbook. Fill in the definitions from the glossary of the textbook as a first step to understanding the material.*

— **heat of reaction (ΔH)**—____

____

— **enthalpy change (ΔH)**—______________________________________

____

— **enthalpy (H)**—____

____

Summary: *Read through this section and fill in the missing words and equations. Reference the textbook when you are not confident in your answers.*

Reactions can transfer energy through heat (q) or PV work. These terms can be combined to give the overall energy change of a system:

$\Delta E =$

which can also be rearranged so that we can solve for q:

$q =$

If a reaction is run in a closed container, the volume is constant, and no PV work can be done. In this case, $\Delta E = q_v$.

If instead the reaction is carried out at constant pressure (such as the atmospheric pressure in the lab), *PV* work can be done, so $q_p = \Delta E + P\Delta V = \Delta H$.

Because reactions at constant pressure are so common, the change of heat in these systems is called the **heat of reaction,** or **enthalpy,** and abbreviated ΔH. Enthalpy is also a state function: it depends only on the initial and final states of the reaction, not the path the reaction takes.

$$\Delta H = H_{\text{products}} - H_{\text{reactants}}.$$

When the *PV* work done by a reaction at constant pressure is calculated, we find that it generally makes only a very small contribution to the overall energy change of a reaction. As a result, $\Delta H \approx \Delta E$. If there is no volume change as a result of the reaction, then $\Delta H = \Delta E$.

Workbook Example 9.4

PROBLEM:
The equation for the reaction of nitrogen and hydrogen to form ammonia is:

$$N_2(g) + 3\ H_2(g) \rightarrow 2\ NH_3(g)$$

If 2.0 L of nitrogen and 6.0 L of hydrogen are reacted together at 1.0 atm, 4.0 L of ammonia is formed. At the same time, 8.23 kJ of heat is released by the reaction. What is the overall energy change of the reaction, and what is the percent contribution of *PV* work?

SOLUTION:
Solve for $P\Delta V$:

$$P\Delta V = (1.0\ atm)(4.0\ L - 8.0\ L) = -4.0\ L \cdot atm$$

$$-4.0\ L \cdot atm \times \frac{101\ J}{1\ L \cdot atm} = -404\ J = -0.404\ kJ$$

Using the equation $\Delta E = q - P\Delta V$

$$\Delta E = -8.23\ kJ - (-0.404\ kJ) = -7.826\ kJ = -7.8\ kJ$$

The percent contribution of the *PV* work is

$$\frac{0.404\ kJ}{7.826\ kJ} = 0.516\ or\ 5.2\%$$

Workbook Problem 9.2

When 7.4 L of hydrogen gas is reacted with 3.7 L of oxygen gas at 13.0 atm, 7.4 L of water vapor is produced, and 79.9 kJ of heat is released. What is the total energy change of the system?

Strategy: Use the process from the previous workbook problem, substituting into $\Delta E = q - P\Delta V$.

Step 1:	Solve for $P\Delta V$
Step 2:	Substitute values into $\Delta E = q - P\Delta V$

POST-LECTURE:

Work Problems: *One of the best ways to learn chemistry is through solving problems. Go to the textbook and work through the Worked Examples for this section, then go to the end of chapter and work the problems there. When you feel confident that you can correctly answer problems like the ones listed below for each objective, check off each of the Learning Objectives for this section above.*

LO 9.6: Worked Example 9.2; Problems 9.62, 9.64, 9.65

Section 9.5 Thermochemical Equations and the Thermodynamic Standard State

PRE-LECTURE:

Learning Objectives: *These are the skills you should be proficient in after reading this section and practicing the associated problems. Check off the boxes below once you have gone through this section, answered the associated questions in your textbook, and feel confident you have mastered the skill.*

☐ **LO 9.7** Calculate the amount of heat transferred given a thermochemical equation and the amount of reactant or product.

Definitions: *To learn, you must do. This is your workbook. Fill in the definitions from the glossary of the textbook as a first step to understanding the material.*

— **thermochemical equation**—____

____

— **thermodynamic standard state**— ___________________________________

____

— **standard enthalpy of reaction ($\Delta H°$)**—______________________________

____

Summary: *Read through this section and fill in the missing words and equations. Reference the textbook when you are not confident in your answers.*

The value of the enthalpy change of a reaction stated in the thermochemical equation is given per stoichiometric equivalent of the reactants and products. For example, when ammonia is created from hydrogen and nitrogen, you get the following thermochemical equation:

$$N_2(g) + 3\ H_2(g) \rightarrow 2\ NH_3(g) \qquad \Delta H = -92.2\ \text{kJ}$$

It says that when 3 moles of hydrogen gas react with 1 mole of nitrogen gas, 2 moles of ammonia gas and 92.2 kJ of energy are produced. This gives us a way to convert between the amount of substance and the accompanying energy change. Thus, for every 3 moles of hydrogen used, there is an energy change of −92.2 kJ.

Workbook Example 9.5

PROBLEM:
What is the change in heat when 25 g of ammonia is created? The equation for the reaction of nitrogen and hydrogen to form ammonia is below.

$$N_2(g) + 3\ H_2(g) \rightarrow 2\ NH_3(g) \qquad \Delta H = -92.2\ \text{kJ}$$

SOLUTION:
Use stoichiometry to convert the amount of ammonia to heat.

$$25\ g\ NH_3 \times \frac{1\ mol\ NH_3}{17.03052\ g} \times \frac{-92.2\ kJ}{2\ mol\ NH_3} = -68\ kJ$$

Since the question asked for heat change, and did not specify a direction, it is important to include the sign of heat change.

In calculating heat and energy change, it is essential to know the phases of all the reactants and products: for instance, the difference in energy change for producing liquid water is more than for producing gaseous water vapor because there is an energy associated with phase changes.

To ensure a common understanding regarding the amounts and phases of products and reactants, a *thermodynamic standard state* has been defined as a substance in its most stable form at ________ pressure and _____ °C. For solutions, the standard state is a solution under the above conditions and at _____ concentration. Measurements made at standard state are identified with a superscript (°) and can be used to calculate other quantities at standard state.

Workbook Example 9.6

PROBLEM:
Identify the standard state for each of the following:

bromine   oxygen   iron   mercury   sodium

SOLUTION:
At 25 °C and 1 atm, bromine is a diatomic liquid, $Br_2(l)$; oxygen is a diatomic gas, $O_2(g)$; iron is a metallic solid, Fe(*s*); mercury is a metallic liquid, Hg(*l*); and sodium is a metallic solid, Na(*s*).

Workbook Example 9.7

PROBLEM:
What is the value of $\Delta E°$ if a reaction that has $\Delta H° = 157.3$ kJ is carried out at a constant pressure of 12.3 atm and its volume change is 453 L?

SOLUTION:

$$\Delta H° = \Delta E° + P\Delta V$$

Replacing all of the known terms with the numbers from the problem, we have

$$157.3\ kJ = \Delta E° + (12.3\ atm \times 453\ L)$$

$$157.3\ kJ = \Delta E° + (5571.9\ L \cdot atm)$$

Converting 5571.9 $\text{L} \cdot \text{atm}$ to J (remember that 1 $\text{L} \cdot \text{atm} = 101$ J), we have

$$157.3\ kJ = \Delta E° + (562.8\ kJ)$$

Solving for $\Delta E°$:

$$\Delta E° = 157.3\ kJ - 562.8\ kJ = -405.5\ kJ$$

Workbook Problem 9.3

When 0.50 mol of aqueous carbonic acid dissociates into water and carbon dioxide, the volume change at 1 atm is 11.2 L. If $\Delta H°$ for this reaction is −20.7 kJ/mol, what is ΔE?

Strategy: Use the equation $\Delta H° = \Delta E° + P\Delta V$.

Step 1: Determine $\Delta H°$ for the reaction given the amount of carbonic acid that dissociated.

Step 2: Determine $P\Delta V$ and convert from $\text{L} \cdot \text{atm}$ to kJ.

Step 3: Determine $\Delta E°$ from the equation $\Delta H° = \Delta E° + P\Delta V$.

POST-LECTURE:

Work Problems: *One of the best ways to learn chemistry is through solving problems. Go to the textbook and work through the Worked Examples for this section, then go to the end of chapter and work the problems there. When you feel confident that you can correctly answer problems like the ones listed below for each objective, check off each of the Learning Objectives for this section above.*

LO 9.7: Worked Example 9.3; Problems 9.68–9.72

Section 9.6 Enthalpies of Chemical and Physical Changes

PRE-LECTURE:

Learning Objectives: *These are the skills you should be proficient in after reading this section and practicing the associated problems. Check off the boxes below once you have gone through this section, answered the associated questions in your textbook, and feel confident you have mastered the skill.*

☐ **LO 9.8** Classify reaction as endothermic or exothermic and give the sign of ΔH.

Definitions: *To learn, you must do. This is your workbook. Fill in the definitions from the glossary of the textbook as a first step to understanding the material.*

— **endothermic—**____

____

— **exothermic—**____

____

— **sublimation—** ____

____

Summary: *Read through this section and fill in the missing words and equations. Reference the textbook when you are not confident in your answers.*

There are enthalpies associated with all physical changes or phase changes. It is important to realize that because all of these changes are reversible, the energies associated with the changes are also reversible: the heat required to melt a certain amount of a substance is the same amount of heat that will be released when it freezes again. The two heats will be equal in magnitude, but opposite in sign. So $\Delta H°$ for the forward reaction = $-\Delta H°$ for the reverse reaction.

Enthalpies of Physical Reactions:

- Heat of fusion ($\Delta H°_{fus}$) is the energy required to ________ a substance at its normal melting/freezing point (for water, this is 32 °F or 0 °C). The heat of fusion is also the heat a substance releases when it freezes.
- Heat of vaporization ($\Delta H°_{vap}$) is the energy required to convert a substance from ________ to ________ at its normal boiling point. This energy tends to be large because

all of the intermolecular interactions must be broken to vaporize a substance. Heat of vaporization is also the energy released by a substance when it condenses from gas to liquid, which is why a steam burn can be so severe.

- Heat of sublimation (ΔH°_{subl}): when a substance is converted straight from a ________ to a ________, it is said to have sublimed. The heat of sublimation is the sum of the heats of fusion and vaporization for a substance, because both of these phase changes are contained in sublimation.

Chemical changes also have associated enthalpies. At constant pressure, heat can flow into or out of a reaction system. Reactions that emit heat into the surroundings are exothermic: the enthalpy of the products is lower than that of the reactants, so $H_{products} - H_{reactants}$ is negative. $\Delta H < 0$ is **exothermic.** A reaction that absorbs heat from the surroundings is endothermic: in this case, the enthalpy of the products is higher than that of the reactants, so $H_{products} - H_{reactants}$ is positive. $\Delta H > 0$ is **endothermic.**

- $-\Delta H =$ ________________
- $+\Delta H =$ ________________

Workbook Example 9.8

PROBLEM:
The standard enthalpy for the combustion of propane (C_3H_8) is −2044 kJ. If a propane tank contains 3.5 kg of propane, how much energy will be released or absorbed when the propane is burned?

SOLUTION:
The ΔH° reported is for the combustion of 1 mol of propane. First, it is necessary to determine the number of moles of propane present. From the formula, the molecular weight of propane is 44 g/mol. 3.5 kg is 3500 g.

$$3500\ g\ propane \times \frac{1\ mol\ propane}{44\ g\ propane} = 79.5\ mol\ propane$$

The amount of heat involved in the combustion of this amount of propane is

$$79.5\ mol\ propane \times \frac{-2044\ kJ}{1\ mol\ propane} = -162{,}500\ kJ$$

This is an exothermic reaction. Two things make this clear: the first is that it is a combustion reaction; common sense says that heat is released when things are burned, and this is no exception. The primary reason that it is obviously an exothermic reaction is the sign on the enthalpy change: when the enthalpy change is negative, heat is being lost by the reactants and gained by the surroundings—an exothermic reaction.

Workbook Example 9.9

PROBLEM:
The reaction of boron and hydrogen proceeds by the following reaction:

$$2\ B(s) + 3\ H_2(g) \rightarrow B_2H_6(g) \qquad \Delta H^\circ = 36.4\ kJ/mol$$

If 25.4 g of boron react, what is the energy change? Is this reaction endothermic or exothermic?

SOLUTION:
The stoichiometric ratios here must be considered very carefully. The reported $\Delta H°$ is per mole of boron hydride, but two moles of boron are required per mole of boron hydride.

$$25.4\ g\ B \times \frac{1\ mol\ B}{10.81\ g\ B} \times \frac{1\ mol\ B_2H_6}{2\ mol\ B} \times \frac{36.4\ kJ}{1\ mol\ B_2H_6} = 42.8\ kJ$$

No calculations are required to determine that this is an endothermic reaction—this can be seen immediately from the positive value of ΔH.

Workbook Problem 9.4

How much heat is released or required when 50.0 g of water is converted from liquid to gas at 100 °C? ΔH_{vap} for water is 40.7 kJ/mol.

Strategy: Use the heat of vaporization to determine the amount of heat involved.

Step 1: Convert the grams of water to moles.

Step 2: Multiply by the heat of vaporization.

Step 3: Make sure the sign of the heat transfer makes sense: water is the system.

Workbook Problem 9.5

For the reaction

$$2\ Ca(s) + O_2(g) \rightarrow 2\ CaO(s) \qquad \Delta H = -634.9\ kJ/mol,$$

how much heat is involved if 3.24 g of calcium metal reacts with an excess of oxygen gas? Is this an exothermic or an endothermic reaction?

Strategy: Convert grams Ca to moles and use the stoichiometry of the reaction.

Step 1: Determine the number of moles of calcium.

Step 2: Using the stoichiometric ratio, determine the heat involved.

Step 3: Based on the sign of the heat transfer, determine whether the reaction is exothermic or endothermic.

POST-LECTURE:

Work Problems: *One of the best ways to learn chemistry is through solving problems. Go to the textbook and work through the Worked Examples for this section, then go to the end of chapter and work the problems there. When you feel confident that you can correctly answer problems like the ones listed below for each objective, check off each of the Learning Objectives for this section above.*

LO 9.8: Worked Example 9.4; Problems 9.66, 9.67, 9.74, 9.75

Section 9.7 Calorimetry and Heat Capacity

PRE-LECTURE:

Learning Objectives: *These are the skills you should be proficient in after reading this section and practicing the associated problems. Check off the boxes below once you have gone through this section, answered the associated questions in your textbook, and feel confident you have mastered the skill.*

☐ **LO 9.9** Calculate heat capacities, temperature changes, or heat transfer using the equation for heat capacity (C), specific heat (c), or molar heat capacity (C_m).

☐ **LO 9.10** Calculate enthalpy changes in a calorimetry experiment.

Definitions: *To learn, you must do. This is your workbook. Fill in the definitions from the glossary of the textbook as a first step to understanding the material.*

— **heat capacity (C)**— ______________________________

____

— **specific heat (c)**— ______________________________

____

— **molar heat capacity (C_m)**— ______________________________

____

Summary: *Read through this section and fill in the missing words and equations. Reference the textbook when you are not confident in your answers.*

Calorimetry is an experimental method for measuring ΔH for a reaction. It can be carried out at either constant pressure or constant volume.

A constant-pressure calorimeter can be as simple as an insulated container with a loosely fitting lid to maintain the contents at atmospheric pressure, a stirrer, and a thermometer. This setup measures ________.

A constant-volume calorimeter—also called a bomb calorimeter—is more complex, and it measures ________ of combustion reactions. A sample of a substance is placed in an oxygen environment, and the contents are electrically ignited. The heat change is calculated from the increase in temperature of the surrounding water. In calorimetry, the heat gained or lost by the water is the heat lost or gained by whatever is in the calorimeter. In other words, the water in the calorimeter is the *surroundings*.

To calculate a heat change from a temperature change, it is necessary to know the relationship between the two. The heat capacity (C) is calculated using the following formula:

C =

where q is heat (measured in joules), and ΔT is the temperature change (remember to calculate it as: final − initial temperature). This is an extensive property: the amount of heat required will change based on the amount of the material present.

The **specific heat, *c*,** is the amount of heat required to increase 1 gram of a material by 1 °C. For water, 4.184 J will raise the temperature of 1 g of water by 1 °C. The units of this heat capacity are therefore $\text{J/g}\cdot{}^\circ\text{C}$. We can calculate specific heat by using the following equation:

$q =$

Molar heat capacities are also used sometimes. The units for these are $\text{J/mol}\cdot{}^\circ\text{C}$. This is an intrinsic property, and can be used to identify unknown samples:

$C_m =$

Workbook Example 9.10

PROBLEM:
Calculate the amount of heat lost when the temperature of 14.3 g of iron drops 17.0 °C. The specific heat of iron is 0.449 J/g·°C.

SOLUTION:
Use the formula $q = mc\Delta T$ to determine the amount of heat involved.

$$q = (14.3\ g)\left(0.449\frac{J}{g\cdot{}^\circ\text{C}}\right)(17.0\ {}^\circ\text{C}) = 109\ J$$

Workbook Example 9.11

PROBLEM:
If a 7.3-g sample of an unknown metal at 100 °C is dropped into 25 g of water at 20.0 °C, and the final temperature of the water is 22.1 °C, calculate the specific heat of the metal. The specific heat of water is 4.184 J/g · °C.

SOLUTION:
Use the formula $q = mc\Delta T$ to determine the amount of heat absorbed by the water. This is also the heat given off by the metal, which will allow the specific heat to be calculated.

$$q(water) = (25\ g)\left(4.184 \frac{J}{g \cdot °C}\right)(22.1\ °C - 20.0°C) = 220\ J$$

220 J is absorbed by the water, so 220 J was given off by the metal:

$$-220\ J = (7.3\ g)c(22.1\ °C - 100°C)$$

$$c = \frac{-220\ J}{(7.3\ g)(22.1\ °C - 100\ °C)} = 0.386 \frac{J}{g \cdot °C}$$

If you were to consult a data table, you would see that the specific heat indicates that the metal is copper.

Workbook Problem 9.6

A student mixes 1.50 g of NaOH with 50.0 mL of water at 22.73 °C in a coffee-cup calorimeter. The final temperature of the reaction is 30.70 °C. Assuming that the calorimeter absorbs a negligible amount of heat, and the density of the solution is the same as that of water, calculate the amount of heat evolved in this dissociation reaction. What is the heat of dissociation of NaOH in kJ/mol?

Strategy: Determine the heat absorbed by the water, then reverse the sign to determine the number of joules evolved by the dissociation. Convert the grams of NaOH to moles, and divide the heat evolved by the number of moles of NaOH to determine the molar heat of dissociation.

Step 1: Determine the amount of heat gained by the water using the equation $q = mc\Delta T$.

Step 2: Determine the amount of heat released by the reaction.

Step 3: Determine the number of moles of NaOH involved.

Step 4: Determine the heat of dissociation per mole of NaOH.

POST-LECTURE:

Work Problems: *One of the best ways to learn chemistry is through solving problems. Go to the textbook and work through the Worked Examples for this section, then go to the end of chapter and work the problems there. When you feel confident that you can correctly answer problems like the ones listed below for each objective, check off each of the Learning Objectives for this section above.*

LO 9.9: Worked Example 9.5; Problems 9.78–9.82
LO 9.10: Worked Examples 9.6, 9.7; Problems 9.82, 9.84, 9.86, 9.88

Section 9.8 Hess's Law

PRE-LECTURE:

Learning Objectives: *These are the skills you should be proficient in after reading this section and practicing the associated problems. Check off the boxes below once you have gone through this section, answered the associated questions in your textbook, and feel confident you have mastered the skill.*

☐ **LO 9.11** Use Hess's Law to find ΔH for an overall reaction, given reaction steps and their ΔH values.

Definitions: *To learn, you must do. This is your workbook. Fill in the definitions from the glossary of the textbook as a first step to understanding the material.*

— **Hess's law—**____

____

Summary: *Read through this section and fill in the missing words and equations. Reference the textbook when you are not confident in your answers.*

Reactions do not always go cleanly, and often produce a mixture of products. This makes a direct measurement of the energetics of these reactions problematic. With enthalpy being a state function, the reactions can be broken down into steps and added together to determine the energetics of a difficult-to-measure step. This is Hess's law: The overall enthalpy change for a reaction is equal to the sum of the enthalpy changes for the individual steps in the reaction.

Here are the rules for calculating the overall enthalpy change:

- Reactants must all appear on the left. It is acceptable to reverse reactions to make this happen.
- Products must all appear on the right. Again, ________ reactions, if necessary.
- Intermediates in the reaction must appear on both sides so they cancel out.
- When a reaction is reversed, the sign on that reaction's ΔH must also be ________.
- Reactions and their energies can be multiplied by any necessary factor.
- All the reactions and their energies are added to get the desired total reaction and its energy.

Workbook Example 9.12

PROBLEM:
Calculate $\Delta H°$ for the reaction $3\ C(s) + 4\ H_2(g) \rightarrow C_3H_8(g)$ given the following information:

$C_3H_8(g) + 5\ O_2(g) \rightarrow 3\ CO_2(g) + 4\ H_2O(g)$ $\quad \Delta H° = -2043$ kJ
$C(s) + O_2(g) \rightarrow CO_2(g)$ $\quad \Delta H° = -393.5$ kJ
$2\ H_2(g) + O_2(g) \rightarrow 2\ H_2O(g)$ $\quad \Delta H° = -483.6$ kJ

SOLUTION:
Reactants
Place the reactants on the left and the products on the right, flipping reactions around if needed, and multiplying, if necessary, to get the equation as written.

$C(s) + O_2(g) \rightarrow CO_2(g)$ $\quad \Delta H° = -393.5$ kJ
$2\ H_2(g) + O_2(g) \rightarrow 2\ H_2O(g)$ $\quad \Delta H° = -483.6$ Kj

These reactions both need to be multiplied by coefficients, the first equation by 3, and the second equation by 2, to give the coefficients stated in the problem:

$3\ C(s) + 3\ O_2(g) \rightarrow 3\ CO_2(g)$ $\quad \Delta H° = -1180.5$ kJ
$4\ H_2(g) + 2\ O_2(g) \rightarrow 4\ H_2O(g)$ $\quad \Delta H° = -967.2$ kJ

Products
Take the first reaction in the given list and turn it around, remembering to also reverse the sign on $\Delta H°$.

$3\ CO_2(g) + 4\ H_2O(g) \rightarrow C_3H_8(g) + 5\ O_2(g)$ $\quad \Delta H° = 2043$ kJ

Stack the final reactions from the steps above, and add them together:

$3\ C(s) + 3\ O_2(g) \rightarrow 3\ CO_2(g)$ $\quad \Delta H° = -1180.5$ kJ
$4\ H_2(g) + 2\ O_2(g) \rightarrow 4\ H_2O(g)$ $\quad \Delta H° = -967.2$ kJ
$3\ CO_2(g) + 4\ H_2O(g) \rightarrow C_3H_8(g) + 5O_2(g)$ $\quad \Delta H° = \ 2043$ kJ

$3\ C(s) + 3\ O_2(g) + 4\ H_2(g) + 2\ O_2(g) + 3\ CO_2(g) + 4\ H_2O(g) \rightarrow$
$3\ CO_2(g) + 4\ H_2O(g) + C_3H_8(g) + 5O_2(g)$

Cancel terms that are the same on the reactant and product sides of the equation, which leaves the equation that was stated in the problem:

$3\ C(s) + 4\ H_2(g) \rightarrow C_3H_8(g)$ $\Delta H° = -1180.5\text{ kJ} + (-967.2\text{ kJ}) + 2043\text{ kJ}$
$3\ C(s) + 4\ H_2(g) \rightarrow C_3H_8(g)$ $\Delta H° = -104.7\text{ kJ}$

Workbook Problem 9.7

Calculate ΔH for the reaction

$$Fe_2O_3(s) + 3\ CO(g) \rightarrow 2\ Fe(s) + 3\ CO_2(g)$$

given the following information:

$4\ Fe(s) + 3\ O_2(g) \rightarrow 2\ Fe_2O_3(s)$ $\Delta H° = -1648.4\text{ kJ}$
$2\ CO(g) + O_2(g) \rightarrow 2\ CO_2(g)$ $\Delta H° = -565.4\text{ kJ}$

Strategy: Place the reactants on the left and the products on the right, flipping reactions, if needed, and multiplying by coefficients, if necessary, to get the equation as written.

Step 1: Place reactants on the left, multiplying, if necessary.

Step 2: Place products on the right, multiplying, if necessary.

Step 3: Stack up the rewritten reactions and add them together.

Step 4: Cancel terms that appear on both sides.

POST-LECTURE:

Work Problems: *One of the best ways to learn chemistry is through solving problems. Go to the textbook and work through the Worked Examples for this section, then go to the end of chapter and work the problems there. When you feel confident that you can correctly answer problems like the ones listed below for each objective, check off each of the Learning Objectives for this section above.*

LO 9.11: Worked Example 9.8; Problems 9.38, 9.39, 9.93, 9.94, 9.96

Section 9.9 Standard Heats of Formation

PRE-LECTURE:

Learning Objectives: *These are the skills you should be proficient in after reading this section and practicing the associated problems. Check off the boxes below once you have gone through this section, answered the associated questions in your textbook, and feel confident you have mastered the skill.*

- ☐ **LO 9.12** Identify standard states of elements.
- ☐ **LO 9.13** Write standard enthalpy of formation reactions (ΔH°_f) for compounds from their elements.
- ☐ **LO 9.14** Use values of (ΔH°_f) for elements and compounds to calculate ΔH° for a reaction.

Definitions: *To learn, you must do. This is your workbook. Fill in the definitions from the glossary of the textbook as a first step to understanding the material.*

— **standard heats of formation (ΔH°_f)**— ____________________________________

____

Summary: *Read through this section and fill in the missing words and equations. Reference the textbook when you are not confident in your answers.*

Another method can be used to determine the energy changes for any reaction, whether actual or theoretical. The **standard heat of formation** for a compound is the enthalpy change for the hypothetical formation of one mole of that compound in its standard state from its constituent elements in their standard states. The standard enthalpy of formation of an element in its standard state is ________.

Workbook Example 9.13

PROBLEM:
Write the equation for the reaction that corresponds to the standard heat of formation of acetone, C_3H_6O.

SOLUTION:
The standard state of carbon is graphite, while hydrogen and oxygen are diatomic gases:

$$3\ C(\textit{graphite}) + 3\ H_2(g) + \tfrac{1}{2}\ O_2(g) \rightarrow C_3H_6O(l)$$

It's as simple as that. You will see fractional coefficients in thermochemical equations, because they are written to provide a coefficient of 1 for the product.

There are extensive tables of heats of formation (one of which is in the appendix of your textbook), and they can be used to calculate heats of reaction in a variation on final − initial: The sum of the heats of formation of the products minus the sum of the heats of formation of the reactants will be the overall heat of reaction. This is summarized in the following formula:

$$\Delta H° = \sum \Delta H°_{\text{products}} - \sum \Delta H°_{\text{reactants}}$$

It is important to note that the standard heats of formation must be multiplied by the coefficients in the balanced equation.

Workbook Example 9.14

PROBLEM:
Calculate $\Delta H°$ for the reaction $C_3H_6O(l) + 4\ O_2(g) \rightarrow 3\ CO_2(g) + 3\ H_2O(l)$ using the following information:

$\Delta H_f^o(C_3H_6O(l)) = -248.4\ kJ/mol$

$\Delta H_f^o(O_2(g)) = -0\ kJ/mol$

$\Delta H_f^o(CO_2(g)) = -393.5\ kJ/mol$

$\Delta H_f^o(H_2O(l)) = -285.8\ kJ/mol$

SOLUTION:
Subtract the sum of the heats of formation of the reactants from the sum of the heats of formation of the products.

$$\Delta H° = [3(-393.5\ kJ/mol) + 3(-285.8\ kJ/mol)] - [1(-248.4\ kJ/mol) + 4(0\ kJ/mol)]$$

$$\Delta H° = [-1180.5\ kJ - 857.4\ kJ + 248.4\ kJ = -1789.5\ kJ]$$

Workbook Problem 9.8

During glycolysis, glucose ($C_6H_{12}O_6$) is taken through a series of reactions during which it is combined with oxygen, and carbon dioxide and water are released. How much energy is released through the slow combustion of 10.0 grams of glucose in this process? What would be the temperature change if this energy was used to heat 500 g of water?

You may use the following information:

$\Delta H_f^o(C_6H_{12}O_6(s)) = -1273.3\ kJ/mol$

$\Delta H_f^o(O_2(g)) = -0\ kJ/mol$

$\Delta H_f^o(CO_2(g)) = -393.5\ kJ/mol$

$\Delta H_f^o(H_2O(l)) = -285.8\ kJ/mol$

Strategy: Write a balanced equation for the reaction and use the standard heats of formation to determine the overall heat of reaction per mole of glucose, then convert grams to moles, and determine the heat released by 10 grams of glucose. Use this heat and the equation $q = mc\Delta T$ to determine the temperature change.

Step 1: Write and balance the equation.

Step 2:	Determine the overall heat of reaction using $\Delta H° = \sum \Delta H°_{\text{products}} - \sum \Delta H°_{\text{reactants}}$.
Step 3:	Convert the grams of glucose to moles.
Step 4:	Determine the energy released by 10 g of glucose.
Step 5:	Using $q = mc\Delta T$, determine the temperature change of 500 g of water.

POST-LECTURE:

Work Problems: *One of the best ways to learn chemistry is through solving problems. Go to the textbook and work through the Worked Examples for this section, then go to the end of chapter and work the problems there. When you feel confident that you can correctly answer problems like the ones listed below for each objective, check off each of the Learning Objectives for this section above.*

LO 9.12: Problems 9.100, 9.101
LO 9.13: Problems 9.102, 9.103
LO 9.14: Worked Example 9.9; Problems 9.106, 9.108, 9.110

Section 9.10 Bond Dissociation Energies

PRE-LECTURE:

Learning Objectives: *These are the skills you should be proficient in after reading this section and practicing the associated problems. Check off the boxes below once you have gone through this section, answered the associated questions in your textbook, and feel confident you have mastered the skill.*

☐ **LO 9.15** Use bond dissociation energies to estimate $\Delta H°$ for a reaction.

Summary: *Read through this section and fill in the missing words and equations. Reference the textbook when you are not confident in your answers.*

Heats of formation are not available for every compound. When there is no data for the heat of formation of a compound, it can be estimated using the known bond dissociation energies for every bond in the compound. Bond dissociation energies are positive: Energy must always be absorbed to break a bond. Bond formation energies will be equal in magnitude and opposite in sign to the bond dissociation energies.

To determine the approximate enthalpy of any reaction, subtract the sum of the bond dissociation energies for the products from the bond dissociation energies in the:

$$\Delta H^\circ = D(\_\_\_\_\_\_\_\_\_\_\_\_) - D(\_\_\_\_\_\_\_\_\_\_\_\_)$$

Workbook Example 9.15

PROBLEM:
Using bond dissociation energies, estimate the heat of reaction of hydrogen and oxygen to form water.

Type of Bond	Bond Energy
H–H	436 kJ/mol
O=O	498 kJ/mol
O–H	464 kJ/mol

SOLUTION:

$$2\ H_2(g) + O_2(g) \rightarrow 2\ H_2O(g)$$

On the product side, four H–O bonds are formed:

$$4 \times 464\ \text{kJ/mol} = 1856\ \text{kJ/mol released}$$

On the reactant side, two H–H bonds are broken, and one O=O bond is broken:

$$(2 \times 436\ \text{kJ/mol}) + 498\ \text{kJ/mol} = 1370\ \text{kJ/mol absorbed}$$

$$\text{Energy of reactants} - \text{Energy of products} = -486\ \text{kJ/mol}$$

This is the answer for the problem as written—for the production of two moles of water. For one mole of water, the answer will be −486 kJ/mol × 0.5 = −243 kJ/mol. This is very close to the heat of formation of water vapor.

Workbook Problem 9.9

Predict the energy change for combustion of methane (CH_4) using bond dissociation energies. Use the following information:

Type of Bond	Bond Energy
C–H	414 kJ/mol
O=O	498 kJ/mol
O–H	464 kJ/mol
C=O	799 kJ/mol

Strategy:	Write a balanced equation for the reaction and determine the bond energies for both products and reactants. Subtract energy of products from energy of reactants to determine the overall energy change of the reaction.
Step 1:	Write and balance the equation.
Step 2:	Determine the energy of the bonds on the reactant side.
Step 3:	Determine the energy of the bonds on the product side.
Step 4:	Determine the energy released by the reaction.

POST-LECTURE:

Work Problems: *One of the best ways to learn chemistry is through solving problems. Go to the textbook and work through the Worked Examples for this section, then go to the end of chapter and work the problems there. When you feel confident that you can correctly answer problems like the ones listed below for each objective, check off each of the Learning Objectives for this section above.*

LO 9.15: Worked Example 9.10; Problems 9.19, 9.20, 9.118–9.121

Section 9.11 An Introduction to Entropy

PRE-LECTURE:

Learning Objectives: *These are the skills you should be proficient in after reading this section and practicing the associated problems. Check off the boxes below once you have gone through this section, answered the associated questions in your textbook, and feel confident you have mastered the skill.*

☐ **LO 9.16** Predict the sign of the entropy change (ΔS) given the chemical equation or a molecular diagram.

Definitions: *To learn, you must do. This is your workbook. Fill in the definitions from the glossary of the textbook as a first step to understanding the material.*

— **spontaneous process—** ____

____

— **entropy (S)—** ____

____

Summary: *Read through this section and fill in the missing words and equations. Reference the textbook when you are not confident in your answers.*

Chemical reactions generally ________ energy when proceeding from higher-energy reactants to lower-energy products. But this is not always the case: there are endothermic processes that occur even though energy must be added to them. For an ice cube to melt, it must absorb energy from the surroundings the release of energy cannot be the only factor in reactions.

The other factor to consider is the entropy (S) of the system. **Entropy** is an increase in molecular disorder or randomness. When an ice cube melts, the rigid crystalline structure of the solid water becomes the far-less-orderly liquid water. The units for entropy are J/K: larger values indicate a greater degree of disorder. As with all change measurements, $\Delta S = S_{\text{final}} - S_{\text{initial}}$. In contrast to ΔH, a positive ΔS is favorable for a reaction, which indicates an ________ in disorder.

The freedom to move and the ability to adopt numerous conformations are all related to entropy. Phase also plays a major role: solids have less entropy than liquids; gases have the most entropy of all. Solid iodine has an entropy of 116 J/K, while gaseous iodine has an entropy of 261 J/K.

Spontaneous processes are those that occur on their own without external influence. Nonspontaneous processes require constant inputs of energy. For example, to roll a marble down a hill, it simply requires that you get it started; rolling down the hill is a ________ process. Rolling it back up the hill requires a constant input of energy and is ________. To summarize the favorable conditions for spontaneous and nonspontaneous reactions:

- spontaneous processes:
  - favored by a _____ΔG
  - favored by a ____ΔS
- nonspontaneous processes:
  - favored by a ____ΔG
  - favored by a _____ΔS

Workbook Example 9.16

PROBLEM:
Predict whether ΔS° is likely to be positive or negative for the following processes:

a. $H_2O(s) \rightarrow H_2O(g)$
b. $Na(s) + \frac{1}{2}\,Cl_2(g) \rightarrow NaCl(s)$
c. $CH_3CH_2OH(l) + 3\,O_2(g) \rightarrow 2\,CO_2(g) + 3\,H_2O(g)$

SOLUTION:

a. Converting water from a crystalline solid to a gas will generate a large positive ΔS.
b. Going from a solid and a gas to a crystal lattice will cause a large negative ΔS.
c. Four moles of reactants (one a liquid) being converted to 5 moles of gas will increase the entropy, which gives a positive ΔS.

POST-LECTURE:

Work Problems: *One of the best ways to learn chemistry is through solving problems. Go to the textbook and work through the Worked Examples for this section, then go to the end of chapter and work the problems there. When you feel confident that you can correctly answer problems like the ones listed below for each objective, check off each of the Learning Objectives for this section above.*

LO 9.16: Worked Example 9.11; Problems 9.21, 9.22, 9.126, 9.127, 9.130

Section 9.12 An Introduction to Free Energy

PRE-LECTURE:

Learning Objectives: *These are the skills you should be proficient in after reading this section and practicing the associated problems. Check off the boxes below once you have gone through this section, answered the associated questions in your textbook, and feel confident you have mastered the skill.*

- ☐ **LO 9.17** Predict the sign of ΔG, ΔH, and ΔS, using the equation for Gibbs free-energy change (ΔG).
- ☐ **LO 9.18** Use the Gibbs free-energy equation to calculate the temperature at which a reaction crosses over from spontaneous to nonspontaneous.

Definitions: *To learn, you must do. This is your workbook. Fill in the definitions from the glossary of the textbook as a first step to understanding the material.*

— **Gibbs free-energy change (ΔG)**— ____

____

Summary: *Read through this section and fill in the missing words and equations. Reference the textbook when you are not confident in your answers.*

To determine whether a reaction will be spontaneous at a certain temperature, both entropy and enthalpy need to be considered. The Gibbs free-energy change (ΔG) can be calculated for any system to show the balance:

$$\Delta G =$$

The sign of ΔG states whether a process is spontaneous. A negative ΔG indicates a ________ process in which free energy is being released; a positive ΔG indicates a ________ process in which free energy is being absorbed. When $\Delta G = 0$, the reaction is at equilibrium.

Below is a chart that helps to see when a reaction is spontaneous.

ΔH (spontaneity)	**ΔS (spontaneity)**	**ΔG**
$-\Delta H$ *(favored)*	$+\Delta S$ *(favored)*	Spontaneous at all temperatures
$+\Delta H$ *(disfavored)*	$-\Delta S$ *(disfavored)*	Nonspontaneous at all temperatures
$-\Delta H$ *(favored)*	$-\Delta S$ *(disfavored)*	Since $\Delta G = \Delta H - T\Delta S$ decrease T to make the reaction spontaneous
$+\Delta H$ *(disfavored)*	$+\Delta S$ *(favored)*	Since $\Delta G = \Delta H - T\Delta S$ increase T to make the reaction spontaneous

Entropy is temperature dependent. If the free-energy equation is set to zero (equilibrium), it is possible to solve for the temperature at which the process will become spontaneous (or nonspontaneous) by using the following equation:

T =

Workbook Example 9.17

PROBLEM:
At what temperature is the vaporization of methanol at equilibrium, given the following values:

ΔH_{vap} methanol = 38.2 kJ/mol

ΔS_{vap} methanol = 112.8 J/K·mol

SOLUTION:

$$T = \frac{\Delta H}{\Delta S} = \frac{38.2\frac{kJ}{mol}}{0.1128\frac{kJ}{K \cdot mol}} = 338.65\ K = 65\ °C$$

This is the boiling point of methanol.

Workbook Problem 9.10

Given the following data, is the reaction of hydrogen and chlorine gas spontaneous at room temperature?

ΔH_f HCl(*g*) = −92.3 kJ/mol $\qquad$ $\Delta S°$ = 20. J/K·mol

Strategy: Write a balanced equation for the reaction and determine ΔH and ΔS for the reaction. Using the Gibbs free-energy equation, determine ΔG for the reaction at 25 °C, remembering to convert from °C to kelvin first.

Step 1: Write and balance the equation.

Step 2:	Determine ΔH.
Step 3:	Write the value for ΔS.
Step 4:	Determine the free energy of the reaction at 25 °C, first converting the temperature to K.
Step 5:	Determine the spontaneity of the reaction.

POST-LECTURE:

Work Problems: *One of the best ways to learn chemistry is through solving problems. Go to the textbook and work through the Worked Examples for this section, then go to the end of chapter and work the problems there. When you feel confident that you can correctly answer problems like the ones listed below for each objective, check off each of the Learning Objectives for this section above.*

LO 9.17: Worked Example 9.12; Problems 9.23, 9.24, 9.40, 9.42, 9.129, 9.132
LO 9.18: Worked Example 9.13; Problems 9.134–9.138

Self-Test

This section is intended to test your knowledge of the material covered in this chapter. Think through these problems and make certain you understand what the problem is about—ask yourself if the answer makes sense. Many of these questions are linked to the chapter learning goals. Therefore, successful completion of these problems indicates that you have achieved the learning goals for this chapter. You will receive the greatest benefit from this section if you use it as a mock exam. Do not use your notes, book, Internet, or friends when taking the test. There is no grade and you only shortchange yourself by getting help with this activity. As you take the test, mark the problems with your level of confidence with the answer.

★ *Use a star for problems you know you got right.*
? *Use a question mark for those you think you got right.*
x *Use an x for those you are guessing on.*

If you get a question right that you marked with a star, you can be confident that you know the associated LO.

If you get a question right that you marked with a question mark, go back and work a few more problems and test yourself again tomorrow.

If you get a question wrong or mark it with an x, go back and read the section again, work a few problems today, a few tomorrow, and a few the next day before testing yourself again.

This will help you to discover which topics you have learned thoroughly and which topics you need to study in more detail.

True–False

1. Energy can be converted from one form to another.
2. Changes in energies are always calculated *initial* − *final*.
3. The energy changes in a state function depend on the path.
4. When a system expands against pressure, it is doing work.
5. The thermodynamic standard state of an element is its most stable form at 0 °C.
6. A reaction can do work or transfer heat, but not both.
7. Melting is an exothermic process.
8. Specific heat is an extensive property, and it cannot be used to identify a material.
9. Fractional coefficients can be used in thermochemical equations.
10. Standard heats of formation can be used to calculate reaction enthalpies.
11. Depending on the needs of the end user, different units can be used for heats of combustion of fuels.
12. A negative ΔS is favorable for a reaction.
13. Gases have lower entropy than liquids.
14. Free-energy changes can be used to predict the spontaneity of reactions.
15. Entropy is a temperature-dependent property.

Matching

16. conservation of energy — a. heat released upon reaction with oxygen

17. energy — b. a process that proceeds without external influence

18. temperature	c. going directly from solid to gas
19. system	d. a measure of molecular kinetic energy
20. entropy	e. energy can be transferred between forms, but cannot be created or destroyed
21. state function	f. absorbing energy
22. heat	g. a measure of randomness or disorder
23. work	h. capacity to do work
24. enthalpy	i. heat required to convert a substance from solid to liquid
25. heat of fusion	j. releasing energy
26. sublimation	k. the amount of energy required to raise the temperature of one gram of a substance by 1 °C
27. specific heat	l. a measurement that is independent of the path
28. exothermic	m. the focus of a thermochemical process
29. Hess's law	n. exerting a force over a distance
30. heat of combustion	o. the overall enthalpy of a reaction can be determined by combining any number of theoretical steps
31. spontaneous	p. $E + P\Delta V$
32. endothermic	q. energy transferred as the result of a temperature difference

Fill in the Blanks

33. Energy can be either ______________________, the energy of motion, or ____________________, stored energy.

34. Molecular kinetic energy is measured as ____________________. When this property differs between objects, the energy transferred is called ____________.

35. Energy that flows out of a system is given a ______________________ sign. Reactions in which this occurs are called ______________________.

36. It is possible for state functions to return to their original condition. As a result, they are said to be completely ________________.

37. If there is no change in the ________________ of a system, no ________ is done.

38. The most stable form of a substance at 25 °C and 1 atm is referred to as its ________________________________.

39. When calorimetry is used to measure heat flows, changes in the temperature of the water can be converted to changes in heat by using the mass of the water and its ________________________.

40. During an endothermic process, heat flows from the ________________ to the ______________ and ΔH is ______________.

41. The energy required to break a chemical bond is its ________________________. It is always ______________________.

42. The greater the disorder in a system, the higher the ____________________. If the temperature of a system is increased, this property will ____________________.

43. The ________________________ can determine the spontaneity of a process. When it is zero, the system is at ____________________. When it is ________________ than zero, the reaction will proceed spontaneously. When it is ____________________ than zero, the reaction will be nonspontaneous.

Problems

44. *Kinetic energy* is defined as 1/2 mv^2. How fast would a 2-gram marble need to be moving to have the same kinetic energy as a 1.3 kg bowling ball that is moving at 5 m/s?

45. If 950 mL of a gas is compressed to 550 mL under a constant external pressure of 8.00 atm, and the gas absorbs 12 kJ, what are the values of q, w, and ΔE for the gas? Is work being done on the system or by the system? What is the value of ΔE for the surroundings?

46. If the specific heat of mercury is 0.140 J/g·°C, how much will the temperature of 15 grams of mercury change, if 150 joules of heat is added to it?

47. If 2.00 L of chlorine gas react stoichiometrically with 1.00 L of oxygen at 1 atm to form 2.00 L of Cl_2O gas, according to the following reaction, what is the overall energy change of the system (the density of chlorine gas is 3.17 g/L)? Is this an endothermic or an exothermic reaction?

$$Cl_2(g) + \tfrac{1}{2}\, O_2(g) \rightarrow Cl_2O(g) \qquad \Delta H = 80.3 \text{ kJ/mol}$$

48. The enthalpy of solution for $MgSO_4 \cdot 7\ H_2O$ (Epsom salts) is +16.11 kJ/mol. If 15 grams of Epsom salts is dissolved in 100 grams of water at 24.36 °C, what is the final temperature of the solution? (Assume that the specific heat and density of the solution are the same as that of pure water.)

49. How much energy, in kJ, is required to take 50 grams of water from −10 °C to steam at 100 °C? Use the following values:

c_{ice} = 2.108 J/g·°C c_{water} = 4.184 J/g·°C
ΔH_{fus} = 6.01 kJ/mol ΔH_{vap} = 40.7 kJ/mol

What is the largest contributor to the energy consumption? Why is this?

50. A 6.73 gram chunk of metal, known to be a mixture of copper (c = 0.385 J/g·°C) and gold (c = 0.129 J/g·°C), is taken from boiling water and dropped into 25 grams of water at 25.00 °C. The final temperature is 26.65 °C. How many grams of copper and how many grams of gold are present in the sample?

51. Use Hess's law to determine the value of ΔH for the following reaction:

$$2\ Fe(s) + 3/2\ O_2(g) \rightarrow Fe_2O_3(s)$$

from the following reactions:

$Fe_2O_3(s) + 3\ CO(g) \rightarrow 2\ Fe(s) + 3\ CO_2(g)$ $\Delta H° = -26.7$ kJ
$CO(g) + ½\ O_2(g) \rightarrow 3\ CO_2(g)$ $\Delta H° = -283.0$ kJ

52. Spraying water on a kitchen grease fire can spread it, so in commercial kitchens, a large box of baking soda is often kept nearby in case of fire. When sodium bicarbonate is thrown on the flames, it physically puts out the fire, but it also decomposes to form carbon dioxide and water by the following reaction:

$$2\ NaHCO_3(s) \rightarrow Na_2CO_3(s) + H_2O(l) + CO_2(g)$$

Given the following heats of formation, calculate ΔH of this decomposition reaction?

$$\Delta H_f^\circ NaHCO_3 = -947.7 \text{ kJ/mol}$$

$$\Delta H_f^\circ Na_2CO_3 = -1131 \text{ kJ/mol}$$

$$\Delta H_f^\circ H_2O = -285.9 \text{ kJ/mol}$$

$$\Delta H_f^\circ CO_2 = -393.5 \text{ kJ/mol}$$

53. Using the bond dissociation energies given, calculate the approximate enthalpy of the following reaction:

$$2\ Cl_2(g) + CH_4(g) \rightarrow 2\ H_2(g) + CCl_4(g)$$

Cl–Cl = 243 kJ/mol   H–H = 436 kJ/mol
C–H = 410 kJ/mol   C–Cl = 330 kJ/mol

54. The entropy value for the equation above is $-158.8\ J/mol \cdot K$, at approximately what temperature will the reverse of the above reaction become spontaneous?

55. Using the following standard heats of formation, calculate the heat of combustion of pentane (C_5H_{12}) in kJ/mol and kJ/g. The density of pentane is 0.626 g/cm^3. What is the heat of combustion per mL?

$$\Delta H_f^o \text{pentane} = -146.3 \text{ kJ/mol}$$
$$\Delta H_f^o H_2O = -285.9 \text{ kJ/mol}$$
$$\Delta H_f^o CO_2 = -393.5 \text{ kJ/mol}$$

Test Yourself: *Take the Practice Test on pages 363 and 364 of your textbook. Again, use a testing environment using only those items that your instructor will allow you to have access to during the test and mark each question with a star, question mark, or x to denote your confidence level. Once you have finished you can check off the LO at the beginning of this chapter based on which questions you starred and got correct. Go back and spend more time with the topics from the other learning objectives.*

CHAPTER TEN

Gases: Their Properties and Behavior

Learning Objectives

As a result of reading and studying this chapter, you should be able to:

- ☐ **LO 10.1** Convert between different units of pressure.
- ☐ **LO 10.2** Determine the pressure from the height of a liquid mercury column in a barometer or manometer.
- ☐ **LO 10.3** Use the individual gas laws to calculate pressure, volume, number of moles, or temperature of a gas sample when conditions change.
- ☐ **LO 10.4** Use the ideal gas law to calculate pressure, volume, number of moles, or temperature of a gas sample.
- ☐ **LO 10.5** Calculate volumes of gases in chemical reactions.
- ☐ **LO 10.6** Calculate the density or molar mass of a gas using the formula for gas density.
- ☐ **LO 10.7** Calculate the partial pressure, mole fraction, or amount of each gas in a mixture.
- ☐ **LO 10.8** Use the assumptions of kinetic–molecular theory to predict gas behavior.
- ☐ **LO 10.9** Calculate the average molecular speed of a gas particle at a given temperature.
- ☐ **LO 10.10** Interpret a molecular picture of effusion and diffusion.
- ☐ **LO 10.11** Use Graham's law to estimate relative rates of diffusion for two gases.
- ☐ **LO 10.12** Identify temperature and pressure conditions that cause significant deviations from ideal behavior.
- ☐ **LO 10.13** Use the van der Waals equation to calculate the properties of real gases.
- ☐ **LO 10.14** Explain the principle of the greenhouse effect.
- ☐ **LO 10.15** Predict if a molecule is a greenhouse gas by examining molecular vibrations.
- ☐ **LO 10.16** Describe the trends in greenhouse gas concentrations over time and measured and predicted effects of climate change.

Chapter Summary

Gases behave very differently than solids or liquids. They are compressible, because they are mostly empty space, and they exert pressure as a result of collisions between their rapidly moving particles and the walls of any container in which they are placed. Extensive experimental work in the early days of the science of chemistry worked out the gas laws: simple equations that can be used to predict how gases will behave when conditions are varied. The four parameters that are relevant to gases are temperature, pressure, volume, and amount. Knowing the mathematical relationships between these variables makes it possible to perform stoichiometric calculations for reactions that involve gases. Mixtures of gases can also be dealt with by using the gas laws, in particular, Dalton's law of partial pressures. Kinetic–molecular theory provides a model for gas behavior that explains the gas laws and is a basis for understanding effusion and diffusion. Generally, it is assumed that gases will behave in an "ideal" way, that is, they will adhere closely to the tenets of kinetic–molecular theory. At high pressures, this ceases to be true, so additional strategies are needed to determine the behavior of gases under these conditions. The chapter ends with a brief look at the Earth's atmosphere and climate change.

The Chapter in Detail

Section 10.1 Gases and Gas Pressure

PRE-LECTURE:

Learning Objectives: *These are the skills you should be proficient in after reading this section and practicing the associated problems. Check off the boxes below once you have gone through this section, answered the associated questions in your textbook, and feel confident you have mastered the skill.*

- ☐ **LO 10.1** Convert between different units of pressure.
- ☐ **LO 10.2** Determine the pressure from the height of a liquid mercury column in a barometer or manometer.

Definitions: *To learn, you must do. This is your workbook. Fill in the definitions from the glossary of the textbook as a first step to understanding the material.*

— **pressure (P)**— ____

____

— **newton (N)**— ____

____

— **pascal (Pa)**— ____

____

— **millimeter of mercury (mm Hg)**— ______________________________

____

— **atmosphere (atm)**— ____________________________________

____

— **bar**— ____

____

— **nanometer**— ____

____

Summary: *Read through this section and fill in the missing words and equations. Reference the textbook when you are not confident in your answers.*

Gas molecules exist very far apart from each other and interact very little. Since they have ___________ interactions, they will mix to form homogeneous mixtures.

Also, as a result of the distance between molecules—the volume of a gas is 99.9% empty space—gases are ___________ compressible.

Gases exert a measurable pressure on the walls of their container. Pressure is measured as force per unit area, and force is measured as mass (m) multiplied by acceleration (a).

The SI unit for force is the newton (N), and the unit for pressure is the pascal (Pa).

$$P = \frac{force}{area} = \frac{m \cdot a}{area} = \frac{kg \cdot \frac{m}{s^2}}{m^2} = \frac{N}{m^2} = Pa$$

The pascal is very small and not convenient to use in chemistry measurements. There is a number of alternative pressure units that are more practical. The millimeter of mercury (mm Hg), also called a torr after the inventor of the barometer, Evangelista Torricelli, is the most well-known. A barometer has an open dish of mercury with a long, thin, glass tube full of mercury in it. The tube is closed at the top end and open at the bottom. An increase in atmospheric pressure forces mercury higher into the tube, and a drop in atmospheric pressure allows the mercury to drop. *Atmospheric pressure* is defined as _____ mm Hg, which is about __________Pa, or _____ kPa. In addition to millimeters of mercury, the atmosphere (atm) and the bar (1 bar = 100,000 Pa) are commonly used in chemistry.

To measure the pressure of a gas, an open-ended manometer can be used, which compares the pressure of the gas in the container to atmospheric pressure. If the liquid level in both is equal, the pressure of the contained gas is __________to atmospheric pressure. If the liquid is lower on the container side, the pressure in the container is __________ than atmospheric pressure, and if the liquid is higher on the container side, the pressure is __________ in the container. The height differences can be measured to quantify the pressure in the container.

With problems that involve a manometer, when in doubt, draw a picture!

Workbook Example 10.1

PROBLEM:
If a gas is contained in an open-ended manometer, and the mercury on the side of the gas is 37 mm higher than the mercury in the side open to the air, what is the pressure of the contained gas?

SOLUTION:
The gas is at less than atmospheric pressure, because the pressure of the air on the open end of the tube pushes the mercury toward the contained gas. Atmospheric pressure is 760 mm Hg, so the pressure of the gas must be 760 mm Hg − 37 mm Hg = 723 mm Hg.

Workbook Problem 10.1

If the difference in height between the two sides of a manometer is 73 mm Hg, with the mercury higher on the open side of the manometer, what is the pressure of the contained gas in atmospheres?

Strategy: Determine the pressure of the gas in mm Hg; convert to atmospheres.

Step 1: Determine the pressure of the gas in mm Hg.

Step 2: Convert from mm Hg to atmospheres.

POST-LECTURE:

Work Problems: *One of the best ways to learn chemistry is through solving problems. Go to the textbook and work through the Worked Examples for this section, then go to the end of chapter and work the problems there. When you feel confident that you can correctly answer problems like the ones listed below for each objective, check off each of the Learning Objectives for this section above.*

LO 10.1: Worked Example 10.1; Problems 10.38, 10.39
LO 10.2: Worked Example 10.2; Problems 10.23–10.25, 10.40, 10.42

Section 10.2 The Gas Laws

PRE-LECTURE:

Learning Objectives: *These are the skills you should be proficient in after reading this section and practicing the associated problems. Check off the boxes below once you have gone through this section, answered the associated questions in your textbook, and feel confident you have mastered the skill.*

☐ **LO 10.3** Use the individual gas laws to calculate pressure, volume, number of moles, or temperature of a gas sample when conditions change.

Definitions: *To learn, you must do. This is your workbook. Fill in the definitions from the glossary of the textbook as a first step to understanding the material.*

— **gas laws—** ______________________________

— **ideal gas—** ______________________________

— **Boyle's law—** ______________________________

— **Charles's law—** ______________________________

— **Avogadro's law—** ______________________________

— **standard molar volume—** ______________________________

Summary: *Read through this section and fill in the missing words and equations. Reference the textbook when you are not confident in your answers.*

Although they have very different chemical properties, gases have very similar physical properties that can be defined by four variables: ___________, ___________, ___________, and number of ___________ (amount). The relationships among these variables form the gas laws, and a substance that follows them exactly is known as an **ideal gas.**

Boyle's law—examines volume and pressure when ___________ and ___________ are held constant. Volume and pressure are inversely related: when volume increases, pressure ___________. Mathematically, Boyle's law is stated in a few ways:

$$V \propto \frac{1}{P} \qquad PV = k \text{ (where } k \text{ is a constant)} \qquad P_1V_1 = P_2V_2$$

Charles's law—considers volume and temperature when ___________ and ___________ are held constant. They are directly related: when temperature drops, the volume ___________ as well. Temperatures *must* be expressed in kelvin for the proportionality to be seen. Mathematically, Charles's law is stated as:

$$V \propto T \qquad \frac{V}{T} = k\left(\text{where } k \text{ is a constant}\right) \qquad \frac{V_1}{T_1} = \frac{V_2}{T_2}$$

Charles's law is also the basis of the kelvin temperature scale: if volume is plotted against temperature, there is a point at which the volume drops (theoretically) to zero. This occurs at −273.15 °C, a temperature which is known as *absolute zero*. This is the lowest possible temperature, and therefore, it is 0 K.

Avogadro's law—studies amount and volume while holding ___________ and ___________ constant. They are directly related: more moles of gas will always mean ___________ volume, with all else being equal. Mathematically, Avogadro's law is stated as:

$$V \propto n \qquad \frac{V}{n} = k\left(\text{where } k \text{ is a constant}\right) \qquad \frac{V_1}{n_1} = \frac{V_2}{n_2}$$

It can also be seen from Avogadro's law that one mole of any gas will have the same volume at the same temperature and pressure. In fact, 1 mol of any gas at 0 °C and 1 atm will have a volume of 22.4 L. These conditions (0 °C and 1 atm) are referred to as STP—standard ___________ and ___________—for gas law calculations.

Workbook Example 10.2

PROBLEM:
If 2.0 mol of a gas at 150 K and 1 atm occupy 24.6 L, how many liters will 4.3 mol of gas occupy under the same conditions?

SOLUTION:
There are a lot of numbers in this problem, making the primary challenge to figure out what is important and what is not. The temperature and pressure of the gas are given, but they do not vary

and so need not be considered. The important numbers are the moles of gas and the one volume that is given. This is enough information to solve for the other volume by using Avogadro's law:

$$\frac{24.6\ L}{2.0\ mol} = \frac{V_2}{4.3\ mol}$$

$$V_2 = \frac{(24.6\ L)(4.3\ mol)}{2.0\ mol} = 53\ L$$

Workbook Problem 10.2

If a quantity of oxygen occupies 27 L at 37 °C, how many liters will it occupy at 350 °C?

Strategy: Determine which gas law is involved, convert temperatures, solve for the final volume.

Step 1: Determine which gas law to use.

Step 2: Convert the temperatures to K.

Step 3: Solve for the final volume.

POST-LECTURE:

Work Problems: *One of the best ways to learn chemistry is through solving problems. Go to the textbook and work through the Worked Examples for this section, then go to the end of chapter and work the problems there. When you feel confident that you can correctly answer problems like the ones listed below for each objective, check off each of the Learning Objectives for this section above.*

LO 10.3: Worked Example 10.3; Problems 10.26, 10.48

Section 10.3 The Ideal Gas Law

PRE-LECTURE:

Learning Objectives: *These are the skills you should be proficient in after reading this section and practicing the associated problems. Check off the boxes below once you have gone through this section, answered the associated questions in your textbook, and feel confident you have mastered the skill.*

☐ **LO 10.4** Use the ideal gas law to calculate pressure, volume, number of moles, or temperature of a gas sample.

Definitions: *To learn, you must do. This is your workbook. Fill in the definitions from the glossary of the textbook as a first step to understanding the material.*

— **ideal gas law**—____

____

— **gas constant (R)**—______________________________________

____

— **standard temperature and pressure (STP)**—________________________

____

Summary: *Read through this section and fill in the missing words and equations. Reference the textbook when you are not confident in your answers.*

All three gas laws can be combined to form the **ideal gas law:**

$$PV =$$

where R is the ideal gas law constant. R can be calculated from the molar volume of gas at STP:

$$R = \frac{PV}{RT} = \frac{(1\,atm)(22.4\,L)}{(1\,mol)(273.15\,K)} = 0.08206 \frac{L \cdot atm}{mol \cdot K}$$

The ideal gas law can be used to calculate any one of the four variables that describe gases if the other three are known, and it can also be rearranged to solve for any of the four variables:

$PV = nRT$ | $P =$ | $V =$ | $n =$ | $T =$

If the problem involves a change in one of these parameters, and you are unsure of which gas law to use, try the following trick: set your two conditions equal to each other like this:

$$\frac{P_1V_1}{n_1RT_1} = \frac{P_2V_2}{n_2RT_2}$$

Then cancel anything that is not changing. For instance, if the experiment is being done at constant pressure, $P_1 = P_2$, and they can be cancelled. R is always the same on both sides, so it can always be cancelled.

Note: the standard state for gas laws is not the same as the thermodynamic standard state (apparently gas law chemists like to be cold).

It's important to note that there are no really "ideal" gases—all gases deviate from ideal behavior to a certain extent. However, the deviations are usually so small as to be negligible.

Workbook Example 10.3

PROBLEM:
What is the volume of 7.62 g of Cl_2 at 45.7 °C and 0.257 atm?

SOLUTION:
First, solve for the moles of chlorine and the temperature in K:

$$n = 7.62\ g\ Cl_2 \times \frac{1\ mol\ Cl_2}{70.906\ g} = 0.1074662\ mol\ Cl_2$$

$$45.7\ °C + 273.15 = 318.85\ K$$

Then solve for the volume:

$$V = \frac{nRT}{P} = \frac{(0.1074662\ mol)\left(0.08206 \frac{L \cdot atm}{K \cdot mol}\right)(318.85\ K)}{0.257\ atm} = 10.9\ L$$

Workbook Problem 10.3

If 6.7 g of fluorine were placed in a 3.0 L container at 25.0 °C, what would be the pressure in the container?

Strategy: Rearrange the ideal gas law to solve for pressure, convert grams to moles, and degrees Celsius to kelvins; solve.

Step 1: Rearrange the ideal gas law.

Step 2: Convert grams to moles and degrees Celsius to kelvin.

Step 3: Solve for the final pressure.

Workbook Problem 10.4

If 12.3 g of neon were taken from 3.2 L at 1.0 atm and 25 °C to 18 atm and 750 °C, what would the final volume of the gas be?

Strategy: Determine what form of the gas laws is needed, convert temperatures and moles, if necessary, and solve for final pressure.

Step 1: Determine the form of gas laws needed.

Step 2: Convert grams to moles and degrees Celsius to kelvin if needed.

Step 3: Solve for the final pressure.

POST-LECTURE:

Work Problems: *One of the best ways to learn chemistry is through solving problems. Go to the textbook and work through the Worked Examples for this section, then go to the end of chapter and work the problems there. When you feel confident that you can correctly answer problems like the ones listed below for each objective, check off each of the Learning Objectives for this section above.*

LO 10.4: Worked Examples 10.4, 10.5; Problems 10.52–10.58

Section 10.4 Stoichiometric Relationships with Gases

PRE-LECTURE:

Learning Objectives: *These are the skills you should be proficient in after reading this section and practicing the associated problems. Check off the boxes below once you have gone through this section, answered the associated questions in your textbook, and feel confident you have mastered the skill.*

- ☐ **LO 10.5** Calculate volumes of gases in chemical reactions.
- ☐ **LO 10.6** Calculate the density or molar mass of a gas using the formula for gas density.

Summary: *Read through this section and fill in the missing words and equations. Reference the textbook when you are not confident in your answers.*

Many chemical reactions involve gases as either reactants or products. ___________ in combination with the gas laws allows for calculations between these substances.

Workbook Example 10.4

PROBLEM:
If 2.52 g of zinc is dropped into concentrated HCl, how many liters of hydrogen gas will be produced at 33 °C and 1.00 atm? The chemical equation is:

$$Zn(s) + 2HCl(aq) \rightarrow ZnCl_2(aq) + H_2(g)$$

SOLUTION:
First, solve for the moles of hydrogen gas produced, then use the ideal gas law to calculate the volume of the gas:

$$2.52\ g\ Zn \times \frac{1.00\ mol\ Zn}{65.30\ g} \times \frac{1\ mol\ H_2}{1\ mol\ Zn} = 0.0385439\ mol\ H_2$$

$$V = \frac{nRT}{P} = \frac{(0.0385439\ mol)\left(0.08206\frac{L \cdot atm}{K \cdot mol}\right)(33 + 273.15\ K)}{1.00\ atm} = 0.968\ L$$

Workbook Problem 10.5

In the following reaction, how many liters of oxygen will be needed to react with 7.45 L of NO? If the temperature of the reaction does not change, how many liters of NO_2 will be produced?

$$2\ NO(g) + O_2(g) \rightarrow 2\ NO_2(g)$$

Strategy: Determine which gas law is involved; solve for the volumes of oxygen and nitrogen dioxide.

Step 1: Determine the gas law to use.

Step 2: Solve for the volume of oxygen.

Step 3: Solve for the volume of nitrogen dioxide.

Density of a gas can be found by weighing a known volume of gas. For consistency, densities are reported at STP; if the volume isn't measured at STP, the volume at STP must be calculated. Since the amount of gas does not change, the relationship below can be used to calculate the volume at STP:

$$\left(\frac{PV}{RT}\right)_{measured} = n = \left(\frac{PV}{RT}\right)_{STP}$$

Rearranged to solve for V_{STP}:

V_{STP}:

Workbook Example 10.5

PROBLEM:
What is the density of chlorine gas at 85 °C and 1.5 atm?

SOLUTION:
It does not look like there is enough information here, but there is! Determine the volume of 1 mol of gas under the given conditions. The mass of 1 mol of Cl_2 is known, so the density can be calculated.

$$V = \frac{nRT}{P} = \frac{(1\,mol)\left(0.08206\,\frac{L \cdot atm}{mol \cdot K}\right)(85 + 273.15\,K)}{1.5\,atm} = 19.5931927\,L$$

1 mol of Cl_2 = 70.906 g, so the density of chlorine under these conditions is

$$\frac{70.906\,g}{19.5931927\,L} = 3.6\,g/L$$

Workbook Problem 10.6

What is the density of xenon at 300 mm Hg and 100 °C?

Strategy: Convert temperatures and pressures to make it possible to solve for the volume of 1 mol under these conditions. Divide the mass of 1 mol of xenon by the volume to determine the density.

Step 1: Convert the temperature and pressure.

Step 2:	Use the ideal gas law to calculate the volume of the gas.
Step 3:	Determine the density of the gas.

If the identity of the gas is not known, density can be used to calculate the molar mass and therefore help to identify the substance. To do this, mass must be introduced into the ideal gas law equation. This is done by multiplying both sides by molar mass; moles (n), already in the equation, multiplied by molar mass (M) equals mass.

$$PV(M) = n(M)RT \quad \text{or} \quad PVM = mRT$$

Since density is equal to the mass per volume, $\frac{m}{V}$, the equation above can be rearranged:

$$d = \frac{m}{V} =$$

Workbook Example 10.6

PROBLEM:
If a sample of gas that weighs 11.2 g has a volume of 3.13 L at 300 K and 2.0 atm, what is the molar mass of the gas? What is a possible identity for the gas?

SOLUTION:
To figure this out, first solve for the number of moles of gas present:

$$n = \frac{PV}{RT} = \frac{(2.0\ atm)(3.13\ L)}{\left(0.08206\ \frac{L \cdot atm}{mol \cdot K}\right)(300\ K)} = 0.254285\ mol$$

So 0.254285 mol of the gas weighs 11.2 g. This makes it possible to solve for the molar mass:

$$\frac{11.2\ g}{0.254285\ mol} = 44\ g/mol$$

It is likely that the gas is carbon dioxide.

Workbook Problem 10.7

The density of a gas was found to be 0.798 g/L at 700 °C and 3.75 atm. What is the molar mass of the gas? Can you suggest a possible identity for the gas?

Strategy: Decide on a volume of gas, then solve for n. Using the density makes it possible to solve for the mass for that volume, and then for the molar mass.

Step 1: Determine the number of moles in 1 L of gas.

Step 2: Solve for the mass in 1 L of gas by using the gas's density.

Step 3: Determine the molar mass by dividing grams by moles.

POST-LECTURE:

Work Problems: *One of the best ways to learn chemistry is through solving problems. Go to the textbook and work through the Worked Examples for this section, then go to the end of chapter and work the problems there. When you feel confident that you can correctly answer problems like the ones listed below for each objective, check off each of the Learning Objectives for this section above.*

LO 10.5: Worked Example 10.6; Problems 10.70, 10.72, 10.74
LO 10.6: Worked Example 10.7; Problems 10.66, 10.68

Section 10.5 Mixture of Gases: Partial Pressure and Dalton's Law

PRE-LECTURE:

Learning Objectives: *These are the skills you should be proficient in after reading this section and practicing the associated problems. Check off the boxes below once you have gone through this section, answered the associated questions in your textbook, and feel confident you have mastered the skill.*

☐ **LO 10.7** Calculate the partial pressure, mole fraction, or amount of each gas in a mixture.

Definitions: *To learn, you must do. This is your workbook. Fill in the definitions from the glossary of the textbook as a first step to understanding the material.*

— **Dalton's law of partial pressures—** ______________________________

____

— **mole fraction (χ)—** ______________________________

____

— **internal energy (E)—** ______________________________

____

— **state function—** ______________________________

____

Summary: *Read through this section and fill in the missing words and equations. Reference the textbook when you are not confident in your answers.*

Because gas particles do not interact much with one another, each gas in a mixture remains independent. This makes dealing with mixtures less complex than might be expected.

Dalton's law of partial pressures states that the total pressure of a mixture of gases is the sum of the pressures of the individual gases:

$$P_{total} = P_1 + P_2 + P_3 + \cdots + P_n$$

Since all the gases in a mixture are by definition at the same temperature and pressure, the pressure exerted by a mixture of gases is purely dependent on the number of moles in the mixture.

To determine the pressures of individual components of the mixture, the mole fraction (χ) is used. This is calculated by dividing the number of moles of the component of interest by the total number of moles in the mixture:

$$mole\ fraction\ (\chi) = \frac{moles\ of\ component}{total\ number\ of\ moles\ in\ mixture}$$

The pressure of the individual component is then calculated as the total pressure of the gas times the mole fraction of the individual component:

$P_1 =$

This equation can be rearranged to determine the mole fraction of a gas based on its partial pressure.

Workbook Example 10.7

PROBLEM:
A 3.0 L flask at 25 °C contains Ar at a partial pressure of 0.76 atm, He at a partial pressure of 0.32 atm, and Ne at a partial pressure of 0.42 atm. What is the total pressure of the mixture? What is the mole fraction of each gas?

SOLUTION:

$$P_{total} = 0.76\ atm + 0.32\ atm + 0.42\ atm = 1.50\ atm$$

The mole fraction of each gas can be calculated from the partial pressure:

$$\chi_{Ar} = \frac{0.76\ atm}{1.50\ atm} = 0.51 \qquad \chi_{He} = \frac{0.32\ atm}{1.50\ atm} = 0.21 \qquad \chi_{Ne} = \frac{0.42\ atm}{1.50\ atm} = 0.28$$

Note: Mole fractions are unitless, and will always add up to 1, just as percentages always add up to 100.

Workbook Problem 10.8

A 30 L tank is filled with the following: 18.3 g CO_2, 23.1 g Ne, and 5.26 g H_2. If the container is at 25 °C, what is the total pressure in the container, and what are the partial pressures of the individual gases?

Strategy: Determine the number of moles of each gas, the total number of moles, and the mole fraction of each component. Using these, calculate the total pressure and the partial pressures using the ideal gas law and Dalton's law.

Step 1: Determine the number of moles of each gas.

Step 2: Use the total number of moles to calculate the total pressure.

Step 3: Use the mole fraction of each component to determine the partial pressures.

POST-LECTURE:

Work Problems: *One of the best ways to learn chemistry is through solving problems. Go to the textbook and work through the Worked Examples for this section, then go to the end of chapter and work the problems there. When you feel confident that you can correctly answer problems like the ones listed below for each objective, check off each of the Learning Objectives for this section above.*

LO 10.7: Worked Example 10.8; Problems 10.82, 10.86, 10.90

Section 10.6 The Kinetic–Molecular Theory of Gases

PRE-LECTURE:

Learning Objectives: *These are the skills you should be proficient in after reading this section and practicing the associated problems. Check off the boxes below once you have gone through this section, answered the associated questions in your textbook, and feel confident you have mastered the skill.*

☐ **LO 10.8** Use the assumptions of kinetic–molecular theory to predict gas behavior.
☐ **LO 10.9** Calculate the average molecular speed of a gas particle at a given temperature.

Definitions: *To learn, you must do. This is your workbook. Fill in the definitions from the glossary of the textbook as a first step to understanding the material.*

— **kinetic–molecular theory—** ____

____

Summary: *Read through this section and fill in the missing words and equations. Reference the textbook when you are not confident in your answers.*

The kinetic–molecular theory is a model developed over a century ago that describes the macroscopic behavior of gases based on a series of assumptions:

1. Gas particles are ___________ , and they move___________ .
2. The volume of the gas particles themselves is ___________ .
3. The gas particles are ___________ of each other.
4. Collisions of gas particles, with one another or with the walls of the container, are ___________ . That is, no energy is lost in these collisions, so the total kinetic energy of a gas is constant at constant temperature.
5. The average kinetic energy is proportional to the kelvin ___________ of the gas.

These assumptions can be used to explain the individual gas laws quite effectively:

- Boyle's law: pressure increases as volume decreases because the smaller the space, the greater the number of collisions, and thus the ___________ the pressure.
- Charles's law: since temperature is a measure of kinetic energy, as temperature increases the amount of kinetic energy increases and the particles move ___________ . If volume was held constant, this would lead to an increase in pressure because the particles would

be colliding with the walls more often. However, in his experiments, Charles held pressure constant, not volume. Thus, volume needs to be increased to keep the number of collisions and therefore the pressure the same.

- Avogadro's law: volume increases as the amount increases at constant pressure because adding more particles to the mixture will __________ the number of collisions unless the space is also increased.
- Dalton's law: because the identity of the particles is irrelevant and they do not interact, the pressure of a mixture of gases depends only on the temperature and amount of gas present. The pressure of any individual component will depend only on the __________ __________ of that component.

From the fifth assumption, it is possible to derive the average speed of the particles of a gas:

$$u = \sqrt{\frac{3RT}{M}}$$

In this equation, u is the speed (m/s), M is the molar mass in kg/mol (note the difference in units of molar mass from what you are used to using), and a different form of R must be used: $8.314\ \frac{\text{J}}{\text{K}\cdot\text{mol}}$. All of these changes are required to make the units work out to m/s—remember from the last chapter that joules are derived from the definition of *kinetic energy* $\left(\text{kg}\frac{\text{m}^2}{s^2}\right)$.

Gas particles move amazingly quickly! Larger molecules move more __________ than smaller molecules, but it is important to remember that this is an average: in any gas, there will be a distribution of actual speeds. However, the total kinetic energy in the gas will remain the same. Also, gas particles do not move very far before colliding either with another gas particle or with something else, so they proceed in a zigzag fashion. The *mean free path*, the average distance between collisions, is longer for a smaller, faster particle and shorter for a larger, slower-moving particle.

Workbook Example 10.8

PROBLEM:
Calculate the average speed of a xenon atom at 398 K.

SOLUTION:

$$u = \sqrt{\frac{3RT}{M}} = \sqrt{\frac{(3)\left(8.314\frac{\frac{kg\cdot m^2}{s^2}}{K\cdot mol}\right)(398\,K)}{0.131293\frac{kg}{mol}}} = \sqrt{75608\frac{m^2}{s^2}} = 275\frac{m}{s}$$

Workbook Problem 10.9

What temperature is required to get gaseous iodine molecules moving at the same average speed as hydrogen molecules at room temperature?

Strategy:	Determine the speed of H_2 molecules at 25 °C, then solve the equation again for the temperature needed for the much larger I_2 molecules to move at that same speed.
Step 1:	Determine the speed of H_2 at 25 °C.
Step 2:	Determine the temperature required for the iodine molecules to achieve the same speed.

POST-LECTURE:

Work Problems: *One of the best ways to learn chemistry is through solving problems. Go to the textbook and work through the Worked Examples for this section, then go to the end of chapter and work the problems there. When you feel confident that you can correctly answer problems like the ones listed below for each objective, check off each of the Learning Objectives for this section above.*

LO 10.8: Problems 10.27, 10.28, 10.32
LO 10.9: Problems 10.94, 10.98

Section 10.7 Gas Diffusion and Effusion: Graham's Law

PRE-LECTURE:

Learning Objectives: *These are the skills you should be proficient in after reading this section and practicing the associated problems. Check off the boxes below once you have gone through this section, answered the associated questions in your textbook, and feel confident you have mastered the skill.*

- ☐ **LO 10.10** Interpret a molecular picture of effusion and diffusion.
- ☐ **LO 10.11** Use Graham's law to estimate relative rates of diffusion for two gases.

Definitions: *To learn, you must do. This is your workbook. Fill in the definitions from the glossary of the textbook as a first step to understanding the material.*

— **diffusion**— ______________________________

____

— **effusion**— ______________________________

____

— **Graham's law**— ______________________________

____

Summary: *Read through this section and fill in the missing words and equations. Reference the textbook when you are not confident in your answers.*

Because of the high speeds at which gas molecules move, and their frequent collisions, gases mix rapidly with one another (this is why you can smell baking cookies from quite a distance). This rapid mixing of gases with frequent collisions is called ___________ . It is difficult to mathematically model diffusion because of the random nature of the collisions.

More readily characterized mathematically is ___________ : the progress of gas particles through a tiny hole into a vacuum. Not surprisingly, speed is inversely proportional to molar mass: a bigger, and thus slower-moving gas will effuse more ___________ than a small, faster-moving gas.

Graham's law describes this property mathematically:

$$Rate \propto \frac{1}{\sqrt{M}}$$

This equation can be used most simply to calculate the relative rates of effusion in a mixture of gases:

$$\frac{Rate_1}{Rate_2} = \sqrt{\frac{M_2}{M_1}}$$

Workbook Example 10.9

PROBLEM:
Calculate the ratio of effusion rates of Ne and Xe from the same container at the same temperature and pressure.

SOLUTION:

$$\frac{Rate\ of\ effusion\ of\ Ne}{Rate\ of\ effusion\ of\ Xe} = \sqrt{\frac{131.293\frac{g\ Xe}{mol}}{20.1797\frac{g\ Ne}{mol}}} = 2.55072$$

Workbook Problem 10.10

Oxygen-16 has an atomic mass of 15.995, and oxygen-18 has an atomic mass of 17.999. What will be their relative rates of effusion?

Strategy: Solve using Graham's law.

Step 1: Solve using Graham's law.

POST-LECTURE:

Work Problems: *One of the best ways to learn chemistry is through solving problems. Go to the textbook and work through the Worked Examples for this section, then go to the end of chapter and work the problems there. When you feel confident that you can correctly answer problems like the ones listed below for each objective, check off each of the Learning Objectives for this section above.*

LO 10.10: Problem 10.33
LO 10.11: Worked Example 10.9; Problems 10.100–10.103, 10.143

Section 10.8 The Behavior of Real Gases

PRE-LECTURE:

Learning Objectives: *These are the skills you should be proficient in after reading this section and practicing the associated problems. Check off the boxes below once you have gone through this section, answered the associated questions in your textbook, and feel confident you have mastered the skill.*

- ☐ **LO 10.12** Identify temperature and pressure conditions that cause significant deviations from ideal behavior.
- ☐ **LO 10.13** Use the van der Waals equation to calculate the properties of real gases.

Definitions: *To learn, you must do. This is your workbook. Fill in the definitions from the glossary of the textbook as a first step to understanding the material.*

— **van der Waals equation—** ____

____

Summary: *Read through this section and fill in the missing words and equations. Reference the textbook when you are not confident in your answers.*

The ideal gas law is a good approximation of the behavior of gases under normal conditions but begins to fail when pressure is very ___________.

The actual volume of gas particles can no longer be neglected when they are very close together. As a result, the volume of a gas at high pressure is _________ than predicted by the ideal gas law.

Attractive forces between gas particles will also become a factor at high pressures: these tend to ___________ the volume of the gas.

Volume and attractive forces negate each other at moderate pressures, but above 300 atm, the volume of the gas particles becomes the primary factor.

These factors have been considered mathematically in the van der Waals equation:

$$\left(P+\frac{an^2}{V^2}\right)(V-nb)=nRT$$

where the a term is a correction for intermolecular attractions and the b term is a correction for molecular volume. Solving this equation for pressure can be done fairly simply but solving for volume is not trivial.

Workbook Example 10.10

PROBLEM:
Calculate the pressure of 1.00 mol of N_2 gas in 50.0 mL at 300.0 K using the ideal gas law and the van der Waals equation. The van der Waals constants for nitrogen are $a = 1.35\ (L^2 \cdot atm)/mol^2$ and $b = 0.0387$ L/mol.

SOLUTION:
Ideal gas law:

$$P=\frac{nRT}{V}=\frac{(1.00\ mol)\left(0.08206\ \frac{L\cdot atm}{mol\cdot K}\right)(300.0\ K)}{0.0500\ L}=492\ atm$$

van der Waals equation:

$$P=\frac{nRT}{V-nb}-\frac{an^2}{V^2}=\frac{(1.00\ mol)\left(0.08206\ \frac{L\cdot atm}{mol\cdot K}\right)(300.0\ K)}{0.0500\ L-(1\ mol)\left(0.0387\ \frac{L}{mol}\right)}-\frac{\left(1.35\ \frac{L^2\cdot atm}{mol^2}\right)}{(0.0500\ L)^2}$$

$$=\frac{24.618\ L\cdot atm}{0.0113\ L}-\frac{1.35\ \frac{L^2\cdot atm}{mol^2}}{0.00250\ L^2}=2178.584\ atm-540\ atm=1640\ atm$$

At very high pressures, the ideal gas law fails rather dramatically.

Workbook Problem 10.11

Calculate the pressure of 1.00 mol of O_2 gas in 100 mL at 300 K by using the ideal gas law and the van der Waals equation. The van der Waals constants for oxygen are $a = 1.38\ (L^2 \cdot atm)/mol^2$ and $b = 0.0318$ L/mol. Does the ideal gas law approximate the pressure acceptably under these conditions?

Strategy: Solve using the ideal gas law and the van der Waals equation.

Step 1: Solve using the ideal gas law.

Step 2: Solve using the van der Waals equation.

Step 3: Compare the two values.

POST-LECTURE:

Work Problems: *One of the best ways to learn chemistry is through solving problems. Go to the textbook and work through the Worked Examples for this section, then go to the end of chapter and work the problems there. When you feel confident that you can correctly answer problems like the ones listed below for each objective, check off each of the Learning Objectives for this section above.*

LO 10.12: Problems 10.110, 10.111
LO 10.13: Problems 10.112, 10.114

Section 10.9 The Earth's Atmosphere and the Greenhouse Effect

PRE-LECTURE:

Learning Objectives: *These are the skills you should be proficient in after reading this section and practicing the associated problems. Check off the boxes below once you have gone through this section, answered the associated questions in your textbook, and feel confident you have mastered the skill.*

☐ **LO 10.14** Explain the principle of the greenhouse effect.

Definitions: *To learn, you must do. This is your workbook. Fill in the definitions from the glossary of the textbook as a first step to understanding the material.*

— **greenhouse effect—**____

____

— **greenhouse gas—**____

____

Summary: *Read through this section and fill in the missing words and equations. Reference the textbook when you are not confident in your answers.*

The atmosphere has four major regions. The ___________ is closest to the Earth's surface, and has both the greatest effect on the surface and is most affected by human activity. Above the troposphere is the ___________ , the ___________ , and then the ___________ .

The *greenhouse effect* refers to the absorption of ___________ , or IR, radiation by gases in the atmosphere. It is a natural phenomenon that helps to regulate Earth's temperature. Without it, the average temperature on Earth would be 0 °F instead of approximately 60 °F.

POST-LECTURE:

Work Problems: *One of the best ways to learn chemistry is through solving problems. Go to the textbook and work through the Worked Examples for this section, then go to the end of chapter and work the problems there. When you feel confident that you can correctly answer problems like the ones listed below for each objective, check off each of the Learning Objectives for this section above.*

LO 10.14: Problems 10.120, 10.122

Section 10.10 Greenhouse Gases

PRE-LECTURE:

Learning Objectives: *These are the skills you should be proficient in after reading this section and practicing the associated problems. Check off the boxes below once you have gone through this section, answered the associated questions in your textbook, and feel confident you have mastered the skill.*

☐ **LO 10.15** Predict if a molecule is a greenhouse gas by examining molecular vibrations.

Summary: *Read through this section and fill in the missing words and equations. Reference the textbook when you are not confident in your answers.*

A few gases in our atmosphere can absorb IR radiation, and are thus referred to as greenhouse gases. These include ___________, ___________ ___________, and ___________. The absorbed radiation makes the bonds in the molecules vibrate. However, the gas molecules will only absorb IR radiation if two conditions are met:

- The energy difference between the lower vibrational state and the excited vibrational state exactly matches the energy of the IR photon.
- The vibration results in a change in dipole moment.

Once these molecules absorb the IR energy, it "traps" the energy within our atmosphere and increases the temperature of our planet—very similar to how a greenhouse works!

POST-LECTURE:

Work Problems: *One of the best ways to learn chemistry is through solving problems. Go to the textbook and work through the Worked Examples for this section, then go to the end of chapter and work the problems there. When you feel confident that you can correctly answer problems like the ones listed below for each objective, check off each of the Learning Objectives for this section above.*

LO 10.15: Problems 10.123–10.126

Section 10.11 Climate Change

PRE-LECTURE:

Learning Objectives: *These are the skills you should be proficient in after reading this section and practicing the associated problems. Check off the boxes below once you have gone through this section, answered the associated questions in your textbook, and feel confident you have mastered the skill.*

☐ **LO 10.16** Describe the trends in greenhouse gas concentrations over time and measured and predicted effects of climate change.

Definitions: *To learn, you must do. This is your workbook. Fill in the definitions from the glossary of the textbook as a first step to understanding the material.*

— **global warming—** ____

____

— **climate change—** ____

____

Summary: *Read through this section and fill in the missing words and equations. Reference the textbook when you are not confident in your answers.*

Two phrases are frequently used when we talk about pollution and our environment:

- **Global** ___________ refers to the theory that increasing concentrations of greenhouse gases will upset the delicate thermal balance of incoming and outgoing radiation on Earth (the greenhouse effect).
- **Climate** ___________ is a newer hypothesis that suggests that there will not be a uniform increase in temperature throughout the planet; some areas will experience an increase in temperature, while some areas may stay the same or experience some cooling.

Climate science is complex because there are numerous factors that influence climate such as cloud cover, particulate matter, solar energy, changing surface reflection due to melting polar ice caps, and deforestation.

POST-LECTURE:

Work Problems: *One of the best ways to learn chemistry is through solving problems. Go to the textbook and work through the Worked Examples for this section, then go to the end of chapter and work the problems there. When you feel confident that you can correctly answer problems like the ones listed below for each objective, check off each of the Learning Objectives for this section above.*

LO 10.16: Problems 10.127–10.133

Self-Test

This section is intended to test your knowledge of the material covered in this chapter. Think through these problems and make certain you understand what the problem is about—ask yourself if the answer makes sense. Many of these questions are linked to the chapter learning goals. Therefore, successful completion of these problems indicates that you have achieved the learning goals for this chapter. You will receive the greatest benefit from this section if you use it as a mock exam. Do not use your notes, book, Internet, or friends when taking the test. There is no grade and you only shortchange yourself by getting help with this activity. As you take the test, mark the problems with your level of confidence with the answer.

★ *Use a star for problems you know you got right.*
? *Use a question mark for those you think you got right.*
x *Use an x for those you are guessing on.*

If you get a question right that you marked with a star, you can be confident that you know the associated LO.

If you get a question right that you marked with a question mark, go back and work a few more problems and test yourself again tomorrow.

If you get a question wrong or mark it with an x, go back and read the section again, work a few problems today, a few tomorrow, and a few the next day before testing yourself again.

This will help you to discover which topics you have learned thoroughly and which topics you need to study in more detail.

True–False

1. The atmosphere is about 70% oxygen.

2. Gases are more compressible than liquids.

3. Pascal, atmosphere, bar, and millimeters of mercury are all accepted units of pressure.

4. If the mercury in a manometer is lower on the side that is open to the air, then the pressure of the gas is lower than the pressure of the air.

5. Stoichiometric ratios can be determined by volume in gas reactions.

6. The smaller a gas particle, the slower it travels.

7. The ratio of effusion rates of two gases is directly proportional to the square roots of their masses.

8. The ideal gas law begins to fail at high temperatures.

9. When you inhale, you are breathing stratosphere.

10. Incomplete combustion of hydrocarbons is a major factor in air pollution.

Matching

11. barometer	a. force per unit area
12. pressure	b. the pressure of one component in a gas mixture is related to its mole fraction in the mixture
13. atmospheric pressure	c. the rate at which a gas effuses is inversely proportional to the square root of its molar mass
14. Boyle's law	d. the escape of gas particles through a tiny hole in a membrane into a vacuum
15. Charles's law	e the pressure exerted by one component in a gaseous mixture
16. Avogadro's law	f. the chaotic mixing of gases through random collisions
17. Ideal gas law	g. number of moles of a component divided by the total moles in a mixture
18. Dalton's law	h. volume and temperature vary directly when pressure and amount are held constant
19. partial pressure	i. volume and pressure are inversely proportional when temperature and amount are constant
20. kinetic–molecular theory	j. a piece of equipment used to measure atmospheric pressure
21. Graham's Law	k. equipment for measuring gas pressures
22. diffusion	l. pressure generated by the atmosphere pressing down on the Earth's surface
23. effusion	m. allows any of the four variables to be calculated—*T*, *P*, *V*, or *n*—if the others are known
24. ideal gas	n. a model that accounts for the behavior of gases
25. STP	o. moles and volume vary directly
26. standard molar volume	p. gas that precisely follows the tenets of kinetic–molecular theory
27. mole fraction	q. 22.4 L at STP
28. manometer	r. 0 °C and 1 atmosphere pressure

Fill in the Blanks

29. Gases exert ________________ on the walls of their container as a result of ________________ with the walls.

30. The physical properties of a gas can be defined by four variables: ________________, ________________, ________________, and ________________.

31. The concentration of a component in a gas mixture is typically calculated as its ________________ ________________. This can also be used to determine the ________________ ________________ of the component.

32. In kinetic–molecular theory, the ________________ of individual gas particles is said to be ________________. The assumption fails at high ________________.

33. Effusion rates are proportional to the square root of molar mass. Gases with __________ molar masses move significantly ________________ than gases with high molar masses.

Problems

34. If atmospheric pressure is 0.983 atm, how high a column of silicone oil (density = 0.970 g/mL) will this pressure support (the density of mercury is 13.6 g/mL)?

35. If you are using an open-ended manometer filled with ethyl alcohol rather than mercury, what is the gas pressure in mm Hg if the level of ethyl alcohol in the arm connected to the bulb is a) 340 mm lower and P_{atm} = 760 mm Hg; and b) 257 mm higher and P_{atm} = 760 mm Hg? The density of ethyl alcohol is 0.789 g/mL, and the density of mercury is 13.6 g/mL.

36. How many moles of gas are present in a 50 mL flask at 5.78 atm and −20 °C? If the mass of the gas is 0.281 g, what is the gas?

37. What is the density of methane (CH_4) at 100 °C and 4 atm?

38. A student carried out a reaction in the lab in which one of the products was a gas. She collected the gas for analysis and found that it contained 7.69% hydrogen and 92.3% carbon. She also observed that 250 mL of the gas at 35 °C and 763 mm Hg had a mass of 0.258 g. a) What is the empirical formula of the gas? b) What is the molar mass of the gas? c) What is its molecular formula?

39. What is the molar mass of a gas with a density of 14.04 g/L at 15 atm and 300 °C?

40. When 25 mL of butane (C_4H_{10}, density 0.6014 g/mL) are burned, how many liters of oxygen are consumed, and how many liters of carbon dioxide and water vapor are produced at 30 °C and atmospheric pressure? What is the overall change in volume for the reaction? Is work being done as well as heat given off?

41. If 7.03 g Ne, 11.7 g N_2, and 15.7 g He are placed in a 5000-mL container at room temperature (25 °C), what is the total pressure, and what is the pressure of each component?

42. What are the relative effusion rates of H_2 and He?

43. If a 50-mL aerosol can at room temperature (25 °C) with an internal pressure of 7.8 atm is emptied into a 375-mL container at atmospheric pressure, what is the final temperature of the gas in Celsius?

Test Yourself: *Take the Practice Test on pages 411 and 412 of your textbook. Again, use a testing environment using only those items that your instructor will allow you to have access to during the test and mark each question with a star, question mark, or x to denote your confidence level. Once you have finished you can check off the LO at the beginning of this chapter based on which questions you starred and got correct. Go back and spend more time with the topics from the other learning objectives.*

CHAPTER ELEVEN

Liquids and Phase Changes

Learning Objectives

As a result of reading and studying this chapter, you should be able to:

- ☐ **LO 11.1** Match an observation of a physical phenomenon with the correct liquid property.
- ☐ **LO 11.2** Predict which substance has higher viscosity or surface tension based on its molecular structure.
- ☐ **LO 11.3** Use the Clausius–Clapeyron equation to calculate ΔH_{vap} or vapor pressure at varying temperatures.
- ☐ **LO 11.4** Use the equation for free energy to calculate the temperature at which a phase change becomes spontaneous or ΔS for a phase change.
- ☐ **LO 11.5** Draw or interpret a heating curve for a substance.
- ☐ **LO 11.6** Calculate the amount of heat associated with temperature and phase changes of substances.
- ☐ **LO 11.7** Use a phase diagram to identify the phase of a substance at a given temperature and pressure or the phase transition that occurs when the temperature and pressure are changed.
- ☐ **LO 11.8** Sketch a phase diagram given the appropriate data.
- ☐ **LO 11.9** Predict which molecules will exhibit a liquid crystal phase based on molecular structure.
- ☐ **LO 11.10** Classify the types of intermolecular forces between molecules in a liquid crystal.
- ☐ **LO 11.11** Match a molecular image with different states of matter, including solid, liquid, and common liquid crystal phases.

Chapter Summary

Unlike the molecules in gases, the molecules in liquids interact considerably. This chapter enhances the previous discussion on intermolecular forces by discussing the effects they have on the properties of liquids and phase transitions. Free-energy changes accompany all physical and chemical changes: in this chapter, we will examine the contributions of entropy and enthalpy to phase changes as well as the contributions of changes in pressure and temperature.

The Chapter in Detail

Section 11.1 Properties of Liquids

PRE-LECTURE:

Learning Objectives: *These are the skills you should be proficient in after reading this section and practicing the associated problems. Check off the boxes below once you have gone through this section, answered the associated questions in your textbook, and feel confident you have mastered the skill.*

☐ **LO 11.1** Match an observation of a physical phenomenon with the correct liquid property.
☐ **LO 11.2** Predict which substance has higher viscosity or surface tension based on its molecular structure.

Definitions: *To learn, you must do. This is your workbook. Fill in the definitions from the glossary of the textbook as a first step to understanding the material.*

— **viscosity—** ____

____

— **surface tension—**____

____

Review: *Before going through the section make sure you are familiar with these concepts:*

- Section 8.6 intermolecular forces
- Section 8.6 strength of intermolecular forces in relation to physical properties

Summary: *Read through this section and fill in the missing words and equations. Reference the textbook when you are not confident in your answers.*

Viscosity is the measure of a liquid's resistance to flow: the greater the intermolecular interactions, the stronger molecules hold on to one another, and the ___________the resistance to flow. Long-chain polar molecules like those found in honey will not flow as readily as those in a nonpolar solvent held together only by dispersion forces.

Intermolecular forces also cause surface tension. Since there is no molecule on top of a molecule at the surface of a liquid to balance out the force of the molecules pulling from underneath, these surface molecules are pulled in tightly. It is this phenomenon that makes it possible for insects to walk on water; it also causes water to bead up on a nonpolar surface.

Both viscosity and surface tension are temperature dependent: increasing kinetic energy will ___________ both properties.

POST-LECTURE:

Work Problems: *One of the best ways to learn chemistry is through solving problems. Go to the textbook and work through the Worked Examples for this section, then go to the end of chapter and work the problems there. When you feel confident that you can correctly answer problems like the ones listed below for each objective, check off each of the Learning Objectives for this section above.*

LO 11.1: Problems 11.20, 11.21
LO 11.2: Problems 11.22–11.25

Section 11.2 Vapor Pressure and Boiling Point

PRE-LECTURE:

Learning Objectives: *These are the skills you should be proficient in after reading this section and practicing the associated problems. Check off the boxes below once you have gone through this section, answered the associated questions in your textbook, and feel confident you have mastered the skill.*

☐ **LO 11.3** Use the Clausius–Clapeyron equation to calculate ΔH_{vap} or vapor pressure at varying temperatures.

Definitions: *To learn, you must do. This is your workbook. Fill in the definitions from the glossary of the textbook as a first step to understanding the material.*

— **vapor pressure—**____

____

— **heat of vaporization (ΔH_{vap})—**______________________________

____

— **Clausius–Clapeyron equation—** ______________________________

____

— **normal boiling point—** ____________________________________

____

Summary: *Read through this section and fill in the missing words and equations. Reference the textbook when you are not confident in your answers.*

Liquid molecules at a certain temperature all have the same average kinetic energy, but a group of individual molecules has a range of energies. If a molecule with a high kinetic energy reaches the surface, it can escape, or evaporate. The amount of energy needed to escape varies between substances and is related to the intermolecular forces that are present. Weaker intermolecular forces give rise to ___________ boiling points because less energy is needed to escape the surface of the liquid.

When the molecules escape the surface of a liquid in an open container, they will drift away, and the liquid will evaporate completely. In a closed container, a dynamic equilibrium will develop: the same number of molecules are escaping into the air as are simultaneously returning to the liquid. As a result, at a specific temperature, a liquid will have a characteristic vapor pressure.

When the vapor pressure is equal to the external pressure, the liquid will ___________.The boiling point that is measured at 1 atm is called the normal boiling point. If the external pressure is higher or lower than 1 atm, the boiling point will change correspondingly: the boiling point at lower pressures will be a ___________temperature, the boiling point at a higher pressure will be a ___________ temperature.

When the natural log of the vapor pressure is plotted against the inverse of the temperature, the graph looks like a linear relationship (it fits the form y = mx + b), known as the Clausius–Clapeyron equation:

$$\ln P_{\text{vap}} = \left(\frac{-\Delta H_{\text{vap}}}{R}\right)\frac{1}{T} + C$$

This form of the Clausius–Clapeyron equation resembles the equation for a straight line (y = mx + b) where *y* is ln P_{vap}, *m* (the slope) is equal to $-\Delta H_{\text{vap}}/R$, *x* is 1/T and *b* (the *y*-intercept) is *C*, where *C* is a constant that is characteristic of each substance.

This equation can also be rearranged to solve for the vapor pressures and corresponding temperatures, provided that ΔH_{vap} is known:

$$\ln\left(\frac{P_1}{P_2}\right) = \frac{\Delta H_{\text{vap}}}{R}\left(\frac{1}{T_2} - \frac{1}{T_1}\right)$$

In all of these equations, R = 8.3145 J/K · mol.

Workbook Example 11.1

PROBLEM:
Methanol has a normal boiling point of 64.7 °C and a vapor pressure of 400 mm Hg at 49.9 °C. What is ΔH_{vap} for methanol?

SOLUTION:
There is a great deal of information here, but some of it is hidden. What you have is two vapor pressures and two temperatures with which to solve for ΔH. The vapor pressure of a liquid is always 760 mm Hg at its normal boiling point.

Using the two-point form of the Clausius–Clapeyron equation from above, we get:

$$\ln\left(\frac{400\ mm\ Hg}{760\ mm\ Hg}\right) = \frac{\Delta H_{vap}}{8.3145\ J/_{mol\cdot K}}\left(\frac{1}{64.7+273.15} - \frac{1}{49.9+273.15}\right)$$

$$\ln\left(\frac{400\ mm\ Hg}{760\ mm\ Hg}\right) = \frac{\Delta H_{vap}}{8.3145\ J/_{mol\cdot K}}\left(\frac{1}{337.85\ K} - \frac{1}{323.05\ K}\right)$$

$$-0.641853886 = \frac{\Delta H_{vap}}{8.3145\ J/_{mol\cdot K}}\left(\frac{0.002959893}{K} - \frac{0.003095496}{K}\right)$$

$$-0.641853886 = \frac{\Delta H_{vap}}{8.3145\ J/_{mol\cdot K}}\left(-\frac{0.000135603}{K}\right)$$

$$\frac{-0.641853886}{-0.000135603} = \frac{\Delta H_{vap}}{8.3145\ J/mol}$$

$$4733.34465 = \frac{\Delta H_{vap}}{8.3145\ J/mol}$$

$$\Delta H_{vap} = (4733.34465)(8.3145\ J/mol) = 39{,}355.3941\ J/mol = 39.4\ kJ/mol$$

Workbook Example 11.2

PROBLEM:
Given the information in Workbook Example 11.3, what is the vapor pressure of methanol at 25 °C?

SOLUTION:
With a variety of temperatures and vapor pressures to choose from, and having solved for ΔH_{vap} above, this is a fairly straightforward problem. Here, the equation above is used with 25 °C on one side and the normal boiling point on the other:

$$\ln\left(\frac{P_1}{760\ mm\ Hg}\right) = \frac{39355.3941\ J/mol}{8.3145\ J/mol \cdot K}\left(\frac{1}{64.7 + 273.15} - \frac{1}{25 + 273.15}\right)$$

$$\ln\left(\frac{P_1}{760\ mm\ Hg}\right) = \frac{39355.3941\ J/mol}{8.3145\ J/mol \cdot K}\left(\frac{1}{337.85\ K} - \frac{1}{298.15\ K}\right)$$

$$\ln\left(\frac{P_1}{760\ mm\ Hg}\right) = 4733.345\ K\left(\frac{0.002959893}{K} - \frac{0.00335402}{K}\right)$$

$$\ln\left(\frac{P_1}{760\ mm\ Hg}\right) = 4733.345\ K\left(\frac{-0.000394123}{K}\right)$$

$$\ln\left(\frac{P_1}{760\ mm\ Hg}\right) = -1.86552$$

$$e^{\ln\left(\frac{P_1}{760\ mm\ Hg}\right)} = e^{-1.86552}$$

$$\frac{P_1}{760\ mm\ Hg} = 0.1548157$$

$$P_1 = 117.6599\ mm\ Hg = 118\ mm\ Hg$$

Workbook Problem 11.1

Ethanol has a normal boiling point of 78.4 °C and a vapor pressure of 400 mm Hg at 63.5 °C. What is ΔH_{vap} for ethanol?

Strategy: Use the Clausius–Clapeyron equation to solve for ΔH.

Step 1: Convert temperatures to kelvin.

Step 2: Solve for ΔH.

Workbook Problem 11.2

Given the information in the previous problem, what is the vapor pressure of ethanol at 25 °C?

Strategy: Use the Clausius–Clapeyron equation to solve for the vapor pressure.

Step 1: Use the Clausius–Clapeyron equation to solve for the vapor pressure.

POST-LECTURE:

Work Problems: *One of the best ways to learn chemistry is through solving problems. Go to the textbook and work through the Worked Examples for this section, then go to the end of chapter and work the problems there. When you feel confident that you can correctly answer problems like the ones listed below for each objective, check off each of the Learning Objectives for this section above.*

LO 11.3: Worked Example 11.1; Problems 11.26, 11.28, 11.36

Section 11.3 Phase Changes between Solid, Liquids, and Gases

PRE-LECTURE:

Learning Objectives: *These are the skills you should be proficient in after reading this section and practicing the associated problems. Check off the boxes below once you have gone through this section, answered the associated questions in your textbook, and feel confident you have mastered the skill.*

☐ **LO 11.4** Use the equation for free energy to calculate the temperature at which a phase change becomes spontaneous or ΔS for a phase change.

Definitions: *To learn, you must do. This is your workbook. Fill in the definitions from the glossary of the textbook as a first step to understanding the material.*

— **phase change—** ____

____

— **phase—** ____

____

Summary: *Read through this section and fill in the missing words and equations. Reference the textbook when you are not confident in your answers.*

There are six possible phase changes between solids, liquids, and gases, all associated with a change in free energy, ΔG.

- fusion (melting): __________ → __________
- freezing: __________ → __________
- evaporation: __________ → __________
- condensation: __________ → __________
- sublimation: __________ → __________
- deposition: __________ → __________

The free-energy change can be described by using the equation $\Delta G = \Delta H - T\Delta S$, and it has two portions. The enthalpy change is related to the breaking or forming of intermolecular interactions, and the entropy change is associated with the increase or decrease in ___________that is associated with breaking or forming intermolecular interactions.

For melting, sublimation, and vaporization, heat must be added, but disorder increases: ΔH and ΔS are both ________. For the opposite phase changes, freezing, condensation, and deposition, heat is released, but order increases: ΔH and ΔS are both _________. Both changes in enthalpy and entropy are larger for the change between liquid and gas than for the change between solid and liquid.

At equilibrium, seen here as a phase change, $\Delta G = 0$. This makes it possible to calculate freezing and boiling points, or, more usefully, to calculate entropy changes from measured freezing or boiling points, using the following equations:

$T =$	$\Delta S =$

Workbook Example 11.3

PROBLEM:
Methanol boils at 64.7 °C and has an enthalpy of vaporization of 35.3 kJ/mol. What is ΔS_{vap} for this phase transition?

SOLUTION:
Using the equation above, we calculate ΔS_{vap} to be

$$\Delta S = \frac{\Delta H}{T} = \frac{35300\ ^{J}/_{mol}}{(64.7 + 273.15)\,K} = 104\ ^{J}/_{K \cdot mol}$$

Workbook Example 11.4

PROBLEM:
A theoretical liquid has an enthalpy of fusion of −15.7 kJ/mol and an entropy of fusion of −142 J/mol · K. What is the freezing point of this liquid?

SOLUTION:
Using the above equation, we calculate T to be:

$$T = \frac{\Delta H}{\Delta S} = \frac{-15700\ ^{J}/_{mol}}{-142\ ^{J}/_{mol \cdot K}} = 111\,K = -163\ ^{\circ}\mathrm{C}$$

Workbook Problem 11.3

If the enthalpy of fusion of ammonia is 5.97 kJ/mol, and the normal freezing point of ammonia is −107 °C, what is the change in entropy associated with the freezing of ammonia?

Strategy: Rearrange the equation $\Delta G = \Delta H - T\Delta S$ to solve for the entropy change.

Step 1: Rearrange to solve for ΔS.

Step 2: Solve for ΔS, being careful to maintain the correct sign.

POST-LECTURE:

Work Problems: *One of the best ways to learn chemistry is through solving problems. Go to the textbook and work through the Worked Examples for this section, then go to the end of chapter and work the problems there. When you feel confident that you can correctly answer problems like the ones listed below for each objective, check off each of the Learning Objectives for this section above.*

LO 11.4: Worked Example 11.2; Problems 11.48–11.57

Section 11.4 Energy Changes during Phase Transitions

PRE-LECTURE:

Learning Objectives: *These are the skills you should be proficient in after reading this section and practicing the associated problems. Check off the boxes below once you have gone through this section, answered the associated questions in your textbook, and feel confident you have mastered the skill.*

- ☐ **LO 11.5** Draw or interpret a heating curve for a substance.
- ☐ **LO 11.6** Calculate the amount of heat associated with temperature and phase changes of substances.

Definitions: *To learn, you must do. This is your workbook. Fill in the definitions from the glossary of the textbook as a first step to understanding the material.*

— **heat of fusion (ΔH_{fusion})**— ____

____

Summary: *Read through this section and fill in the missing words and equations. Reference the textbook when you are not confident in your answers.*

Heating curves are graphical representations of phase changes. The vertical axis shows __________, and the horizontal axis shows the __________ added. In heating curves, we see that the temperature increases up to the normal melting point, then it stops increasing as the phase change occurs. The same happens at the boiling point.

Heating curves have a number of notable features. The slopes when solids are being warmed to their melting point, liquids are being warmed to their boiling point, and gases are being heated are equivalent to the molar heat capacities of these phases. The amount of kJ/mol that is added in the times that the temperature does not change is the heat of fusion (melting) and the heat of vaporization.

The heat of vaporization will always be greater than the heat of fusion: the flat line that occurs while a liquid is being vaporized is always longer than the flat portion of the line that corresponds to melting. This is due to the fact that in melting, the molecules only need to overcome the intermolecular forces enough to be able to move around each other, while in vaporization, they must completely overcome the intermolecular force so the molecules can be fully separated from one another.

POST-LECTURE:

Work Problems: *One of the best ways to learn chemistry is through solving problems. Go to the textbook and work through the Worked Examples for this section, then go to the end of chapter and work the problems there. When you feel confident that you can correctly answer problems like the ones listed below for each objective, check off each of the Learning Objectives for this section above.*

LO 11.5: Problems 11.42, 11.44, 11.50
LO 11.6: Worked Example 11.3; Problems 11.46, 11.48

Section 11.5 Phase Diagrams

PRE-LECTURE:

Learning Objectives: *These are the skills you should be proficient in after reading this section and practicing the associated problems. Check off the boxes below once you have gone through this section, answered the associated questions in your textbook, and feel confident you have mastered the skill.*

- ☐ **LO 11.7** Use a phase diagram to identify the phase of a substance at a given temperature and pressure or the phase transition that occurs when the temperature and pressure are changed.
- ☐ **LO 11.8** Sketch a phase diagram given the appropriate data.

Definitions: *To learn, you must do. This is your workbook. Fill in the definitions from the glossary of the textbook as a first step to understanding the material.*

— **phase diagram—** ____

____

— **triple point—** ____

____

— **normal melting point—** ____

____

— **critical point—** ____

____

— **supercritical fluid—** ____

____

Summary: *Read through this section and fill in the missing words and equations. Reference the textbook when you are not confident in your answers.*

Temperature can cause a spontaneous phase change, but pressure can as well. Graphs of pressure (vertical axis) versus temperature (horizontal axis) are called **phase diagrams.**

The lines on a phase diagram indicate interfaces between ___________. It is fairly straightforward to remember which phase is which by thinking of water, a substance familiar in all three phases. As the temperature increases, the substance first becomes ___________, then ___________.

A similar pattern is seen on the pressure axis: at the highest pressures, most substances are ___________. The very few exceptions are substances, such as water and gallium, in which the liquid form is more dense than the solid. These are described with a backward-sloping line between the solid and liquid phases, which indicates that increasing pressure favors the liquid form.

There are a few features of phase diagrams that should be noted.

- *triple point*: this is the temperature and pressure at which all three phases exist in equilibrium.
- *critical point*: this is the temperature and pressure above which gases and liquids behave indistinguishably as *supercritical fluids*: gas and liquid phases have the same density and are miscible.
- *critical pressure*: this is the pressure above which a liquid cannot be vaporized.
- *critical temperature*: this is the temperature above which a gas cannot be liquefied.
- *normal melting/boiling point*: this is the phase transition at 1 atm pressure.

POST-LECTURE:

Work Problems: *One of the best ways to learn chemistry is through solving problems. Go to the textbook and work through the Worked Examples for this section, then go to the end of chapter and work the problems there. When you feel confident that you can correctly answer problems like the ones listed below for each objective, check off each of the Learning Objectives for this section above.*

LO 11.7: Worked Example 11.4; Problems 11.16, 11.17, 11.55, 11.61
LO 11.8: Problems 11.56, 11.57, 11.62

Section 11.6 Liquid Crystals

PRE-LECTURE:

Learning Objectives: *These are the skills you should be proficient in after reading this section and practicing the associated problems. Check off the boxes below once you have gone through this section, answered the associated questions in your textbook, and feel confident you have mastered the skill.*

☐ **LO 11.9** Predict which molecules will exhibit a liquid crystal phase based on molecular structure.

☐ **LO 11.10** Classify the types of intermolecular forces between molecules in a liquid crystal.

☐ **LO 11.11** Match a molecular image with different states of matter, including solid, liquid, and common liquid crystal phases.

Definitions: *To learn, you must do. This is your workbook. Fill in the definitions from the glossary of the textbook as a first step to understanding the material.*

— **liquid crystal—** ____________________

____

— **nematic liquid crystal phase—** ____________________

____

— **smectic liquid crystal phase—** ____________________

____

— **cholesteric liquid crystal phase—** ____

____

— **anisotropic properties—**____

____

Summary: *Read through this section and fill in the missing words and equations. Reference the textbook when you are not confident in your answers.*

A liquid crystal has particles that are ordered like a solid, but that can still flow like those in a ___________. There are a variety of ways the particles can align. In the ___________ liquid crystal phase, the long ends of the crystals align. In the ___________ liquid crystal phase, the long axes are parallel and the ends are aligned. In the ___________ liquid crystal phase, the molecules are arranged in layers in which the long axes are slightly rotated to the layers above and below it. The properties of the liquid depends on the arrangement of its particles.

Self-Test

This section is intended to test your knowledge of the material covered in this chapter. Think through these problems and make certain you understand what the problem is about—ask yourself if the answer makes sense. Many of these questions are linked to the chapter learning goals. Therefore, successful completion of these problems indicates that you have achieved the learning goals for this chapter. You will receive the greatest benefit from this section if you use it as a mock exam. Do not use your notes, book, Internet, or friends when taking the test. There is no grade and you only shortchange yourself by getting help with this activity. As you take the test, mark the problems with your level of confidence with the answer.

★ *Use a star for problems you know you got right.*
? *Use a question mark for those you think you got right.*
x *Use an x for those you are guessing on.*

If you get a question right that you marked with a star, you can be confident that you know the associated LO.

If you get a question right that you marked with a question mark, go back and work a few more problems and test yourself again tomorrow.

If you get a question wrong or mark it with an x, go back and read the section again, work a few problems today, a few tomorrow, and a few the next day before testing yourself again.

This will help you to discover which topics you have learned thoroughly and which topics you need to study in more detail.

True–False

1. Freezing causes a positive entropy change and a negative enthalpy change.

2. Vapor pressure does not vary with temperature.

3. Anisotropic properties depend on the temperature of the liquid.

4. A supercritical fluid is extra slippery.

5. It always takes more energy to vaporize a substance than it does to melt it.

6. Heat of vaporization is always greater than the heat of fusion for the same substance.

7. Increasing pressure increases the boiling point of a liquid.

8. Stronger intermolecular forces lead to lower vapor pressures.

9. Some bugs can walk on water because of viscosity.

Matching

10. critical point
11. deposition
12. liquid crystal
13. normal boiling point
14. sublimation
15. triple point
16. vapor pressure
17. viscosity

a. temperature and pressure at which the solid, liquid, and gas phases of the same substance are at equilibrium

b. point at which any increase in temperature or pressure produces a supercritical fluid

c. resistance to flow

d. the process in which a solid directly becomes a gas

e. substance with particles that are ordered like a solid, but that flow around one another

f. temperature at which the phase spontaneously changes from liquid to gas at 1 atm pressure

g. the process in which a gas directly becomes a solid

h. the partial pressure of a gas in equilibrium with a liquid in a closed container.

Multiple Choice

18. Which of the following liquids has the highest boiling point?
   a. $CH_3CH_2CH_3$
   b. $CH_3(CH_2)_2CH_3$
   c. $CH_3(CH_2)_4CH_3$
   d. $CH_3(CH_2)_6CH_3$

19. Which of the following liquids has the lowest vapor pressure?
   (A) CH_3CH_3 (B) CH_3OCH_3 (C) CH_3CH_2OH
   a. A
   b. B
   c. C
   d. B and C will have comparable values

20. If the external pressure is more than 1 atm, a liquid boils at
    a. the normal boiling point.
    b. a temperature below the normal boiling point.
    c. a temperature above the normal boiling point.
    d. there is not enough information to say

21. The phase change where a liquid directly becomes a solid.
    a. deposition
    b. fusion
    c. sublimation
    d. vaporization

22. Which liquid crystal phase occurs when the long axes of the molecules are aligned and the ends line up?
    a. nematic
    b. smectic
    c. cholesteric
    d. anisotropic

23. The point at which the phase transition from liquid to solid occurs at 1 atm is
    a. the normal freezing point
    b. the triple point
    c. the critical point
    d. the critical pressure

Fill in the Blanks

24. Phase changes have an associated change in free energy. The _______________ portion of this change is involved in disrupting or forming intermolecular forces.

25. In a heating curve, the temperature increase stops during ___________ _____________.

26. In a phase diagram, the phase change represented by the line between solid and gas is called _______________________.

27. As temperature decreases, vapor pressure _______________________, and viscosity and surface tension ___________________.

Problems

28. What is the most important intermolecular force in the following molecules?
    $CH_3CH_2CH_2OH$ $\quad$ $CH_3(CH_2)_4CH_3$ $\quad$ PCl_3

29. Which of the following would you expect to have the highest boiling point and why?
    a. Br_2 or Cl_2 $\quad$ b. CH_3OH or CH_3CH_2OH $\quad$ c. CH_4 or CH_3OH

30. For platinum, $\Delta H_{vap} = 565.3$ kJ/mol and $\Delta S_{vap} = 150.8$ J/ mol · K. What is the normal boiling point of platinum?

31. If 0.237 g of water condensed on a 50.0 g block of iron at 25.0 °C, what would the final temperature of the iron be? $\Delta H_{vap} = 40.7$ kJ/mol, $c_{iron} = 0.449$ J/g · °C.

32. If a prankster were to take a 45-g ice cube out of a household freezer at −20 °C and drop it down the back of an unsuspecting sibling, how much heat would be absorbed from the hapless sibling if she decided to pretend she did not notice? Normal body temperature is 37 °C, the molar heat capacity of ice is 36.57 J/mol · °C, ΔH_{fus} = 6.01 kJ/mol, and the specific heat of water is 4.184 J/g · °C.

33. The normal boiling point of benzene is 80.1 °C and ΔH_{vap} = 30.8 kJ/mol. What is the vapor pressure of benzene at 75.0 °C?

34. For the phase diagram shown, identify the normal freezing point, the triple point, the critical point, and the lines that indicate boiling, sublimation, and melting. What does the slope of the central line indicate about the substance whose data is shown?

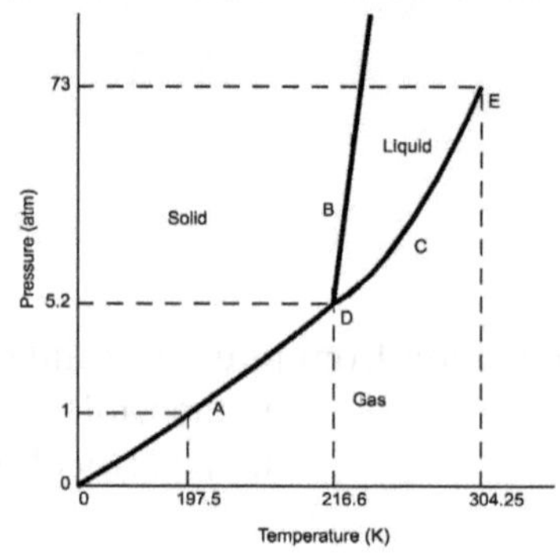

Challenge Problem

35. How many grams of water can be taken from −20 °C to 115 °C by burning 320 g of propane (C_3H_8)? Assume that all of the energy of combustion goes into the water.

Necessary information:

ΔH_{comb} for propane	= 2220 kJ/mol	Specific heat for ice	= 2.03 J/g · °C
Specific heat for water	= 4.184 J/g · °C	Specific heat of steam	= 2.08 J/g · °C
ΔH_{fus} for water	= 6.01 kJ/mol	ΔH_{vap} for water	= 40.67 kJ/mol

Test Yourself: *Take the Practice Test on pages 442–444 of your textbook. Again, use a testing environment using only those items that your instructor will allow you to have access to during the test and mark each question with a star, question mark, or x to denote your confidence level. Once you have finished you can check off the LO at the beginning of this chapter based on which questions you starred and got correct. Go back and spend more time with the topics from the other learning objectives.*

CHAPTER TWELVE

Solids and Solid-State Materials

Learning Objectives

As a result of reading and studying this chapter, you should be able to:

- ☐ **LO 12.1** Classify types of solids based on their chemical composition and properties.
- ☐ **LO 12.2** Use the Bragg equation to calculate spacing between atomic layers in a crystal.
- ☐ **LO 12.3** Identify the four kinds of spherical packing arrangements in crystalline solids and the three kinds of cubic unit cells.
- ☐ **LO 12.4** Calculate the density of a substance, atomic radii of its atoms, or molecular mass given its unit cell dimensions.
- ☐ **LO 12.5** Use an image of a unit cell to determine the formula and geometry of ionic compounds.
- ☐ **LO 12.6** Describe the electron-sea model of metals, and explain how it accounts for their properties.
- ☐ **LO 12.7** Describe the molecular orbital theory for metals (band theory), draw MO energy-level diagrams for metals, and use band theory to account for their properties.
- ☐ **LO 12.8** Draw MO diagrams for insulators, semiconductors, and conductors. Use band theory to explain why different substances have different conductivities.
- ☐ **LO 12.9** Classify doped semiconductors as *n*-type or *p*-type and draw their MO diagrams.
- ☐ **LO 12.10** Predict relative band gap energies based on periodic trends in atomic size and electronegativity.
- ☐ **LO 12.11** Relate the wavelength of light emitted or absorbed by a semiconductor to the band gap in its band theory energy-level diagram.
- ☐ **LO 12.12** Describe some applications of semiconductors.
- ☐ **LO 12.13** Describe the appearance of a plot of electrical resistance versus temperature for a superconductor.
- ☐ **LO 12.14** Describe the bonding in ceramics and account for their properties in terms of the bonding model.
- ☐ **LO 12.15** Describe the synthesis of ceramics, and write balanced equations for the reactions that occur in the sol–gel synthesis method.

Chapter Summary

Solids take a number of forms. Those that are crystalline can be characterized by X-ray crystallography, which provides data about unit cells. This information can be used to gain insights into packing and allows for calculations of atomic radius and density. The properties of solids, such as melting points and hardness, depend on the forces holding the particles together. Whether a substance acts an insulator, semiconductor, or metal depends on how easily electrons can flow within it.

The Chapter in Detail

Section 12.1 Types of Solids

PRE-LECTURE:

Learning Objectives: *These are the skills you should be proficient in after reading this section and practicing the associated problems. Check off the boxes below once you have gone through this section, answered the associated questions in your textbook, and feel confident you have mastered the skill.*

☐ **LO 12.1** Classify types of solids based on their chemical composition and properties.

Definitions: *To learn, you must do. This is your workbook. Fill in the definitions from the glossary of the textbook as a first step to understanding the material.*

— **crystalline solid**— ____

____

— **amorphous solid**— ____

____

— **ionic solid**— ____

____

— **molecular solid**— ____

____

— **covalent network solid**— ____

____

— **metallic solid**— ____

____

Review: *Before going through the section make sure you are familiar with these concepts:*

- Section 2.11 molecules
- Section 2.12 ions
- Section 8.6 intermolecular forces
- Section 8.6 relative strength of intermolecular forces

Summary: *Read through this section and fill in the missing words and equations. Reference the textbook when you are not confident in your answers.*

Amorphous solids have randomly arranged constituent particles. Crystalline solids have an ordered, long-range structure. This order is visible macroscopically as sharp edges and flat faces.

Fill in the chart below about the four types of crystalline solids are:

Type of Solid	Intermolecular Forces	Properties
Ionic		
Molecular		
Covalent Network		
Metallic		

Workbook Example 12.1

PROBLEM:
Classify the type of solid for the following examples: MgO, $C_{10}H_8$(naphthalene), SiO_2(quartz), and Au.

SOLUTION:
MgO: Since it is composed of the ions Mg^{2+} and O^{2-}, it is held together by ionic bonds and is thus an ionic solid.
$C_{10}H_8$: Discrete molecules held together by intermolecular forces such as dispersion forces, dipole–dipole interactions, or hydrogen bonds are molecular solids.
SiO_2: Substances that have a long-range order held together by covalent bonds are covalent network solids.
Au: As a metal, it will form a metallic solid.

POST-LECTURE:

Work Problems: *One of the best ways to learn chemistry is through solving problems. Go to the textbook and work through the Worked Examples for this section, then go to the end of chapter and work the problems there. When you feel confident that you can correctly answer problems like the ones listed below for each objective, check off each of the Learning Objectives for this section above.*

LO 12.1: Problems 12.28–12.31

Section 12.2 Probing the Structure of Solids: X-Ray Crystallography

PRE-LECTURE:

Learning Objectives: *These are the skills you should be proficient in after reading this section and practicing the associated problems. Check off the boxes below once you have gone through this section, answered the associated questions in your textbook, and feel confident you have mastered the skill.*

☐ **LO 12.2** Use the Bragg equation to calculate spacing between atomic layers in a crystal.

Definitions: *To learn, you must do. This is your workbook. Fill in the definitions from the glossary of the textbook as a first step to understanding the material.*

— **diffraction—** ____

____

— **Bragg equation—** ____

____

Summary: *Read through this section and fill in the missing words and equations. Reference the textbook when you are not confident in your answers.*

It is a principle of optics that the wavelength of light must be no more than twice the length of an object for it to be visible. Atoms are too small to see with visible light, and so, the short wavelengths of X-rays are used to "see" atoms.

When X-rays are directed at a crystal with regularly spaced atoms, the electromagnetic radiation is bent, or diffracted. The diffracted X-rays can interact constructively or destructively giving rise to light and dark spots on a detector. The diffraction pattern can then be interpreted to provide data about interatomic distances.

- __________ interference—wave peaks and troughs line up
  — results in an amplified wave
- __________ interference—wave peaks and troughs do not line up
  — cancels the waves

The basis for this interpretation is the Bragg equation, for which a father and son shared the Nobel Prize in physics in 1915:

$$d = \qquad$$

The wavelength, λ, is known; the angle of reflection, θ, can be measured, and n is a small whole number, usually 1. Use d, λ, and θ to fill in the five boxes in the figure below.

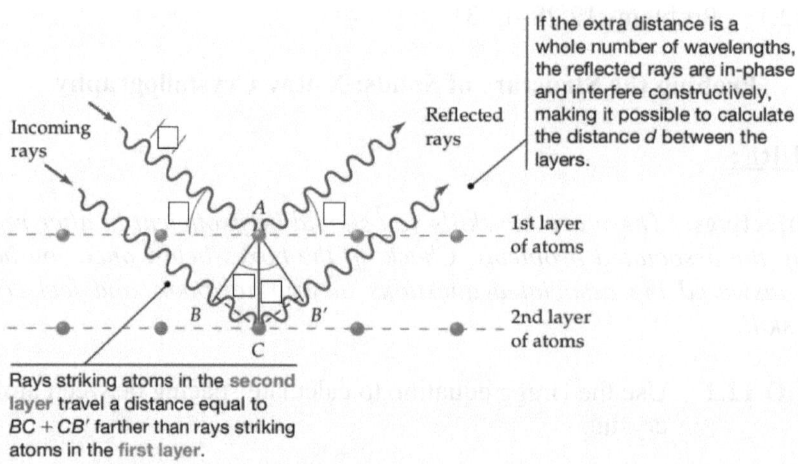

Computer-controlled diffractometers are now used for both data collection and analysis, making it possible to investigate the structures of even the largest macromolecules.

POST-LECTURE:

Work Problems: *One of the best ways to learn chemistry is through solving problems. Go to the textbook and work through the Worked Examples for this section, then go to the end of chapter and work the problems there. When you feel confident that you can correctly answer problems like the ones listed below for each objective, check off each of the Learning Objectives for this section above.*

LO 12.2: Problems 12.32, 12.33

Section 12.3 The Packing of Spheres in Crystalline Solids: Unit Cells

PRE-LECTURE:

Learning Objectives: *These are the skills you should be proficient in after reading this section and practicing the associated problems. Check off the boxes below once you have gone through this section, answered the associated questions in your textbook, and feel confident you have mastered the skill.*

☐ **LO 12.3** Identify the four kinds of spherical packing arrangements in crystalline solids and the three kinds of cubic unit cells.

Definitions: *To learn, you must do. This is your workbook. Fill in the definitions from the glossary of the textbook as a first step to understanding the material.*

— **simple cubic packing—** ____________________

____

— **coordination number—** ____________________

____

— **body-centered cubic packing—** ____________________

____

— **hexagonal closest-packing—** ____________________

____

— **cubic closest-packing—** ____________________

____

— **unit cell—** ____________________

____

— **primitive-cubic unit cell—** ____________________

____

— **body-centered cubic unit—** ______________________________

____

— **face-centered cubic unit cell—** ______________________________

____

Summary: *Read through this section and fill in the missing words and equations. Reference the textbook when you are not confident in your answers.*

Particles pack as closely as possible to maximize intermolecular forces. The coordination number is the number of atoms touching the atom being considered. Unit cells are the small repeating units of a crystal. There are four ways for spheres to pack. Fill in the chart below.

Structure	**Stacking Pattern**	**Coordination Number**	**Space Used (%)**	**Unit Cell**
simple cubic				
body-centered cubic				
hexagonal closest-packing				
cubic closest-packing				

- **simple cubic packing:** each atom touches six others, 1/8 of eight atoms are found on the corners of the unit cell. There is one atom per unit cell with an atomic diameter = edge length.
- **body-centered cubic packing:** the central atom has four neighbors above and four below. There are two atoms per unit cell.
- **hexagonal closest-packing:** this is a noncubic arrangement with the atoms arranged in a hexagonal pattern in two offset layers.
- **cubic closest-packing:** three different hexagonally arranged layers. Atomic radius = $\sqrt{\frac{d^2}{8}}$ where d is the edge length of a unit cell.

Workbook Example 12.2

PROBLEM:
Aluminum has a face-centered cubic unit cell. If the edge of a unit cell is 404.5 pm, what is the density of aluminum in g/cm^3?

SOLUTION:
A face-centered cubic unit cell has a total of four atoms.

Based on this, it is possible to calculate the mass of a unit cell from the molar mass of aluminum and Avogadro's number:

$$4\ atoms\ Al \times \frac{1\ mol\ Al}{6.022\times10^{23}\ atoms\ Al} \times \frac{26.98\ g\ Al}{1\ mole\ Al} = 1.792\times10^{-22}\ g\ Al$$

The volume of a unit cell can be calculated from the edge length, being sure to convert from picometers to centimeters:

$$\left(404.5\times10^{-12}\ m \times \frac{100\ cm}{1\ m}\right)^3 = 6.618\times10^{-23}\ cm^3$$

Since we know the mass and the volume, we can calculate the density:

$$d = \frac{m}{V} = \frac{1.792\times10^{-22}\ g}{6.618\times10^{-23}\ cm^3} = 2.71\ {}^{g}\!/_{cm^3}$$

This agrees closely with the calculated density of aluminum.

Workbook Example 12.3

PROBLEM:
Nickel has a face-centered cubic unit cell. If the edge of a unit cell is 352.4 pm, what is the diameter of a nickel atom in pm?

SOLUTION:
The side of a face-centered cubic unit cell has one full atom and two radii as the diagonal, making the diagonal equal to 4*r* or 2*d*.

Using the Pythagorean theorem:

$$(352.4\times10^{-12}\ m)^2 + (352.4\times10^{-12}\ m)^2 = (2d)^2$$

$$2.484\times10^{-19}\ m^2 = 4d^2$$

$$\frac{2.484\times10^{-19}\ m^2}{4} = d^2$$

$$\sqrt{6.209\times10^{-20}\ m^2} = d$$

$$2.492\times10^{-10}\ m = d$$

The diameter of a nickel atom is approximately 249 pm.

Workbook Problem 12.1

Polonium has a density of 9.3 g/cm^3 and a simple cubic structure. Estimate the radius of a polonium atom.

Strategy: Use the number of atoms per unit cell and the molecular mass to determine the mass of a unit cell, then use the density to solve for the volume of a cell. With the volume, determine the length of one side, then determine the atomic radius.

Step 1: Determine the mass of a unit cell.

Step 2: Solve for the volume of a unit cell.

Step 3: Determine the edge length of a unit cell.

Step 4: Determine the atomic radius.

POST-LECTURE:

Work Problems: *One of the best ways to learn chemistry is through solving problems. Go to the textbook and work through the Worked Examples for this section, then go to the end of chapter and work the problems there. When you feel confident that you can correctly answer problems like the ones listed below for each objective, check off each of the Learning Objectives for this section above.*

LO 12.3: Problem 12.18

Section 12.4 Structures of Some Ionic Solids

PRE-LECTURE:

Learning Objectives: *These are the skills you should be proficient in after reading this section and practicing the associated problems. Check off the boxes below once you have gone through this section, answered the associated questions in your textbook, and feel confident you have mastered the skill.*

☐ **LO 12.4** Calculate the density of a substance, atomic radii of its atoms, or molecular mass given its unit cell dimensions.

Review: *Before going through the section make sure you are familiar with these concepts:*

- Section 1.7 density
- Section 3.3 molecular weight
- Section 12.3 unit cell

Summary: *Read through this section and fill in the missing words and equations. Reference the textbook when you are not confident in your answers.*

Like the metals discussed above, when ionic solids crystallize, they adopt a unit cell structure. Unlike metals, these unit cells must accommodate ions of different sizes: cations are ___________ than their neutral atoms, and anions are ___________. The unit cell of an ionic solid is always electrically ___________.

POST-LECTURE:

Work Problems: *One of the best ways to learn chemistry is through solving problems. Go to the textbook and work through the Worked Examples for this section, then go to the end of chapter and work the problems there. When you feel confident that you can correctly answer problems like the ones listed below for each objective, check off each of the Learning Objectives for this section above.*

LO 12.4: Worked Examples 12.1, 12.2; Problems 12.36, 12.38, 12.42, 12.44

Section 12.5 Structures of Some Covalent Network Solids

PRE-LECTURE:

Learning Objectives: *These are the skills you should be proficient in after reading this section and practicing the associated problems. Check off the boxes below once you have gone through this section, answered the associated questions in your textbook, and feel confident you have mastered the skill.*

☐ **LO 12.5** Use an image of a unit cell to determine the formula and geometry of ionic compounds.

Definitions: *To learn, you must do. This is your workbook. Fill in the definitions from the glossary of the textbook as a first step to understanding the material.*

— **allotrope**—____

____

Summary: *Read through this section and fill in the missing words and equations. Reference the textbook when you are not confident in your answers.*

Covalent network solids are comprised of a giant three-dimensional array of ___________ bonded atoms. Since covalent bonds are so strong, these solids are hard and have ________ melting points.

Carbon has more than 40 allotropes—different structural forms of the element with varying physical and chemical properties. Most are amorphous, but there are three famous exceptions:

- *graphite*: the most common and most stable allotrope, graphite consists of two-dimensional sheets of fused, sp^2-hybridized, six-membered rings. Graphite is the "lead" in pencils, an electrode in batteries, and a lubricant for locks. All of these functions rely on the fact that the layers of graphite can slide over one another when air and water adsorb onto them. A closely related structure is carbon *nanotubes*, which are like sheets of graphite rolled into tubes. They are being investigated as structural composites because they have 50 to 60 times the tensile strength of steel.
- *diamond*: a covalent network solid in which sp^3-hybridized carbon atoms bond to one another in a vast network. Diamond is the hardest known substance, and because of this, it is used in industrial saw blades and drill bits, as well as in jewelry. It is an electrical insulator and has a melting point over 3550 °C.
- *fullerene*: discovered in 1985 as a component of soot, fullerene is a soccer ball-shaped molecule of 60 sp^2-hybridized carbon atoms in alternating hexagons and pentagons. When reacted with rubidium, a superconducting material called rubidium fulleride (Rb_3C_{60}) is formed.

Silicon has a similar ability to form network solids. It forms four single bonds with oxygen that are used to bridge to other silicon atoms. Silica has the empirical formula SiO_2. Most minerals and rocks are formed by silicates: these account for 75% of the earth's crust by mass.

When silica is heated above 1600 °C, many of the Si–O bonds break, thus converting the silica into a viscous liquid. When it cools, the bonds reform into a random arrangement and an amorphous solid—quartz glass—is formed. Addition of transition metal ions lends color, and the addition of B_2O_3 adds heat resistance. This *borosilicate* glass is sold as Pyrex.

POST-LECTURE:

Work Problems: *One of the best ways to learn chemistry is through solving problems. Go to the textbook and work through the Worked Examples for this section, then go to the end of chapter and work the problems there. When you feel confident that you can correctly answer problems like the ones listed below for each objective, check off each of the Learning Objectives for this section above.*

LO 12.5: Worked Example 12.3; Problems 12.19, 12.20

Section 12.6 Bonding in Metals

PRE-LECTURE:

Learning Objectives: *These are the skills you should be proficient in after reading this section and practicing the associated problems. Check off the boxes below once you have gone through this section, answered the associated questions in your textbook, and feel confident you have mastered the skill.*

- ☐ **LO 12.6** Describe the electron-sea model of metals, and explain how it accounts for their properties.
- ☐ **LO 12.7** Describe the molecular orbital theory for metals (band theory), draw MO energy-level diagrams for metals, and use band theory to account for their properties.

Definitions: *To learn, you must do. This is your workbook. Fill in the definitions from the glossary of the textbook as a first step to understanding the material.*

— **electron-sea model—**____

____

— **band—** ____

____

— **band theory—** ____

____

Review: *Before going through the section make sure you are familiar with these concepts:*

- Section 2.12 cations
- Section 5.12 valence electrons
- Section 8.7 MO theory
- Section 5.10 degenerate energy levels

Summary: *Read through this section and fill in the missing words and equations. Reference the textbook when you are not confident in your answers.*

Metals have some universal characteristics—they are ___________, ___________, ___________, and good ___________ of heat and electricity.

Two theoretical models are used to explain these properties: the *electron-sea model* and the *molecular orbital theory.*

Electron-Sea Model of Metals

Metals do not have enough valence electrons to bond to adjacent molecules; their crystal structures contain delocalized electrons that belong to the structure as a whole. This acts as an electrostatic glue that holds the metal cations together.

This is a simple qualitative explanation for the properties of metals: electrons are mobile and therefore free to move from a negative electrode to a positive electrode when subjected to an electrical potential. Mobile electrons are also able to carry kinetic energy (heat) from one part of the crystal to another. Metals are malleable and ductile because no local bonds are broken when the solid is deformed. The more valence electrons there are, the harder the metal.

Molecular Orbital Theory for Metals

The number of molecular orbitals (MOs) formed is the same as the number of atomic orbitals combined. As the number of atoms increases, the difference in energy between successive MOs __________. Since metals form an extended lattice of a very large number of atoms, the orbitals merge into an almost __________ band of energy levels.

For this reason, molecular orbital theory in metals is referred to as **band theory.** The bottom half of the band consists of ______ bonding MOs, while the top half consists of _______ antibonding MOs.

Energy levels within a band occur in pairs and are said to be degenerate. In the absence of a current, electrons are equally likely to be traveling in one direction as another. One set of energy levels applies to electrons moving one way and the other applies to electrons moving in the opposite direction within the metal. If an electrical potential is applied, the electrons moving toward the positive terminal are accelerated, while those moving toward the negative terminal slow down, and, if they are going slow enough, may even end up changing directions. When this happens, there are more electrons moving toward the positive terminal than the negative, and there is a net electric current.

If there are no vacant orbitals in the band, there are no orbitals available for electrons to pass through, which make the material an electrical insulator.

Main group metals have *s* and *p* subshells that are sufficiently close in energy that the *s* and *p* bands overlap, which results in a partially filled composite band and electrical conductivity.

Transition metals have overlapping *d* and *s* bands, which allows six MOs per metal atom, three bonding and three antibonding. As a result, the metals with six valence electrons (bonding orbitals completely filled) are the__________ and have the __________ melting points.

Workbook Problem 12.2

Rank Mo, Cd, and Y in order of their expected melting points based on band theory.

Strategy: Determine the number of bonding and antibonding electrons; rank in order by maximum bonding electrons and minimum antibonding electrons.

Step 1: Determine the numbers of electrons of each type.

Step 2: Rank in order of expected bonding strength.

POST-LECTURE:

Work Problems: *One of the best ways to learn chemistry is through solving problems. Go to the textbook and work through the Worked Examples for this section, then go to the end of chapter and work the problems there. When you feel confident that you can correctly answer problems like the ones listed below for each objective, check off each of the Learning Objectives for this section above.*

LO 12.6: Problems 12.60–12.64
LO 12.7: Worked Example 12.4; Problems 12.21, 12.68–12.72

Section 12.7 Semiconductors

PRE-LECTURE:

Learning Objectives: *These are the skills you should be proficient in after reading this section and practicing the associated problems. Check off the boxes below once you have gone through this section, answered the associated questions in your textbook, and feel confident you have mastered the skill.*

- ☐ **LO 12.8** Draw MO diagrams for insulators, semiconductors, and conductors. Use band theory to explain why different substances have different conductivities.
- ☐ **LO 12.9** Classify doped semiconductors as *n*-type or *p*-type and draw their MO diagrams.
- ☐ **LO 12.10** Predict relative band gap energies based on periodic trends in atomic size and electronegativity.
- ☐ **LO 12.11** Relate the wavelength of light emitted or absorbed by a semiconductor to the band gap in its band theory energy-level diagram.

Definitions: *To learn, you must do. This is your workbook. Fill in the definitions from the glossary of the textbook as a first step to understanding the material.*

— **semiconductor**— ______________________________

____

— **valence band**— ______________________________

____

— **conduction band**— ______________________________

____

— **band gap**— ______________________________

____

— **doping**— ______________________________

____

— ***n*-type semiconductor**— ______________________________

____

— ***p*-type semiconductor—** ____

____

Summary: *Read through this section and fill in the missing words and equations. Reference the textbook when you are not confident in your answers.*

A semiconductor has electrical conductivity between that of a ___________ and that of an ___________. In metals, there is no energy gap between the highest occupied orbitals and the lowest unoccupied orbital: electrons can move freely between these energy levels. In insulators, there is a large energy gap between the highest occupied orbitals (the valence band) and the lowest unoccupied orbital (the conduction band): electrons cannot move between these energy levels.

The difference in energy between the valence and conduction bands is the **band gap.** Metals have no band gap, insulators have an insurmountable band gap, and semiconductors fall between the two.

In a semiconductor, a few electrons will have the energy that is needed to jump across the band gap. Conductivity increases with increasing temperature as the number of high-energy electrons increases. This is the opposite of metals: as their temperature increases, conductivity drops, because the increased vibrational motion of the atoms disrupts the flow of electrons. This provides a convenient way to distinguish between the two types of material.

Doping is a process in which the conductivity of a semiconductor is increased by adding small amounts of ___________.

If an impurity is introduced that has more electrons than the semiconductor, the extra electrons will reside in the conduction band and increase conductivity. Because ___________ electrons are being added, these are called *n*-type semiconductors.

If an impurity is introduced that has fewer electrons than the semiconductor, there will be vacancies in the valence band that allow for increased conductivity. Because these can be thought of as adding ___________ holes in the valence band, these are *p*-type semiconductors.

Workbook Problem 12.3

Identify the following as *n*- or *p*-type semiconductors:

a. silicon doped with gallium
b. germanium doped with antimony
c. silicon doped with arsenic
d. germanium doped with phosphorus

Strategy: Determine the number of valence electrons in the semiconductor and compare it to the number of valence electrons in the dopant. Because all of the elements are main group elements, their group numbers can be used.

Step 1: Determine the numbers of valence electrons in each element.

Step 2:	Determine whether the dopant has more or fewer valence electrons than the semiconductor.
Step 3:	Identify the doped semiconductors as *n*- or *p*-type.

POST-LECTURE:

Work Problems: *One of the best ways to learn chemistry is through solving problems. Go to the textbook and work through the Worked Examples for this section, then go to the end of chapter and work the problems there. When you feel confident that you can correctly answer problems like the ones listed below for each objective, check off each of the Learning Objectives for this section above.*

LO 12.8: Problems 12.22–12.24, 12.78
LO 12.9: Worked Example 12.5; Problems 12.82, 12.84, 12.86
LO 12.10: Problems 12.98–12.101
LO 12.11: Problems 12.25, 12.96, 12.105

Section 12.8 Semiconductor Applications

PRE-LECTURE:

Learning Objectives: *These are the skills you should be proficient in after reading this section and practicing the associated problems. Check off the boxes below once you have gone through this section, answered the associated questions in your textbook, and feel confident you have mastered the skill.*

☐ **LO 12.12** Describe some applications of semiconductors.

Definitions: *To learn, you must do. This is your workbook. Fill in the definitions from the glossary of the textbook as a first step to understanding the material.*

Summary: *Read through this section and fill in the missing words and equations. Reference the textbook when you are not confident in your answers.*

Doped semiconductors are common in modern solid-state electronic devices.

Diodes

Diodes permit electron flow in one direction while ___________ electron flow in the other direction. They can be used to convert alternating current to direct current. They consist of a *p*-type semiconductor in contact with an *n*-type semiconductor to create what is known as a *p–n* junction.

If the *n*-type side is connected to the negative terminal of a battery, current will flow through the junction; the electrons can find their way into the conductance band and across the gap into the valence band. If the *p*-type side is connected to the negative terminal of a battery, the positive holes move toward the battery, and the electrons in the conductance band across the gap are repelled. In this case, no current will flow.

Light-Emitting Diodes (LEDs)
If there is a difference in energy levels between the conduction band and the valence band, electrons will be able to fall from one to the other, releasing energy as __________. The energy of the light emitted is roughly equivalent to the size of the band gap ($E_g = h\nu = hc/\lambda$). The band gap can be tuned by varying the composition of the semiconductor.

LEDs are smaller, brighter, longer lived, more energy efficient, and have faster switching times than incandescent bulbs.

Photovoltaic (Solar) Cells
Photovoltaic cells work in the opposite direction of an LED. Whereas an LED converts the energy difference of the band gap to light, a photovoltaic cell converts light to __________.

When light shines on the *p–n* junctions, the photons from the light excite electrons from the valence band of the *p*-type semiconductor into the conduction band of the *n*-type semiconductor.

If the photovoltaic cell is part of an electrical circuit, light can flow in, charging the system, and the resultant energy can be used to power a device.

Current cells are only 20% efficient, so research is focusing on increasing efficiency and reducing cost.

Transistors
Transistors consist of *n–p–n* or *p–n–p* junctions that control or amplify electrical signals in modern integrated circuits. Vast numbers of these devices can be packed into tiny spaces, thus increasing the speed and decreasing the size of modern electronics.

Workbook Problem 12.4

If an LED has a band gap of 178 kJ/mol, what is the wavelength of the resulting light?

Strategy: Convert the band gap energy to photon energy, and then to wavelength.

Step 1: Determine the energy of an individual photon.

Step 2: Convert the photon energy to wavelength.

POST-LECTURE:

Work Problems: *One of the best ways to learn chemistry is through solving problems. Go to the textbook and work through the Worked Examples for this section, then go to the end of chapter and work the problems there. When you feel confident that you can correctly answer problems like the ones listed below for each objective, check off each of the Learning Objectives for this section above.*

LO 12.12: Problems 12.92–12.94

Section 12.9 Superconductors

PRE-LECTURE:

Learning Objectives: *These are the skills you should be proficient in after reading this section and practicing the associated problems. Check off the boxes below once you have gone through this section, answered the associated questions in your textbook, and feel confident you have mastered the skill.*

☐ **LO 12.13** Describe the appearance of a plot of electrical resistance versus temperature for a superconductor.

Definitions: *To learn, you must do. This is your workbook. Fill in the definitions from the glossary of the textbook as a first step to understanding the material.*

— **superconductor—** ____

____

— **superconducting transition temperature, T_c—** ______________________

____

Summary: *Read through this section and fill in the missing words and equations. Reference the textbook when you are not confident in your answers.*

A **superconductor** is a material that loses all electrical ___________ below a characteristic temperature, called the **superconducting transition temperature** (T_c). Below that temperature, the material becomes a perfect conductor: once an electric current is started, it will flow indefinitely without loss of energy.

Superconductors were discovered in 1911, but only superconductors with relatively low temperatures were known. In 1986, $Ba_xLa_{2-x}CuO_4$ was demonstrated to have a T_c of 35 K, and soon after that, other copper-containing oxides were found to have even higher superconducting transition temperatures. The record at the time of this writing is 138 K.

This was unexpected, because most metal oxides, nonmetallic inorganic solids called *ceramics,* are insulators. In this field, experimental data exceeds the ability of scientists to explain it: there is no generally accepted theory of superconductivity in ceramic superconductors.

If a superconductor is cooled to below T_c and a magnet is lowered toward it, the magnet and superconductor repel each other, causing the magnet to levitate. Magnets induce a supercurrent in the superconductor that generates a magnetic field opposite to that of the magnet. This is called the *Meissner effect*, and it is already being used to operate a high-speed train in Shanghai, China.

Superconducting magnets are also used in MRI instruments and particle accelerators, but these uses currently require cooling with liquid helium, which is expensive and requires cryogenic equipment.

Once the T_c of a material goes above 77 K, it can be cooled with liquid nitrogen, an abundant refrigerant that is cheaper than milk.

Currently available high-temperature superconductors are brittle powders with high melting points, so they cannot be made easily into wires and coils needed for electrical equipment. However, superconducting thin films are being used as microwave filters, and superconducting wires as long as 1 km are now available.

High-temperature superconductors are also being developed based on fullerene (C_{60}) molecules. K_3C_{60} is a metallic conductor at room temperature and a superconductor at 18 K.

Fluorine-doped lanthanum oxide iron arsenide also shows promise, superconducting at 26 K, with substitution of the lanthanum by other elements, raising the temperature to 77 K.

POST-LECTURE:

Work Problems: *One of the best ways to learn chemistry is through solving problems. Go to the textbook and work through the Worked Examples for this section, then go to the end of chapter and work the problems there. When you feel confident that you can correctly answer problems like the ones listed below for each objective, check off each of the Learning Objectives for this section above.*

LO 12.13: Problem 12.111

Section 12.10 Ceramics and Composites

PRE-LECTURE:

Learning Objectives: *These are the skills you should be proficient in after reading this section and practicing the associated problems. Check off the boxes below once you have gone through this section, answered the associated questions in your textbook, and feel confident you have mastered the skill.*

- ☐ **LO 12.14** Describe the bonding in ceramics and account for their properties in terms of the bonding model.
- ☐ **LO 12.15** Describe the synthesis of ceramics, and write balanced equations for the reactions that occur in the sol–gel synthesis method.

Definitions: *To learn, you must do. This is your workbook. Fill in the definitions from the glossary of the textbook as a first step to understanding the material.*

— **ceramic—**____

— **advanced ceramic—** ____

____

— **sintering—** ____

____

— **sol–gel method—** ____

____

— **ceramic composite—** ____

____

— **ceramic–metal composite (cermet)—** ______________________________

____

— **ceramic–polymer composite—** ____________________________________

____

Summary: *Read through this section and fill in the missing words and equations. Reference the textbook when you are not confident in your answers.*

Ceramics are inorganic, ___________, ___________ solids, including both crystalline and amorphous materials, such as glasses. Traditional silicate ceramics are made by heating aluminosilicate clays to high temperatures, while **advanced ceramics**—materials that have high-tech engineering, electronic, and biomedical applications—include oxide ceramics such as alumina (Al_2O_3) and nonoxide ceramics such as silicon carbide and silicon nitride.

Oxide ceramics are named by adding an *–a* suffix in place of the *–um* of the metal.

Compared to metals, ceramics have ___________ melting points; they are stiffer, harder, and more resistant to wear and corrosion, maintain their strength at high temperatures, and are less dense than steel. This makes them attractive lightweight, high-temperature materials for replacing metal components in aircraft, space vehicles, and automobiles.

Nonoxide ceramics are covalent network solids with highly directional covalent bonds. This prevents planes of atoms from sliding over one another when the solid is subjected to the stress of a load or an impact. Because the solid cannot change shape as a metal would, the bonds instead give way. This brittleness is the primary drawback of these materials. The bonding is largely ionic in oxide ceramics, but the material is still brittle.

Ceramic processing, the series of steps that leads from raw material to the finished ceramic object, determines the strength and the resistance to fracture of the product.

Sintering is a process done below the melting point of a material, in which the particles of the powder are welded together without completely melting. During the process, crystal grains grow larger and the density of the material increases as the void spaces between particles disappear. High-purity, fine powders that are tightly compacted prior to sintering are needed to prevent impurities and leftover voids that lead to cracking and material failure.

In the **sol–gel method** for preparing high-purity fine powders, a metal oxide powder is synthesized from a metal alkoxide (a compound derived from a metal and an alcohol). This forms a colloidal dispersion called a *sol*, which consists of extremely fine particles that can then be dehydrated and linked with oxygen bridges. As this cross-linking occurs, the sol becomes a more rigid material called a *gel*.

Oxide ceramics have many uses. Alumina is used in spark plugs, dental crowns, and the heads of artificial hips as well as being the substrate for circuit boards. Silica is used to make the heat-resistant tiles that protect the space shuttle on re-entry into the atmosphere. Military armor is now also ceramic rather than steel, thus reducing its weight and allowing the armor to shatter and deflect bullets.

To counteract the brittleness of ceramic, a ceramic powder can be mixed, prior to sintering, with a second ceramic material to combine the properties of both components. This second compound is often in the form of *whiskers* (tiny fiber-shaped particles) and fibers.

Fibers and whiskers increase the strength and fracture-resistance of composite materials, because most of the chemical bonds are aligned along the fiber axis, which gives the fibers great strength. The fibers can deflect cracks, preventing them from moving cleanly in one direction, or bridging cracks to hold them together.

Composites can also be made of different materials, for example, **ceramic–metal composites,** or **cermets,** and **ceramic–polymer composites.** These materials have a high strength-to-weight ratio, which makes them ideal for aerospace and other applications.

Ceramic fibers used in composites are usually made by high-temperature methods. For example, carbon fiber is made by heating polyacrylonitrile to 1400–2500 °C, converting it to graphite. Silicon carbide chains are made using a similar method.

POST-LECTURE:

Work Problems: *One of the best ways to learn chemistry is through solving problems. Go to the textbook and work through the Worked Examples for this section, then go to the end of chapter and work the problems there. When you feel confident that you can correctly answer problems like the ones listed below for each objective, check off each of the Learning Objectives for this section above.*

LO 12.14: Problems 12.116, 12.118, 12.120
LO 12.15: Worked Example 12.6; Problems 12.126, 12.128

Self-Test

This section is intended to test your knowledge of the material covered in this chapter. Think through these problems and make certain you understand what the problem is about—ask yourself if the answer makes sense. Many of these questions are linked to the chapter learning goals. Therefore, successful completion of these problems indicates that you have achieved the learning goals for this chapter. You will receive the greatest benefit from this section if you use it as a mock exam. Do not use your notes, book, Internet, or friends when taking the test. There is no grade and you only shortchange yourself by getting help with this activity. As you take the test, mark the problems with your level of confidence with the answer.

★ *Use a star for problems you know you got right.*
? *Use a question mark for those you think you got right.*
x *Use an x for those you are guessing on.*

If you get a question right that you marked with a star, you can be confident that you know the associated LO.

If you get a question right that you marked with a question mark, go back and work a few more problems and test yourself again tomorrow.

If you get a question wrong or mark it with an x, go back and read the section again, work a few problems today, a few tomorrow, and a few the next day before testing yourself again.

This will help you to discover which topics you have learned thoroughly and which topics you need to study in more detail.

True–False

1. Covalent network solids are extremely large molecules.

2. X-rays can be used to determine the strength of intermolecular forces in liquids.

3. The coordination number of a packing arrangement is the number of atoms in the unit cell.

4. Fullerene and graphite are both isotopes of carbon.

5. The electron-sea model of metal bonding accounts for the superconductivity of some materials.

6. Valence band theory accounts for the different conductivities of metals and semiconductors.

7. Doping of semiconductors increases their conductivity.

8. Heating a semiconductor reduces its conductivity.

9. Diodes can convert alternating current to direct current.

10. Ceramics are ideal for aerospace applications because they are strong, malleable, and light.

Matching

11. allotrope
12. amorphous solid
13. band gap
14. band theory

a. solid in which the constituent particles are randomly arranged
b. solid held together by intermolecular forces
c. antibonding molecular orbitals in a semiconductor
d. solid in which the constituent particles have long-range order

15. ceramic		e.	different form of a pure element: can have different physical and chemical properties
16. conduction band		f.	molecular orbital theory for metals
17. crystalline solid		g.	material that has conductivity between that of metals and insulators
18. doping		h.	energy difference between bonding MOs in the valence band and the nonbonding MOs in the conductance band
19. electron-sea model		i.	process in which the particles of a powder are welded together without completely melting
20. molecular solid		j.	addition of a small amount of an impurity to increase the conductivity of a semiconductor
21. *p*-type semiconductor		k.	an inorganic, nonmetallic, nonmolecular solid
22. semiconductor		l.	material that loses all electrical resistance below a certain temperature
23. sintering		m.	model that visualizes metals as cations surrounded by delocalized electrons
24. sol–gel method		n.	temperature below which all electrical resistance is lost
25. superconductor		o.	semiconductor doped with a material that has fewer electrons than necessary for bonding
26. superconducting temperature		p.	bonding molecular orbitals in a semiconductor
27. valence band		q.	method of preparing ceramics that involves synthesis of metal oxide powder from metal alkoxide

Multiple Choice

28. A solid whose constituent particles are held together by charge interactions is a(n)
    a. ionic solid
    b. molecular solid
    c. covalent network solid
    d. metallic solid

29. The type of packing present when the unit cell contains a complete central atom is
    a. simple cubic packing
    b. body-centered cubic
    c. hexagonal closest packed
    d. cubic closest packed

30. Most of the allotropes of carbon are amorphous, but diamond, fullerene, and graphite are all
    a. ionic solids
    b. covalent network solids
    c. forms of silica
    d. metallic

31. Models of metallic bonding must account for metals
    a. malleability and ductility
    b. dull appearance
    c. electrical conductivity
    d. a and c

32. Semiconducting metals conduct electricity, but not efficiently. This is as a result of
    a. impurities
    b. a large band gap
    c. paramagnetism
    d. no available *p* orbitals

33. Doping silicon with which of the following would produce a *p*-type semiconductor?
    a. boron
    b. phosphorus
    c. nitrogen
    d. arsenic

34. Light-emitting diodes
    a. can be tuned to different wavelengths
    b. are energy efficient
    c. a and b
    d. none of the above

35. Superconductors
    a. are strongly magnetic
    b. operate best at high temperatures
    c. have no electrical resistance below T_c
    d. are easily made into wires

36. Most ceramics are
    a. malleable
    b. insulators
    c. shiny
    d. paramagnetic

Fill in the Blanks

37. A brittle, crystalline solid with sharp edges and a very high melting point is a(n) ____________________________. A relatively soft, low-melting-point crystalline solid is most likely a ____________________________, and an extremely hard, extremely high-melting-point solid is probably a ________________________________.

38. The electron-sea model of metal bonding accounts for the ______________________ and ____________________ of metals. These two abilities of metals to be shaped come from the fact that bonds and electrons are not ______________________. As a result, metals are not ______________: they will deform rather than shatter.

39. In a metal the bonding and antibonding orbitals are ____________________ in energy. However, in semiconductors and insulators a ____________________ gap separates the __________________ and ____________________ bands. This gap is ____________________ in insulators than in semiconductors.

40. Superconductors are fascinating, poorly understood materials with some interesting properties. ______________ the T_c, the material loses all electrical ____________________. The challenge for researchers is to find materials that can be cooled using liquid ________________, a common and inexpensive coolant that requires no specialized equipment.

41. Advanced ceramics have properties that make them superior to _______________ for many applications. They have a lower density than metals and so products made from ceramics are _______________.They are electrical _______________ and resistant to _______________. Their durability is limited in one respect: they are not _______________ and will shatter under extreme stress.

42. Composite materials combine the _______________ properties of more than one material. For example, brittle ceramics can be _______________ by addition of fibers.

Problems

43. Gold has a density of 19.3 g/cm^3 and a face-centered cubic unit cell. What is the atomic radius of gold?

44. Palladium has a face-centered cubic unit cell. What is the density of palladium if the atomic radius is 137 pm?

45. Cobalt, chromium, and copper are all period 4 transition metals. Using molecular orbital theory, rank them in order of melting point.

46. Silicon and germanium are both semiconductors. Phosphorus, arsenic, and boron are all doping agents. Determine which type of semiconductor is formed by all of the combinations.

Test Yourself: *Take the Practice Test on pages 485–487 of your textbook. Again, use a testing environment using only those items that your instructor will allow you to have access to during the test and mark each question with a star, question mark, or x to denote your confidence level. Once you have finished you can check off the LO at the beginning of this chapter based on which questions you starred and got correct. Go back and spend more time with the topics from the other learning objectives.*

CHAPTER THIRTEEN

Solutions and Their Properties

Learning Objectives

As a result of reading and studying this chapter, you should be able to:

- ☐ **LO 13.1** Classify the types of intermolecular forces involved in the process of forming a solution.
- ☐ **LO 13.2** Determine the enthalpy of solution from the enthalpy changes of steps in the solution-making process.
- ☐ **LO 13.3** Predict the solubility of a substance based on the chemical structure of the solute and solvent.
- ☐ **LO 13.4** Calculate the concentration of a solution in units of mass percent.
- ☐ **LO 13.5** Calculate the concentration of a solution in units of parts per million (ppm) or parts per billion (ppb).
- ☐ **LO 13.6** Calculate the concentration of a solution in units of molarity or molality.
- ☐ **LO 13.7** Convert from one unit of concentration to another.
- ☐ **LO 13.8** Determine the solubility of a solute at a given temperature by reading a solubility graph.
- ☐ **LO 13.9** Use Henry's law to calculate the solubility of a gas.
- ☐ **LO 13.10** Qualitatively predict changes to the vapor pressure of a solution when a solute is added.
- ☐ **LO 13.11** Calculate the vapor pressure of a solution containing a nonvolatile solute.
- ☐ **LO 13.12** Calculate the vapor pressure of a solution containing a volatile solute.
- ☐ **LO 13.13** Calculate the amount of boiling-point elevation and freezing-point depression for a solution.
- ☐ **LO 13.14** Calculate osmotic pressure for a solution.
- ☐ **LO 13.15** Use osmotic pressure or another colligative property to calculate the molecular weight of a solute.

Chapter Summary

This chapter focuses on homogeneous mixtures, primarily solutions. The energetics of solution formation, the free-energy changes involved, and factors that affect solubility are covered, as are the great variety of mass- and mole-based concentration units, along with calculations involving these units and the ways that they are interconverted. The effects of pressure and temperature on the solubility of gases are explored as the four colligative properties: lowering of vapor pressure, raising of boiling point, lowering of freezing point, and production of osmotic pressure; methods of their calculation are also explained.

The Chapter in Detail

Section 13.1 Solutions

PRE-LECTURE:

Learning Objectives: *These are the skills you should be proficient in after reading this section and practicing the associated problems. Check off the boxes below once you have gone through this section, answered the associated questions in your textbook, and feel confident you have mastered the skill.*

☐ **LO 13.1** Classify the types of intermolecular forces involved in the process of forming a solution.

Definitions: *To learn, you must do. This is your workbook. Fill in the definitions from the glossary of the textbook as a first step to understanding the material.*

— **mixture—** ______________________________

____

— **solution—** ______________________________

____

— **colloid—** ______________________________

____

— **solute—** ______________________________

____

— **solvent—** ______________________________

____

Summary: *Read through this section and fill in the missing words and equations. Reference the textbook when you are not confident in your answers.*

Mixtures can be classified as *homogeneous* or *heterogeneous*. In heterogeneous mixtures, the components are visibly nonuniform; homogeneous mixtures appear the same throughout.

Homogeneous mixtures are classified according to the size and behavior of the constituent particles:

- *Solutions* contain particles the size of ______________________.

  They are ____________ and ________ separate on standing.
- *Colloids* contain particles with diameters ______________________.

  They scatter light and may be murky or opaque, but they ________ separate on standing.
- *Suspensions* are not truly homogenous mixtures because______________________

  and ______________________.

Any state of matter can form a solution with another state. Seven possible solution types are known: gases can dissolve in solids, liquids, or other gases; liquids can dissolve in other liquids or in solids; and solids can dissolve in liquids or in one another.

When a gas or solid is dissolved in a liquid, the liquid is referred to as the ____________ and the dissolved substance is the ____________.

When two liquids are dissolved in one another, the main component is the solvent and the lesser component is the solute.

Workbook Example 13.1

PROBLEM:
Drugstore "rubbing alcohol" can be purchased as 90% isopropyl alcohol or as 70% isopropyl alcohol. For each of these, what is the solvent and what is the solute?

SOLUTION:
In both of these cases, the isopropyl alcohol is the solvent. The minor component, or solute, is water.

Workbook Problem 13.1

A grocery store bottle of hydrogen peroxide is labeled as 3% solution. What is the solute and what is the solvent?

Strategy: Determine which is the major component and which is the minor component.

POST-LECTURE:

Work Problems: *One of the best ways to learn chemistry is through solving problems. Go to the textbook and work through the Worked Examples for this section, then go to the end of chapter and work the problems there. When you feel confident that you can correctly answer problems like the ones listed below for each objective, check off each of the Learning Objectives for this section above.*

LO 13.1: Problems 13.40, 13.41, 13.48

Section 13.2 Enthalpy Changes and the Solution Process

PRE-LECTURE:

Learning Objectives: *These are the skills you should be proficient in after reading this section and practicing the associated problems. Check off the boxes below once you have gone through this section, answered the associated questions in your textbook, and feel confident you have mastered the skill.*

☐ **LO 13.2** Determine the enthalpy of solution from the enthalpy changes of steps in the solution-making process.

Definitions: *To learn, you must do. This is your workbook. Fill in the definitions from the glossary of the textbook as a first step to understanding the material.*

— **enthalpy of solution (ΔH_{soln})**—____

____

Review: *Before going through the section make sure you are familiar with these concepts:*

- Section 8.6 Intermolecular forces

Summary: *Read through this section and fill in the missing words and equations. Reference the textbook when you are not confident in your answers.*

Except for mixtures of gases, intermolecular forces are at work in solutions as well as in pure substances, but the situation is made somewhat more complex by the fact that there are solvent–solvent interactions, solvent–solute interactions, and solute–solute interactions that all need to be considered.

The enthalpy change can be broken into three components, the sum of which will be the overall enthalpy change:

- Solvent–solvent interactions: these have a _________ ΔH because solvent molecules must be separated and their intermolecular forces disrupted to make room for solute particles.
- Solute–solute interactions: these also have a _________ ΔH, and for the same reason: the interactions between the solute particles have to be overcome. For ionic solids, this is proportional to the lattice energy.
- Solvent–solute interactions: these have a _________ ΔH because the solvent molecules solvate the solute particles. This portion of the energy has a larger contribution in the case of smaller ions and larger charges.

When ionic substances dissolve in water, water molecules surround the ions and neutralize their charges. When this occurs, the ions are said to be *solvated*, or *hydrated*, when water is the solvent.

Workbook Example 13.2

PROBLEM:
Which of the following would you expect to be soluble in hexane, a nonpolar solvent? Rank them in order from most soluble to least soluble.

KCl $\quad$ C_2H_6 $\quad$ CH_3OH

SOLUTION:
The polarity of the solutes will determine their solubility. Using "like dissolves like," the most nonpolar solute will be the most soluble. The least polar solute here is C_2H_6, which has no dipole. Methanol (CH_3OH) is the next most soluble. It is able to hydrogen bond, but has a nonpolar component that may provide some solubility. The most polar solute is KCl, which is unlikely to dissolve at all.

C_2H_6 > CH_3OH > KCl

Workbook Problem 13.2

Rank the following in order of increasing hydration energy (least negative to most negative):

Mg^{2+} Cl^{-} Li^{+}

Strategy: Charge and molecular size determine the hydration energy.

POST-LECTURE:

Work Problems: *One of the best ways to learn chemistry is through solving problems. Go to the textbook and work through the Worked Examples for this section, then go to the end of chapter and work the problems there. When you feel confident that you can correctly answer problems like the ones listed below for each objective, check off each of the Learning Objectives for this section above.*

LO 13.2: Problems 13.42, 13.43

Section 13.3 Predicting Solubility

PRE-LECTURE:

Learning Objectives: *These are the skills you should be proficient in after reading this section and practicing the associated problems. Check off the boxes below once you have gone through this section, answered the associated questions in your textbook, and feel confident you have mastered the skill.*

☐ **LO 13.3** Predict the solubility of a substance based on the chemical structure of the solute and solvent.

Definitions: *To learn, you must do. This is your workbook. Fill in the definitions from the glossary of the textbook as a first step to understanding the material.*

— **entropy of solution (ΔS_{soln})**— ____

____

Review: *Before going through the section make sure you are familiar with these concepts:*

- Section 9.13 Gibb's free-energy change

Summary: *Read through this section and fill in the missing words and equations. Reference the textbook when you are not confident in your answers.*

The Gibb's free-energy change helps predict solubility.

- $-\Delta G$—the process is ______________________________
- $+\Delta G$—the process is ______________________________

Entropies of solution are usually ____________________ because the particles randomize as they mix with another substance.

Enthalpies of solution are more difficult to predict because they take into consideration the solvent–solvent, solute–solute, and solvent–solute interactions discussed in the last section.

When all three sets of interactions are similar, whether polar or nonpolar, dissolution will occur. This is often summarized simply in the expression "like dissolves like." Polar substances will dissolve in polar solvents, and nonpolar substances will dissolve in nonpolar solvents.

POST-LECTURE:

Work Problems: *One of the best ways to learn chemistry is through solving problems. Go to the textbook and work through the Worked Examples for this section, then go to the end of chapter and work the problems there. When you feel confident that you can correctly answer problems like the ones listed below for each objective, check off each of the Learning Objectives for this section above.*

LO 13.3: Worked Example 13.1; Problems 13.27, 13.44–13.49, 13.52

Section 13.4 Concentration Units for Solutions

PRE-LECTURE:

Learning Objectives: *These are the skills you should be proficient in after reading this section and practicing the associated problems. Check off the boxes below once you have gone through this section, answered the associated questions in your textbook, and feel confident you have mastered the skill.*

- ☐ **LO 13.4** Calculate the concentration of a solution in units of mass percent.
- ☐ **LO 13.5** Calculate the concentration of a solution in units of parts per million (ppm) or parts per billion (ppb).
- ☐ **LO 13.6** Calculate the concentration of a solution in units of molarity or molality.
- ☐ **LO 13.7** Convert from one unit of concentration to another.

Definitions: *To learn, you must do. This is your workbook. Fill in the definitions from the glossary of the textbook as a first step to understanding the material.*

— **mass percent (mass %)**— ______________________________

____

— **parts per million (ppm)**— ______________________________

____

— **parts per billion (ppb)—** ______________________________

____

— **molality (*m*)—** ______________________________

____

Review: *Before going through the section make sure you are familiar with these concepts:*

- Section 10.5 mole fraction (χ)

Summary: *Read through this section and fill in the missing words and equations. Reference the textbook when you are not confident in your answers.*

There are four main methods for expressing the concentration of a solution, that is, the amount of solute per unit of solvent. Below we will look at a few units of concentration measurement along with their advantages (+) and disadvantages (−).

MOLARITY (M):

M =

(+) stoichiometry calculations and titrations are greatly simplified by using moles.
(+) amounts are easily obtained from volume and substances do not need to be weighed.
(−) temperature dependence (the volume of solvent will change slightly with increases and decreases in temperature) and the inability to determine solvent quantities without density.

MOLE FRACTION (χ):

$\chi =$

(+) it is temperature independent.
(+) convenient when with calculations involving gases.
(−) this method is rarely used for liquid solutions because other methods are easier for calculations involving liquid solutions.

MASS PERCENT (mass %):

mass % =

(+) it is temperature independent.
(−) volume is easier to measure than mass for most solutions.

Two variants on this method are often used for dilute solution:

Parts per million (ppm) =

Parts per billion (ppb) =

These are often used for mixtures in water. Since the density of water is 1 g/mL, it follows that 1 kg solution = 1 L solution. So:

$$1\ ppm = \frac{1\ mg\ solute}{1\ L\ solution} \qquad 1\ ppb = \frac{1\ \mu g\ solute}{1\ L\ solution}$$

Molality (m):

X =

(+) temperature independent.
(−) it is difficult to use this method to measure the mass of liquids, and the density of the solution must be known in order to convert to molarity.

Workbook Example 13.3

PROBLEM:
A sample of water is found to have 65.0 ppm of arsenic. How many grams of arsenic will 1 L contain? What is the molarity of this solution?

SOLUTION:
A concentration of 1 ppm means that each liter of an aqueous solution contains 1 mg of solute. Therefore, the mass of arsenic in 1 L of this water is 65 mg. To calculate the molarity of this solution, we need to convert milligrams to moles.

$$65\ mg\ As \times \frac{1\ g}{1000\ mg} \times \frac{1\ mole\ As}{74.92\ g\ As} = 0.000868\ mol\ As = 868\ \mu mol\ As$$

Since we have 1 L of solution, the molarity is equal to 868 μM. Density is not needed to convert such dilute solutions from mass-based measures to molarity, because the solution will have the same density as water.

Workbook Example 13.4

PROBLEM:
A solution is made by dissolving 35.0 g of NaCl in 100 g of water. The resulting solution has a density of 1.16 g/mL. Determine the mass percent and molarity of the solution.

SOLUTION:
The mass percent of the solution is the mass of the solute divided by the total mass of the solution:

$$mass\ percent = \frac{35.0\ g}{135\ g} \times 100\% = 25.9\%\ NaCl$$

To convert to molarity, the mass of solute must be converted to moles and the solution converted to volume using the density:

$$35.0\ g \times \frac{1\ mol\ NaCl}{58.44\ g\ NaCl} = 0.599\ mol\ NaCl$$

$$\frac{135\ g}{1.16\ {}^{g}/_{mL}} = 116\ mL$$

$$M = \frac{0.599\ mol\ NaCl}{0.116\ L} = 5.16\ M\ NaCl$$

Workbook Problem 13.3

A solution is prepared by dissolving 15.43 g of glucose ($C_6H_{12}O_6$) in 300 g of water. The density of this solution is 1.09 g/mL. Calculate the mass percent, molality, and molarity of the solution.

Strategy: Calculate mass percent of solution, volume of solution, and moles of solute to provide all units.

Step 1: Determine mass percent.

Step 2: Calculate the moles of solute and the molality of the solution.

Step 3: Calculate the volume of the solution and the molarity of the solution.

POST-LECTURE:

Work Problems: *One of the best ways to learn chemistry is through solving problems. Go to the textbook and work through the Worked Examples for this section, then go to the end of chapter and work the problems there. When you feel confident that you can correctly answer problems*

like the ones listed below for each objective, check off each of the Learning Objectives for this section above.

LO 13.4: Worked Example 13.2; Problem 13.66
LO 13.5: Worked Example 13.3; Problems 13.58, 13.69
LO 13.6: Worked Example 13.4; Problems 13.60, 13.62, 13.70, 13.76
LO 13.7: Worked Example 13.5; Problems 13.56, 13.64, 13.72, 13.74

Section 13.5 Some Factors That Affect Solubility

PRE-LECTURE:

Learning Objectives: *These are the skills you should be proficient in after reading this section and practicing the associated problems. Check off the boxes below once you have gone through this section, answered the associated questions in your textbook, and feel confident you have mastered the skill.*

☐ **LO 13.8** Determine the solubility of a solute at a given temperature by reading a solubility graph.
☐ **LO 13.9** Use Henry's law to calculate the solubility of a gas.

Definitions: *To learn, you must do. This is your workbook. Fill in the definitions from the glossary of the textbook as a first step to understanding the material.*

— **saturated—** ____

____

— **supersaturated—**____

____

— **miscible—** ____

____

— **Henry's law—** ____

____

Summary: *Read through this section and fill in the missing words and equations. Reference the textbook when you are not confident in your answers.*

Solubility has natural limits: at some point, the solute molecules will be in equilibrium, meaning that they are going in and out of solution at an equal rate. When this occurs, the solution is said to be ___________.

Some solutes are more soluble at high temperatures. In this case, it is possible to produce a **supersaturated** solution with a greater-than-equilibrium amount of solute by creating a saturated solution at a high temperature and allowing it to cool undisturbed. These solutions are unstable.

The amount of solute that can be dissolved in a solvent to form a saturated solution at a given temperature is the **solubility** of that substance. Sometimes two substances are soluble in one another regardless of proportions. When this happens, the substances are ___________.

The temperature effect on solubility of solids is unpredictable; however, the solubility of most solids increases as temperature ___________. This occurs because the positive ΔS associated with a solid dissolving in a liquid increases as temperature increases. Since it is associated with $-T\Delta S$ in the Gibb's free-energy equation, it causes ΔG to become more negative.

Gases tend to become less soluble as the temperature of the solution ___________. Entropy of gases decrease as they are trapped in a liquid, lowering the entropy of the gas. A $-\Delta S$ becomes a positive term when put into $-T\Delta S$ in the equation and so a lower temperature is needed to overcome this decrease in entropy.

Pressure does not affect the solubility of solids and liquids, but has a large effect on gases, which is described by **Henry's law:**

$$solubility = k \times P$$

where k is a constant related to the gas (with usual units of mol/L · atm), and P is the partial pressure of the gas. In contrast to other gas calculations, these measurements are reported at 25 °C.

Solubility of a gas increases as pressure ___________ because an increase in pressure forces more particles into the solution.

Workbook Example 13.5

PROBLEM:
Determine the solubility of CO_2 (*g*) at 25 °C and a partial pressure of 33.5 mm Hg. Note: the k constant is equal to 0.0032 mol/L · atm.

SOLUTION:
Use Henry's law, being careful to convert units as needed.

$$solubility = 0.0032 \frac{mol}{L \cdot atm} \times 33.5\ mm\ Hg \times \frac{1\ atm}{760\ mm\ Hg} = 1.41 \times 10^{-4} \frac{mol}{L}$$

Workbook Problem 13.4

Given the Henry's law constant in the last problem (0.0032 mol/L · atm), how many liters of CO_2 will bubble out of a 500 mL soda after it is opened at 25 °C if it was previously under 2.3 atm of pressure with 100% CO_2 in the air, and it comes to equilibrium with the atmosphere where the partial pressure of CO_2 is only 0.3 mm Hg?

Strategy: Calculate the initial and final amounts of CO_2 in a saturated solution; determine the number of moles and then the liters of CO_2.

Step 1: Calculate the moles of dissolved CO_2 under pressure.

Step 2: Calculate the moles of dissolved CO_2 after pressure is released and equilibrium with the atmosphere is reached.

Step 3: Calculate the change in dissolved CO_2—this is the CO_2 that has been released.

Step 4: Determine the volume of gas at 25 °C and 1 atm pressure.

POST-LECTURE:

Work Problems: *One of the best ways to learn chemistry is through solving problems. Go to the textbook and work through the Worked Examples for this section, then go to the end of chapter and work the problems there. When you feel confident that you can correctly answer problems like the ones listed below for each objective, check off each of the Learning Objectives for this section above.*

LO 13.8: Problems 13.80, 13.81
LO 13.9: Worked Example 13.6; Problems 13.82–13.86

Section 13.6 Physical Behavior of Solutions: Colligative Properties

PRE-LECTURE:

Learning Objectives: *There are no specific learning objectives for this section; the information in this section is used as a basis for understanding subsequent sections.*

Definitions: *To learn, you must do. This is your workbook. Fill in the definitions from the glossary of the textbook as a first step to understanding the material.*

— **colligative properties—**____

____

Summary: *Read through this section and fill in the missing words and equations. Reference the textbook when you are not confident in your answers.*

Solutions behave in some ways differently than pure solvents. Properties that depend on the amount of dissolved solute but not its chemical identity are called ___________. They include:

- ____
- ____
- ____
- ____

These properties will be examined in detail in the following sections.

Section 13.7 Vapor-Pressure Lowering of Solutions: Raoult's Law

PRE-LECTURE:

Learning Objectives: *These are the skills you should be proficient in after reading this section and practicing the associated problems. Check off the boxes below once you have gone through this section, answered the associated questions in your textbook, and feel confident you have mastered the skill.*

☐ **LO 13.10** Qualitatively predict changes to the vapor pressure of a solution when a solute is added.
☐ **LO 13.11** Calculate the vapor pressure of a solution containing a nonvolatile solute.
☐ **LO 13.12** Calculate the vapor pressure of a solution containing a volatile solute.

Definitions: *To learn, you must do. This is your workbook. Fill in the definitions from the glossary of the textbook as a first step to understanding the material.*

— **Raoult's law—**____

____

— **van't Hoff factor (i)—**____

____

— **ion pair**—____

____

Review: *Before going through the section make sure you are familiar with these concepts:*

- Section 11.2 vapor pressure
- Section 10.5 mole fraction
- Section 10.5 Dalton's law of partial pressure

Summary: *Read through this section and fill in the missing words and equations. Reference the textbook when you are not confident in your answers.*

In a solution with a nonvolatile solute (one that has no vapor pressure of its own), the vapor pressure is ___________ than it would be in the pure solvent. Since the solution has a higher entropy than the pure solvent, the entropy gain in evaporating is reduced, and the vapor pressure is reduced.

Use Raoult's law to calculate the vapor pressure of the solution:

$$P_{soln} =$$

where X is the mole fraction of the solvent in the mixture.

When calculating the mole fraction, the total number of moles of particles must be used for the solute. For a molecular solute, like glucose, this will be the same as the number of moles of molecules. For an ionic solute, this will be the number of moles of ions generated. Unfortunately, ionic compounds don't often dissociate completely. Moles of ions generated are mathematically described by the **van't Hoff factor,** i.

$$i =$$

Workbook Example 13.6

PROBLEM:
Determine the vapor pressure of a solution made by dissolving 25.7 g of LiBr in 1.00 kg of water at 25 °C. The vapor pressure of pure water at 25 °C is 23.76 mm Hg. Assume complete dissociation.

SOLUTION:
Use Raoult's law. First, determine the mole fraction of water in the solution:

$$moles\ LiBr = 25.7\ g \times \frac{1\ mol\ LiBr}{86.845\ g} \times \frac{2\ mol\ ions}{1\ mol\ LiBr} = 0.591859\ mol\ ions$$

$$moles\ H_2O = 1000\ g \times \frac{1\ mol\ H_2O}{18.015\ g} = 55.096419\ moles\ H_2O$$

$$\chi_{H_2O} = \frac{55.096419\ moles\ H_2O}{55.096419\ moles\ H_2O + 0.591859\ mol\ ions} = 0.98937$$

Then multiply by the vapor pressure of pure solvent:

$$P_{solution} = 23.76\ mm\ Hg \times 0.98937 = 23.5\ mm\ Hg$$

Workbook Example 13.7

PROBLEM:
How much LiBr would need to be dissolved in 1.00 kg of water to lower the vapor pressure to 20 mm Hg, assuming complete dissociation?

SOLUTION:
First, calculate the mole fraction of water:

$$\frac{20.0\ mm\ Hg}{23.76\ mm\ Hg} = \chi_{H_2O} = 0.842$$

Next, determine the number of moles of solute needed for this mole fraction:

$$\chi_{H_2O} = \frac{55.56\ mol\ H_2O}{(55.56\ mol\ H_2O + x\ mol\ LiBr)} = 0.842$$

$$55.56\ mol = 0.842(55.56\ mol + x\ mol)$$

$$55.56 - 46.78 = 0.842x$$

$$x = 10.43\ mole\ ions = 5.21\ mol\ LiBr$$

$$5.21\ mol \times \frac{86.85\ g\ LiBr}{1\ mol\ LiBr} = 453\ g\ LiBr$$

Workbook Problem 13.5

Assuming complete dissociation for both, calculate the quantity of $CaCl_2$ and LiCl that are needed to lower the vapor pressure of 500 g of water by 3 mm Hg at 70 °C. The vapor pressure of water at this temperature is 233.7 mm Hg.

Strategy: Determine the mole fraction of ions needed to lower the vapor pressure by this amount, then convert to grams of each ionic substance.

Step 1:	Calculate the mole fraction of water in a solution under these conditions.
Step 2:	Calculate the moles of dissolved ions that will lead to this mole fraction.
Step 3:	Convert to moles, then grams, of $CaCl_2$.
Step 4:	Convert to moles, then grams, of LiCl.

In solutions with a volatile solute, the vapor pressure of the solution will be ___________ between that of the two pure solvents.

$$P_{total} = P_A + P_B =$$

Workbook Example 13.8

PROBLEM:
The vapor pressure of water at 25 °C is 23.8 mm Hg. The vapor pressure of ethyl alcohol (C_2H_5OH) at 25°C is 61.2 mm Hg. What is the vapor pressure of a solution prepared by combining 50 g of each liquid?

SOLUTION:
Calculate the moles of each to determine the mole fractions:

$$50\ g\ C_2H_5OH \times \frac{1\ mol\ C_2H_5OH}{46.0\ g\ C_2H_5OH} = 1.09\ mol\ C_2H_5OH$$

$$50\ g\ H_2O \times \frac{1\ mol\ H_2O}{18.0\ g\ H_2O} = 2.78\ mol\ H_2O$$

$$\chi_{C_2H_5OH} = \frac{1.09\ mol\ C_2H_5OH}{(1.09\ mol\ C_2H_5OH + 2.78\ mol\ H_2O)} = 0.282$$

$$\chi_{H_2O} = \frac{2.78\ mol\ H_2O}{(1.09\ mol\ C_2H_5OH + 2.78\ mol\ H_2O)} = 0.718$$

$$P_{\text{total}} = (0.282 \times 61.2\ mmHg) + (0.718 \times 23.8\ mmHg) = 34.35\ mmHg$$

Workbook Problem 13.6

The label on a bottle of wine states that it is 13% ethanol. Using the vapor pressures listed above, what is the pressure above a 150-g glass of wine, ignoring the fact that other volatile components are present in the wine.

Strategy: Determine the mole fraction of each component, and calculate the partial pressure of each.

Step 1: Solve for the mass and moles of each component.

Step 2: Determine the mole fraction of each component.

Step 3: Determine the partial pressure of each component and the total pressure.

Just like the ideal gas law, Raoult's law applies to "ideal" solutions. It works best at low solute concentrations.

POST-LECTURE:

Work Problems: *One of the best ways to learn chemistry is through solving problems. Go to the textbook and work through the Worked Examples for this section, then go to the end of chapter and work the problems there. When you feel confident that you can correctly answer problems like the ones listed below for each objective, check off each of the Learning Objectives for this section above.*

LO 13.10: Problems 13.29–13.33, 13.88
LO 13.11: Worked Example 13.7; Problems 13.92, 13.102
LO 13.12: Worked Example 13.8; Problems 13.108, 13.110

Section 13.8 Boiling-Point Elevation and Freezing-Point Depression of Solutions

PRE-LECTURE:

Learning Objectives: *These are the skills you should be proficient in after reading this section and practicing the associated problems. Check off the boxes below once you have gone through this section, answered the associated questions in your textbook, and feel confident you have mastered the skill.*

☐ **LO 13.13** Calculate the amount of boiling-point elevation and freezing-point depression for a solution.

Definitions: *To learn, you must do. This is your workbook. Fill in the definitions from the glossary of the textbook as a first step to understanding the material.*

— **molal boiling-point elevation constant (K_b)**—______________________________

____

— **molal freezing-point elevation constant (K_f)**—______________________________

____

Review: *Before going through the section make sure you are familiar with these concepts:*

- Section 9.11 entropy of phases of matter
- Section 11.3 boiling point

Summary: *Read through this section and fill in the missing words and equations. Reference the textbook when you are not confident in your answers.*

As seen with vapor pressure in the last section, boiling-point elevation and freezing-point depression depend on the number of particles dissolved not the type of substance. All three, boiling-point elevation, freezing-point depression, and vapor-pressure lowering, are due to the entropy difference between the pure solvent and the solvent in solution. For each of these, the solution has more molecular randomness than the pure solvent.

- *Boiling-point elevation:* The solution, with more molecular randomness than the pure solvent, has less of an entropy change to convert to a more random gas. Thus, ΔS is

___________ for the solution, and the temperature must be ___________ to boil the solution.

- *Freezing-point depression:* The solution, still with more molecular randomness than the pure solvent, has more of an entropy change to convert to a more ordered solid. Thus, ΔS is ___________ for the solution, and the temperature must be ___________ to freeze the solution.

Mathematically, these are described with the following equations:

$\Delta T_b =$	$\Delta T_f =$

For these equations, K_b is the **molal boiling-point elevation constant** and K_f is the **molal freezing-point depression constant.** Both are dependent on the *solvent*, not the identity of the particles. m is the molal concentration of solute particles, whether they are ions or molecules. Molality rather than molarity is used for these calculations because temperature extremes affect the volume of solutions. The actual dissociation of ionic compounds is taken into account with the van't Hoff factor for the compound.

Workbook Example 13.9

PROBLEM:
What will be the freezing point and boiling point of an aqueous solution containing 55.0 g of KCl in 250 g of water? $K_b(H_2O) = 0.51\ °C/m$ and $K_f = 1.86\ °C/m$. Assume complete dissociation.

SOLUTION:
Determine the molality of the solution, then calculate the boiling-point elevation and the freezing-point depression:

$$55.0\ g\ KCl \times \frac{1\ mol\ KCl}{74.55\ g\ KCl} \times \frac{2\ mol\ ions}{1\ mol\ KCl} = 1.48\ mol\ ions$$

$$\frac{1.48\ mol\ ions}{0.250\ kg} = 5.92\ m$$

$$\Delta T_b = \left(0.51\ {}^{°C}\!/\!_{m}\right)(5.92\ m) = 3.02 \qquad 100\ °C + 3.02\ °C = 103\ °C$$

$$\Delta T_f = \left(1.86\ {}^{°C}\!/\!_{m}\right)(5.92\ m) = 11.0 \qquad 0\ °C - 11.0\ °C = -11\ °C$$

Workbook Problem 13.7

How many grams of $(NH_4)_2SO_4$ need to be added to 300 g of H_2O so that the freezing point of the solution is lowered to −15.0 °C? $K_f = 1.86\ °C/m$. Assume complete dissociation.

Strategy: Using the equation for freezing-point depression, solve for the molality of ions needed, then convert to moles and grams of ammonium sulfate.

Step 1:	Determine the molality of ions needed to generate this freezing-point depression.
Step 2:	Determine the moles of $(NH_4)_2SO_4$ needed to generate this quantity of ions.
Step 3:	Convert to grams of ammonium sulfate needed.

POST-LECTURE:

Work Problems: *One of the best ways to learn chemistry is through solving problems. Go to the textbook and work through the Worked Examples for this section, then go to the end of chapter and work the problems there. When you feel confident that you can correctly answer problems like the ones listed below for each objective, check off each of the Learning Objectives for this section above.*

LO 13.13: Worked Example 13.9; Problems 13.90, 13.94, 13.114, 13.116, 13.132

Section 13.9 Osmosis and Osmotic Pressure

PRE-LECTURE:

Learning Objectives: *These are the skills you should be proficient in after reading this section and practicing the associated problems. Check off the boxes below once you have gone through this section, answered the associated questions in your textbook, and feel confident you have mastered the skill.*

- ☐ **LO 13.14** Calculate osmotic pressure for a solution.
- ☐ **LO 13.15** Use osmotic pressure or another colligative property to calculate the molecular weight of a solute.

Definitions: *To learn, you must do. This is your workbook. Fill in the definitions from the glossary of the textbook as a first step to understanding the material.*

— **osmosis—** ______________________________

____

— **osmotic pressure (Π)**—____

____

Review: *Before going through the section make sure you are familiar with these concepts:*

- Particles travel from a region of higher concentration to a region of lower concentration.

Summary: *Read through this section and fill in the missing words and equations. Reference the textbook when you are not confident in your answers.*

Cell membranes, along with some manufactured membranes, are ______________, meaning that solvent molecules and some very small molecules can pass freely through the membrane, but large molecules and ions are blocked.

When solutions of different concentrations are allowed to mix, they act to form a uniform solution. When a semipermeable membrane separates these solutions, the large particles are not able to pass through to equilibrate the concentrations; however, the small solvent particles can. The small solvent particles travel from a region of higher concentration (for the solvent particles this is the more dilute solution) to the region of lower concentration (for the solvent particles this is the more concentrated solution). As a result, the level of the low concentration or pure solvent will ___________, and the level of the high-concentration solution will ___________.

This is an entropy-driven process: the dilution of a solution increases the disorder of the solution. The process will continue until the external pressure and the **osmotic pressure** are equal.

Osmotic pressure is calculated with the following equation:

$$\Pi = \qquad\qquad$$

Where Π is the osmotic pressure, *M* is the molarity of solute particles in solution, *R* is the gas law constant, and T is the temperature in kelvins. Molarity is used for these calculations because the temperature is a variable as well.

Workbook Example 13.10

PROBLEM:
Determine the osmotic pressure of a 0.050 M solution of sodium chloride at 24.3 °C.

SOLUTION:

$$\Pi = \left(\frac{0.050\ mol}{L} \times \frac{2\ mol\ ions}{1\ mol\ NaCl}\right)\left(0.08206\frac{L \cdot atm}{K \cdot mol}\right)(273.15 + 24.3\,K) = 2.4\ atm$$

Workbook Problem 13.8

A solution containing an unknown substance has an osmotic pressure of 7.34 atm at 25 °C. What is the molarity of the solution?

Strategy: Using the equation for osmotic pressure, solve for the molarity of the solution.

Any of the colligative properties can be used to determine molecular weights, but osmotic pressure, as the effect is so large, can be used most accurately.

Workbook Problem 13.9

A solution is prepared from 23 mg of a compound in 100 mL of aqueous solution. The osmotic pressure of the solution is 8.428 mm Hg at 25 °C. What is the approximate molecular weight of the compound?

Strategy: Using the equation for osmotic pressure, calculate the molarity of the solution, and from that the molecular weight of the compound.

Step 1: Calculate the molarity of the solution.

Step 2: Calculate the number of moles in the solution.

Step 3: Determine the molecular weight.

POST-LECTURE:

Work Problems: *One of the best ways to learn chemistry is through solving problems. Go to the textbook and work through the Worked Examples for this section, then go to the end of chapter and work the problems there. When you feel confident that you can correctly answer problems*

like the ones listed below for each objective, check off each of the Learning Objectives for this section above.

LO 13.14: Worked Example 13.10; Problems 13.34, 13.118, 13.120
LO 13.15: Worked Example 13.11; Problems 13.124, 13.143

Self-Test

This section is intended to test your knowledge of the material covered in this chapter. Think through these problems and make certain you understand what the problem is about—ask yourself if the answer makes sense. Many of these questions are linked to the chapter learning goals. Therefore, successful completion of these problems indicates that you have achieved the learning goals for this chapter. You will receive the greatest benefit from this section if you use it as a mock exam. Do not use your notes, book, Internet, or friends when taking the test. There is no grade and you only shortchange yourself by getting help with this activity. As you take the test, mark the problems with your level of confidence with the answer.

★ *Use a star for problems you <u>know</u> you got right.*
? *Use a question mark for those you <u>think</u> you got right.*
x *Use an x for those you are guessing on.*

If you get a question right that you marked with a star, you can be confident that you know the associated LO.

If you get a question right that you marked with a question mark, go back and work a few more problems and test yourself again tomorrow.

If you get a question wrong <u>or</u> mark it with an x, go back and read the section again, work a few problems today, a few tomorrow, and a few the next day before testing yourself again.

This will help you to discover which topics you have learned thoroughly and which topics you need to study in more detail.

<u>*True–False*</u>

1. A solution of metals is called an *alloy*.

2. Entropy always increases when a solution is formed, but enthalpy can increase or decrease.

3. Molality is the most practical unit of concentration for environmental measurements.

4. Potassium nitrate will dissolve readily in oil.

5. The density of a solution is needed to convert from mass percent to molarity.

6. Supersaturated solutions are stable at low temperatures.

7. Increasing pressure causes ionic solids to become more soluble.

8. The more concentrated a solution, the lower the vapor pressure.

9. The more concentrated a solution, the higher the freezing point.

10. Osmotic pressure can be overcome with high external pressures.

Multiple Choice

11. When a gas or solid is dissolved in a liquid, the solute is
    a. the dissolved substance
    b. the liquid
    c. the major component
    d. the minor component

12. ΔH_{soln} will be exothermic if
    a. ΔH for solvent–solvent interactions is negative
    b. ΔH for solute–solute interactions is negative
    c. ΔH for solute–solvent interactions is negative
    d. the sum of the three types of interactions leads to a negative ΔH

13. Benzene, a nonpolar organic compound, is most likely to dissolve in
    a. water
    b. CH_3CH_2OH
    c. CCl_4
    d. NH_3

14. When the temperature of a solvent is lowered, the solubility of gases
    a. decreases
    b. increases
    c. is not affected
    d. cannot be generally predicted

15. A mixture that contains particles large enough to be visible with a low-power microscope is a
    a. heterogeneous mixture
    b. solution
    c. suspension
    d. colloid

Matching

16. colligative property
17. colloid
18. mass percent
19. heat of solution
20. miscible
21. molality
22. osmosis
23. semipermeable membrane

a. measure of concentration that uses mass of component per mass of solution
b. passage of solvent from a less-concentrated to a more-concentrated solution
c. a solution holding the maximum quantity of solute
d. able to mix together in any proportions
e. the liquid portion of a solution
f. a homogeneous mixture containing large particles that do not settle out
g. allows passage of small molecules only, not ions or large molecules
h. a solid or gas that is dissolved in a liquid

24. saturated
25. solubility
26. solute
27. solution
28. solvent
29. supersaturated

i. homogeneous mixture containing very small molecules or ions
j. a solution that has been heated to dissolve more solute, then cooled: unstable
k. property that depends on the amount of dissolved solute, not its identity
l. measure of concentration that is moles per kilogram of solvent
m. enthalpy change, positive or negative, that occurs when a solution forms
n. the amount of solute that a solution can contain at a specific temperature

Fill in the Blanks

30. Solution formation has two endothermic elements: disrupting ____________________ interactions and disrupting ____________________ interactions both require energy.

31. To calculate molarity you need to find moles of ______________ per liters of ______________. For molality you need moles of ______________ per kilograms of ______________.

32. To convert between mass-based units and mole-based units requires that the ______________ of a solution be known.

33. Gases become ______________ soluble as the temperature decreases.

34. Adding a nonvolatile solute to a solution causes the vapor pressure to ______________, the boiling point to ______________, the freezing point to ______________, and ____________ ______________ to develop across semipermeable membranes.

35. Colligative properties can be used to determine the ______________ ______________ of an unknown solute.

36. Dissolved ions that are surrounded and stabilized by a shell of solvent molecules are said to be ________________. If water is the solvent, the term used is ______________.

37. The spontaneous dissolution of ionic solids can be either ________________ or ________________, depending on the energetics of solution formation.

Problems

38. Rank the following in order of increasing solubility in benzene, a nonpolar solvent.

I_2 NaCl $CH_3(CH_2)_2CH_2OH$

39. If 28.4 g of sucrose ($C_{12}H_{22}O_{11}$) is dissolved in 350 g of water, the final volume is 374 mL. Calculate the concentration of the solution in terms of: (a) mass percent, (b) mole fraction, (c) molality, and (d) molarity.

40. Household hydrogen peroxide is listed on the label as 3% H_2O_2. What is the molarity of this solution if the density of the solution is 1.01 g/mL?

41. A sample of industrial effluent is found to have an arsenic concentration of 15 ppb. What is the molarity of this solution? How many grams of arsenic would be found in 100,000 L of this effluent?

42. Goldfish prefer cold water to warm. If the mole fraction of oxygen in air is 0.21, what is the oxygen concentration in water at 25 °C? The Henry's law constant for oxygen is 1.32×10^{-3} mol/L · atm.

43. The Henry's law constant for nitrogen is 6.25×10^{-4} mol/L · atm. Calculate the solubility of nitrogen if its partial pressure above water is 976 mm Hg.

44. The vapor pressure of pure water is 42.175 mm Hg at 35 °C. Calculate the vapor pressure of a solution made with 4.5 g of glucose ($C_6H_{12}O_6$) and 55 g of H_2O.

45. The vapor pressure of water at 25 °C is 23.8 mm Hg. Calculate the vapor pressure of an aqueous solution that contains 5.8 g $FeCl_3$ and 72.1 g of water at this temperature. The van't Hoff factor for $FeCl_3$ is 3.4.

46. The vapor pressure of benzene (C_6H_6) at 25 °C is 93.4 mm Hg. The vapor pressure of toluene ($C_6H_5CH_3$) is 26.9 mm Hg at 25 °C. Calculate the vapor pressure of a solution prepared from 10.0 g of benzene and 10.0 g of toluene.

47. The normal boiling point of carbon tetrachloride (CCl_4) is 76.7 °C. How many grams of naphthalene ($C_{10}H_8$) need to be added to 1.0 kg of carbon tetrachloride to raise the boiling point of the solution to 80.0 °C? K_b for CCl_4 is 5.03 °C/m. Assume complete dissociation.

48. Calculate the boiling point and freezing point of a solution prepared by mixing 10.0 g NaCl with 90.0 g of water. Assume complete dissociation. $K_b = 0.51$ °C/m, $K_f = 1.86$ °C/m.

49. A solution was prepared by mixing 75.0 g of benzene (freezing point = 5.5 °C) and 2.50 g of an unknown substance. The freezing point of the solution was 3.5 °C. Calculate the molar mass of the unknown substance. $K_f = 5.12$ °C/m.

50. Calculate the osmotic pressure of a solution at 15 °C containing 18.0 g of glucose ($C_6H_{12}O_6$) in 1.2 L of solution.

51. Calculate the molar mass of 0.250 g of a peptide in 1 L of water at 15 °C that has an osmotic pressure of 16.4 mm Hg.

Challenge Problem

52. The vapor above a room-temperature mixture of pentane and hexane is 43.2% pentane by mass. What is the mass percent composition of the solution if the vapor pressure of pure pentane is 425 mm Hg and the vapor pressure of hexane is 151 mm Hg at room temperature? What is the total pressure of the solution?

Test Yourself: *Take the Practice Test on pages 529–530 of your textbook. Again, use a testing environment using only those items that your instructor will allow you to have access to during the test and mark each question with a star, question mark, or x to denote your confidence level. Once you have finished you can check off the LO at the beginning of this chapter based on which questions you starred and got correct. Go back and spend more time with the topics from the other learning objectives.*

CHAPTER FOURTEEN

Chemical Kinetics

Learning Objectives

As a result of reading and studying this chapter, you should be able to:

- ☐ **LO 14.1** Determine the average reaction rate over a specified period of time and estimate instantaneous reaction rate from a graph of concentration versus time.
- ☐ **LO 14.2** Relate the rate of consumption of any reactant to the rate of formation of any product using reaction stoichiometry.
- ☐ **LO 14.3** Find the order of a reaction with respect to each reactant, the overall order of a reaction, and the units of the rate constant, given the rate law.
- ☐ **LO 14.4** Predict the change in reaction rate when the concentration of a reactant changes by a specified amount.
- ☐ **LO 14.5** Determine the rate law and rate constant using initial rate and concentration data.
- ☐ **LO 14.6** Use the integrated rate law to determine the half-life and the concentrations remaining at various times for a zeroth-order reaction.
- ☐ **LO 14.7** Use the integrated rate law to determine the half-life and the concentrations remaining at various times for a first-order reaction.
- ☐ **LO 14.8** Use the integrated rate law to determine the half-life and the concentrations remaining at various times for a second-order reaction.
- ☐ **LO 14.9** Determine the reaction order and rate constant graphically.
- ☐ **LO 14.10** Interpret potential energy diagrams and suggest geometries for successful collisions and transition states using collision theory.
- ☐ **LO 14.11** Use the Arrhenius equation to calculate the activation energy of a reaction given rate constants at different temperatures.
- ☐ **LO 14.12** Given a reaction mechanism, write the overall reaction, identify intermediates, and determine the molecularity for each elementary step.
- ☐ **LO 14.13** For mechanisms with an initial slow step (a) predict the rate law for the overall reaction or (b) propose a mechanism that agrees with the experimental rate law.
- ☐ **LO 14.14** For mechanisms with an initial fast equilibrium step, predict the rate law.
- ☐ **LO 14.15** Use a molecular diagram for a catalyzed reaction to determine the rate law and propose a mechanism consistent with the rate law.
- ☐ **LO 14.16** Identify catalysts and intermediates in the mechanism of a catalyzed reaction.
- ☐ **LO 14.17** Interpret a potential energy diagram representing the mechanism in a catalyzed reaction.

Chapter Summary

In this chapter, you will learn about chemical kinetics: the study of reaction rates and mechanisms. The appearance of products can be related to the disappearance of reactants, and reaction orders can be determined from experimental data and plots of concentrations versus time in both logarithmic and nonlogarithmic forms. Integrated rate laws can be used to calculate the concentration of a reactant at any given time point, the fraction of reactant remaining, or the time required for the reaction to progress to a certain point. Radioactive decay follows first-order kinetics with a decay constant in place of a reaction constant. This makes it possible to determine how long an isotope has been decaying, determine how long it will take for a specific amount or proportion of decay to occur, or solve for either half-life or decay constant given experimental data. Kinetics make it possible to identify a rate-determining step and propose a reaction mechanism. You will also learn how to use plots of concentration versus time to determine a reaction order. Kinetic molecular theory and the Arrhenius equation make it possible to derive equations to determine activation energy and rate constants at different temperatures. Catalysts, both heterogeneous and homogeneous, and their mechanisms are also discussed.

The Chapter in Detail

Section 14.1 Reaction Rates

PRE-LECTURE:

Learning Objectives: *These are the skills you should be proficient in after reading this section and practicing the associated problems. Check off the boxes below once you have gone through this section, answered the associated questions in your textbook, and feel confident you have mastered the skill.*

- ☐ **LO 14.1** Determine the average reaction rate over a specified period of time and estimate instantaneous reaction rate from a graph of concentration versus time.
- ☐ **LO 14.2** Relate the rate of consumption of any reactant to the rate of formation of any product using reaction stoichiometry.

Definitions: *To learn, you must do. This is your workbook. Fill in the definitions from the glossary of the textbook as a first step to understanding the material.*

— **chemical kinetics—** ______________________________

____

— **reaction rate—** ______________________________

____

— **instantaneous rate—** ______________________________

____

— **initial rate—** ______________________________

____

Summary: *Read through this section and fill in the missing words and equations. Reference the textbook when you are not confident in your answers.*

Rate is determined by measuring the change in concentration over a period of time:

rate =

As the concentration of reactants decreases, the concentration of products increases. As always, changes are calculated as final–initial, so for a disappearing reactant, the sign on the rate will be negative and for an appearing product, the rate will be positive.

The units associated with these rates are M/s or mol/(L · s), which allows the calculated rate to be independent of the scale of the reaction. The reactant or product measured must also be specified as rates differ based on the coefficients in the balanced equation. To use a very simple example:

$$2\ H_2(g) + O_2(g) \rightarrow 2\ H_2O(l)$$

In this reaction, hydrogen gas will disappear at the same rate that water is formed, but oxygen will disappear at half the rate of hydrogen disappearance and water formation. This leads to a number of possible rates. To avoid this ambiguity, chemists define a *general rate* as:

$$general\ rate\ of\ reaction = \frac{change\ in\ concentration}{coefficient}$$

So, for the above reaction:

$$general\ rate\ of\ reaction = -\frac{1}{2} \times \frac{\Delta[H_2]}{\Delta t} = -\frac{\Delta[O_2]}{\Delta t} = \frac{1}{2} \times \frac{\Delta[H_2O]}{\Delta t}$$

When concentration is plotted versus time, we see average rates over a specific time period, but *instantaneous rates* can also be seen. An ______________________ is defined by the slope of a tangent to a curve of concentration versus time. The instantaneous rate when time = 0, the very beginning of the reaction, is called the ___________.

Workbook Example 14.1

PROBLEM:
In the following reaction:

$$4\ NH_3(g) + 3\ O_2(g) \rightarrow 2\ N_2(g) + 6\ H_2O(g)$$

the rate of formation of H_2O is 0.73 M · s^{-1} at 300 s. What is the rate of disappearance of NH_3?

SOLUTION:
Along with the rate of formation of water, the stoichiometry of the balanced equation can be used to determine the rate of disappearance of NH_3.

$$\frac{0.73\ mol\ H_2O}{L \cdot s} \times \frac{4\ mol\ NH_3}{6\ mol\ H_2O} = \frac{0.49\ mol\ NH_3}{L \cdot s}$$

This rate is reported with a minus sign in front of it, $-0.49\ M \cdot s^{-1}$, because we are reporting the rate of *disappearance* of the reactant NH_3.

Workbook Problem 14.1

Given the following unbalanced equation:

$$HBr(g) \rightarrow H_2(g) + Br_2(g)$$

express the general rate of reaction with respect to all reactants and products. If the initial rate of disappearance of HBr is -3.26×10^{-3} M/s, what is the rate of appearance of bromine gas?

Strategy: The stoichiometry of the reaction allows relative reaction rates to be determined.

Step 1: Balance the equation.

Step 2: Write the relative reaction rates.

Step 3: Determine the rate of formation of bromine from the rate of disappearance of HBr.

POST-LECTURE:

Work Problems: *One of the best ways to learn chemistry is through solving problems. Go to the textbook and work through the Worked Examples for this section, then go to the end of chapter and work the problems there. When you feel confident that you can correctly answer problems like the ones listed below for each objective, check off each of the Learning Objectives for this section above.*

LO 14.1: Problems 14.39, 14.50, 14.52, 14.54
LO 14.2: Worked Example 14.1; Problems 14.56, 14.58

Section 14.2 Rate Laws and Reaction Order

PRE-LECTURE:

Learning Objectives: *These are the skills you should be proficient in after reading this section and practicing the associated problems. Check off the boxes below once you have gone through this section, answered the associated questions in your textbook, and feel confident you have mastered the skill.*

- ☐ **LO 14.3** Find the order of a reaction with respect to each reactant, the overall order of a reaction, and the units of the rate constant, given the rate law.
- ☐ **LO 14.4** Predict the change in reaction rate when the concentration of a reactant changes by a specified amount.

Definitions: *To learn, you must do. This is your workbook. Fill in the definitions from the glossary of the textbook as a first step to understanding the material.*

— **rate law—**____

____

— **rate constant—** ____________________________________

____

— **reaction order—** ___________________________________

____

Summary: *Read through this section and fill in the missing words and equations. Reference the textbook when you are not confident in your answers.*

The rate law for a reaction indicates its dependence on the concentration of each reactant. For a general reaction:

$$a\,\mathrm{A} + b\,\mathrm{B} \rightarrow \text{products},$$

rate =

In a **rate law,** k is a proportionality constant called the **rate constant.**

The exponents m and n must be experimentally determined because they are unrelated to the coefficients in the chemical reaction. These coefficients determine the **reaction order,** which is the sum of the exponents. So, a reaction in which $m + n = 2$ is said to be second order, a reaction in which $m + n = 3$ is said to be third order, and so on.

The exponents themselves are usually small positive integers, but in complex reactions, they can be negative, zero, or even fractions.

- $m = -1$: if the concentration is doubled, the rate __________ by a factor of __________.
- $m = 0$: the rate is __________ of concentration.
- $m = 1$: if the concentration is doubled, the rate ________.
- $m = 2$: if the concentration is doubled, the rate ________ by a factor of ________.

POST-LECTURE:

Work Problems: *One of the best ways to learn chemistry is through solving problems. Go to the textbook and work through the Worked Examples for this section, then go to the end of chapter and work the problems there. When you feel confident that you can correctly answer problems like the ones listed below for each objective, check off each of the Learning Objectives for this section above.*

LO 14.3: Worked Example 14.2; Problems 14.60, 14.61
LO 14.4: Problems 14.40, 14.62, 14.64

Section 14.3 Method of Initial Rates: Experimental Determination of a Rate Law

PRE-LECTURE:

Learning Objectives: *These are the skills you should be proficient in after reading this section and practicing the associated problems. Check off the boxes below once you have gone through this section, answered the associated questions in your textbook, and feel confident you have mastered the skill.*

☐ **LO 14.5** Determine the rate law and rate constant using initial rate and concentration data.

Summary: *Read through this section and fill in the missing words and equations. Reference the textbook when you are not confident in your answers.*

Determining the reaction order experimentally is accomplished by measuring the initial rate of a reaction with differing initial concentrations. This is done carefully, so as to make it possible to see what effect changing each reaction has independently.

__________ rates are always used for these determinations because the reverse reaction will begin to occur, even if only slightly, as the reaction continues.

Only reactants and catalysts appear in the rate law.

Once the rate law is determined, it can be used to solve for the rate constant. Rate constants are temperature-dependent and are a characteristic of a reaction. The units of k depend on the order of the reaction.

Workbook Problem 14.2

What are the units for zeroth-order, first-order, and second-order rate constants if the rate is expressed as M/s?

Strategy: Determine the units needed to cancel out to the rate of each reaction.

Step 1: Zeroth-order reaction:

Step 2: First-order reaction:

Step 3: Second-order reaction:

Workbook Example 14.2

PROBLEM:
For the following theoretical reaction:

$$2\ A + 3\ B \rightarrow A_2B_3$$

the following measurements were made:

Initial [A] (M)	Initial [B] (M)	Rate of formation of A_2B_3
0.1	0.1	$1\ M \cdot s^{-1}$
0.2	0.1	$2\ M \cdot s^{-1}$
0.1	0.2	$1\ M \cdot s^{-1}$

Determine the rate law, reaction order, and rate constant.

SOLUTION:
What this shows is that doubling the concentration of A doubles the rate, while doubling the concentration of B does not affect the rate.

From this, the rate law can be determined as

$$rate = k[A]^1[B]^0 = k[A]$$

This is a first-order reaction.

To solve for *k*, the rate expression must be rearranged:

$$k = \frac{rate}{[A]} = \frac{1M \cdot s^{-1}}{0.1M} = 10\, s^{-1}$$

The unit of s^{-1} is characteristic of a first-order reaction.

Workbook Example 14.3

PROBLEM:
The following reaction

$$2\ NO(g) + 2\ H_2(g) \rightarrow N_2(g) + 2\ H_2O(g)$$

is first order in H_2 and second order in NO. Determine the rate law and the reaction order. How does the reaction rate change if the concentration of NO is doubled and the concentration of H_2 is held constant? How does the reaction rate change if the concentration of H_2 is cut in half and the concentration of NO is held constant?

SOLUTION:
Based on the information given, the rate law is

$$rate = k[NO]^2[H_2]$$

The reaction is third order.

If the concentration of NO is doubled while the concentration of H_2 is held constant, the rate will quadruple. If the concentration of H_2 is halved, the rate will also be halved.

Workbook Problem 14.3

The following data were collected for this reaction:

$$2\ NO_2(g) + F_2(g) \rightarrow 2NO_2F(g)$$

Exp	$[NO_2]$ (M)	$[F_2]$ (M)	Rate $(M \cdot s^{-1})$
1	0.10	0.10	0.026
2	0.20	0.10	0.051
3	0.20	0.20	0.103

Determine the rate law from this data. What is the order of the reaction with respect to each reactant? What is the overall order of the reaction? Calculate the value of *k*.

Strategy: The rate law for the reaction is

$$rate = k[NO_2]^m[F_2]^n$$

To find *m* and *n*, compare the change in concentration with the change in rate. Verify your answer by using the data from any of the three experiments.

Step 1: Determine the order of the reaction with respect to NO_2.

Step 2:	Determine the order of the reaction with respect to F_2.
Step 3:	Write the equation for the rate law.
Step 4:	Calculate the value of *k*.

POST-LECTURE:

Work Problems: *One of the best ways to learn chemistry is through solving problems. Go to the textbook and work through the Worked Examples for this section, then go to the end of chapter and work the problems there. When you feel confident that you can correctly answer problems like the ones listed below for each objective, check off each of the Learning Objectives for this section above.*

LO 14.5: Worked Examples 14.3, 14.4; Problems 14.66–14.68

Section 14.4 Integrated Rate Law: Zeroth-Order Reactions

PRE-LECTURE:

Learning Objectives: *These are the skills you should be proficient in after reading this section and practicing the associated problems. Check off the boxes below once you have gone through this section, answered the associated questions in your textbook, and feel confident you have mastered the skill.*

☐ **LO 14.6** Use the integrated rate law to determine the half-life and the concentrations remaining at various times for a zeroth-order reaction.

☐ **LO 14.9** Determine the reaction order and rate constant graphically. (Sections 14.4–14.6)

Definitions: *To learn, you must do. This is your workbook. Fill in the definitions from the glossary of the textbook as a first step to understanding the material.*

— **integrated rate law—** ______________________________

____

— **zeroth-order reaction—** ______________________________

____

Summary: *Read through this section and fill in the missing words and equations. Reference the textbook when you are not confident in your answers.*

Using calculus, it is possible to convert a zeroth-order rate law:

$$rate = -\frac{\Delta[A]}{\Delta t} = k[A]^0 = k$$

to an integrated form:

[ ]

A plot of [A] versus time is a straight line for a zeroth-order reaction where the slope = $-k$, and the y-intercept = $[A]_0$.

Workbook Problem 14.4

A general zeroth-order reaction gave the following rate data:

Exp	[A] (M)	Rate ($M \cdot s^{-1}$)
1	0.10	0.026
2	0.20	0.026
3	0.40	0.026

Determine the rate constant and the amount of reactant remaining after 20 seconds with an initial concentration of 1.0 M.

Strategy: Substitute into the rate law for zeroth-order reactions, solve for [A] after 20 seconds.

Step 1: Solve for k.

Step 2: Substitute into the integrated rate law and solve for [A] after 20 seconds.

POST-LECTURE:

Work Problems: *One of the best ways to learn chemistry is through solving problems. Go to the textbook and work through the Worked Examples for this section, then go to the end of chapter and work the problems there. When you feel confident that you can correctly answer problems*

like the ones listed below for each objective, check off each of the Learning Objectives for this section above.

LO 14.6: Worked Example 14.5; Problems 14.9, 14.10

Section 14.5 Integrated Rate Law: First-Order Reactions

PRE-LECTURE:

Learning Objectives: *These are the skills you should be proficient in after reading this section and practicing the associated problems. Check off the boxes below once you have gone through this section, answered the associated questions in your textbook, and feel confident you have mastered the skill.*

- ☐ **LO 14.7** Use the integrated rate law to determine the half-life and the concentrations remaining at various times for a first-order reaction.
- ☐ **LO 14.9** Determine the reaction order and rate constant graphically. (Sections 14.4–14.6)

Definitions: *To learn, you must do. This is your workbook. Fill in the definitions from the glossary of the textbook as a first step to understanding the material.*

— **first-order reaction—** ______________________________

____

— **half-life—** ______________________________

____

Summary: *Read through this section and fill in the missing words and equations. Reference the textbook when you are not confident in your answers.*

A first-order rate law:

$$rate = -\frac{\Delta[A]}{\Delta t} = k[A]$$

converted to an integrated form:

[ ]

can be rearranged to the $y = mx + b$ form:

[ ]

Graphically, a plot of ln [A] versus time gives a straight line if the reaction is first order in A. The slope is again −k. This is indicative of a first-order reaction: a concentration versus time plot will be curved, but a plot of ln A versus time will be straight.

Workbook Example 14.4

PROBLEM:
When sucrose reacts with water, glucose is formed according to the reaction

$$C_{12}H_{22}O_{11} + H_2O \rightarrow 2\ C_6H_{12}O_6$$

This reaction follows first-order kinetics with respect to the sucrose. Calculate the value of *k* if it takes 3.00 hr for the concentration of sucrose to decrease from 0.0500 M to 0.0442 M. Determine the amount of time required for the reaction to go to 95% completion.

SOLUTION:
To calculate the value of *k*, we simply substitute the data given into the first-order integrated rate equation.

$$\ln = \frac{0.0442}{0.0500} = -k(3.00\ h)$$
$$k = 4.11 \times 10^{-2}\ h^{-1}$$

To determine the amount of time needed for the reaction to go to 95% completion determine the concentration of sucrose at that point. The amount of sucrose remaining at 95% completion is 5% of the initial concentration:

$$0.05 \times 0.0500\ M = 0.0025\ M$$

This can be substituted into the integrated rate law with the rate constant previously calculated:

$$\ln \frac{0.0025}{0.0500} = -\left(4.11 \times 10^{-2}\ h^{-1}\right)t$$
$$t = 72.9\ h$$

Workbook Problem 14.5

In a first-order reaction, the concentration of the reactant goes from 0.100 M to 0.030 M over the course of 73 min. What is *k*, and how long will it take for the reaction to go to 80% completion?

Strategy: Substitute into the integrated rate law for a first-order reaction, then use *k* to solve for the time to 80% completion.

Step 1: Solve for *k*.

Step 2: Determine the concentration of the reactant at 80% completion.

Step 3: Solve for t at 80% completion.

The half-life of a reaction is the time it takes for the reactant concentration to drop to one-half its initial value. From the integrated rate law:

$$\ln\frac{1}{2} = -kt_{1/2}$$
$$-0.693 = -kt_{1/2}$$
$$t_{1/2} = \frac{0.693}{k}$$

For a first-order reaction, half-life is a constant.

Workbook Example 14.5

PROBLEM:
If the reaction of sucrose and water has a rate constant of -4.11×10^{-2}, what is the half-life for the reaction?

SOLUTION:
Substitute the rate constant into the equation for the half-life of a first-order equation.

$$t_{1/2} = \frac{0.693}{4.11\times10^{-2}\ h^{-1}} = 16.9\ h$$

Workbook Problem 14.6

What is the half-life for the reaction described in workbook problem 13.5?

Strategy: Substitute into the half-life equation.

POST-LECTURE:

Work Problems: *One of the best ways to learn chemistry is through solving problems. Go to the textbook and work through the Worked Examples for this section, then go to the end of chapter and work the problems there. When you feel confident that you can correctly answer problems like the ones listed below for each objective, check off each of the Learning Objectives for this section above.*

LO 14.7: Worked Examples 14.6, 14.8; Problems 14.41, 14.42, 14.70, 14.72, 14.79, 14.85

Section 14.6 Integrated Rate Law: Second-Order Reactions

PRE-LECTURE:

Learning Objectives: *These are the skills you should be proficient in after reading this section and practicing the associated problems. Check off the boxes below once you have gone through this section, answered the associated questions in your textbook, and feel confident you have mastered the skill.*

- ☐ **LO 14.8** Use the integrated rate law to determine the half-life and the concentrations remaining at various times for a second-order reaction.
- ☐ **LO 14.9** Determine the reaction order and rate constant graphically.

Definitions: *To learn, you must do. This is your workbook. Fill in the definitions from the glossary of the textbook as a first step to understanding the material.*

— **second-order reaction—**____

____

Summary: *Read through this section and fill in the missing words and equations. Reference the textbook when you are not confident in your answers.*

Second-order reactions are dependent either on the concentration of a single reactant raised to the second power, $[A]^2$, or on two first-order reactants, [A][B].

The rate law for a reaction of the simpler type is:

$$rate = -\frac{\Delta[A]}{\Delta t} = k[A]^2$$

The integrated rate law for a second-order reaction is:

[ ]

A second order reaction can be identified from a plot of ___________ versus ___________. The slope is equal to the rate constant (k), and the y-intercept is equal to $1/[A]_0$.

The half-life equation (taking the integrated rate law and set $[A]_t = 1/2\ [A]_0$) is:

$$t_{1/2} = \frac{1}{k[A]_0}$$

The half-life for a second-order reaction depends on both the rate constant and the initial concentration: as a result, each half-life for a second-order reaction is twice as long as the previous half-life.

Workbook Example 14.6

PROLBEM:
The reaction 2 NOBr(*g*) → 2 NO(*g*) + Br_2(*g*) is a second-order reaction with respect to NOBr. The rate constant for this reaction is k = 0.810 $M^{-1} \cdot s^{-1}$ when the reaction is carried out at a temperature of 10 °C. If the initial concentration of NOBr = 8.3 × 10^{-3} M, how much NOBr will be left after a reaction time of 50 min? What is the half-life of this reaction?

SOLUTION:
We can solve for the amount of NOBr after 50 min by substituting into the integrated rate law for a second-order reaction.

$$\frac{1}{[NOBr]} = \left(0.81\,M^{-1}\,s^{-1}\right)\left(50\text{ min} \times \frac{60\,s}{1\,min}\right) + \frac{1}{8.3 \times 10^{-3}\,M} = 2550\,M^{-1}$$

$$[NOBr] = 3.92 \times 10^{-4}\,M$$

To determine the half-life, again substitute into the equation for the half-life of a second-order reaction:

$$t_{1/2} = \frac{1}{\left(0.810\,M^{-1}\,s^{-1}\right)\left(8.3 \times 10^{-3}\right)} = 149\,s$$

Workbook Problem 14.7

The following data were collected for the second-order reaction NO_2(*g*) → NO(*g*) + O(*g*)

Time(s)	0	100	200	300	400
[NO_2]	0.1000	0.0797	0.0662	0.0567	0.0495

Determine the rate constant, the time required for the reaction to reach 50% completion, and the time required to reach 95% completion.

Strategy: Use the integrated second-order rate law and experimental data to determine the rate constant, then solve for the first half-life (50% completion) and then for 95% completion using the integrated second-order rate law.

Step 1: Solve for k.

Step 2:	Use the equation for half-life to determine the time required for 50% completion.
Step 3:	Determine the concentration of NO_2 at 95% completion (5% of NO_2 remaining).
Step 4:	Substitute the concentrations into the integrated rate law, and solve for *t*.

POST-LECTURE:

Work Problems: *One of the best ways to learn chemistry is through solving problems. Go to the textbook and work through the Worked Examples for this section, then go to the end of chapter and work the problems there. When you feel confident that you can correctly answer problems like the ones listed below for each objective, check off each of the Learning Objectives for this section above.*

LO 14.8: Worked Example 14.9; Problems 14.74–14.77
LO 14.9: Worked Example 14.9; Problems 14.81–14.83

Section 14.7 Reaction Rates and Temperature: The Arrhenius Equation

PRE-LECTURE:

Learning Objectives: *These are the skills you should be proficient in after reading this section and practicing the associated problems. Check off the boxes below once you have gone through this section, answered the associated questions in your textbook, and feel confident you have mastered the skill.*

☐ **LO 14.10** Interpret potential energy diagrams and suggest geometries for successful collisions and transition states using collision theory.

Definitions: *To learn, you must do. This is your workbook. Fill in the definitions from the glossary of the textbook as a first step to understanding the material.*

— **collision theory—** ____

____

— **activation energy (E_a)—** ____

____

— **transition state—**____

____

— **steric factor (*p*)—** ____

____

— **Arrhenius equation—**____

____

— **frequency factor (*A*)—** ____

____

Summary: *Read through this section and fill in the missing words and equations. Reference the textbook when you are not confident in your answers.*

Reaction rates tend to double when reaction temperature is increased by 10 °C. This is consistent with the **collision theory** model of reactions. Increasing the kinetic energy of a reaction system by raising the temperature increases the energy with which the molecules collide, increasing the chances of a successful reaction.

Molecules must be correctly oriented at collision in order to produce a reaction, and an intermediate must form that is an aggregate of all the reactants. This aggregate is known as the ___________ **state,** and the energy required to form it is called the ___________ **energy.**

Evidence for the concept of an activation energy barrier comes from kinetic–molecular theory: only a tiny proportion of gas-phase collisions result in reactions. The proportion of the collisions that do result in reactions is described by the formula:

$$f = e^{-\frac{E_a}{RT}}$$

As the temperature increases in a reaction, more molecules are available to collide at the energies necessary to generate a reaction. In fact, the distribution of these molecules according to both the kinetic–molecular theory and the dependence of reaction rate on temperature are exponential.

Also involved is the orientation of the reactants when they collide: the proportion of molecules colliding with the proper orientation is the **steric factor (*p*).**

These factors are all brought together in the Arrhenius equation:

$$\boxed{k = \qquad\qquad}$$

The parameter A is called the **frequency factor,** or the pre-exponential factor. From this equation, the rate constant decreases as the activation energy (E_a) increases and increases as T increases.

<u>POST-LECTURE:</u>

Work Problems: *One of the best ways to learn chemistry is through solving problems. Go to the textbook and work through the Worked Examples for this section, then go to the end of chapter*

and work the problems there. When you feel confident that you can correctly answer problems like the ones listed below for each objective, check off each of the Learning Objectives for this section above.

LO 14.10: Worked Example 14.10; Problems 14.90–14.92

Section 14.8 Using the Arrhenius Equation

PRE-LECTURE:

Learning Objectives: *These are the skills you should be proficient in after reading this section and practicing the associated problems. Check off the boxes below once you have gone through this section, answered the associated questions in your textbook, and feel confident you have mastered the skill.*

☐ **LO 14.11** Use the Arrhenius equation to calculate the activation energy of a reaction given rate constants at different temperatures.

Definitions: *To learn, you must do. This is your workbook. Fill in the definitions from the glossary of the textbook as a first step to understanding the material.*

— **Arrhenius plot—** ______________________________

____

Summary: *Read through this section and fill in the missing words and equations. Reference the textbook when you are not confident in your answers.*

If the rate constant is known at two different temperatures, the natural log of the equation can be taken:

$$\ln k = \ln A - E_a/RT$$

A plot of ln k versus $1/T$ gives a straight line, with a slope equal to $-E_a/R$ and intercept b = ln A.

It is possible to calculate the activation energy from just two temperatures. To do this, subtract the Arrhenius equation at T_1 from the Arrhenius equation at T_2. This equation can be rearranged to give the two-point form of the Arrhenius equation:

$$\ln\left(\frac{k_2}{k_1}\right) =$$

Workbook Example 14.7

PROBLEM:
The activation energy for the reaction $ClO_2F(g) \rightarrow ClOF(g) + O(g)$ is 186 kJ/mol. If the value of k is $6.76 \times 10^{-4}\ s^{-1}$ at 322 °C, what is the value of k at 150 °C?

SOLUTION:
Convert all temperatures to kelvins, and substitute into the Arrhenius equation:

$$\ln\frac{k_2}{6.76\times10^{-4}\ s^{-1}} = -\frac{1.86\times10^{5}\ J/mol}{8.314\ J/K\cdot mol}\left(\frac{1}{423\ K}-\frac{1}{595\ K}\right)$$

$$\ln\frac{k_2}{6.76\times10^{-4}\ s^{-1}} = -15.3$$

Taking the antilog of both sides gives us

$$e^{\ln\frac{k_2}{6.76\times10^{-4}\ s^{-1}}} = e^{-15.3}$$

$$\ln\frac{k_2}{6.67\times10^{-4}\ s^{-1}} = 2.27\times10^{-7}$$

$$k_2 = 1.51\times10^{-10}\ s^{-1}$$

Workbook Problem 14.8

If a reaction has a rate constant of 0.0123 s^{-1} at 700 K and 0.00342 s^{-1} at 250 K, what is the activation energy of the reaction, and what will the rate constant be at 500 K?

Strategy: Solve for E_a using the two-point Arrhenius equation, then use the same equation to solve for the rate constant at 500 K.

Step 1: Solve for E_a using the two-point Arrhenius equation.

Step 2: Solve for k at 500 K.

POST-LECTURE:

Work Problems: *One of the best ways to learn chemistry is through solving problems. Go to the textbook and work through the Worked Examples for this section, then go to the end of chapter and work the problems there. When you feel confident that you can correctly answer problems like the ones listed below for each objective, check off each of the Learning Objectives for this section above.*

LO 14.11: Worked Example 14.11; Problems 14.94, 14.96, 14.98, 14.102

Section 14.9 Reaction Mechanisms

PRE-LECTURE:

Learning Objectives: *These are the skills you should be proficient in after reading this section and practicing the associated problems. Check off the boxes below once you have gone through this section, answered the associated questions in your textbook, and feel confident you have mastered the skill.*

☐ **LO 14.12** Given a reaction mechanism, write the overall reaction, identify intermediates, and determine the molecularity for each elementary step.

Definitions: *To learn, you must do. This is your workbook. Fill in the definitions from the glossary of the textbook as a first step to understanding the material.*

— **reaction mechanism—** ____

____

— **elementary reaction—** ____

____

— **elementary step—** ____

____

— **reaction intermediate—** ____

____

— **molecularity—**____

____

— **unimolecular reaction—**____

____

— **bimolecular reaction—**____

____

— **termolecular reaction—** ____

____

Summary: *Read through this section and fill in the missing words and equations. Reference the textbook when you are not confident in your answers.*

The ________________is the sequence of molecular events that leads from reactants to products. Each step is known as an ________________or an ________________. Knowing reaction mechanisms makes it possible to increase the efficiency of industrial reactions.

Elementary reaction describes ________________________________.
Overall reaction describes ________________________________.

A species that is formed then consumed in the course of reaction is known as ____________. These do not appear in the overall equation.

Elementary reactions are classified based on their molecularity: the number of molecules or atoms on the reactant side. A unimolecular reaction involves ________reactant molecule, a bimolecular reaction involves energetic collisions between ________ reactant molecules, and a termolecular reaction (rare) involves collisions of ________ molecules or atoms.

The elementary steps must sum to give the overall equation.

POST-LECTURE:

Work Problems: *One of the best ways to learn chemistry is through solving problems. Go to the textbook and work through the Worked Examples for this section, then go to the end of chapter and work the problems there. When you feel confident that you can correctly answer problems like the ones listed below for each objective, check off each of the Learning Objectives for this section above.*

LO 14.12: Worked Examples 14.12, 14.13; Problems 14.106, 14.108, 14.110

Section 14.10 Rate Laws for Elementary Reactions

PRE-LECTURE:

Learning Objectives: *There are no specific learning objectives for this section; the information is a continuation of the previous section.*

Summary: *Read through this section and fill in the missing words and equations. Reference the textbook when you are not confident in your answers.*

Although overall rate laws must be determined __________, the rate law for an elementary reaction will contain the concentration of each reactant raised to an exponent equal to its coefficient in the chemical equation.

For a unimolecular reaction, the rate will always be __________ order in the concentration of reactant, because the single molecule is being converted to products.

For a bimolecular reaction, the rate depends on the frequency of collisions between the reactant molecules, and thus on the concentration of those molecules. These will be __________ order reactions.

This applies only to elementary reactions, not overall reactions.

Workbook Example 14.8

PROBLEM:
Write the rate law for the following elementary reaction:

$$2\ H_2(g) + O_2(g) \rightarrow 2\ H_2O$$

SOLUTION:
The rate law for an elementary reaction is equal to the concentration of the reactants raised to their stoichiometric coefficients:

$$rate = k[H_2]^2[O_2]$$

Section 14.11 Rate Laws for Overall Reactions

PRE-LECTURE:

Learning Objectives: *These are the skills you should be proficient in after reading this section and practicing the associated problems. Check off the boxes below once you have gone through this section, answered the associated questions in your textbook, and feel confident you have mastered the skill.*

- ☐ **LO 14.13** For mechanisms with an initial slow step (a) predict the rate law for the overall reaction or (b) propose a mechanism that agrees with the experimental rate law.
- ☐ **LO 14.14** For mechanisms with an initial fast equilibrium step, predict the rate law.

Definitions: *To learn, you must do. This is your workbook. Fill in the definitions from the glossary of the textbook as a first step to understanding the material.*

— **rate-determining step—** ____

____

Summary: *Read through this section and fill in the missing words and equations. Reference the textbook when you are not confident in your answers.*

The experimentally observed rate law for an overall reaction depends on the reaction mechanism.

When the overall reaction occurs in two or more steps, the slowest step will determine the rate of the overall reaction. This is known as the ________________.

The rate-determining step will determine the rate law for the overall reaction and will match the predicted rate law for the elementary reaction.

When a rate law for an overall reaction is proposed, the plausible mechanism must meet two criteria:

- ____
- ____

To establish a reaction mechanism, the overall rate law must first be experimentally determined, then a series of elementary steps devised. The rate law can then be predicted based on the rate-determining step. If the observed and predicted rate laws agree, the proposed method is a possible one.

Workbook Example 14.9

PROBLEM:
Given the following reaction mechanism:

$$A_2 \rightarrow 2\,A$$
$$2\,A + 3\,H_2O \rightarrow 2\,AH_3 + 3/2\,O_2$$
$$2\,AH_3 + 4\,O_2 \rightarrow 2\,HAO_3 + 2\,H_2O$$

1. Determine the overall reaction.
2. Identify the reaction intermediates, and calculate the molecularity of each step.
3. Identify the rate law if the second step is the rate-determining step.

SOLUTION:
To determine the overall reaction, sum up the elementary reactions and cancel out the species that occur on both sides:

$$A_2 + 2\,A + 3\,H_2O + 2\,AH_3 + 4\,O_2 \rightarrow 2\,A + 2\,AH_3 + 3/2\,O_2 + 2\,HAO_3 + 2\,H_2O$$

Overall reaction: $A_2 + H_2O + 5/2\,O_2 \rightarrow 2\,HAO_3$

The reaction intermediates are the species that appear in the reaction mechanism but not in the overall reaction: A, AH_3.

The molecularity of the first step = 1 (unimolecular).
The molecularity of the second step = 5.
The molecularity of the third step = 6.
The rate law is

$$rate = k[A]^2[H_2O]^3$$

Workbook Problem 14.9

What rate law and overall reaction would be expected from the following mechanism:

$2\,A \rightarrow A_2$	rate-determining step
$A_2 + B \rightarrow A_2B$	fast step

Strategy: Write the expected rate law based on the rate-determining step, and sum the reactions for the overall reaction.

Step 1: Write the expected rate law.

Step 2: Sum for the overall reaction.

POST-LECTURE:

Work Problems: *One of the best ways to learn chemistry is through solving problems. Go to the textbook and work through the Worked Examples for this section, then go to the end of chapter and work the problems there. When you feel confident that you can correctly answer problems like the ones listed below for each objective, check off each of the Learning Objectives for this section above.*

LO 14.13: Worked Example 14.14; Problems 14.46, 14.111, 14.112
LO 14.14: Worked Example 14.15; Problems 14.113–14.115

Section 14.12 Catalysis

PRE-LECTURE:

Learning Objectives: *These are the skills you should be proficient in after reading this section and practicing the associated problems. Check off the boxes below once you have gone through this section, answered the associated questions in your textbook, and feel confident you have mastered the skill.*

- ☐ **LO 14.15** Use a molecular diagram for a catalyzed reaction to determine the rate law and propose a mechanism consistent with the rate law.
- ☐ **LO 14.16** Identify catalysts and intermediates in the mechanism of a catalyzed reaction.
- ☐ **LO 14.17** Interpret a potential energy diagram representing the mechanism in a catalyzed reaction.

Definitions: *To learn, you must do. This is your workbook. Fill in the definitions from the glossary of the textbook as a first step to understanding the material.*

— **catalyst—** ____

____

Summary: *Read through this section and fill in the missing words and equations. Reference the textbook when you are not confident in your answers.*

A ___________ is a substance that increases the rate of a reaction without being consumed in the reaction. Catalysts are very important in the chemical industry, where they increase reaction efficiencies, and in living organisms where enzymes (biological catalysts) facilitate biologically essential reactions.

Catalysts make a lower-energy mechanism available for reaction, either by increasing the frequency factor (for instance, by ensuring that reactants are brought together in the proper configuration to react) or, more commonly, by reducing the activation energy.

POST-LECTURE:

Work Problems: *One of the best ways to learn chemistry is through solving problems. Go to the textbook and work through the Worked Examples for this section, then go to the end of chapter*

and work the problems there. When you feel confident that you can correctly answer problems like the ones listed below for each objective, check off each of the Learning Objectives for this section above.

LO 14.15: Worked Example 14.16; Problems 14.31, 14.32, 14.47
LO 14.16: Problems 14.118, 14.119, 14.121
LO 14.17: Problem 14.48

Section 14.13 Homogeneous and Heterogeneous Catalysts

PRE-LECTURE:

Learning Objectives: *There are no specific learning objectives for this section; the information is a continuation of the previous section.*

Definitions: *To learn, you must do. This is your workbook. Fill in the definitions from the glossary of the textbook as a first step to understanding the material.*

— **homogeneous catalyst—** ____

____

— **heterogeneous catalyst—** ____

____

Summary: *Read through this section and fill in the missing words and equations. Reference the textbook when you are not confident in your answers.*

___________ catalysts exist in the same phase as the reactants—an aqueous enzyme, for example.

___________ catalysts are in a different phase than the reactants: a metal mesh that provides a reaction surface, for example. The mechanism for heterogeneous catalysis is not well understood, but seems to involve three steps:

- **Adsorption** of the reactants to the reaction surface
- Facilitation of the reaction
- **Desorption** of products from the reaction surface

Heterogeneous catalysts are the most important for industrial processes, in part because they are easily separated from the reactants and products. The catalytic converters in the exhaust systems of newer automobiles use heterogeneous catalysts to convert pollutants to water, carbon dioxide, nitrogen, and oxygen.

Self-Test

This section is intended to test your knowledge of the material covered in this chapter. Think through these problems and make certain you understand what the problem is about—ask yourself if the answer makes sense. Many of these questions are linked to the chapter learning goals. Therefore, successful completion of these problems indicates that you have achieved the learning goals for this chapter. You will receive the greatest benefit from this section if you use it as a mock exam. Do not use your notes, book, Internet, or friends when taking the test. There is no grade and you only shortchange yourself by getting help with this activity. As you take the test, mark the problems with your level of confidence with the answer.

★ *Use a star for problems you know you got right.*
? *Use a question mark for those you think you got right.*
x *Use an x for those you are guessing on.*

If you get a question right that you marked with a star, you can be confident that you know the associated LO.

If you get a question right that you marked with a question mark, go back and work a few more problems and test yourself again tomorrow.

If you get a question wrong or mark it with an x, go back and read the section again, work a few problems today, a few tomorrow, and a few the next day before testing yourself again.

This will help you to discover which topics you have learned thoroughly and which topics you need to study in more detail.

True–False

1. The rate of a reaction can change as the reaction proceeds.
2. The exponents in rate laws cannot be experimentally determined.
3. The order of a reaction is determined by adding up the coefficients in the balanced equation.
4. The rate of a reaction is not the same as the rate constant unless the reaction is of the zeroth order.
5. The rate of a first-order reaction is dependent on the concentration of a single reactant raised to the first power, $[A]^1$.
6. The half-life of a reaction is half of the original amount of a reactant.
7. The half-life of a second-order reaction is dependent on the initial concentration.
8. Zeroth-order reactions are fairly common.
9. Reactions can occur in more than one step, and all the steps occur at the same rate.
10. High temperatures increase the likelihood that reactants will collide with sufficient energy to react.
11. Activation energy can be negative for exothermic reactions.
12. A catalyst is unchanged at the end of a reaction.
13. Biological reactions rarely involve catalysts.
14. Catalysts can be in a different phase than the rest of the reactant mixture.

Multiple Choice

15. The overall reaction order for the rate law, rate = $k[A]^2[B]$ is
   a. two b. three c. five d. six

16. A reaction has the rate law, rate = $k[A]^2$. If the initial rate for this reaction is 0.15 mol/L · sec, and the concentration of A is doubled, the new initial rate for this reaction will be
   a. 0.45 mol/L · sec
   b. 1.20 mol/L · sec
   c. 0.60 mol/L · sec
   d. 1.35 mol/L · sec

17. If a plot of ln [A] versus time gives a straight line for a particular reaction, that reaction is
   a. first order b. ½ order c. second order d. zeroth order

18. The reaction order for the elementary reaction step, A + B → products, is
   a. second order
   b. third order
   c. unable to be determined from the balanced equation
   d. first order

19. For every collision that occurs in a reaction
   a. products are formed
   b. the particles have enough kinetic energy to overcome the potential energy barrier
   c. a small fraction of molecules will overcome the potential energy barrier to reaction
   d. none of the above

20. The exponents in a rate law indicate the
   a. dependence of the rate on time
   b. sum of the coefficients in a balanced equation
   c. time required for a certain amount of reactant to disappear
   d. order of the reaction

21. The half-life of a first-order reaction with $k = 3.23 \times 10^{-3}\ s^{-1}$ is
   a. unable to be determined
   b. 1.62×10^{-3} s
   c. 3.10×10^{2}
   d. 2.15×10^{2} s

22. The half-life of a second-order reaction depends on
   a. the initial concentration
   b. only the rate constant
   c. the rate constant and the initial concentration
   d. none of the above

23. Reaction rates tend to double when the temperature is
   a. doubled
   b. increased by 10 °C
   c. increased by 100 K
   d. increased by 100 °C

24. An activated complex
   a. has more energy than either the reactants or products
   b. is the configuration of the atoms at the maximum of the potential energy barrier
   c. is a weak linking of all the atoms involved in reaction
   d. all of the above

Matching

25. activation energy
26. bimolecular reaction
27. catalyst
28. kinetics
29. decay constant
30. enzyme
31. half-life
32. heterogeneous catalyst
33. homogeneous catalyst
34. initial rate
35. molecularity
36. rate-determining step
37. steric factor
38. termolecular reaction
39. transition state
40. zeroth-order reaction

a. biological catalyst
b. study of rates and mechanisms of chemical reactions
c. having a different phase than the reaction
d. involving two molecules
e. not dependent on the concentration of reactants
f. first-order rate constant for radiodecay
g. slowest elementary reaction in an overall reaction mechanism
h. factor related to orientation of molecules in collisions and their ability to react
i. occurring when the concentration of products is zero
j. involving three particles
k. the potential energy barrier to reaction
l. reaction intermediate with the maximum potential energy
m. time required for the concentration of a reactant to drop to one-half its initial value
n. species that lowers the activation energy without being consumed in the reaction
o. in the same phase as the reaction
p. the number of particles, whether atoms or molecules, involved in a reaction

Fill in the Blanks

41. The ______________ ___________ is defined as either the _____________________ in concentration of a reactant or the ___________________ in the concentration of a ___________________.

42. Rate laws depend on the concentration(s) of reactants and/or a proportionality constant known as the _____________ ________________, abbreviated _______.

43. In a first-order reaction, doubling the concentration of a reactant multiplies the rate by a factor of ___. In a second-order reaction, doubling the concentration of a reactant multiplies the rate by a factor of ___________. In a third-order reaction, doubling the concentration of a reactant multiplies the rate by a factor of _____________. In a zeroth-order reaction, doubling the concentration of a reactant multiplies the rate by a factor of _______________.

44. In experimental determinations of rate laws, initial rates of reaction are used because when the products reach a certain concentration, it is likely that the _________________ _________________ will begin to occur.

45. If a plot of log [A] versus time gives a straight line, the reaction is ___________________ with respect to A.

46. Reaction mechanisms can be made up of individual ______________________ _________, each of which will have its own rate. The overall rate of the reaction will be determined by the ________ __________________ __________, which is the ______________ step.

47. The Arrhenius equation can be used to calculate the __________________ _____________ of a reaction, the energy required to form a ________________ __________ and from there to form products.

48. A catalyst can either increase the efficiency of collisions or make a ______________ _________ reaction mechanism available.

49. When the temperature increases by 10 °C, reaction rates tend to ____________________.

50. Heterogeneous catalysts ______________ reactants, facilitate reactions, then ______________ products.

Problems

51. The following data for the reaction A + B → C was collected. Calculate an estimate of the initial rate and the rate between 30 and 75 sec.

[C] (mol/L)	Time (s)
0	0
0.005	15
0.010	30
0.015	45
0.017	60
0.019	75
0.020	90
0.021	105
0.022	120

52. The rate of disappearance of PH_3 for the reaction $4\ PH_3(g) \rightarrow P_4(g) + 6\ H_2(g)$ is 1.67×10^{-2} mol/L · sec. What is the rate of appearance of H_2?

53. The following data were collected for the reaction $2\ NO(g) + Cl_2(g) \rightarrow 2\ NOCl(g)$

[NO] (mol/L)	[Cl_2] (mol/L)	Initial Rate mol/(L·sec)
0.025	0.025	15.9
0.050	0.025	63.6
0.025	0.050	31.8

Determine the rate law from this data. What is the order of the reaction with respect to each reactant? What is the overall order of the reaction? Provide the rate law, and solve for k.

54. The following data were collected for the reaction 3 A + 2 B → 2 D

[A] (mol/L)	[B] (mol/L)	Initial Rate mol/(L·sec)
0.150	0.150	6.70×10^{-2}
0.300	0.150	0.134
0.150	0.300	6.70×10^{-2}

Determine the rate law from this data. What is the order of the reaction with respect to each reactant? What is the overall order of the reaction? What is the rate law? Calculate the value of k. If the initial concentration of A = 0.542 M and the initial concentration of B = 0.830 M, what would the initial rate be?

55. The rate law for the decomposition of O_3 to O_2 is second order in ozone. The rate constant for this reaction equals 1.40×10^{-2} L/(mol · sec). If the initial concentration of ozone is 2.37 M, how much O_3 will be present after 12 hr?

56. The following data were collected for the reaction 2 HI(g) → H_2(g) + I_2(g)

[HI] (mol/L)	Time (min)
2.50	0
1.45	3
1.02	6
0.788	9
0.641	12
0.541	15
0.468	18
0.412	21
0.368	24
0.332	27
0.303	30

Determine the order of this reaction. Calculate the value of k and the half-life from the initial reaction.

57. The proposed reaction mechanism for a reaction is

$$NO_2(g) + Cl_2(g) \rightarrow NO_2Cl(g) + Cl(g)$$
$$Cl(g) + NO_2(g) \rightarrow NO_2Cl(g)$$

What is the overall reaction? What is the molecularity of each step? What intermediates are there? If the first step is the slow step in the mechanism, write the expected rate law. What will the order of the reaction be?

58. If the rate constant for a reaction is $k = 3.79 \times 10^{-5}$ L/mol · sec at 108 °C and the activation energy for the reaction is 11.8 kJ/mol, what is the value of A?

59. The activation energy is E_a = 98.62 kJ/mol for the reaction 2 NOCl → 2 NO + Cl_2. If $k = 7.2 \times 10^{-5}$ s^{-1} at 67 °C, at what temperature will $k = 3.2 \times 10^{-3}$ s^{-1}?

<u>*Challenge Problem*</u>

60. For the reaction $C_2H_5I + OH^- \rightarrow C_2H_5OH + I^-$, E_a = 79.7 kJ/mol and A = 1.09×10^9 $M^{-1}s^{-1}$at 45°C. If the concentration of C_2H_5I is doubled while [OH^-] remains constant, the rate doubles. If the [OH^-] is doubled while [C_2H_5I] remains constant, the rate doubles. If a 250 mL solution of 0.475 g KOH in ethanol is mixed with a 250 mL solution of 1.378 g C_2H_5I in ethanol, what is the initial rate at 35 °C?

Test Yourself: *Take the Practice Test on pages 587–589 of your textbook. Again, use a testing environment using only those items that your instructor will allow you to have access to during the test and mark each question with a star, question mark, or x to denote your confidence level. Once you have finished you can check off the LO at the beginning of this chapter based on which questions you starred and got correct. Go back and spend more time with the topics from the other learning objectives.*

CHAPTER FIFTEEN

Chemical Equilibrium

Learning Objectives

As a result of reading and studying this chapter, you should be able to:

- ☐ **LO 15.1** Describe the characteristics of a reaction in chemical equilibrium.
- ☐ **LO 15.2** Write an equilibrium constant expression K_c for gas- and solution-phase reactions.
- ☐ **LO 15.3** Determine a new equilibrium constant expression when reactions are manipulated or added.
- ☐ **LO 15.4** Calculate the value of the equilibrium constant K_c given concentrations of reactant and products.
- ☐ **LO 15.5** Calculate the equilibrium constant from a molecular representation of a reaction mixture.
- ☐ **LO 15.6** Write the equilibrium expression K_p in terms of partial pressures of reactants and products. Relate the equilibrium constants K_p and K_c.
- ☐ **LO 15.7** Write equilibrium constant expressions for reactions involving heterogeneous equilibria.
- ☐ **LO 15.8** Determine the extent of a reaction given the equilibrium constant.
- ☐ **LO 15.9** Predict the direction a reaction will shift to reach equilibrium given nonequilibrium concentrations.
- ☐ **LO 15.10** Calculate concentrations or partial pressures of products and reactants in equilibrium.
- ☐ **LO 15.11** Calculate concentrations of reactants and products in equilibrium using the quadratic equation to solve the mathematical operations.
- ☐ **LO 15.12** Predict the direction a reaction in equilibrium will shift when concentrations of reactants or products are changed.
- ☐ **LO 15.13** Predict the direction a reaction in equilibrium will shift when volume or pressure is changed.
- ☐ **LO 15.14** Predict the direction a reaction in equilibrium will shift when temperature is changed.
- ☐ **LO 15.15** Describe the effect of a catalyst on the equilibrium constant for a reaction.
- ☐ **LO 15.16** Relate the equilibrium constant to the rate of the forward and reverse reactions.

Chapter Summary

This chapter introduces the concept of chemical equilibrium, the state reached when the concentration of reactants and products remains constant over time despite both forward and backward reactions continuing to occur. Equilibrium constants can be related to either concentrations or partial pressures; these constants can be used to determine the final concentrations of reactants and products in an equilibrium mixture. When equilibria are stressed through changing conditions, the equilibrium will shift to minimize the stress. Changes in temperature, pressure/volume, and concentrations all shift the equilibrium mixture. Adding a catalyst will decrease the time it takes to reach equilibrium, but does not change the equilibrium concentrations. At the end of the chapter, the link between kinetics

(Chapter 14) and equilibrium is explained along with the reason that equilibrium constants are temperature-dependent.

The Chapter in Detail

Section 15.1 The Equilibrium State

PRE-LECTURE:

Learning Objectives: *These are the skills you should be proficient in after reading this section and practicing the associated problems. Check off the boxes below once you have gone through this section, answered the associated questions in your textbook, and feel confident you have mastered the skill.*

☐ **LO 15.1** Describe the characteristics of a reaction in chemical equilibrium.

Definitions: *To learn, you must do. This is your workbook. Fill in the definitions from the glossary of the textbook as a first step to understanding the material.*

— **chemical equilibrium—** ____

____

— **equilibrium mixture—** ____

____

Review: *Before going through the section make sure you are familiar with these concepts:*

- Section 14.2 reaction rates

Summary: *Read through this section and fill in the missing words and equations. Reference the textbook when you are not confident in your answers.*

Reactions do not always go to completion, but instead reach a ___________ ___________ at which the concentrations of reactants and products no longer change. To indicate that a reaction can proceed in either direction, a double arrow is used. To avoid confusion, species on the left of an equation are always called *reactants* and those on the right are always called *products*, even though the direction of the reaction may be unclear.

Remember from Chapter 14 that as concentration decreases the rate ___________. So, if you start with all reactants, as they react their concentration decreases and thus, the forward rate decreases. As this occurs, the products accumulate, and so, the rate of the reverse begins to increase. As a reaction proceeds, there comes a point when the reverse reaction rate increases enough and the forward reaction rate decreases enough that they proceed at the same rate.

When equilibrium is reached, the reactions, forward and backward, continue to occur, but have the same rates: there is no net conversion of reactants to products.

POST-LECTURE:

Work Problems: *One of the best ways to learn chemistry is through solving problems. Go to the textbook and work through the Worked Examples for this section, then go to the end of chapter and work the problems there. When you feel confident that you can correctly answer problems like the ones listed below for each objective, check off each of the Learning Objectives for this section above.*

LO 15.1: Problems 15.41, 15.52, 15.53

Section 15.2 The Equilibrium Constant K_c

PRE-LECTURE:

Learning Objectives: *These are the skills you should be proficient in after reading this section and practicing the associated problems. Check off the boxes below once you have gone through this section, answered the associated questions in your textbook, and feel confident you have mastered the skill.*

- ☐ **LO 15.2** Write an equilibrium constant expression K_c for gas- and solution-phase reactions.
- ☐ **LO 15.3** Determine a new equilibrium constant expression when reactions are manipulated or added.
- ☐ **LO 15.4** Calculate the value of the equilibrium constant K_c given concentrations of reactant and products.
- ☐ **LO 15.5** Calculate the equilibrium constant from a molecular representation of a reaction mixture.

Definitions: *To learn, you must do. This is your workbook. Fill in the definitions from the glossary of the textbook as a first step to understanding the material.*

— **equilibrium constant, K_c—** ____

____

— **equilibrium equation—** ____

____

Summary: *Read through this section and fill in the missing words and equations. Reference the textbook when you are not confident in your answers.*

When reactions are permitted to go to equilibrium, the concentrations of reactants and products are related by the equilibrium constant, K_c.

For the generic reaction $a\text{ A} + b\text{ B} \rightleftharpoons c\text{ C} + d\text{ D}$, the equilibrium constant is:

$K_c =$

The equilibrium constant is temperature-dependent and unitless: all of the concentrations are considered to be ratios and so are divided by their standard state concentrations (1 M), which eliminates the units from the concentration values. Pure solids and pure liquids are excluded from the equilibrium expression.

For the reverse reaction, the equilibrium constant will be the inverse of the constant for the forward reaction. This makes it important that the form of the equation be given.

When chemical equations are added, the equilibrium expressions are multiplied.

$$\begin{array}{l} a\,A \rightleftharpoons b\,B \\ \underline{+\; b\,B \rightleftharpoons c\,C} \\ a\,A \rightleftharpoons c\,C \end{array} \qquad K_c \text{ overall} = [B]^b/[A]^a = [C]^c/[B]^b = [C]^c/[A]^a$$

Workbook Example 15.1

PROBLEM:
For the reaction $CO(g) + 2\,H_2(g) \rightleftharpoons CH_3OH(g)$, the equilibrium concentrations are $[CO] = 2.38$ M; $[H_2] = 0.260$ M; $[CH_3OH] = 3.15$ M at a temperature of 110 °C. Calculate the equilibrium constant for this reaction.

SOLUTION:
Write an equilibrium equation for the balanced chemical reaction:

$$K_c = \frac{[CH_3OH]}{[CO][H_2]^2}$$

Substitute the equilibrium concentrations and solve for K_c:

$$K_c = \frac{(3.15)}{(2.38)(0.260)^2} = 19.6$$

Workbook Problem 15.1

A weak acid dissociates according to the following equation:

$$HA(aq) \rightleftharpoons H^+(aq) + A^-(aq)$$

What is the equilibrium constant at 20 °C if a 0.250 M solution is 1.73% ionized?

Strategy: Write the equilibrium equation for K_c, calculate concentrations of all species, and solve.

Step 1: Write the equation for the equilibrium constant.

Step 2:	Determine the concentration of each species.
Step 3:	Solve for K_c.

POST-LECTURE:

Work Problems: *One of the best ways to learn chemistry is through solving problems. Go to the textbook and work through the Worked Examples for this section, then go to the end of chapter and work the problems there. When you feel confident that you can correctly answer problems like the ones listed below for each objective, check off each of the Learning Objectives for this section above.*

LO 15.2: Worked Example 15.1; Problems 15.54, 15.56
LO 15.3: Problems 15.60, 15.62, 15.64
LO 15.4: Worked Example 15.2; Problems 15.66, 15.72
LO 15.5: Worked Example 15.3; Problem 15.43

Section 15.3 The Equilibrium Constant K_p

PRE-LECTURE:

Learning Objectives: *These are the skills you should be proficient in after reading this section and practicing the associated problems. Check off the boxes below once you have gone through this section, answered the associated questions in your textbook, and feel confident you have mastered the skill.*

☐ **LO 15.6** Write the equilibrium expression K_p in terms of partial pressures of reactants and products. Relate the equilibrium constants K_p and K_c.

Definitions: *To learn, you must do. This is your workbook. Fill in the definitions from the glossary of the textbook as a first step to understanding the material.*

— **equilibrium constant, K_p**—____

____

Review: *Before going through the section make sure you are familiar with these concepts:*

- Section 10.3 Ideal gas law
- Section 10.5 Dalton's law of partial pressure
- Section 10.5 $P_a = n_aRT/V$

Summary: *Read through this section and fill in the missing words and equations. Reference the textbook when you are not confident in your answers.*

For gas-phase reactions, ___________ ___________can be used instead of molar concentrations. The ideal gas law can be rearranged to show the relationship between concentration, [A] or (n/V), and pressure:

$$P_A = [A]RT$$

Then these terms are substituted into the equation for an equilibrium constant, the relationship between K_c and K_p can be shown to be:

$$K_P = K_c(RT)^{\Delta n}$$

There Δn is the sum of the coefficients of the gaseous ___________ minus the sum of the coefficients of the gaseous ___________.

Workbook Example 15.2

PROBLEM:
Calculate K_p for the reaction in the previous example.

SOLUTION:
The coefficients of the balanced equation can be used to calculate Δn. The sum of the coefficients on the products side is 1, while the sum of the coefficients on the reactant side is 3. Δn = products − reactants = −2.

Substitute Δn and K_c into the equation, relating K_p to K_c:

$$K_P = (19.6)\left[\left(0.0821\frac{L\cdot atm}{K\cdot mol}\right)(383\,K)\right]^{-2} = 1.98\times10^{-2}$$

Workbook Problem 15.2

Determine the value of K_c and K_p for the reaction

$$PCl_5(g) \rightleftharpoons PCl_3(g) + Cl_2(g)$$

At equilibrium at 250 K, the concentrations measured for each reactant are $[PCl_5]$ = 2.35 M and $[PCl_3] = [Cl_2]$ = 1.56 M.

Strategy: Write the equilibrium equation for K_c.

Step 1: Solve for K_c using the molar concentrations given.

Step 2:	Determine Δn.
Step 3:	Solve for K_p, using the equation that expresses the relationship between K_c and K_p.

POST-LECTURE:

Work Problems: *One of the best ways to learn chemistry is through solving problems. Go to the textbook and work through the Worked Examples for this section, then go to the end of chapter and work the problems there. When you feel confident that you can correctly answer problems like the ones listed below for each objective, check off each of the Learning Objectives for this section above.*

LO 15.6: Worked Example 15.4, 15.5; Problems 15.68, 15.70

Section 15.4 Heterogeneous Equilibria

PRE-LECTURE:

Learning Objectives: *These are the skills you should be proficient in after reading this section and practicing the associated problems. Check off the boxes below once you have gone through this section, answered the associated questions in your textbook, and feel confident you have mastered the skill.*

☐ **LO 15.7** Write equilibrium constant expressions for reactions involving heterogeneous equilibria.

Definitions: *To learn, you must do. This is your workbook. Fill in the definitions from the glossary of the textbook as a first step to understanding the material.*

— **Homogenous equilibria—**____

____

— **Heterogeneous equilibria—** ____

____

Summary: *Read through this section and fill in the missing words and equations. Reference the textbook when you are not confident in your answers.*

In ___________ ___________, all the reactants and products are in the same phase; in ___________ ___________, reactants and products are in a mixture of phases.
Because the concentrations of solids and liquids are constant, they are not included in the equilibrium expression.

Workbook Example 15.3

PROBLEM:
Write the equilibrium expression for the reaction $H_2CO_3(s) \rightleftharpoons CO_2(g) + H_2O(l)$.

SOLUTION: This reaction generates only one product that is not a pure solid or liquid. As a result, the equilibrium constant will be:

$$K_c = [CO_2]$$

Because this is a gas, K_p should be used:

$$K_P = K_c RT$$

POST-LECTURE:

Work Problems: *One of the best ways to learn chemistry is through solving problems. Go to the textbook and work through the Worked Examples for this section, then go to the end of chapter and work the problems there. When you feel confident that you can correctly answer problems like the ones listed below for each objective, check off each of the Learning Objectives for this section above.*

LO 15.7: Worked Example 15.6; Problems 15.74, 15.75

Section 15.5 Using the Equilibrium Constant

PRE-LECTURE:

Learning Objectives: *These are the skills you should be proficient in after reading this section and practicing the associated problems. Check off the boxes below once you have gone through this section, answered the associated questions in your textbook, and feel confident you have mastered the skill.*

- ☐ **LO 15.8** Determine the extent of a reaction given the equilibrium constant.
- ☐ **LO 15.9** Predict the direction a reaction will shift to reach equilibrium given nonequilibrium concentrations.
- ☐ **LO 15.10** Calculate concentrations or partial pressures or products and reactants in equilibrium.
- ☐ **LO 15.11** Calculate concentrations of reactants and products in equilibrium using the quadratic equation to solve the mathematical operations.

Definitions: *To learn, you must do. This is your workbook. Fill in the definitions from the glossary of the textbook as a first step to understanding the material.*

— **Reactant quotient, Q_c—** ____

____

Review: *Before going through the section make sure you are familiar with these concepts:*

- Quadratic equation

Summary: *Read through this section and fill in the missing words and equations. Reference the textbook when you are not confident in your answers.*

The K_c value provides information about the way the reaction proceeds at a specific temperature:

- If K_c is large, over 10^3, the reaction proceeds almost to completion. __________ predominate over __________________.
- If K_c is small, under 10^{-3}, the reaction proceeds very little. __________ predominate over __________________.
- If the value of K_c is intermediate, there will be appreciable quantities of both reactants and products in an equilibrium mixture.

The equilibrium constant can also be used to predict the direction of reaction given the concentrations of reactants and products. The reaction quotient, Q_c, is calculated in exactly the same way as the equilibrium constant expression, but uses concentrations that may or may not be equilibrium values. Comparing Q and K makes it possible to predict which way the reaction will go (pg. 615):

- $Q_c < K_c$; reaction goes from left to right, more __________ will be formed.
- $Q_c > K_c$; reaction goes from right to left, more __________ will be formed.
- $Q_c = K_c$; reaction is at equilibrium, there will be ______________.

If all but one of the equilibrium concentrations is known, the last can be calculated algebraically using the equilibrium constant.

It is also possible to calculate equilibrium concentrations from initial concentrations using the equilibrium coefficient and the balanced chemical equation. To do this, the following steps are required:

- Compare the values of Q_c to K_c to determine the direction of the reaction.
- Write and balance the chemical equation. Make an I.C.E. table that lists:
  - I: the initial concentrations of all reactants and products in the reaction
  - C: the change expected when going to equilibrium; for the change, define the concentration of one of the species as x and use the balanced equation to define the changes in the other species in terms of x.
  - E: the concentration at equilibrium
- Substitute the equilibrium concentrations into the equation and solve for x. If this requires solving a quadratic equation, choose the answer that makes chemical sense (there are no negative concentrations).
- Calculate the other concentrations from the calculated value of x.
- Check your results by substituting the values back into the equilibrium equation.

Workbook Example 15.4

PROBLEM:
The following reaction has a K_c value of 4.18×10^{-9} at 425 °C:

$$2\ HBr(g) \rightleftharpoons H_2(g) + Br_2(g)$$

What is the position of the equilibrium? If the concentrations of all species present are [HBr] = 0.35 M; [H_2] = [Br_2] = 3.2 × 10^{-3} M, is the reaction mixture at equilibrium? If not, in which direction will the reaction proceed?

SOLUTION:
We can determine the position of the equilibrium by looking at the value of K_c. Since $K_c < 10^{-3}$, we know that reactants predominate over products and that the position of the equilibrium is to the left. This is reflected in the concentrations provided, but to determine if the reaction mixture is at equilibrium, we need to calculate Q_c and compare that value to K_c.

$$Q_c = \frac{[H_2][Br_2]}{[HBr]^2} = \frac{(3.2\times10^{-3}\,M)(3.2\times10^{-3}\,M)}{(0.35\,M)^2} = 8.4\times10^{-5}$$

Since $Q_c > K_c$, the reaction is not at equilibrium and will proceed from right to left.

Workbook Example 15.5

PROBLEM:
The reaction, $PCl_5(g) \rightleftharpoons PCl_3(g) + Cl_2(g)$, has $K_c = 85.0$ at a temperature of 760 °C. Calculate the equilibrium concentrations of PCl_5, PCl_3, and Cl_2 if the initial concentration of PCl_5 is 3.50 M. Assume a volume of 1.00 L.

SOLUTION:
To solve this problem, follow the procedure outlined above:

The balanced equation is given.

The initial concentration of PCl_5 is 3.50 M. Let x be the concentration of PCl_5 that reacts on going to the equilibrium state. Since the mole-to-mole ratios are 1:1:1, if x amount of PCl_5 reacts, then x mol/L of PCl_3 and Cl_2 will be present at equilibrium. This can be summarized in the following table:

	PCl_5 (g)	$\rightleftharpoons$	PCl_3 (g)	+	Cl_2 (g)
Initial concentration (M)	3.5		0		0
Change (M)	$-x$		$+x$		$+x$
Equilibrium concentration (M)	$3.5 - x$		x		x

The equilibrium expression for the reaction is

$$K_{eq} = \frac{[PCl_3][Cl_2]}{[PCl_5]}$$

The equilibrium concentrations from the above table can be substituted into this expression, and the equation rearranged to facilitate the use of the quadratic equation.

$$85.0 = \frac{x^2}{3.5 - x}$$
$$x^2 + 85x - 297.5 = 0$$

Use the quadratic equation to solve for x:

$$x = \frac{-85 \pm \sqrt{(85)^2 - 4(1)(-297.5)}}{2}$$

$$x = 3.37 - 88.4$$

A negative number does not make physical sense: the answer is 3.37.

Calculate the equilibrium concentrations:

$$[PCl_5] = 3.5 - 3.37 = 0.13 \text{ M}$$
$$[PCl_3] = [Cl_2] = 3.37 \text{ M}$$

Workbook Problem 15.3

The K_c value for the following reaction is 3.0×10^{-2}. Determine the equilibrium concentration if the initial concentration of water is 11.75 M.

$$C(s) + H_2O(g) \rightleftharpoons CO(g) + H_2(g)$$

Strategy: Follow the previously outlined steps in your textbook.

Step 1: Write the balanced equation for the reaction.

Step 2: Make a table listing the initial concentration, the change in concentration, and the equilibrium concentration. (Let x = the amount of substance that reacts.)

Initial concentration (M)
Change (M)
Equilibrium concentration (M)

Step 3:	From the balanced equation, write the equilibrium equation. Substitute the algebraic expressions for the equilibrium concentrations into the equilibrium equation, and solve for x. Use the quadratic equation if necessary.

POST-LECTURE:

Work Problems: *One of the best ways to learn chemistry is through solving problems. Go to the textbook and work through the Worked Examples for this section, then go to the end of chapter and work the problems there. When you feel confident that you can correctly answer problems like the ones listed below for each objective, check off each of the Learning Objectives for this section above.*

LO 15.8: Worked Example 15.7; Problems 15.76, 15.78
LO 15.9: Worked Example 15.8; Problems 15.80, 15.82
LO 15.10: Worked Examples 15.9, 15.11; Problems 15.84, 15.86, 15.90, 15.116
LO 15.11: Worked Examples 15.10, 15.12; Problems 15.88, 15.92

Section 15.6 Factors That Alter the Composition of an Equilibrium Mixture: Le Châtelier's Principle

PRE-LECTURE:

Learning Objectives: *There are no specific learning objectives for this section; the information in this section is used as a basis for understanding subsequent sections.*

Definitions: *To learn, you must do. This is your workbook. Fill in the definitions from the glossary of the textbook as a first step to understanding the material.*

— **Le Châtelier's principle—** ____

____

Summary: *Read through this section and fill in the missing words and equations. Reference the textbook when you are not confident in your answers.*

If a reaction does not proceed at room temperature and atmospheric pressure, the reaction conditions can be altered to force the reaction to proceed. There are four factors that alter the composition of an equilibrium mixture:

- ______________________________
- ______________________________
- ______________________________
- ______________________________

The fourth (above), addition of a catalyst, only affects the rate at which equilibrium is reached, not the composition of the equilibrium mixture.

Le Châtelier's principle states that if a stress is applied to a reaction mixture at equilibrium, the equilibrium will shift to minimize the stress. There are four ways to "stress" a reaction. They are a change in:

- ______________________
- ______________________
- ______________________
- ______________________

Section 15.7 Altering an Equilibrium Mixture: Changes in Concentration

PRE-LECTURE:

Learning Objectives: *These are the skills you should be proficient in after reading this section and practicing the associated problems. Check off the boxes below once you have gone through this section, answered the associated questions in your textbook, and feel confident you have mastered the skill.*

☐ **LO 15.12** Predict the direction a reaction in equilibrium will shift when concentrations of reactants or products are changed.

Review: *Before going through the section make sure you are familiar with these concepts:*

- Section 14.1 reaction rates

Summary: *Read through this section and fill in the missing words and equations. Reference the textbook when you are not confident in your answers.*

When equilibrium is disturbed a reaction will occur to bring the system back to equilibrium.

- If you add a reactant or product, the reaction shifts to ______________.
- If you remove a reactant or product, the reaction shifts to ______________.

Mathematically, this works by changing Q_c: when a reactant is added or a product removed, Q_c becomes smaller: the reaction must shift right toward products, increasing the numerator again. When a reactant is removed, or a product added, Q_c becomes larger. To return to equilibrium, the reaction must shift left: products will be converted to reactants.

Workbook Example 15.6

PROBLEM:
Predict the results of adding $CaCO_3$, adding CO_2, and removing CO_2 to the following reaction:

$$CaCO_3(s) \rightleftharpoons CaO(s) + CO_2(g)$$

SOLUTION:
Adding $CaCO_3$ will not affect the equilibrium: solids do not appear in either K or Q.

Adding CO_2 will push the reaction to shift left toward reactants, thus reducing the amount of CO_2.

Removing CO_2 will push the reaction to shift right toward products, thereby increasing the amount of CO_2.

Workbook Problem 15.4

The following reaction is at equilibrium:

$$2\ BrNO(g) \rightleftharpoons 2\ NO(g) + Br_2(g)$$

What would be the effect on the equilibrium if (1) BrNO was added, (2) BrNO was removed, (3) NO was added, and (4) Br_2 was removed.

Strategy: Use Le Châtelier's principle.

Step 1: Determine the direction of reaction needed to minimize each stress.

POST-LECTURE:

Work Problems: *One of the best ways to learn chemistry is through solving problems. Go to the textbook and work through the Worked Examples for this section, then go to the end of chapter and work the problems there. When you feel confident that you can correctly answer problems like the ones listed below for each objective, check off each of the Learning Objectives for this section above.*

LO 15.12: Worked Example 15.13; Problems 15.45, 15.124, 15.128

Section 15.8 Altering an Equilibrium Mixture: Changes in Pressure and Volume

PRE-LECTURE:

Learning Objectives: *These are the skills you should be proficient in after reading this section and practicing the associated problems. Check off the boxes below once you have gone through this section, answered the associated questions in your textbook, and feel confident you have mastered the skill.*

☐ **LO 15.13** Predict the direction a reaction in equilibrium will shift when volume or pressure is changed.

Review: *Before going through the section make sure you are familiar with these concepts:*

- Section 10.3 the pressure volume relationship for gases

Summary: *Read through this section and fill in the missing words and equations. Reference the textbook when you are not confident in your answers.*

When changing the volume and/or pressure, it can disturb the equilibrium and the reaction will shift to bring the system back to equilibrium.

- An increase in the pressure or decrease in volume will shift the reaction to the side with ________ moles of gas.
- A decrease in the pressure or increase in volume will shift the reaction in the direction of ________ moles of gas.

Increases in pressure (due to decrease in volume) are akin to increases in concentration for gases, as n/V increases.

- If a reaction has the same number of moles of gaseous products and reactants, changing the pressure will not affect the reaction.
- In heterogeneous equilibria, changes in pressure do not affect solids or liquids, because their volume is essentially independent of pressure.

These effects only are seen when pressure is changed as a result of changing volume. If pressure is increased by adding a nonreacting (inert) gas to the mixture, the equilibrium will not change.

Workbook Problem 15.5

The following reaction is at equilibrium:

$$N_2O_4(g) \rightleftharpoons 2\ NO_2(g)$$

What would be the effect on the equilibrium if the pressure was increased? If the pressure was decreased? If 10 L of neon were added to the reaction mixture at constant volume?

Strategy: Use Le Châtelier's principle.

Step 1: Determine the direction of reaction needed to minimize each stress.

POST-LECTURE:

Work Problems: *One of the best ways to learn chemistry is through solving problems. Go to the textbook and work through the Worked Examples for this section, then go to the end of chapter and work the problems there. When you feel confident that you can correctly answer problems like the ones listed below for each objective, check off each of the Learning Objectives for this section above.*

LO 15.13: Worked Example 15.14; Problems 15.44, 15.120, 15.131

Section 15.9 Altering an Equilibrium Mixture: Changes in Temperature

PRE-LECTURE:

Learning Objectives: *These are the skills you should be proficient in after reading this section and practicing the associated problems. Check off the boxes below once you have gone through this section, answered the associated questions in your textbook, and feel confident you have mastered the skill.*

☐ **LO 15.14** Predict the direction a reaction in equilibrium will shift when temperature is changed.

Review: *Before going through the section make sure you are familiar with these concepts:*

- Section 9.6 heat flow in endothermic and exothermic reactions
- Heat that flows in can be thought of as a reactant
- Heat that flows out can be thought of as a product

Summary: *Read through this section and fill in the missing words and equations. Reference the textbook when you are not confident in your answers.*

When changing the concentration, volume, or pressure of a reaction, the equilibrium constant does not change, only the reaction quotient changes. When the temperature changes, the equilibrium constant itself is altered:

- In an exothermic reaction, the equilibrium constant decreases as temperature increases. To help you remember this, you can use Le Châtelier's principle that states that the system will shift to get rid of added stress. Since in an exothermic reaction, heat can be considered a product, the reaction will shift to the ____________ to get rid of the heat.
- In an endothermic reaction, the equilibrium constant increases as temperature increases. Again, use Le Châtelier's to help you remember. Since in an endothermic reaction, heat can be considered a reactant, the reaction will shift to the ________ to get rid of the heat.

Workbook Example 15.7

PROBLEM:
How will the following changes alter the equilibrium for the reaction?

$$\text{C(graphite)} + CO_2(g) \rightleftharpoons 2CO(g) \qquad \Delta H^\circ = +172.5 \text{ kJ}$$

a. CO is removed from the system.
b. CO_2 is removed from the system.
c. The volume of the container is increased.
d. Graphite is added to the system.
e. The temperature is raised.

SOLUTION:
a. When CO is removed from the system, the equilibrium shifts right.
b. When CO_2 is removed from the system, the equilibrium shifts left.

c. Increasing the volume of the container decreases the pressure in the container. The equilibrium will shift to the right to minimize the change.
d. Adding a solid to the system does not affect the equilibrium since solids are not included in the equilibrium expression.
e. Raising the temperature causes the equilibrium to shift right because the reaction absorbs temperature.

POST-LECTURE:

Work Problems: *One of the best ways to learn chemistry is through solving problems. Go to the textbook and work through the Worked Examples for this section, then go to the end of chapter and work the problems there. When you feel confident that you can correctly answer problems like the ones listed below for each objective, check off each of the Learning Objectives for this section above.*

LO 15.14: Worked Examples 15.15, 15.16; Problems 15.46, 15.51, 15.122, 15.156

Section 15.10 The Link between Chemical Equilibrium and Chemical Kinetics

PRE-LECTURE:

Learning Objectives: *These are the skills you should be proficient in after reading this section and practicing the associated problems. Check off the boxes below once you have gone through this section, answered the associated questions in your textbook, and feel confident you have mastered the skill.*

☐ **LO 15.15** Describe the effect of a catalyst on the equilibrium constant for a reaction.
☐ **LO 15.16** Relate the equilibrium constant to the rate of the forward and reverse reactions.

Review: *Before going through the section make sure you are familiar with these concepts:*

- Section 14.1 reaction rates
- Section 14.8 Arrhenius equation

Summary: *Read through this section and fill in the missing words and equations. Reference the textbook when you are not confident in your answers.*

If the reaction $A + B \rightleftharpoons C + D$ is a simple reaction with an elementary forward and reverse reaction, it has two rate constants:

- Rate of ___________ reaction = $k_f[A][B]$
- Rate of ___________ reaction = $k_r[C][D]$

At equilibrium, these rates are equal to one another, and they can be related to the equilibrium constant:

$$k_f[A][B] = k_r[C][D] \text{ or } \frac{k_f}{k_r} = \frac{[C][D]}{[A][B]} = K_c$$

Thus, the equilibrium constant depends on the reaction rates, and it is the relative values of k_f and k_r that determine the composition of the equilibrium mixture. Looking at the rates helps us to calculate the K_c value and to determine where the equilibrium lies.

If the rate constant for the forward reaction is much larger than that of the reverse reaction, K_c is very ___________, and the reaction goes to completion. Because the reverse reaction is too slow to be detected, this is an irreversible reaction.

If instead the values of k_f and k_r are close to one another, there will be an equilibrium in which all species—reactants and products—will be present.

The temperature dependence of equilibrium constants is also explained by the relationship derived above:

$$\frac{k_f}{k_r} = K_c$$

Because the kinetic constants are temperature-dependent and increase by different amounts when the temperature increases (as described by the Arrhenius equation), equilibrium is also temperature-dependent.

Workbook Example 15.8

PROBLEM:
Calculate the equilibrium constant K_c for the reaction

$$H_2O(l) \rightleftharpoons H^+(aq) + OH^-(aq)$$

given that $k_f = 2.4 \times 10^{-5}\ s^{-1}$ and $k_r = 1.3 \times 10^{11}\ M^{-1} \cdot s^{-1}$ and that the equilibrium expression is

$$K_c = [H^+][OH^-]$$

SOLUTION:
To calculate K_c, use the equation

$$K_c = \frac{k_f}{k_r} = \frac{2.4\times10^{-5}\, s^{-1}}{1.3\times10^{11}\, s^{-1}} = 1.8\times10^{-16}$$

Adding a catalyst ___________ the activation energy for both the forward and reverse reactions. As a result, the reaction reaches equilibrium faster, but the equilibrium constant remains the same, and the equilibrium concentrations will be unaffected.

Although a catalyst does not affect the equilibrium, it can affect the reaction conditions chosen: if good yields are obtained slowly at a certain temperature, finding a catalyst can make the reaction more practical.

Workbook Problem 15.6

Consider the following reaction, which occurs in a catalytic converter in the presence of a palladium catalyst:

$$2CO(g) + O_2(g) \xrightarrow{Pd} 2CO_2(g) \qquad \Delta H° = -566 \text{ kJ}$$

What would be the effect on the concentration of CO_2 if the pressure was increased? If the pressure was decreased? If the palladium catalyst was removed? If the temperature was lowered? If the concentration of O_2 was increased?

Strategy: Use Le Châtelier's principle.

Step 1: Determine the direction of reaction needed to minimize each stress and its effect on $[CO_2]$.

POST-LECTURE:

Work Problems: *One of the best ways to learn chemistry is through solving problems. Go to the textbook and work through the Worked Examples for this section, then go to the end of chapter and work the problems there. When you feel confident that you can correctly answer problems like the ones listed below for each objective, check off each of the Learning Objectives for this section above.*

LO 15.15: Problems 15.50, 15.132
LO 15.16: Worked Example 15.17; Problems 15.134, 15.137, 15.138

Self-Test

This section is intended to test your knowledge of the material covered in this chapter. Think through these problems and make certain you understand what the problem is about—ask yourself if the answer makes sense. Many of these questions are linked to the chapter learning goals. Therefore, successful completion of these problems indicates that you have achieved the learning goals for this chapter. You will receive the greatest benefit from this section if you use it as a mock exam. Do not use your notes, book, Internet, or friends when taking the test. There is no grade and you only shortchange yourself by getting help with this activity. As you take the test, mark the problems with your level of confidence with the answer.

★ *Use a star for problems you know you got right.*
? *Use a question mark for those you think you got right.*
x *Use an x for those you are guessing on.*

If you get a question right that you marked with a star, you can be confident that you know the associated LO.

If you get a question right that you marked with a question mark, go back and work a few more problems and test yourself again tomorrow.

If you get a question wrong or mark it with an x, go back and read the section again, work a few problems today, a few tomorrow, and a few the next day before testing yourself again.

This will help you to discover which topics you have learned thoroughly and which topics you need to study in more detail.

True–False

1. All chemical reactions are reversible to some extent.

2. The equilibrium constant can be determined by running the same reaction at different temperatures.

3. The equilibrium constant has units of M^{-1}.

4. Equilibrium constants can be related to concentrations or to partial pressures.

5. A large equilibrium constant indicates a reaction with more reactants than products at equilibrium.

6. Increasing the pressure on a gas-phase reaction will drive the reaction in the direction of more moles of gas.

7. Adding more of a reactant pushes the reaction to the right toward products.

8. Catalysts speed the rate at which equilibrium is achieved.

9. The relative values of k_f and k_r determine the composition of the equilibrium mixture.

10. The value of K makes it possible to predict the direction of a reaction.

Multiple Choice

11. A state of chemical equilibrium is reached when
    a. the rate of the forward reaction is the same as the rate of the reverse reaction
    b. $Q = K$
    c. the concentrations of the products and reactants have reached constant value
    d. all of the above

12. The equilibrium constant is calculated as
    a. concentration of products over concentration of reactants
    b. concentration of products raised to stoichiometric coefficients over concentration of reactants raised to stoichiometric coefficients
    c. concentration of reactants over concentration of products
    d. concentration of reactants raised to stoichiometric coefficients over concentration of products raised to stoichiometric coefficients

13. In heterogeneous equilibria, the following are included in the equilibrium constant:
   a. solids, liquids, solutions, and gases
   b. liquids, solutions, and gases
   c. solutions and gases
   d. only gases

14. An equilibrium constant in the range of 10^{-3} to 10^{3} indicates that
   a. both products and reactants will be appreciably present at equilibrium
   b. only products will be appreciably present at equilibrium
   c. only reactants will be appreciably present at equilibrium
   d. no information about final concentrations can be gained from this

15. The following stresses will result in a change in equilibrium concentrations:
   a. change of temperature
   b. change of concentrations
   c. change in volume and pressure
   d. all of the above

16. Which of the following is an expected response to stress on a reaction?
   a. Lowering the temperature on an exothermic reaction increases the concentration of products.
   b. Lowering the temperature on an endothermic reaction increases the concentration of products.
   c. Raising the temperature on an exothermic reaction increases the concentration of products.
   d. Raising the temperature on an endothermic reaction has no effect on the concentration of products.

17. A catalyst
   a. increases the concentration of products
   b. increases the percent yield of a reaction
   c. makes a reaction go to completion
   d. increases the speed at which equilibrium is obtained

18. K_p
   a. is the same as K_c
   b. is the reciprocal of K_c
   c. is the equilibrium constant calculated using partial pressures
   d. includes the concentration of pure liquids in the equilibrium expression

19. The reaction quotient, Q_c, is
   a. calculated in the same manner as K_c, but not necessarily at equilibrium
   b. less than K_c at equilibrium
   c. greater than K_c at equilibrium
   d. equal to K_p

20. When a reactant is removed from a system at equilibrium
   a. the reaction shifts right
   b. the reaction shifts left
   c. the reactants are converted to products
   d. the denominator in the K_c expression becomes larger

Matching

21. chemical equilibrium

22. equilibrium constant, K_c

23. equilibrium constant, K_p

24. equilibrium mixture

25. homogeneous equilibria

26. heterogeneous equilibria

27. reaction quotient, Q_c

28. reversible reaction

29. dynamic state

30. Le Châtelier's principle

a. constant in the equilibrium equation for gas-phase reactions

b. equilibrium in which all reactants and products are in the same phase

c. calculated like K_c, but with nonequilibrium concentrations

d. constant in the equilibrium equation for solutions

e. if a stress is applied to a reaction, the reaction occurs in the direction that relieves the stress

f. any state in which change in forward and reverse directions is equal in rate, leading to no net change in concentration of reactants and products

g. a state in which the concentration of products and reactants in a reaction no longer shows any net change

h. equilibrium in which reactants and products are in different phases

i. a reaction that can go forward toward products or back to reactants, depending on reaction conditions

j. combination of reactants and products with characteristic proportions

Fill in the Blanks

31. Chemical equilibrium is a ____________________ state in which the ____________________ of reactants and products is ____________________.

32. The equilibrium equation for a ____________________ equilibrium does not include concentrations of pure ______________ or ______________.

33. When an equilibrium mixture is subjected to a stress, it responds to ______________ that stress. Adding more reactant will shift a reaction toward ______________, while adding more products will shift a reaction toward ____________________. Changing the ______________ affects the mixture by changing the ____________________ constants.

Adding a ____________________ does not alter the equilibrium mixture, simply ______________ the rate at which it is achieved.

34. Comparing K and Q makes it possible to determine the ________________ of a reaction. When $K = Q$, the reaction is at ______________________. When $K < Q$, the reaction will proceed from ____________________ to ______________________. When $K > Q$, the reaction will proceed from ____________________ to ______________________.

35. Increasing the pressure on a gas-phase reaction will shift the equilibrium away from small molecules toward larger molecules. This has the result of _________________ the pressure.

Problems

36. Write equilibrium expressions for the following balanced equations.

a. $O_3(g) + Cl(g) \rightleftharpoons O_2(g) + ClO(g)$

b. $2\ NO(g) + Br_2(g) \rightleftharpoons 2\ NOBr(g)$

c. $2\ Cu(s) + O_2(g) \rightleftharpoons 2\ CuO(s)$

d. $2\ N_2O(g) \rightleftharpoons 2\ N_2(g) + O_2(g)$

37. The partial equilibrium pressures for N_2, O_2, and NO in the reaction

$$N_2(g) + O_2(g) \rightleftharpoons 2\ NO(g)$$

are $p(N_2) = 0.27$ atm, $p(O_2) = 0.187$ atm, and $p(NO) = 0.045$ atm. Calculate K_p for this reaction.

38. Calculate K_c for the reaction in Question 37.

39. From the value of K_c, determine whether mainly products or mainly reactants exist at equilibrium for the following balanced equations:

a. $CH_3Cl(aq) + OH^-(aq) \rightleftharpoons CH_3OH(aq) + Cl^-(aq)$ $\quad K_c = 1 \times 10^{16}$

b. $2SO_2(g) + O_2(g) \rightleftharpoons 2\ SO_3(g)$ $\quad K_p = 3.3$

c. $2\ HBr(g) \rightleftharpoons H_2(g) + Br_2(g)$ $\quad K = 2 \times 10^{-19}$

40. For the reaction

$$PCl_5(g) \rightleftharpoons PCl_3(g) + Cl_2(g)$$

$K_c = 33.3$ at 760 °C. Is this system at equilibrium if $[PCl_5] = 2.43 \times 10^{-3}$ M, $[PCl_3] = [Cl_2] = 0.830$ M? If not, determine the direction the reaction must go to reach equilibrium.

41. Nitrogen dioxide is produced from the reaction of dinitrogen oxide and oxygen according to the equation

$$2\ N_2O(g) + 3\ O_2(g) \rightleftharpoons 4\ NO_2(g)$$

At 25 °C, 0.0342 moles of N_2O and 0.0415 moles of O_2 are placed in a 1.00-L container and allowed to react. The $[NO_2]$ at equilibrium is 0.0298 moles. What are the equilibrium concentrations of N_2O and O_2? What is the value of K_c?

42. The equilibrium constant for the reaction

$$SO_2(g) + NO_2(g) \rightleftharpoons NO(g) + SO_3(g)$$

is $K_c = 85.0$ at 460 °C. Calculate the concentrations of all species at equilibrium when the initial reaction mixture contains 0.100 M SO_2 and 0.100 M NO_2.

43. Consider the reaction

$$2\ SO_2(g) + O_2(g) \rightleftharpoons 2\ SO_3(g) \qquad \Delta H^\circ = +197\ \text{kJ}$$

What will be the direction of the reaction when the following stress is applied?

a. addition of SO_3
b. decrease in temperature
c. increase in volume
d. addition of N_2 gas
e. addition of SO_2 gas

44. For the reaction $A + B \rightleftharpoons 2\ C$, $k_f = 6.8 \times 10^{-5}\ s^{-1}$ and $k_r = 7.2 \times 10^{-7}\ s^{-1}$. Calculate K.

45. Calculate the equilibrium concentration for all species present in the reaction

$$HCONH_2(g) \rightleftharpoons NH_3(g) + CO(g)$$

if the initial concentration of formamide ($HCONH_2$) is 0.125 M. $K_c = 4.84$ at 127 °C.

46. For the reaction

$$2\ H_2S(g) + CH_4(g) \rightleftharpoons 4\ H_2(g) + CS_2(g)$$

$K_c = 5.27 \times 10^{-8}$ at 700 °C. What is the position of equilibrium for this reaction? Is the reaction mixture at equilibrium when the concentrations of the reactants and products are $[CH_4] = 1.75$ M, $[H_2S] = 2.00$ M, $[H_2] = 0.450$ M, and $[CS_2] = 0.085$ M? If not, in which direction will the reaction proceed?

Challenge Problem

47. The following reaction was carried out in a 4.00-L vessel:

$$NO_2(g) + NO(g) \rightleftharpoons N_2O(g) + O_2(g)$$

Determine the number of moles of reactants and products present at equilibrium if the initial number of moles are 0.100 mol NO_2, 0.050 mol NO, 0.0125 mol N_2O, and 0.00125 mol O_2. Use a value of $K_c = 0.914$.

Test Yourself: *Take the Practice Test on pages 641 and 642 of your textbook. Again, use a testing environment using only those items that your instructor will allow you to have access to during the test and mark each question with a star, question mark, or x to denote your confidence level. Once you have finished you can check off the LO at the beginning of this chapter based on which questions you starred and got correct. Go back and spend more time with the topics from the other learning objectives.*

CHAPTER SIXTEEN

Aqueous Equilibria: Acids and Bases

Learning Objectives

As a result of reading and studying this chapter, you should be able to:

- ☐ **LO 16.1** Identify Arrhenius acids and bases and Brønsted–Lowry acids and bases and conjugate acid–base pairs.
- ☐ **LO 16.2** Rank acids in order of increasing strength given a molecular image representing dissociation in solution.
- ☐ **LO 16.3** Predict the direction of a reaction based upon relative strengths of the conjugate acid–base pairs.
- ☐ **LO 16.4** Predict the relative strengths of binary acids (HA) and oxoacids (H_nYO_m) based on their chemical structure.
- ☐ **LO 16.5** Calculate the concentration of $[H_3O^+]$ or $[OH^-]$ using the value of K_w.
- ☐ **LO 16.6** Calculate the pH of a solution given $[H_3O^+]$ or $[OH^-]$, or calculate $[H_3O^+]$ or $[OH^-]$ given the pH of a solution.
- ☐ **LO 16.7** Calculate the pH of a strong acid or base solution.
- ☐ **LO 16.8** Calculate the value for K_a of a weak acid given the initial concentration of the acid and its equilibrium pH.
- ☐ **LO 16.9** Given the K_a value of a weak acid and its initial concentration, calculate the pH, percent dissociation, and the concentration of all species present in solution.
- ☐ **LO 16.10** Calculate the pH and the concentration of all species present in a solution of a diprotic acid.
- ☐ **LO 16.11** Calculate the pH and equilibrium concentrations in a solution of a weak base.
- ☐ **LO 16.12** Relate K_a, K_b, pK_a, and pK_b for a conjugate acid–base pair.
- ☐ **LO 16.13** Predict whether a salt is acidic, basic, or neutral and calculate the pH of the salt solution.
- ☐ **LO 16.14** Identify the Lewis acid and Lewis base and use curved arrow notation to indicate donation of a lone pair of elections in Lewis acid–base reaction.

Chapter Summary

The study of equilibrium carries over into the study of acids and bases. In this chapter, we discuss the three definitions of *acids* and *bases* and *conjugate acids* and *base pairs*. We also introduce reactions of strong and weak acids and bases and the concepts of pH and pK_a. We explain polyprotic acids and their reactions, as well as the identifications of acidic salts and basic salts and the pH of their solutions. Finally, we look at periodic variations in acid strength and how they can be used to predict the strength of various acids.

The Chapter in Detail

Section 16.1 Acid–Base Concepts: The Brønsted–Lowry Theory

PRE-LECTURE:

Learning Objectives: *These are the skills you should be proficient in after reading this section and practicing the associated problems. Check off the boxes below once you have gone through this section, answered the associated questions in your textbook, and feel confident you have mastered the skill.*

☐ **LO 16.1** Identify Arrhenius acids and bases and Brønsted–Lowry acids and bases and conjugate acid–base pairs.

Definitions: *To learn, you must do. This is your workbook. Fill in the definitions from the glossary of the textbook as a first step to understanding the material.*

— **Arrhenius acid**—____

____

— **Arrhenius base**— ____

____

— **Brønsted–Lowry theory**— ____________________________________

____

— **Brønsted–Lowry acid**— ____________________________________

____

— **Brønsted–Lowry base**—____________________________________

____

— **conjugate acid–base pair**— __________________________________

____

— **conjugate base**—____

____

— **conjugate acid**— ____

____

— **hydronium ion, H_3O^+**— ____________________________________

____

Summary: *Read through this section and fill in the missing words and equations. Reference the textbook when you are not confident in your answers.*

The Arrhenius definition of *acids* and *bases* is the simplest, and we have used it in this book thus far. In the Arrhenius definition, acids are described as dissociating to form hydrogen ions and bases as dissociating to form hydroxide ions. This theory is limited to aqueous solutions, and it is unable to account for the basicity of substances—such as ammonia (NH_3)—that are bases without having any hydroxide ions to contribute.

The Brønsted–Lowry theory of acids and bases solves these problems by describing an acid as a *proton* ___________ and a base as a *proton* ___________, with acid–base reactions being *proton transfers*.

Chemical species that differ only by a proton are called ______________________.

HA indicates the acid and A^- the conjugate base. A conjugate base has one less proton than the acid.

BH^+ is the conjugate acid of B; the conjugate acid has one more proton than the base.

Acid dissociation is when a proton is transferred to a solvent, for example:

$$HA + H_2O(l) \leftrightarrows A^-(aq) + H_3O^+(aq)$$

Base dissociation is when a proton is removed from a solvent, for example:

$$B + H_2O(l) \leftrightarrows BH^+(aq) + OH^-(aq)$$

All Brønsted–Lowry bases have one or more lone pairs of electrons. This makes it possible for them to pull protons off solvent molecules.

A generic reaction for an acid–base reaction is given. Identify each species.

$$HA \quad + \quad B \quad \leftrightarrows \quad BH^+ \quad + \quad A^-$$

_____ _____ _____ _____

Workbook Example 16.1

PROBLEM:
Write the proton-transfer equilibria for the following acids or bases in aqueous solution, and identify the conjugate acid–base pairs in each one: CH_3COOH (acetic acid), PO_4^{3-} (phosphate ion–base)

SOLUTION:

$$CH_3COOH(l) + H_2O(l) \leftrightarrows CH_3COO^-(aq) + H_3O^+(aq)$$

Acid = CH_3COOH conjugate base = CH_3COO^-

$$PO_4^{3-}(aq) + H_2O(l) \leftrightarrows HPO_4^{2-}(aq) + OH^-(aq)$$

Base = PO_4^{3-} conjugate acid = HPO_4^{2-}

Workbook Problem 16.1

Write the proton-transfer reactions for nitrous acid (HNO_2) and basic sulfate (SO_4^{2-}). Identify the acid or base and its conjugate.

Strategy: Write the reaction with water, in which the acid protonates the water, and the base deprotonates the water.

Step 1: Write the reactions in the forward direction.

Step 2: Identify the conjugate acid and the conjugate base.

POST-LECTURE:

Work Problems: *One of the best ways to learn chemistry is through solving problems. Go to the textbook and work through the Worked Examples for this section, then go to the end of chapter and work the problems there. When you feel confident that you can correctly answer problems like the ones listed below for each objective, check off each of the Learning Objectives for this section above.*

LO 16.1: Worked Examples 16.1, 16.2; Problems 16.38, 16.48, 16.52

Section 16.2 Acid Strength and Base Strength

PRE-LECTURE:

Learning Objectives: *These are the skills you should be proficient in after reading this section and practicing the associated problems. Check off the boxes below once you have gone through this section, answered the associated questions in your textbook, and feel confident you have mastered the skill.*

- ☐ **LO 16.2** Rank acids in order of increasing strength given a molecular image representing dissociation in solution.
- ☐ **LO 16.3** Predict the direction of a reaction based upon relative strengths of the conjugate acid–base pairs.

Definitions: *To learn, you must do. This is your workbook. Fill in the definitions from the glossary of the textbook as a first step to understanding the material.*

— **strong acid—** ____

____

— **weak acid—** ____

____

Review: *Before going through the section make sure you are familiar with these concepts:*

- Section 4.3 electrolytes

Summary: *Read through this section and fill in the missing words and equations. Reference the textbook when you are not confident in your answers.*

A strong acid is like a strong electrolyte, it dissociates to a ___________ extent.

A weak acid is like a weak electrolyte, it dissociates to a ___________ extent.

Strong acids:

- ______________________
- ______________________
- ______________________
- ______________________
- ______________________
- ______________________

Acids other than your strong acids are weak acids.

There is a ___________ relationship between the strength of an acid and its conjugate base.

The weaker the acid, the stronger its conjugate base.
The stronger the acid, the weaker its conjugate base.

Do not confuse the concept of strong acid/weak acid with strong<u>er</u> acid/weak<u>er</u> acid!

The strong acid/weak acid concept deals with the extent of dissociation of the acid in water. Stronger acid/weaker acid is a comparison between two different acids (both of which could be "weak" acids; one is just a stronger weak acid than the other).

When you start with an equal concentration of reactants and products, the proton will always be transferred to the ___________ base. Thus, the stronger acid gives the stronger base its proton forming the weaker acid and the weaker base.

Workbook Example 16.2

PROBLEM:
If you mix equal concentrations of reactants and products, will the following reaction proceed to the right or the left?

$$HCl + CH_3COO^-(aq) \leftrightarrows CH_3COOH(aq) + Cl^-(aq)$$

SOLUTION:
Hydrochloric acid is a strong acid, making Cl^- a very weak base. Acetic acid is a weak acid, so acetate ions are a relatively strong conjugate base. As a result, hydrochloric acid will transfer protons to acetate ions. The reaction will proceed to the right.

Workbook Problem 16.2

Write the proton-transfer reaction between HCl and CO_3^{2-} and the proton-transfer reaction between CN^- and H_3PO_4.

Strategy: Determine which species is the stronger acid and which is the stronger base, then transfer protons accordingly.

Step 1: Write the reactions, transferring protons from the stronger acid to the stronger base.

POST-LECTURE:

Work Problems: *One of the best ways to learn chemistry is through solving problems. Go to the textbook and work through the Worked Examples for this section, then go to the end of chapter and work the problems there. When you feel confident that you can correctly answer problems like the ones listed below for each objective, check off each of the Learning Objectives for this section above.*

LO 16.2: Problem 16.39
LO 16.3: Worked Example 16.3; Problems 16.56, 16.57

Section 16.3 Factors That Affect Acid Strength

PRE-LECTURE:

Learning Objectives: *These are the skills you should be proficient in after reading this section and practicing the associated problems. Check off the boxes below once you have gone through this section, answered the associated questions in your textbook, and feel confident you have mastered the skill.*

☐ **LO 16.4** Predict the relative strengths of binary acids (HA) and oxoacids (H_nYO_m) based on their chemical structure.

Review: *Before going through the section make sure you are familiar with these concepts:*

- Section 7.2 dissociation energy

Summary: *Read through this section and fill in the missing words and equations. Reference the textbook when you are not confident in your answers.*

The extent of dissociation of an acid is often determined by the ___________ and ___________ of the HA bond: the weaker or more polar the HA bond is, the stronger the acid.

For binary acids, as the size of A^- ___________ down the periodic table, the strength of the HA bond is weakened; therefore, acidity ___________ down the groups.

Within a row, the acidity will ___________ with increasing polarity. As A^- becomes more electronegative, acidity increases.

The dissociation of an oxoacid involves the breaking of an ___________ bond, so anything that weakens these bonds or increases their polarity will ___________ the strength of the acid.

- As Y becomes more electronegative, the strength of oxoacids ___________.
- As the number of oxygens changes in an oxoacid (sulfuric acid to sulfurous acid, for example), an increase in the oxidation number of Y will ___________ acid strength. In other words, the more oxygens the central atom is trying to hold on to, the stronger the acid will be.

Add the trend of acid strength to your periodic table.

Workbook Example 16.3

PROBLEM:
Order the following by increasing acid strength:

a) HF, NH_3, H_2O
b) NH_3, AsH_3, PH_3
c) HIO_3, HIO, HIO_2, HIO_4
d) H_2SeO_4, H_2TeO_4, H_2SO_4

SOLUTION:

a) Across a period, an increase in electronegativity gives rise to an increase in acidity: $NH_3 < H_2O < HF$.

b) Down a group, an increase in size gives rise to an increase in acid strength: $NH_3 < PH_3 < AsH_3$.

c) For oxoacids, acid strength increases with an increasing number of oxygen atoms: $HIO < HIO_2 < HIO_3 < HIO_4$.

d) For oxoacids that contain the same number of OH groups and the same number of O atoms, acid strength increases with increasing electronegativity of Y: $H_2TeO_4 < H_2SeO_4 < H_2SO_4$.

POST-LECTURE:

Work Problems: *One of the best ways to learn chemistry is through solving problems. Go to the textbook and work through the Worked Examples for this section, then go to the end of chapter and work the problems there. When you feel confident that you can correctly answer problems like the ones listed below for each objective, check off each of the Learning Objectives for this section above.*

LO 16.4: Worked Example 16.4; Problems 16.40, 16.60, 16.62, 16.66

Section 16.4 Dissociation of Water

PRE-LECTURE:

Learning Objectives: *These are the skills you should be proficient in after reading this section and practicing the associated problems. Check off the boxes below once you have gone through this section, answered the associated questions in your textbook, and feel confident you have mastered the skill.*

☐ **LO 16.5** Calculate the concentration of $[H_3O^+]$ or $[OH^-]$ using the value of K_w.

Definitions: *To learn, you must do. This is your workbook. Fill in the definitions from the glossary of the textbook as a first step to understanding the material.*

— **dissociation of water—** ____

____

— **ion product constant for water (K_w)—** ______________________________

____

Review: *Before going through the section make sure you are familiar with these concepts:*

- Section 15.2 writing a K_c
- Section 15.4 heterogeneous equilibria

Summary: *Read through this section and fill in the missing words and equations. Reference the textbook when you are not confident in your answers.*

Water can act as an ___________ or as a ___________. When in the presence of an acid, water is the base. In the presence of a base, water is the acid. It can even react in this way with another water molecule; one acting as the acid and the other the base.

$$H_2O(l) + H_2O(l) \leftrightarrows H_3O^+(aq) + OH^-(aq)$$

The equilibrium constant, K_C, for this equation is called the ion product of water and abbreviated K_W.

$K_W =$

The equilibrium lies far to the left and ___________ water molecules are dissociated.

$$[H_3O^+] = [OH^-] =$$

In aqueous solutions, the relative values of $[H_3O^+]$ and $[OH^-]$ determine whether the solution is acidic, basic, or neutral:

- $[H_3O^+] = [OH^-]$ ___________ solution $[H_3O^+] = 1.0 \times 10^{-7}$ M.
- $[H_3O^+] > [OH^-]$ ___________ solution $[H_3O^+] > 1.0 \times 10^{-7}$ M.
- $[H_3O^+] < [OH^-]$ ___________ solution $[H_3O^+] < 1.0 \times 10^{-7}$ M.

Workbook Example 16.4

PROBLEM:
Calculate the molarity of OH^- in a solution with an H_3O^+ concentration of 0.0042 M. Is this solution acidic, basic, or neutral? What is the $[H_3O^+]$ in a solution with $[OH^-]$ of 9.35×10^{-7} M? Will the solution be acidic, basic, or neutral?

SOLUTION:
If we know the concentration of either the $[H_3O^+]$ or the $[OH^-]$, we can calculate the other from the relationship $K_w = [H_3O^+][OH^-]$.

$$\left[OH^-\right] = \frac{1 \times 10^{-14}}{0.0042\ M} = 2.4 \times 10^{-12}$$

$[H_3O^+] > [OH^-]$, so this solution is acidic.

$$\left[H_3O^+\right] = \frac{1.0 \times 10^{-14}}{9.35 \times 10^{-7}\ M} = 1.07 \times 10^{-8}$$

$[H_3O^+] < [OH^-]$; this solution is basic.

Workbook Problem 16.3

Determine the $[OH^-]$ in a solution with $[H_3O^+] = 2.3 \times 10^{-5}$ M, and the $[H_3O^+]$ of a solution with $[OH^-] = 5.73 \times 10^{-9}$ M. Are these solutions acidic, basic, or neutral?

Strategy: Use the water dissociation constant to determine the concentrations. A basic solution will have $[OH^-] > [H_3O^+]$, and an acid solution will have $[H_3O^+] > [OH^-]$.

Step 1: Use K_w to calculate the ion concentrations.

POST-LECTURE:

Work Problems: *One of the best ways to learn chemistry is through solving problems. Go to the textbook and work through the Worked Examples for this section, then go to the end of chapter and work the problems there. When you feel confident that you can correctly answer problems like the ones listed below for each objective, check off each of the Learning Objectives for this section above.*

LO 16.5: Worked Example 16.5; Problems 16.68, 16.70

Section 16.5 The pH Scale

PRE-LECTURE:

Learning Objectives: *These are the skills you should be proficient in after reading this section and practicing the associated problems. Check off the boxes below once you have gone through this section, answered the associated questions in your textbook, and feel confident you have mastered the skill.*

☐ **LO 16.6** Calculate the pH of a solution given $[H_3O^+]$ or $[OH^-]$, or calculate $[H_3O^+]$ or $[OH^-]$ given the pH of a solution.

Definitions: *To learn, you must do. This is your workbook. Fill in the definitions from the glossary of the textbook as a first step to understanding the material.*

— **pH—**____

____

Summary: *Read through this section and fill in the missing words and equations. Reference the textbook when you are not confident in your answers.*

The pH scale is a logarithmic scale used to express the hydronium ion concentration in a solution:

pH =

$[H_3O^+]$

Note that only the digits to the right of the decimal point are significant in a logarithm: the number to the left of the decimal point is an exact number related to the integral power of 10.

As $[H_3O^+]$ increases, the pH decreases:

- Acidic solutions: pH < 7.
- Basic solutions: pH > 7.
- Neutral solutions: pH = 7.

A less-often-used scale is pOH:

pOH =

Just as $K_w = [H_3O^+][OH^-] = 1 \times 10^{-14}$, the sum of pH + pOH always equals 14.

Workbook Example 16.5

PROBLEM:
Calculate the $[H_3O^+]$ and $[OH^-]$ for orange juice, which has a pH of 3.87.

SOLUTION:

$$\left[H_3O^+\right] = antilog(-pH) = 10^{-3.87} = 1.35 \times 10^{-4}\ M$$

$$\left[OH^-\right] = \frac{1 \times 10^{-14}}{1.35 \times 10^{-4}} = 7.41 \times 10^{-11}\ M$$

Workbook Example 16.6

PROBLEM:
Calculate the pH of a solution with a $[H_3O^+]$ of 3.56×10^{-7} M. Is this solution acidic, basic, or neutral?

SOLUTION:

$$pH = -\log\left(3.56 \times 10^{-7}\ M\right) = 6.45$$

The pH is less than 7: this solution is slightly acidic.

Workbook Problem 16.4

Calculate the pH of a solution whose $[OH^-] = 2.35 \times 10^{-6}$ M. Is the solution acidic, basic, or neutral?

Strategy: Calculate the $[H_3O^+]$, then determine pH.

Step 1: Calculate the $[H_3O^+]$, using the expression for the dissociation of water.

Step 2: Calculate the pH.

Step 3: Determine if the solution is acidic, basic, or neutral.

POST-LECTURE:

Work Problems: *One of the best ways to learn chemistry is through solving problems. Go to the textbook and work through the Worked Examples for this section, then go to the end of chapter and work the problems there. When you feel confident that you can correctly answer problems like the ones listed below for each objective, check off each of the Learning Objectives for this section above.*

LO 16.6: Worked Examples 16.6, 16.7; Problems 16.74, 16.76

Section 16.6 Measuring pH

PRE-LECTURE:

Learning Objectives: *There are no specific learning objectives for this section; the information is a continuation of the previous section.*

Definitions: *To learn, you must do. This is your workbook. Fill in the definitions from the glossary of the textbook as a first step to understanding the material.*

— **acid–base indicator—** ____

____

Summary: *Read through this section and fill in the missing words and equations. Reference the textbook when you are not confident in your answers.*

To measure pH, there are a number of acid–base indicators: compounds that change color in a specific pH range. They are weak acids with different colors in their acid and conjugate base forms and change color over a range of 2 pH units.

Particularly useful is a mixture of indicators that change color over the entire range of pH: this is known as a ______________________.

For greater precision than is possible with such indicators, pH meters are electronic instruments that measure the pH-dependent electrical potential of the test solution.

Section 16.7 The pH in Solutions of Strong Acids and Strong Bases

PRE-LECTURE:

Learning Objectives: *These are the skills you should be proficient in after reading this section and practicing the associated problems. Check off the boxes below once you have gone through*

this section, answered the associated questions in your textbook, and feel confident you have mastered the skill.

☐ **LO 16.7** Calculate the pH of a strong acid or base solution.

Definitions: *To learn, you must do. This is your workbook. Fill in the definitions from the glossary of the textbook as a first step to understanding the material.*

— **strong bases**— ____

____

Summary: *Read through this section and fill in the missing words and equations. Reference the textbook when you are not confident in your answers.*

Three of the strong acids are ___________, meaning that they dissociate completely to form one proton. Because these dissociate completely, $[H_3O^+] = [A^-]$ = initial concentration of the acid, and $[HA] = 0$. This makes the pH of a strong acid solution very straightforward to calculate:

$$pH = -\log[acid]$$

The most common strong bases are alkali metal hydroxides with the formula MOH. These also dissociate completely to form metal ions and hydroxide ions. The pH can be calculated from the $[OH^-]$.

Alkaline earth hydroxides have the formula $M(OH)_2$. When dissolved, they dissociate completely to form one metal ion and two hydroxide ions, but they are not as soluble as the alkali metal hydroxides.

Alkaline earth oxides are very strong bases because the O^{2-} ion is an even stronger base than the hydroxide ion. An O^{2-} ion will immediately pull a proton off a water molecule, forming two hydroxide ions: one from the oxide ion, now bound to a hydrogen, and the other from the split water molecule that is now missing a hydrogen:

$$O^{2-} + H_2O \rightarrow 2\ OH^-$$

Workbook Example 16.7

PROBLEM:
Calculate the pH of a 2.37×10^{-2} M HNO_3 solution.

SOLUTION:
HNO_3 is a strong acid that completely dissociates in water. Therefore, $[H_3O^+]$ = initial concentration of the undissociated acid.

$$pH = -\log\left[H_3O^+\right] = -\log\left[2.37 \times 10^{-2}\right] = 1.63$$

Workbook Example 16.8

PROBLEM:
Calculate the pH of a 2.37×10^{-6} M $Mg(OH)_2$ solution.

SOLUTION:
$Mg(OH)_2$ is a strong base that completely dissociates in water to form a metal cation and two hydroxide anions. Therefore, $[OH^-] = 2 \times$ initial concentration of the undissociated base.

$$\left[OH^-\right]=2\left(2.37\times10^{-6}\ M\right)=4.74\times10^{-6}\ M$$

in dilute aqueous solution, $[OH^-][H_3O^+] = 1 \times 10^{-14}$

$$\left[H_3O^+\right]=\frac{1\times10^{-14}}{4.74\times10^{-6}}=2.11\times10^{-9}\ M$$

$$pH=-\log\left[H_3O^+\right]=-\log\left(2.11\times10^{-9}\ M\right)=8.68$$

Workbook Problem 16.5

Calculate the pH of a 7.2×10^{-3} M KOH solution.

Strategy: Determine the $[H^+]$ from the K_w expression and then calculate pH.

Step 1: Determine the concentration of $[OH^-]$.

Step 2: Calculate the $[H_3O^+]$.

Step 3: Calculate the pH.

POST-LECTURE:

Work Problems: *One of the best ways to learn chemistry is through solving problems. Go to the textbook and work through the Worked Examples for this section, then go to the end of chapter and work the problems there. When you feel confident that you can correctly answer problems*

like the ones listed below for each objective, check off each of the Learning Objectives for this section above.

LO 16.7: Worked Example 16.8; Problems 16.82, 16.84

Section 16.8 Equilibria in Solutions of Weak Acids

Learning Objectives: *There are no specific learning objectives for this section; the information in this section is used as a basis for understanding subsequent sections.*

Definitions: *To learn, you must do. This is your workbook. Fill in the definitions from the glossary of the textbook as a first step to understanding the material.*

— **acid-dissociation constant (K_a)**—____

____

Review: *Before going through the section make sure you are familiar with these concepts:*

- Section 15.2 writing a K_c

Summary: *Read through this section and fill in the missing words and equations. Reference the textbook when you are not confident in your answers.*

A weak acid only partially dissociates in solution. The equilibrium expression for an acid in water is called the ___________ and abbreviated K_a:

$$HA(aq) + H_2O(l) \leftrightarrows H_3O^+(aq) + A^-(aq)$$

$K_a =$

Because it is a pure liquid, water is omitted from the equilibrium expression.

K_a values are experimentally determined: the higher the value, the stronger the acid.

Also sometimes used are pK_a values. We calculate pK_a in the same manner as we do for pH:

p$K_a =$

The higher the K_a is, the ___________ the pK_a will be. It is fair to note that the pK_a scale is used very heavily in organic chemistry, so some of you will be seeing this again!

Workbook Example 16.9

PROBLEM:
The pH of a 0.400 M solution of nicotinic acid ($HC_6H_4NO_2$) is 2.63. Determine the value of K_a for this acid.

SOLUTION:
To determine K_a, first write the balanced equation for the dissociation equilibrium and the equilibrium equation that defines K_a.

$$HC_6H_4NO_2(aq) + H_2O(l) \leftrightarrows H_3O^+(aq) + C_6H_4NO_2^-(aq)$$

$$K_a = \frac{[H_3O^+][C_6H_4NO_2^-]}{[HC_6H_4NO_2]}$$

To determine K_a, we need to know the concentrations of the species in the equilibrium mixture. We can determine the concentration of H_3O^+ from the pH.

$$[H_3O^+] = antilog(-pH) = 10^{-2.63} = 2.34 \times 10^{-3}\ M$$

Dissociation of one $HC_6H_4NO_2$ molecule produces one H_3O^+ and $C_6H_4NO_2^-$ ion; the H_3O^+ and $C_6H_4NO_2^-$ concentrations are equal.

$$[H_3O^+] = [C_6H_4NO_2^-] = 2.34 \times 10^{-3}\ M$$

The $[HC_6H_4NO_2]$ concentration at equilibrium is equal to the initial concentration minus the amount of $HC_6H_4NO_2$ that dissociates.

$$[HC_6H_4NO_2] = 0.400 - (2.34 \times 10^{-3}) = 0.398\ M$$

$$K_a = \frac{[H_3O^+][C_6H_4NO_2^-]}{[HC_6H_4NO_2]} = \frac{(2.34\times10^{-3})(2.34\times10^{-3})}{0.398} = 1.38\times10^{-5}$$

Section 16.9 Calculating Equilibrium Concentrations in Solutions of Weak Acid

PRE-LECTURE:

Learning Objectives: *These are the tasks you should be able to do after reading this section and practicing the associated problems. Check off the boxes below once you have gone through this section, answered the associated questions in your textbook, and feel confident you have mastered the task.*

☐ **LO 16.8** Calculate the value for K_a of a weak acid given the initial concentration of the acid and its equilibrium pH.

Definitions: *To learn, you must do. This is your workbook. Fill in the definitions from the glossary of the textbook as a first step to understanding the material.*

— **principal acid–base reaction**— ______________________________

____

— **subsidiary reaction**— ______________________________

____

Review: *Before going through the section make sure you are familiar with these concepts:*

- quadratic equation

Summary: *Read through this section and fill in the missing words and equations. Reference the textbook when you are not confident in your answers.*

After experimentally calculating the K_a value of a weak acid, it can be used to calculate equilibrium concentrations or pH.

Steps to solve acid–base equilibrium problems:

- STEP 1: ____
  - List the species initially present
  - Classify them as acids or bases
  - Choose the one with the __________ dissociation constant
- STEP 2: ____
- STEP 3: ____
  - Check the validity of your approximation
  - If invalid, use the quadratic equation to solve for x
- STEP 4: ____
- STEP 5: ____

Workbook Example 16.10

PROBLEM:
Calculate the pH and the concentration of all species present (H_3O^+, $C_4H_7O_2^-$, $HC_4H_7O_2$, and OH^-) in 0.355 M butyric acid ($HC_4H_7O_2$), which has $K_a = 1.5 \times 10^{-5}$.

SOLUTION:
Because $K_a >> K_w$, it is possible to disregard the contribution of autoionization of water. The dissociation reaction is

$$HC_4H_7O_2(aq) + H_2O(l) \leftrightharpoons H_3O^+(aq) + C_4H_7O_2^-(aq)$$

Principal reaction	$HC_4H_7O_2(aq)$ $\leftrightharpoons$	H_3O^+ +	$C_4H_7O_2^-$
Initial concentration (M)	0.355	0	0
Change (M)	$-x$	$+x$	$+x$
Equilibrium concentration (M)	$0.355 - x$	$+x$	$+x$

Substitute the equilibrium concentrations into the equilibrium expression:

$$K_a = 1.5 \times 10^{-5} = \frac{\left[H_3O^+\right]\left[C_4H_7O_2^-\right]}{\left[HC_4H_7O_2\right]} = \frac{(x)(x)}{0.355 - x}$$

Assume that x is negligible compared with the initial concentration of the acid, so $0.355 - x \approx 0.355$. Using this value in the denominator, solve for x.

$$x^2 = \left(1.5 \times 10^{-5}\right)(0.355)$$

$$x^2 = 5.33 \times 10^{-6}\ M$$

$$x = \sqrt{5.33 \times 10^{-6}\ M} = 2.31 \times 10^{-3}\ M$$

The equilibrium concentrations are

$$[HC_4H_7O_2] = 0.355 - 0.00231 = 0.353\text{ M}$$
$$[H_3O^+] = [C_4H_7O_2^-] = 2.31 \times 10^{-3}\text{ M}$$

$[OH^-]$ is obtained from the dissociation of water.

$$\left[OH^-\right] = \frac{K_w}{\left[H_3O^+\right]} = \frac{1.0 \times 10^{-14}}{2.31 \times 10^{-3}\ M} = 4.33 \times 10^{-12}\ M$$

$$pH = -\log\left(2.31 \times 10^{-3}\ M\right) = 2.64$$

Workbook Problem 16.6

Determine the pH of a 1.63 M lactic acid ($HC_3H_5O_3$) solution. ($K_a = 1.4 \times 10^{-4}$).

Strategy: Follow the steps outlined previously in the textbook.

Step 1: Determine the species present initially.

Step 2: Write the proton-transfer reaction.

Step 3: Make a table showing the principal reaction and the initial and equilibrium concentrations.

Step 4:	Substitute the equilibrium concentrations into the equilibrium equation, and solve for x. (Since $K_a < 1.0 \times 10^{-3}$, assume that x is negligible and that $1.63 - x \cong 1.63$.)
Step 5:	Calculate the equilibrium concentrations.
Step 6:	Calculate the pH.

POST-LECTURE:

Work Problems: *One of the best ways to learn chemistry is through solving problems. Go to the textbook and work through the Worked Examples for this section, then go to the end of chapter and work the problems there. When you feel confident that you can correctly answer problems like the ones listed below for each objective, check off each of the Learning Objectives for this section above.*

LO 16.8: Worked Example 16.9; Problems 16.90, 16.92

Section 16.10 Percent Dissociation in Solutions of Weak Acids

PRE-LECTURE:

Learning Objectives: *These are the skills you should be proficient in after reading this section and practicing the associated problems. Check off the boxes below once you have gone through this section, answered the associated questions in your textbook, and feel confident you have mastered the skill.*

☐ **LO 16.9** Given the K_a value of a weak acid and its initial concentration, calculate the pH, percent dissociation, and the concentration of all species present in solution.

Definitions: *To learn, you must do. This is your workbook. Fill in the definitions from the glossary of the textbook as a first step to understanding the material.*

— **percent dissociation—**____

____

Summary: *Read through this section and fill in the missing words and equations. Reference the textbook when you are not confident in your answers.*

Another way of expressing acid strength is in terms of ______________________. This is calculated as:

Percent dissociation =

The value depends on the concentration and type of acid: in general, the percent dissociation ___________ with an increasing K_a, and ___________ in more dilute solutions.

Workbook Example 16.11

PROBLEM:
Calculate the percent dissociation of 0.100 and 1.00 M vitamin C solutions. The molecular formula for ascorbic acid is $C_6H_8O_6$; the $K_a = 8.0 \times 10^{-5}$.

SOLUTION:
We need to know what the concentration is of the conjugate base of ascorbic acid; namely, $C_6H_7O_6^-$. To do this, we need to write the acid-dissociation constant expression and solve for $C_6H_7O_6^-$:

$$K_a = \frac{\left[H_3O^+\right]\left[C_6H_7O_6^-\right]}{\left[C_6H_8O_6\right]} = 8.0 \times 10^{-5}$$

Working with the first solution ($[C_6H_8O_6] = 0.100$ M):

$$K_a = \frac{x^2}{0.100}$$
$$x^2 = (0.100)\left(8.0 \times 10^{-5}\right) = 8.0 \times 10^{-6}$$
$$x = 0.00283\,M$$

This tells us that the concentration of dissociated ascorbic acid is 0.00283 M. Substituting this into the percent dissociation equation:

$$\textit{percent dissociated} = \frac{0.00283\,M}{0.100\,M} \times 100\% = 2.83\%$$

For the next solution, the concentration of ascorbic acid is 1.00 M:

$$K_a = \frac{x^2}{1.00}$$
$$x^2 = (1.00)\left(8.0 \times 10^{-5}\right) = 8.0 \times 10^{-5}$$
$$x = 0.00894\,M$$

Substituting this value into the percent dissociated equation gives

$$percent\ dissociated = \frac{0.00894\,M}{1.00\,M} \times 100\% = 0.89\%$$

What this tells us is this: the more dilute the solution, the more we'll see molecules dissociate.

Workbook Problem 16.7

Calculate the percent dissociation of a 0.79 M benzoic acid solution. Benzoic acid has the molecular formula C_6H_5COOH and has a K_a value of 6.5×10^{-5}.

Strategy: To determine the percent dissociation, we need to first determine the concentration of the dissociated HA. This requires a determination of the equilibrium concentration of H_3O^+ and $C_6H_5COO^-$.

Step 1: Determine the species present initially, and write the proton-transfer reaction.

Step 2: Make a table showing the principal reaction, initial concentration, change in concentration, and the equilibrium concentration.

Step 3: Substitute the equilibrium concentrations into the equilibrium expression, and solve for *x*.

Step 4: Calculate the concentrations of the major species present.

Step 5: Calculate the percent dissociation.

POST-LECTURE:

Work Problems: *One of the best ways to learn chemistry is through solving problems. Go to the textbook and work through the Worked Examples for this section, then go to the end of chapter and work the problems there. When you feel confident that you can correctly answer problems*

like the ones listed below for each objective, check off each of the Learning Objectives for this section above.

LO 16.9: Worked Example 16.10; Problems 16.94, 16.96

Section 16.11 Polyprotic Acids

PRE-LECTURE:

Learning Objectives: *These are the skills you should be proficient in after reading this section and practicing the associated problems. Check off the boxes below once you have gone through this section, answered the associated questions in your textbook, and feel confident you have mastered the skill.*

☐ **LO 16.10** Calculate the pH and the concentration of all species present in a solution of a diprotic acid.

Definitions: *To learn, you must do. This is your workbook. Fill in the definitions from the glossary of the textbook as a first step to understanding the material.*

— **polyprotic acid—**____

____

Review: *Before going through the section make sure you are familiar with these concepts:*

- Section 16.9 solve acid–base equilibrium problems

Summary: *Read through this section and fill in the missing words and equations. Reference the textbook when you are not confident in your answers.*

Some acids, such as sulfuric acid (strong) and phosphoric acid (weak), have more than one proton than can dissociate. The acids are called ___________ acids. As a result, they have a stepwise dissociation, and each step has its own characteristic K_a.

It is more difficult to pull a hydrogen off of an already negatively charged ion, so the dissociation constants decrease dramatically:

$$K_{a1} > K_{a2} > K_{a3}$$

A diprotic acid solution contains a mixture of acids: H_2A, HA^-, and H_2O. The diprotic species is by far the strongest, and as a rule, the first dissociation step is the only one that needs to be considered, because the hydronium ions contributed by the second dissociation (or the third in the case of phosphoric acid) are too few to change the pH calculated from the first dissociation.

The exception to the above rule is sulfuric acid, the only strong polyprotic acid. Sulfuric acid dissociates completely at the first dissociation step and has a $K_{a2} = 1.2 \times 10^{-2}$, stronger than the K_a of many weak acids. In this case, it is necessary to calculate the $[H_3O^+]$ generated by both dissociations.

Use the same five steps from section 16.9 to solve acid–base equilibrium problems.

Workbook Example 16.12

PROBLEM:
Calculate the pH of a 0.145 M sulfuric acid solution. K_{a1} = complete, $K_{a2} = 1.2 \times 10^{-2}$.

SOLUTION:
Because K_{a1} is complete, it means that $[H_3O^+]$ is equal to the concentration of $[HSO_4^-]$, which is 0.145 M.

To determine the concentration of H_3O^+ for the second dissociation:

$$K_{a2} = \frac{[H_3O^+][SO_4^{2-}]}{[HSO_4^-]} = \frac{x^2}{0.145 - x} = 1.2 \times 10^{-2}$$

With a K_{a2} of this magnitude, it is unlikely that the value of x will be negligible. We need to solve for x using the quadratic formula:

$$x^2 + 0.012x - 0.00174 = 0$$

$$x = \frac{-0.012 \pm \sqrt{(0.012)^2 - 4(1)(-0.001740)}}{2(1)}$$

$$x = 0.036\ M$$

The total concentration of H_3O^+ is: 0.145 M + 0.036 M = 0.181 M.

$$pH = -\log(0.181\ M) = 0.74$$

Workbook Problem 16.8

Write the stepwise dissociation for arsenic acid (H_3AsO_4) and determine the pH of a 1.69×10^{-3} M solution. ($K_{a1} = 5.62 \times 10^{-3}$; $K_{a2} = 1.70 \times 10^{-7}$; $K_{a3} = 3.95 \times 10^{-12}$)

Step 1: Write the stepwise dissociation for arsenic acid.

Step 2: Determine the principal reaction.

Step 3: Make a table showing the principal reaction, initial concentration, change in concentration, and equilibrium concentration of the reactant and products.

Step 4: Substitute the equilibrium concentrations into the equilibrium expression.

Step 5: Calculate the pH of the solution.

POST-LECTURE:

Work Problems: *One of the best ways to learn chemistry is through solving problems. Go to the textbook and work through the Worked Examples for this section, then go to the end of chapter and work the problems there. When you feel confident that you can correctly answer problems like the ones listed below for each objective, check off each of the Learning Objectives for this section above.*

LO 16.10: Worked Example 16.11; Problems 16.106, 16.108

Section 16.12 Equilibria in Solutions of Weak Bases

PRE-LECTURE:

Learning Objectives: *These are the skills you should be proficient in after reading this section and practicing the associated problems. Check off the boxes below once you have gone through this section, answered the associated questions in your textbook, and feel confident you have mastered the skill.*

☐ **LO 16.11** Calculate the pH and equilibrium concentrations in a solution of a weak base.

Definitions: *To learn, you must do. This is your workbook. Fill in the definitions from the glossary of the textbook as a first step to understanding the material.*

— **base-dissociation constant, K_b**— ______________________________

____

Summary: *Read through this section and fill in the missing words and equations. Reference the textbook when you are not confident in your answers.*

Weak bases will accept a proton from water, leaving OH^- ions and the protonated conjugate acid of the base:

$$B(aq) + H_2O(l) \leftrightarrows BH^+(aq) + OH^-(aq)$$

The position of the equilibrium is characterized by the ________________________, K_b.

$K_b =$

Many weak bases are amines, ammonia derivatives in which one of the hydrogen atoms is replaced by another chemical group: the basicity of these compounds is due to the lone pair of electrons on the nitrogen atom, which can be used to bind a proton.

Problems involving weak bases are solved just like those involving weak acids. The only difference is that $[OH^-]$ is being solved for in the equilibrium equation. To obtain the $[H^+]$ and the pH of the solution requires using the autoionization constant of water.

Workbook Example 16.13

PROBLEM:
Calculate the pH of a 0.10 M solution of methylamine, CH_3NH_2; $K_b = 3.7 \times 10^{-4}$.

SOLUTION:

$$K_b = 3.7\times10^{-4} = \frac{[CH_3NH_3^+][OH^-]}{[CH_3NH_2]} = \frac{x^2}{0.10-x}$$

K_b is within 3% of the concentration of the base; use the quadratic equation:

$$x^2 + 3.7\times10^{-4}x - 3.7\times10^{-5} = 0$$

$$x = \frac{-3.7\times10^{-4} \pm \sqrt{(3.7\times10^{-4})^2 - 4(1)(-3.7\times10^{-5})}}{2(1)} = 0.0059\,M$$

To solve for pH, calculate the concentration of H^+ using the autoionization constant of water:

$$[H^+] = \frac{1\times10^{-14}}{[OH^-]} = \frac{1\times10^{-14}}{0.0059\,M} = 1.69\times10^{-12}\,M$$

$$pH = -\log(1.69\times10^{-12}) = 11.8$$

Workbook Problem 16.9

Determine the pH of a 0.975 M trimethylamine, $(CH_3)_3N$, solution. $K_b = 6.5 \times 10^{-5}$.

Strategy: Solve using a table of equilibrium values.

Step 1: Write the principal reaction.

Step 2: Construct a table with concentrations of the reactant and products.

Step 3: Substitute the equilibrium concentrations into the equilibrium expression, and solve for x.

Step 4: Determine the equilibrium concentrations of the species present.

Step 5: Calculate the pH of the solution.

POST-LECTURE:

Work Problems: *One of the best ways to learn chemistry is through solving problems. Go to the textbook and work through the Worked Examples for this section, then go to the end of chapter and work the problems there. When you feel confident that you can correctly answer problems like the ones listed below for each objective, check off each of the Learning Objectives for this section above.*

LO 16.11: Worked Example 16.12; Problems 16.114, 16.116, 16.118

Section 16.13 Relation between K_a and K_b

PRE-LECTURE:

Learning Objectives: *These are the skills you should be proficient in after reading this section and practicing the associated problems. Check off the boxes below once you have gone through this section, answered the associated questions in your textbook, and feel confident you have mastered the skill.*

- ☐ **LO 16.12** Relate K_a, K_b, pK_a, and pK_b for a conjugate acid–base pair.

Summary: *Read through this section and fill in the missing words and equations. Reference the textbook when you are not confident in your answers.*

For a conjugate acid–base pair, K_a and K_b can be related through the ionization constant of water:

$$K_W =$$

As the strength of an acid increases, the strength of its conjugate base ____________; likewise, as the strength of a base increases, the strength of its conjugate acid ____________. This makes sense, especially since these values are always proportional to K_w. This gives another useful relationship:

$$= pK_W = 14$$

This can be rearranged to solve for K_a or K_b if either are unknown:

$K_a =$ $\qquad$ $K_b =$

Workbook Example 16.14

PROBLEM:
Calculate K_a for methylamine, where $K_b = 1.4 \times 10^{-4}$.

SOLUTION:

$$K_a = \frac{K_W}{K_b} = \frac{1 \times 10^{-14}}{1.4 \times 10^{-4}} = 7.14 \times 10^{-11}$$

POST-LECTURE:

Work Problems: *One of the best ways to learn chemistry is through solving problems. Go to the textbook and work through the Worked Examples for this section, then go to the end of chapter and work the problems there. When you feel confident that you can correctly answer problems*

like the ones listed below for each objective, check off each of the Learning Objectives for this section above.

LO 16.12: Worked Example 16.13; Problems 16.120, 16.122

Section 16.14 Acid–Base Properties of Salts

PRE-LECTURE:

Learning Objectives: *These are the skills you should be proficient in after reading this section and practicing the associated problems. Check off the boxes below once you have gone through this section, answered the associated questions in your textbook, and feel confident you have mastered the skill.*

☐ **LO 16.13** Predict whether a salt is acidic, basic, or neutral and calculate the pH of the salt solution.

Summary: *Read through this section and fill in the missing words and equations. Reference the textbook when you are not confident in your answers.*

The pH of a salt solution is determined by the acid–base properties of the cations and anions that make up that solution. The salt(s) that are formed in an acid–base reaction will contain the conjugate acid and the conjugate base of the reactants. The stronger of the two will predominate:

- Strong acid + Strong base → ___________ solution
- Strong acid + Weak base → ___________ solution
- Weak acid + Strong base → ___________ solution

Let's look at these one by one.

1) Salts that yield neutral solutions:
   - ___________ acid combined with a ___________ base
   - pH = ______
   - K_a (for the cation) ≈ K_b (for the anion)
2) Salts that yield acidic solutions:
   - ___________ acid combined with a ___________ base
   - pH ______7
   - K_a (for the cation) > K_b (for the anion)
3) Salts that yield basic solutions:
   - ___________ acid combined with a ___________ base
   - pH ______7
   - K_a (for the cation) < K_b (for the anion)

Workbook Example 16.15

PROBLEM:
Determine whether aqueous solutions of the following salts are acidic, neutral, or basic. Write the hydrolysis reaction for those solutions that are acidic or basic. NH_4ClO_4, RbCl, $NaCH_3CO_2$, NH_4CO_3.

SOLUTION:
To determine the acidity of an aqueous solution of a salt, we must determine if the salt is derived from a (1) strong acid/strong base reaction, (2) weak acid/strong base reaction, (3) strong acid/weak base reaction, or (4) weak acid/weak base reaction.

NH_4ClO_4: derived from the weak base $NH_3(aq)$ and the strong acid $HClO_4(aq)$. Acidic (NH_4^+ is the weak conjugate acid of NH_3).

$$NH_4^+(aq) + H_2O \leftrightarrows NH_3(aq) + H_3O^+(aq)$$

RbCl: derived from the strong base RbOH(*aq*) and the strong acid HCl(*aq*). Neutral.

$NaCH_3CO_2$: derived from the strong base NaOH and the weak acid CH_3COOH. Basic.

($CH_3CO_2^-$ is the weak conjugate base of CH_3COOH.)

$$CH_3CO_2^-(aq) + H_2O \leftrightarrows CH_3COOH(aq) + OH^-(aq)$$

NH_4CO_3: derived from the weak base NH_3 and the weak acid $HCO_3^-(aq)$. To determine the acidity of this solution, we must calculate the K_a of the cation and the K_b of the anion of the salt and compare the two.

$$NH_4^+\ K_a = \frac{1\times10^{-14}}{K_b\ NH_3} = \frac{1\times10^{-14}}{1.8\times10^{-5}} = 5.6\times10^{-10}$$

$$CO_3^{2-}\ K_b = \frac{1\times10^{-14}}{K_a\ HCO_3^-} = \frac{1\times10^{-14}}{5.6\times10^{-11}} = 1.8\times10^{-4}$$

Since the $K_b > K_a$, this solution has to be basic. The hydrolysis reaction is

$$CO_3^{2-}(aq) + H_2O(l) \leftrightarrows HCO_3^-(aq) + OH^-(aq)$$

Workbook Example 16.16

PROBLEM:
Determine the pH of a 0.150 M solution of NH_4Cl.

SOLUTION:
To determine the acidity of an aqueous solution of a salt, the K_a for the acidic portion must be used.

From the last example problem, the K_a value of the ammonium cation is 5.6×10^{-10}.

$$5.6\times10^{-10} = \frac{[NH_3][H_3O^+]}{[NH_4^+]} = \frac{x^2}{0.150}$$

$$x = 9.17\times10^{-6}$$

$$\text{pH} = -\log(9.17\times10^{-6}) = 5.04$$

Workbook Problem 16.10

Calculate the pH of a 0.137 M KOCl solution. The K_a for HOCl is 3.5×10^{-8}.

Strategy: Determine if KOCl produces an acidic, basic, or neutral aqueous solution, write the hydrolysis reaction for this salt, and determine the pH.

Step 1: Write the hydrolysis reaction for this salt.

Step 2: Write an equilibrium expression for this reaction.

Step 3: Construct a table.

Step 4: Substitute the equilibrium concentrations into the equilibrium expression.

Step 5: Calculate the pH of the solution.

POST-LECTURE:

Work Problems: *One of the best ways to learn chemistry is through solving problems. Go to the textbook and work through the Worked Examples for this section, then go to the end of chapter and work the problems there. When you feel confident that you can correctly answer problems like the ones listed below for each objective, check off each of the Learning Objectives for this section above.*

LO 16.13: Worked Example 16.14; Problems 16.44, 16.124, 16.128, 16.133, 16.134

Section 16.15 Lewis Acids and Bases

PRE-LECTURE:

Learning Objectives: *These are the skills you should be proficient in after reading this section and practicing the associated problems. Check off the boxes below once you have gone through this section, answered the associated questions in your textbook, and feel confident you have mastered the skill.*

☐ **LO 16.14** Identify the Lewis acid and Lewis base and use curved arrow notation to indicate donation of a lone pair of elections in Lewis acid–base reaction.

Definitions: *To learn, you must do. This is your workbook. Fill in the definitions from the glossary of the textbook as a first step to understanding the material.*

— **Lewis acid—** ____

____

— **Lewis base—** ____

____

Summary: *Read through this section and fill in the missing words and equations. Reference the textbook when you are not confident in your answers.*

At this point in the story of acids and bases, we have been introduced to the Arrhenius theory of acids and bases, and we have worked extensively with the Brønsted–Lowry theory of acids and base. Another theory of acids and bases was developed by G.N. Lewis, of the Lewis dot structure fame. According to the Lewis theory a:

- base is an electron-pair ___________
- acid is an electron-pair ___________

For bases, the Lewis and Brønsted–Lowry definitions agree, but for acids, the Lewis definition accommodates several situations that are not addressed by the Brønsted–Lowry definition, including cations and neutral molecules with vacant valence orbitals. The Lewis definition explains the acidity of Al^{3+} and CO_2, for example.

This theory is very useful when dealing with organic compounds. If you continue on to organic chemistry (which I hope you will!), you will be working with this in a lot more detail.

POST-LECTURE:

Work Problems: *One of the best ways to learn chemistry is through solving problems. Go to the textbook and work through the Worked Examples for this section, then go to the end of chapter and work the problems there. When you feel confident that you can correctly answer problems like the ones listed below for each objective, check off each of the Learning Objectives for this section above.*

LO 16.14: Worked Example 16.15; Problems 16.45–16.47, 16.142

Self-Test

This section is intended to test your knowledge of the material covered in this chapter. Think through these problems and make certain you understand what the problem is about—ask yourself if the answer makes sense. Many of these questions are linked to the chapter learning goals. Therefore, successful completion of these problems indicates that you have achieved the learning goals for this chapter. You will receive the greatest benefit from this section if you use it as a mock exam. Do not use your notes, book, Internet, or friends when taking the test. There is no grade and you only shortchange yourself by getting help with this activity. As you take the test, mark the problems with your level of confidence with the answer.

- ★ *Use a star for problems you know you got right.*
- ? *Use a question mark for those you think you got right.*
- x *Use an x for those you are guessing on.*

If you get a question right that you marked with a star, you can be confident that you know the associated LO.

If you get a question right that you marked with a question mark, go back and work a few more problems and test yourself again tomorrow.

If you get a question wrong or mark it with an x, go back and read the section again, work a few problems today, a few tomorrow, and a few the next day before testing yourself again.

This will help you to discover which topics you have learned thoroughly and which topics you need to study in more detail.

True–False

1. The Brønsted–Lowry definition states that acids are electron donors and bases are electron acceptors.
2. A strong acid is more corrosive than a weak acid.
3. H^+ is unstable in solution and attaches to a water molecule to form a hydronium (H_3O^+) ion.
4. A solution with a pH of 7.9 is slightly basic.
5. The strength of an acid and the strength of its conjugate base are directly related: the stronger the acid, the stronger its conjugate base.
6. Dilute solutions of weak acids will have a higher percent dissociation.
7. For most polyprotic acids, two or more dissociations must be considered when calculating the hydronium ion concentration, $[H_3O^+]$.
8. Salts can be acidic, basic, or neutral.
9. K_a can be used to calculate the K_b of a weak base.
10. $Mg(OH)_2$ is a weak base.

Multiple Choice

11. In the Arrhenius, Brønsted–Lowry, and Lewis definitions of an acid, acids are
    a. hydrogen ion generators, proton acceptors, and electron acceptors
    b. hydroxide acceptors, proton donors, and electron acceptors
    c. hydrogen ion generators, proton donors, and electron acceptors
    d. hydroxide acceptors, proton acceptors, and electron donors

12. The pH of a salt solution
    a. = 7 for all salts
    b. > 7 for salts derived from a strong acid and a strong base
    c. = 7 for salts derived from a weak acid and a weak base
    d. < 7 for salts derived from a strong acid and a weak base

13. A weak acid or base
    a. dissociates incompletely in solution
    b. has a pH close to 7
    c. dissociates fully in solution
    d. is always less corrosive than a strong acid or base

14. The concentrations of H_3O^+ and OH^- in pure water at 25 °C are
    a. 1×10^{-7} M, 1×10^{-14} M
    b. 1×10^{-7} M, 1×10^{-7} M
    c. 1×10^{-14} M, 1×10^{-7} M
    d. 1×10^{-14} M, 1×10^{-7} M

15. Solutions with $[H_3O^+] = 1 \times 10^{-5}$ M have a pH of
    a. 1×10^{-5}
    b. −5
    c. 5
    d. 9

16. The strength of a binary acid is affected by
    a. polarity of the HA bond
    b. the size of A–
    c. the strength of the HA bond
    d. all of the above

17. Acid–base indicators
    a. are strong acids
    b. are substances that change color in a specific pH range
    c. have the same colors in their acid (HIn) and conjugate base (In^-) forms
    d. can be used to determine the exact pH of a solution

18. Which of the following is the strongest acid?
    a. $HClO_4$
    b. $HClO_3$
    c. $HClO_2$
    d. HClO

Matching

19. acid–base indicator	a. K_a
20. acid-dissociation constant	b. H^+ donor
21. Arrhenius acid	c. chemical species that differ only by one proton
22. Arrhenius base	d. molecule that dissociates in water to form OH^- ions
23. base-dissociation constant	e. electron acceptor

24. Brønsted–Lowry acid	f. a molecule with different colored forms at different pH values
25. Brønsted–Lowry base	g. K_b
26. conjugate acid–base pair	h. dissociated/undissociated ×100
27. hydronium ion	i. $-\log [H_3O^+]$
28. ion product for water ions	j. molecule that dissociates in water to release H^+
29. Lewis acid	k. H^+ acceptor
30. Lewis base	l. acid that is capable of more than one dissociation
31. percent dissociation	m. ion in which H^+ is bonded to the oxygen atom of a solvent water molecule
32. pH	n. electron donor
33. polyprotic acid	o. given by $K_w = \left[H_3O^+\right]\left[OH^-\right]$
34. strong acid	p. acid that dissociates incompletely
35. weak acid	q. acid that dissociates completely

Fill in the Blanks

36. Only a small fraction of __________ acid molecules transfer a ____________ to water.

37. In acidic solutions, the concentration of ______________ ions is greater than the concentration of _______________ ions. In neutral solutions, these concentrations are _______________.

38. In general, the percent dissociation of an acid _______________ with increasing K_a, and _______________ with increasing concentration.

39. Polyprotic acids have a _________________ dissociation in which each step has a much ______________ K_a.

40. Amines are weak bases. They have ____________________ that can be donated to a bond with ________________.

Problems

41. What is the $[OH^-]$ for a solution of $Ca(OH)_2$ whose pH = 11.78?

42. Calculate the pH of a solution prepared by dissolving 1.83 g of $Ba(OH)_2$ in 150 mL of water.

43. The percent dissociation for a 7.35×10^{-3} M weak acid solution is 0.51%. Calculate the pH of this solution and determine the K_a for the acid.

44. Determine the K_a of histidine, a weak organic acid, if the pH of a 1.30×10^{-3} M solution is 6.03.

45. What is the pH of a 1.24 M solution of pyridine, a weak base whose formula is C_6H_5N? The base-dissociation constant is 1.8×10^{-9}.

46. Calculate the concentration of all species present and the pH of a 0.125 M oxalic acid ($H_2C_2O_4$) solution. The acid-dissociation constants are $K_{a1} = 5.9 \times 10^{-2}$, $K_{a2} = 6.4 \times 10^{-5}$.

47. Calculate the pH of a 0.350 M solution of benzylamine ($K_b = 2.14 \times 10^{-5}$).

48. Determine (a) K_b for the conjugate base of ascorbic acid ($K_a = 8.0 \times 10^{-5}$) and (b) the K_a for the conjugate acid of hydrazine ($K_b = 8.9 \times 10^{-7}$).

49. 1.75 g of KCN is dissolved in 150 mL of water. What is the concentration of HCN(*aq*) at equilibrium? What is the pH of the solution? The acid-dissociation constant is $K_a = 4.9 \times 10^{-10}$.

50. Determine which acid is stronger and explain why.
    a. HBr(*aq*) or HI(*aq*)
    b. H_3PO_4 or H_3PO_3

Challenge Problem

51. A 25.0% by mass solution of H_3PO_4 has a density of 1.1667 g/mL. Calculate the pH and concentrations of all phosphate-containing species present.

Test Yourself: *Take the Practice Test on pages 698 and 699 of your textbook. Again, use a testing environment using only those items that your instructor will allow you to have access to during the test and mark each question with a star, question mark, or x to denote your confidence level. Once you have finished you can check off the LO at the beginning of this chapter based on which questions you starred and got correct. Go back and spend more time with the topics from the other learning objectives.*

CHAPTER SEVENTEEN

Applications of Aqueous Equilibria

Learning Objectives

As a result of reading and studying this chapter, you should be able to:

- ☐ **LO 17.1** Write a balanced equation for a neutralization reaction and predict whether the pH after neutralization is greater than, equal to, or less than 7.
- ☐ **LO 17.2** Calculate the equilibrium concentrations of all species, pH, and percent dissociation in a solution of a weak acid when a common ion is present.
- ☐ **LO 17.3** Calculate the initial pH of a buffer solution and the pH after the addition of a strong acid or a strong base.
- ☐ **LO 17.4** Use the Henderson–Hasselbalch equation to calculate the pH of a buffer solution.
- ☐ **LO 17.5** Calculate the pH at various points in a strong acid–strong base titration.
- ☐ **LO 17.6** Calculate the pH at various points in a weak acid–strong base titration or weak base–strong acid titration.
- ☐ **LO 17.7** Calculate the pH at various points in a diprotic acid–strong base titration.
- ☐ **LO 17.8** Visualize the molecular species present during a titration and interpret titration curves.
- ☐ **LO 17.9** Write the equilibrium-constant expression for dissolution of an ionic compound, and calculate the value of its K_{sp}.
- ☐ **LO 17.10** Calculate equilibrium concentrations of ions and the solubility of an ionic compound from its K_{sp}.
- ☐ **LO 17.11** Predict how the presence of molecular and ionic species will affect the solubility of an ionic compound.
- ☐ **LO 17.12** Calculate solubility in a solution that contains a common ion.
- ☐ **LO 17.13** Use the formation constant K_f to calculate ion concentrations in a solution that contains a complex ion.
- ☐ **LO 17.14** Use the formation constant K_f to calculate the solubility of an ionic compound when the cation forms a complex ion.
- ☐ **LO 17.15** Calculate the ion product IP for an ionic compound, and determine whether a precipitate will form when various solutions are mixed.
- ☐ **LO 17.16** Design a scheme for separating ions in a mixture based on selective precipitation.

Chapter Summary

Concepts of equilibrium can be applied to neutralization reactions between strong and weak acids and bases. The titration curves of each of these reactions have distinctive features that make it possible to determine the relative strength of the reactants. The common-ion effect also affects equilibrium mixtures. It is this corollary of Le Châtelier's principle that explains the functioning of buffers: solutions of weak acids and their conjugate bases that resist changes in pH. The pH of a buffer system will be close to the pK_a of the weak acid, and it can be calculated using the Henderson–Hasselbalch equation. The solubility product of an ionic compound, K_{sp}, is the equilibrium constant for its dissolution. This equilibrium constant can be converted to molar solubility and can be altered by changing the conditions surrounding the compound. Using knowledge of these constants and how they can be changed enables us to separate ions in solution through selective precipitation.

The Chapter in Detail

Section 17.1 Neutralization Reactions

PRE-LECTURE:

Learning Objectives: *These are the skills you should be proficient in after reading this section and practicing the associated problems. Check off the boxes below once you have gone through this section, answered the associated questions in your textbook, and feel confident you have mastered the skill.*

☐ **LO 17.1** Write a balanced equation for a neutralization reaction and predict whether the pH after neutralization is greater than, equal to, or less than 7.

Review: *Before going through the section make sure you are familiar with these concepts:*

- Section 4.4 neutralization reactions
- Section 16.4 ion product constant for water

Summary: *Read through this section and fill in the missing words and equations. Reference the textbook when you are not confident in your answers.*

STRONG ACID–STRONG BASE

When a strong acid and a strong base react, the net ionic equation for the reaction is:

When equal numbers of moles of acid and base are mixed together, $[H_3O^+]$ and $[OH^-]$ = ______ M, the solution is neutral, and the reaction proceeds to completion. The equilibrium constant is the reciprocal of the ion product constant for water, K_W.

$K_n =$

The reaction leaves behind a neutral salt, and the pH = ________.

WEAK ACID–STRONG BASE

When a weak acid and a strong base react, the net ionic reaction involves proton transfer from HA to the strong base. Write the net ionic reaction below:

The equilibrium constant for this reaction is obtained by combining the reactions for the dissociation of the weak acid and formation of water (reverse of the dissociation of water). As a result, the equilibrium constant for these reactions is:

$K_n =$

The neutralization reaction will go to completion, but it will leave behind the conjugate base of the weak acid; therefore, the pH will be ________ than 7.

STRONG ACID–WEAK BASE

When a strong acid and a weak base react, the net ionic equation involves proton transfer from the strong acid to the weak base. Write it below:

The equilibrium constant for this reaction is obtained by multiplying the equilibrium constant for the dissociation of the base with the formation of water (this is the reverse of the dissociation of water).

$K_n =$

The neutralization reaction will go to completion but will leave behind the conjugate acid of the base. As a result, the pH will be ________ than 7.

WEAK ACID–WEAK BASE

When a weak acid and a weak base react, the reaction involves proton transfer from the weak acid to the weak base:

The equilibrium constant for the overall reaction will be the combination of the dissociation of the weak acid, the protonation of the weak base, and the formation of water:

$K_n =$

This type of reaction has far less tendency to proceed to completion than neutralizations involving strong acids or strong bases.

Workbook Example 17.1

PROBLEM:
Write the net ionic equation and predict whether the pH after neutralization is greater than, equal to, or less than 7:

$HF(aq) + KOH$ $HCl(aq) + CH_3NH_2$ (a weak base)

SOLUTION:

The first equation is the reaction of a weak acid with a strong base. The net ionic equation is

$$HF(aq) + OH^-(aq) \leftrightarrows H_2O(l) + F^-(aq)$$

The pH of this solution will be higher than 7 (basic) because a weak base is left in solution.

The second equation is the reaction of a strong acid with a weak base. The net ionic equation is

$$H_3O^+(aq) + CH_3NH_2 \leftrightarrows H_2O(l) + CH_3NH_3^+(aq)$$

The pH of this solution is expected to be less than 7 (acidic), since the conjugate acid of a weak base is a weak acid.

Workbook Problem 17.1

Write a balanced net ionic equation for the neutralization of nitrous acid by sodium hydroxide. Determine K_n and the position of equilibrium for this neutralization reaction. Predict whether the pH after neutralization is greater than, equal to, or less than 7. The K_a for nitrous acid is 4.5×10^{-4}.

Strategy: Write the net reaction and determine K_n.

Step 1: Write the net neutralization reaction by writing individual reactions.

Step 2: Calculate K_n based on the equilibrium constants for the individual reactions in step 1.

Step 3: Determine the position of equilibrium from the value of K_n.

Step 4:	Predict the pH.

POST-LECTURE:

Work Problems: *One of the best ways to learn chemistry is through solving problems. Go to the textbook and work through the Worked Examples for this section, then go to the end of chapter and work the problems there. When you feel confident that you can correctly answer problems like the ones listed below for each objective, check off each of the Learning Objectives for this section above.*

LO 17.1: Worked Example 17.1; Problems 17.38, 17.47, 17.49, 17.51

Section 17.2 The Common-Ion Effect

PRE-LECTURE:

Learning Objectives: *These are the skills you should be proficient in after reading this section and practicing the associated problems. Check off the boxes below once you have gone through this section, answered the associated questions in your textbook, and feel confident you have mastered the skill.*

☐ **LO 17.2** Calculate the equilibrium concentrations of all species, pH, and percent dissociation in a solution of a weak acid when a common ion is present.

Definitions: *To learn, you must do. This is your workbook. Fill in the definitions from the glossary of the textbook as a first step to understanding the material.*

— **common ion—** ____

____

— **common-ion effect—** ____

____

Review: *Before going through the section make sure you are familiar with these concepts:*

- Section 15.6 Le Châtelier's principle

Summary: *Read through this section and fill in the missing words and equations. Reference the textbook when you are not confident in your answers.*

A solution of a weak acid and a salt of its conjugate base is an important mixture because this type of mixture regulates the pH of biological systems.

The ________effect is a shift in equilibrium caused by adding more of an ion that is involved in an equilibrium. It is an example of Le Châtelier's principle: adding the salt of the conjugate base of an acid will push the equilibrium toward less dissociation of the acid.

To determine the properties of a solution prepared from a weak acid and a salt of its conjugate base, we need to know the properties of the various species. The principal reaction is the dissociation of the weak acid. The equilibrium calculations must be adjusted for the initial concentration of the conjugate base—this provides an initial concentration of A^-, and the dissociation of the weak acid will provide $[H_3O^+]$ and the change in $[A^-]$.

Workbook Example 17.2

PROBLEM:
Calculate the concentration of all species present, the pH, and the percent dissociation of nitrous acid in a solution that is 0.150 M HNO_2 and 0.075 M $NaNO_2$.

SOLUTION:
Because the salt is 100% dissociated, the species present initially are HNO_2, NO_2^-, Na^+, and H_2O. Na^+ is inert; HNO_2 is a weak acid ($K_a = 4.5 \times 10^{-4}$); NO_2^- is the conjugate base of a weak acid; and H_2O can be either an acid or a base. The principal reaction is transfer of a proton from HNO_2 to H_2O.

Principal reaction	$HNO_2(aq) + H_2O(l) \leftrightarrows$	$H_3O^+(aq)$	$NO_2^-(aq)$
Initial concentration (M)	0.150	0	0.075
Change (M)	$-x$	$+x$	$+x$
Eq. concentration (M)	$0.150 - x$	$+x$	$0.075 + x$

The common ion in this problem is NO_2^-. The equilibrium equation for the principal reaction is

$$K_a = 4.5 \times 10^{-4} = \frac{[H_3O^+][NO_2^-]}{[HNO_2]} = \frac{(x)(0.075+x)}{(0.150-x)} = \frac{(x)(0.075)}{(0.150)}$$

Because K_a is so small, x is assumed to be negligible. The equilibrium is shifted to the left due to the common-ion effect.

$$x = [H_3O^+] = \frac{(4.5\times10^{-4})(0.150)}{(0.075)} = 9.0\times10^{-4}$$

Note that the assumption concerning the size of x is justified.

$$pH = -\log(9.0\times10^{-4}) = 3.05$$

The percent dissociation of HNO_2 is

$$\text{percent dissociation} = \frac{[HNO_2]_{\text{dissociated}}}{[HNO_2]_{\text{initial}}} \times 100\% = \frac{9.0\times10^{-4}}{0.150} \times 100\% = 0.60\%$$

Workbook Problem 17.2

Calculate the pH and the percent dissociation of phosphorous acid in a solution that is 0.250 M H_3PO_3 and 0.175 M NaH_2PO_3. The K_a of phosphorous acid is $K_a = 1.0 \times 10^{-2}$.

Strategy: Write the reaction; determine the concentration of H_3O^+, $H_2PO_3^-$.

Step 1: Write the overall reaction.

Step 2: Make a chart of initial and equilibrium concentrations.

Step 3: Solve for the equilibrium concentrations of H_3O^+, $H_2PO_2^-$.

Step 4: Determine the pH of the solution.

Step 5: Determine the percent dissociation of the acid.

POST-LECTURE:

Work Problems: *One of the best ways to learn chemistry is through solving problems. Go to the textbook and work through the Worked Examples for this section, then go to the end of chapter and work the problems there. When you feel confident that you can correctly answer problems like the ones listed below for each objective, check off each of the Learning Objectives for this section above.*

LO 17.2: Worked Examples 17.2, 17.3; Problems 17.57, 17.59, 17.61

Section 17.3 Buffer Solutions

PRE-LECTURE:

Learning Objectives: *These are the skills you should be proficient in after reading this section and practicing the associated problems. Check off the boxes below once you have gone through this section, answered the associated questions in your textbook, and feel confident you have mastered the skill.*

☐ **LO 17.3** Calculate the initial pH of a buffer solution and the pH after the addition of a strong acid or a strong base.

Definitions: *To learn, you must do. This is your workbook. Fill in the definitions from the glossary of the textbook as a first step to understanding the material.*

— **buffer solutions—** ____

____

— **buffer capacity—**____

____

Summary: *Read through this section and fill in the missing words and equations. Reference the textbook when you are not confident in your answers.*

Solutions like these, which contain a weak acid and its conjugate base or a weak base and its conjugate acid, are called ________________. They are important to biological systems because they are able to resist changes in pH.

If a small amount of a base is added to one of these solutions, it will be neutralized by the ________________. If a small amount of acid is added, it will be neutralized by the ________________.

For a buffer prepared from a weak acid, HA, and its conjugate base, B:

$$\left[H_3O^+\right]=K_a\frac{[HA]}{[B]}$$

Because the dissociations involved in these reactions are so small, these calculations can be done using initial concentrations.

Buffer capacity is a measure of how much acid or base a solution can absorb before the pH begins to be seriously affected. The take-home message of buffer capacity is this: the higher the concentration of the acid and base, the ________ the buffer capacity. For the same concentration, a larger volume of solution will also have a ________ buffer capacity.

Workbook Example 17.3

PROBLEM:
Which of these solutions will have the greatest buffer capacity? 1 M acetic acid, 0.1 M sodium acetate; 0.1M acetic acid, 1 M sodium acetate; 1 M acetic acid, 1 M sodium acetate; or 0.1 M acetic acid and 0.1 M sodium acetate?

SOLUTION:
The higher concentrations will have the greatest buffer capacities, so the solution that is 1 M in both species will have the highest buffer capacity.

Workbook Problem 17.3

Calculate the pH of a 1.0 L buffer solution containing 0.35 M HCOOH and 0.25 M HCOONa. Determine the pH of this solution after the addition of 0.10 mol HCl (assume no volume change). K_a for formic acid equals 1.8×10^{-4}.

Strategy: The pH of the initial solution is determined from K_a and the equilibrium expression. The pH of the solution after addition of HCl is determined knowing the changes in concentration that occur after the addition of $[H_3O^+]$.

Step 1: Determine the principal reaction and equilibrium concentrations.

Step 2: Using the equilibrium equation, solve for $[H_3O^+]$.

Step 3: Solve for the pH.

Step 4: Write the neutralization reaction that occurs upon addition of HCl and set up a table showing the number of moles present before and after the addition of HCl.

Step 5: Determine the concentrations of the buffer components after neutralization occurs.

Step 6:	Substitute the concentrations of the buffer components into the equilibrium expression and calculate the pH.

POST-LECTURE:

Work Problems: *One of the best ways to learn chemistry is through solving problems. Go to the textbook and work through the Worked Examples for this section, then go to the end of chapter and work the problems there. When you feel confident that you can correctly answer problems like the ones listed below for each objective, check off each of the Learning Objectives for this section above.*

LO 17.3: Worked Example 17.4; Problems 17.67, 17.69

Section 17.4 The Henderson–Hasselbalch Equation

PRE-LECTURE:

Learning Objectives: *These are the skills you should be proficient in after reading this section and practicing the associated problems. Check off the boxes below once you have gone through this section, answered the associated questions in your textbook, and feel confident you have mastered the skill.*

☐ **LO 17.4** Use the Henderson–Hasselbalch equation to calculate the pH of a buffer solution.

Definitions: *To learn, you must do. This is your workbook. Fill in the definitions from the glossary of the textbook as a first step to understanding the material.*

— **Henderson–Hasselbalch equation—** ____

____

Summary: *Read through this section and fill in the missing words and equations. Reference the textbook when you are not confident in your answers.*

The equation developed above can be rearranged in logarithmic form called the Henderson–Hasselbalch equation:

pH =

The Henderson–Hasselbalch equation helps us:

☐ determine the pH of a solution at certain moments during a titration.

☐ Helps calculate the percent dissociation of a weak acid.

The Henderson–Hasselbalch equation gives the following relationships:

- At pH = pK_a + 2.00, there is ________% dissociation
- At pH = pK_a + 1.00, there is ________% dissociation
- At pH = pK_a + 0.00, there is ________% dissociation
- At pH = pK_a − 1.00, there is ________% dissociation
- At pH = pK_a − 2.00, there is ________% dissociation

This equation is also extremely useful for preparing buffer solutions. To do this, first choose a weak acid with a pK_a close to the desired pH of the solution. Then adjust the ratio of base to acid according the Henderson–Hasselbalch equation to reach the desired pH.

The pH of the final solution is determined only by the mole ratio of base to acid, not by the volume, but the volume will also determine the buffer capacity.

Workbook Example 17.4

PROBLEM:
You are performing an experiment that requires your solution be buffered at a pH of 4.32. Benzoic acid has a K_a of 6.5×10^{-5}. What proportion of benzoic acid to sodium benzoate would give a pH of 4.32?

SOLUTION:
The pK_a of benzoic acid is $-\log(6.5 \times 10^{-5}) = 4.19$.

$$\log\frac{[base]}{[acid]} = pH - pK_a = 4.32 - 4.19 = 0.13$$

$$\frac{[base]}{[acid]} = 10^{0.13} = 1.35$$

A good buffer for this pH would contain 1.35 moles of sodium benzoate for every mole of benzoic acid.

Workbook Problem 17.4

Determine the [sodium acetate]/[acetic acid] ratio for a buffer system with a pH = 5.75. The K_a of acetic acid is 1.8×10^{-5}. How much sodium acetate would you need to add to 1 L of a 0.01 M solution of acetic acid to make a buffer at this pH?

Strategy: Use the Henderson–Hasselbalch equation to calculate the [base]/[acid] ratio. This information provides the mole ratio of acetate to acetic acid.

Step 1: Calculate the pK_a of acetic acid.

Step 2: Use the Henderson–Hasselbalch equation to calculate the [base]/[acid] ratio.

Step 3: Determine the mass of sodium acetate needed.

Workbook Problem 17.5

The pK_a of asparagine is 8.8. At what pH is asparagine 25% dissociated?

Strategy: Using the Henderson–Hasselbalch equation, calculate the log of the [base]/[acid] ratio, then determine the pH at which this value is obtained.

Step 1: Use the Henderson–Hasselbalch equation.

Step 2: Calculate pH.

POST-LECTURE:

Work Problems: *One of the best ways to learn chemistry is through solving problems. Go to the textbook and work through the Worked Examples for this section, then go to the end of chapter and work the problems there. When you feel confident that you can correctly answer problems like the ones listed below for each objective, check off each of the Learning Objectives for this section above.*

LO 17.4: Worked Examples 17.5, 17.6; Problems 17.71, 17.73, 17.75, 17.77

Section 17.5 pH Titration Curves

PRE-LECTURE:

Learning Objectives: *These are the skills you should be proficient in after reading this section and practicing the associated problems. Check off the boxes below once you have gone through this section, answered the associated questions in your textbook, and feel confident you have mastered the skill.*

☐ **LO 17.8** Visualize the molecular species present during a titration and interpret titration curves.

Definitions: *To learn, you must do. This is your workbook. Fill in the definitions from the glossary of the textbook as a first step to understanding the material.*

— **pH titration curve—** ____

____

— **equivalence point—**____

____

Review: *Before going through the section make sure you are familiar with these concepts:*

- Section 4.9 titrations

Summary: *Read through this section and fill in the missing words and equations. Reference the textbook when you are not confident in your answers.*

In an acid–base titration, a solution of a known concentration of base (or acid) is slowly added from a buret to a solution containing an unknown concentration of acid (or base).

This method is used to determine the equivalence point—the point at which stoichiometrically equivalent quantities of acid and base have been mixed together.

Titrations can be graphically represented as ________ curves: a plot of the pH of the solution versus the volume of added titrant. These are useful in determining the equivalence point, which allows the experimenter to choose a suitable acid–base indicator.

Small volumes are typically used for titrations. To make calculations easy, molarity is calculated in mL. Thus, the equation for molarity looks like this:

$$\mathrm{M} = \frac{\_\_\_\_\_\_ \textit{ of solute}}{\textit{mL of solvent}}$$

POST-LECTURE:

Work Problems: *One of the best ways to learn chemistry is through solving problems. Go to the textbook and work through the Worked Examples for this section, then go to the end of chapter and work the problems there. When you feel confident that you can correctly answer problems*

like the ones listed below for each objective, check off each of the Learning Objectives for this section above.

LO 17.8: Problems 17.41–17.44

Section 17.6 Strong Acid–Strong Base Titrations

PRE-LECTURE:

Learning Objectives: *These are the skills you should be proficient in after reading this section and practicing the associated problems. Check off the boxes below once you have gone through this section, answered the associated questions in your textbook, and feel confident you have mastered the skill.*

- ☐ **LO 17.5** Calculate the pH at various points in a strong acid–strong base titration.
- ☐ **LO 17.8** Visualize the molecular species present during a titration and interpret titration curves.

Summary: *Read through this section and fill in the missing words and equations. Reference the textbook when you are not confident in your answers.*

The process of titrating a strong acid with a strong base is outlined below. The process is similar, but you see a curve in the opposite direction, when titrating a strong base with a strong acid.

BEFORE ANY BASE IS ADDED:

- The $[H_3O^+]$ = concentration of strong acid.

BEFORE THE EQUIVALENCE POINT:

- As base is added incrementally, it will neutralize the acid, leading to a slow rise in pH.
- It is easier to calculate amount of acid using moles rather than molarity because every drop of added base also increases volume.

$$mmol\ H_3O^+\ after\ neutralization = mmol\ H_3O^+\ initial - mmol\ OH^-\ added$$

$$\left[H_3O^+\right] after\ neutralization = \frac{mmol\ H_3O^+\ after\ neutralization}{total\ volume\ of\ acid\ and\ base}$$

- Just before the equivalence point pH changes drastically.

AT THE EQUIVALENCE POINT:

- At the equivalence point, all of the initial acid is neutralized, and the solution contains a ________.
- The pH will be ________.

BEYOND THE EQUIVALENCE POINT:

- Just after the equivalence point, the pH will increase ________.
- Outside of this region, the pH change slows.

- To calculate excess amount of OH^- present:

$$mmol\ of\ excess\ OH^- = mmol\ of\ OH^-\ added - mmol\ of\ acid\ initially\ present$$

$$\left[OH^-\right]\ after\ neutralization = \frac{mmol\ excess\ of\ OH^-}{total\ volume\ of\ acid\ and\ base}$$

Workbook Problem 17.6

What is the pH that results when 50 mL of 0.1 M NaOH are added to 15 mL of 1.0 M HCl?

Strategy: Determine the number of moles of H_3O^+ and OH^- present in the mixture and determine the pH based on the excess.

Step 1: Calculate the moles of OH^- and the moles of H_3O^+.

Step 2: Determine the excess quantity and the resultant $[H_3O^+]$.

Step 3: Determine the pH of the solution.

POST-LECTURE:

Work Problems: *One of the best ways to learn chemistry is through solving problems. Go to the textbook and work through the Worked Examples for this section, then go to the end of chapter and work the problems there. When you feel confident that you can correctly answer problems like the ones listed below for each objective, check off each of the Learning Objectives for this section above.*

LO 17.5: Worked Example 17.7; Problems 17.81, 17.83

Section 17.7 Weak Acid–Strong Base Titrations

PRE-LECTURE:

Learning Objectives: *These are the skills you should be proficient in after reading this section and practicing the associated problems. Check off the boxes below once you have gone through this section, answered the associated questions in your textbook, and feel confident you have mastered the skill.*

☐ **LO 17.6** Calculate the pH at various points in a weak acid–strong base titration or weak base–strong acid titration.

☐ **LO 17.8** Visualize the molecular species present during a titration and interpret titration curves.

Review: *Before going through the section make sure you are familiar with these concepts:*

- Section 16.7 weak acid equilibrium problems
- Section 16.14 basic salts

Summary: *Read through this section and fill in the missing words and equations. Reference the textbook when you are not confident in your answers.*

BEFORE ANY BASE IS ADDED:

- Calculate pH as usual for a solution of weak acid.

BEFORE THE EQUIVALENCE POINT:

- The weak acid is largely undissociated and will react with all the OH^- from the strong base to form the conjugate base of the weak acid.
- Set up a table to determine amounts as neutralization occurs.
- Use the Henderson–Hasselbalch equation to calculate pH.

HALFWAY TO THE EQUIVALENCE POINT:

- pH = ___________ of the weak acid

AT THE EQUIVALENCE POINT:

- pH ___ 7 because the anion of the weak acid is a base.

AFTER THE EQUIVALENCE POINT:

- In general, the anion created from the weak acid can be neglected after the equivalence point and pH is calculated from excess $[OH^-]$.

$$\left[H_3O^+\right] = \frac{K_w}{\left[OH^-\right]}$$

The pH titration curve for a strong base and a weak acid has a different appearance than that of a strong base and a strong acid. Because of the buffering action of the weak acid/conjugate base mixture that develops before the equivalence point, the pH initially rises quickly, then levels off. Where pH = pK_a, the curve is flattest. Near the equivalence point, the pH increase is smaller than in a strong acid/strong base reaction, and the pH at the equivalence point is greater than 7.

As the K_a of a weak acid increases, the equivalence point gets more difficult to detect: in other words, the weaker the acid, the flatter the curve, because the initial pH is so close to neutral.

Workbook Example 17.5

PROBLEM:
25.0 mL of 0.200 M HF is titrated with 0.100 M NaOH. How many mL of base are required to reach the equivalence point? Calculate the pH at each of the following points: (a) after addition of

10.0 mL of base, (b) halfway to the equivalence point, (c) at the equivalence point, and (d) after addition of 65.0 mL of base.

SOLUTION:
To determine the number of mL of base needed to reach the equivalence point, we need to first calculate the number of mmol of HF present.

$$mmol\ HF = 25.0\ mL \times \frac{0.200\ mmol\ HF}{1\ mL} = 5.00\ mmol$$

We need 5.0 mmol of NaOH to reach the equivalence point, which means we need 50 mL of 0.100 M NaOH.

To calculate the pH at the different points in the titration, we need to use the steps that are outlined above.

(i) After addition of 10.0 mL of NaOH:

$$mmol\ F^- = 10.0\ mL \times \frac{0.100\ mmol}{1\ mL} = 1.00\ mmol$$

$$\left[F^-\right] = \frac{1.00\ mmol}{35.0\ mL} = 2.86 \times 10^{-2}\ M$$

$$mmol\ HF = 5.00\ mmol - 1.00\ mmol = 4.00\ mmol$$

$$[HF] = \frac{4.00\ mmol}{35.0\ mL} = 0.114\ M$$

$$pH = pK_a + \log\frac{\left[F^-\right]}{[HF]}$$

The K_a for HF is 3.5 × 10^{-4}; the pK_a is 3.46. Plugging this information into the Henderson–Hasselbalch equation:

$$pH = pK_a + \log\frac{\left[F^-\right]}{[HF]}$$

$$pH = 3.46 + \log\frac{\left[2.82 \times 10^{-2}\ M\right]}{[0.114 M]} = 2.85$$

(ii) Halfway to the equivalence point:
At this point, 25.0 mL of NaOH has been added to the solution (because 50 mL is required to reach the equivalence point). Following the same steps as before:

$$mmol\ F^- = 25.0\ mL \times \frac{0.100\ mmol}{1\ mL} = 2.50\ mmol$$

$$\left[F^-\right] = \frac{2.50\ mmol}{50.0\ mL} = 0.0500\ M$$

$$mmol\ HF = 5.00\,mmol - 2.50\,mmol = 2.50\,mmol$$

$$[HF] = \frac{2.50\,mmol}{50.0\,mL} = 0.0500\,M$$

$$pH = 3.46 + \log\frac{[0.0500\,M]}{[0.0500\,M]} = 3.46$$

(iii) At the equivalence point:
At this point, all the HF has been neutralized, and the pH is determined by [F⁻].

$$mmol\ F^- = 50.0\,mL \times \frac{0.100\ mmol}{1\ mL} = 5.00\ mmol$$

$$[F^-] = \frac{5.00\,mmol}{75.0\,mL} = 0.0667\,M$$

Remember that F⁻ is the anion of a weak acid and gives a basic solution. The principal reaction is

$$F^-(aq) + H_2O(aq) \rightleftharpoons HF(aq) + OH^-(aq)$$

We can calculate the K_b for this reaction from the K_a:

$$K_b = \frac{K_w}{K_a} = \frac{1.0\times10^{-14}}{3.5\times10^{-4}} = 2.9\times10^{-11}$$

We can also solve for the [OH⁻]:

$$K_b = \frac{[HF][OH^-]}{[F^-]} = \frac{x^2}{6.67\times10^{-2}}$$

$$x = [OH^-] = 1.4\times10^{-6}\ M$$

$$[H_3O^+] = \frac{1.0\times10^{-14}}{1.4\times10^{-6}} = 7.1\times10^{-9}$$

$$pH = -\log(7.1\times10^{-9}) = 8.15$$

(iv) After addition of 65.0 mL of NaOH:

$$mmol\ OH^-\ added = 65.0\ mL \times \frac{0.100\ mmol}{1\ mL} = 6.50\ mmol$$

We calculated in the last part that the amount of fluoride ion present was 5.00 mmol. Using this in the equation:

$$mmol\ OH^-\ present = 6.50\ mmol - 5.00\ mmol = 1.50\ mmol$$

$$\left[OH^-\right] = \frac{1.5\ mmol}{90.0\ mL} = 1.67 \times 10^{-2}\ M$$

$$\left[H_3O^+\right] = \frac{1.0 \times 10^{-14}}{1.67 \times 10^{-2}} = 6.0 \times 10^{-13}$$

$$pH = -\log\left(6.0 \times 10^{-13}\right) = 12.22$$

Workbook Problem 17.7

What quantity of 0.10 M NaOH is required to titrate 100 mL of 0.050 M phenol ($K_a = 1.3 \times 10^{-10}$) to the equivalence point? Calculate the pH at the equivalence point.

Strategy: Determine the volume of NaOH required to provide equimolar amounts of base and acid; determine the concentration of conjugate base and pH.

Step 1: Calculate the moles of phenol present and thus the moles of OH^- needed.

Step 2: Determine the volume of sodium hydroxide solution required.

Step 3: Determine the total volume of the resultant solution.

Step 4: Solve for the concentration of salt.

Step 5: Set up an equilibrium table.

Step 6: Solve for K_b and $[OH^-]$.

Step 7: Solve for $[H_3O^+]$ and pH.

POST-LECTURE:

Work Problems: *One of the best ways to learn chemistry is through solving problems. Go to the textbook and work through the Worked Examples for this section, then go to the end of chapter and work the problems there. When you feel confident that you can correctly answer problems like the ones listed below for each objective, check off each of the Learning Objectives for this section above.*

LO 17.6: Worked Example 17.8; Problems 17.41, 17.85, 17.89

Section 17.8 Weak Base–Strong Acid Titrations

PRE-LECTURE:

Learning Objectives: *These are the skills you should be proficient in after reading this section and practicing the associated problems. Check off the boxes below once you have gone through this section, answered the associated questions in your textbook, and feel confident you have mastered the skill.*

☐ **LO 17.6** Calculate the pH at various points in a weak acid–strong base titration or weak base–strong acid titration.

☐ **LO 17.8** Visualize the molecular species present during a titration and interpret titration curves.

Review: *Before going through the section make sure you are familiar with these concepts:*

- Section 16.12 weak base equilibrium problems

Summary: *Read through this section and fill in the missing words and equations. Reference the textbook when you are not confident in your answers.*

BEFORE ANY ACID IS ADDED:

- Calculate pH as usual for a solution of weak base.

BEFORE THE EQUIVALENCE POINT:

- The neutralization goes to completion, but there is still some weak base left. This and the salt from the weak base's conjugate acid form a buffer solution.
- Set up a table to determine amounts as neutralization occurs.
- Use the Henderson–Hasselbalch equation to calculate pH.

AT THE EQUIVALENCE POINT:

- pH ___ 7 because the anion of the weak base is an acid.

AFTER THE EQUIVALENCE POINT:

- In general, the anion created from the weak base can be neglected after the equivalence point and pH is calculated from excess $[H_3O^+]$.

Workbook Problem 17.8

A 40.00 mL solution of 0.375 M trimethylamine, $(CH_3)_3N$, is titrated with 0.100 M HCl. (a) How many mL of acid are required to reach the equivalence point? (b) Calculate the initial pH and the pH, (c) after the addition of 10.00 mL of acid, (d) halfway to the equivalence point, (e) at the equivalence point, and (f) after addition of 100.00 mL of acid. The K_b of trimethylamine is 6.5×10^{-5}.

Strategy: Use the method outlined above to calculate the pH at various points in the titration.

a) mL of acid required to reach the equivalence point:

Step 1: Write the neutralization reaction that occurs and determine the number of mmoles of $(CH_3)_3N$ in the initial solution.

Step 2: Determine the mmol of HCl needed to react with the $(CH_3)_3N$ and the volume of 0.100 M HCl needed.

b) Initial pH of the $(CH_3)_3N$ solution:

Step 1: Determine the initial pH of the solution.

c) pH of the solution after addition of 10.00 mL of acid.

Step 1: Calculate the concentration of $(CH_3)_3N$ and its conjugate acid after the addition of 10.00 mL of HCl.

Step 2: Determine the K_a and pK_a of the conjugate acid of $(CH_3)_3N$.

Step 3: Using the Henderson–Hasselbalch equation, calculate the pH of the solution.

d) pH halfway to the equivalence point:

Step 1: Use the Henderson–Hasselbalch equation to calculate the pH. Halfway to the equivalence point, [base] = [acid].

e) pH at the equivalence point:

Step 1: Calculate the concentration of the conjugate acid, using the initial moles of base, and the total volume of solution.

Step 2: Determine the pH of the solution.

f) pH after the addition of 200.00 mL of 0.100 M HCl:

Step 1: Determine the number of mmoles of excess HCl.

Step 2: Determine the concentration of excess HCl, and calculate the pH.

POST-LECTURE:

Work Problems: *One of the best ways to learn chemistry is through solving problems. Go to the textbook and work through the Worked Examples for this section, then go to the end of chapter and work the problems there. When you feel confident that you can correctly answer problems like the ones listed below for each objective, check off each of the Learning Objectives for this section above.*

LO 17.6: Worked Example 17.8; Problems 17.41, 17.85, 17.89

Section 17.9 Polyprotic Acid–Strong Base Titrations

PRE-LECTURE:

Learning Objectives: *These are the skills you should be proficient in after reading this section and practicing the associated problems. Check off the boxes below once you have gone through this section, answered the associated questions in your textbook, and feel confident you have mastered the skill.*

- ☐ **LO 17.7** Calculate the pH at various points in a diprotic acid–strong base titration.
- ☐ **LO 17.8** Visualize the molecular species present during a titration and interpret titration curves.

Review: *Before going through the section make sure you are familiar with these concepts:*

- Section 16.11 polyprotic acids
- Section 16.11 calculation of pH of a diprotic acid
- Section 16.5 calculate pH from $[OH^-]$

Summary: *Read through this section and fill in the missing words and equations. Reference the textbook when you are not confident in your answers.*

These calculations share the same features of other calculations except that there are two equivalence points. Amino acids have two dissociable protons, and so they will react with two molar amounts of strong base. These titrations are done with protonated amino acids, so they begin at a low pH with the form H_2A^+. To simplify calculations, solid NaOH will be added as the base. This leads to negligible volume changes during the reaction.

BEFORE ANY BASE IS ADDED:

- Calculate pH of the diprotic acid as in Section 16.11.

HALFWAY TO THE FIRST EQUIVALENCE POINT:

- As NaOH is added, the diprotic acid, H_2A^+, is converted to HA. Write the equation for the neutralization reaction below:

- There is now an H_2A^+-HA buffer and the ____________________ equation is used to calculate pH.

AT THE FIRST EQUIVALENCE POINT:

- All of the H_2A^+ has been converted to HA.
- The principal reaction at the first equivalence point is:

- The pH at this point is equal to the average of the two pK_a values.

For an amino acids, this is also the *isoelectric point*—the neutral form [HA] is at a maximum and the [H_2A^+] and [A^-] are both equal and very small. In biochemistry, the isoelectric point is useful for separating amino acids and proteins.

HALFWAY BETWEEN THE FIRST AND SECOND EQUIVALENCE POINTS:

- The primary acid is now HA and the principal reaction is:

- There is now an HA-A^- buffer and the ______________________ equation is used to calculate pH.

AT THE SECOND EQUIVALENCE POINT:

- Enough NaOH has now been added to convert all of the acid to A^-: leaving a solution of a basic salt. Write the principal reaction:

- The equilibrium constant is:

K_b =

- pH is calculated as usual:

K_b =

BEYOND THE SECOND EQUIVALENCE POINT:

- pH is calculated from amount of OH^- present.

Workbook Problem 17.9

What would the pH be if 25 mg of NaOH was added to 100 mL of a 0.050 M solution of the amino acid proline? (pK_{a1} = 1.952, pK_{a2} = 10.64) What is the isoelectric point of proline?

Strategy: Determine how far this amount of sodium hydroxide takes the titration: use the Henderson–Hasselbalch equation to determine pH. Calculate the isoelectric point from pK_{a1} and pK_{a2}.

Step 1:	Determine how many moles of proline and how many moles of sodium hydroxide are present.
Step 2:	Use the Henderson–Hasselbalch equation to determine the pH.
Step 3:	Calculate the isoelectric point.

POST-LECTURE:

Work Problems: *One of the best ways to learn chemistry is through solving problems. Go to the textbook and work through the Worked Examples for this section, then go to the end of chapter and work the problems there. When you feel confident that you can correctly answer problems like the ones listed below for each objective, check off each of the Learning Objectives for this section above.*

LO 17.7: Worked Example 17.9; Problem 17.95
LO 17.8: Problems 17.41–17.44

Section 17.10 Solubility Equilibria for Ionic Compounds

PRE-LECTURE:

Learning Objectives: *These are the skills you should be proficient in after reading this section and practicing the associated problems. Check off the boxes below once you have gone through this section, answered the associated questions in your textbook, and feel confident you have mastered the skill.*

☐ **LO 17.9** Write the equilibrium-constant expression for dissolution of an ionic compound, and calculate the value of its K_{sp}.

Definitions: *To learn, you must do. This is your workbook. Fill in the definitions from the glossary of the textbook as a first step to understanding the material.*

— **solubility product K_{sp}**— ____

____

Summary: *Read through this section and fill in the missing words and equations. Reference the textbook when you are not confident in your answers.*

When a sparingly soluble ionic compound dissolves, an equilibrium is set up between the solid and the dissolved ions in solution. The equilibrium constant for solubility is called the **solubility product constant,** or K_{sp}.

For the generic reaction:

$$M_aX_b(s) \leftrightarrows a\,M^{b+}(aq) + b\,X^{a-}(aq)$$

$K_{sp} =$

POST-LECTURE:

Work Problems: *One of the best ways to learn chemistry is through solving problems. Go to the textbook and work through the Worked Examples for this section, then go to the end of chapter and work the problems there. When you feel confident that you can correctly answer problems like the ones listed below for each objective, check off each of the Learning Objectives for this section above.*

LO 17.9: Worked Example 17.10; Problems 17.99, 17.101, 17.103

Section 17.11 Measuring K_{sp} and Calculating Solubility from K_{sp}

PRE-LECTURE:

Learning Objectives: *These are the skills you should be proficient in after reading this section and practicing the associated problems. Check off the boxes below once you have gone through this section, answered the associated questions in your textbook, and feel confident you have mastered the skill.*

☐ **LO 17.10** Calculate equilibrium concentrations of ions and the solubility of an ionic compound from its K_{sp}.

Summary: *Read through this section and fill in the missing words and equations. Reference the textbook when you are not confident in your answers.*

The solubility constant is measured ________: the concentrations of an equilibrium mixture of species are measured, either by allowing the solid to come to equilibrium or by generating a solution by adding together soluble components to create a precipitate. The solubility constant is temperature-dependent and can be used to calculate molar solubility.

Workbook Example 17.6

PROBLEM:
Calculate the molar solubility of AgCl if the K_{sp} for AgCl is 1.8×10^{-10}.

SOLUTION:
The solubility equilibrium for AgCl is

$$AgCl \leftrightarrows Ag^{+}(aq) + Cl^{-}(aq)$$

$$K_{sp} = \left[Ag^+\right]\left[Cl^-\right]$$

$$1.8 \times 10^{-10} = x^2$$

$$x = 1.34 \times 10^{-5}\ M$$

The molar solubility of AgCl is 1.34×10^{-5} M.

Workbook Example 17.7

PROBLEM:
The molar solubility of copper (I) chloride is 1.095×10^{-3} M. Calculate K_{sp} for CuCl.

SOLUTION:
The solubility equilibrium for AgCl is

$$CuCl \leftrightarrows Cu^+(aq) + Cl^-(aq)$$

$$K_{sp} = \left[Cu^+\right]\left[Cl^-\right]$$

$$K_{sp} = \left(1.095 \times 10^{-3}\right)^2$$

$$K_{sp} = 1.20 \times 10^{-6}$$

Workbook Problem 17.10

Calculate the K_{sp} for a solution of $CdCl_2$ prepared in pure water, given that the $[Cd^{2+}] = 1.10 \times 10^{-5}$ M.

Strategy: Write the balanced chemical equation for the solubility of $CdCl_2$, and use the stoichiometry and information given to calculate K_{sp}.

Step 1: Write the solubility equilibrium expression for $CdCl_2$.

Step 2: Determine the concentration of Cd^{2+} and Cl^- and calculate the K_{sp}.

Workbook Problem 17.11

The K_{sp} of $Cu(OH)_2$ is 2.18×10^{-20}. Calculate the molar solubility of $Cu(OH)_2$.

Strategy: Write the solubility equilibrium for $Cu(OH)_2$ and the equilibrium expression. Calculate the molar solubility of $Cu(OH)_2$.

Step 1: Write the equilibrium for $Cu(OH)_2$.

Step 2: Write the equilibrium expression.

Step 3: Let x be the number of mol/L of $Cu(OH)_2$ that dissolves. The saturated solution then contains x mol/L of Cu^{2+} and $2x$ mol/L of OH^-. Solve for K_{sp}.

POST-LECTURE:

Work Problems: *One of the best ways to learn chemistry is through solving problems. Go to the textbook and work through the Worked Examples for this section, then go to the end of chapter and work the problems there. When you feel confident that you can correctly answer problems like the ones listed below for each objective, check off each of the Learning Objectives for this section above.*

LO 17.10: Worked Examples 17.11, 17.12; Problem 17.107

Section 17.12 Factors That Affect Solubility

PRE-LECTURE:

Learning Objectives: *These are the skills you should be proficient in after reading this section and practicing the associated problems. Check off the boxes below once you have gone through this section, answered the associated questions in your textbook, and feel confident you have mastered the skill.*

- ☐ **LO 17.11** Predict how the presence of molecular and ionic species will affect the solubility of an ionic compound.
- ☐ **LO 17.12** Calculate solubility in a solution that contains a common ion.

☐ **LO 17.13** Use the formation constant K_f to calculate ion concentrations in a solution that contains a complex ion.

☐ **LO 17.14** Use the formation constant K_f to calculate the solubility of an ionic compound when the cation forms a complex ion.

Definitions: *To learn, you must do. This is your workbook. Fill in the definitions from the glossary of the textbook as a first step to understanding the material.*

— **complex ion—** ____

____

— **formation constant (K_f)—**____________________________________

____

— **amphoteric—**____

____

Summary: *Read through this section and fill in the missing words and equations. Reference the textbook when you are not confident in your answers.*

There are three factors that affect solubility:

1. Common-ion effect
   - Addition of a compound that shares an ion with the slightly soluble salt in question will ___________ solubility of the ionic compound causing more of it to precipitate out.
   - K_{sp} is a constant equal to the concentrations of the ionic species that go into solution for a particular compound. Since it is a constant, if the concentration of one species increases the other must _________.
2. pH
   - Addition of a compound that contains a basic anion will ___________ solubility as the pH decreases.
3. Formation of complex ions
   - The formation of complex ions, in which a Lewis base forms a coordinate covalent bond with the metal cation, will push the equilibrium in the direction of increased solubility by removing dissolved metal cations by forming complex ions.
   - The formation of complex ions is a stepwise process, but the equilibrium constant for the ion formation incorporates all of the steps. A high K_f indicates the formation of a stable complex ion.

Amphoteric oxides are metal hydroxides that will dissolve in both strongly basic and strongly acidic solutions. This occurs in basic solutions when excess OH^- ions convert the hydroxide into a complex ion in basic solution. In acid solutions, the OH^- is neutralized, and the metal cation becomes soluble.

Only oxides that can be converted to more soluble complex ions exhibit this behavior.

Workbook Example 17.8

PROBLEM:
Determine the molar solubility of $Cu(C_2O_4)$ ($K_{sp} = 2.87 \times 10^{-8}$ at 25 °C) in a 0.25 M $CuCl_2$ solution.

SOLUTION:
The solubility equilibrium expression is

$$Cu(C_2O_4) \leftrightarrows Cu^{2+}(aq) + C_2O_4^{2-}(aq)$$

The equilibrium expression for this reaction is

$$K_{sp} = \left[Cu^{2+}\right]\left[C_2O_4^{2-}\right]$$

Let x be the number of mol/L of $Cu(C_2O_4)$ that dissolves. We can now construct a table showing the equilibrium concentrations:

Solubility equilibrium	$Cu(C_2O_4)_2(s)$	$\leftrightarrows$	$Cu^{2+}(aq)$ +	$C_2O_4^{2-}(aq)$
Initial concentration (M)			0.25	0
Equilibrium concentration (M)			$0.25 + x$	$+x$

Given the value of K_{sp}, we can assume that x is negligible; it means that the equilibrium concentration of Cu^{2+} is approximately 0.25 M. Substituting these values into the equilibrium expression gives:

$$2.87 \times 10^{-8} = (0.25)(x)$$

$$x = 1.15 \times 10^{-7}\ M$$

Workbook Problem 17.12

The initial pH of a solution containing $Sn(OH)_2$ was 9.45 after addition of excess NH_3. Determine the molar solubility of $Sn(OH)_2$ in this solution. For $Sn(OH)_2$, $K_{sp} = 5.4 \times 10^{-27}$.

Step 1: Write the solubility equation and solubility expression.

Step 2: Using the pH, determine the $[OH^-]$ concentration.

Step 3: Let x be the number of mol/L of $Sn(OH)_2$ that dissolves. Construct a table showing the equilibrium concentrations.

Step 4: Calculate the solubility of $Sn(OH)_2$.

Workbook Example 17.9

PROBLEM:
Write a balanced net ionic equation for the dissolution reaction between AgBr and $Na_2S_2O_3$, and calculate the equilibrium constant given K_{sp} (AgBr) = 5.4×10^{-13} and K_f ($Ag(S_2O_3)_2^{3-}$) = 4.7×10^{13}.

SOLUTION:
The net ionic equation for the dissolution of AgBr in $Na_2S_2O_3$ is obtained by combining the solubility expression for AgBr and the formation expression for $Ag(S_2O_3)_2^{3-}$.

$$AgBr(s) \leftrightarrows Ag^+(aq) + Br^-(aq) \qquad K_{sp} = 5.4 \times 10^{-13}$$

$$Ag^+(aq) + 2\,S_2O_3^{2-}(aq) \leftrightarrows Ag(S_2O_3)_2^{3-}(aq) \qquad K_f = 4.7 \times 10^{13}$$

Adding the two equations together and canceling out what is common on both the reactant and product sides gives

$$AgBr(s) \leftrightarrows \cancel{Ag^+}(aq) + Br^-(aq) \qquad K_{sp} = 5.4 \times 10^{-13}$$
$$\cancel{Ag^+}(aq) + 2\,S_2O_3^{2-}(aq) \leftrightarrows Ag(S_2O_3)_2^{3-}(aq) \qquad K_f = 4.7 \times 10^{13}$$
$$AgBr(s) + 2\,S_2O_3^{2-}(aq) \leftrightarrows Ag(S_2O_3)_2^{3-}(aq) + Br^-(aq) \qquad K = K_{sp} \times K_f = 25.4$$

POST-LECTURE:

Work Problems: *One of the best ways to learn chemistry is through solving problems. Go to the textbook and work through the Worked Examples for this section, then go to the end of chapter and work the problems there. When you feel confident that you can correctly answer problems like the ones listed below for each objective, check off each of the Learning Objectives for this section above.*

LO 17.11: Problems 17.109, 17.113, 17.115, 17.117
LO 17.12: Worked Example 17.13; Problem 17.111
LO 17.13: Worked Example 17.14; Problem 17.119
LO 17.14: Worked Example 17.15; Problems 17.123, 17.125

Section 17.13 Precipitation of Ionic Compounds

PRE-LECTURE:

Learning Objectives: *These are the skills you should be proficient in after reading this section and practicing the associated problems. Check off the boxes below once you have gone through this section, answered the associated questions in your textbook, and feel confident you have mastered the skill.*

☐ **LO 17.15** Calculate the ion product IP for an ionic compound, and determine whether a precipitate will form when various solutions are mixed.

Definitions: *To learn, you must do. This is your workbook. Fill in the definitions from the glossary of the textbook as a first step to understanding the material.*

— **ion product (IP)**—____

____

Review: *Before going through the section make sure you are familiar with these concepts:*

- Section 15.5 reaction quotient

Summary: *Read through this section and fill in the missing words and equations. Reference the textbook when you are not confident in your answers.*

Solubility guidelines from Section 4.6 begin to describe the behavior of ions in solution, but a far more accurate method is the use of the **ion product (IP).**

IP =

The ion product is calculated using the initial concentration, not equilibrium concentrations: this is a reaction quotient, and it will determine which direction the reaction will go using the following guidelines:

- If IP > K_{sp}, the solution is ____________________________.
- If IP = K_{sp}, the solution is ____________________________.
- If IP < K_{sp}, the solution is ____________________________.

Workbook Problem 17.13

Will a precipitate form when 100 mL of 0.75 M $Zn(NO_3)_2$ is mixed with 250 mL of 1.50 M Na_2CO_3?

Strategy: Write a metathesis (exchange) reaction for the reaction between $Zn(NO_3)_2$ and Na_2CO_3, calculate the IP and compare its value to the K_{sp}

Step 1:	Write the solubility equation and IP expression for the precipitate.
Step 2:	Calculate the IP and compare its value to the K_{sp}. Will a precipitate form?

POST-LECTURE:

Work Problems: *One of the best ways to learn chemistry is through solving problems. Go to the textbook and work through the Worked Examples for this section, then go to the end of chapter and work the problems there. When you feel confident that you can correctly answer problems like the ones listed below for each objective, check off each of the Learning Objectives for this section above.*

LO 17.15: Worked Example 17.16; Problems 17.127, 17.130

Section 17.14 Separation of Ions by Selective Precipitation

PRE-LECTURE:

Learning Objectives: *These are the skills you should be proficient in after reading this section and practicing the associated problems. Check off the boxes below once you have gone through this section, answered the associated questions in your textbook, and feel confident you have mastered the skill.*

☐ **LO 17.16** Design a scheme for separating ions in a mixture based on selective precipitation.

Summary: *Read through this section and fill in the missing words and equations. Reference the textbook when you are not confident in your answers.*

It is possible to separate a mixture of ions in solution by adding a reagent that will precipitate some of the ions but not others.

Using an acid solution, it is possible to separate insoluble and soluble metal sulfides. The equilibrium constant for this reaction is K_{spa}—the solubility constant in acid solution.

For the general equation

$$MS(s) + 2\ H_3O^+(aq) \leftrightarrows M^{2+}(aq) + H_2S(aq) + 2\ H_2O(l)$$

$K_{spa} =$

Using a reaction quotient:

$$Q_c = \frac{[M^{2+}][H_2S]}{[H_3O^+]}$$

It is possible to adjust $[H_3O^+]$ so that $Q_c > K_{spa}$ for the more insoluble metal sulfide. The more insoluble sulfide will precipitate out, and the more soluble sulfide will stay in solution.

POST-LECTURE:

Work Problems: *One of the best ways to learn chemistry is through solving problems. Go to the textbook and work through the Worked Examples for this section, then go to the end of chapter and work the problems there. When you feel confident that you can correctly answer problems like the ones listed below for each objective, check off each of the Learning Objectives for this section above.*

LO 17.16: Problems 17.136, 17.137

Section 17.15 Qualitative Analysis

PRE-LECTURE:

Learning Objectives: *These are the skills you should be proficient in after reading this section and practicing the associated problems. Check off the boxes below once you have gone through this section, answered the associated questions in your textbook, and feel confident you have mastered the skill.*

☐ **LO 17.16** Design a scheme for separating ions in a mixture based on selective precipitation.

Definitions: *To learn, you must do. This is your workbook. Fill in the definitions from the glossary of the textbook as a first step to understanding the material.*

— **qualitative analysis—** ____

____

Summary: *Read through this section and fill in the missing words and equations. Reference the textbook when you are not confident in your answers.*

Qualitative analysis is a procedure for identifying the ions that are present in an unknown solution.

In the traditional scheme of analysis for metal cations, 20 cations are separated into five groups for selective precipitation.

- ________________will remove Ag, Hg, and Pb as insoluble chlorides (some Pb will remain).
- Bubbling ________ through the acid solution will precipitate out the remaining Pb, as well as Cu, Hg, Cd, Bi, and Sn.

- Addition of ________ neutralizes the solution, precipitating out the remaining sulfides: Mn, Fe, Co, Ni, and Zn. Insoluble hydroxides of Al and Cr also precipitate out.
- ________ is now added to precipitate out Ca and Ba as carbonates.
- ________ will precipitate out Mg.

All that can remain in solution are ________. These can be identified using flame testing: Na imparts a persistent yellow color to a flame, while K imparts a transient violet flame.

The ions in each group can be separated with further analysis. Although there are now more sophisticated techniques for the analysis of metal ions in solution, this method is still an excellent technique for learning laboratory skills and learning about acid–base, solubility, and complex-ion equilibria.

POST-LECTURE:

Work Problems: *One of the best ways to learn chemistry is through solving problems. Go to the textbook and work through the Worked Examples for this section, then go to the end of chapter and work the problems there. When you feel confident that you can correctly answer problems like the ones listed below for each objective, check off each of the Learning Objectives for this section above.*

LO 17.16: Problems 17.136, 17.137

Self-Test

This section is intended to test your knowledge of the material covered in this chapter. Think through these problems and make certain you understand what the problem is about—ask yourself if the answer makes sense. Many of these questions are linked to the chapter learning goals. Therefore, successful completion of these problems indicates that you have achieved the learning goals for this chapter. You will receive the greatest benefit from this section if you use it as a mock exam. Do not use your notes, book, Internet, or friends when taking the test. There is no grade and you only shortchange yourself by getting help with this activity. As you take the test, mark the problems with your level of confidence with the answer.

★ *Use a star for problems you know you got right.*
? *Use a question mark for those you think you got right.*
x *Use an x for those you are guessing on.*

If you get a question right that you marked with a star, you can be confident that you know the associated LO.

If you get a question right that you marked with a question mark, go back and work a few more problems and test yourself again tomorrow.

If you get a question wrong or mark it with an x, go back and read the section again, work a few problems today, a few tomorrow, and a few the next day before testing yourself again.

This will help you to discover which topics you have learned thoroughly and which topics you need to study in more detail.

True–False

1. Mixing 100 mL of 0.1 M HCl with 100 mL of 0.1 M NaOH will yield a solution with pH 7.
2. Mixing 100 mL of 0.1 M acetic acid with 100 mL of 0.1 M NaOH will yield a solution with pH < 7.
3. Adding a common ion will reduce the dissociation of a weak acid.
4. When OH^- ions are added to a buffer system, the equilibrium shifts toward undissociated acid molecules.
5. When pH = pK_a, all of the weak acid in a solution is dissociated.
6. A strong acid–strong base titration has an equivalence point at pH 7.
7. A strong acid–weak base titration has an equivalence point above pH 7.
8. Molar solubility and K_{sp} must be experimentally determined separately.
9. Metal ions in solution can be separated based on reactivity.
10. Na and K ions in solution can be separated by adding HCl to solution.

Matching

11. amphoteric	a. metal cation bonded to one or more small molecules
12. buffer capacity	b. solution of weak acid and conjugate base that resists changes in pH
13. buffer solution	c. point in a titration at which molar equivalents have been added to solution
14. common-ion effect	d. exhibiting both acidic and basic properties
15. complex ion	e. relationship between pH and pK_a of a weak acid, calculated using the amount of dissociation
16. equivalence point	f. procedure for identifying ions in solution
17. formation constant	g. shift in equilibrium caused by adding an ion already involved in the equilibrium
18. Henderson–Hasselbalch equation	h. K_{eq} for complex ions
19. ion product	i. The amount of acid or base a buffer can absorb without a significant change in pH
20. qualitative analysis	j. Q for K_{sp}

Fill in the Blanks

21. The neutralization of a weak acid with a __________ base, will go to ______________ because OH^- has a great affinity for ______________.

22. Solutions which contain a __________ acid and ______________ base resist changes in __________ through shifts in __________________. These solutions are called ________________.

23. Complex ions are formed when ________________ cations form ________________ covalent bonds with small ______________ like ammonia or water.

Problems

24. Write balanced net ionic equations, and predict the pH for reactions between
   a. HNO_3 and NaOH
   b. HCl and NH_2NH_2
   c. CH_3COOH (acetic acid) and NaOH

25. Calculate the pH of a solution containing 0.10 mol of NH_3 and 0.25 mol of NH_4Cl in 1.0 L. For ammonia, $K_b = 1.8 \times 10^{-5}$.

26. Determine the pH and concentration of all species present in a buffer solution containing 0.45 mol HClO and 0.25 mol NaClO in 1.0 L of solution. Determine the change in pH upon addition of 0.10 mol NaOH. Determine the change in pH on addition of 0.10 mol HCl. K_a for HClO is 3.5×10^{-8}.

27. Determine the ratio of lactic acid to lactate ion required for preparing a buffer solution whose pH is 4.75. The K_a for lactic acid is 1.4×10^{-4}. If the buffer is to be 0.750 M overall in these compounds, how many moles of each will be added per liter of solution?

28. Determine the change in pH that will occur on addition of 0.300 mol HCl (assuming no volume change) to a 1 L buffer solution prepared from 0.50 M HCOOH and 0.25 M HCO_2Na. The pK_a of formic acid is 3.74.

29. Calculate the pH of 75.0 mL of 0.0400 M HCl after addition of the following volumes of 0.1000M NaOH: (a) 0.0 mL, (b) 10.0 mL, (c) 30.0 mL, and (d) 50.0 mL.

30. Calculate the pH of 50.0 mL of 0.0500 M acetic acid ($K_a = 1.8 \times 10^{-5}$) after addition of the following volumes of 0.1000 M NaOH: (a) 0.0 mL, (b) 10.0 mL, (c) 25.0 mL, and (d) 30.0 mL.

31. The molar solubility of $CaCO_3$ is 7.07×10^{-5} M. Calculate the value of K_{sp}.

32. Calculate the molar solubility for $Ca_3(PO_4)_2$, given that the $K_{sp} = 2.1 \times 10^{-33}$.

33. Calculate the solubility of a solution of $CaSO_4$ that contains 0.50 M Na_2SO_4. ($K_{sp} = 7.1 \times 10^{-5}$)

34. Which of the following compounds are more soluble in acidic solution than in pure water? Why?
   a. AgBr b. Na_2S
   c. LiCN d. CaF_2

35. What are the concentrations of Zn^{2+} and $Zn(NH_3)_4{}^{2+}$ in a solution prepared by adding 0.50 mol of $Zn(NO_3)_2$ to 1.0 L of 4.0 M NH_3? ($K_f = 7.8 \times 10^8$)

36. Will a precipitate form when 350 mL of 0.75 M $CaCl_2$ is mixed with 200 mL of 1.50 M Na_3PO_4? $K_{sp} = 2.1 \times 10^{-33}$ for calcium phosphate.

37. Is it possible to separate Cu^{2+} from Fe^{2+} by bubbling H_2S through a 0.20 M HCl solution that contains 0.007 M Cu^{2+} and 0.007 M Fe^{2+}? K_{spa} for CuS = 6×10^{-16}, K_{spa} for FeS = 6×10^2, and $[H_2S] = 0.10$ M.

Challenge Problem

38. CuS has $K_{spa} = 6.0 \times 10^{-16}$. Determine the solubility of CuS in a buffer prepared from 0.45 M formic acid and 0.25 M sodium formate. $K_a = 1.8 \times 10^{-4}$ for formic acid.

Test Yourself: *Take the Practice Test on pages 758 and 759 of your textbook. Again, use a testing environment using only those items that your instructor will allow you to have access to during the test and mark each question with a star, question mark, or x to denote your confidence level. Once you have finished you can check off the LO at the beginning of this chapter based on which questions you starred and got correct. Go back and spend more time with the topics from the other learning objectives.*

CHAPTER EIGHTEEN

Thermodynamics: Entropy, Free Energy, and Spontaneity

Learning Objectives

As a result of reading and studying this chapter, you should be able to:

- ☐ **LO 18.1** Define a *spontaneous process* and classify various physical processes and chemical reactions as spontaneous or nonspontaneous.
- ☐ **LO 18.2** Predict the sign of ΔS for physical processes and chemical reactions.
- ☐ **LO 18.3** Determine the number of arrangements in a system and use Boltzmann's equation to calculate the entropy.
- ☐ **LO 18.4** Calculate ΔS for the expansion or compression of an ideal gas at constant temperature.
- ☐ **LO 18.5** Predict how the phase of matter, or conditions such as temperature, volume, and pressure, affect the entropy of a substance.
- ☐ **LO 18.6** Predict how atomic structure affects the entropy of a substance.
- ☐ **LO 18.7** Use standard molar entropies ($S°$) to calculate the standard entropy of reaction ($\Delta S°$).
- ☐ **LO 18.8** Describe the factors that affect the spontaneity of a reaction.
- ☐ **LO 18.9** Calculate values of ΔS_{sys}, ΔS_{surr}, and ΔS_{total} for a reaction, and use these values to determine whether the reaction is spontaneous.
- ☐ **LO 18.10** Describe how the sign and magnitude of ΔS, ΔH, and temperature affect the value of ΔG and the spontaneity of the reaction.
- ☐ **LO 18.11** Estimate the temperature at which a reaction changes between spontaneous and nonspontaneous.
- ☐ **LO 18.12** Calculate the standard free-energy change ($\Delta G°$) from $\Delta H°$ and $\Delta S°$ and interpret the meaning of the sign of $\Delta G°$.
- ☐ **LO 18.13** Calculate $\Delta G°$ for a reaction from values of $\Delta G°_f$ and determine if the reaction is spontaneous under standard-state conditions and how temperature will affect spontaneity.
- ☐ **LO 18.14** Calculate ΔG for a reaction under nonstandard-state conditions.
- ☐ **LO 18.15** Use the relationship between $\Delta G°$ and the equilibrium constant for a reaction, and be able to calculate one from the other.
- ☐ **LO 18.16** Interpret the shape of the curve in a plot of free energy versus reaction progress.

Chapter Summary

Spontaneous processes are those that proceed toward equilibrium without any outside influence. Spontaneity is related to entropy, the amount of molecular randomness. The standard molar entropy of substances can be used to calculate the standard entropy of reaction, and the second law of thermodynamics states that in any spontaneous process, the total entropy of the system and its surroundings increases. Free energy is related to entropy through the equation $\Delta G = \Delta H - T\Delta S$. It is a state function that determines the spontaneity of a reaction. The relative importance of the enthalpy (ΔH) and entropy (ΔS) terms is determined by the temperature of a reaction. Free energy of reaction can be calculated either using the equation above or the equation for free energies of formation. The free energy of a reaction is also mathematically related to the equilibrium constant, *K*, through the equation $\Delta G° = -RT \ln K$.

Section 18.1 Spontaneous Processes

PRE-LECTURE:

Learning Objectives: *These are the skills you should be proficient in after reading this section and practicing the associated problems. Check off the boxes below once you have gone through this section, answered the associated questions in your textbook, and feel confident you have mastered the skill.*

☐ **LO 18.1** Define a *spontaneous process* and classify various physical processes and chemical reactions as spontaneous or nonspontaneous.

Definitions: *To learn, you must do. This is your workbook. Fill in the definitions from the glossary of the textbook as a first step to understanding the material.*

— **thermodynamics**— ____

____

— **spontaneous process**— ____

____

Summary: *Read through this section and fill in the missing words and equations. Reference the textbook when you are not confident in your answers.*

A ___________ reaction is one that proceeds on its own without any external influence or input of energy.

Spontaneous reactions always move toward equilibrium. Because of this, their spontaneity is dependent on the same things that the equilibrium is dependent on: temperature, pressure, and the composition of the reaction mixture.

When $Q < K$, the reaction proceeds in the ___________ direction. When $Q > K$, the reaction proceeds in the ___________ direction.

Spontaneity of a reaction is not an indication of the speed of the reaction. The speed of a reaction is a function of kinetics and the height of the ____________________ barrier.

Workbook Example 18.1

PROBLEM:
Determine which of the following processes are spontaneous and which are nonspontaneous.

a. The cooling of a cup of tea.
b. The smell of baking cookies being detectable by sensors throughout a house.
c. The decomposition of table salt into sodium metal and chlorine gas.
d. The reaction of S and H_2 to produce SH_2 if the concentration of H_2 is 1 M and the concentration of SH_2 is 5 M; the equilibrium constant is 7.8×10^5.

SOLUTION:
Both a. and b. are spontaneous processes and c. is a nonspontaneous process. To determine if d. is spontaneous, we need to determine Q for the reaction. If $Q < K$, then the reaction is spontaneous.

If $Q > K$, then the reaction is nonspontaneous in the forward direction, but will be spontaneous in the reverse direction. The equilibrium expression for this reaction is

$$S(s) + H_2(g) \leftrightarrows SH_2(g)$$

$$Q = \frac{[SH_2]}{[H_2]} = \frac{5\,M}{1\,M} = 5$$

$Q < K$; the reaction is spontaneous in the forward direction.

POST-LECTURE:

Work Problems: *One of the best ways to learn chemistry is through solving problems. Go to the textbook and work through the Worked Examples for this section, then go to the end of chapter and work the problems there. When you feel confident that you can correctly answer problems like the ones listed below for each objective, check off each of the Learning Objectives for this section above.*

LO 18.1: Problems 18.38, 18.40

Section 18.2 Enthalpy, Entropy, and Spontaneous Processes

PRE-LECTURE:

Learning Objectives: *These are the skills you should be proficient in after reading this section and practicing the associated problems. Check off the boxes below once you have gone through this section, answered the associated questions in your textbook, and feel confident you have mastered the skill.*

☐ **LO 18.2** Predict the sign of ΔS for physical processes and chemical reactions.

Definitions: *To learn, you must do. This is your workbook. Fill in the definitions from the glossary of the textbook as a first step to understanding the material.*

— **entropy (S)**—____

____

Review: *Before going through the section make sure you are familiar with these concepts:*

- Section 9.4 enthalpy
- Section 9.11 entropy
- Section 9.12 free energy

Summary: *Read through this section and fill in the missing words and equations. Reference the textbook when you are not confident in your answers.*

Remember from Chapter 9 that spontaneity of a reaction depends on both entropy and enthalpy. Although many spontaneous reactions give off heat, some occur because of the increase in molecular randomness.

When ΔS is positive, disorder has ___________. When ΔS is negative, order has ___________.

Workbook Example 18.2

PROBLEM:
Predict the sign of ΔS for

a. $Br_2(l) \rightarrow Br_2(g)$
b. $H_2O(l) \rightarrow H_2O(s)$
c. $NaCl(aq) \rightarrow NaCl(s)$
d. $Zn(s) + 2\ HCl(aq) \rightarrow H_2(g) + ZnCl_2(aq)$

SOLUTION:

a. ΔS is positive; gas molecules are less orderly than liquid molecules.
b. ΔS is negative; liquid molecules decrease in randomness when they freeze into a solid.
c. ΔS is negative; aqueous ions lose their randomness when they condense into a crystal.
d. ΔS is positive; gas is generated, which greatly increases the disorder of the system.

POST-LECTURE:

Work Problems: *One of the best ways to learn chemistry is through solving problems. Go to the textbook and work through the Worked Examples for this section, then go to the end of chapter and work the problems there. When you feel confident that you can correctly answer problems like the ones listed below for each objective, check off each of the Learning Objectives for this section above.*

LO 18.2: Worked Example 18.1; Problems 18.26, 18.44, 18.45

Section 18.3 Entropy and Probability

PRE-LECTURE:

Learning Objectives: *These are the skills you should be proficient in after reading this section and practicing the associated problems. Check off the boxes below once you have gone through this section, answered the associated questions in your textbook, and feel confident you have mastered the skill.*

- ☐ **LO 18.3** Determine the number of arrangements in a system and use Boltzmann's equation to calculate the entropy.
- ☐ **LO 18.4** Calculate ΔS for the expansion or compression of an ideal gas at constant temperature.
- ☐ **LO 18.5** Predict how the phase of matter, or conditions such as temperature, volume, and pressure, affect the entropy of a substance.

Summary: *Read through this section and fill in the missing words and equations. Reference the textbook when you are not confident in your answers.*

A random state is more likely than an ordered one. Although we are discussing here the state of things at the molecular level, we can see this in everyday life as well. It would be a rare thing indeed, for instance, to open a clothes dryer and find that the clothes have spontaneously folded themselves. It is much more likely that they are found in one of the multitude of disordered states that are available to them.

Ludwig Boltzmann proposed that the entropy of a particular state is related to the number of ways that this state can be achieved, and it can be calculated by the following formula:

$$S = k \ln W$$

where $k = R/N_A = 1.38 \times 10^{-23}$ J/K, and $\ln W$ = the number of available states.

For a perfect crystal, entropy will be 0, because there is only one way to achieve that state, and $\ln 1 = 0$. Molecules with strong **dipole moments** are more likely to form more orderly crystals. As a result, their entropy will be lower.

Gas expands spontaneously because the state of greater volume is more ___________: the larger the volume, the higher the number of disordered states available to the gas particles.

The Boltzmann equation can be manipulated to calculate the entropy change due to a change in volume:

$\Delta S =$

Or from a change in pressure:

$\Delta S =$

From these equations, it is possible to see that increasing the volume or decreasing the pressure both ___________ the entropy of a gas.

Workbook Problem 18.1

What is the entropy change if the volume of 3 g of hydrogen gas increases from 5 L to 20 L at a constant temperature? (R = 8.314 J/K)

Strategy: Determine the number of moles of hydrogen; calculate ΔS.

Step 1: Calculate the moles of hydrogen.

Step 2: Calculate ΔS.

POST-LECTURE:

Work Problems: *One of the best ways to learn chemistry is through solving problems. Go to the textbook and work through the Worked Examples for this section, then go to the end of chapter and work the problems there. When you feel confident that you can correctly answer problems like the ones listed below for each objective, check off each of the Learning Objectives for this section above.*

LO 18.3: Worked Example 18.2; Problems 18.49, 18.50
LO 18.4: Problem 18.58
LO 18.5: Problem 18.54

Section 18.4 Entropy and Temperature

PRE-LECTURE:

Learning Objectives: *These are the skills you should be proficient in after reading this section and practicing the associated problems. Check off the boxes below once you have gone through this section, answered the associated questions in your textbook, and feel confident you have mastered the skill.*

☐ **LO 18.5** Predict how the phase of matter, or conditions such as temperature, volume, and pressure, affect the entropy of a substance.

Definitions: *To learn, you must do. This is your workbook. Fill in the definitions from the glossary of the textbook as a first step to understanding the material.*

— **third law of thermodynamics—**____

____

Summary: *Read through this section and fill in the missing words and equations. Reference the textbook when you are not confident in your answers.*

As the temperature of a substance increases, the overall kinetic energy of the system also ________.

This affects entropy in several ways. When temperature increases, molecular motion increases, randomness increases, and individual molecular energies occur across a wider spectrum, introducing another area of disorder. All of these events increase the entropy of a system.

The third law of thermodynamics states ____

____.

A plot of entropy versus temperature will show a steady increase in entropy according to temperature, with ___________ jumps for phase changes.

POST-LECTURE:

Work Problems: *One of the best ways to learn chemistry is through solving problems. Go to the textbook and work through the Worked Examples for this section, then go to the end of chapter*

and work the problems there. When you feel confident that you can correctly answer problems like the ones listed below for each objective, check off each of the Learning Objectives for this section above.

LO 18.5: Problem 18.56

Section 18.5 Standard Molar Entropies and Standard Entropies of Reaction

PRE-LECTURE:

Learning Objectives: *These are the skills you should be proficient in after reading this section and practicing the associated problems. Check off the boxes below once you have gone through this section, answered the associated questions in your textbook, and feel confident you have mastered the skill.*

- ☐ **LO 18.6** Predict how atomic structure affects the entropy of a substance.
- ☐ **LO 18.7** Use standard molar entropies ($S°$) to calculate the standard entropy of reaction ($\Delta S°$).

Definitions: *To learn, you must do. This is your workbook. Fill in the definitions from the glossary of the textbook as a first step to understanding the material.*

— **standard molar entropy ($S°$)**— ______________________________________

____

— **standard entropy of reaction ($\Delta S°$)**— ______________________________

____

Summary: *Read through this section and fill in the missing words and equations. Reference the textbook when you are not confident in your answers.*

Standard molar entropies, $S°$ are available for many materials. A standard molar entropy is the entropy of ___ mole of the pure substance at ___________ pressure and a specified temperature, usually _____°C. The units of these values are in J/(K · mol).

For the same substance, the gas-phase form will have the ___________ entropy, the solid phase the ___________ entropy.

Within the same phase, larger molecules—which have the ability to adopt different conformations and thus can have internal entropy—have ___________ entropy than smaller molecules.

Entropies of reaction can be calculated from the standard entropies of the reactants and products by subtracting the sum of the entropies of the reactants from the sum of the entropies of the products.

$$\Delta S° = S°(\text{products}) - S°(\text{reactants})$$

Just as with enthalpies of reaction, the individual values must be multiplied by the stoichiometric coefficients. Write the standard entropy change of reaction for the general reaction

$$a\,A + b\,B \rightarrow c\,C + d\,D$$

$\Delta S° =$

Workbook Example 18.3

PROBLEM:
Calculate the $\Delta S°_{rxn}$ for

$$2\,NO(g) + O_2(g) \rightarrow 2\,NO_2(g)$$

SOLUTION:
Remember that the equation we need to use is

$$\Delta S°_{rxn} = \Delta S°_{products} - \Delta S°_{reactants}$$

By looking up each of the compounds (keep in mind that states of matter do matter!) in Appendix B of your text, we see that we have the following entropy values for the reactants and products:

$$S°\,(NO) = 210.7\ J/mol \cdot K$$
$$S°\,(O_2) = 205.0\ J/mol \cdot K$$
$$S°\,(NO_2) = 240.0\ J/mol \cdot K$$

Substituting these values into the equation above, we get

$$\Delta S°_{rxn} = \left[2\left(240.0\ {}^{J}/_{K \cdot mol}\right)\right] - \left[\left(205.0\ {}^{J}/_{K \cdot mol}\right) + 2\left(210.7\ {}^{J}/_{K \cdot mol}\right)\right] = -146.4\ {}^{J}/_{K \cdot mol}$$

Workbook Problem 18.2

Calculate the $\Delta S°_{rxn}$ for the combustion of hydrogen to form water vapor.

$$S°\,(H_2) = 130.6\ J/mol \cdot K$$
$$S°\,(O_2) = 205.0\ J/mol \cdot K$$
$$S°\,(H_2O) = 188.7\ J/mol \cdot K$$

Strategy: Write the formula for the reaction; calculate ΔS.

Step 1: Write the chemical equation.

Step 2: Calculate ΔS.

POST-LECTURE:

Work Problems: *One of the best ways to learn chemistry is through solving problems. Go to the textbook and work through the Worked Examples for this section, then go to the end of chapter and work the problems there. When you feel confident that you can correctly answer problems like the ones listed below for each objective, check off each of the Learning Objectives for this section above.*

LO 18.6: Problems 18.62, 18.63
LO 18.7: Worked Example 18.3; Problems 18.64, 18.66

Section 18.6 Entropy and the Second Law of Thermodynamics

PRE-LECTURE:

Learning Objectives: *These are the skills you should be proficient in after reading this section and practicing the associated problems. Check off the boxes below once you have gone through this section, answered the associated questions in your textbook, and feel confident you have mastered the skill.*

- ☐ **LO 18.8** Describe the factors that affect the spontaneity of a reaction.
- ☐ **LO 18.9** Calculate values of ΔS_{sys}, ΔS_{surr}, and ΔS_{total} for a reaction, and use these values to determine whether the reaction is spontaneous.

Definitions: *To learn, you must do. This is your workbook. Fill in the definitions from the glossary of the textbook as a first step to understanding the material.*

— **first law of thermodynamics—** ____________________________________

____

— **second law of thermodynamics—** ____________________________________

____

Review: *Before going through the section make sure you are familiar with these concepts:*

- Section 9.1 the first law of thermodynamics
- Section 9.13 free-energy change

Summary: *Read through this section and fill in the missing words and equations. Reference the textbook when you are not confident in your answers.*

Free-energy changes determine the spontaneity of a reaction; we calculate free energy by using the Gibb's free-energy equation:

$\Delta G =$

The value of the free energy can be interpreted this way:

- when the free energy (ΔG) > than 0, the reaction is ____________________;
- when $\Delta G = 0$, the reaction is at ____________________;
- and when $\Delta G < 0$, the reaction is ____________________.

Remember that the first law of thermodynamics states that in any process, whether spontaneous or nonspontaneous, the total energy of a system and its surroundings is constant. You may remember that the first law of thermodynamics is really just a restatement of the law of conservation of energy, which says nothing about the spontaneity of a chemical reaction.

The second law of thermodynamics states that in any spontaneous process, the total entropy of a system and its surroundings will always ____________. Therefore, we can use the equation for total entropy change to determine spontaneity:

$\Delta S_{total} =$

The value of ΔS_{total} can be interpreted this way:

- when ΔS_{total} is positive, the reaction is ____________________;
- when ΔS_{total} is equal to 0, the reaction is ____________________;
- and when ΔS_{total} is less than 0, the reaction is ____________________.

All reactions proceed spontaneously in the direction that ____________ the total entropy of the system plus its surroundings.

To determine ΔS_{system} use the method outlined in Section 18.5 for finding enthalpy of reaction.

To determine entropy change in the surroundings use the following equation:

$\Delta S_{surr} =$

Using the equation above, it is possible to rewrite the equation for total entropy change as:

$$\Delta S_{total} = \Delta S_{system} - \frac{\Delta H_{rxn}}{T}$$

Workbook Example 18.4

PROBLEM:
Determine whether the following reaction is spontaneous at 25 °C by determining ΔS_{total}:

$$H_2(g) + Cl_2(g) \rightarrow 2\ HCl(g)$$

SOLUTION:
We need to calculate $\Delta S°$, ΔH_{rxn}, and ΔS_{total}. Appendix B of the textbook provides the following entropy values:

$S°(H_2) = 130.6$ J/mol · K
$S°(Cl_2) = 223.0$ J/mol · K
$S°(HCl) = 186.8$ J/mol · K
$\Delta H°_f(HCl) = -92.3$ kJ/mol

$$\Delta S°_{rxn} = \left[2\left(186.8\ {}^{J}\!/_{K\cdot mol}\right)\right] - \left[\left(130.6\ {}^{J}\!/_{K\cdot mol}\right) + \left(223.0\ {}^{J}\!/_{K\cdot mol}\right)\right] = 20.0\ {}^{J}\!/_{K\cdot mol}$$

$$\Delta H°_{rxn} = \left[2\left(-92.3\ {}^{kJ}\!/_{mol}\right)\right] - \left[\left(0\ {}^{kJ}\!/_{mol}\right) + \left(0\ {}^{kJ}\!/_{mol}\right)\right] = -184.6\ {}^{kJ}\!/_{mol}$$

$$\Delta S_{total} = \Delta S°_{rxn} - \frac{\Delta H_{rxn}}{T} = 20.0\ {}^{J}\!/_{K\cdot mol} - \frac{-184{,}600\ {}^{J}\!/_{mol}}{298\ K} = 639\ {}^{J}\!/_{K\cdot mol}$$

The overall entropy change is positive, so the reaction is spontaneous.

Workbook Problem 18.3

Calculate ΔS_{total} for the following reaction at 298 K, and determine if the reaction is spontaneous.

$$AgNO_3(aq) + NaBr(aq) \rightarrow AgBr(s) + NaNO_3(aq)$$

The following values may be used in this problem:

$S°(Ag^+) = 72.7$ J/mol · K
$\Delta H°(Ag^+) = 105.6$ kJ/mol
$S°(AgBr) = 107.1$ J/mol · K
$\Delta H°(AgBr) = -100.4$ kJ/mol
$S°(Br^-) = 82.4$ J/mol · K
$\Delta H°(Br^-) = -121.5$ kJ/mol

Strategy: Calculate the values of ΔS_{system} and ΔS_{surr}.

Step 1: Write the net ionic equation for this reaction.

Step 2: Calculate ΔS_{system}.

Step 3: Calculate ΔH_{rxn}.

Step 4: Calculate ΔS_{total}.

POST-LECTURE:

Work Problems: *One of the best ways to learn chemistry is through solving problems. Go to the textbook and work through the Worked Examples for this section, then go to the end of chapter and work the problems there. When you feel confident that you can correctly answer problems like the ones listed below for each objective, check off each of the Learning Objectives for this section above.*

LO 18.8: Problems 18.68, 18.70
LO 18.9: Worked Example 18.4; Problems 18.74, 18.76

Section 18.7 Free Energy and the Spontaneity of Chemical Reactions

PRE-LECTURE:

Learning Objectives: *These are the skills you should be proficient in after reading this section and practicing the associated problems. Check off the boxes below once you have gone through this section, answered the associated questions in your textbook, and feel confident you have mastered the skill.*

- ☐ **LO 18.10** Describe how the sign and magnitude of ΔS, ΔH, and temperature affect the value of ΔG and the spontaneity of the reaction.
- ☐ **LO 18.11** Estimate the temperature at which a reaction changes between spontaneous and nonspontaneous.

Definitions: *To learn, you must do. This is your workbook. Fill in the definitions from the glossary of the textbook as a first step to understanding the material.*

— **free energy (G)**— ______________________________

____

Summary: *Read through this section and fill in the missing words and equations. Reference the textbook when you are not confident in your answers.*

By combining and rearranging the equations for Gibb's free-energy change:

$$\Delta G = \Delta H - T\Delta S$$

and total entropy change:

$$\Delta S_{\text{total}} = \Delta S_{\text{system}} + \Delta S_{\text{surroundings}}$$

you can get an equation that shows the relationship between entropy and free energy:

$$-T\Delta S_{\text{total}} = \Delta G$$

The derivation of the final equation, shown above, is in your book.

Based on this last equation, we can now conclude that if ΔG is:

- less than 0 ($\Delta G < 0$), the reaction will be ________________;
- greater than 0 ($\Delta G > 0$), the reaction will be ________________;
- equal to 0 ($\Delta G = 0$), the reaction mixture is ________________.

In any spontaneous process at constant temperature and pressure, the free energy of the system ________. Temperature can determine the relative importance of the enthalpy and entropy changes.

A chart can help see how enthalpy and entropy interact to predict the direction of the free-energy change and how that corresponds to reaction spontaneity.

ΔH	ΔS	ΔG	**Spontaneity**
$-\Delta H$	$+\Delta S$		________________ at all temperatures
$+\Delta H$	$-\Delta S$		________________ at all temperatures
$-\Delta H$	$-\Delta S$		Since $\Delta G = \Delta H - T\Delta S$ ________T to make the reaction spontaneous
$+\Delta H$	$+\Delta S$		Since $\Delta G = \Delta H - T\Delta S$ ________ T to make the reaction spontaneous

Using the free-energy equation, it is possible to determine the temperature at which a reaction becomes spontaneous by finding the temperature at which the reaction is at equilibrium. Keeping in mind that a reaction is at equilibrium when ΔG is equal to 0, we can rearrange the equation to find the crossover temperature:

$$T =$$

Workbook Example 18.5

PROBLEM:
Pure chromium is obtained by reducing Cr_2O_3 with aluminum.

$$Cr_2O_3(s) + 2\ Al(s) \rightarrow 2\ Cr(s) + Al_2O_3(s)$$

Determine $\Delta H°$ and $\Delta S°$ for the reaction. Is the reaction spontaneous at 350 °C?

SOLUTION:
Using Appendix B in your textbook to get the $\Delta S°$ and $\Delta H°_f$ values that we need, we can solve for $\Delta H°_{\text{rxn}}$ and $\Delta S°_{\text{rxn}}$:

$$\Delta H°_{\text{rxn}} = \left[\Delta H°_f(Al_2O_3) + \Delta H°_f(Cr)\right] - \left[\Delta H°_f(Cr_2O_3) + \Delta H°_f(Al)\right]$$

$$\Delta H°_{\text{rxn}} = \left[1\,mol\left(-1676\ {}^{kJ}\!/\!_{mol}\right)\right] - \left[1\,mol\left(-1140\ {}^{kJ}\!/\!_{mol}\right)\right]$$

$$\Delta H°_{\text{rxn}} = -536\ kJ$$

$$\Delta S°_{\text{rxn}} = \left[\Delta S°_f(Al_2O_3) + \Delta S°_f(Cr)\right] - \left[\Delta S°_f(Cr_2O_3) + \Delta S°_f(Al)\right]$$

$$\Delta S°_{rxn} = \left[\left(1\ mol\ Al_2O_3 \times 50.9\ {}^{J}\!/_{K \cdot mol}\right) + \left(2\ mol\ Cr \times 23.8\ {}^{J}\!/_{K \cdot mol}\right)\right]$$
$$-\left[\left(1\ mol\ Cr_2O_3 \times 81.2\ {}^{J}\!/_{K \cdot mol}\right) + \left(2\ mol\ Al \times 28.3\ {}^{J}\!/_{K \cdot mol}\right)\right]$$
$$\Delta S°_{rxn} = -39.3\ {}^{J}\!/_{K}$$

To determine if the reaction is spontaneous, we need to calculate ΔG.

$$\Delta G° = \Delta H° - T\Delta S°$$
$$\Delta G° = -536\ kJ - (623.15\ K)\left(-0.0393\ {}^{kJ}\!/_{K}\right) = -512\ kJ$$

Since the value for $\Delta G°$ is negative, the reaction is spontaneous.

Workbook Problem 18.4

Given the following information, what is the normal boiling point of methanol?

$$CH_3OH(l) \rightarrow CH_3OH(g)$$

$S°(l)$ = 127 J/mol · K　　　$S°(g)$ = 239.8 J/mol · K
$\Delta H°_f(l)$ = −239.2 kJ/mol　　　$\Delta H°_f(g)$ = −201.0 kJ/mol

Strategy: Solve for ΔS and ΔH; solve for the temperature at which $\Delta G = 0$.

Step 1: Calculate ΔS.

Step 2: Calculate ΔH.

Step 3: Calculate T.

POST-LECTURE:

Work Problems: *One of the best ways to learn chemistry is through solving problems. Go to the textbook and work through the Worked Examples for this section, then go to the end of chapter and work the problems there. When you feel confident that you can correctly answer problems like the ones listed below for each objective, check off each of the Learning Objectives for this section above.*

LO 18.10: Problems 18.27–18.31, 18.82, 18.84, 18.86
LO 18.11: Worked Example 18.5; Problems 18.88, 18.90

Section 18.8 Standard Free-Energy Changes for Reactions

PRE-LECTURE:

Learning Objectives: *These are the skills you should be proficient in after reading this section and practicing the associated problems. Check off the boxes below once you have gone through this section, answered the associated questions in your textbook, and feel confident you have mastered the skill.*

☐ **LO 18.12** Calculate the standard free-energy change ($\Delta G°$) from $\Delta H°$ and $\Delta S°$ and interpret the meaning of the sign of $\Delta G°$.

Definitions: *To learn, you must do. This is your workbook. Fill in the definitions from the glossary of the textbook as a first step to understanding the material.*

— **standard free-energy change ($\Delta G°$)**—____________________________________

____

Summary: *Read through this section and fill in the missing words and equations. Reference the textbook when you are not confident in your answers.*

The ________________, $\Delta G°$, is the change in free energy that occurs when reactants in their standard states are converted to products in their standard states. It is an extensive property, and refers to the number of moles in the equation.

The standard state conditions are:

- the pure form of a solid, liquid, or gas at ___________ pressure
- solutes at _____ concentration and ___________ pressure
- and a specified temperature, usually _____ °C.

$\Delta G°$ can be calculated from the standard enthalpy change, $\Delta H°$, and the standard entropy change, $\Delta S°$, using the equation $\Delta G° = \Delta H° - T\Delta S°$.

Workbook Problem 18.5

Determine $\Delta G°$ for the reaction

$$NaCl(aq) + AgNO_3(aq) \rightarrow NaNO_3(aq) + AgCl(s)$$

given the following information:

	$\Delta H°$ (kJ/mol)	$S°$ (J/mol · K)
$AgCl(s)$	−127.1	96.2
$AgNO_3(aq)$	−101.8	219.1
$NaNO_3(aq)$	−447.5	205.4
$NaCl(aq)$	−407.3	115.5

Strategy: From the information given, you can determine $\Delta G°$ from the equation $\Delta G° = \Delta H° - T\Delta S°$.

Step 1: Determine $\Delta H°_{rxn}$.

Step 2: Determine $\Delta S°_{rxn}$.

Step 3: Calculate $\Delta G°_{rxn}$.

POST-LECTURE:

Work Problems: *One of the best ways to learn chemistry is through solving problems. Go to the textbook and work through the Worked Examples for this section, then go to the end of chapter and work the problems there. When you feel confident that you can correctly answer problems like the ones listed below for each objective, check off each of the Learning Objectives for this section above.*

LO 18.12: Worked Example 18.6; Problems 18.94, 18.95

Section 18.9 Standard Free Energies of Formation

PRE-LECTURE:

Learning Objectives: *These are the skills you should be proficient in after reading this section and practicing the associated problems. Check off the boxes below once you have gone through this section, answered the associated questions in your textbook, and feel confident you have mastered the skill.*

☐ **LO 18.13** Calculate $\Delta G°$ for a reaction from values of $\Delta G°_f$ and determine if the reaction is spontaneous under standard-state conditions and how temperature will affect spontaneity.

Definitions: *To learn, you must do. This is your workbook. Fill in the definitions from the glossary of the textbook as a first step to understanding the material.*

— **standard free energy of formation ($\Delta G°_f$)**— ______________________________

____

Summary: *Read through this section and fill in the missing words and equations. Reference the textbook when you are not confident in your answers.*

The **standard free energy of formation,** ΔG°_f, of a substance is the free-energy change of formation of ___________ of the substance in its standard state from the most stable form of its constituent elements in their standard states. For an element in its most stable form at 25 °C, $\Delta G^\circ_f = 0$.

The standard free energy of formation of a substance measures its thermodynamic stability with respect to its constituent elements.

- If ΔG°_f is negative, the substance is ________ and ________ decompose back into its elements.
- If ΔG°_f is positive, the substance is ________ and are ________ to decompose. This says nothing about the rate of decomposition, however.

We can use ΔG°_f to calculate the standard free-energy changes for reactions:

$$\Delta G^\circ = \Delta G^\circ_f(\_\_\_\_\_\_\_\_\_\_) - \Delta G^\circ_f(\_\_\_\_\_\_\_\_\_\_)$$

For the general reaction

$$a\,A + b\,B \rightarrow c\,C + d\,D$$

the free-energy change is

$$\Delta G^\circ =$$

Workbook Example 18.6

PROBLEM:
Calculate ΔG°_{rxn} for the following reaction:

$$C_2H_4(g) + Cl_2(g) \rightarrow C_2H_3Cl(g) + HCl(g)$$

given the following information:

	ΔG°_f(kJ/mol)
$C_2H_4(g)$	68.1
$Cl_2(g)$	0
$HCl(g)$	−95.3
C_2H_3Cl	51.9

SOLUTION:

$$\Delta G^\circ_{rxn} = \left[51.9\ {kJ}/{mol} + \left(-95.3\ {kJ}/{mol}\right)\right] - \left[\left(68.1\ {kJ}/{mol}\right) + 0\ {kJ}/{mol}\right] = -111.5\ kJ$$

Workbook Problem 18.6

Determine $\Delta G°_{rxn}$ from $\Delta G°_f$ for the reaction below. Is the reaction spontaneous? If not, determine the temperature at which the reaction becomes spontaneous. Will the reaction be spontaneous at 400 °C? 45 °C? −20 °C?

$$CH_4(g) + 2\ Cl_2(g) \rightleftharpoons CCl_4(l) + 2\ H_2(g)$$

Strategy: Determine $\Delta G°_{rxn}$, then calculate $\Delta H°$ and $\Delta S°$ to determine the temperature at which the reaction becomes spontaneous.

	$\Delta G°$ (kJ/mol)	$\Delta H°$ (kJ/mol)	$S°$ (J/K · mol)
$CH_4(g)$	−50.8	−74.8	186.2
$Cl_2(g)$	0	0	223.0
$CCl_4(l)$	−65.3	−135.4	216.4
$H_2(g)$	0	0	130.6

Step 1: Calculate $\Delta G°_{rxn}$; determine if the reaction is spontaneous.

Step 2: Determine the temperature at which the reaction will be spontaneous.

Step 3: Determine the spontaneity at the temperatures listed.

POST-LECTURE:

Work Problems: *One of the best ways to learn chemistry is through solving problems. Go to the textbook and work through the Worked Examples for this section, then go to the end of chapter and work the problems there. When you feel confident that you can correctly answer problems like the ones listed below for each objective, check off each of the Learning Objectives for this section above.*

LO 18.13: Worked Example 18.7; Problems 18.96, 18.98, 18.100, 18.104

Section 18.10 Free-Energy Changes for Reactions Under Nonstandard-State Conditions

PRE-LECTURE:

Learning Objectives: *These are the skills you should be proficient in after reading this section and practicing the associated problems. Check off the boxes below once you have gone through this section, answered the associated questions in your textbook, and feel confident you have mastered the skill.*

☐ **LO 18.14** Calculate ΔG for a reaction under nonstandard-state conditions.

Review: *Before going through the section make sure you are familiar with these concepts:*

- Section 15.5 reaction quotient, Q_c

Summary: *Read through this section and fill in the missing words and equations. Reference the textbook when you are not confident in your answers.*

When reactants and products are present at nonstandard-state pressures and concentrations we use the following equation:

$\Delta G =$

where ΔG is the ___________ under nonstandard-state conditions, R is the ___________, T is the ___________ in kelvin, and Q is the reaction quotient and takes the same form as the equilibrium constant expression.

Workbook Example 18.7

PROBLEM:
Calculate ΔG for the formation of acetylene (C_2H_2) if the pressure of hydrogen is 415 atm, the pressure of acetylene is 0.1 atm, and the temperature is 25 °C.

SOLUTION:
The formula for this reaction is

$$2\,C(s) + H_2(g) \rightarrow C_2H_4(g)$$

$$\Delta G^\circ = 209.2\ ^{kJ}\!/_{mol}$$

$$\Delta G = 209200\ ^{J}\!/_{mol} + 8.314\ ^{J}\!/_{K\cdot mol}(298\,K)\ln\frac{0.1}{415} = 188600\ ^{J}\!/_{mol} = 189\ ^{kJ}\!/_{mol}$$

Workbook Problem 18.7

Consider the following reaction:

$$2\,NO(g) + O_2(g) \rightleftharpoons 2NO_2(g)$$

Determine the spontaneity under standard-state conditions and again under the following conditions: $P_{NO} = 0.150\ atm,\ P_{O_2} = 0.250\ atm,\ P_{NO_2} = 0.001\ atm.$ Under which conditions is the reaction more spontaneous?

Strategy: Determine $\Delta G°_{rxn}$ and determine which reaction is more spontaneous.

	$\Delta G°$ (kJ/mol)
NO(*g*)	87.6
$O_2(g)$	0
$NO_2(g)$	51.3

Step 1: Calculate $\Delta G°_{rxn}$, determine if the reaction is spontaneous.

Step 2: Calculate ΔG.

POST-LECTURE:

Work Problems: *One of the best ways to learn chemistry is through solving problems. Go to the textbook and work through the Worked Examples for this section, then go to the end of chapter and work the problems there. When you feel confident that you can correctly answer problems like the ones listed below for each objective, check off each of the Learning Objectives for this section above.*

LO 18.14: Worked Example 18.8; Problems 18.108, 18.110

Section 18.11 Free Energy and Chemical Equilibrium

PRE-LECTURE:

Learning Objectives: *These are the skills you should be proficient in after reading this section and practicing the associated problems. Check off the boxes below once you have gone through this section, answered the associated questions in your textbook, and feel confident you have mastered the skill.*

☐ **LO 18.15** Use the relationship between $\Delta G°$ and the equilibrium constant for a reaction, and be able to calculate one from the other.

☐ **LO 18.16** Interpret the shape of the curve in a plot of free energy versus reaction progress.

Summary: *Read through this section and fill in the missing words and equations. Reference the textbook when you are not confident in your answers.*

As a reaction proceeds toward equilibrium, the value of Q changes:

- when the mixture is mostly reactants,

  $Q << 1$ RT $\ln Q << 0$ $\Delta G < 0$

  the total free energy decreases as the reaction proceeds spontaneously in the __________ direction.
- when the mixture is mostly products,

  $Q >> 1$ RT $\ln Q >> 0$ $\Delta G > 0$

  the total free energy decreases as the reaction proceeds spontaneously in the __________ direction.

Free energy can be plotted against reaction progress: there will be a minimum between pure reactants and pure products; this minimum is the equilibrium mixture.

At equilibrium, $\Delta G° = 0$, and $Q = K$, allowing the following relationship to be derived:

$$\Delta G° = -RT \ln K$$

The relationships between the variables in the equation above can be summarized in the chart below:

$\Delta G°$	$\ln K$	K	Spontaneity
< 0	> 0	> 1	The equilibrium mixture is mostly __________
> 0	< 0	< 1	The equilibrium mixture is mostly __________
$= 0$	$= 0$	$= 1$	The equilibrium mixture contains __________

Workbook Example 18.8

PROBLEM:
Calculate K for the following reaction at 25 °C:

$$C_2H_6(g) + H_2(g) \rightarrow 2\ CH_4(g)$$

with the following data:

$\Delta G°$ for $CH_4 = -50.8$ kJ/mol $\qquad$ $\Delta G°$ for $C_2H_6 = -32.9$ kJ/mol

SOLUTION:
Calculate $\Delta G°_{rxn}$:

$$\Delta G°_{rxn} = 2\left(-50.8\ {kJ}/{mol}\right) - \left(-32.9\ {kJ}/{mol}\right) = -68.7\ {kJ}/{mol}$$

Calculate K:

$$\ln K = -\frac{\Delta G^\circ}{RT} = -\frac{-68{,}700\ J/mol}{\left[\left(8.314\ J/K\cdot mol\right)\left(298\ K\right)\right]} = 27.7$$

$$e^{\ln K} = e^{27.7} = 1.07 \times 10^{12}$$

Workbook Problem 18.8

Consider the following reaction:

$$I_2(g) + Cl_2(g) \rightleftharpoons 2ICl(g)$$

The equilibrium constant, K_p, for this reaction is 81.9 at 25 °C. What is the standard free-energy change for this reaction?

Strategy: Use the equation $\Delta G^\circ = -RT \ln K$.

Step 1: Solve for ΔG°.

POST-LECTURE:

Work Problems: *One of the best ways to learn chemistry is through solving problems. Go to the textbook and work through the Worked Examples for this section, then go to the end of chapter and work the problems there. When you feel confident that you can correctly answer problems like the ones listed below for each objective, check off each of the Learning Objectives for this section above.*

LO 18.15: Worked Examples 18.9, 18.10; Problems 18.37, 18.112, 18.118
LO 18.16: Problem 18.34

Self-Test

This section is intended to test your knowledge of the material covered in this chapter. Think through these problems and make certain you understand what the problem is about—ask yourself if the answer makes sense. Many of these questions are linked to the chapter learning goals. Therefore, successful completion of these problems indicates that you have achieved the learning goals for this chapter. You will receive the greatest benefit from this section if you use it as a mock exam. Do not use your notes, book, Internet, or friends when taking the test. There is no grade and you only shortchange yourself by getting help with this activity. As you take the test, mark the problems with your level of confidence with the answer.

★ *Use a star for problems you know you got right.*
? *Use a question mark for those you think you got right.*
x *Use an x for those you are guessing on.*

If you get a question right that you marked with a star, you can be confident that you know the associated LO.

If you get a question right that you marked with a question mark, go back and work a few more problems and test yourself again tomorrow.

If you get a question wrong or mark it with an x, go back and read the section again, work a few problems today, a few tomorrow, and a few the next day before testing yourself again.

This will help you to discover which topics you have learned thoroughly and which topics you need to study in more detail.

True–False

1. A spontaneous process occurs without any external energy and occurs quickly.

2. A reaction that is spontaneous and endothermic will have a large positive ΔS.

3. An increase in temperature always leads to an increase in entropy.

4. Standard entropies of reaction cannot be calculated without knowing the enthalpy of reaction.

5. A spontaneous reaction always increases the total entropy of the system plus the surroundings.

6. If ΔH is negative and ΔS is also negative, the reaction will be spontaneous at low temperatures.

7. The temperature of a reaction system determines the relative importance of the enthalpy and entropy terms.

8. The standard enthalpy of formation of a pure element is always negative.

9. ΔH determines whether a reaction is spontaneous or nonspontaneous.

10. The standard molar entropy of a pure substance is always positive.

Matching

11. entropy	a. the total energy of a system and its surroundings is constant
12. first law of thermodynamics	b. the study of energy changes in chemical reactions
13. free energy	c. the amount of molecular randomness in a system
14. second law of thermodynamics	d. a process that proceeds on its own without any external influence

15. spontaneous process
16. standard state
17. thermodynamics
18. third law of thermodynamics

e. in any spontaneous process, the total entropy of a system and its surroundings always increases
f. the capacity of a system to continue a chemical reaction
g. the entropy of a perfectly ordered crystalline substance at 0 K is zero
h. most stable form at 25 °C and 1 atm, 1 M for solutions

Fill in the Blanks

19. The expansion of a gas is a(n) ________________ process that ________________ the entropy of the gas.

20. When an ionic solid dissolves, the entropy of the solid ________________, but the entropy of the solvent ________________.

21. As the temperature of a system increases, molecular ________________ and ________________ increase.

22. Standard molar entropies show that ________________ molecules have higher entropy than ______ ________________ molecules, ________________ have higher entropy than liquids, and liquids have higher entropy than ________________.

23. Reactions can be made spontaneous by the ________________ term, the ________________ term, or both. What matters is their relative contributions to ______ ________________. The relative contributions of the two terms are determined by the ________________.

24. If a substance has a ________________ free energy of formation, it is thermodynamically ________________. However, it may decompose imperceptibly slowly.

25. A reaction with a positive free energy will be spontaneous in the ________________ direction.

Problems

26. Which has the higher entropy?
    a. An orderly dishwasher or a toy box
    b. Sugar crystals in a sugar bowl or sweetened tea
    c. 32 g O_2 gas at STP or 1 mol of O_2 gas at 273 K in a volume of 35.5 L

27. Predict the sign of ΔS for the following processes or reactions:
    a. an increase in the volume of a gas at constant temperature
    b. formation of solid products from aqueous reactants
    c. $MgO(s) + 2\ NH_4Cl(s) \rightarrow 2\ NH_3(g) + MgCl_2(s) + H_2O(l)$
    d. $N_2(g) + H_2(g) \rightarrow N_2H_4(l)$
    e. separation of $^{235}_{92}U$ from a mixture of $^{235}_{92}U$ and $^{238}_{92}U$.

28. Calculate the standard entropy of reaction for
   a. $2\ NH_4NO_3(s) \rightarrow 2\ N_2(g) + O_2(g) + 4\ H_2O(g)$
      $S°\ NH_4NO_3(s) = 151.1$ J/(mol · K)
      $S°\ N_2(g) = 191.5$ J/(mol · K)
      $S°\ O_2(g) = 205.0$ J/(mol · K)
      $S°\ H_2O(g) = 188.7$ J/(mol · K)

   b. $2\ S(s) + 3\ O_2(g) \rightarrow 2\ SO_3(g)$
      $S°\ S(s) = 31.8$ J/(mol · K)
      $S°\ O_2(g) = 205.0$ J/(mol · K)
      $S°\ SO_3(g) = 256.6$ J/(mol · K)

29. For a given reaction at room temperature, if $\Delta S°_{\text{total}} = 1.957 \times 10^3\ J/K$ and $\Delta S°_{\text{rxn}} = 515.7\ J/K$. Calculate $\Delta H°_{\text{rxn}}$.

30. Calculate ΔS if 1 mol of hydrogen gas at STP expands to a volume of 45.3 L at 273 K and 1 atm of pressure.

31. Determine the normal boiling point of ethanol given the following information:

   $S°(l) = 161$ J/mol · K
   $S°(g) = 282.6$ J/mol · K
   $\Delta H°_f(l) = -277.7$ kJ/mol
   $\Delta H°_f(g) = -235.1$ kJ/mol

32. Calculate ΔS_{total}, ΔS_{sys}, and ΔS_{surr} for the following reaction at 25 °C:

$$H_2(g) + Cl_2(g) \rightarrow 2\ HCl(g)$$

   $S°\ H_2(g) = 130.6$ J/(mol · K)
   $\Delta H°_f\ H_2(g) = 0$ kJ/mol
   $S°\ Cl_2(g) = 223.0$ J/(mol · K)
   $\Delta H°_f\ Cl_2(g) = 0$ kJ/mol
   $S°\ HCl(g) = 186.8$ J/(mol · K)
   $\Delta H°_f\ HCl(g) = -92.3$ kJ/mol

33. Calculate the standard free energy of reaction from the standard free energies of formation for the following reaction:

$$C_2H_4(g) + 3\ O_2(g) \rightarrow 2\ CO_2(g) + 2\ H_2O(l)$$

   $\Delta G°_f\ C_2H_4(g) = 68.1$ kJ/mol
   $\Delta G°_f\ O_2(g) = 0$ kJ/mol
   $\Delta G°_f\ CO_2(g) = -394.4$ kJ/mol
   $\Delta G°_f\ H_2O(l) = -237.2$ kJ/mol

34. Calculate the free-energy change for the following reaction if the partial pressures are 0.5 atm for CO, 15 atm for CO_2, and 0.1 atm for O_2 at 25 °C.

$$2\ CO(g) + O_2(g) \leftrightarrows 2\ CO_2(g)$$

   $\Delta G°_f\ CO(g) = -137.2$ kJ/mol
   $\Delta G°_f\ O_2(g) = 0$ kJ/mol
   $\Delta G°_f\ CO_2(g) = -394.4$ kJ/mol

35. Determine the equilibrium constant for the following reaction at 37 °C.

$$H_2CO_3(aq) \rightleftharpoons CO_2(aq) + H_2O(l)$$

$\Delta G^\circ_f\ H_2CO_3(aq) = -623$ kJ/mol $\quad$ $\Delta G_f^\circ\ CO_2(aq) = -386.0$ kJ/mol
$\Delta G^\circ_f\ H_2O(l) = -237.2$ kJ/mol

36. Alkali metals are refined at high temperatures. What is ΔG_{rxn} for this process? At what temperature does it become spontaneous? What is the equilibrium constant for the reaction at this temperature?

$$2\ NaCl(s) \leftrightarrows 2\ Na(s) + Cl_2(g)$$

	ΔG° (kJ/mol)	ΔH° (kJ/mol)	S° (J/mol · K)
$NaCl(s)$	−384.2	−411.2	72.1
$Cl_2(g)$	0	0	223.0
$Na(s)$	0	0	51.2

Challenge Problem

37. Using the information below, (a) determine the value of K at 25 °C, (b) determine the value of K at 773 K, and (c) solve for the partial pressures of each substance at equilibrium if 3 moles of each of the reactants were allowed to react in a 15 L vessel at 773 K.

$$N_2(g) + O_2(g) \leftrightarrows 2\ NO(g)$$

	ΔG° (kJ/mol)	ΔH° (kJ/mol)	S° (J/mol · K)
$N_2(g)$	0	0	191.5
$O_2(g)$	0	0	205.0
$NO(g)$	86.6	90.2	210.7

Test Yourself: *Take the Practice Test on pages 802–804 of your textbook. Again, use a testing environment using only those items that your instructor will allow you to have access to during the test and mark each question with a star, question mark, or x to denote your confidence level. Once you have finished you can check off the LO at the beginning of this chapter based on which questions you starred and got correct. Go back and spend more time with the topics from the other learning objectives.*

CHAPTER NINETEEN

Electrochemistry

Learning Objectives

As a result of reading and studying this chapter, you should be able to:

- ☐ **LO 19.1** Assign oxidation numbers and classify half-reactions as oxidation or reduction reactions.
- ☐ **LO 19.2** Use the half-reaction method to balance a redox reaction in acidic solution.
- ☐ **LO 19.3** Use the half-reaction method to balance a redox reaction in basic solution.
- ☐ **LO 19.4** Draw a galvanic cell. Label the anode and cathode, indicate the direction of electron and ion flow, and write balanced equations for the half-reactions and overall cell reaction.
- ☐ **LO 19.5** Use shorthand notation to represent a galvanic cell.
- ☐ **LO 19.6** Interconvert between cell potentials and standard free-energy changes for electrochemical reactions.
- ☐ **LO 19.7** Use tabulated values of standard reduction potentials for half-reactions (Appendix D) to order substances by increasing strength as an oxidizing or reducing agent.
- ☐ **LO 19.8** Use standard reduction potentials for half-reactions to calculate the potential of a redox reaction and determine spontaneity.
- ☐ **LO 19.9** Use standard reduction potentials for half-reactions to calculate the voltage of a galvanic cell.
- ☐ **LO 19.10** Calculate the cell potential under nonstandard-state conditions using the Nernst equation.
- ☐ **LO 19.11** Use the cell potential to calculate solution concentration.
- ☐ **LO 19.12** Use standard reduction potentials to calculate the equilibrium constant.
- ☐ **LO 19.13** Describe batteries and fuel cells, and calculate values of $E°$, $\Delta G°$, and K for the cell reaction.
- ☐ **LO 19.14** Describe how corrosion occurs and methods used to prevent it.
- ☐ **LO 19.15** Draw an electrolytic cell. Label the anode and cathode, indicate the direction of electron and ion flow, and write balanced equations for the electrode and overall cell reactions.
- ☐ **LO 19.16** Predict the products formed and the electrode and overall cell reactions when an aqueous solution of an ionic compound is electrolyzed.
- ☐ **LO 19.17** Relate the current, time, and amount of product produced in an electrolytic cell.

Chapter Summary

Electrochemistry is the area of chemistry concerned with the interconversion of chemical and electrical energy. Galvanic cells convert chemical energy to electrical energy. The standard cell potential of these reactions can be calculated from tables of standard half-reactions, and potentials at nonstandard-state conditions can be calculated using the Nernst equation. These potentials can also be used to calculate the free-energy changes and equilibrium constants of the reactions. The most common applications of these reactions are in batteries and fuel cells, with the same processes causing the much-less-productive process of corrosion. When the process is reversed, electrical energy can be used to drive nonspontaneous chemical changes. This process, known as *electrolysis,* is used to produce sodium, chlorine, and sodium hydroxide; it is also used in electrorefining and electroplating. The products obtained during electrolysis depend on the overvoltage required to drive the processes, not only on the standard voltages of the reactions involved. Quantitative calculations involving electrolytic processes require that the amount of current, and thus the number of electrons that flow into the reaction, be known.

The Chapter in Detail

Section 19.1 Balancing Redox Reactions by the Half-Reaction Method

PRE-LECTURE:

Learning Objectives: *These are the skills you should be proficient in after reading this section and practicing the associated problems. Check off the boxes below once you have gone through this section, answered the associated questions in your textbook, and feel confident you have mastered the skill.*

- ☐ **LO 19.1** Assign oxidation numbers and classify half-reactions as oxidation or reduction reactions.
- ☐ **LO 19.2** Use the half-reaction method to balance a redox reaction in acidic solution.
- ☐ **LO 19.3** Use the half-reaction method to balance a redox reaction in basic solution.

Definitions: *To learn, you must do. This is your workbook. Fill in the definitions from the glossary of the textbook as a first step to understanding the material.*

— **electrochemistry—** ____

____

— **half-reaction method—**____

____

— **half-reactions—**____

____

Review: *Before going through the section make sure you are familiar with these concepts:*

- Section 4.10 redox reactions
- Section 4.10 oxidation numbers

Summary: *Read through this section and fill in the missing words and equations. Reference the textbook when you are not confident in your answers.*

Electrons are transferred during redox reactions. During a redox reaction both an oxidation and a reduction are occurring.

- Oxidation is the ___________ of one or more electrons.
- Reduction is the ___________ of one or more electrons.

Separating the reaction into two half-reactions provides another way to balance these complex equations. The half-reaction method is outlined below:

- STEP 1—____
- STEP 2—____
- STEP 3—____
- STEP 4—____
- STEP 5—____
- STEP 6—____
- STEP 7—____

Workbook Example 19.1

PROBLEM:
Using the half-reaction method, balance the following reaction, which takes place in an acidic solution:

$$Mn^{2+}(aq) + NaBiO_3(s) \rightarrow Bi^{3+}(aq) + MnO_4^-(aq)$$

SOLUTION:

1. Write two unbalanced half-reactions:

   Oxidation: $Mn^{2+}(aq) \rightarrow MnO_4^-(aq)$ (Mn goes from +2 to +7)

   Reduction: $NaBiO_3(s) \rightarrow Bi^{3+}(aq)$ (Bi goes from +5 to +3)

2. Balance each half-reaction for atoms other than H and O.

   $Mn^{2+}(aq) \rightarrow MnO_4^-(aq)$ $\quad$ $NaBiO_3(s) \rightarrow Bi^{3+}(aq) + Na^+(aq)$

3. Add H_2O to balance for oxygen and H^+ to balance for hydrogen.

   $4\ H_2O(l) + Mn^{2+}(aq) \rightarrow MnO_4^-(aq) + 8\ H^+(aq)$

   $6\ H^+(aq) + NaBiO_3(s) \rightarrow Bi^{3+}(aq) + Na^+(aq) + 3\ H_2O(l)$

4. Balance each reaction for charge.

   $4\ H_2O(l) + Mn^{2+}(aq) \rightarrow MnO_4^-(aq) + 8\ H^+(aq) + 5e^-$

   $2\ e^- + 6\ H^+(aq) + NaBiO_3(s) \rightarrow Bi^{3+}(aq) + Na^+(aq) + 3\ H_2O(l)$

5. Make the electron count the same in both reactions.

   $(4\ H_2O(l) + Mn^{2+}(aq) \rightarrow MnO_4^-(aq) + 8\ H^+(aq) + 5e^-) \times 2$

   $(2\ e^- + 6\ H^+(aq) + NaBiO_3(s) \rightarrow Bi^{3+}(aq) + Na^+(aq) + 3\ H_2O(l)) \times 5$

   $8\ H_2O(l) + 2\ Mn^{2+}(aq) \rightarrow 2\ MnO_4^-(aq) + 16\ H^+(aq) + 10\ e^-$

   $10\ e^- + 30\ H^+(aq) + 5\ NaBiO_3(s) \rightarrow 5\ Bi^{3+}(aq) + 5\ Na^+(aq) + 15\ H_2O(l)$

6. Add the two half-reactions together, canceling anything that appears on both sides of the equation.

~~$8\ H_2O(l)$~~ + $2\ Mn^{2+}(aq)$ + ~~$10\ e^-$~~ + 14 ~~30~~$H^+(aq)$ + $5\ NaBiO_3(s) \rightarrow$
$2\ MnO_4^-(aq)$ + ~~$16\ H^+$~~(aq) + ~~$10e^-$~~ + $5Bi^{3+}(aq) + 5\ Na^+(aq)$ + 7 ~~15~~$H_2O(l)$

$2\ Mn^{2+}(aq) + 14\ H^+(aq) + 5\ NaBiO_3(s) \rightarrow$
$2\ MnO_4^-(aq) + 5\ Bi^{3+}(aq) + 5\ Na^+(aq) + 7\ H_2O(l)$

7. Check to make sure all atoms and charges are balanced.

Reactant side:	$2Mn^{2+}$	14H	5Na	5Bi	15O	net charge: +18
Product Side:	2Mn	14H	5Na	5Bi	15O	net charge: +18

Workbook Problem 19.1

Using the half-reaction method, balance the following reaction, which takes place in a basic solution:

$$KMnO_4 + Na_2SO_3 \rightarrow MnO_2 + Na_2SO_4$$

Strategy: Follow the steps in the preceding worked example.

Step 1: Write two unbalanced half-reactions.

Step 2: Balance each half-reaction for atoms other than H and O.

Step 3: Add H_2O to balance oxygen and H^+ to balance hydrogen.

Step 4: Balance each reaction for charge.

Step 5: Make the electron count the same in both reactions.

Step 6: Add the two half-reactions together, canceling anything that appears on both sides of the equation.

Step 7: Make the solution basic by adding 1 OH^- to each side for every H^+.

Step 8: Cancel water molecules that appear on both sides of the equation.

Step 9: Check your answer to make sure both atoms and charges are balanced.

POST-LECTURE:

Work Problems: *One of the best ways to learn chemistry is through solving problems. Go to the textbook and work through the Worked Examples for this section, then go to the end of chapter and work the problems there. When you feel confident that you can correctly answer problems like the ones listed below for each objective, check off each of the Learning Objectives for this section above.*

LO 19.1: Problems 19.42, 19.46
LO 19.2: Worked Example 19.1; Problems 19.50, 19.51
LO 19.3: Worked Example 19.2; Problems 19.52, 19.53

Section 19.2 Galvanic Cells

PRE-LECTURE:

Learning Objectives: *These are the skills you should be proficient in after reading this section and practicing the associated problems. Check off the boxes below once you have gone through this section, answered the associated questions in your textbook, and feel confident you have mastered the skill.*

☐ **LO 19.4** Draw a galvanic cell. Label the anode and cathode, indicate the direction of electron and ion flow, and write balanced equations for the half-reactions and overall cell reaction.

Definitions: *To learn, you must do. This is your workbook. Fill in the definitions from the glossary of the textbook as a first step to understanding the material.*

— **electrochemical cell—** ____

____

— **galvanic cell—** ____

____

— **electrolytic cell—**____

____

— **electrode—**____

____

— **salt bridge—** ____

____

— **anode—** ____

____

— **cathode—**____

____

Summary: *Read through this section and fill in the missing words and equations. Reference the textbook when you are not confident in your answers.*

In ________ cells, a spontaneous chemical reaction generates an electric current. In ________ cells, an electric current drives a nonspontaneous reaction.

In redox reactions, recall that oxidation is a loss of electrons (an increase in oxidation number), and reduction is a gain of electrons (a decrease in oxidation number). To make it easier to see what is happening in these types of reactions, they are typically broken into half-reactions: one for the oxidation portion (with electrons as products) and another for the reduction portion (with electrons as reactants). The species that causes the oxidation and accepts the electrons is referred to as the ________ *agent*, while the species that causes the reduction and gives up the electrons is referred to as the ________ *agent.*

These reactions can be carried out directly, as when a strip of zinc is placed in a beaker that contains copper ions; in this case, the copper is reduced by accepting electrons directly off the zinc metal, while the zinc is oxidized by donating electrons to the copper ions. Any enthalpy of reaction is lost into the solution.

Redox reactions can also be separated and run in a Daniell or other type of galvanic cell. In this situation, the oxidation and reduction portions of the reaction are run separately. The metals are immersed in a solution that contains a common anion and are connected by a wire through which electrons flow from the reducing agent to the oxidizing agent.

Specific names are given to the two metals, which are known as ________.

- Anode
  - the electrode at which ________ occurs
  - is where electrons are ________
  - is what ________ migrate toward
  - has a ________ sign
- Cathode
  - the electrode at which ________ occurs
  - is where electrons are ________
  - is what ________ migrate toward
  - has a ________ sign

The solutions are connected by a ________, a U-shaped tube containing a gel permeated with an electrolyte that does not participate in the reaction. Neutrality of the solutions is maintained by the electrolyte in the salt bridge. In the galvanic cell, ________ move toward the anode, and ________ move toward the cathode.

The half-reaction at the anode and the half-reaction at the cathode must add together to give the overall reaction.

Workbook Example 19.2

PROBLEM:
Describe the construction of a galvanic cell based on the following reaction:

$$Ni^{2+}(aq) + Mg(s) \rightarrow Ni(s) + Mg^{2+}(aq)$$

SOLUTION:
Start by taking the overall cell reaction and breaking it into two half-reactions.

$$Ni^{2+}(aq) + 2\ e^- \rightarrow Ni(s)$$

$$Mg(s) \rightarrow Mg^{2+}(aq) + 2\ e^-$$

Looking at the two half-reactions, we find that the Ni^{2+} is being reduced and the Mg is being oxidized. Therefore, the anode compartment of our cell consists of a strip of magnesium metal immersed in a solution containing Mg^{2+} ions (such as magnesium nitrate). The cathode compartment consists of a strip of nickel immersed in a solution containing Ni^{2+} ions (such as nickel(II) nitrate). The two half-cells are connected to each other with a salt bridge and an external wire. Electrons flow through the wire from the magnesium anode to the nickel cathode. Anions move from the cathode compartment toward the anode, while cations migrate from the anode compartment toward the cathode.

Workbook Problem 19.2

Describe the construction of a galvanic cell based on the following reaction:

$$Pb^{2+}(aq) + Cd(s) \rightarrow Pb(s) + Cd^{2+}(aq)$$

Strategy: Break the overall reaction into half-reactions and design a galvanic cell, identifying the cathode and the anode.

Step 1: Write the half-reactions, identifying the oxidizing agent and the reducing agent.

Step 2: Identify solutions for each half-cell and a suitable inert electrolyte for the salt bridge.

POST-LECTURE:

Work Problems: *One of the best ways to learn chemistry is through solving problems. Go to the textbook and work through the Worked Examples for this section, then go to the end of chapter and work the problems there. When you feel confident that you can correctly answer problems like the ones listed below for each objective, check off each of the Learning Objectives for this section above.*

LO 19.4: Worked Example 19.3; Problems 19.34, 19.56

Section 19.3 Shorthand Notation for Galvanic Cells

PRE-LECTURE:

Learning Objectives: *These are the skills you should be proficient in after reading this section and practicing the associated problems. Check off the boxes below once you have gone through this section, answered the associated questions in your textbook, and feel confident you have mastered the skill.*

☐ **LO 19.5** Use shorthand notation to represent a galvanic cell.

Summary: *Read through this section and fill in the missing words and equations. Reference the textbook when you are not confident in your answers.*

Shorthand notation is used to quickly see what is happening in galvanic cell. There is a lot of information represented by the shorthand notation. Fill in the boxes in the image below to help you remember what the notations mean.

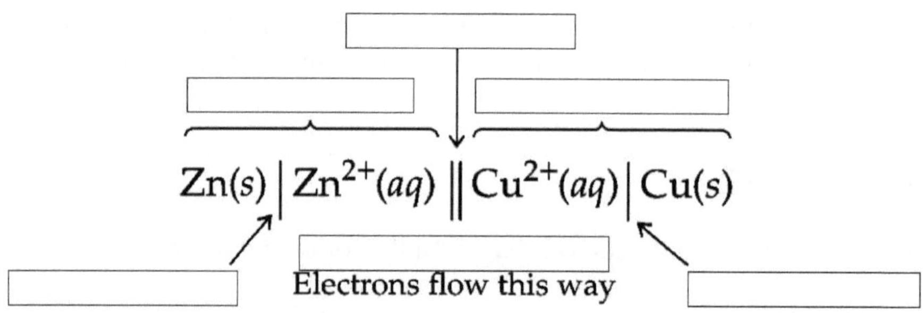

A few notes:

- __________ are written first, followed by __________.
- Commas are to separate species in the __________ phase.
- An additional vertical line is used to represent additional __________.

A very detailed version of the shorthand notation also includes ion concentrations and gas pressures.

Workbook Example 19.3

PROBLEM:
Give the shorthand notation for a galvanic cell that employs the overall reaction

$$Ni(NO_3)_2(aq) + Cd(s) \rightarrow Ni(s) + Cd(NO_3)_2(aq)$$

Give a brief description of the cell.

SOLUTION:
The two half-reactions for this overall reaction are

$$Ni^{2+}(aq) + 2\ e^{-} \rightarrow Ni(s)$$

$$Cd(s) \rightarrow Cd^{2+}(aq) + 2\ e^{-}$$

From these half-reactions, we know that nickel is being reduced and cadmium is being oxidized. Therefore, cadmium is the anode and nickel is the cathode. The cell notation is

$$Cd(s) \mid Cd^{2+}(aq) \parallel Ni^{2+}(aq) \mid Ni(s)$$

This cell consists of a strip of cadmium metal as the anode dipping into an aqueous solution of $Cd(NO_3)_2$ and a strip of nickel metal as the cathode dipping into an aqueous solution of $Ni(NO_3)_2$. The two half-cells are connected by a salt bridge and a wire.

Workbook Problem 19.3

Given the shorthand notation below, determine the half-reactions occurring at the anode and cathode. Write the overall reaction for the cell and give a brief description of the cell's construction.

$$Mg(s) \mid Mg^{2+}(aq) \parallel Cr^{3+}(aq) \mid Cr(s)$$

Strategy: Based on the shorthand notation above, determine which species is the cathode and which is the anode, and write the half-reaction that occurs at each electrode.

Step 1: Write the half-reaction occurring at the anode. Remember, the reactant in each half-cell is written first, followed by the product.

Step 2: Write the half-reaction occurring at the cathode.

Step 3: Combine the two half-reactions to obtain the overall cell reaction (recall that the equation must be balanced in electrons).

Step 4: Describe the cell indicated in the shorthand notation.

POST-LECTURE:

Work Problems: *One of the best ways to learn chemistry is through solving problems. Go to the textbook and work through the Worked Examples for this section, then go to the end of chapter and work the problems there. When you feel confident that you can correctly answer problems like the ones listed below for each objective, check off each of the Learning Objectives for this section above.*

LO 19.5: Worked Example 19.4; Problems 19.58, 19.60, 19.62

Section 19.4 Cell Potentials and Free-Energy Changes for Cell Reactions

PRE-LECTURE:

Learning Objectives: *These are the skills you should be proficient in after reading this section and practicing the associated problems. Check off the boxes below once you have gone through this section, answered the associated questions in your textbook, and feel confident you have mastered the skill.*

☐ **LO 19.6** Interconvert between cell potentials and standard free-energy changes for electrochemical reactions.

Definitions: *To learn, you must do. This is your workbook. Fill in the definitions from the glossary of the textbook as a first step to understanding the material.*

— **electromotive force (emf)**— ______________________________

____

— **cell potential (E)**— ______________________________

____

— **cell voltage**— ______________________________

____

— **faraday**— ______________________________

____

— **standard cell potential (*E*°)**— ______________________________

____

Summary: *Read through this section and fill in the missing words and equations. Reference the textbook when you are not confident in your answers.*

The chemical potential that pushes electrons away from the anode and toward the cathode is known as the **electromotive force,** the **cell potential** (*E*), or the **cell voltage.** The potential of a galvanic cell is defined as a *positive quantity with units of volts (V).*

The following relationships are found among the units of energy, electrical charge, and current:

$$1\ watt = 1\frac{Joule}{sec} = \frac{1\ Coulomb \times 1\ volt}{sec} = 1\ ampere \times 1\ volt$$

Watts are a measure of power, coulombs are a measure of charge, volts are a measure of potential, and amperes are a measure of current.

Cell potential is measured with a voltmeter: when the + and − terminals of the voltmeter are connected to cathode (+) and anode (−), a positive value will be obtained for the voltage. This makes it possible to use a voltmeter to determine which electrode is the cathode and which is the anode.

Write the equation that shows the relationship between cell potential and free energy:

$\Delta G =$

where $n =$ ________ of ________ transferred in the reaction and F is the ________. The Faraday constant is the electrical charge per mole of electrons and is equal to 96,500 C/mol e^-.

Note that ΔG and E have opposite signs: a spontaneous reaction is indicated by a ________ free-energy change or a ________ cell potential.

Standard cell potential, $E°$, is the cell potential when both reactants and products are in their standard states: solutes at 1 M concentration, gases at a partial pressure of 1 atm, solids and liquids in pure form, at 25 °C. At standard state, the equation becomes:

$\Delta G° =$

Workbook Example 19.4

PROBLEM:
If the standard cell potential for the following reaction is 2.37 *V*, what is the standard free-energy change? Write the shorthand notation for this reaction.

$$Al(s) + Fe^{3+}(aq) \rightarrow Al^{3+}(aq) + Fe(s)$$

SOLUTION:
Use the equation $\Delta G° = -nFE°$

F is 96,500 *C/mol*, and n can be inferred from the balanced equation. Three electrons are being transferred per mole of reactants, so $n = 3$.

$$\Delta G° = -3\left(96500\ \frac{C}{mol}\right)(2.37\ V)\left(1\frac{J}{C \cdot V}\right) = -686\ {}^{kJ}\!/_{mol}$$

The shorthand notation for this reaction is $Al(s) | Al^{3+}(aq) || Fe^{3+}(aq) | Fe(s)$.

Workbook Problem 19.4

The reaction shown below in shorthand notation has a standard cell potential of 1.61 V. Write the equation for this reaction and calculate the standard free-energy change.

$$Mg(s)\,|\,Mg^{2+}(aq)\,||\,Zn^{2+}(aq)\,|\,Zn(s)$$

Strategy: Based on the shorthand notation above, write the equation. Using the formula $\Delta G° = -nFE°$, calculate the standard free-energy change.

Step 1: Write the balanced equation for the reaction.

Step 2: Identify n and use $\Delta G° = -nFE°$ to calculate $\Delta G°$.

POST-LECTURE:

Work Problems: *One of the best ways to learn chemistry is through solving problems. Go to the textbook and work through the Worked Examples for this section, then go to the end of chapter and work the problems there. When you feel confident that you can correctly answer problems like the ones listed below for each objective, check off each of the Learning Objectives for this section above.*

LO 19.6: Worked Example 19.5; Problems 19.68, 19.70, 19.72

Section 19.5 Standard Reduction Potentials

PRE-LECTURE:

Learning Objectives: *These are the skills you should be proficient in after reading this section and practicing the associated problems. Check off the boxes below once you have gone through this section, answered the associated questions in your textbook, and feel confident you have mastered the skill.*

☐ **LO 19.7** Use tabulated values of standard reduction potentials for half-reactions (Appendix D) to order substances by increasing strength as an oxidizing or reducing agent.

Definitions: *To learn, you must do. This is your workbook. Fill in the definitions from the glossary of the textbook as a first step to understanding the material.*

— **standard hydrogen electrode (S.H.E.)**—________________________________

____

— **standard reduction potential—** ______________________________

____

— **standard electrode potential—** ______________________________

____

Summary: *Read through this section and fill in the missing words and equations. Reference the textbook when you are not confident in your answers.*

The standard potential of a galvanic cell is the sum of the half-cell potentials:

$$E^\circ_{cell} =$$

Only a potential *difference* can be measured, so to determine half-reaction potentials, a standard electrode must be used against which all other reactions can be measured. The standard hydrogen electrode (S.H.E.) is assigned an arbitrary potential of exactly 0 V, and all other voltages are relative to this reaction:

$$2\,H^+(aq, 1\,M) + 2\,e^- \rightarrow H_2(g, 1\,atm) \qquad E^\circ = 0.00\,V$$

The shorthand notation for S.H.E. is $H^+\ (1\ M)\,|\,H_2\ (1\ atm)\,|\,Pt(s)$, where the Pt electrode is in contact with both H_2 gas and $H^+(aq)$.

Standard potentials for half-cells can be determined by constructing a galvanic cell in which the half-cell of interest is paired up with the standard hydrogen electrode: standard oxidation potential is the half-cell potential for an oxidation half-reaction, and standard reduction potential is the half-cell potential for a reduction half-reaction.

As with free-energy changes, when the direction of a half-reaction is reversed, the sign of E° must be ________.

In a table of half-cell reaction, all the reactions are written as ________. Oxidizing agents are on the ________ side, and reducing agents are on the ________ side. Reactions are listed in decreasing order of standard reduction potential, with the ________ oxidizing agents at the upper left and the ________ reducing agents at the lower right.

POST-LECTURE:

Work Problems: *One of the best ways to learn chemistry is through solving problems. Go to the textbook and work through the Worked Examples for this section, then go to the end of chapter and work the problems there. When you feel confident that you can correctly answer problems like the ones listed below for each objective, check off each of the Learning Objectives for this section above.*

.

LO 19.7: Worked Example 19.6; Problems 19.74, 19.76, 19.78

Section 19.6 Using Standard Reduction Potentials

PRE-LECTURE:

Learning Objectives: *These are the skills you should be proficient in after reading this section and practicing the associated problems. Check off the boxes below once you have gone through this section, answered the associated questions in your textbook, and feel confident you have mastered the skill.*

- ☐ **LO 19.8** Use standard reduction potentials for half-reactions to calculate the potential of a redox reaction and determine spontaneity.
- ☐ **LO 19.9** Use standard reduction potentials for half-reactions to calculate the voltage of a galvanic cell.

Summary: *Read through this section and fill in the missing words and equations. Reference the textbook when you are not confident in your answers.*

Tables of standard reduction potentials summarize an enormous quantity of chemical information in a small space. They make it possible to arrange oxidizing or reducing agents in order of strength and predict the spontaneity of thousands of redox reactions.

To balance the equations in electrons, it may be necessary to place coefficients on the half-reactions. What this means is that it may be necessary to multiply half-reactions by some factor to ensure that electrons cancel. In this case, the values of $E°$ are ________ multiplied by the same factor: the voltage and free energy depend on the number of electrons transferred, not on the stoichiometry of the atoms.

An oxidizing agent can oxidize any reducing agent that lies ________ it in the table. For the reaction to be spontaneous, $E°$ must be ________.

Workbook Example 19.5

PROBLEM:
Write the balanced net ionic equation, and calculate $E°$ for the following galvanic cell:

$$Al(s)\ |\ Al^{3+}(aq)\ ||\ Cl_2(g)\ |\ Cl^-(aq)$$

SOLUTION:
$Al(s)$ is the anode and therefore undergoes oxidation, while $Cl_2(s)$ is the cathode and undergoes reduction. The half-reactions and their cell potentials are

$$Al(s) \rightarrow Al^{3+}(aq) + 3\ e^- \qquad E° = +1.66\ V$$

$$Cl_2(g) + 2\ e^- \rightarrow Cl^-(aq) \qquad E° = +1.36\ V$$

(Notice that the sign for $E°$ for the Al/Al^{3+} half-reactions has been reversed.) To write the balanced net ionic equation, we need to make sure that the electrons cancel out on both sides. Therefore, we need to multiply the top reaction by 2 and the bottom reaction by 3.

$$2\ Al(s) \rightarrow 2\ Al^{3+}(aq) + 6\ e^- \qquad E° = +1.66\ V$$

$$3\ Cl_2(aq) + 6\ e^- \rightarrow 6\ Cl^-(aq) \qquad E° = +1.36\ V$$

Remember that $E°$ values are independent of the amount of reaction and are not multiplied by stoichiometric coefficients.

$$E°_{\text{cell}} = E°_{Al \to Al^{3+}} + E°_{Cl_2 \to Cl^-} = +1.66\,V + 1.36\,V = 3.02\,V$$

Workbook Problem 19.5

Using the table of standard reduction potentials and without calculating the cell potential, determine if the following reactions are spontaneous.

$$Cu(s) + 2\ Ag^+(aq) \to 2\ Ag(aq) + Cu^{2+}(aq)$$

$$2\ MnO_4^-(aq) + 16\ H^+(aq) + 10\ F^-(aq) \to 2\ Mn^{2+}(aq) + 8\ H_2O(l) + 5\ F_2(g)$$

Strategy: Determine the position of the reducing agent relative to the position of the oxidizing agent, and based on those results, determine the spontaneity of the reactions.

Step 1: Determine the position of the reducing agent relative to the position of the oxidizing agent.

Step 2: Determine the spontaneity of the reactions.

POST-LECTURE:

Work Problems: *One of the best ways to learn chemistry is through solving problems. Go to the textbook and work through the Worked Examples for this section, then go to the end of chapter and work the problems there. When you feel confident that you can correctly answer problems like the ones listed below for each objective, check off each of the Learning Objectives for this section above.*

LO 19.8: Worked Example 19.7; Problems 19.80, 19.82
LO 19.9: Worked Example 19.8; Problems 19.84, 19.86

Section 19.7 Cell Potentials Under Nonstandard-State Conditions: The Nernst Equation

PRE-LECTURE:

Learning Objectives: *These are the skills you should be proficient in after reading this section and practicing the associated problems. Check off the boxes below once you have gone through this section, answered the associated questions in your textbook, and feel confident you have mastered the skill.*

☐ **LO 19.10** Calculate the cell potential under nonstandard-state conditions using the Nernst equation.

Definitions: *To learn, you must do. This is your workbook. Fill in the definitions from the glossary of the textbook as a first step to understanding the material.*

— **Nernst equation—**____

____

Review: *Before going through the section make sure you are familiar with these concepts:*

- Section 18.10 $\Delta G = \Delta G° + RT \ln Q$

Summary: *Read through this section and fill in the missing words and equations. Reference the textbook when you are not confident in your answers.*

Like free-energy changes, cell potentials depend on temperature and on the composition of the reaction mixture. The Nernst equation makes it possible to calculate cell potentials under nonstandard-state conditions. Write the Nernst equation below:

$E =$

This equation works at 25 °C, and the unit of E is in volts. Its derivation comes from combining $\Delta G = \Delta G° + RT \ln Q$ (Section 18.10) and $\Delta G = -nFE$ (Section 19.4) and can be found in your book.

Workbook Example 19.6

PROBLEM:
Calculate E_{cell} for the following reaction:

$$2\ Fe^{3+}(aq) + Pb(s) \rightarrow 2\ Fe^{+}(aq) + Pb^{2+}(aq)$$

with the following concentrations: $[Pb^{2+}] = 0.15$ M, $[Fe^{3+}] = 0.50$ M, and $[Fe^{+}] = 0.10$ M.

SOLUTION:
The half-reactions for this equation are

$$Pb(s) \rightarrow Pb^{2+} + 2\ e^{-} \qquad E° = +0.13\ V$$

$$Fe^{3+}(aq) + 2\ e^{-} \rightarrow Fe^{+}(aq) \qquad E° = +0.77\ V$$

$$E°_{cell} = 0.13\ V + 0.77\ V = 0.90\ V$$

The Nernst equation for this reaction is

$$E_{cell} = E°_{cell} - \frac{0.0592}{2} \log \frac{\left[Pb^{2+}\right]\left[Fe^{+}\right]^2}{\left[Fe^{3+}\right]^2}$$

Substituting the information given and solving for E_{cell} gives:

$$E_{\text{cell}} = 0.90\,V - \frac{0.0592}{2}\log\frac{(0.15\,M)(0.10\,M)^2}{(0.50\,M)^2} = 0.97\,V$$

Workbook Problem 19.6

If E_{cell} = 0.17 V for the following reaction, what is the concentration of $[Sn^{4+}]$ ions? The concentrations are $[Cu^{2+}]$ = 0.63 M, $[Sn^{2+}]$ = 0.72 M.

$$Sn^{2+}(aq) + Cu^{2+}(aq) \rightarrow Sn^{4+}(aq) + Cu(s)$$

Strategy: Calculate the $E°_{cell}$ for the cell reaction, substitute the information given into the Nernst equation, and solve for $[Sn^{4+}]$.

Step 1: Calculate $E°_{cell}$.

Step 2: Using $E°_{cell}$ and E, solve for $[Sn^{4+}]$.

POST-LECTURE:

Work Problems: *One of the best ways to learn chemistry is through solving problems. Go to the textbook and work through the Worked Examples for this section, then go to the end of chapter and work the problems there. When you feel confident that you can correctly answer problems like the ones listed below for each objective, check off each of the Learning Objectives for this section above.*

LO 19.10: Worked Examples 19.9, 19.10; Problems 19.41 19.94, 19.98

Section 19.8 Electrochemical Determination of pH

PRE-LECTURE:

Learning Objectives: *These are the skills you should be proficient in after reading this section and practicing the associated problems. Check off the boxes below once you have gone through this section, answered the associated questions in your textbook, and feel confident you have mastered the skill.*

☐ **LO 19.11** Use the cell potential to calculate solution concentration.

Summary: *Read through this section and fill in the missing words and equations. Reference the textbook when you are not confident in your answers.*

Using the Nernst equation, it is possible to derive an equation to calculate the pH of a solution.

When the following reaction is considered with a hydrogen electrode in a solution of unknown pH

$$Pt(s) \mid H_2\ (1\ atm) \mid H^+\ (?\ M) \parallel \text{reference cathode}$$

the difference between E_{cell} and E_{ref} can be directly related to pH, by the following derived equation:

pH =

As a result of this equation, the pH of a solution can be measured by measuring E_{cell}.

The awkwardness of the hydrogen electrode has resulted in its substitution with a glass electrode consisting of a silver wire coated with silver chloride immersed in a reference solution of dilute hydrochloric acid. The reference electrode is known as a *calomel* electrode: Hg_2Cl_2 in contact with liquid mercury and aqueous KCl.

Workbook Example 19.7

PROBLEM:
The following cell has a potential of 0.65 V. Calculate the pH of the solution in the anode compartment.

$$Pt(s) \mid H_2(g)\ (1\ atm) \mid H^+\ (pH = ?) \parallel Cl^-(aq)\ (1\ M) \mid Hg_2Cl_2(s) \mid Hg(l)$$

SOLUTION:
The cell reaction is

$$Hg_2Cl_2(s) + H_2(g) \rightarrow 2\ Hg(l) + 2\ Cl^-(aq) + 2\ H^+(aq)$$

$E°$ for the calomel electrode is 0.28 V:

$$E° = E°_{H_2 \rightarrow H^+} + E°_{\mathrm{Hg_2Cl_2 \rightarrow Hg, Cl^-}} = 0.00\ V + 0.28\ V = 0.28\ V$$

The pH can be calculated using the equation

$$pH = \frac{E_{\text{cell}} - E_{\text{ref}}}{0.0592}$$

$$pH = \frac{0.65\ V - 0.28\ V}{0.0592} = 6.25$$

Workbook Problem 19.7

If $E_{cell} = 0.62$ V for the following reaction, what is the pH in the anode compartment?

$$Cu^{2+}(aq) + H_2(g) \rightarrow 2\ H^+(aq) + Cu(s)$$

Strategy: Determine the half-reactions occurring and calculate E_{ref}. Substitute into the modified Nernst equation to determine pH of the solution ($Cu^{2+}(aq) + 2\ e^- \rightarrow Cu(s)\ E° = 0.34$).

Step 1: Determine E_{ref}.

Step 2: Calculate the pH of the anode solution.

POST-LECTURE:

Work Problems: *One of the best ways to learn chemistry is through solving problems. Go to the textbook and work through the Worked Examples for this section, then go to the end of chapter and work the problems there. When you feel confident that you can correctly answer problems like the ones listed below for each objective, check off each of the Learning Objectives for this section above.*

LO 19.11: Worked Example 19.11; Problems 19.104, 19.106

Section 19.9 Standard Cell Potentials and Equilibrium Constants

PRE-LECTURE:

Learning Objectives: *These are the skills you should be proficient in after reading this section and practicing the associated problems. Check off the boxes below once you have gone through this section, answered the associated questions in your textbook, and feel confident you have mastered the skill.*

☐ **LO 19.12** Use standard reduction potentials to calculate the equilibrium constant.

Review: *Before going through the section make sure you are familiar with these concepts:*

- Section 18.10 $\Delta G = \Delta G° + RT \ln Q$
- Section 15.2 determination of the equilibrium constant K from concentration data
- Section 18.11 determination of the equilibrium constant K from thermochemical data

Summary: *Read through this section and fill in the missing words and equations. Reference the textbook when you are not confident in your answers.*

The standard cell potential is related to the standard free-energy change for the reaction. Derivation of the equation showing this relationship can be found in your textbook. Write it below:

$E° =$

This makes it possible to calculate equilibrium constants from cell potentials. This is convenient, as potentials are easily measured, but small concentrations are not as easily measured.

Compared with the equilibrium constants for acid–base reactions, those for redox reactions tend to be either very large or very small. In other words, redox reactions either go to completion or do not proceed at all. A positive value of $E°$ corresponds to $K > 1$, while a negative value of $E°$ corresponds to $K < 1$.

With the addition of this method, there are three different ways to determine the value of an equilibrium constant K.

- K from concentration data: $K =$
- K from thermochemical data: $\Delta G° =$ or $\ln K =$
- K from electrochemical data: $E° =$ or $\ln K =$

Workbook Example 19.8

PROBLEM:
Calculate the equilibrium constant for the following reaction at 25 °C.

$$2\ Fe^{3+}(aq) + 3\ Pb(s) \rightarrow 2\ Fe(s) + 3\ Pb^{2+}(aq)$$

SOLUTION:
The half-reactions for this reaction are:

$$Fe^{3+}(aq) + 3\ e^{-} \rightarrow Fe(s) \qquad E° = 0.77 - 0.45 = 0.32\ V$$

$$Pb(s) \rightarrow Pb^{2+}(aq) + 2\ e^{-} \qquad E° = 0.13\ V$$

$$E°_{cell} = 0.32 + 0.13 = +0.45\ V$$

The value of n for this reaction is 6. We can now solve for K.

$$\log K = \frac{nE^\circ}{0.0592\ V} = \frac{(6)(0.45\ V)}{0.0592\ V} = 45.6$$

$$K = 10^{45.6} = 3.98 \times 10^{45}$$

Workbook Problem 19.8

Find K for the reaction:

$$H_2O_2(aq) + 2\ H^+(aq) + 2\ Br^-(aq) \rightarrow 2\ H_2O(l) + Br_2(l)$$

Strategy: Determine the half-reactions occurring and calculate E° from the standard potentials of the half-cell reactions. Calculate K from the equation $\log K = E^\circ \frac{n}{0.0592}$.

Step 1: Determine E°.

Step 2: Calculate K from E°.

POST-LECTURE:

Work Problems: *One of the best ways to learn chemistry is through solving problems. Go to the textbook and work through the Worked Examples for this section, then go to the end of chapter and work the problems there. When you feel confident that you can correctly answer problems like the ones listed below for each objective, check off each of the Learning Objectives for this section above.*

LO 19.12: Worked Example 19.12; Problems 19.110, 19.112, 19.116

Section 19.10 Batteries

PRE-LECTURE:

Learning Objectives: *These are the skills you should be proficient in after reading this section and practicing the associated problems. Check off the boxes below once you have gone through this section, answered the associated questions in your textbook, and feel confident you have mastered the skill.*

☐ **LO 19.13** Describe batteries and fuel cells, and calculate values of E°, ΔG°, and K for the cell reaction.

Summary: *Read through this section and fill in the missing words and equations. Reference the textbook when you are not confident in your answers.*

Galvanic cells can be linked in series, with the final voltage the sum of all the individual voltages. Battery features depend on the application, but in general, batteries should be rugged, compact, lightweight, and inexpensive and should provide stable power for a relatively long time.

_______________ batteries have been generating power for automobiles for nearly a century. Six 2 V cells are connected to generate 12 V of power. Inside the battery, the anode is a series of grids packed with spongy lead, while the cathode is a series of grids packed with lead dioxide; both are dipped in a 38% solution of sulfuric acid.

Anode:	$Pb(s) + HSO_4^-(aq) \rightarrow PbSO_4(s) + H^+(aq) + 2e^-$	$E° = 0.296$ V
Cathode:	$PbO_2(s) + 3H^+(aq) + HSO_4^-(aq) + 2e^- \rightarrow PbSO_4(s) + 2H_2O(l)$	$E° = 1.628$ V
Overall:	$Pb(s) + PbO_2(s) + 2H^+(aq) + 2HSO_4^-(aq) \rightarrow 2PbSO_4(s) + 2H_2O(l)$	$E° = 1.924$ V

As the reaction proceeds, $PbSO_4$ adheres to the surface of the electrodes. By running power back through the battery, the $PbSO_4$ is driven off the surfaces. Eventually, this is no longer possible, and the battery must be recycled.

_______________ batteries are common household batteries. They originated with *Leclanché cells*, which were patented in 1866. Leclanché cells consist of a zinc metal can (the anode), an inert graphite rod surrounded by a paste of solid MnO_2, and carbon black (cathode), and a paste of NH_4Cl and $ZnCl_2$ in starch (electrolyte).

Leclanché cells have been displaced by alkaline dry cells, in which the acidic NH_4Cl solution (which corrodes the zinc) is replaced by NaOH or KOH. This slows the corrosion, which increases battery life, power, and stability.

Anode: $Zn(s) + 2\ OH^-(aq) \rightarrow Zn(OH)_2(aq) + 2\ e^-$

Cathode: $2\ MnO_2(s) + H_2O(l) + 2\ e^- \rightarrow Mn_2O_3(s) + 2\ OH^-(aq)$

_______________, or "*ni-cad*," batteries are rechargeable: like lead storage batteries, the solid products of the reactions adhere to the surface of the electrodes. The anode is Cd, and the cathode is NiO(OH).

Anode: $Cd(s) + 2\ OH^-(aq) \rightarrow Cd(OH)_2(s) + 2\ e^-$

Cathode: $NiO(OH)(s) + H_2O(l) + e^- \rightarrow Ni(OH)_2(s) + OH^-(aq)$

_______________ *batteries*, or *NiMH*, batteries have replaced ni-cad batteries in many applications because cadmium is expensive and toxic—*NiMH* batteries are more environmentally friendly. They have the same voltage as ni-cad batteries, but have twice the energy density: half as much battery mass is required for the same voltage. The cathode is the same as in ni-cad batteries, but the anode is a special metal alloy that can absorb and release large amounts of H_2 at ordinary temperature:

Anode: $MH_{ab}(s) + OH^-(aq) \rightarrow M(s) + H_2O(l) + e^-$

Cathode: $NiO(OH)(s) + H_2O(l) + e^- \rightarrow Ni(OH)_2(s) + OH^-(aq)$

The overall reaction transfers hydrogen from the anode to the cathode.

NiMH batteries are used in many consumer products and also in hybrid automobiles, where they supply energy to the electric motor and are recharged when the brakes are applied, using some of the kinetic energy of the vehicle.

________________ and ________________ batteries are lightweight, high-voltage, and rechargeable. They are used in cell phones, laptops, cameras, tools, and Tesla electric cars. With an atomic weight of 6.97 g/mol, less than 7 g of lithium can provide 1 mol of electrons. In addition, lithium has the highest standard oxidation potential of any metal.

The anode is lithium metal, the cathode is typically MnO_2, and the electrolyte is a lithium salt such as $LiClO_4$ in an organic solvent.

Anode: $x\,Li(s) \rightarrow x\,Li^{+}(soln) + x\,e^{-}$

Cathode: $MnO_2(s) + Li^{+}(soln) + x\,e^{-} \rightarrow Li_xMnO_2(s)$

Lithium ion batteries contain a lithiated graphite anode. Abbreviated Li_xC_6, this is graphite with lithium atoms inserted between the layers. The cathode is CoO_2, which is also able to incorporate lithium ions into its structure. The electrolyte is a solution of lithium salt in an organic solvent or a solid-state polymer that can transport lithium ions.

Anode: $Li_xC_6(s) \rightarrow x\,Li^{+}(soln) + 6\,C(s) + x\,e^{-}$

Cathode: $Li_{1-x}CoO_2(s) + x\,Li^{+}(soln) + x\,e^{-} \rightarrow LiCoO_2(s)$

There are safety concerns about lithium ion batteries that are being addressed by the development of alternative electrode materials.

POST-LECTURE:

Work Problems: *One of the best ways to learn chemistry is through solving problems. Go to the textbook and work through the Worked Examples for this section, then go to the end of chapter and work the problems there. When you feel confident that you can correctly answer problems like the ones listed below for each objective, check off each of the Learning Objectives for this section above.*

LO 19.13: Problems 19.118, 19.120, 19.122

Section 19.11 Corrosion

PRE-LECTURE:

Learning Objectives: *These are the skills you should be proficient in after reading this section and practicing the associated problems. Check off the boxes below once you have gone through this section, answered the associated questions in your textbook, and feel confident you have mastered the skill.*

☐ **LO 19.14** Describe how corrosion occurs and methods used to prevent it.

Definitions: *To learn, you must do. This is your workbook. Fill in the definitions from the glossary of the textbook as a first step to understanding the material.*

— **corrosion—** ______________________________

____

— **galvanizing—** ______________________________

____

— **cathodic protection—** ______________________________

____

Summary: *Read through this section and fill in the missing words and equations. Reference the textbook when you are not confident in your answers.*

________ is the oxidative deterioration of a metal. The conversion of iron to rust is a well-known example; it requires both oxygen and water and involves pitting of the metal surface and deposition of hydrated iron(III) oxide at a location separate from the pit.

A possible mechanism involves an electrochemical process in which iron is oxidized in one region of the surface, while oxygen is reduced in another region. Step 1 is

Anode region:	$Fe(s) \rightarrow Fe^{2+}(aq) + 2\ e^-$	$E° = 0.45$ V
Cathode region:	$O_2(g) + 4\ H^+(aq) + 4\ e^- \rightarrow 2\ H_2O(l)$	$E° = 1.23$ V

Soluble Fe^{2+} ions migrate through the water droplets and encounter dissolved oxygen, leading to the further oxidation of Fe^{2+} to Fe^{3+}. Fe^{3+} then reacts with water in the final step, producing $Fe_2O_3 \cdot H_2O$, which is rust.

This mechanism also explains the increased corrosion of cars in the presence of road salt—salt in the water greatly increases its conductivity, allowing the reaction to proceed more quickly.

O_2 is able to oxidize all metals except a few, as can be seen by the fact that the O_2/H_2O half-reaction lies above the M^{n+}/M half-reaction for all but a very few metals, such as gold. In the case of many metals, including aluminum, magnesium, chromium, titanium, and zinc, the oxidation forms a hard coating that prevents further oxidation.

Corrosion of iron can be avoided by shielding it from oxygen and water. This can be done with paint, but is more effective with another metal. Coating steel with molten zinc is known as ________, and it provides protection even if the zinc layer is scratched, because the zinc is preferentially oxidized and will reduce any oxidized iron.

________________, a similar method, uses a *sacrificial anode*: a metal that is more easily oxidized is electrically connected to the steel or iron and corrodes instead of the iron. This method is used to protect ships and large buildings from corrosion.

POST-LECTURE:

Work Problems: *One of the best ways to learn chemistry is through solving problems. Go to the textbook and work through the Worked Examples for this section, then go to the end of chapter and work the problems there. When you feel confident that you can correctly answer problems like the ones listed below for each objective, check off each of the Learning Objectives for this section above.*

LO 19.14: Problems 19.127, 19.128, 19.130

Section 19.12 Electrolysis and Electrolytic Cells

PRE-LECTURE:

Learning Objectives: *These are the skills you should be proficient in after reading this section and practicing the associated problems. Check off the boxes below once you have gone through this section, answered the associated questions in your textbook, and feel confident you have mastered the skill.*

☐ **LO 19.15** Draw an electrolytic cell. Label the anode and cathode, indicate the direction of electron and ion flow, and write balanced equations for the electrode and overall cell reactions.

☐ **LO 19.16** Predict the products formed and the electrode and overall cell reactions when an aqueous solution of an ionic compound is electrolyzed.

Definitions: *To learn, you must do. This is your workbook. Fill in the definitions from the glossary of the textbook as a first step to understanding the material.*

— **electrolysis**— ____

____

— **overpotential**— ____

____

Summary: *Read through this section and fill in the missing words and equations. Reference the textbook when you are not confident in your answers.*

In a galvanic cell, a spontaneous reaction generates a current. In an **electrolytic cell,** a current is used to drive a nonspontaneous reaction. The process of using an electric current to bring about chemical change is called ________. The relationship between these is summarized below:

Reaction Type	E	ΔG	Cell Type
	+	−	
	−	+	
	0	0	

An electrolytic cell has two electrodes connected to a source of current and dipped into an electrolyte. The battery acts as an electron pump, removing electrons from the anode, moving electrons through the electrolyte, and pulling electrons into the cathode. The signs of the electrodes are switched in galvanic and electrolytic cells.

Type of Cell	**Anode**	**Cathode**
	Oxidation takes place (−)	Reduction takes place (+)
	Oxidation takes place (+)	Reduction takes place (−)

Electrolysis of molted NaCl: the negative electrode (cathode) attracts Na^+, which is reduced to $Na(l)$. The positive electrode (anode) attracts Cl^-, which is oxidized to $Cl_2(g)$.

Anode: $2\ Cl^-(l) \rightarrow Cl_2(g) + e^-$

Cathode: $Na^+(l) + e^- \rightarrow Na(l)$

Electrolysis of aqueous NaCl: when an aqueous solution is used, the reactions may differ because water may be involved.

Cathode: $Na^+(aq) + e^- \rightarrow Na(s)$ $E° = -2.71$ V *or*
$2\ H_2O + 2\ e^- \rightarrow H_2(g) + 2\ OH^-(aq)$ $E° = -0.83$ V

Anode: $2\ Cl^- \rightarrow Cl_2(g) + e^-$ $E° = -1.36$ V *or*
$2\ H_2O \rightarrow O_2(g) + 4\ H^+(aq) + 4\ e^-$ $E° = -1.23$ V

At the cathode, water is split because that reaction is far less negative than the reduction of sodium, and hydrogen bubbles are seen. At the anode, however, chlorine gas is formed. This is due to a phenomenon known as *overvoltage.* A basic solution of NaOH is formed.

________ is the amount of voltage needed above the calculated standard reduction (or oxidation) potential for electrolysis to occur. It is needed when the rate of electron transfer is very slow. Only a small overvoltage is needed for the solution or deposition of metals—they are good conductors of electrons—but a large overvoltage is required for the formation of gases, especially O_2 or H_2. Current theories do not allow this to be predicted easily, so experimental evidence is needed when cell potentials are similar.

Electrolysis of water: if the electrolyte in solution is less easily oxidized and reduced than water, water will react at both electrodes:

Anode: $2\ H_2O(l) \rightarrow O_2(g) + 4\ H^+(aq) + 4\ e^-$

Cathode: $4\ H_2O(l) + 4\ e^- \rightarrow 2\ H_2(g) + 4\ OH^-(aq)$

Equal quantities of $H^+(aq)$ and $OH^-(aq)$ are formed, so the solution will remain neutral.

Overall cell reaction: $2\ H_2O(l) \rightarrow O_2(g) + 2\ H_2(g)$

Workbook Example 19.9

PROBLEM:
Predict the half-cell reactions when a solution of aqueous KCl is electrolyzed. Write the overall cell reaction.

SOLUTION:
The possible half-reactions are

Cathode:	$K^+(aq) + e^- \rightarrow K(s)$	$E° = -2.93$ V *or*
	$2\ H_2O(l) + 2\ e^- \rightarrow H_2(g) + 2\ OH^-(aq)$	$E° = -0.83$ V
Anode:	$2\ Cl^-(aq) \rightarrow Cl_2(g) + 2\ e^-$	$E° = -1.36$ V *or*
	$2\ H_2O(l) \rightarrow O_2(g) + 4\ H^+(aq) + 4\ e^-$	$E° = -1.23$ V

At the cathode, water will be reduced, generating hydrogen gas. Remember that at the anode, the overvoltage needed to generate chlorine gas is less than that needed to generate oxygen gas, so chlorine gas will be produced, despite its higher oxidation potential.

Overall reaction: $2\ H_2O(l) + 2\ Cl^-(aq) \rightarrow H_2(g) + Cl_2(g) + 2\ OH^-(aq)$

Workbook Problem 19.9

Predict the half-cell and overall reaction for the electrolysis of an aqueous Li_2SO_4 solution. The half-reaction for sulfate is shown below, all others can be found in your textbook.

$$2\ SO_4^{2-}(aq) \rightarrow S_2O_8^{2-}(aq) + 2e^- \qquad E° = -2.01\ V$$

Strategy: Determine the possible half-reactions and their standard voltages to predict the reactions at the cathode and anode, then determine the overall reaction.

Step 1: Determine the possible reactions at the cathode and their standard voltages.

Step 2: Determine the possible reactions at the anode and their standard voltages.

Step 3: Predict which reactions will occur and calculate the overall reaction.

POST-LECTURE:

Work Problems: *One of the best ways to learn chemistry is through solving problems. Go to the textbook and work through the Worked Examples for this section, then go to the end of chapter and work the problems there. When you feel confident that you can correctly answer problems like the ones listed below for each objective, check off each of the Learning Objectives for this section above.*

LO 19.15: Problems 19.38, 19.39, 19.134
LO 19.16: Problems 19.136, 19.137

Section 19.13 Commercial Applications of Electrolysis

PRE-LECTURE:

Learning Objectives: *There are no specific learning objectives for this section; the information is a continuation of the previous section.*

Definitions: *To learn, you must do. This is your workbook. Fill in the definitions from the glossary of the textbook as a first step to understanding the material.*

— **Hall–Héroult process—** ____

____

— **electrorefining—** ____

____

— **electroplating—** ____

____

Summary: *Read through this section and fill in the missing words and equations. Reference the textbook when you are not confident in your answers.*

Sodium is produced commercially in a *Downs cell* by electrolysis of a molten mixture of NaCl and $CaCl_2$. The addition of the $CaCl_2$ lowers the melting point of the mixture by more than 200 °C. The liquid Na produced at the cathode is less dense than the mixture and can be drawn off from the top. The Downs cell process requires extremely high currents, so manufacture is often near hydroelectric power plants.

Chlorine and sodium hydroxide are produced by the electrolysis of aqueous sodium chloride in what is known as the *chlor–alkali industry*. The anode reaction, which produces Cl_2 gas, is separated from the cathode compartment by a membrane that permits the passage of sodium ions. When water is hydrolyzed at the cathode, a sodium hydroxide solution is produced.

Aluminum is produced in the *Hall–Héroult* process, which involves the electrolysis of a molten mixture of Al_2O_3 and cryolite (Na_3AlF_6) at 1000 °C in a cell with graphite electrodes. At the cathode, molten aluminum is produced and sinks to the bottom of the cell. At the anode, O_2 gas is formed. This reacts with the electrode to generate CO_2 gas, so frequent replacement of the anode is necessary. The reactions at the electrodes are not well understood, and a great deal of current is required for the process, making aluminum production the largest single consumer of electricity in the United States.

________________ is the purification of a metal by means of electrolysis. For example, copper can be purified using an impure copper anode and a pure copper cathode. Copper ions move away from the impure electrode and deposit on the pure electrode, leaving other metals either dissolved in the electrode solution or on the bottom of the container as *anode mud*.

________________ is a similar process in which one metal is coated on the surface of another using electrolysis. The cathode is the carefully cleaned object to be plated, while the solution in the electrolytic cell contains ions of the metal to be deposited.

Section 19.14 Quantitative Aspects of Electrolysis

PRE-LECTURE:

Learning Objectives: *These are the skills you should be proficient in after reading this section and practicing the associated problems. Check off the boxes below once you have gone through this section, answered the associated questions in your textbook, and feel confident you have mastered the skill.*

☐ **LO 19.17** Relate the current, time, and amount of product produced in an electrolytic cell.

Summary: *Read through this section and fill in the missing words and equations. Reference the textbook when you are not confident in your answers.*

The amount of substance produced at an electrode by electrolysis depends on the quantity of electrons passed through the cell. This is directly related to the stoichiometry of the reaction: 3 mol of electrons are needed to reduce 1 mol of aluminum ions to aluminum metal.

$$Al^{3+} + 3\,e^- \rightarrow Al$$

Moles of electrons passed through a cell can be calculated from electric current and time by the following formulas:

Charge (C) =

Moles of e^- =

The sequence of conversions below can be used to calculate the amount of product produced by passing a current through an electrolytic cell for a fixed period of time. The figure is reproduced below.

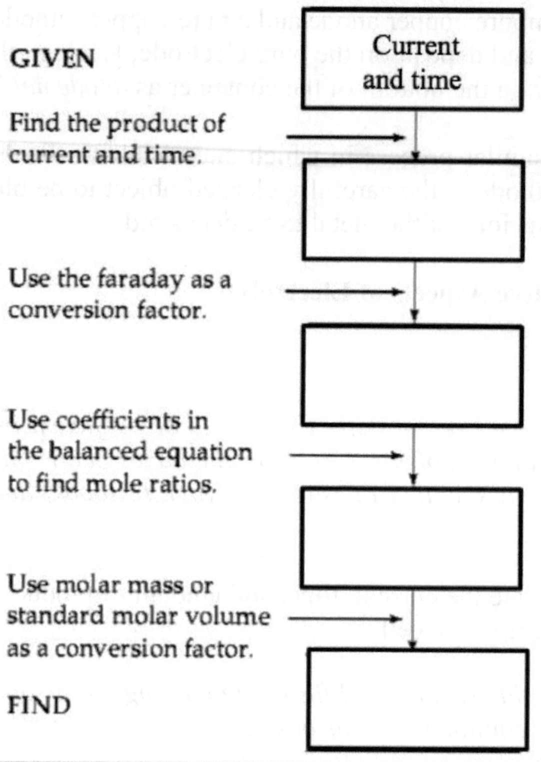

Workbook Example 19.10

PROBLEM:
How many grams of Cl_2 gas will be produced in the electrolysis of molten NaCl by a current of 5.75 A for 50.0 min? How many grams of sodium will be produced in the same time period?

SOLUTION:
(Remember that a coulomb is an A · s and that an ampere is C/s.)

$$2\ Cl^- \rightarrow Cl_2 + 2\ e^-$$

Moles of electrons = 2

$$\frac{5.75\ C}{s} \times 50.0\ min \times \frac{60\ s}{1\ min} \times \frac{1\ mol\ e^-}{96500\ C} \times \frac{1\ mol\ Cl_2}{2\ mol\ e^-} \times \frac{70.9\ g\ Cl_2}{1\ mole\ Cl_2} = 6.34\ g\ Cl_2$$

$$\frac{5.75\ C}{s} \times 50.0\ min \times \frac{60\ s}{1\ min} \times \frac{1\ mol\ e^-}{96500\ C} \times \frac{1\ mol\ Na}{1\ mol\ e^-} \times \frac{23.0\ g\ Na}{1\ mole\ Na} = 4.11\ g\ Na$$

Workbook Problem 19.10

The Dow process isolates Mg metal from seawater. The final step in this process involves the electrolysis of molten $MgCl_2$ to metal. How long would it take to produce 12 kg of Mg(*s*) at a current of 15 A? How many liters of Cl_2 gas would be produced (at STP) in the same amount of time?

Strategy: Write the electrolysis reaction and consider the conversion process to calculate time.

Step 1: Treat the electrons as reactants in the chemical equation, and solve for the time required to produce 12 kg of magnesium. Remember, the conversion factor of 96,500 C (or A·s) per mole of electrons.

Step 2: Use the information calculated above to determine the volume of Cl_2 gas.

POST-LECTURE:

Work Problems: *One of the best ways to learn chemistry is through solving problems. Go to the textbook and work through the Worked Examples for this section, then go to the end of chapter and work the problems there. When you feel confident that you can correctly answer problems like the ones listed below for each objective, check off each of the Learning Objectives for this section above.*

LO 19.17: Worked Example 19.13; Problems 19.138, 19.142, 19.144

Self-Test

This section is intended to test your knowledge of the material covered in this chapter. Think through these problems and make certain you understand what the problem is about—ask yourself if the answer makes sense. Many of these questions are linked to the chapter learning goals. Therefore, successful completion of these problems indicates that you have achieved the learning goals for this chapter. You will receive the greatest benefit from this section if you use it as a mock exam. Do not use your notes, book, Internet, or friends when taking the test. There is no grade and you only shortchange yourself by getting help with this activity. As you take the test, mark the problems with your level of confidence with the answer.

★ *Use a star for problems you know you got right.*
? *Use a question mark for those you think you got right.*
x *Use an x for those you are guessing on.*

If you get a question right that you marked with a star, you can be confident that you know the associated LO.

If you get a question right that you marked with a question mark, go back and work a few more problems and test yourself again tomorrow.

If you get a question wrong or mark it with an x, go back and read the section again, work a few problems today, a few tomorrow, and a few the next day before testing yourself again.

This will help you to discover which topics you have learned thoroughly and which topics you need to study in more detail.

True–False

1. In galvanic or voltaic cells, a spontaneous reaction occurs.

2. A salt bridge allows electrons to flow between the two halves of a galvanic cell.

3. If the cell potential of a reaction is positive, the free-energy change of the reaction is also positive.

4. When balancing electrochemical equations, $E°$ is never multiplied by the stoichiometric coefficients.

5. E for a reaction at nonstandard conditions can only be determined by experiment.

6. Cell potential and pH have a linear relationship.

7. The voltage of galvanic cells is added when they are connected in series to form a battery.

8. Direct methanol fuel cells have been proposed for use in small consumer electronics.

9. Corrosion of any metal can be prevented by contact with another metal above it in the list of standard reduction potentials.

10. The results of the electrolysis of an aqueous salt solution can be predicted using only standard half-cell voltages.

Multiple Choice

11. For the following galvanic cell, $Ni(s) \mid Ni^{2+}(aq) \parallel Br^{-}(aq) \mid Br_2(l) \mid Pt(s)$,
   a. the cathode is $Ni(s)$
   b. the electrons flow from the $Ni(s)$ electrode to the $Pt(s)$ electrode
   c. the Ni^{2+} ions flow to the anode
   d. electrons flow from the $Pt(s)$ electrode to the $Ni(s)$ electrode

12. The reaction carried out in a galvanic cell is
   a. spontaneous, and therefore has a positive $E°$ value
   b. nonspontaneous, and therefore has a positive $E°$ value
   c. spontaneous, and therefore has a negative $E°$ value
   d. nonspontaneous, and therefore has a negative $E°$ value

13. For the net reaction $Cr(s) + Fe^{3+}(aq) \rightarrow Cr^{3+}(aq) + Fe(s)$
    a. $Cr(s)$ is the reducing agent
    b. $Fe(s)$ is the anode
    c. $Fe^{3+}(aq)$ is the reducing agent
    d. $Cr(s)$ undergoes reduction

14. Given the following reduction half-cells:

$PbO_2(s) + 3\ H^+(aq) + HSO_4^-(aq) + 2\ e^- \rightarrow PbSO_4(s) + 2\ H_2O(l)$ $E° = 1.628$ V

$Cr_2O_7^{2-}(aq) + 14\ H^+(aq) + 6\ e^- \rightarrow 2\ Cr^{3+}(aq) + 7\ H_2O(l)$ $E° = 1.33$ V

$SO_4^{2-}(aq) + 4\ H^+(aq) + 2\ e^- \rightarrow H_2SO_3(aq) + H_2O(l)$ $E° = 0.17$ V

$2\ CO_2(g) + 2\ H^+(aq) + 2\ e^- \rightarrow H_2C_2O_4(aq)$ $E° = -0.49$ V

    a. the strongest oxidizing agent is $PbSO_4(s)$
    b. $PbSO_4(s)$ will spontaneously react with $CO_2(g)$
    c. the strongest oxidizing agent is $PbO_2(s)$
    d. $H_2C_2O_4(aq)$ will not spontaneously react with $PbO_2(s)$

15. In the electrolysis of molten BaI_2,
    a. the Ba^{2+} ions migrate toward the anode
    b. the I^- ions migrate toward the anode
    c. water undergoes oxidation at the anode
    d. the system becomes increasingly basic

16. In the table of standard reduction potentials,
    a. the strongest reducing agents are located in the bottom left of the table
    b. the strongest reducing agents are located in the top left of the table
    c. the strongest reducing agents are located in the lower right of the table
    d. the strongest oxidizing agents are located in the bottom left of the table

17. In an electrolytic cell,
    a. the anode has a positive sign
    b. reduction occurs at the anode
    c. the anode has a negative sign
    d. oxidation occurs at the cathode

18. In an electrolytic cell,
    a. anions migrate toward the cathode
    b. cations migrate toward the anode
    c. ions migrate through a salt bridge
    d. a current forces nonspontaneous reactions to occur

19. Alkaline dry cells
    a. contain an electrolyte of a moist paste of NaOH and $ZnCl_2$
    b. have a cathode in which steel is in contact with HgO in an alkaline medium
    c. contain an electrolyte of a moist paste of NH_4Cl and $ZnCl_2$
    d. are rechargeable nickel–cadmium batteries

20. The amount of substance produced at an electrode by electrolysis depends
    a. on the quantity of reactant present
    b. on the quantity of charge passed through the cell
    c. on the spontaneity of the reaction
    d. on the size of the electrolytic cell

Fill in the Blanks

21. Car batteries are ____________________ batteries, while common household batteries are ____________________.

22. The oxidative deterioration of a metal is called ____________________. In the case of steel, this can be prevented by coating with zinc, also called ________________ for small objects, or with the use of another metal as a(n) __________ ________________ for buildings or ships.

23. The refining of ________________ and _____________ metals require very large quantities of ________________. As a result, these processes are generally located near ________________.

24. Standard half-cell potentials are written as ______________________, and are listed in ___________ tendency to occur as written, and _____________ tendency to occur in reverse. The "zero" on the scale is the ______________________________________.

25. The ________________________ of a reaction can be calculated from the standard ____________________, but these tend to be either very __________ or very ____________, indicating that these reactions either go to ________________, or not at all.

Matching

26. anode
27. cathode
28. corrosion
29. electrolysis
30. electroplating
31. electrorefining
32. galvanizing
33. overvoltage
34. salt bridge

a. the process of using an electric current to bring about a chemical change
b. using current to deposit a thin layer of one metal on another
c. electrode at which reduction occurs
d. covering one metal with another to prevent corrosion
e. the voltage above the standard potential needed to speed a reaction
f. electrode at which oxidation occurs
g. mechanism for allowing ion flow in a galvanic cell
h. oxidative deterioration of a metal
i. using current to purify impure metals

Problems

35. Describe the galvanic cell that uses the following reaction:

$$Ni^{2+}(aq) + Mg(s) \rightarrow Ni(s) + Mg^{2+}(aq)$$

Write the half-reactions, identify the anode and cathode, and provide the shorthand notation.

36. Determine standard reaction potentials for the following unbalanced reactions and indicate whether they are spontaneous in the forward or reverse direction:
   a. $2\ Al(s) + 3\ Pb^{2+}(aq) \rightarrow 2\ Al^{3+}(aq) + 3\ Pb(s)$
   b. $Br_2(l) + 2\ Cl^-(aq) \rightarrow 2\ Br^-(aq) + Cl_2(g)$
   c. $Ni(s) + Pb^{2+}(aq) \rightarrow Ni^{2+}(aq) + Pb(s)$
   d. $Fe(s) + HNO_3 \rightarrow Fe^{2+}(aq) + NO_3^-(aq) + H_2(g)$

37. Calculate E for the following cell, $Mg(s) \mid Mg^{2+}(aq) \parallel Fe^{3+}(aq) \mid Fe(s)$, under the following conditions:
   a. standard conditions
   b. $[Mg^{2+}] = 0.1$ M, $[Fe^{3+}] = 2.0$ M
   c. $[Mg^{2+}] = 2.0$ M, $[Fe^{3+}] = 0.1$ M

38. Given the following reaction:

$Cu^{2+}(aq) + 2\ Ag(s) + 2\ Br^-(aq) \rightarrow Cu(s) + 2\ AgBr(s)$ $\qquad E° = 0.27$ V

Calculate $\Delta G°$ and K.

39. $E°$ for a galvanic cell in which Zn^{2+} is reduced to $Zn(s)$ is +1.61 V. What is the potential at the anode? What metal is oxidized at the anode? Write the half-reaction that occurs at the anode.

40. If the following cell has a potential of 0.14 V, what is the pH in the anode compartment?

$$Pt(s) \mid H_2\ (1\ atm) \mid H^+(aq) \parallel Ni^{2+}\ (1\ M) \mid Ni(s)$$

41. Which of these metals, if coated onto iron, would prevent corrosion? Explain your answer.
   a. Pb
   b. Mg
   c. Cu

42. Calculate the equilibrium constant for the reaction of iodine and aluminum using the table of standard reduction potentials.

$$3\ I_2(s) + 2\ Al(s) \rightarrow 6\ I^-(aq) + 2\ Al^{3+}(aq)$$

43. Write the reactions that occur at the anode and cathode in the electrolysis of molten NaBr. How much current would be required to produce sodium metal at the rate of 3 g/hour?

44. If 3 A of current is applied to a copper object immersed in a solution of $AgNO_3$ and copper, how much time will be required to deposit 15 g of silver?

45. An acidic solution of $NaMnO_4$ can be used to oxidize zinc metal. If 3.27 g of zinc is to be oxidized in 250 mL of solution, what is the minimum number of grams of $NaMnO_4$ that need to be dissolved?

46. How many liters of O_2 are produced at 356 mm Hg when 2.97×10^3 C is passed through water at 25 °C?

Challenge Problem

47. The K_{sp} for lead(II) sulfate is 1.8×10^{-8}. The standard reduction potential for Pb^{2+}/Pb is $E° = -0.126$ V. Use this information to determine $E°$ for the reaction below.

$$PbSO_4 + 2\ e^- \rightarrow Pb(s) + SO_4^{2-}(aq)$$

Test Yourself: *Take the Practice Test on pages 858 and 859 of your textbook. Again, use a testing environment using only those items that your instructor will allow you to have access to during the test and mark each question with a star, question mark, or x to denote your confidence level. Once you have finished you can check off the LO at the beginning of this chapter based on which questions you starred and got correct. Go back and spend more time with the topics from the other learning objectives.*

CHAPTER TWENTY

Nuclear Chemistry

Learning Objectives

As a result of reading and studying this chapter, you should be able to

- ☐ **LO 20.1** Write balanced nuclear reactions for different types of radioactive decay.
- ☐ **LO 20.2** Predict the type of radioactive decay for a given isotope.
- ☐ **LO 20.3** Write a balanced nuclear reaction for a radioactive decay series.
- ☐ **LO 20.4** Relate half-life and decay constant.
- ☐ **LO 20.5** Calculate the amount of radioactive isotope remaining after a given amount of time.
- ☐ **LO 20.6** Relate decay rates to decay constant, half-life, or amount remaining.
- ☐ **LO 20.7** Calculate the age of an object using radioisotopic dating.
- ☐ **LO 20.8** Calculate the mass defect and binding energy of a nucleus.
- ☐ **LO 20.9** Calculate the mass and energy change of a chemical reaction.
- ☐ **LO 20.10** Identify key aspects of fission and fusion reactions and their role in nuclear power and weapons.
- ☐ **LO 20.11** Calculate the energy change associated with a fission or fusion reaction.
- ☐ **LO 20.12** Write balanced equations for nuclear transmutation reactions.
- ☐ **LO 20.13** Calculate the radiation intensity in a radioactive sample and convert between units of radiation intensity.

Chapter Summary

Nuclear chemistry is the study of the properties and reactions of atomic nuclei. The stability of an element is related to its proton/neutron ratio, and the binding energy of the nucleus can be calculated from this ratio. Some elements can undergo fission, either spontaneously or by being induced, as in nuclear reactors. Other elements can fuse together under extreme conditions to form larger nuclei. All of the transuranium elements and many other isotopes have been made by nuclear transmutation, a process by which the nuclei of one element are bombarded with particles to increase their size. Radiation varies in its ability to damage tissues; a number of different units are used for radiation exposures depending on the reason for the measurement. There are numerous practical uses of nuclear chemistry, from radiocarbon dating of ancient objects to therapeutic and imaging uses in medicine.

The Chapter in Detail

Section 20.1 Nuclear Reactions and Their Characteristics

PRE-LECTURE:

Learning Objectives: *There are no specific learning objectives for this section; the information in this section is used as a basis for understanding subsequent sections.*

Definitions: *To learn, you must do. This is your workbook. Fill in the definitions from the glossary of the textbook as a first step to understanding the material.*

— **nuclear chemistry—** ______________________________

____

— **nuclear reaction—** ______________________________

____

— **nuclear equation—** ______________________________

____

— **nucleon—** ______________________________

____

— **nuclide—** ______________________________

____

Review: *Before going through the section make sure you are familiar with these concepts:*

- Section 2.8 isotopes
- Section 2.8 isotope symbols

Summary: *Read through this section and fill in the missing words and equations. Reference the textbook when you are not confident in your answers.*

Isotope symbols are written using the following conventions, where *A* is the __________ number or number of nucleons, *Z* is the __________ number, and *X* indicates the __________:

$$^{A}_{Z}X$$

Nuclear equations use these isotope symbols to represent the changes occurring in the nucleus of the reacting atoms. A nuclear equation is balanced when the sum of the mass numbers of the reactants equals the sum of the __________ of the __________ and the sum of the atomic numbers of the reactants equals the sum of the __________ of the __________.

Nuclear reactions differ from chemical reactions in several ways:

- __________ reactions involve a change in an atom's nucleus and usually create a different element. A __________ reaction does not create a different element, but combines elements in different ways.
- Different isotopes of an element exhibit essentially the __________ chemical behavior, but often completely __________ behavior in nuclear reactions.
- The energy change in nuclear reactions is __________ than that from chemical reactions.

Workbook Example 20.1

PROBLEM:
Determine the number of protons, neutrons, and electrons in the following isotopes:

$$^{71}_{31}Ga \qquad ^{191}_{77}Ir \qquad ^{88}_{39}Y$$

SOLUTION:
The subscript is the atomic number (Z), and the superscript is the mass number (A).
Z = the number of protons = the number of electrons, and $A - Z$ = the number of neutrons.

$^{71}_{31}Ga$: $Z = 31$; number of protons = number of electrons = 31
$A - Z = 71 - 31 = 40$; number of neutrons = 40

$^{191}_{77}Ir$: $Z = 77$; number of protons = number of electrons = 77
$A - Z = 191 - 77 = 114$; number of neutrons = 114

$^{88}_{39}Y$: $Z = 39$; number of protons = number of electrons = 39
$A - Z = 88 - 39 = 49$; number of neutrons = 49

Workbook Problem 20.1

Oxygen is the most abundant element by mass in the biosphere. All classes of biomolecules, proteins, sugars, fats, and nucleic acids contain oxygen, and oxygen makes up almost 90% of the mass of the world's oceans. Oxygen is made up of three stable isotopes, with mass numbers 16, 17, and 18. Write the standard symbols for these isotopes.

Strategy: Use the periodic table to determine the atomic number and the number of neutrons in each isotope. Give the standard symbol for each.

Step 1: First, find the chemical symbol for oxygen.

Step 2: Use the periodic table to determine the atomic number for oxygen.

Step 3: Determine the number of neutrons from the definition for mass number.

Step 4: The standard symbol is written with the mass number as a superscript and the atomic number as a subscript, both to the left of the symbol.

Workbook Problem 20.2

The metalloid X has been used as a cosmetic since at least 3000 B.C. When bound to sulfur atoms in a 2:3 ratio, it is known as *kohl.* An atom of the metal contains 51 protons, and its most stable isotope contains 70 neutrons. Identify the metal and write the symbol in standard format.

Strategy: Determine the atomic number from the information given.

Step 1: From the atomic number, identify the element by using the periodic table.

Step 2: The standard symbol is written with the mass number as a superscript and the atomic number as a subscript, both to the left of the symbol.

Section 20.2 Radioactivity

PRE-LECTURE:

Learning Objectives: *These are the skills you should be proficient in after reading this section and practicing the associated problems. Check off the boxes below once you have gone through this section, answered the associated questions in your textbook, and feel confident you have mastered the skill.*

☐ **LO 20.1** Write balanced nuclear reactions for different types of radioactive decay.

Definitions: *To learn, you must do. This is your workbook. Fill in the definitions from the glossary of the textbook as a first step to understanding the material.*

— **radioactive—** ____

____

— **radioisotopes—** ____

____

— **alpha (α) radiation—** ____

____

— **beta (β) radiation—**____

____

— **gamma (γ) radiation—** ____

____

— **positron emission—**____

____

— **electron capture—**____

____

Summary: *Read through this section and fill in the missing words and equations. Reference the textbook when you are not confident in your answers.*

Many nuclei are radioactive and undergo spontaneous decay to emit some form of radiation.

- ________ radiation consists of a stream of helium nuclei.
- ________ radiation occurs when electrons are ejected as a neutron spontaneously decays into a proton and electron.
- ________ radiation is the release of high-energy electromagnetic radiation.
- ________ emission occurs as a proton decays into a neutron and ejects a positron (positive electron).
- ________occurs as an electron is captured by the nucleus converting a proton into a neutron.

These types of radiation are summarized below:

Process	Symbol	Change in Atomic Number	Change in Mass Number	Change in Neutron Number
Alpha emission		−2	−4	−2
Beta emission		+1	0	−1
Gamma emission		0	0	0
Positron emission		−1	0	+1
Electron capture		−1	0	+1

Workbook Example 20.2

PROBLEM:
Smoke detectors used in homes and apartments contain americium-241, which undergoes alpha decay. When alpha particles collide with air molecules, charged particles are produced that generate an electrical current. If smoke particles enter the detector, they interfere with the formation of charged particles in the air, and the electrical current is interrupted. This causes the alarm to sound and warns the occupants of a potential fire. Complete the following nuclear equation for the decay of americium-241:

$$^{241}_{95}\text{Am} \rightarrow \_\_\_\_\_ + {}^{4}_{2}\text{He}$$

SOLUTION:
We need to figure out the missing atomic number and atomic mass. The atomic number tells us the number of protons, which, in turn, tells us the identity of the resulting nucleus.

Looking at the nuclear equation, we have a mass number of 241 on the left side and a mass number of 4 on the right-hand side. Subtracting 4 from 241 gives us a mass number of 237; this is the mass number of the new nucleus.

$$241 = x + 4$$
$$x = 237$$

Doing the same thing with atomic numbers, we have 95 on the left side and 2 on the right side. Subtracting 2 from 95 gives us 93; the new nucleus has an atomic number of 93. Looking up 93 on the periodic table tells us the new nucleus (nuclide) is neptunium, $^{237}_{93}Np$.

$$95 = x + 2$$
$$x = 93$$

Inserting the nuclide into the equation gives us:

$$^{241}_{95}\text{Am} \rightarrow {}^{237}_{93}\text{Np} + {}^{4}_{2}\text{He}$$

Workbook Problem 20.3

Write a balanced nuclear equation for the alpha decay of polonium-214.

Strategy: Use the periodic table to write the nuclide symbol of polonium-214 and the alpha particle. Using this information, write an incomplete nuclear equation and solve for the missing particle.

Step 1: Write the nuclide symbols for polonium-214 and the alpha particle.

Step 2: Write an incomplete nuclear equation representing the alpha decay of polonium-214.

Step 3: Determine the missing mass number.

Step 4: Determine the missing atomic number and determine the symbol of the new nucleus.

Step 5: Write the complete balanced nuclear equation.

Workbook Problem 20.4

Write the balanced nuclear equation for the bombardment of nickel-58 by a proton, which produces a radioactive isotope and an alpha particle.

Strategy: Use the periodic table to write the nuclide symbols given in the problem. Using this information, write an incomplete nuclear equation and solve for the missing particle.

Step 1: Write the incomplete nuclear equation.

Step 2: Determine the missing atomic mass.

Step 3: Determine the missing atomic number and the identity of the new nucleus.

Step 4:	Write the complete balanced nuclear equation.

POST-LECTURE:

Work Problems: *One of the best ways to learn chemistry is through solving problems. Go to the textbook and work through the Worked Examples for this section, then go to the end of chapter and work the problems there. When you feel confident that you can correctly answer problems like the ones listed below for each objective, check off each of the Learning Objectives for this section above.*

LO 20.1: Worked Example 20.1; Problems 20.30, 20.34, 20.36

Section 20.3 Nuclear Stability

PRE-LECTURE:

Learning Objectives: *These are the skills you should be proficient in after reading this section and practicing the associated problems. Check off the boxes below once you have gone through this section, answered the associated questions in your textbook, and feel confident you have mastered the skill.*

☐ **LO 20.2** Predict the type of radioactive decay for a given isotope.
☐ **LO 20.3** Write a balanced nuclear reaction for a radioactive decay series.

Definitions: *To learn, you must do. This is your workbook. Fill in the definitions from the glossary of the textbook as a first step to understanding the material.*

— **transuranium element**— ______________________________________

____

— **radioactive decay chain**— ______________________________________

____

Summary: *Read through this section and fill in the missing words and equations. Reference the textbook when you are not confident in your answers.*

When the 3600 known isotopes of all the elements are plotted on a graph with the number of neutrons on the *y*-axis and the number of protons on the *x*-axis, we get something called the *band of* ________________.

A few generalizations can be made about this band of nuclear stability:

- Every element in the periodic table has at least one radioactive isotope.
- Hydrogen is the only element whose most abundant isotope (hydrogen-1) contains more protons (1) than neutrons (0).

- The ratio of neutrons to protons gradually increases, giving a curved appearance to the band of stability.
- All isotopes heavier than bismuth-209 are radioactive, even though they may decay slowly and be stable enough to occur naturally.

Nuclei at the top of the band of stability tend to undergo ___________ decay.
Nuclei at the bottom of the band of stability tend to undergo ___________ decay.

It turns out that that many radioactive nuclei cannot reach a stable nucleus with just one decay reaction; a series of nuclear decays, called ________________, will occur until an unstable nucleus becomes stable.

POST-LECTURE:

Work Problems: *One of the best ways to learn chemistry is through solving problems. Go to the textbook and work through the Worked Examples for this section, then go to the end of chapter and work the problems there. When you feel confident that you can correctly answer problems like the ones listed below for each objective, check off each of the Learning Objectives for this section above.*

LO 20.2: Problems 20.40, 20.42
LO 20.3: Problems 20.44, 20.46, 20.48

Section 20.4 Radioactive Decay Rates

PRE-LECTURE:

Learning Objectives: *These are the skills you should be proficient in after reading this section and practicing the associated problems. Check off the boxes below once you have gone through this section, answered the associated questions in your textbook, and feel confident you have mastered the skill.*

☐ **LO 20.4** Relate half-life and decay constant.
☐ **LO 20.5** Calculate the amount of radioactive isotope remaining after a given amount of time.
☐ **LO 20.6** Relate decay rates to decay constant, half-life, or amount remaining.

Definitions: *To learn, you must do. This is your workbook. Fill in the definitions from the glossary of the textbook as a first step to understanding the material.*

— **decay constant—** ____

____

Review: *Before going through the section make sure you are familiar with these concepts:*

- Section 14.5 first-order process
- Section 14.5 integrated rate law

Summary: *Read through this section and fill in the missing words and equations. Reference the textbook when you are not confident in your answers.*

The rate at which a radioactive nucleus decays is characterized by the time it takes for half of the sample to disappear or its ___________. The decay of radioactive nuclei is a first-order process and as such each half-life is the same length of time. With each half-life, the number of remaining radioactive nuclei is divided by __________.

Radioactive decay is a first-order process in which the number of radioactive nuclei in a sample can be used in lieu of the concentration of a reactant. The rate constant, k, is here called the **decay constant.**

Decay rate constant =

The integrated rate law for radioactive decay is:

where N_t is the ______________________________ at a time, t, and N_0 is the ____________________ at time t = 0.

The half-life and decay constant are related by the following equations:

$$t_{1/2} = \frac{\ln 2}{k} \qquad k = \frac{\ln 2}{t_{1/2}}$$

Workbook Example 20.3

PROBLEM:
If the decay constant of technetium-93 is 0.254 h^{-1}, what is its half-life?

SOLUTION:
Substitute the decay constant into the above equation:

$$t_{1/2} = \frac{\ln 2}{0.254\ h^{-1}} = 2.73\ h$$

Workbook Example 20.4

PROBLEM:
If the half-life of technetium-96 is 4.3 days, what is its decay constant?

SOLUTION:
Substitute the half-life into the above equation:

$$k = \frac{\ln 2}{4.3\,days} = 0.161\,days^{-1}$$

Workbook Problem 20.5

The half-life for the decay of C-14 is 5730 years. What is the decay constant? Since this is a first-order reaction, how old is a piece of wood in which 23% of the C-14 has decayed?

Strategy: Substitute into the equation relating half-life to the decay constant; use the integrated first-order rate law.

Step 1: Solve for *k*.

Step 2: Substitute into the integrated rate law and solve for *t* after 23% decay.

POST-LECTURE:

Work Problems: *One of the best ways to learn chemistry is through solving problems. Go to the textbook and work through the Worked Examples for this section, then go to the end of chapter and work the problems there. When you feel confident that you can correctly answer problems like the ones listed below for each objective, check off each of the Learning Objectives for this section above.*

LO 20.4: Worked Example 20.2; Problems 20.50, 20.56
LO 20.5: Worked Example 20.3; Problems 20.52, 20.54, 20.62
LO 20.6: Worked Example 20.4; Problems 20.58, 20.59

Section 20.5 Dating with Radioisotopes

PRE-LECTURE:

Learning Objectives: *These are the skills you should be proficient in after reading this section and practicing the associated problems. Check off the boxes below once you have gone through this section, answered the associated questions in your textbook, and feel confident you have mastered the skill.*

☐ **LO 20.7** Calculate the age of an object using radioisotopic dating.

Summary: *Read through this section and fill in the missing words and equations. Reference the textbook when you are not confident in your answers.*

________________ depends on the slow production of C-14 in the upper atmosphere by neutron bombardment of nitrogen. From this reaction, $^{14}CO_2$ diffuses into the lower atmosphere, where it is taken up by plants and eventually makes it way to animals. While the organisms are living, the ratio of ^{14}C to ^{12}C is constant and is the same as that in the atmosphere. When an organism dies, the ^{14}C decays and is not replenished. By measuring the $^{14}C/^{12}C$ ratio, the approximate age of biological remains can be determined using the half-life of ^{14}C, which is 5715 years.

Variations on this technique use the fact that uranium-238 decays to lead-206 with a half-life of 4.47×10^9 years, making the dating of uranium-containing rocks possible. Potassium-40 decays with a half-life of 1.25×10^9 years to yield argon-40. The age of a rock can be estimated by crushing it and comparing the amount of ^{40}Ar with the amount of ^{40}K.

Workbook Example 20.5

PROBLEM:
A skull hoped to be ancient was found to have a ^{14}C decay rate of 4.7 disintegrations/min. What age is implied by this if living organisms have a decay rate of 15.3 disintegrations/min, and the half-life of ^{14}C is 5715 years?

SOLUTION:
Radioactive decay is a first-order process, so the time since the beginning of the reaction can be determined by calculating using the integrated rate law:

$$\ln\frac{N_t}{N_0} = -0.693\left(\frac{t}{t_{1/2}}\right)$$

Substituting:

$$\ln\frac{4.7}{15.3} = -0.693\left(\frac{t}{5715\ years}\right)$$

$$\frac{\left(\ln\frac{4.7}{15.3}\right)(5715\ years)}{-0.693} = 9734\ years$$

The skull is approximately 10,000 years old.

POST-LECTURE:

Work Problems: *One of the best ways to learn chemistry is through solving problems. Go to the textbook and work through the Worked Examples for this section, then go to the end of chapter and work the problems there. When you feel confident that you can correctly answer problems like the ones listed below for each objective, check off each of the Learning Objectives for this section above.*

LO 20.7: Worked Example 20.5; Problems 20.64, 20.66, 20.70

Section 20.6 Energy Changes during Nuclear Reactions

PRE-LECTURE:

Learning Objectives: *These are the skills you should be proficient in after reading this section and practicing the associated problems. Check off the boxes below once you have gone through this section, answered the associated questions in your textbook, and feel confident you have mastered the skill.*

- ☐ **LO 20.8** Calculate the mass defect and binding energy of a nucleus.
- ☐ **LO 20.9** Calculate the mass and energy change of a chemical reaction.

Definitions: *To learn, you must do. This is your workbook. Fill in the definitions from the glossary of the textbook as a first step to understanding the material.*

— **mass defect—**____

____

— **binding energy—**____

____

Review: *Before going through the section make sure you are familiar with these concepts:*

- Section 9.1 the law of conservation of energy

Summary: *Read through this section and fill in the missing words and equations. Reference the textbook when you are not confident in your answers.*

The ratio of neutrons to protons required for stability varies with the mass of the element. Lighter elements require a 1:1 ratio of neutrons to protons, but ________ neutrons are needed for heavier elements.

The more massive an atom, the more neutrons are needed to minimize the repulsions of the protons: neutrons serve as nuclear "glue."

The energy released when a nucleus is created cannot be measured, because the activation energy required to force the particles together is so high that temperatures rivaling those in the interior of the Sun are needed.

Although the energy changes cannot be measured, they can be calculated using Einstein's famous equation:

$\Delta E =$

Using the known masses of neutrons and protons, it is possible to calculate the **mass defect** of a nucleus. The mass of a nucleus is always less than the mass of the individual particles. This lost mass is a direct measure of the **binding energy** holding the nucleus together. The ________ the binding energy, the more stable the nucleus.

The energy changes involved in nuclear reactions are millions of times greater than those involved in typical processes.

To make it easier to compare the stability of different nuclei, energies are usually expressed per **nucleon** in mega-electron volts, MeV. To covert mega-electron volts to joules, the conversion factor is 1 $MeV = 1.6 \times 10^{-13}$ J.

If binding energy is plotted against atomic number, a stability peak is found at Fe-56. Nuclear stability is increased when lighter nuclei fuse together or heavier nuclei break apart.

The result of these calculations is the discovery that mass and energy are not independently conserved, but the combination of the two is conserved. For non-nuclear reactions, the effect is too small to be measured and can thus be disregarded.

Workbook Example 20.6

PROBLEM:
Determine the mass defect and binding energy for chlorine-35 in both J/mol and MeV/nucleon. The mass of ^{35}Cl is 34.968 85 u.

SOLUTION:
First, calculate the total mass of the nucleons and the mass defect:

Mass of 17 protons = (17)(1.007 28 u)	= 17.123 76 u
Mass of 18 neutrons = (18)(1.008 67 u)	= 18.156 06 u
Mass of 17 protons and 18 neutrons	= 35.279 82 u

Mass of chlorine-35 atom	= 34.968 85 u
Mass of electrons = (17) (5.486×10^{-4})	= 0.009 33 u
Mass of chlorine-35 *nucleus*	= 34.978 18 u

Mass defect: 35.279 82 u − 34. 978 18 u = 0.301 64 u

Convert this mass to grams:

$$0.30164\, u \times \frac{1.6605 \times 10^{-24}}{1\, u} = 5.00873 \times 10^{-25}\ g$$

Use the Einstein equation to convert the mass defect to binding energy:

$$\Delta E = 5.00873 \times 10^{-25}\ g \times \frac{1\ kg}{10^3\ g} \times \left(\frac{3.00 \times 10^8\ m}{s}\right)^2 = 4.50786 \times 10^{-11}\ J$$

Convert to J/mol:

$$\frac{4.50786 \times 10^{-11}\ J}{1\, nucleus} \times \frac{6.022 \times 10^{23}\ nuclei}{1\ mol} = 2.71463 \times 10^{13}\ J/mol$$

Convert to MeV/nucleon:

$$\frac{4.50786\times10^{-11}\ J}{1\ nucleus}\times\frac{1\ MeV}{1.60\times10^{-13}\ J}\times\frac{1\ nucleus}{35\ nucleons}=7.95\ {}^{MeV}\!/_{nucleon}$$

Workbook Example 20.7

PROBLEM:
Calculate the mass change in g/mol when N atoms combine to form N_2 molecules.

$$2\,N \rightarrow N_2 \qquad\qquad \Delta E = -945.4\ kJ/mol$$

SOLUTION:
With a known energy change, rearrange the Einstein equation to solve for Δm:

$$\Delta m = \frac{\Delta E}{c^2}$$

$$\Delta m = \frac{945.4\ kJ/mol}{\left(3.00\times10^{8}\ m/s\right)^2}\times\frac{10^3\ J}{kJ}=\frac{1.05044\times10^{-11}\ {}^{J}\!/_{mol}}{{}^{m^2}\!/_{s^2}}$$

$$\frac{1.05044\times10^{-11}\ {}^{J}\!/_{mol}}{{}^{m^2}\!/_{s^2}}\times\frac{{}^{kg\cdot m^2}\!/_{s^2}}{J}=\frac{1.05044\times10^{-11}\ kg}{mol}$$

$$\frac{1.05044\times10^{-11}\ kg}{mol}\times\frac{10^3\ g}{kg}=1.05044\times10^{-8}\ {}^{g}\!/_{mol}$$

This is an amount far too small to be measured.

Workbook Problem 20.6

Calculate the binding energy of a Xe-130 nucleus in Mev/nucleon. The mass of a ^{130}Xe nucleus is 129.903 *u*.

Strategy: Calculate the mass defect and convert to energy using the Einstein equation.

Step 1: Determine the number of nucleons and their mass, subtract the mass of the electrons from the atomic mass, and calculate the mass defect in *u*.

Step 2: Convert the mass defect to grams.

Step 3: Determine ΔE and convert to *MeV/nucleon.*

POST-LECTURE:

Work Problems: *One of the best ways to learn chemistry is through solving problems. Go to the textbook and work through the Worked Examples for this section, then go to the end of chapter and work the problems there. When you feel confident that you can correctly answer problems like the ones listed below for each objective, check off each of the Learning Objectives for this section above.*

LO 20.8: Worked Example 20.6; Problems 20.76, 20.78, 20.86
LO 20.9: Worked Example 20.7; Problems 20.84, 20.85

Section 20.7 Nuclear Fission and Fusion

PRE-LECTURE:

Learning Objectives: *These are the skills you should be proficient in after reading this section and practicing the associated problems. Check off the boxes below once you have gone through this section, answered the associated questions in your textbook, and feel confident you have mastered the skill.*

- ☐ **LO 20.10** Identify key aspects of fission and fusion reactions and their role in nuclear power and weapons.
- ☐ **LO 20.11** Calculate the energy change associated with a fission or fusion reaction.

Definitions: *To learn, you must do. This is your workbook. Fill in the definitions from the glossary of the textbook as a first step to understanding the material.*

— **fusion—** ____

____

— **fission—** ____

____

— **chain reaction—** ____

____

— **critical mass—** ____

____

Summary: *Read through this section and fill in the missing words and equations. Reference the textbook when you are not confident in your answers.*

Because the maximum stability is found in the middle of the periodic table, there is stability to be gained by the ________ of light nuclei and the ________ of heavy nuclei.

NUCLEAR FISSION

- Certain large nuclei shatter into smaller pieces when bombarded with ________.
- When neutrons are released as part of a nuclear fission in addition to those used to begin the reaction, a ________________ can occur in which the reaction continues even after the external supply of neutrons is cut off.

- In a small sample, many of the neutrons will escape without initiating further fission, but if there is a sufficient amount of reaction—known as a ________—it will become a self-perpetuating chain reaction.

Workbook Example 20.8

PROBLEM:
How much energy is released in both J and kJ/mol for the neutron-induced fission of U-235 (atomic mass = 235.043 92 *u*) to form Te-140 (atomic mass = 139.938 85 *u*) and Kr-93 (atomic mass = 92.931 27 *u*)? The mass of a neutron is 1.008 67 *u*.

SOLUTION:
To calculate the energy difference, the mass difference must be calculated and converted into energy.

Write and balance the equation: the overall masses need to balance!

$$^{235}_{92}U + ^{1}_{0}n \rightarrow ^{140}_{52}Te + ^{93}_{36}Kr + 2\,^{1}_{0}n$$

mass of reactants = 235.043 92 *u* + 1.008 67 *u* = 236.052 59 *u*
mass of products = 139.938 85 *u* + 92.931 27 *u* + 2 (1.008 67 *u*) = 234.887 46 *u*

Calculate the mass difference:

$$\text{reactants} - \text{products} = 1.16513\ u$$

Convert mass to kg:

$$1.16513\,u \times \frac{1.6605 \times 10^{-24}\ g}{1\,u} \times \frac{1\,kg}{1000\ g} = 1.93470 \times 10^{-27}\ kg$$

Use the Einstein equation to convert the mass difference to energy:

$$\Delta E = \left(1.93470 \times 10^{-27}\ kg\right)\left(\frac{3.00 \times 10^{8}\ m}{1\,s}\right)^{2} = 1.74123 \times 10^{-10}\ J$$

Convert to J/mol:

$$\frac{1.74123 \times 10^{-10}\ J}{1\,nucleus} \times \frac{6.022 \times 10^{23}\ nuclei}{1\,mol} = 1.049 \times 10^{14}\ \frac{J}{mol}$$

NUCLEAR REACTORS

Controlled fission reactions can be used to generate ________. Uranium fuel is placed in a pressurized containment vessel surrounded by water and *control rods* made of boron and cadmium are added. The water slows the escape of neutrons, and the control rods absorb neutrons to regulate the speed of the fission reaction: they are raised to speed the reaction, allowing more neutrons into the fuel, and lowered to slow the reaction. The energy generated is used to heat circulating coolant, which produces steam that drives a turbine to produce electricity.

Naturally occurring uranium is a mixture of 99.3% nonfissionable ^{238}U and 0.07% fissionable ^{235}U. Reactor fuel is compressed UO_2 pellets that have been enriched to 3% ^{235}U and encased in zirconium.

The amount and concentration of nuclear fuel in a reactor is not enough to sustain a nuclear explosion: the worst-case scenario is when uncontrolled fission melts the reactor and the containment vessel, releasing radioactivity into the environment. Newer reactors have *passive safety* designs that automatically slow runaway reactors.

Thirty countries now use nuclear energy to supply some of their electricity needs. The primary issue holding back further use of nuclear energy is safe disposal of the wastes: ^{90}Sr takes 600 years to decay to safe levels, and ^{239}Pu takes 20,000 years to decay to safe levels.

NUCLEAR FUSION

Very light nuclei also release enormous amounts of energy when they undergo ________: it is hydrogen fusion reactions that power the Sun and other stars. Hydrogen is cheap and plentiful, and fusion products are nonradioactive and nonpolluting, but with a temperature of approximately 40,000,000 K required to initiate the process, the technical barriers to making this a viable source of power are formidable, to say the least.

Workbook Problem 20.7

In Oklo, Gabon, a natural nuclear reactor operated in uranium ore for hundreds of thousands of years. Part of the evidence for this is the presence of ^{136}Xe and ^{99}Ru in the surrounding rocks. Write and balance an equation for the fission of ^{235}U resulting in these isotopes, and calculate the energy generated by this reaction in J/mol. The mass of ^{235}U is 235.043 92 *u*, the mass of ^{136}Xe is 135.907 219 5 *u*, and the mass of ^{99}Ru is 98.905 939 *u*. One neutron is needed to induce the fission, and one neutron is produced by the fission.

Strategy: Balance the equation, calculate the mass defect, and convert to energy using the Einstein equation.

Step 1: Write and balance the equation.

Step 2: Calculate the mass defect in *u* and convert to kilograms.

Step 3: Determine ΔE and convert to J/mol.

Workbook Problem 20.8

Hydrogen bombs utilize the reaction between deuterium (^{2}H – 2.014 10 *u*) and tritium (^{3}H – 3.016 05 *u*) that produces ^{4}He (4.002 60 *u*) and a neutron (1.008 67 *u*).
Calculate the energy generated by the reaction per mol of ^{4}He produced.

Strategy: Calculate the mass defect and convert to energy using the Einstein equation.

Step 1: Write and balance the equation.

Step 2: Calculate the mass defect in *u* and convert to kilograms.

Step 3: Determine ΔE and convert to J/mol.

POST-LECTURE:

Work Problems: *One of the best ways to learn chemistry is through solving problems. Go to the textbook and work through the Worked Examples for this section, then go to the end of chapter and work the problems there. When you feel confident that you can correctly answer problems like the ones listed below for each objective, check off each of the Learning Objectives for this section above.*

LO 20.10: Problems 20.88, 20.93
LO 10.11: Worked Example 20.8; Problems 20.94–20.96

Section 20.8 Nuclear Transmutation

PRE-LECTURE:

Learning Objectives: *These are the skills you should be proficient in after reading this section and practicing the associated problems. Check off the boxes below once you have gone through*

this section, answered the associated questions in your textbook, and feel confident you have mastered the skill.

☐ **LO 20.12** Write balanced equations for nuclear transmutation reactions.

Definitions: *To learn, you must do. This is your workbook. Fill in the definitions from the glossary of the textbook as a first step to understanding the material.*

— **nuclear transmutation—** ____

____

Summary: *Read through this section and fill in the missing words and equations. Reference the textbook when you are not confident in your answers.*

Of the 3600 known isotopes, only ________ occur naturally. The remainder have been made by ________________, the change of one element into another. This is often accomplished by bombardment of an atom with a high-energy particle such as a proton, a neutron, or an α-particle. In the collision, an unstable nucleus is momentarily generated, and a new element can be produced.

All of the transuranium elements—those with atomic numbers above 92—have been produced using transmutation.

The reactions and subsequent decays can be quite complex: because the elements can be produced in a variety of ways, they can also decay in a variety of ways.

Workbook Example 20.9

PROBLEM:
Ernest Rutherford was the first to accomplish nuclear transmutation, but the husband and wife team of Frederic and Irene Joliot-Curie were not far behind; they bombarded Al-27 with α-particles to produce P-30. Write and balance an equation for this reaction. What particle is also produced in this reaction?

SOLUTION:
According to the periodic table, Al has an atomic number of 13, and phosphorus has an atomic number of 15. An α-particle is a helium nucleus: it has a mass of 4, and an atomic number of 2.

$$^{27}_{13}\text{Al} + ^{4}_{2}\text{He} \rightarrow ^{30}_{15}\text{P} + \_\_\_\_\_$$

The number of protons adds up, but the number of neutrons does not: a neutron must also be produced by the reaction:

$$^{27}_{13}\text{Al} + ^{4}_{2}\text{He} \rightarrow ^{30}_{15}\text{P} + ^{1}_{0}\text{n}$$

Workbook Problem 20.9

If U-238 is bombarded with C-12, the result is a new element and six neutrons. Write and balance the equation and determine the identity of the element produced using the periodic table.

Strategy: Determine the overall mass, subtract out the neutrons, and determine the identity and mass of the element produced.

Step 1: Determine the number of protons in the new element.

Step 2: Determine the mass number of the new element.

Step 3: Determine the identity of the new element using the number of protons.

Step 4: Write the balanced nuclear equation.

POST-LECTURE:

Work Problems: *One of the best ways to learn chemistry is through solving problems. Go to the textbook and work through the Worked Examples for this section, then go to the end of chapter and work the problems there. When you feel confident that you can correctly answer problems like the ones listed below for each objective, check off each of the Learning Objectives for this section above.*

LO 20.12: Worked Example 20.9; Problems 20.98, 20.100, 20.104

Section 20.9 Detecting and Measuring Radioactivity

PRE-LECTURE:

Learning Objectives: *These are the skills you should be proficient in after reading this section and practicing the associated problems. Check off the boxes below once you have gone through this section, answered the associated questions in your textbook, and feel confident you have mastered the skill.*

☐ **LO 20.13** Calculate the radiation intensity in a radioactive sample and convert between units of radiation intensity.

Summary: *Read through this section and fill in the missing words and equations. Reference the textbook when you are not confident in your answers.*

Radiation is invisible, but it can be measured and detected by instruments, and its effect on biological systems can be readily observed.

Radiation intensity is measured in a number of different ways. These are summarized below (from Table 20.2 in the textbook).

Unit	Quantity Measured	Description
Becquerel (Bq)		1 Bq = 1 disintegration/s
Curie (Ci)		1 Ci = 3.7×10^{10} disintegration/s
Gray (Gy)		1 Gy = 1 J/kg tissue
Rad		1 rad = 0.01 Gy
Sievert (Sv)		1 Sv = 1 J/kg
Rem		1 rem = 0.01 Sv

The seriousness of radiation exposures varies with the type of radiation, the energy of the radiation, the length of exposure, and whether the source is external or internal.

Coming from an external source, X-rays and γ-rays are the ________ harmful to the human body because they penetrate clothing and skin. They move at the speed of light and have 1000 times the penetrating power of α particles. ________________are required to stop them.

Coming from an internal source, α particles and β particles are more dangerous. Alpha particles move at one-tenth the speed of light and can be stopped by ______________________________.
β particles are lighter and faster—they move at nine-tenths the speed of light and require ________________________________ to stop them from penetrating the body.

The biological effects of radiation exposure can be dire, but the usual exposure of an average person is only about 120 mrem annually. Transient health effects begin to be seen between 25 and 100 rem, where white cell counts begin to be temporarily decreased. The lethal dose of radiation is 600 rem—5000 times higher than most people's annual exposure.

About 70% of yearly exposures are natural, from rocks and cosmic rays, with the remaining 30% coming from medical procedures such as X-rays.

POST-LECTURE:

Work Problems: *One of the best ways to learn chemistry is through solving problems. Go to the textbook and work through the Worked Examples for this section, then go to the end of chapter and work the problems there. When you feel confident that you can correctly answer problems like the ones listed below for each objective, check off each of the Learning Objectives for this section above.*

LO 20.13: Problems 20.108, 20.110, 20.112

Self-Test

This section is intended to test your knowledge of the material covered in this chapter. Think through these problems and make certain you understand what the problem is about—ask yourself if the answer makes sense. Many of these questions are linked to the chapter learning goals. Therefore, successful completion of these problems indicates that you have achieved the learning goals for this chapter. You will receive the greatest benefit from this section if you use it as a mock exam. Do not use your notes, book, Internet, or friends when taking the test. There is no grade and you only shortchange yourself by getting help with this activity. As you take the test, mark the problems with your level of confidence with the answer.

★ *Use a star for problems you know you got right.*
? *Use a question mark for those you think you got right.*
x *Use an x for those you are guessing on.*

If you get a question right that you marked with a star, you can be confident that you know the associated LO.

If you get a question right that you marked with a question mark, go back and work a few more problems and test yourself again tomorrow.

If you get a question wrong or mark it with an x, go back and read the section again, work a few problems today, a few tomorrow, and a few the next day before testing yourself again.

This will help you to discover which topics you have learned thoroughly and which topics you need to study in more detail.

True–False

1. Radioactive decay is a second-order process.
2. Lighter elements require more neutrons for stability, but heavier elements have a 1:1 ratio of protons and neutrons.
3. When nucleons come together to form a nucleus, a small amount of mass is gained.
4. When elements undergo neutron-induced fission, they tend to break into pieces of roughly equivalent size.
5. The control rods in a reactor absorb excess neutrons to slow the reaction.
6. Nuclear fusion has no toxic by-products but is unlikely to be a practical source of power in the near future.
7. Given their positions in the periodic table, a nuclear transmutation reaction is more likely to turn gold into lead than lead into gold.
8. Alpha particles are most damaging from an internal source.
9. The curie measures only disintegrations per second, and it says nothing about tissue damage.
10. Radioisotopes can be used for disease treatment, imaging, and in vivo studies.

Multiple Choice

11. All of the following are nuclear processes *except*
    a. alpha emission
    b. electron capture
    c. beta modification
    d. transmutation

12. Which of the following statements about nuclear stability is true?
    a. Heavy nuclei are most stable with a 1:1 ratio of protons and neutrons.
    b. Nuclei tend to be more stable when they have less neutrons than protons.
    c. Nuclei tend to be more stable when they have more neutrons than protons.
    d. Nuclei tend to be more stable when they have no neutrons.

13. The mass defect in nuclei is the result of
    a. inaccurate masses of subatomic particles
    b. the conversion of mass to energy to hold the nucleus together
    c. the law of conservation of mass
    d. nuclear transmutation

14. The process of nuclear fission
    a. can self-perpetuate as a chain reaction
    b. generates very little energy
    c. generates only ^{4}He as a product
    d. b and c

15. In a reactor, nuclear fission is used to generate power by
    a. generating electrons
    b. converting water to steam, which turns a turbine
    c. driving a piston with repeated small explosions
    d. none of the above

16. Which of the following could effectively shield you from γ radiation?
    a. paper
    b. thick clothing
    c. a lead block several inches thick
    d. none of these

17. *Nuclear transmutation* is a process that can be used to
    a. generate new elements and isotopes
    b. generate electricity
    c. diagnose disease
    d. date ancient objects

18. Which of the following provides information about tissue damage?
    a. becquerel
    b. rem
    c. sievert
    d. b and c

19. Which of the following correctly ranks these types of emissions by speed, fastest to slowest?
    a. alpha particle, gamma ray, beta particle
    b. beta particle, alpha particle, X-ray
    c. gamma ray, beta particle, alpha particle
    d. X-ray, alpha particle, gamma ray

20. Radiocarbon dating can only be used to determine the ages of
    a. meteorites
    b. formerly living objects
    c. uranium-containing rocks
    d. all of the above

Fill in the Blanks

21. Neutrons appear to function as a nuclear ________ that overcomes the ___________ to ____________ repulsions that might otherwise cause a(n) _______________ to fly apart.

22. The energy to hold the nucleus together cannot be directly measured, but can be calculated using the _________ _____________ and the equation ________________.

23. Binding energy per nucleon reaches a(n) ________________ in the middle of the periodic table. Therefore, small nuclei can increase their stability through ______________, and large nuclei can increase their stability through ________________ or radioactive _____________.

24. In nuclear ______________________, an atom is bombarded with any of a number of high-energy_______________. In the collision, an unstable _______________ is momentarily formed, and a new _________________ is produced.

25. The effects of radiation on a biological system vary according to a large number of variables. The most damage comes from large, slow particles such as _______________ and ________________ particles as a(n) ________________ source. Also damaging are the fastest moving radiation, ________________ rays and _________________ rays, which cause a lot of damage even as ________________ sources. Large, slow particles are _______________ by skin. The fast-moving rays require several inches of _____________ to stop them.

26. Radiopharmaceuticals are used in a number of ways. They can be used in ______________ procedures in which they are administered and monitored, in therapeutic procedures in which they are used to target __________________ tissue (_______________ radiation therapy is always more specific and targeted than _________________ radiation therapy), and in ____________, in which the isotope's location or absence is used to assist in diagnosis.

Matching

27. binding energy
28. chain reaction
29. critical mass
30. fission
31. fusion

a. a reaction in which the nucleus of an element splits into two nearly equal pieces

b. the loss of mass when nucleons come together to form a nucleus

c. the joining of nuclei in a reaction

d. the amount of material necessary for a nuclear reaction to be self-sustaining

e. the energy that holds nucleons together

32. mass defect — f. change of one element into another

33. nuclear transmutation — g. a self-sustaining reaction whose product initiates further reaction

Problems

34. Calculate the binding energy of europium-130 (mass = 129.963 57 *u*), europium-153 (mass = 152.921 23 *u*), and europium-167 (mass = 166.953 21 *u*) in MeV/nucleon. From these data, which is the most stable isotope? Which is the least stable?

35. Calculate the change in mass for the following reaction in g/mol:

$$2\,O \rightarrow O_2 \qquad \Delta E = -498.4 \text{ kJ/mol}$$

36. How much energy in J/mol is released in the following fission reaction?

$$^{235}U + {}^{1}n \rightarrow {}^{92}Kr + {}^{141}Ba + 3\,{}^{1}n$$

The mass of ^{235}U is 235.0439 *u*, the mass of ^{92}Kr is 91.9262 *u*, the mass of ^{141}Ba is 140.9144 *u*, and the mass of a neutron is 1.008 67 *u*.

37. How much energy is released in the following fusion reaction?

$$^{3}H + {}^{3}H \rightarrow {}^{4}He + 2\,{}^{1}n$$

The mass of tritium is 3.016 049 *u*, the mass of ^{4}He is 4.002 60 *u*, and the mass of a neutron is 1.008 67 *u*.

38. Write chemical reactions for the following nuclear transmutations:
   a. Be-9 is bombarded with hydrogen to form Li-6 and an alpha particle.
   b. Pu-239 reacts with an alpha particle to produce an unknown element and a neutron.
   c. Pb-208 is collided with another element to form Hs-265 and a neutron.

39. The half-life for zirconium-96 is 6.3×10^{26} s. What is the decay constant for this radioactive element?

40. What is the age of a bone fragment that shows an average of 3.5 disintegrations per minute per gram of carbon? The carbon in living organisms undergoes an average of 15.3 disintegrations per minute per gram, and the half-life of C-14 is 5715 y.

41. The decay of U-235 to Pb-207 takes place with a half-life of 704 million years. If an asteroid that originally contained no lead is found to contain 56 g of Pb-207 and 27 g of U-235, how old is it?

<u>*Challenge Problem*</u>

42. In an experimental form of cancer treatment known as boron neutron capture therapy (BNCT), boron-10 is injected into a cancer patient, where it binds preferentially to tumor cells. A beam of neutrons is then directed at the tumor, and the boron-10 disintegrates into lithium. Only one type of radiation is produced by the reaction, and only one neutron is absorbed in the reaction. Write and balance the equation for this reaction and explain the rationale behind this therapeutic method.

Test Yourself: *Take the Practice Test on page 898 of your textbook. Again, use a testing environment using only those items that your instructor will allow you to have access to during the test and mark each question with a star, question mark, or x to denote your confidence level. Once you have finished you can check off the LO at the beginning of this chapter based on which questions you starred and got correct. Go back and spend more time with the topics from the other learning objectives.*

CHAPTER TWENTY-ONE

Transition Elements and Coordination Chemistry

Learning Objectives

As a result of reading and studying this chapter, you should be able to:

- ☐ **LO 21.1** Write electron configurations for transition metals and predict the number of unpaired electrons in transition metal ions.
- ☐ **LO 21.2** Describe the trends in physical and chemical properties of transition metal elements across a period. Explain the reasons for trends.
- ☐ **LO 21.3** Relate the oxidizing and reducing strength of a transition metal ion to its oxidation state and location in the periodic table.
- ☐ **LO 21.4** Identify Lewis acids, Lewis bases, ligands, and donor atoms in a coordination complex.
- ☐ **LO 21.5** Determine the formula of a coordination complex, identify the oxidation state and coordination number of the metal atom, and draw the structure of the complex.
- ☐ **LO 21.6** Classify a ligand as mono, bi, tri, tetra, or hexadentate based on its chemical structure.
- ☐ **LO 21.7** Write the systematic name for a coordination complex.
- ☐ **LO 21.8** Classify isomers according to the scheme given in Figure 21.11.
- ☐ **LO 21.9** Draw diastereomers for octahedral and square planar complexes.
- ☐ **LO 21.10** Classify a coordination complex as chiral or achiral and draw enantiomers for chiral compounds.
- ☐ **LO 21.11** Relate the color of a metal complex, the wavelength of light it absorbs, and the energy difference between its ground and excited state.
- ☐ **LO 21.12** Draw crystal field energy-level diagrams for octahedral, tetrahedral, and square complexes. Classify ligands as strong- or weak-field and predict (or account for) color and magnetic properties of complexes in terms of crystal field theory.
- ☐ **LO 21.13** Describe the bonding in a metal complex in terms of valence bond theory, including orbital diagrams, the hybrid orbitals used by the metal atom, and the number of unpaired electrons.

Chapter Summary

This chapter examines the properties and chemical behavior of transition metal compounds, with particular emphasis on coordination compounds. Beginning with a review of the electronic configurations of the transition elements and how these configurations affect the properties and oxidation states of the ions formed, the chemistry of chromium, iron, and copper are examined in detail. Coordination chemistry is then explored, including structural properties, isomerization, color, and magnetism of the compounds. Two different theories, valence bond theory and crystal field theory that account for these properties, are then discussed.

The Chapter in Detail

Section 21.1 Electron Configurations

PRE-LECTURE:

Learning Objectives: *These are the skills you should be proficient in after reading this section and practicing the associated problems. Check off the boxes below once you have gone through this section, answered the associated questions in your textbook, and feel confident you have mastered the skill.*

☐ **LO 21.1** Write electron configurations for transition metals and predict the number of unpaired electrons in transition metal ions.

Review: *Before going through the section make sure you are familiar with these concepts:*

- Section 5.10 electron configurations
- Section 5.11 anomalous electron configurations
- Section 6.1 electron configurations of transition metal ions

Summary: *Read through this section and fill in the missing words and equations. Reference the textbook when you are not confident in your answers.*

Note: For convenience in this chapter, the orbitals will be listed by principal quantum number, because the *ns* electrons are lost first when transition metals ionize. However, the filling still proceeds according to Hund's rule, with the *ns* shells filling first, followed by one electron in each of the $(n-1)d$ orbitals, before doubling up the electrons.

There are exceptions to this pattern: ____ and ____ are both able to gain stability by varying from this rule. Cr places one electron in each of the orbitals for a completely half-filled configuration: $4s^1\ 3d^5$. Copper pulls one electron from the 4*s* orbital to gain a fully filled 3*d* orbital and leaves a half-filled *s* orbital: $4s^1\ 3d^{10}$.

Exceptions from the expected orbital-filling pattern result in completely half-filled subshells or completely filled shells. The $(n-1)d$ and *ns* orbitals have very similar energies, allowing electrons to move between them if it is energetically favorable to do so.

Two half-filled subshells minimize electron–electron repulsions, as does a completely filled *d* subshell with a half-filled *s* subshell.

When electrons are lost, the effective nuclear charge ___________. The *d* orbitals experience a steeper drop in energy as the effective nuclear charge increases, making them quickly ___________ in energy than the *s* orbital. As a result, the *s* orbital empties first.

The designation of transition metals as groups 1B–8B was developed because these groups have analogous electron structures to the elements in groups 1A–8A. Copper, in group 1B, has Cu^+ as its most common ion, while Zn, in group 2B, will most commonly form a Zn^{2+} ion, like the main-group metals in group 2A. However, these designations are not as straightforward as those of the main-group elements, and they should be used with care.

Workbook Example 21.1

PROBLEM:
Write the electron configuration of Cu^{3+}. Would you expect this to be a stable species?

SOLUTION:
Writing the electron configuration for neutral copper gives

$$[Ar]\ 3d^{10}\ 4s^1$$

Removing three electrons to make Cu^{3+} gives

$$[Ar]\ 3d^8$$

because the *s* orbital will empty first. This will not be a stable species, as it has a high effective nuclear charge and two half-filled *d* orbitals. The most stable ion for copper will be Cu^+, with electron configuration $[Ar]\ 3d^{10}$.

Workbook Problem 21.1

Write the electron configurations for the following species:

Ti   Mn in MnO_4^- Ni^{2+}

Strategy: Write the expected electron configuration for the neutral element, determine the charge, and remove electrons accordingly.

Step 1: Write the electron configuration of the neutral atom.

Step 2: Determine the oxidation state of the metal.

Step 3: Remove the necessary electrons.

POST-LECTURE:

Work Problems: *One of the best ways to learn chemistry is through solving problems. Go to the textbook and work through the Worked Examples for this section, then go to the end of chapter and work the problems there. When you feel confident that you can correctly answer problems like the ones listed below for each objective, check off each of the Learning Objectives for this section above.*

LO 21.1: Worked Example 21.1; Problems 21.16, 21.17, 21.37, 21.38

Section 21.2 Properties of Transition Elements

PRE-LECTURE:

Learning Objectives: *These are the skills you should be proficient in after reading this section and practicing the associated problems. Check off the boxes below once you have gone through this section, answered the associated questions in your textbook, and feel confident you have mastered the skill.*

☐ **LO 21.2** Describe the trends in physical and chemical properties of transition metal elements across a period. Explain the reasons for trends.

Definitions: *To learn, you must do. This is your workbook. Fill in the definitions from the glossary of the textbook as a first step to understanding the material.*

— **lanthanide contraction—** ____

____

Summary: *Read through this section and fill in the missing words and equations. Reference the textbook when you are not confident in your answers.*

METALLIC PROPERTIES

As metals, all these elements are malleable, ductile, lustrous, and ___________ conductors of heat and electricity. The sharing of *d*, as well as *s*, electrons gives rise to stronger metallic bonding in these elements than is seen in the group 1A and 2A metals, which makes those metals harder and ___________ dense, with ___________melting and boiling points.

Melting points ___________ as the number of unpaired *d* electrons available for metallic bonding increases and then ___________ as the *d* electrons pair up and become less available for bonding, meaning that the melting points reach a maximum in the middle of each series.

ATOMIC RADII AND DENSITIES

Atomic radii and densities demonstrate a similar pattern: as the unpaired *d* electrons increase to a half-filled orbital, the atomic radii ___________ and densities ___________. As the electrons begin to pair up, the electron repulsions ___________ and the effective nuclear charge ___________, so the atomic radii will ___________ and the density will ___________.

The effects on atomic radii are more subtle in the transition metals than they are in the main-group elements. The similar atomic radii among the transition metals account for their ability to blend with one another into an alloy.

Across the *f*-block lanthanide elements, the increase in effective nuclear charge is almost exactly balanced by the increase in size expected from adding an entire quantum shell, resulting in smaller-than-expected atomic radii. Because of this **lanthanide contraction,** the elements in the third transition series have almost the same radii as the elements in the second transition series. This means that the third series, which includes Os, the most dense metal, are unusually dense.

IONIZATION ENERGIES AND OXIDATION POTENTIALS

The ionization energies of transition metals generally ___________ from left to right across a series, due to an increase in the effective nuclear charge and the corresponding decrease in atomic radius.

The $E°$ for the **oxidation** potential of the first transition series is positive for every metal except Cu, meaning that these metals are oxidized ___________ than H_2 gas is oxidized to H^+. These metals can be oxidized by simple acids that lack an oxidizing anion, but the oxidation of Cu(*s*) requires a stronger oxidizing agent such as HNO_3.

The general trend for $E°$ is the same as that for ionization energies. The easier a metal is to oxidize, the easier it is to ionize.

POST-LECTURE:

Work Problems: *One of the best ways to learn chemistry is through solving problems. Go to the textbook and work through the Worked Examples for this section, then go to the end of chapter and work the problems there. When you feel confident that you can correctly answer problems like the ones listed below for each objective, check off each of the Learning Objectives for this section above.*

LO 21.2: Problems 21.28, 21.40, 21.42, 21.45, 21.48

Section 21.3 Oxidation States of Transition Elements

PRE-LECTURE:

Learning Objectives: *These are the skills you should be proficient in after reading this section and practicing the associated problems. Check off the boxes below once you have gone through this section, answered the associated questions in your textbook, and feel confident you have mastered the skill.*

☐ **LO 21.3** Relate the oxidizing and reducing strength of a transition metal ion to its oxidation state and location in the periodic table.

Summary: *Read through this section and fill in the missing words and equations. Reference the textbook when you are not confident in your answers.*

Transition metals exhibit a ___________ of oxidation states, which can be ___________ than their group number. All of the first series except scandium will form +2 cations, corresponding to the loss of the ____ shell electrons.

The first series can also lose 3*d* electrons to form ions with greater charges. The increased energy required to remove additional electrons is balanced by the ___________ $\Delta G°$ of hydration of the more highly charged cations. The most highly charged species are found in combination with the most electronegative elements, typically O.

For the group 3B–7B metals, the group number is the highest possible oxidation state, corresponding to the loss or sharing of ___________ valence *s* and *d* electrons. This is not true for the 8B elements, however, as the loss of all the valence electrons is energetically prohibitive.

Transition metal ions in a high oxidation state are good __________ agents, while early transition metal ions in a low oxidation state are good __________ agents, but divalent ions of the later metals are poor reducing agents because of the larger effective nuclear charge.

POST-LECTURE:

Work Problems: *One of the best ways to learn chemistry is through solving problems. Go to the textbook and work through the Worked Examples for this section, then go to the end of chapter and work the problems there. When you feel confident that you can correctly answer problems like the ones listed below for each objective, check off each of the Learning Objectives for this section above.*

LO 21.3: Problems 21.56, 21.58, 21.60

Section 21.4 Coordination Compounds

PRE-LECTURE:

Learning Objectives: *These are the skills you should be proficient in after reading this section and practicing the associated problems. Check off the boxes below once you have gone through this section, answered the associated questions in your textbook, and feel confident you have mastered the skill.*

- ☐ **LO 21.4** Identify Lewis acids, Lewis bases, ligands, and donor atoms in a coordination complex.
- ☐ **LO 21.5** Determine the formula of a coordination complex, identify the oxidation state and coordination number of the metal atom, and draw the structure of the complex.

Definitions: *To learn, you must do. This is your workbook. Fill in the definitions from the glossary of the textbook as a first step to understanding the material.*

— **coordination compound—** ____

____

— **ligand—** ____

____

— **ligand donor atom—**____

____

— **metal complex—** ____

____

— **coordination number—** ____

____

Review: *Before going through the section make sure you are familiar with these concepts:*

- Section 16.5 Lewis acids and bases

Summary: *Read through this section and fill in the missing words and equations. Reference the textbook when you are not confident in your answers.*

A **coordination compound** is a compound in which a central metal ion is attached to a group of surrounding molecules or ions by coordinate covalent bonds. The molecules or ions surrounding the central metal ion are called __________, while the atoms that are directly attached to the metal ion are the **ligand donor atoms.** Formation of these compounds is a Lewis acid–base interaction: the electron __________ (Lewis bases) are the ligands, which always have lone pairs, and the electron __________ (acting as a Lewis acid) is the central metal ion.

Some coordination compounds are salts containing a complex cation or anion along with enough ions of opposite charge to give a compound that is electrically neutral overall. The complex ion is enclosed in brackets when it is written to show that the other ions are simply balancing charges, while the complex ion is a discrete structural unit. The term **metal complex** refers to both neutral coordination compounds and the salts of ionic coordination compounds.

The number of ligand donor atoms surrounding the metal ion is the __________ **number.** The most common are 4 and 6, but others are also well known. The coordination number depends on the __________, __________, and ____________________ of the central atom, as well as the __________ and __________ of the ligands.

The characteristic shape of a metal complex depends on the metal ion's coordination number. A coordination number of 2 indicates a __________, while a coordination number of 4 can be either __________ or __________. A coordination number of 6 will be __________.

The charge on a metal complex is equal to the oxidation state of the __________ plus the sum of the charges on the __________. If two are known, the third can be found.

Workbook Example 21.2

PROBLEM:
Silver forms a neutral complex with two ammonia molecules and a nitrate ion. What is the oxidation state of the metal, and what is the formula of the complex?

SOLUTION:
The charge on the complex is the sum of all the charges. The ammonia molecules are neutral, and nitrate has a charge of −1. For the complex to be neutral, the oxidation state of silver must be +1.

The formula of the complex is $[Ag(NH_3)_2]NO_3$.

Workbook Problem 21.2

Determine the oxidation state and coordination number of the metal in the following complexes:

$AgCl_2^-$ $[MnCl_4]^{2-}$ $[Fe(CN)_6]^{4-}$ $[Co(NH_3)_4Br_2]Br$

Strategy:	Because we are given the charge and ligands, it is possible to determine the oxidation state of the metals. The coordination number is the number of ligand attachments.
Step 1:	Determine the oxidation state of the metals.
Step 2:	Determine the coordination numbers.

POST-LECTURE:

Work Problems: *One of the best ways to learn chemistry is through solving problems. Go to the textbook and work through the Worked Examples for this section, then go to the end of chapter and work the problems there. When you feel confident that you can correctly answer problems like the ones listed below for each objective, check off each of the Learning Objectives for this section above.*

LO 21.4: Problems 21.74, 21.75
LO 21.5: Worked Example 21.2; Problems 21.62, 21.66, 21.68, 21.70

Section 21.5 Ligands

PRE-LECTURE:

Learning Objectives: *These are the skills you should be proficient in after reading this section and practicing the associated problems. Check off the boxes below once you have gone through this section, answered the associated questions in your textbook, and feel confident you have mastered the skill.*

☐ **LO 21.6** Classify a ligand as mono, bi, tri, tetra, or hexadentate based on its chemical structure.

Definitions: *To learn, you must do. This is your workbook. Fill in the definitions from the glossary of the textbook as a first step to understanding the material.*

— **monodentate ligand—**____

____

— **polydentate ligand—**____

____

— **bidentate ligand—**____

____

— **hexadentate ligand**— ______________________________

____

— **chelating agent**— ______________________________

____

— **chelate ring**— ______________________________

____

— **chelate**— ______________________________

____

Summary: *Read through this section and fill in the missing words and equations. Reference the textbook when you are not confident in your answers.*

Ligands can be classified as **monodentate** or **polydentate** depending on the number of electron pairs that bond to the metal atom. Water and ammonia are monodentate ligands, binding through the O and N atoms. Glycinate ($NH_2CH_2COO^-$) is a bidentate ligand that binds through both the N and O.

Chelating agents are polydentate ligands. When the chelating agent is bound to a metal forming a ring, it forms a __________. A complex that contains one or more such rings is called a metal __________.

POST-LECTURE:

Work Problems: *One of the best ways to learn chemistry is through solving problems. Go to the textbook and work through the Worked Examples for this section, then go to the end of chapter and work the problems there. When you feel confident that you can correctly answer problems like the ones listed below for each objective, check off each of the Learning Objectives for this section above.*

LO 21.6: Problems 21.30, 21.31

Section 21.6 Naming Coordination Compounds

PRE-LECTURE:

Learning Objectives: *These are the skills you should be proficient in after reading this section and practicing the associated problems. Check off the boxes below once you have gone through this section, answered the associated questions in your textbook, and feel confident you have mastered the skill.*

☐ **LO 21.7** Write the systematic name for a coordination complex.

Definitions: *To learn, you must do. This is your workbook. Fill in the definitions from the glossary of the textbook as a first step to understanding the material.*

— **hydrate**— ______________________________

____

Summary: *Read through this section and fill in the missing words and equations. Reference the textbook when you are not confident in your answers.*

IUPAC rules for naming coordination compounds:

1. If the compound is a salt, name the __________ first and then the __________.
2. In naming a complex ion or a neutral complex, name the __________ first, in alphabetical order (ignoring prefixes), and then the __________.
    - Anionic ligands end in *–o*. (see chart below)
    - Neutral ligands are specified by their __________ names; some exceptions are listed in the chart below.
    - The complex name is one word with no spaces.

Anionic Ligand	Ligand Name	Neutral Ligand	Ligand Name
Bromide		Ammonia	
Carbonate		Water	
Chloride		Carbon monoxide	
Cyanide		Ethylenediamine, en	
Fluoride			
Glycinate			
Hydroxide			
Oxalate			
Thiocyanate (S ligand)			
Thiocyanate (N ligand)			

3. For more than one ligand of a particular type, use Greek prefixes (*di-*, *tri-*, *tetra,* etc.) to indicate the number of ligands.
    - Ligands are listed in __________ order, but these prefixes are ignored.
4. If the name of a ligand itself contains a Greek prefix, put the ligand name in parentheses and use an alternate prefix (*bis- (___), tris- (___), tetrakis- (___)*).
5. Use a Roman numeral in parentheses immediately following the name of the metal to indicate the __________ state of the metal.
6. A solid that contains water is called a __________.
7. In naming the metal, use the ending ______ if the metal is in an anionic complex.

Metal	Anion Name	Metal	Anion Name
Aluminum		Iron	
Chromium		Manganese	
Cobalt		Nickel	
Copper		Platinum	
Gold		Zinc	

Workbook Example 21.3

PROBLEM:
Name the following compounds:

$[Cr(H_2O)_5Cl](NO_3)_2$ $Na_2[OsCl_4NO_3]^{3-}$ $[Co(gly)NO_3]Cl$

SOLUTION:
Using the rules for naming coordination complexes:

$[Cr(H_2O)_5Cl](NO_3)_2$	The ligands are *aqua* and *chloro*. With 5 H_2O, 1 Cl^-, and 2 charge-balancing NO_3^- ions, the charge on chromium is +3. This is pentaaquachlorochromium(III) nitrate.
$Na_2[OsCl_4NO_3]^{3-}$	The ligands in this compound are *chloro* and *nitrato*. With 2 Na^+, 4 Cl^-, and 1 NO_3^-, and an overall charge of −3, the charge on osmium must be +6. Since the complex is an anion, the ending on osmium is changed to *ate*. This is sodium tetrachloronitratoosmate(VI).
$[Co(gly)NO_3]Cl$	The ligands in this compound are *glycinate* and *nitrato*. With gly^-, NO_3^- and a balancing Cl^-, the charge on cobalt is +3. The compound is glycinonitratocobalt(III) chloride.

Workbook Example 21.4

PROBLEM:
Write formulas for the following compounds: hexaaquanickel(II) chloride, ammonium diaquatribromochlorovanadate(III), sodium hexanitratochromium(III).

SOLUTION:
Determine the formula for the compounds:

The aqua ligand is H_2O. The formula is $[Ni(H_2O)_6]Cl_2$
The ammonium cation precedes the formula. $NH_3[V(H_2O)_2Br_3Cl]$
The nitrato cation is NO_3^-. $Na_3[Cr(NO_3)_6]$

POST-LECTURE:

Work Problems: *One of the best ways to learn chemistry is through solving problems. Go to the textbook and work through the Worked Examples for this section, then go to the end of chapter and work the problems there. When you feel confident that you can correctly answer problems like the ones listed below for each objective, check off each of the Learning Objectives for this section above.*

LO 21.7: Worked Examples 21.3, 21.4; Problems 21.32, 21.78, 21.82

Section 21.7 Isomers

PRE-LECTURE:

Learning Objectives: *These are the skills you should be proficient in after reading this section and practicing the associated problems. Check off the boxes below once you have gone through this section, answered the associated questions in your textbook, and feel confident you have mastered the skill.*

☐ **LO 21.8** Classify isomers according to the scheme given in Figure 21.11.
☐ **LO 21.9** Draw diastereomers for octahedral and square planar complexes.

Definitions: *To learn, you must do. This is your workbook. Fill in the definitions from the glossary of the textbook as a first step to understanding the material.*

— **isomer—** ______________________________

____

— **cis isomer—** ______________________________

____

— **constitutional isomer—** ______________________________

____

— **linkage isomer—** ______________________________

____

— **ionization isomer—** ______________________________

____

— **stereoisomer—** ______________________________

____

— **diastereoisomer—** ______________________________

____

— **trans isomer—** ______________________________

____

Summary: *Read through this section and fill in the missing words and equations. Reference the textbook when you are not confident in your answers.*

In coordination chemistry, it is possible for compounds to have the same formula but a different __________ of constituent atoms. This leads to different __________ and __________ properties.

__________ **isomers** are isomers that have different connections among their constituent atoms.

- __________ **isomers** arise when a ligand can bond to a metal through either of two different donor atoms.
- __________ **isomers** differ in the anion that is bonded to the metal ion.

__________ have the same connections among atoms but have a different arrangement of the atoms in space.

- __________ are nonmirror-image isomers
- __________ are mirror-image isomers

Diastereomers have different properties due to their different geometries, which affect dipoles and polarities in the isomer.

Square planar complexes of the type MA_2B_2 and MA_2BC (M = metal ion; A, B, C = ligands) can exist as *cis–trans* isomers. Tetrahedral complexes cannot have isomers, because all four corners are adjacent.

Octahedral complexes of the type MA_4B_2 or MA_4BC can also exist as diastereoisomers: two B ligands can be either on adjacent or opposite corners of the octahedron.

Workbook Example 21.5

PROBLEM:
Of the following compounds, which have geometric isomers? (Assume square planar or octahedral geometry.)

$Co(H_2O)_2NO_3Cl$ $[Fe(CN)_6]^{4-}$ $Ni(CO)_2Cl_2$

SOLUTION:
Determine the basic structure of the compounds and whether isomers are possible.

$Co(H_2O)_2NO_3Cl$ has the basic structure MA_2BC—it will have isomers
$[Fe(CN)_6]^{4-}$ has six identical ligands—it will not have isomers
$Ni(CO)_2Cl_2$ has the basic structure MA_2B_2—it will have isomers

POST-LECTURE:

Work Problems: *One of the best ways to learn chemistry is through solving problems. Go to the textbook and work through the Worked Examples for this section, then go to the end of chapter and work the problems there. When you feel confident that you can correctly answer problems like the ones listed below for each objective, check off each of the Learning Objectives for this section above.*

LO 21.8: Worked Example 21.5; Problems 21.84, 21.85, 21.94
LO 21.9: Worked Example 21.6; Problems 21.86, 21.87

Section 21.8 Enantiomers and Molecular Handedness

PRE-LECTURE:

Learning Objectives: *These are the skills you should be proficient in after reading this section and practicing the associated problems. Check off the boxes below once you have gone through this section, answered the associated questions in your textbook, and feel confident you have mastered the skill.*

☐ **LO 21.10** Classify a coordination complex as chiral or achiral and draw enantiomers for chiral compounds.

Definitions: *To learn, you must do. This is your workbook. Fill in the definitions from the glossary of the textbook as a first step to understanding the material.*

— **enantiomer**—____

____

— **chiral**— ____

____

— **achiral**— ____

____

— **symmetry plane**— ____

____

— **racemic mixture**— ____

____

Summary: *Read through this section and fill in the missing words and equations. Reference the textbook when you are not confident in your answers.*

Enantiomers are molecules or ions that are nonidentical mirror images of one another; they are said to have different handedness, or *chirality*. They are also called *optical isomers* because of their opposite effects on polarized light.

Chiral objects do not have a symmetry plane; **achiral** objects do have a symmetry plane.

Certain molecules and ions are chiral. If in order of size, the ligands spiral to the right (clockwise), the enantiomer is referred to as "right-handed." If the ligands spiral to the left (counterclockwise), the enantiomer is referred to as "left-handed."

Enantiomers have identical properties except for their __________ with other __________ substances and their effect on plane-polarized light: a solution of one enantiomer will rotate light to either the right or left, and the other enantiomer will rotate the light the same amount in the other direction.

A **racemic mixture** is a 50:50 mixture of the two enantiomers. This will have no net effect on polarized light, because the two rotations cancel each other out.

POST-LECTURE:

Work Problems: *One of the best ways to learn chemistry is through solving problems. Go to the textbook and work through the Worked Examples for this section, then go to the end of chapter and work the problems there. When you feel confident that you can correctly answer problems like the ones listed below for each objective, check off each of the Learning Objectives for this section above.*

LO 21.10: Worked Example 21.7; Problems 21.88, 21.90, 21.96

Section 21.9 Color of Transition Metal Complexes

PRE-LECTURE:

Learning Objectives: *These are the skills you should be proficient in after reading this section and practicing the associated problems. Check off the boxes below once you have gone through this section, answered the associated questions in your textbook, and feel confident you have mastered the skill.*

☐ **LO 21.11** Relate the color of a metal complex, the wavelength of light it absorbs, and the energy difference between its ground and excited state.

Definitions: *To learn, you must do. This is your workbook. Fill in the definitions from the glossary of the textbook as a first step to understanding the material.*

— **absorption spectrum—**____

____

Review: *Before going through the section make sure you are familiar with these concepts:*

- Section 5.3 atomic line spectra
- Section 5.4 atoms can absorb energy to form an excited state

Summary: *Read through this section and fill in the missing words and equations. Reference the textbook when you are not confident in your answers.*

The color of transition metal complexes depends on the identity of the__________ and the __________. The colors will vary for the same metal with different ligands; they will also vary for different metals with the same ligand.

Just like atoms, metal complexes absorb light by undergoing an electronic transition from the lowest energy state (E_1) to a higher energy state (E_2). The wavelength absorbed depends on the energy difference between the two energy levels:

$$\Delta E = E_2 - E_1$$
$$\Delta E = h\nu = \frac{hc}{\lambda}$$

where h is __________, ν is the __________, λ is the __________, and c is ______________.

Absorbance is the measure of the amount of light absorbed by a substance. The **absorption spectrum** is a plot of absorbance versus wavelength. The observed color is __________ to the color absorbed; for instance, if a complex strongly absorbs green light, it will appear orange.

Workbook Example 21.6

PROBLEM:
If a compound strongly absorbs at a frequency of $1.23 \times 10^{15}\ s^{-1}$, what is the difference between the ground state and the excited state of this compound in kJ/mol?

SOLUTION:
Use the formula $\Delta E = h\nu$ to determine the energy of the absorbed photons, then convert to kJ/mol.

$$\Delta E = h\nu = \left(6.626\times10^{-34}\ J\cdot s\right)\left(1.23\times10^{15}\ s^{-1}\right) = 8.15\times10^{-19}\ J/_{photon}$$

$$\frac{8.15\times10^{-19}\ J}{1\ photon}\times\frac{6.02\times10^{23}\ photons}{1\ mole} = 491\ kJ/_{mole}$$

Workbook Problem 21.3

If a coordination compound strongly absorbs light with a wavelength of 550 nm, what is the difference between the ground and excited states of the complex in kJ/mol of complex?

Strategy: Using the formula $\Delta E = \frac{hc}{\lambda}$, determine the energy of the absorbed photons, then convert to kJ/mol.

Step 1: Determine the energy of the photons absorbed.

Step 2: Convert to kJ/mol.

POST-LECTURE:

Work Problems: *One of the best ways to learn chemistry is through solving problems. Go to the textbook and work through the Worked Examples for this section, then go to the end of chapter and work the problems there. When you feel confident that you can correctly answer problems like the ones listed below for each objective, check off each of the Learning Objectives for this section above.*

LO 21.11: Problems 21.100, 21.101

Section 21.10 Crystal Field Theory

PRE-LECTURE:

Learning Objectives: *These are the skills you should be proficient in after reading this section and practicing the associated problems. Check off the boxes below once you have gone through this section, answered the associated questions in your textbook, and feel confident you have mastered the skill.*

☐ **LO 21.12** Draw crystal field energy-level diagrams for octahedral, tetrahedral, and square complexes. Classify ligands as strong- or weak-field and predict (or account for) color and magnetic properties of complexes in terms of crystal field theory.

Definitions: *To learn, you must do. This is your workbook. Fill in the definitions from the glossary of the textbook as a first step to understanding the material.*

— **crystal field theory—** ______________________________

____

— **crystal field splitting (Δ)—** ______________________________

____

— **spectrochemical series—** ______________________________

____

— **weak-field ligand—** ______________________________

____

— **strong-field ligand—** ______________________________

____

— **high-spin complex—** ______________________________

____

— **low-spin complex—** ______________________________

____

Review: *Before going through the section make sure you are familiar with these concepts:*

- Section 5.7 *d* orbital shapes

Summary: *Read through this section and fill in the missing words and equations. Reference the textbook when you are not confident in your answers.*

Crystal field theory views the bonding in complexes as arising from __________ interactions and considers the effect of the ligand charges on the energies of the metal ion *d* orbitals. It is able to account for the colors and magnetic properties of transition metal complexes, as well as the difference between the high- and low-spin complexes.

In crystal field theory, there are no covalent bonds, no sharing of electrons, and no hybrid orbitals. Interactions with neutral ligands are ion–dipole.

In octahedral complexes, metal is __________ charged and the ligands are __________charged. The anions adopt the configuration that allows them to get as far away from each other as possible. The metal–ligand attraction is stronger than the ligand–ligand repulsions, so the complex is stable.

Ligand charges repel __________ charged *d* electrons, causing orbital energies to be higher in the complex: the lobes of the orbitals that are closest to the ligands are higher in energy than those that fall between ligands. This causes **crystal field splitting,** represented by the Greek letter Δ.

Crystal field splitting energy, Δ, corresponds to wavelengths in the visible region of the spectrum, providing an explanation for the colors of the complexes. The value of Δ can be calculated from the wavelength of the absorbed light.

The effect that ligands have on Δ allows ligands to be ranked in terms of the energy changes they cause. This is known as the **spectrochemical series. Weak-field ligands** produce relatively small values of Δ and low-spin complexes, while **strong-field ligands** produce high values of Δ and high-spin complexes.

If the value of Δ is greater than the value of *P*, the energy required to put two electrons in one orbital, it is more likely to be with a strong-field ligand and the electrons will pair up in the lower-energy orbital. If the value of Δ is less than that of *P*, the electrons will spread out through the available orbitals.

This is only an issue for complexes in which there are four to seven *d* electrons, known as d^4-d^7 complexes. For others, only one ground-state configuration is possible, and the compounds are colorless.

In tetrahedral complexes, the energy splitting for the *d* orbitals is different. In tetrahedral complexes, none of the ligands points directly at the *d* orbitals. As a result, *P* is almost always __________ than Δ, and so tetrahedral complexes tend to be high-spin.

In square planar complexes, the energy is similar to that of octahedral complexes, with the *z*-axis ligands missing. This leads to a large energy gap between the *d* orbitals that are pointing toward ligands and those that are not, favoring low-spin complexes with the higher-energy orbitals vacant.

POST-LECTURE:

Work Problems: *One of the best ways to learn chemistry is through solving problems. Go to the textbook and work through the Worked Examples for this section, then go to the end of chapter and work the problems there. When you feel confident that you can correctly answer problems like the ones listed below for each objective, check off each of the Learning Objectives for this section above.*

LO 21.12: Worked Examples 21.8, 21.9; Problems 21.104, 21.106, 21.108, 21.114, 21.116, 21.124

Section 21.11 Bonding in Complexes: Valence Bond Theory

PRE-LECTURE:

Learning Objectives: *These are the skills you should be proficient in after reading this section and practicing the associated problems. Check off the boxes below once you have gone through this section, answered the associated questions in your textbook, and feel confident you have mastered the skill.*

☐ **LO 21.13** Describe the bonding in a metal complex in terms of valence bond theory, including orbital diagrams, the hybrid orbitals used by the metal atom, and the number of unpaired electrons.

Definitions: *To learn, you must do. This is your workbook. Fill in the definitions from the glossary of the textbook as a first step to understanding the material.*

— **low-spin complex**—____

____

— **high-spin complex**—____

____

Review: *Before going through the section make sure you are familiar with these concepts:*

- Section 5.3 atomic line spectra
- Section 5.4 atoms can absorb energy to form an excited state
- Section 8.3 hybrid orbitals

Summary: *Read through this section and fill in the missing words and equations. Reference the textbook when you are not confident in your answers.*

In **valence bond theory,** bonding results when a _________ ligand orbital containing a pair of electrons overlaps a _________ hybrid orbital on the metal ion to produce a _________ _________ bond.

The geometry of these complexes is related to the hybrid orbitals used. Linear complexes have ___ orbitals, tetrahedral complexes use ___ orbitals. Octahedral complexes have six equivalent hybrid orbitals that can be formed in either from ___ or ___; all are ___° from one another. Square planar complexes also use *d* orbitals in ___ hybrids. This is summarized in Table 21.6 in your textbook.

Coordination Number	Geometry	Hybridization
2	Linear	
4	Tetrahedral	
4	Square planar	
6	Octahedral	

Transition metal complexes can be **paramagnetic,** containing _________ electrons and attracted by magnetic fields, or **diamagnetic,** containing only _________ electrons and weakly repelled by magnetic fields. The number of unpaired electrons can be measured by the force exerted on the complex by a magnetic field.

Metals can form complexes described as **high-spin complexes,** in which the *d* electrons are arranged to _________ unpaired electrons and minimize doubly occupied orbitals, or **low-spin complexes,** in which *d* electrons are arranged to _________ unpaired orbitals and _________ unpaired electrons.

Workbook Example 21.7

PROBLEM:
$[Co(NH_3)_6]^{3+}$ is a low-spin complex. Draw an electron diagram to demonstrate this.

SOLUTION:
Determine which hybrid orbitals are being used to minimize spin.

The free Co^{3+} ion has the valence [Ar]$3d^6$. This is an octahedral complex, so will use either d^2sp^3 or sp^3d^2 hybrid orbitals.

	3d	4s	4p	4d	
Co^{3+} :	↑↓ ↑ ↑ ↑ ↑	__	__ __ __		d^6
$[Co(NH_3)_6]^{3+}$:	↑↓ ↑↓ ↑↓ ↑↓ ↑↓	↑↓	↑↓ ↑↓ ↑↓		d^2sp^3

or

	3d	4s	4p	4d	
$[Co(NH_3)_6]^{3+}$:	↑↓ ↑ ↑ ↑ ↑	↑↓	↑↓ ↑↓ ↑↓	↑↓ ↑↓ __ __ __	sp^3d^2

The d^2sp^3 configuration has fewer unpaired electrons and is also lower energy overall. Both support the choice of this configuration.

Workbook Problem 21.4

$[CoCl_6]^{3-}$ is a high-spin complex. Draw an electron diagram to demonstrate this.

Strategy: Draw the orbital diagram and determine how to maximize unpaired electrons.

Step 1: Determine the electron configuration of the metal alone.

Step 2: Identify the possible hybridization schemes.

Step 3: Draw electron diagrams with both hybridizations.

Step 4: Choose the hybridization that maximizes electron spin.

POST-LECTURE:

Work Problems: *One of the best ways to learn chemistry is through solving problems. Go to the textbook and work through the Worked Examples for this section, then go to the end of chapter and work the problems there. When you feel confident that you can correctly answer problems like the ones listed below for each objective, check off each of the Learning Objectives for this section above.*

LO 21.13: Worked Example 21.10; Problems 21.16, 21.128

Self-Test

This section is intended to test your knowledge of the material covered in this chapter. Think through these problems and make certain you understand what the problem is about—ask yourself if the answer makes sense. Many of these questions are linked to the chapter learning goals. Therefore, successful completion of these problems indicates that you have achieved the learning goals for this chapter. You will receive the greatest benefit from this section if you use it as a mock exam. Do not use your notes, book, Internet, or friends when taking the test. There is no grade and you only shortchange yourself by getting help with this activity. As you take the test, mark the problems with your level of confidence with the answer.

★ *Use a star for problems you know you got right.*
? *Use a question mark for those you think you got right.*
x *Use an x for those you are guessing on.*

If you get a question right that you marked with a star, you can be confident that you know the associated LO.

If you get a question right that you marked with a question mark, go back and work a few more problems and test yourself again tomorrow.

If you get a question wrong or mark it with an x, go back and read the section again, work a few problems today, a few tomorrow, and a few the next day before testing yourself again.

This will help you to discover which topics you have learned thoroughly and which topics you need to study in more detail.

True–False

1. Electron configurations are dependent on both orbital energies and electron–electron repulsions.

2. All of the transition elements are metals, and all are solids.

3. Densities of transition elements have a minimum density in the center of the row.

4. Transition metals are rarely found in more than one oxidation state.

5. Atomic size of the transition metals does not vary much as you go down a period.

6. Ligands connect to metal atoms by one or more ligand donor atoms.

7. The oxidation state of the transition metal is not given as part of the name of a coordination compound.

8. Optical activity in an enantiomer is the result of chirality.

9. Complexes containing the SCN (thiocyanate) ion can have linkage isomers.

10. Crystal field theory assumes that all interactions between metal and ligand are covalent.

Multiple Choice

11. In a coordinate covalent bond,
   a. no electrons are shared
   b. a lone pair of electrons becomes a bonding pair
   c. metals bind to other metals
   d. ligands ionize

12 The multiple oxidation states of transition elements originate in
   a. the varying sizes of the elements
   b. increasing metallic character with ionization
   c. the closeness in energy of the *s* orbitals and the $n-1$ *d* orbitals
   d. transition elements do not have multiple oxidation states

13. All of the following are properties of transition metals:
   a. malleable, ductile, good conductors of heat and electricity
   b. malleable, ductile, poor conductors of heat and electricity
   c. brittle, ductile, good conductors of heat and electricity
   d. malleable, dull, good conductors of heat and electricity

14. What shape is common for complexes with a coordinate number of 4?
   a. tetrahedral
   b. see-saw
   c. square planar
   d. a and c

15. Chromium is used to electroplate other metals. It is useful for this because
   a. it does not oxidize
   b. it is dull and grey, discouraging thieves
   c. It forms a hard, invisible layer of oxidation that protects against further corrosion
   d. none of the above

16. Ligands in coordination compounds are always
   a. anions
   b. cations
   c. Lewis bases
   d. Lewis acids

17. Numerous spatial arrangements are possible for coordination compounds. The shape of these compounds is determined by
   a. the hybridization of the *d* orbitals
   b. the coordination number
   c. the polarity of the ligand
   d. the group number of the metal

18. Which of the following will have geometric isomers?
    a. MA_2B_4
    b. MA_6
    c. MA_4B_2
    d. a and c

19. The observed color of transition metal complexes is
    a. the maximum absorbed wavelength
    b. complementary to the absorbed wavelength
    c. the absorbed wavelength rotated 90°
    d. none of the above

20. In contrast to valence bond theory, crystal field theory is able to account for
    a. geometry of complexes
    b. color and spin
    c. paramagnetism
    d. all of the above

Fill in the Blanks

21. Transition elements exhibit a variety of ________________ states. Ions in which this is high are good ________________ agents, while ions in which this is low are good ________________ agents.

22. ________________ compounds are those in which a central metal atom is connected to surrounding ________________ by ____________ ____________ bonds.

23. Many compounds exist as ________________, which have the same formula, but different arrangements of the atoms. These can be ________________ in which the connectivity is different, or ________________ in which the connections are the same, but have a different spatial arrangement.

24. The ________________ of the light absorbed by a transition metal complex can be used to calculate the ________________ required for its most common transition and can also be used to predict its observed ____________.

25. ____________ ____________ ____________ describes metal complexes in terms of overlapping ____________ orbitals. ____________ ____________ ____________ describes metal complexes as purely ________________ in nature.

Matching

26. absorption spectrum

27. achiral

28. bidentate ligand

29. chelate

a. a ligand that binds using two donor atoms

b. having handedness

c. the number of ligand donor atoms surrounding a metal ion

d. isomer in which identical ligands or groups are opposite one another

30. chiral
31. *cis* isomer
32. coordination number
33. crystal field splitting
34. enantiomer
35. high-spin complex
36. ligand
37. ligand donor atom
38. low-spin complex
39. racemic mixture
40. strong-field ligand
41. *trans* isomer
42. weak-field ligand

e. complex in which there is a maximum number of unpaired electrons
f. a cyclic complex formed by a metal and a polydentate ligand
g. isomer in which identical ligands or groups are adjacent to one another
h. a plot of the amount of light absorbed versus wavelength
i. atom attached directly to the metal
j. optically inactive mixture of enantiomers in a 50:50 ratio
k. the energy difference between two sets of *d* orbitals
l. ligand that has strong crystal field splitting
m. lacking handedness
n. stereoisomers that are mirror images of one another
o. ligand that has weak crystal field splitting
p. complex in which there are a maximum number of paired electrons
q. molecule or ion that bonds to a central atom in a complex

Problems

43. Write the valence electron configurations for the following atoms and ions: Mn^{5+}, Mo, Cr^{3+}, Fe^{2+}.

44. Identify the following neutral elements from the following electron configurations: [Ar] $4s^2 3d^5$, [Kr] $5s^2 4d^7$, [Xe]$6s^2 5d^1$.

45. Arrange the following from the strongest oxidizing agent to the strongest reducing agent:
Co   Cu   Mn   Zn

46. Arrange the following from the strongest oxidizing agent to the strongest reducing agent:
Fe_2O_3 CoO MnO_4^- CrO_3^-

47. Identify the coordination number of each of the following compounds:
$[HgCl_3]^-$ $[Mo(CN)_8]^{4-}$ $Fe(CO)_5$ $[CuCl_2]^-$

48. For the following compounds identify the ligands and their donor atoms. Determine the coordination number and the oxidation state of the metal and the charge on any complex ion. Determine if the compound can display geometric, linkage, or optical isomerism.
    a. $[Co(H_2O)_4(NH_3)_2]Cl_3$ b. $[Co(NH_3)_4(H_2O)Br]NO_3$ c. $[Cr(NH_3)_4(SCN)_2]$
    d. $(NH_4)_3[Fe(ox)_3]$ e. $[Co(en)_3]I_3$

49. Name the compounds in question 48.

50. Write formulas for the following compounds:
    a. tetraamminediaquachromium(III) chloride
    b. ethylenediaminedithiocyanatocopper(II)
    c. pentacarbonylchloromanganate(I)
    d. potassium tetracyanodichloroferrate(II) dihydrate
    e. tetraaquacopper(II) sulfate

51. Use valence bond theory to explain the bonding in $[Zn(NH_3)_4]^{2+}$ (tetrahedral complex) and $[Ni(CN)_4]^{2-}$ (square planar). What hybrid orbitals are used by the metal? Are the complexes low-spin or high-spin complexes?

52. Why is the crystal field splitting in a tetrahedral complex approximately half of the crystal field splitting in an octahedral complex?

Challenge Problem

53. Three different isomers with the formula $CrCl_3{\cdot}6\ H_2O$ exist. Two are green in color; one is violet. These isomers are labeled *A, B,* and *C.* When a 0.500 g sample of isomer *A* reacts with a dehydrating agent, the resulting mass is 0.432 g. When 50 mL of a 0.125 M solution of isomer *A* is titrated with excess $AgNO_3$, 0.269 g of AgCl is produced.

    When a 0.500 g sample of isomer *B* reacts with a dehydrating agent, the resulting mass is 0.466 g. Reaction of a 50 mL solution of 0.125 M isomer *B* with excess $AgNO_3$ produced 0.539 g of AgCl. When a 0.500 g sample of isomer *C* reacts with a dehydrating agent, the resulting mass is 0.500 g. When 50 mL of a 0.125 M solution of isomer *C* is titrated with $AgNO_3$, 0.809 g of AgCl is produced. Determine the formula of each isomer.

Test Yourself: *Take the Practice Test on pages 944 and 945 of your textbook. Again, use a testing environment using only those items that your instructor will allow you to have access to during the test and mark each question with a star, question mark, or x to denote your confidence level. Once you have finished you can check off the LO at the beginning of this chapter based on which questions you starred and got correct. Go back and spend more time with the topics from the other learning objectives.*

CHAPTER TWENTY-TWO

The Main-Group Elements

Learning Objectives

As a result of reading and studying this chapter, you should be able to:

- ☐ **LO 22.1** Use the periodic table to predict the chemical and physical properties of the main-group elements.
- ☐ **LO 22.2** Compare the number and types of bonds in compounds with second-row elements and compounds with third- and higher-row elements.
- ☐ **LO 22.3** Write and balance a chemical equation for the reaction of ionic hydrides with water.
- ☐ **LO 22.4** Classify binary hydrides and describe their bonding and properties.
- ☐ **LO 22.5** Use periodic trends to predict chemical and physical properties of Group 1A and 2A elements.
- ☐ **LO 22.6** Write and balance reactions of alkali and alkaline earth metals with halogens, oxygen, and water.
- ☐ **LO 22.7** Write electrode reactions and perform calculations in the electrolysis of molten salts to produce alkali and alkaline earth metals.
- ☐ **LO 22.8** Use periodic trends to predict chemical and physical properties of Group 3A elements.
- ☐ **LO 22.9** Describe the structure and bonding in boron halides and diborane.
- ☐ **LO 22.10** Draw electron-dot structures and predict hybrid orbitals and geometry for molecules and ions in main-group compounds containing elements from Group 4A.
- ☐ **LO 22.11** Use the shorthand notation for a silicate anion to represent its structure and interpret the notation to find the formula and charge of a silicate anion.
- ☐ **LO 22.12** Write electron-dot structures and describe the structure, bonding, and properties of molecules and ions in main-group compounds containing nitrogen and phosphorus.
- ☐ **LO 22.13** Determine oxidation states of nitrogen and phosphorus in acids and write acid–base reactions.
- ☐ **LO 22.14** Classify oxides as ionic or covalent and as acidic, basic, or amphoteric and write balanced equations for the reaction that occurs when oxides dissolve in water.
- ☐ **LO 22.15** Write electron-dot structures and describe the structure, bonding, and properties of compounds containing oxygen and sulfur.
- ☐ **LO 22.16** Write balanced equations for the formation of halogens and reactions of their compounds.
- ☐ **LO 22.17** Determine oxidation states of halogens in oxoacids and predict the relative acid strengths of oxoacids.
- ☐ **LO 22.18** Draw electron-dot structures and predict geometry for molecules and ions containing noble gases.

Chapter Summary

The main-group elements demonstrate strong periodic trends in ionization energy, electronegativity, atomic radius, and metallic character. The second-row elements have some unique properties as a result of their small size and high electronegativities. The families will be examined as groups, with particular attention given to the unique properties of hydrogen, boron, carbon, silicon, nitrogen, phosphorus, oxygen, and sulfur.

The Chapter in Detail

Section 22.1 A Review of General Properties and Periodic Trends

PRE-LECTURE:

Learning Objectives: *These are the skills you should be proficient in after reading this section and practicing the associated problems. Check off the boxes below once you have gone through this section, answered the associated questions in your textbook, and feel confident you have mastered the skill.*

☐ **LO 22.1** Use the periodic table to predict the chemical and physical properties of the main-group elements.

Review: *Before going through the section make sure you are familiar with these concepts:*

- Section 2.3 metals, nonmetals, semimetals
- Section 5.9 effective nuclear charge
- Section 6.3 ionization energy
- Section 5.13 atomic radius
- Section 7.3 electronegativity

Summary: *Read through this section and fill in the missing words and equations. Reference the textbook when you are not confident in your answers.*

Overall, the main group elements are divided into ___________ on the left of the periodic table, ___________ on the right of the periodic table, and ___________ with intermediate properties along a stair-step line dividing them.

Periodic trends:

- These decrease going down the periodic table and increase from left to right
  - ______________________________
  - ______________________________
  - ______________________________
  - ______________________________
- These increase going down the periodic table and decrease from left to right
  - ______________________________
  - ______________________________

From left to right across the periodic table, the effective nuclear charge (Z_{eff}) ___________, because the additional electrons are being added into a shell that does not completely shield the additional nuclear charge. As a result of the increased Z_{eff}, electrons are ___________ strongly attracted to the nucleus. Thus, ionization energy and electronegativity ___________ to the right, while atomic radius and metallic character ___________.

From top to bottom down the periodic table, atomic radius increases as additional ___________ shells are occupied. Because the valence electrons are ___________ from the nucleus, ionization energy and electronegativity decrease and metallic character increases.

Metals

- at 25 °C most are ___________ except Hg which is a liquid
- most ___________ a silvery shine
- ___________ and ___________
- ___________ conductors of heat and electricity
- relatively ___________ ionization energies
- relatively ___________ electronegativities
- ___________ electrons to become cations

Nonmetals

- most are gases, Br is a ___________, and C, P, S, Se, and I are ___________ at 25 °C
- most ___________ a metallic luster
- ___________ and ___________
- ___________ conductors of heat and electricity (except graphite)
- relatively ___________ ionization energies
- relatively ___________ electronegativities
- ___________ electrons to become anions; ___________ electrons to become oxoanions

Workbook Problem 22.1

Predict which element; carbon, nitrogen, or silicon has the following trait:

- the highest ionization energy
- is the most electronegative
- has the biggest atomic radius
- the most metallic character

Strategy: Examine the position of the elements on the periodic table.

Step 1: Predict the traits of the elements based on their position on the periodic table.

POST-LECTURE:

Work Problems: *One of the best ways to learn chemistry is through solving problems. Go to the textbook and work through the Worked Examples for this section, then go to the end of chapter and work the problems there. When you feel confident that you can correctly answer problems like the ones listed below for each objective, check off each of the Learning Objectives for this section above.*

LO 22.1: Worked Example 22.1; Problems 22.18, 22.19, 22.32, 22.38, 22.40

Section 22.2 Distinctive Properties of the Second-Row Elements

PRE-LECTURE:

Learning Objectives: *These are the skills you should be proficient in after reading this section and practicing the associated problems. Check off the boxes below once you have gone through this section, answered the associated questions in your textbook, and feel confident you have mastered the skill.*

☐ **LO 22.2** Compare the number and types of bonds in compounds with second-row elements and compounds with third- and higher-row elements.

Review: *Before going through the section make sure you are familiar with these concepts:*

- Section 7.3 electronegativity
- Section 8.2 sigma bond
- Section 8.4 pi bond

Summary: *Read through this section and fill in the missing words and equations. Reference the textbook when you are not confident in your answers.*

The elements of the second-row are much different than those below them in the same periodic group. This is because they have especially ___________ sizes and ___________ electronegativities. This causes the elements in this row to be ___________ nonmetallic: they form mainly covalent molecular compounds, with a maximum of ___________ covalent bonds, because they do not have access to ___ orbitals.

Hydrogen bonding is limited to three elements ___, ___, and ___.

The small size of second-row elements allows for the formation of multiple bonds when $2p$ orbitals overlap to form ___ bonds. $3p$ orbitals are more diffuse, leading to ___________ bond distances and poor π overlap. For example, O_2 is double-bonded, while S_8 consists of single-bonded, crown-shaped rings.

POST-LECTURE:

Work Problems: *One of the best ways to learn chemistry is through solving problems. Go to the textbook and work through the Worked Examples for this section, then go to the end of chapter and work the problems there. When you feel confident that you can correctly answer problems like the ones listed below for each objective, check off each of the Learning Objectives for this section above.*

LO 22.2: Worked Example 22.2; Problems 22.29, 22.46, 22.48, 22.50, 22.51

Section 22.3 Group 1A: Hydrogen

PRE-LECTURE:

Learning Objectives: *These are the skills you should be proficient in after reading this section and practicing the associated problems. Check off the boxes below once you have gone through*

this section, answered the associated questions in your textbook, and feel confident you have mastered the skill.

☐ **LO 22.3** Write and balance a chemical equation for the reaction of ionic hydrides with water.
☐ **LO 22.4** Classify binary hydrides and describe their bonding and properties.

Definitions: *To learn, you must do. This is your workbook. Fill in the definitions from the textbook as a first step to understanding the material.*

— **binary hydride**—____

____

— **ionic hydride**— ____

____

— **covalent hydride**— ____

____

— **metallic hydride**—____

____

— **interstitial hydride**— ____

____

— **nonstoichiometric compound**—____

____

Review: *Before going through the section make sure you are familiar with these concepts:*

- Section 11.7 face-centered cubic
- Section 16.1 Brønsted–Lowry base

Summary: *Read through this section and fill in the missing words and equations. Reference the textbook when you are not confident in your answers.*

Hydrogen is a nonmetal with properties different from the other group 1A metals.

ELEMENTAL HYDROGEN

At ordinary temperatures and pressures, hydrogen is a colorless, odorless diatomic gas with the formula ___. Its lack of polarity and small size lead to very ___________ intermolecular forces, and ___________ melting and boiling points, but it has the ___________ dissociation energy of any diatomic element.

HYDROGEN COMPOUNDS

Hydrogen is the simplest atom containing ___ 1*s* electron and ___ proton. On the periodic table, hydrogen is usually placed at the top of group 1A because like the rest of the group, it has just ___ electron. However, sometimes it is placed at the top of 7A because like the rest of the group

7A, it only needs ___ electron to complete its valence shell. Thus, it has characteristics similar to both and will form a hydrogen cation, ____, or a hydride anion, ____. Most often it does not completely transfer its electron because of the ___________ ionization energy required to pull of its electron.

BINARY HYDRIDES

Binary hydrides contain hydrogen and just one other element and can be ionic, covalent, or metallic.

Ionic hydrides

- Saltlike
- ___________
- ___________
- ___________
- formed with the ___________ and the heavier ___________
- form ______________________ crystal structures

The hydride ion is a proton ___________—a Brønsted–Lowry base—and therefore a good ___________ agent. When dissolved in water, the hydride ion reduces water to generate H_2 gas and OH^- ions.

Covalent hydrides are molecules in which hydrogen is ___________ bonded to another nonmetal. These tend to be ___________ molecules with relatively ___________ intermolecular forces, and so are ___________ or ___________ at room temperature. Many are quite familiar—H_2O, NH_3, and CH_4 are all covalent hydrides.

Metallic hydrides will also form with large metal atoms, especially the lanthanides, actinides, and certain *d*-block transition metals. The nature of the bonding in these molecules is not well understood, but it seems the hydrogens pack into the interstitial spaces. These compounds can be stoichiometric, for example, UH_3, or not, as in the case of $ZrH_{1.9}$. The properties of these metallic hydrides depend on their composition, which is also dependent on the partial pressure of H_2 gas in the surroundings.

POST-LECTURE:

Work Problems: *One of the best ways to learn chemistry is through solving problems. Go to the textbook and work through the Worked Examples for this section, then go to the end of chapter and work the problems there. When you feel confident that you can correctly answer problems like the ones listed below for each objective, check off each of the Learning Objectives for this section above.*

LO 22.3: Problems 22.66, 22.67
LO 22.4: Worked Example 22.3; Problems 22.21–22.23, 22.58, 22.60

Section 22.4 Group 1A: Alkali Metals and Group 2A: Alkaline Earth Metals

PRE-LECTURE:

Learning Objectives: *These are the skills you should be proficient in after reading this section and practicing the associated problems. Check off the boxes below once you have gone through this section, answered the associated questions in your textbook, and feel confident you have mastered the skill.*

- ☐ **LO 22.5** Use periodic trends to predict chemical and physical properties of Group 1A and 2A elements.
- ☐ **LO 22.6** Write and balance reactions of alkali and alkaline earth metals with halogens, oxygen, and water.
- ☐ **LO 22.7** Write electrode reactions and perform calculations in the electrolysis of molten salts to produce alkali and alkaline earth metals.

Summary: *Read through this section and fill in the missing words and equations. Reference the textbook when you are not confident in your answers.*

ALKALI METALS

Alkali metals:

- group ____
- have the ___________ ionization energies of any elements, because they lose an ____ electron to gain a noble gas configuration.
- are among the most ___________ reducing agents in the periodic table.
- are metallic and so they have ___________ appearance, ___________, and ___________ conductors.
- soft metals (can be cut with a knife)
- ___________ melting points
- ___________ densities

Production:

Lithium and sodium are produced by ___________: the chloride salt is melted and an electric current passed through it while it is held at high temperature. Potassium, rubidium, and cesium are produced by chemical reaction.

Reactions:

Alkali metals react with halogens to form halides with the formula ______. The reactivity of the alkali metals ___________ as the ionization energy decreases: cesium is the most reactive, lithium the least reactive.

Alkali metals react ___________ with oxygen, but produce different kinds of products:

- Li_2O is an oxide
- Na_2O_2 is a ___________
- K, Rb, and Cs for ___________ or ___________ depending on conditions

The ionic solids can form a variety of unit cells depending on ___________ and ___________ of the ion.

Alkali metals take their name from their reaction with water to yield hydrogen gas and an alkali metal hydroxide. These reactions are vigorous enough to split water molecules, and, as with the halide reactions above, they ___________ in vigor with decreasing ionization energy.

ALKALINE EARTH METALS

Alkaline earth metals:

- lose ___ ns^2 electrons to gain a noble gas configuration
- have slightly ___________ ionization energies than the alkali metals.
- follow the same reactivity trend, in which the larger members of the group, with lower ionization energies, are the ___________reactive.
- are harder than the alkali metals, but still soft.
- have ___________ densities and melting points.
- despite their lower reactivity, they are also found in nature only as salts.

Production: Pure alkaline earth metals are produced by chemical or electrolytic reduction.

Reactions: Alkaline earth metals react with halogens to form halides with the formula _____ and with oxygen to form oxides with the expected formula _____, though strontium and barium will also form peroxides. Beryllium and magnesium are relatively ___________, but will burn with a bright flame. The other alkaline earth metals are all reactive enough that they are best protected from air.

With the exception of beryllium, the alkaline earth metals will react with water to form hydroxides with the formula _____, but with much less vigor than the alkali metals. Magnesium will react only at temperatures exceeding 100 °C, calcium and strontium will react slowly at room temperature, and barium, with its low ionization energy, reacts vigorously at room temperature.

Workbook Problem 22.2

What is the predicted outcome of the following reactions? Balance the reactions that occur.

$Sr(s) + O_2(g)$ $Ca(s) + H_2O(l)$ $Be(s) + H_2O(l)$

Strategy: Based on the reactivity of the metals and the expected reactions:

Step 1: Determine what reaction, if any, will occur.

Step 2: If reactions occur, write their balanced equations.

POST-LECTURE:

Work Problems: *One of the best ways to learn chemistry is through solving problems. Go to the textbook and work through the Worked Examples for this section, then go to the end of chapter*

and work the problems there. When you feel confident that you can correctly answer problems like the ones listed below for each objective, check off each of the Learning Objectives for this section above.

LO 22.5: Problem 22.68
LO 22.6: Problems 22.70, 22.72
LO 22.7: Problems 22.76, 22.77

Section 22.5 Group 3A Elements

PRE-LECTURE:

Learning Objectives: *These are the skills you should be proficient in after reading this section and practicing the associated problems. Check off the boxes below once you have gone through this section, answered the associated questions in your textbook, and feel confident you have mastered the skill.*

- ☐ **LO 22.8** Use periodic trends to predict chemical and physical properties of Group 3A elements.
- ☐ **LO 22.9** Describe the structure and bonding in boron halides and diborane.

Definitions: *To learn, you must do. This is your workbook. Fill in the definitions from the textbook as a first step to understanding the material.*

— **boranes—**____

____

— **three-center, two-electron bond—**________________________________

____

Review: *Before going through the section make sure you are familiar with these concepts:*

- Section 16.1 Brønsted–Lowry acid

Summary: *Read through this section and fill in the missing words and equations. Reference the textbook when you are not confident in your answers.*

The group 3A elements are the first of the ___ block elements; they have the valence electron configuration ___________and an oxidation state of ___. As expected, metallic character generally ___________ down the group. Boron (described below) is the only semimetal; the others behave as metals.

___________ is important because of its commercial products.

___________ is a liquid form from 29.2 °C to 2204 °C. This is not only an unusually low melting point, but also a very large range of temperatures for the liquid range. Gallium arsenide is a semiconductor used in making diode lasers for laser printers, CD players, and fiber-optic devices.

___________ is also used in making semiconductor devices such as transistors and *thermistors*—electrical-resistance thermometers.

___________ is extremely toxic.

ELEMENTAL BORON

Boron is relatively rare, but occurs in concentrated deposits of borate minerals. It is a strong, high-melting-point substance that is chemically ___________ at room temperature, making it a desirable component in high-strength composite materials.

Boron halides are ___________ reactive, ___________ covalent compounds that consist of ___________ BX_3 molecules. They have a vacant 2*p* orbital that allows them to behave as Lewis ___________: they will form adducts with Lewis bases by accepting a share of a nonbonded pair of electrons.

Boron hydrides (boranes) are ___________, ___________ compounds with formulas B_nH_m, the simplest of which is diborane, B_2H_6. Diborane is of interest because of its unusual structure: rather than having a structure similar to ethane (C_2H_6), it has an *electron-deficient* structure in which two BH_2 groups are connected by bridging hydrogens; therefore, there are only 12 valence electrons. The unusual B–H–B bonds are much longer than the terminal B–H bonds.

POST-LECTURE:

Work Problems: *One of the best ways to learn chemistry is through solving problems. Go to the textbook and work through the Worked Examples for this section, then go to the end of chapter and work the problems there. When you feel confident that you can correctly answer problems like the ones listed below for each objective, check off each of the Learning Objectives for this section above.*

LO 22.8: Problems 22.80–22.82
LO 22.9: Problems 22.84–22.87

Section 22.6 Group 4A Elements

PRE-LECTURE:

Learning Objectives: *These are the skills you should be proficient in after reading this section and practicing the associated problems. Check off the boxes below once you have gone through this section, answered the associated questions in your textbook, and feel confident you have mastered the skill.*

- ☐ **LO 22.10** Draw electron-dot structures and predict hybrid orbitals and geometry for molecules and ions in main-group compounds containing elements from Group 4A.
- ☐ **LO 22.11** Use the shorthand notation for a silicate anion to represent its structure and interpret the notation to find the formula and charge of a silicate anion.

Definitions: *To learn, you must do. This is your workbook. Fill in the definitions from the textbook as a first step to understanding the material.*

— **carbide—** ____

____

— **zone refining—** ____

____

— **silicate—**____

____

Summary: *Read through this section and fill in the missing words and equations. Reference the textbook when you are not confident in your answers.*

Group 4A elements are important both in industry and in living organisms. Carbon is a nonmetal that is present in all plants and animals; it is an essential part of biological molecules. Silicon is a semimetal that makes up numerous silicate minerals, is the second-most abundant element in the Earth's crust, and with germanium is used to make solid-state electronic devices.

Germanium is a high-melting-point brittle semiconductor with the same structure as diamond and silicon. Tin and lead are both soft, malleable, low-melting-point metals that have been known since ancient times. Tin has two allotropes: malleable, silvery metallic *white tin* and brittle, semiconducting *gray tin*. White tin is the most stable allotrope at room temperature, but when kept below 13 °C for long periods, it crumbles into gray tin, in what is known as "tin disease." Lead is found only in metallic form.

Group 4A has the valence electron configuration ___________, with ____ as the most common oxidation state. The chlorides of the group 4A elements are all sp^3 covalently bonded volatile molecular liquids. An oxidation state of +2 is also possible for Sn and Pb in solution, which generally leads to ionic bonding. There are no +4 aqueous ions: these species instead exist as complex ions such as $Sn(OH)_6^{2-}$.

ELEMENTAL CARBON

Elemental carbon forms a number of allotropes including ___________, ___________, and _________

Graphene is a two-dimensional array of hexagonally arranged carbon atoms just one atom thick. It is very strong and flexible and is an excellent conductor, making it a possible material for electronics and composite materials.

CARBON COMPOUNDS

Carbon forms more than ___________ compounds, most of which are organic. Some of the inorganic compounds of carbon are as follows:

Oxides of Carbon include carbon monoxide and carbon dioxide.

___________ is a colorless, odorless, toxic gas that forms when carbon or hydrocarbons are burned in limited oxygen. The toxicity arises from the fact that it binds 200 times more tightly to hemoglobin than oxygen. Thus, even low levels reduce the blood's ability to deliver oxygen to the tissues.

___________ is a colorless, odorless, nonpoisonous gas. It is produced when fuels burn in an excess of O_2, is an end product of metabolism, is a by-product of the yeast-catalyzed fermentation of sugar in the manufacture of alcoholic beverages, and is produced when metal carbonates are heated or reacted with acids.

Carbon dioxide has numerous uses, including providing the "fizz" in beverages and fighting fires. CO_2 is denser than air, so it settles over small fires like a blanket, thereby smothering them. Solid CO_2 is known as "dry ice." It sublimes at −78 ° and is used primarily as a refrigerant.

Carbonates also come in two forms: carbonates, which contain the _____ ion, and hydrogen carbonates, or bicarbonates, which contain the _____ ion.

Hydrogen cyanide, ____, is a highly toxic volatile substance that is weakly acidic in solution, dissociating to form CN^- ions. These are known as *pseudohalides* because they behave like Cl^-, precipitating out Ag ions. In forming complex ions, cyanide behaves as a Lewis base, bonding through the lone pair of electrons on carbon.

The toxicity of HCN and cyanides is due to the strong and irreversible binding of CN^- to the iron (III) atom contained in cytochrome oxidase, an important enzyme in cellular metabolism. With this enzyme unable to function, cellular energy production ceases and death occurs rapidly.

CN^- ions are able to react with ores containing Au and Ag. When the cyanide salts are reduced with zinc, the precious metal can be recovered.

___________ are binary compounds of carbon in which the carbon atom has a negative oxidation state. They can be ionic carbides of active metals (CaC_2), interstitial carbides of transition metals (Fe_3C), or covalent network carbides (SiC).

Workbook Problem 22.3

Determine the oxidation state of carbon in the following compounds:

CO CO_2 C_2O_3 CO_3^{2-} HCN CaC_2

Strategy: Use the rules for assigning oxidation numbers to determine the oxidation state of carbon.

Step 1: Determine the oxidation state of carbon.

ELEMENTAL SILICON

Silicon is a hard, gray, semiconducting solid that melts at 1414 °C. It has a diamond-like structure, but does not have a graphite-like allotrope because the silicon atoms are too large for effective overlap of ___ orbitals.

In nature, silicon is generally found in silica and silicate minerals. To generate elemental silicon, silica sand is reduced by carbon to form silicon and carbon monoxide. To make the ultrapure silicon needed for electronics, silicon is converted to $SiCl_4$, purified by fractional distillation, and reduced with hydrogen gas to form pure silicon and HCl. This can then be further purified by __________, in which a rod of silicon is slowly melted from top to bottom. Impurities concentrate in the molten zone and are dragged to the bottom of the rod.

__________ are ionic compounds that contain silicon oxoanions and metal cations. The basic building block is the SiO_4 tetrahedron. Alone, it is the orthosilicate ion SiO_4^{4-}, but more commonly, there is an oxygen shared between units, leading to a large number of possible mineral structures, including cyclic anions, chains, layers, and extended three-dimensional structures. This bonding flexibility means that silicates include such varied mineral types as fibrous asbestos, sheets of mica, and emeralds.

Workbook Problem 22.4

The formula unit of garnet has three SiO_4^{4-} units balanced by Ca^{2+} and Al^{3+} cations. How is the formula unit of garnet written?

Strategy: From the information provided, determine the overall silicate structure, remembering that oxygens are shared in the tetrahedral units. Balance the charge on the silicate using the ions in a correct ratio.

Step 1: Determine the silicate structure.

Step 2: Determine how many of each cation is needed to balance the charge on the silicate unit.

Step 3: Write the formula unit.

POST-LECTURE:

Work Problems: *One of the best ways to learn chemistry is through solving problems. Go to the textbook and work through the Worked Examples for this section, then go to the end of chapter and work the problems there. When you feel confident that you can correctly answer problems like the ones listed below for each objective, check off each of the Learning Objectives for this section above.*

LO 22.10: Problems 22.90, 22.92, 22.93
LO 22.11: Worked Example 22.4; Problems 22.31, 22.100, 22.101

Section 22.7 Group 5A Elements

PRE-LECTURE:

Learning Objectives: *These are the skills you should be proficient in after reading this section and practicing the associated problems. Check off the boxes below once you have gone through this section, answered the associated questions in your textbook, and feel confident you have mastered the skill.*

- ☐ **LO 22.12** Write electron-dot structures and describe the structure, bonding, and properties of molecules and ions in main-group compounds containing nitrogen and phosphorus.
- ☐ **LO 22.13** Determine oxidation states of nitrogen and phosphorus in acids and write acid–base reactions.

Summary: *Read through this section and fill in the missing words and equations. Reference the textbook when you are not confident in your answers.*

Nitrogen, phosphorus, arsenic, antimony, and bismuth exhibit the expected trends down the periodic group—___________ atomic size, ___________ ionization energy and electronegativity, and acid–base properties of oxides. N and P are ___________, As and Sb are ___________, and Bi is a ___________.

The valence configuration of the group is ___________, which allows for a variety of oxidation states. Nitrogen and phosphorus will exhibit all oxidation states between −3 and +5, while As and Sb will exhibit both +3 and +5 oxidation states, but the +5 oxidation state is less stable as atomic size increases.

Sb^{3+} and Bi^{3+} are found in salts, but there are no simple cations of N or P.

As, Sb, and Bi are found in sulfide ores and are used in making various metal alloys. Arsenic is used in pesticides and semiconductors, and bismuth is present in some pharmaceuticals, such as Pepto-Bismol.

ELEMENTAL NITROGEN

Nitrogen is a colorless, odorless, tasteless gas that makes up 78% of Earth's atmosphere. It can be separated from liquid air by ______________________.

Nitrogen gas can be used as a protective inert atmosphere in manufacturing processes, while liquid nitrogen is commonly used as a refrigerant.

A ___________ amount of energy is required to break the N≡N bond, so reactions involving N_2 typically have a high activation energy and/or an unfavorable equilibrium constant. As temperatures increase, equilibrium shifts to the right. The many compounds formed by nitrogen are conveniently classified by oxidation state.

NITROGEN COMPOUNDS

___________ (NH_3) is the starting material for industrial synthesis of other nitrogen compounds. It is synthesized by the Haber process, which requires high temperatures, high pressures, and catalysis.

Colorless, strong-smelling, and gaseous at room temperature, NH_3 molecules have a lone pair of electrons, making them polar, trigonal pyramidal, water soluble, and easily condensed, because they are able to hydrogen bond. This ability also makes NH_3 an excellent solvent for ionic compounds. It is a Brønsted–Lowry base and reacts with acids to yield ammonium salts.

___________ ($H_2N–NH_2$) is an ammonia derivative in which one H is replaced by an *amino* (NH_2) group. It is prepared by reacting ammonia with OCl^-.

A poisonous, colorless liquid that smells like ammonia, hydrazine is explosive in the presence of oxidizing agents, including air, and is used as a rocket fuel: when it reacts with $N_2O_4(l)$, the two liquids generate large volumes of nitrogen gas and water vapor. It is safely handled in aqueous solution, where it is a weak base and versatile reducing agent.

Oxides of nitrogen: although many exist, we describe only three here:

- ______, or nitrous oxide, is a colorless, sweet-smelling gas used as a propellant. It is also used as a dental anesthetic, because it is mildly intoxicating in small doses, leading to its common name of "laughing gas." It is produced by gently heating molten ammonium nitrate.
- ______, or nitric oxide, is a colorless gas that can be produced in small quantities by reacting copper metal with dilute nitric acid. It is important in biological processes, where it transmits nerve impulses, kills harmful bacteria, and dilates blood vessels to increase blood flow.
- ______, or nitrogen dioxide, is a toxic, reddish-brown gas that can be produced by reacting copper with concentrated nitric acid. It is paramagnetic and tends to dimerize, forming an N–N bond between two units.

___________, or nitric acid, is a strong acid primarily used to manufacture ammonium nitrate for fertilizers, as well as explosives, plastics, and dyes.

___________ is a 3:1 volume ratio of concentrated HCl and concentrated HNO_3. It is an even stronger oxidizing agent and will oxidize even inactive metals, such as gold, which do not react with either component separately.

Workbook Problem 22.5

Upon heating, ammonium nitrate decomposes to produce nitrous oxide (N_2O) and water. What volume of nitrous oxide, collected over water, at a total pressure of 705 *mm* Hg and 22 °C, can be produced from 5.3 *g* ammonium nitrate? (The vapor pressure of water at 22 °C is 24 *mm* Hg.)

Strategy: After writing the balanced chemical reaction, determine the volume of N_2O from the pressure of N_2O using the ideal gas law. You must use Dalton's law to determine the pressure of N_2O.

Step 1: Write the balanced chemical equation.

Step 2:	Calculate the pressure of N_2O, using the total pressure and the vapor pressure of water.
Step 3:	Calculate the number of moles of N_2O produced from 3.5 *g* NH_4NO_3.
Step 4:	Calculate the volume of N_2O using the ideal gas law.

ELEMENTAL PHOSPHOROUS

Phosphorus is found in phosphate rock as $Ca_3(PO_4)_2$ and in fluorapatite as $Ca_5(PO_4)_3F$. Apatites are minerals with the general formula 3 $Ca_3(PO_4)_2 \cdot CaX_2$ ($X^- = F^-$ or OH^-).

Phosphorus is also important in living systems, and it is the sixth most abundant element in the human body. Tooth enamel is almost pure hydroxyapatite, $Ca_5(PO_4)_3OH$, and bones contain hydroxyapatite and collagen. Phosphate is also an essential part of the backbone of nucleic acids and the phospholipids that are the major components of cell membranes.

Industrial production of phosphorus involves heating phosphate rock, coke, and silica sand at 1500 °C. The gaseous P_4 tetrahedra produced are condensed by passing the mixture through water.

Two main allotropes of phosphorus exist:

- _________ phosphorus is toxic, waxy, white solid with discrete P_4 tetrahedral molecules. It is nonpolar, has a low melting point, and is highly reactive due to an unusual bonding structure that requires "bent" overlap of *p* orbitals. If exposed to air, white phosphorus will burst into flame.
- __________ phosphorus can be produced by heating white phosphorus without air. It is a polymeric form, which is nontoxic, less soluble, and less reactive than white phosphorus.

PHOSPHOROUS COMPOUNDS

Phosphorus will form compounds with all oxidation states from −3 to +5, but +3 and +5 are most common. It is less electronegative than nitrogen, so is more likely to be found in a positive oxidation state.
___________ (PH_3) is a colorless, extremely poisonous gas. It is neutral in solution and is easily oxidized, burning in air to form phosphoric acid, H_3PO_4.

Halides will form when phosphorus reacts with any halogen. When halogen is limited, the halides have the formula _____. When halogen is in excess, _____ is formed. These compounds are gases, volatile liquids, or low-melting-point solids.

Oxides of phosphorus follow a similar pattern. When oxygen is limited, ______ forms, but when oxygen is in excess, ______ is formed.

___________, H_3PO_3, is a weak diprotic acid in which only two of the three hydrogens are attached to oxygen.

___________, H_3PO_4, is a ___________ acid because all three hydrogens are attached to oxygens.

POST-LECTURE:

Work Problems: *One of the best ways to learn chemistry is through solving problems. Go to the textbook and work through the Worked Examples for this section, then go to the end of chapter and work the problems there. When you feel confident that you can correctly answer problems like the ones listed below for each objective, check off each of the Learning Objectives for this section above.*

LO 22.12: Problems 22.28, 22.108, 22.110, 22.114
LO 22.13: Problems 22.106, 22.107, 22.112

Section 22.8 Group 6A Elements

PRE-LECTURE:

Learning Objectives: *These are the skills you should be proficient in after reading this section and practicing the associated problems. Check off the boxes below once you have gone through this section, answered the associated questions in your textbook, and feel confident you have mastered the skill.*

- ☐ **LO 22.14** Classify oxides as ionic or covalent and as acidic, basic, or amphoteric and write balanced equations for the reaction that occurs when oxides dissolve in water.
- ☐ **LO 22.15** Write electron-dot structures and describe the structure, bonding, and properties of compounds containing oxygen and sulfur.

Definitions: *To learn, you must do. This is your workbook. Fill in the definitions from the textbook as a first step to understanding the material.*

— **oxide—** ____

____

— **contact process—** ____

____

Summary: *Read through this section and fill in the missing words and equations. Reference the textbook when you are not confident in your answers.*

Oxygen, sulfur, selenium, tellurium, and polonium exhibit the expected trends down the periodic group. O and S are typical ___________. Se and Te are largely nonmetallic and are both ___________. Polonium is a radioactive ___________ found in trace amounts in uranium ores.

The valence configuration of the group is ___________, which makes the −2 oxidation state the most common. As metallic character ___________, this oxidation state is increasingly unstable, and S, Se, and Te are commonly found in positive oxidation states, especially +4 and +6.

Commercial uses of Se, Te, and Po are limited. Se is used in making red glass and in photocopiers, Te is used in alloys to improve machinability, and Po is a heat source for space equipment and a source of alpha particles.

ELEMENTAL OXYGEN

Oxygen is the most abundant element on the surface of our planet and is crucial to human life. Approximately 25% of all biological molecules contain oxygen.

Oxygen can also be prepared on an industrial scale by undergoing the fractional distillation of liquefied air. More than 66% of the oxygen produced industrially is used in making steel. Oxygen is also used in sewage treatment and in paper bleaching.

OXYGEN COMPOUNDS

Oxygen is reactive as can be predicted by its electron configuration and ___________ electronegativity.

<u>Acid–Base Properties of Oxides</u>
Binary compounds with oxygen (with a −2 oxidation state) are called ___________ and can be acidic, basic, or amphoteric.

ELEMENTAL SULFUR

Sulfur accounts for 0.016% of the Earth's crust by mass, has numerous allotropes, and is found in large underground deposits and as a component of numerous minerals. In biology, it is a component of proteins.

___________ sulfur is the most stable form. It is a yellow crystalline solid that contains crown-shaped S_8 rings. Above 95 °C, monoclinic sulfur is more stable. In this allotrope, the S_8 rings have a different packing arrangement than the crystals of rhombic sulfur.

Molten sulfur is a fluid, straw-colored liquid at just above its melting point, but between 160 °C and 195 °C, the S_8 rings open and form long chains that can contain more than 200,000 S atoms. In this state, the sulfur becomes viscous and dark reddish-brown. Above 195 °C, the chains break, and viscosity declines. If molten sulfur is dropped in water, it freezes in a disordered state known as plastic sulfur.

SULFUR COMPOUNDS

Sulfur forms numerous compounds:

- ___________ (H_2S) is a colorless gas with a strong odor of rotten eggs. It is very toxic and dulls the sense of smell, making exposure all the more dangerous. It can be produced in the lab by treating iron(II) sulfide with dilute sulfuric acid or by hydrolysis of thioacetamide for use in qualitative analysis. It is a weak diprotic acid and a mild reducing agent.

- ___________ and ___________ (SO_2 and SO_3) are the most important oxides of sulfur. Sulfur dioxide is produced by burning sulfur and is a colorless, toxic gas with a pungent odor. It is toxic to microorganisms, so is used to sterilize wine and dried fruit. In the atmosphere, sulfur dioxide is oxidized to sulfur trioxide, which dissolves in rainwater to form sulfuric acid. Both species are acidic, but little sulfurous acid (H_2SO_3) is found in solution: SO_2 will dissolve without reacting.
- Sulfuric acid is the world's most important industrial chemical. It is manufactured by the **contact process:** S is burned to SO_2, SO_2 is oxidized to SO_3 with a catalyst, and the SO_3 is dissolved in sulfuric acid, rather than water, to make more sulfuric acid because SO_3 dissolves slowly in water. Concentrated sulfuric acid is 18 M and 98% H_2SO_4 by mass.
- Sulfuric acid is a strong acid at the first proton and dissociates at the second proton with a K_{a2} of 1.2×10^{-2}. It forms two series of salts: ___________, with a hydrogen as one of the cations ($NaHSO_4$), and ___________ (Na_2SO_4). In dilute solutions at room temperature, sulfuric acid behaves like HCl, oxidizing metals above H in the activity series. Hot, concentrated H_2SO_4 is a stronger oxidant than sulfuric acid and will come apart to re-form SO_2. Sulfuric acid is used to manufacture fertilizers and in many other industrial processes.

POST-LECTURE:

Work Problems: *One of the best ways to learn chemistry is through solving problems. Go to the textbook and work through the Worked Examples for this section, then go to the end of chapter and work the problems there. When you feel confident that you can correctly answer problems like the ones listed below for each objective, check off each of the Learning Objectives for this section above.*

LO 22.14: Worked Example 22.5; Problems 22.24, 22.26, 22.130
LO 22.15: Problems 22.140–22.143

Section 22.9 Group 7A: The Halogens

PRE-LECTURE:

Learning Objectives: *These are the skills you should be proficient in after reading this section and practicing the associated problems. Check off the boxes below once you have gone through this section, answered the associated questions in your textbook, and feel confident you have mastered the skill.*

- ☐ **LO 22.16** Write balanced equations for the formation of halogens and reactions of their compounds.
- ☐ **LO 22.17** Determine oxidation states of halogens in oxoacids and predict the relative acid strengths of oxoacids.

Summary: *Read through this section and fill in the missing words and equations. Reference the textbook when you are not confident in your answers.*

The halogens are too __________ to occur in nature as free elements. They are diatomic __________, characterized by their gaining electrons in chemical reactions. They have large __________ electron affinities and __________ positive ionization energies.

Production: Halogens are produced by __________ of their anions. Fluorine and chlorine are produced by electrolysis, while bromine and iodine are prepared by oxidation of their anions by chlorine.

HALOGEN COMPOUNDS

Halogens are among the __________ reactive elements in the periodic table, with fluorine being the most reactive: it will form compounds with every other element in the periodic table except He, Ne, and Ar. Halogens become __________ reactive as they get larger, for exactly the same reason that metals gain reactivity as they get larger: outer electron shells are more weakly held, so elements that are gaining rather than losing electrons will become less reactive as their size increases.

__________ can be formed from every metal, predictably with main-group metals and less-predictably with transition metals, some of which will form more than one halide.

Hydrogen and halogens react to form hydrogen halides: fluorine reacts explosively and spontaneously, chlorine reacts in the presence of UV light or a spark, and bromine and iodine react very slowly. Hydrogen halides will dissolve in water to form **acids.** Hydrofluoric acid, though it is the only weak acid among the hydrogen halides, is also one of the few substances that can be used to etch glass.

Halogens have a valence configuration of __________. They gain an electron to ionize and share one electron in molecular compounds in which they have an oxidation state of −1.

Numerous positive oxidation states are also available for the larger halogens—Cl, Br, and I—allowing them to share valence electrons with oxygen. The oxidation states available are +1, +3, +5, or +7, and the general formula for an oxoacid is HXO_n, where the value of n depends on the oxidation state of the halogen.

The higher the oxidation state of the halogen, the __________ the acid. Acidic protons are bonded to the oxygen (despite how the formula is typically written), and these are all strong oxidizing agents.

POST-LECTURE:

Work Problems: *One of the best ways to learn chemistry is through solving problems. Go to the textbook and work through the Worked Examples for this section, then go to the end of chapter and work the problems there. When you feel confident that you can correctly answer problems like the ones listed below for each objective, check off each of the Learning Objectives for this section above.*

LO 22.16: Problems 22.146, 22.148, 22.149
LO 22.17: Problems 22.150, 22.154

Section 22.10 Group 8A: Noble Gases

PRE-LECTURE:

Learning Objectives: *These are the skills you should be proficient in after reading this section and practicing the associated problems. Check off the boxes below once you have gone through this section, answered the associated questions in your textbook, and feel confident you have mastered the skill.*

☐ **LO 22.18** Draw electron-dot structures and predict geometry for molecules and ions containing noble gases.

Summary: *Read through this section and fill in the missing words and equations. Reference the textbook when you are not confident in your answers.*

Noble gases are ___________ gases. They have ___________ outer valence shells and complete energy levels. They become ___________ reactive as they become larger, because their outer electrons are less tightly held than those closer to the nucleus. Even the larger noble gases will react only with ___________, the most electronegative of all the elements.

POST-LECTURE:

Work Problems: *One of the best ways to learn chemistry is through solving problems. Go to the textbook and work through the Worked Examples for this section, then go to the end of chapter and work the problems there. When you feel confident that you can correctly answer problems like the ones listed below for each objective, check off each of the Learning Objectives for this section above.*

LO 22.18: Problems 22.157, 22.158

Self-Test

This section is intended to test your knowledge of the material covered in this chapter. Think through these problems and make certain you understand what the problem is about—ask yourself if the answer makes sense. Many of these questions are linked to the chapter learning goals. Therefore, successful completion of these problems indicates that you have achieved the learning goals for this chapter. You will receive the greatest benefit from this section if you use it as a mock exam. Do not use your notes, book, Internet, or friends when taking the test. There is no grade and you only shortchange yourself by getting help with this activity. As you take the test, mark the problems with your level of confidence with the answer.

- ★ *Use a star for problems you know you got right.*
- ? *Use a question mark for those you think you got right.*
- x *Use an x for those you are guessing on.*

If you get a question right that you marked with a star, you can be confident that you know the associated LO.

If you get a question right that you marked with a question mark, go back and work a few more problems and test yourself again tomorrow.

If you get a question wrong or mark it with an x, go back and read the section again, work a few problems today, a few tomorrow, and a few the next day before testing yourself again.

This will help you to discover which topics you have learned thoroughly and which topics you need to study in more detail.

True–False

1. The periodic table is divided into metals on the left, nonmetals on the right, and semimetals in the center.

2. Metallic character decreases from top to bottom in a column.

3. Second-row elements have unique bonding properties because of their small sizes and high electronegativities.

4. It is common to find double (π) bonds in third- and fourth-row elements.

5. Boranes can be connected with bridging hydrogen atoms.

6. Carbon allotropes include nonconducting covalent solids and sheets with delocalized electrons that will conduct electricity.

7. Silicates are major components of the Earth's crust and often consist of lengthy chains of double-bonded SiO_4^{4-} tetrahedra.

8. Nitrogen is obtained in pure form by fractional distillation of liquefied air.

9. White phosphorus is made up of tetrahedra of P atoms with strained, unstable bonds.

10. When it is liquefied, sulfur decreases steadily in viscosity to its boiling point.

Multiple Choice

11. Which of the following are characteristic of metals?
    a. brittle
    b. low ionization energy
    c. form covalent hydrides
    d. readily become anions

12. Because of their small size, second-row elements are able to form a maximum of ____ bonds.
    a. 2
    b. 3
    c. 4
    d. 5

13. Boron halides are able to act as Lewis acids because
    a. the empty *p* orbital can accept a lone pair from another atom
    b. they readily give up hydrogen ions in solution
    c. they are so reactive that they split water molecules
    d. all of the above

14. Nitrogen can exist in which of the following oxidation states?
    a. −5, −4, −3, 0, +1, +2
    b. −3, −2, −1, 0
    c. +1, +2, +3, +4, +5
    d. −3, −2, −1, 0, +1, +2, +3, +4, +5

15. Phosphorus is found in
   a. bones and teeth
   b. cell membranes
   c. DNA and RNA
   d. all of the above

16. 6A elements have the electron configuration
   a. ns^2np^2
   b. ns^2np^3
   c. ns^2np^4
   d. ns^2np^5

17. The most reactive halogen is
   a. F
   b. Cl
   c. Br
   d. I

18. The least reactive elements in the periodic table are the
   a. halogens
   b. noble gases
   c. metals
   d. boranes

19. The only elements that will form triple bonds are
   a. N and P
   b. C and N
   c. Br and I
   d. O and N

20. What type of hydride is TiH_2?
   a. ionic hydride
   b. covalent hydride
   c. polar covalent hydride
   d. interstitial hydride

Fill in the Blanks

21. From left to right across the periodic table, effective nuclear charge_________________, ionization energy _________________, atomic radius _________________, and metallic character _________________.

22. Carbon has a number of allotropes, ranging from the tetrahedral covalent solid _________________, to the sheets of _______________, to the remarkable spheres of _________________.

23. When phosphorus is manufactured, an unstable allotrope known as ____________ ____________ is produced. It is highly reactive because of the ____________ bonds in the P_4 tetrahedra. If it is heated in the absence of air, it is converted to the more stable, less toxic, polymeric form known as ____________ ____________.

24. Silicates are typically found as chains of rings with a basic unit of _______________.

25. The properties of sulfur are dramatically temperature dependent. At room temperature, sulfur is a ________________. When heated, it melts into a ________________, then becomes more viscous as the ______________ break and form ______________. When these begin to break down, the ______________ of the liquid begins to decrease with temperature.

Matching

26. aluminosilicate
27. borane
28. carbide
29. contact process
30. aqua regia
31. silicate
32. zone refining

a. a mixture of concentrated hydrochloric acid and concentrated nitric acid
b. purification technique in which a heater melts a narrow zone at the top of a rod of a material, then sweeps slowly down the rod, carrying impurities
c. silicate mineral in which partial substitution of Al^{3+} for Si^{4+} has occurred
d. any compound of boron and hydrogen
e. an ionic compound containing silicon oxoanions along with other cations
f. commercial process for making sulfuric acid from sulfur
g. binary compound in which carbon has a negative oxidation state

Problems

33. For the following pairs of elements, determine which has the more metallic character.
    a. Na or Rb   b. In or Ga   c. Cs or Ba

34. For the following pairs of elements, determine which has the more ionic hydride.
    a. K or Na   b. Sn or In   c. Sb or As

35. What are the general trends down a group in the periodic table for ionization energy, atomic radius, electronegativity, and basicity of oxides?

36. Describe the unique bonding in diborane.

37. Describe four of the allotropes of carbon in terms of their bonding, structure, and properties.

38. Give an example of a nitrogen-containing compound for each common oxidation state exhibited by nitrogen.

39. Give the name of the halogen oxoacid HXO_n (X = Cl, Br, or I and n = 1, 2, 3, and 4).

40. Give the chemical equation for the production of NO_2 from nitric acid and copper.

Challenge Problem

41. Calcium carbide reacts with water to form acetylene (C_2H_2). If 15.0 g of calcium carbide react with 7.45 g of water, how many grams of acetylene are produced? If the resulting acetylene is burned to form carbon dioxide and water, how much energy is produced? (Assume 100% efficiency for all reactions.)

    Possibly useful information:

    ΔH_f $C_2H_2(g)$ = 227.4 kJ/mol
    ΔH_f $CO_2(g)$ = −393.5 kJ/mol
    ΔH_f $H_2O(g)$ = −241.8 kJ/mol

Test Yourself: *Take the Practice Test on pages* 993 and 994 *of your textbook. Again, use a testing environment using only those items that your instructor will allow you to have access to during the test and mark each question with a star, question mark, or x to denote your confidence level. Once you have finished you can check off the LO at the beginning of this chapter based on which questions you starred and got correct. Go back and spend more time with the topics from the other learning objectives.*

CHAPTER TWENTY-THREE

Organic and Biological Chemistry

Learning Objectives

As a result of reading and studying this chapter, you should be able to:

- ☐ **LO 23.1** Represent the chemical structure of alkanes as condensed structures or line drawings.
- ☐ **LO 23.2** Identify and draw constitutional isomers of alkanes.
- ☐ **LO 23.3** Identify chiral centers in a molecule and classify molecules as chiral or achiral.
- ☐ **LO 23.4** Identify and name functional groups in organic molecules.
- ☐ **LO 23.5** Represent molecules with functional groups using condensed structures and line drawings.
- ☐ **LO 23.6** Classify a monosaccharide as an aldose or ketose and draw a typical open chain structure.
- ☐ **LO 23.7** Classify the type of isomerism in monosaccharides.
- ☐ **LO 23.8** Draw orbital overlap pictures to describe bonding in organic compounds.
- ☐ **LO 23.9** Identify and draw cis–trans isomers.
- ☐ **LO 23.10** Recognize and draw the structure of triacylglycerols.
- ☐ **LO 23.11** Relate lipid structure to physical properties.
- ☐ **LO 23.12** Use curved arrows to draw resonance structures and specify formal charges in organic line drawings.
- ☐ **LO 23.13** Draw orbital overlap pictures for conjugated systems and identify localized and delocalized electrons.
- ☐ **LO 23.14** Identify conjugated systems and the hybridization of atoms included in the system.
- ☐ **LO 23.15** Identify amino acids and draw structures for peptides.
- ☐ **LO 23.16** Predict if an organic compound is aromatic.
- ☐ **LO 23.17** Draw structures showing how different components of the nucleic acid are joined, and write the DNA sequence that is complementary to a specified strand.
- ☐ **LO 23.18** Draw resonance structures to show that DNA bases are aromatic.

Chapter Summary

Organic chemistry is the study of carbon compounds. In this chapter, we introduce you to the vast array of compounds and reactions in organic chemistry. We will describe naming of compounds, varying structures with multiple bonds, cyclic structures and functional groups, and the most basic reactions of these molecules. We will then provide a basic introduction to biochemical energetics and the main classes of biomolecules. Proteins, carbohydrates, lipids, and nucleic acids are discussed in terms of their basic building blocks, functions, and properties.

The Chapter in Detail

Section 23.1 Organic Molecules and Their Structures: Constitutional Isomers

PRE-LECTURE:

Learning Objectives: *These are the skills you should be proficient in after reading this section and practicing the associated problems. Check off the boxes below once you have gone through this section, answered the associated questions in your textbook, and feel confident you have mastered the skill.*

- ☐ **LO 23.1** Represent the chemical structure of alkanes as condensed structures or line drawings.
- ☐ **LO 23.2** Identify and draw constitutional isomers of alkanes.

Definitions: *To learn, you must do. This is your workbook. Fill in the definitions from the textbook as a first step to understanding the material.*

— **biochemistry—**______________________________

____

— **organic chemistry—** ______________________________

____

— **hydrocarbon—** ______________________________

____

— **alkane—**______________________________

____

— **straight-chain alkane—** ______________________________

____

— **constitutional isomer—**______________________________

____

— **branched-chain alkane—** ______________________________

____

— **condensed structure—** ______________________________

____

— **line drawing—**______________________________

____

— **cycloalkane—** ______________________________

____

Review: *Before going through the section make sure you are familiar with these concepts:*

- Section 8.1 VSEPR

Summary: *Read through this section and fill in the missing words and equations. Reference the textbook when you are not confident in your answers.*

There are a multitude of single-bonded compounds that contain only carbon and hydrogen. They are known as ___________ and belong to a family of single-bonded organic molecules called ___________.

Each carbon can form _____ bonds, either to hydrogen or to another carbon. There is only one possible structure each for alkanes with one to three carbons, but at four carbons, ___________-chain hydrocarbons become possible. This greatly increases the number of possible compounds. At six carbons, for example, there is one straight-chain alkane with no branches and four branched-chain alkanes.

Compounds such as these, with the same formula but different arrangements of atoms are called ___________. Despite their identical numbers and types of atoms, their differing structures give them ___________ properties, both physical and chemical.

There are a number of ways to represent these molecules, expanded structures (shown below):

can be condensed. Here, vertical bonds are not shown, but "understood."

CH_3CH_3

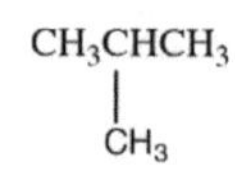

The condensed structure provides information about connectivity only; information about 3-D shape can be predicted from VSEPR. As molecules get larger, they are more and more able to adopt different conformations, but their connectivity remains the same. As a result, all of the following are different representations of the same molecule:

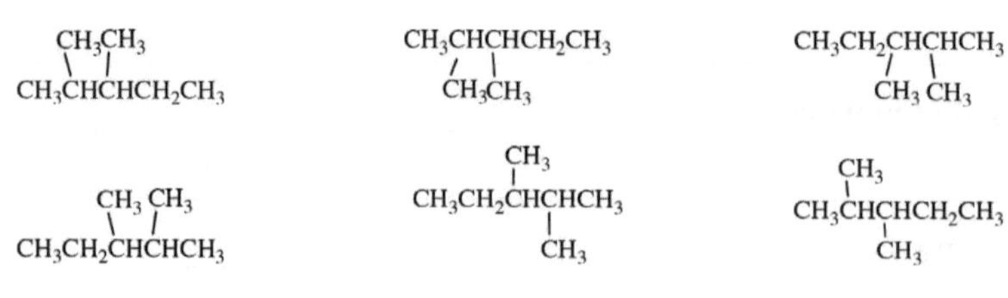

These are all the same molecule, 2,3-dimethylpentane.

Single bonds are free to rotate, so there are an infinite number of possible *conformations* for organic molecules. Most alkanes will have an extended zig-zag shape due to the tetrahedral arrangement of sp^3 orbitals.

Line drawings are another way to represent the structure of organic molecules. Here, each vertex or endpoint represents a ___________ and hydrogen atoms can be determined using the knowledge that each carbon forms ____ bonds.

Workbook Example 23.1

PROBLEM:
Draw the straight-chain alkane C_8H_{18} as a condensed structure. Then draw the line drawing.

SOLUTION:
Remembering that all carbons in an alkane must have four bonds, the structure is:

$$CH_3CH_2CH_2CH_2CH_2CH_2CH_2CH_3$$

To draw the line drawing, a zig-zag line with two endpoints and six vertices is needed.

Ringed molecules also form and are known as cyclic compounds.

POST-LECTURE:

Work Problems: *One of the best ways to learn chemistry is through solving problems. Go to the textbook and work through the Worked Examples for this section, then go to the end of chapter and work the problems there. When you feel confident that you can correctly answer problems like the ones listed below for each objective, check off each of the Learning Objectives for this section above.*

LO 23.1: Worked Example 23.1; Problems 23.42, 23.43
LO 23.2: Worked Example 23.2; Problems 23.44, 23.45

Section 23.2 Stereoisomers: Chiral Molecules

PRE-LECTURE:

Learning Objectives: *These are the skills you should be proficient in after reading this section and practicing the associated problems. Check off the boxes below once you have gone through this section, answered the associated questions in your textbook, and feel confident you have mastered the skill.*

☐ **LO 23.3** Identify chiral centers in a molecule and classify molecules as chiral or achiral.

Review: *Before going through the section make sure you are familiar with these concepts:*

- Section 21.7 stereoisomers
- Section 21.8 chirality

Summary: *Read through this section and fill in the missing words and equations. Reference the textbook when you are not confident in your answers.*

___________ are chiral molecules that are nonidentical mirror images of one another. Chiral centers are atoms that are attached to _____ different groups. Large organic molecules can have a large number of chiral centers; when there are *n* different chiral centers, there are _____ possible isomers.

POST-LECTURE:

Work Problems: *One of the best ways to learn chemistry is through solving problems. Go to the textbook and work through the Worked Examples for this section, then go to the end of chapter and work the problems there. When you feel confident that you can correctly answer problems like the ones listed below for each objective, check off each of the Learning Objectives for this section above.*

LO 23.3: Worked Example 23.3; Problems 23.49–23.52

Section 23.3 Families of Organic Compounds: Functional Groups

PRE-LECTURE:

Learning Objectives: *These are the skills you should be proficient in after reading this section and practicing the associated problems. Check off the boxes below once you have gone through this section, answered the associated questions in your textbook, and feel confident you have mastered the skill.*

- ☐ **LO 23.4** Identify and name functional groups in organic molecules.
- ☐ **LO 23.5** Represent molecules with functional groups using condensed structures and line drawings.

Definitions: *To learn, you must do. This is your workbook. Fill in the definitions from the textbook as a first step to understanding the material.*

— **functional group-** ____

____

— **alkene—** ____

____

— **alkyne—** ____

____

— **carbonyl group—** ____

____

Structures: *Draw an example of each of the following functional groups.*

alkane	alkene	alkyne
aromatic	alcohol	ether
aldehyde	ketone	carboxylic acid
ester	amine	amide

Summary: *Read through this section and fill in the missing words and equations. Reference the textbook when you are not confident in your answers.*

Organic compounds can be classified into families according to ___________ features, and members of a given family also have similar ___________. Alkanes, alkenes, alkynes, and the aromatic compounds contain only carbon–carbon bonds. Others contain bonds to nitrogen or oxygen: some contain bonds to halogens as well. A feature in many functional groups is the C=O or ___________ group.

The following is a condensed version of the much more extensive Table 23.1 in your textbook: for additional information, refer to that chart as well.

Family Name	**Functional Group**	**Name Ending**
	Single bonds	-ane
	One or more C=C bonds	-ene
	One or more C≡C bonds	-yne
Arenes	C_6 ring, alternating C-C and C=C bonds	None
	-C-OH	-ol
	-C-O-C-	-ether
	-C-N-	-amine
	$-C-\overset{\overset{O}{\|}}{C}-H$	-al
	$-C-\overset{\overset{O}{\|}}{C}-C-$	-one
	$-C-\overset{\overset{O}{\|}}{C}-OH$	-oic acid
	$-C-\overset{\overset{O}{\|}}{C}-O-C-$	-oate
	$-C-\overset{\overset{O}{\|}}{C}-N-$	-amide

Workbook Problem 23.1

Identify the functional groups in the following compounds:

a. $CH_3-\overset{\overset{O}{\|}}{C}-OH$ b. $CH_3-\overset{\overset{O}{\|}}{C}-CH_3$ c. $CH_3-\overset{\overset{O}{\|}}{C}-H$ d. CH_3-O-CH_3

Strategy: Look for noncarbon atoms and multiple bonds. Compare with the chart above to name the functional groups.

Step 1: Name the functional groups.

POST-LECTURE:

Work Problems: *One of the best ways to learn chemistry is through solving problems. Go to the textbook and work through the Worked Examples for this section, then go to the end of chapter and work the problems there. When you feel confident that you can correctly answer problems like the ones listed below for each objective, check off each of the Learning Objectives for this section above.*

LO 23.4: Problems 23.33, 23.34
LO 23.5: Worked Example 23.4; Problems 23.58, 23.61, 23.64, 23.66

Section 23.4 Carbohydrates: A Biological Example of Isomers

PRE-LECTURE:

Learning Objectives: *These are the skills you should be proficient in after reading this section and practicing the associated problems. Check off the boxes below once you have gone through this section, answered the associated questions in your textbook, and feel confident you have mastered the skill.*

☐ **LO 23.6** Classify a monosaccharide as an aldose or ketose and draw a typical open chain structure.
☐ **LO 23.7** Classify the type of isomerism in monosaccharides.

Definitions: *To learn, you must do. This is your workbook. Fill in the definitions from the textbook as a first step to understanding the material.*

— **carbohydrate—** ____

____

— **monosaccharides—** ____

____

— **polysaccharides—** ____

____

— **anomer—** ____

____

Summary: *Read through this section and fill in the missing words and equations. Reference the textbook when you are not confident in your answers.*

Carbohydrates occur in all living organisms. The name originated with carbon, with the empirical formula $(CH_2O)_n$, which led to the idea that it was a hydrate of carbon. Although that idea was discarded, the name persisted, and it is now used for the large class of hydroxyl-containing aldehydes and ketones that we commonly call ___________.

MONOSACCHARIDES

______________________, or *simple sugars,* are carbohydrates that cannot be broken into smaller molecules by acid hydrolysis. An *aldose* contains an ___________ carbonyl group, while a *ketose* contains a ___________ carbonyl group. The *–ose* suffix designates a sugar. A prefix; ______, ______, ______, or _____ denotes the number of carbons in the sugar.

Carbohydrates exist in a variety of isomeric forms. ______________________ isomers have the atoms connected differently. ______________________ are the nonsuperimposable mirror images of one another where D means right and L means left. ___________ occur as a ring is formed. The –OH group at C1 can end up on the opposite side of the ring as C6 to form the α-anomer or the same side of the ring as C6 to form the β-anomer.

Workbook Example 23.2

PROBLEM:

Classify the following monosaccharides:

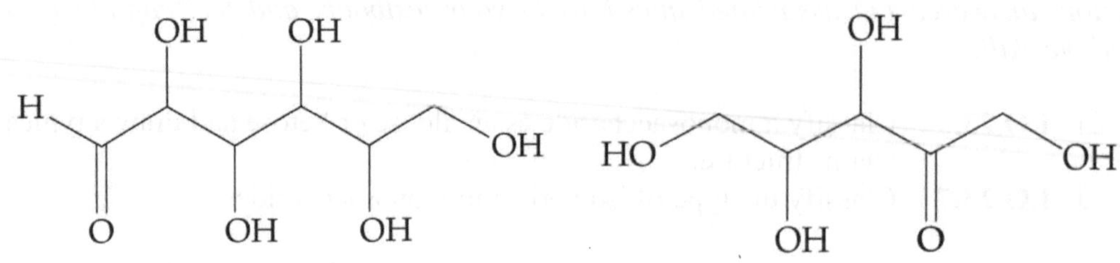

SOLUTION:
The molecule on the left has six carbons and an aldehyde group. This is an aldohexose.
The molecule on the right has five carbons and a keto group. This is a ketopentose.

POLYSACCHARIDES
Polysaccharides are formed when monosaccharides join together. ___________ is a *disaccharide* formed of one glucose and one fructose molecule. Cellulose, the structural material in plants, is made up of thousands of __-glucose molecules joined together in an immensely long chain, while starch is made up of thousands of __-glucose units. This small difference in configuration makes starches—found in rice, beans, and potatoes—digestible to us, and cellulose indigestible.

POST-LECTURE:

Work Problems: *One of the best ways to learn chemistry is through solving problems. Go to the textbook and work through the Worked Examples for this section, then go to the end of chapter and work the problems there. When you feel confident that you can correctly answer problems like the ones listed below for each objective, check off each of the Learning Objectives for this section above.*

LO 23.6: Problems 23.71, 23.72
LO 23.7: Problems 23.74, 23.75

Section 23.5 Valence Bond Theory and Orbital Overlap Pictures

PRE-LECTURE:

Learning Objectives: *These are the skills you should be proficient in after reading this section and practicing the associated problems. Check off the boxes below once you have gone through this section, answered the associated questions in your textbook, and feel confident you have mastered the skill.*

- ☐ **LO 23.8** Draw orbital overlap pictures to describe bonding in organic compounds.
- ☐ **LO 23.9** Identify and draw cis–trans isomers.

Definitions: *To learn, you must do. This is your workbook. Fill in the definitions from the textbook as a first step to understanding the material.*

— **cis–trans isomers—** ____

____

Review: *Before going through the section make sure you are familiar with these concepts:*

- Section 8.2 valence bond theory
- Section 8.2 sigma bonds
- Sections 8.3 and 8.4 hybridization of orbitals
- Section 8.4 pi-bonds

Summary: *Read through this section and fill in the missing words and equations. Reference the textbook when you are not confident in your answers.*

Condensed structure and line drawings are simplistic representations of molecules. When a more complete picture is needed, valence bond theory will help visualize orbital overlap to help interpret structure and predict reactivity.

Workbook Example 23.3

PROBLEM:
Draw the structure of ethanal (CH_3-CHO) and identify the hybridization of each carbon.

SOLUTION:
The -al ending of ethanal indicates that we are dealing with an aldehyde. The ethan-part of the name indicates that we have two carbons. Add hydrogens so that each carbon has four bonds. Putting this together we have:

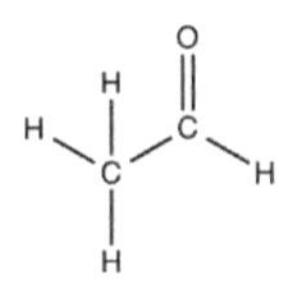

Based on this structure, the carbon on the left is *sp*^{3}-hybridized, and the carbon on the right is *sp*^{2}-hybridized.

Organic molecules can have ____, ____, or ____ hybridized atoms. The *sp*-hybridization is the least common, while *sp*^{2}-hybridization is common and important. Since the π bond __________ rotation around the bond, there can be two isomers almost anytime there is a double bond. The cis isomer has the same groups on the __________ side of the double bond, and the trans isomers have the same groups on __________ sides of the double bond.

Workbook Example 23.4

PROBLEM:
Which of the following molecules can exhibit cis–trans isomerism? For those that do, decide which isomer is shown.

a.  b. 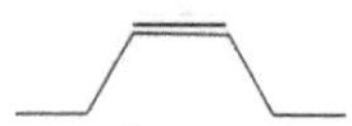 c. 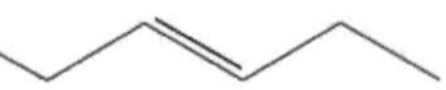

SOLUTION:
Locate the similar groups across the double bonds. If the groups are the on the same side the molecule is cis, if they are on opposite sides the molecule is trans. If there are no similar groups on the carbons across the double bond, cis–trans isomerism is not possible.

a. This molecule does not show cis–trans isomerism because one side of the double bond has two of the same groups (H) attached.
b. A trans-isomer is shown with both of the hydrogen atoms across the double bond, but on the same side.
c. A cis–isomer is shown with both of the hydrogen atoms across the double bond, but on opposite sides.

POST-LECTURE:

Work Problems: *One of the best ways to learn chemistry is through solving problems. Go to the textbook and work through the Worked Examples for this section, then go to the end of chapter and work the problems there. When you feel confident that you can correctly answer problems like the ones listed below for each objective, check off each of the Learning Objectives for this section above.*

LO 23.8: Worked Example 23.6; Problems 23.76, 23.78, 23.79
LO 23.9: Worked Example 23.7; Problems 23.80–23.85

Section 23.6 Lipids: A Biological Example of Cis–Trans Isomerism

PRE-LECTURE:

Learning Objectives: *These are the skills you should be proficient in after reading this section and practicing the associated problems. Check off the boxes below once you have gone through this section, answered the associated questions in your textbook, and feel confident you have mastered the skill.*

☐ **LO 23.10** Recognize and draw the structure of triacylglycerols.
☐ **LO 23.11** Relate lipid structure to physical properties.

Definitions: *To learn, you must do. This is your workbook. Fill in the definitions from the textbook as a first step to understanding the material.*

— **lipid—** ____

____

— **triacylglycerol—** ________________________________

____

— **fatty acid—** ____

____

— **unsaturated—** ____

____

— **saturated—** ____

____

— **hydrogenation—** ____

____

— **polyunsaturated fats—** ____

____

Review: *Before going through the section make sure you are familiar with these concepts:*

- Section 8.6 London dispersion forces

Summary: *Read through this section and fill in the missing words and equations. Reference the textbook when you are not confident in your answers.*

A lipid is an organic molecule that dissolves in nonpolar solvent. The fact that *lipids* are defined by a physical property means that they exist in a variety of forms, but they all contain large ________ portions.

Animal fats and vegetable oils are the most abundant lipids. Fats and oils are both *triacylglycerols*, or *triglycerides*. These are esters of glycerol (1,2,3-propanetriol) with three long-chain carboxylic acids called *fatty acids*.

Fatty acids are classified as:

- ________ if they have the maximum number of hydrogens on each carbon (outside of the carbonyl carbon);

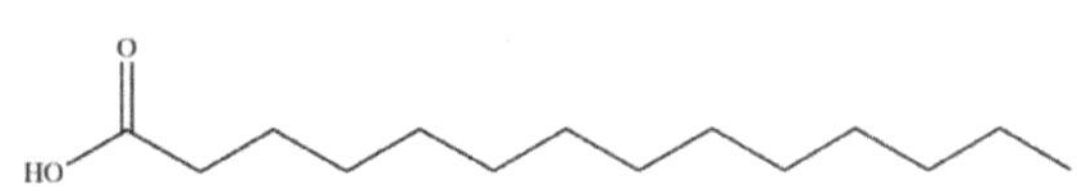

myristic acid

- ________ if they have a carbon to carbon double bond and therefore less than the maximum number of hydrogens on a carbon;

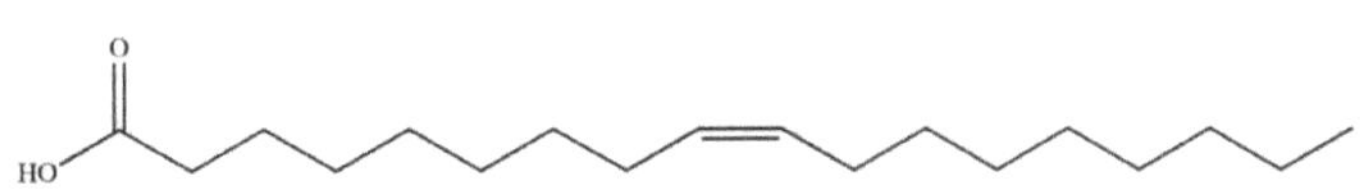

oleic acid

- or ________ if there is more than one carbon to carbon double bond.

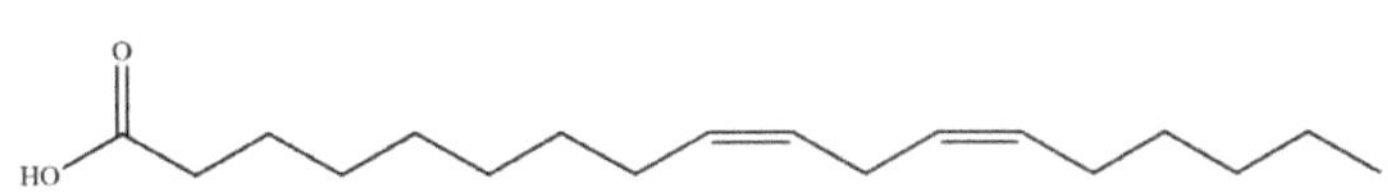

linoleic acid

It is the different properties of these fatty acids that lead to the different properties of fats and oils. Fats have primarily saturated fatty acids that pack together more efficiently increasing the London dispersion forces between the individual molecules and therefore they have higher melting points. Oils contain a higher proportion of unsaturated fatty acids. Unsaturated fatty acids have kinks in the hydrocarbon tail. This interferes with stacking of the molecules and ________ the intermolecular forces lowering melting points.

Unsaturated fats can be chemically processed and converted to saturated fatty acids. To do this, hydrogen atoms are added to the carbons of the double bond through a process called ________.

POST-LECTURE:

Work Problems: *One of the best ways to learn chemistry is through solving problems. Go to the textbook and work through the Worked Examples for this section, then go to the end of chapter and work the problems there. When you feel confident that you can correctly answer problems like the ones listed below for each objective, check off each of the Learning Objectives for this section above.*

LO 23.10: Problems 23.88, 23.89
LO 23.11: Problems 23.92, 23.93

Section 23.7 Formal Charge and Resonance in Organic Compounds

PRE-LECTURE:

Learning Objectives: *These are the skills you should be proficient in after reading this section and practicing the associated problems. Check off the boxes below once you have gone through this section, answered the associated questions in your textbook, and feel confident you have mastered the skill.*

☐ **LO 23.12** Use curved arrows to draw resonance structures and specify formal charges in organic line drawings.

Review: *Before going through the section make sure you are familiar with these concepts:*

- Section 7.10 formal charge
- Section 7.9 resonance structures

Summary: *Read through this section and fill in the missing words and equations. Reference the textbook when you are not confident in your answers.*

Formal charge and resonance help in understanding ________ and ________. Most organic reactions occur at regions of high and low electron density, and this is often shown as formal charge.

Workbook Example 23.5

PROBLEM:
Calculate formal charge on the oxygen and nitrogen atoms in the molecules below.

a. 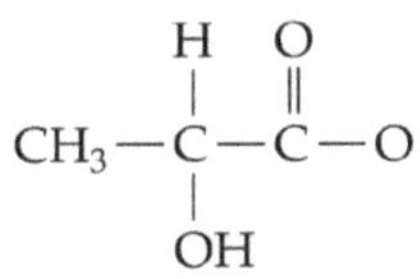

b. 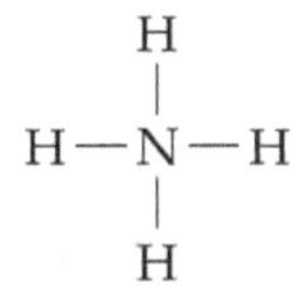

SOLUTION:

$$\text{Formal charge} = \begin{pmatrix} \text{Number of} \\ \text{valence electrons} \\ \text{in free atoms} \end{pmatrix} - \frac{1}{2}\begin{pmatrix} \text{number of} \\ \text{bonding} \\ \text{electrons} \end{pmatrix} - \begin{pmatrix} \text{number of} \\ \text{nonbonding} \\ \text{electrons} \end{pmatrix}$$

a. The oxygen in the carbonyl group (C=O) has a formal charge of 0.

$$(6) - \tfrac{1}{2}(4) - 4 = 0$$

The other oxygen (C—O) has a formal charge of −1.

$$(6) - \tfrac{1}{2}(2) - 6 = -1$$

b. The nitrogen has a formal charge of +1.

$$(5) - \tfrac{1}{2}(8) - 0 = +1$$

There are a few common patterns that indicate a resonance structure can be drawn for a particular molecule:

- A lone pair on an atom that is a sigma-bond away from a pi-bond. (Draw an example of this from your textbook.)

- A polar pi-bond can be rewritten as an extra lone pair on the electronegative atom. (Draw an example of this from your textbook.)

- A positive formal charge on an atom one sigma-bond from a pi-bond. (Draw an example of this from your textbook.)

POST-LECTURE:

Work Problems: *One of the best ways to learn chemistry is through solving problems. Go to the textbook and work through the Worked Examples for this section, then go to the end of chapter and work the problems there. When you feel confident that you can correctly answer problems like the ones listed below for each objective, check off each of the Learning Objectives for this section above.*

LO 23.12: Worked Examples 23.8, 23.9; Problems 23.96, 23.98, 23.100, 23.102

Section 23.8 Conjugated Systems

PRE-LECTURE:

Learning Objectives: *These are the skills you should be proficient in after reading this section and practicing the associated problems. Check off the boxes below once you have gone through this section, answered the associated questions in your textbook, and feel confident you have mastered the skill.*

☐ **LO 23.13** Draw orbital overlap pictures for conjugated systems and identify localized and delocalized electrons.

☐ **LO 23.14** Identify conjugated systems and the hybridization of atoms included in the system.

Definitions: *To learn, you must do. This is your workbook. Fill in the definitions from the textbook as a first step to understanding the material.*

— **pi (π) system—** ____

____

— **conjugated systems—** ____

____

— **delocalized—** ____

____

Summary: *Read through this section and fill in the missing words and equations. Reference the textbook when you are not confident in your answers.*

________ systems occur when *p* orbitals are connected in compounds with alternating single and double bonds. When two pi-bonds are separated by a sigma-bond, that system is conjugated.

In line drawings, electrons appear to be localized in bonds between two atoms, but in a conjugated system, the electrons are ________, or associated with more than one or two atoms in a molecule. Remember that molecular orbital theory describes delocalization very well.

In valence bond theory, the major principle is that atoms will adopt the hybridization that makes the molecule the most stable by ___________ the overall energy. We follow two principles to make the most stable structure of a molecule:

- Conjugation is immensely stabilizing, so always look for conjugated systems first.
- If an atom is not conjugated, it will adopt a stable hybridization according to VSEPR theory.

POST-LECTURE:

Work Problems: *One of the best ways to learn chemistry is through solving problems. Go to the textbook and work through the Worked Examples for this section, then go to the end of chapter and work the problems there. When you feel confident that you can correctly answer problems like the ones listed below for each objective, check off each of the Learning Objectives for this section above.*

LO 23.13: Worked Example 23.10; Problems 23.112, 23.113
LO 23.14: Worked Example 23.11; Problems 23.108, 23.109

Section 23.9 Proteins: A Biological Example of Conjugation

PRE-LECTURE:

Learning Objectives: *These are the skills you should be proficient in after reading this section and practicing the associated problems. Check off the boxes below once you have gone through this section, answered the associated questions in your textbook, and feel confident you have mastered the skill.*

☐ **LO 23.15** Identify amino acids and draw structures for peptides.

Definitions: *To learn, you must do. This is your workbook. Fill in the definitions from the textbook as a first step to understanding the material.*

— **enzyme—** ____

____

— **protein—** ____

____

— **peptide bond—** ____

____

— **peptide—** ____

____

— **alpha-(α-) amino acid—** ____

____

Summary: *Read through this section and fill in the missing words and equations. Reference the textbook when you are not confident in your answers.*

Proteins are major components of biological systems. Fifty percent of the dry weight of a human body is protein: our bodies contain more than 150,000 proteins that perform multiple functions. Proteins can be structural, they can carry messages as hormones, and they can catalyze reactions as enzymes.

Proteins are polymers made up of individual units called ___________. These small molecules contain an amino group, a central carbon, and a carboxylic acid group. They are linked together by amide bonds, which, in proteins, are referred to as ___________ *bonds*.

The amide bond that bonds amino acids together to form a protein has a nitrogen atom with a lone pair of electrons that is one sigma-bond from a pi-bond and thus is part of a conjugated system.

Twenty amino acids occur naturally, all of which have the basic structure below, with a variety of –R groups attached to the central carbon, known as the α-carbon. These are therefore also referred to as α-amino acids. Each is known by its name and a three-letter abbreviation (Serine = Ser, for example). There is a chart of all 20 in Figure 23.13 on page 1036 of the textbook.

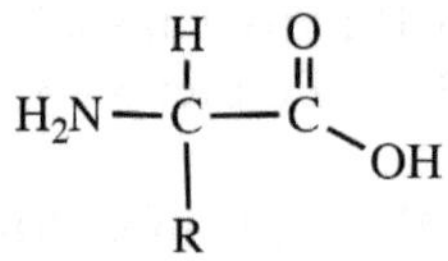

The –R group is also known as a *side chain.* The side chain is used to classify the amino acid. Fifteen of the 20 have neutral side chains, two are acidic, and three are basic. Of the neutral amino acids, these can again be divided into polar and nonpolar residues. The polar side chains, including the acidic and basic residues, are described as ___________ (water loving). The nonpolar side chains are described as ___________ (water fearing).

Regardless of the number and order of amino acids, each protein will have an *N-terminal amino acid* with a free NH_2 group and a *C-terminal amino acid* with a free CO_2H group. By convention, sequences are written N-terminus on the ___________ to C-terminus on the ___________.

POST-LECTURE:

Work Problems: *One of the best ways to learn chemistry is through solving problems. Go to the textbook and work through the Worked Examples for this section, then go to the end of chapter and work the problems there. When you feel confident that you can correctly answer problems like the ones listed below for each objective, check off each of the Learning Objectives for this section above.*

LO 23.15: Worked Example 23.12; Problems 23.115–23.117

Section 23.10 Aromatic Compounds and Molecular Orbital Theory

PRE-LECTURE:

Learning Objectives: *These are the skills you should be proficient in after reading this section and practicing the associated problems. Check off the boxes below once you have gone through this section, answered the associated questions in your textbook, and feel confident you have mastered the skill.*

☐ **LO 23.16** Predict if an organic compound is aromatic.

Definitions: *To learn, you must do. This is your workbook. Fill in the definitions from the textbook as a first step to understanding the material.*

— **aromatic—** ____

____

— **Huckel's rule—** ____

____

Review: *Before going through the section make sure you are familiar with these concepts:*

- Section 8.7 molecular orbital

Summary: *Read through this section and fill in the missing words and equations. Reference the textbook when you are not confident in your answers.*

Aromatic molecules:

- behave ___________ than most organic substances
- are cyclic and ___________
- have adjacent __ orbitals around the ring

Benzene (C_6H_6) is the most common aromatic molecule:

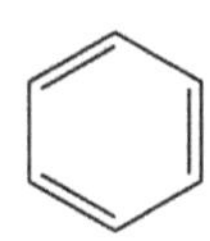

The carbons are sp^2-hybridized, with p orbitals perpendicular to the ring. This leads to a delocalized pi-bond that spreads over the entire molecule, or a conjugated pi-system. MO theory bests help us understand how this conjugation leads to unusual stability. Because of this stability, these molecules are not usually reactive.

POST-LECTURE:

Work Problems: *One of the best ways to learn chemistry is through solving problems. Go to the textbook and work through the Worked Examples for this section, then go to the end of chapter and work the problems there. When you feel confident that you can correctly answer problems like the ones listed below for each objective, check off each of the Learning Objectives for this section above.*

LO 23.16: Worked Example 23.13; Problems 23.120–23.122, 23.124

Section 23.11 Nucleic Acids: A Biological Example of Aromaticity

PRE-LECTURE:

Learning Objectives: *These are the skills you should be proficient in after reading this section and practicing the associated problems. Check off the boxes below once you have gone through*

this section, answered the associated questions in your textbook, and feel confident you have mastered the skill.

- ☐ **LO 23.17** Draw structures showing how different components of the nucleic acid are joined, and write the DNA sequence that is complementary to a specified strand.
- ☐ **LO 23.18** Draw resonance structures to show that DNA bases are aromatic.

Definitions: *To learn, you must do. This is your workbook. Fill in the definitions from the textbook as a first step to understanding the material.*

— **nucleic acid—** ____

____

— **deoxyribonucleic acid (DNA)—**____

____

— **ribonucleic acid (RNA)—**____

____

— **nucleotide—**____

____

— **nucleoside—**____

____

— **Watson–Crick model—** ____

____

— **base stacking—** ____

____

Summary: *Read through this section and fill in the missing words and equations. Reference the textbook when you are not confident in your answers.*

Deoxyribonucleic acid (DNA) and ribonucleic acid (RNA) are the carriers of genetic information. Coded in an organism's nucleic acids is all the information needed to produce the many thousands of proteins required by that organism.

Nucleic acids are polymers built of nucleotide units. Nucleotides are built of nucleosides, which are amine bases linked to aldopentose sugars (___________ for RNA and ___________ for DNA) and phosphoric acid.

DNA and RNA share three amine bases:

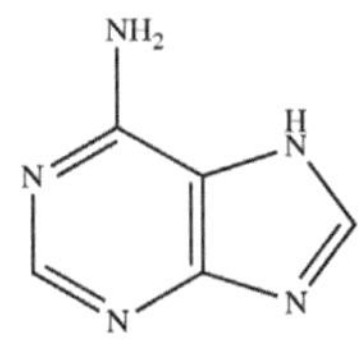

adenine

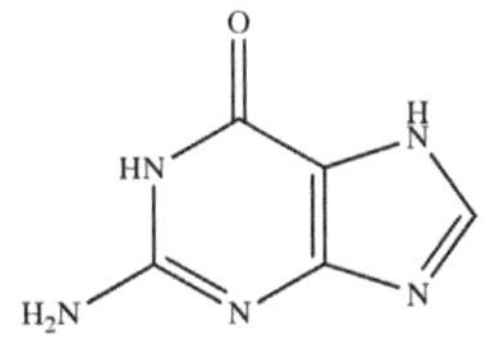

guanine

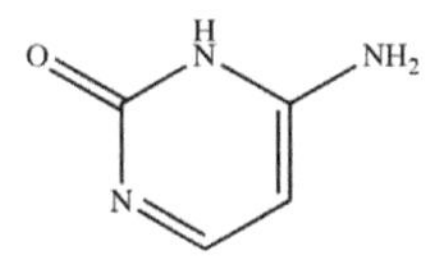

cytosine

Only DNA contains thymine:

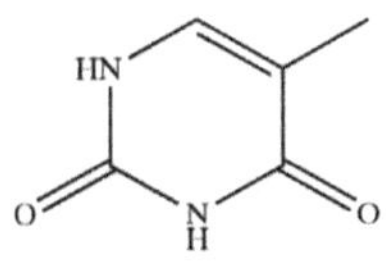

thymine

Only RNA contains uracil:

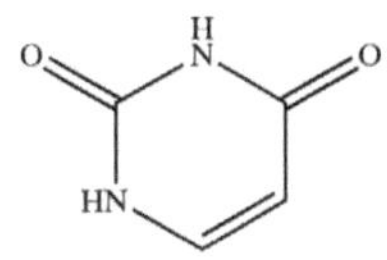

uracil

In both DNA and RNA, the amine base is bonded to the C1′ of the sugar, and the phosphoric acid is bonded to the C5′ carbon of the sugar. The "prime" on the C indicates a sugar carbon. Other numbered carbons are in the amine base. When the nucleotides are joined to make polymers, the phosphates join to the C3′ of the next nucleoside.

DNA sequences are listed beginning at the C5′ end, using the first letters of the nucleotides: TACG. DNA sequences from different tissues of the same species have the same proportions of each nucleotide, but sequences from different species may have very different proportions.

There are always equal amounts of A and ___ and G and ___, regardless of species. In the Watson–Crick model of DNA structure, two complementary strands of DNA hydrogen-bond to each other, with A forming two hydrogen bonds with T, and G forming three hydrogen bonds with C. The two strands then coil around one another to form a double helix. The bases are planar allowing them to stack. This leads to an increase in intermolecular interactions and an ___________ stability.

Workbook Problem 23.2

Determine the complementary DNA sequence that would be produced from the following DNA sequence:

AGCTTGACACGTGTCACTAGA

Strategy: Determine the complementary bases.

POST-LECTURE:

Work Problems: *One of the best ways to learn chemistry is through solving problems. Go to the textbook and work through the Worked Examples for this section, then go to the end of chapter*

and work the problems there. When you feel confident that you can correctly answer problems like the ones listed below for each objective, check off each of the Learning Objectives for this section above.

LO 23.17: Problems 23.128–23.130
LO 23.18: Problems 23.132, 23.133

Self-Test

This section is intended to test your knowledge of the material covered in this chapter. Think through these problems and make certain you understand what the problem is about—ask yourself if the answer makes sense. Many of these questions are linked to the chapter learning goals. Therefore, successful completion of these problems indicates that you have achieved the learning goals for this chapter. You will receive the greatest benefit from this section if you use it as a mock exam. Do not use your notes, book, Internet, or friends when taking the test. There is no grade and you only shortchange yourself by getting help with this activity. As you take the test, mark the problems with your level of confidence with the answer.

★ *Use a star for problems you know you got right.*
? *Use a question mark for those you think you got right.*
x *Use an x for those you are guessing on.*

If you get a question right that you marked with a star, you can be confident that you know the associated LO.

If you get a question right that you marked with a question mark, go back and work a few more problems and test yourself again tomorrow.

If you get a question wrong or mark it with an x, go back and read the section again, work a few problems today, a few tomorrow, and a few the next day before testing yourself again.

This will help you to discover which topics you have learned thoroughly and which topics you need to study in more detail.

True–False

1. Different isomers will have different physical and chemical properties.

2. A double or triple bond is not considered a functional group.

3. Whether an alkene is cis or trans has very little practical significance.

4. Hands are chiral.

5. Organic carboxylic acids are strong acids.

6. The links between amino acids are amide bonds, but when they are in peptides, they are referred to as peptide bonds.

7. *Lipids* are defined by their physical properties, not by their structure.

8. Monosaccharides can be aldoses or ketoses.

9. All polysaccharides are readily digested to form glucose monomers.

10. Nucleic acids are the primary structural components of plant cell walls.

Multiple Choice

11. In naming alkanes, the parent chain is
    a. the longest straight chain
    b. the longest continuous chain
    c. numbered to give the maximum numbers for substituents
    d. b and c

12. Molecules of which functional group can hydrogen bond in pure samples?
    a. alcohols
    b. amines
    c. both a and b
    d. neither a nor b

13. A three-carbon alkane is called
    a. propane b. triane c. propene d. butane

14. The smallest cycloalkane without stress in the bonds is
    a. cyclopropane b. cyclobutane c. cyclopentane d. cyclohexane

15. Aromatic rings
    a. contain six-membered carbon rings
    b. have delocalized *p* electrons
    c. are flat
    d. all of the above

16. Alcohols and ethers can both be considered water derivatives, but they differ in that
    a. ethers have double bonds to oxygen
    b. alcohols have one hydrogen of the water substituted, ethers have both
    c. ethers contain nitrogens
    d. alcohols have double bonds to oxygen

17. Which of the following are carbonyl compounds?
    a. aldehydes and ketones
    b. amides
    c. carboxylic acids
    d. all of the above

18. Which of the following places the molecules in order from smallest to largest?
    a. protein, tripeptide, peptide
    b. amino acid, tripeptide, dipeptide
    c. amino acid, dipeptide, protein
    d. peptide, protein, amino acid

19. Complex carbohydrates are
    a. nonpolar
    b. long chains
    c. all digestible by humans
    d. a and b

20. Lipids
    a. are soluble in nonpolar solvents
    b. are important components of cell membranes
    c. often contain fatty acids bonded to a glycerol backbone
    d. all of the above

Fill in the Blanks

21. Because of carbon's flexible _______________ there are more than 40 million organic compounds. These are organized into families according to the _______________ they contain.

22. _______________ contain only hydrogen and _______________. _______________ have only single bonds, _______________ have at least one double bond, and _______________ have at least one _______________ bond.

23. When six-membered rings alternate single and _______________ bonds, the result is a(n) _______________ ring.

24. _______________ are both structural components of plants and important sources of _______________ for animals.

25. Proteins are made up of long chains of _______________ _______________. When they catalyze reactions in biological systems, they are called _______________.

26. The nucleic acid bases _______________, _______________, and _______________ are found in DNA and RNA. While _______________ is only found in RNA and _______________ is only found in DNA.

Matching

27. alcohol	a. protein monomer containing an $-NH_2$ group and a $-COOH$ group
28. aldehyde	b. a group of biological molecules that have importance to all living organisms
29. alkane	c. large class of polyhydroxylated aldehydes and ketones
30. alkene	d. carbonyl group with hydrogen on one side and an alkyl group (carbon containing group) on the other
31. alkyne	e. hydrocarbon containing only single bonds that contain a ring structure
32. amino acid	f. C=O

33. amide	g. molecule that contains two alkyl groups each single bonded to oxygen
34. amine	h. organic derivative of ammonia
35. protein	i. hydrocarbon containing at least one double bond
36. aromatic	j. carboxylic acid with a long hydrocarbon tail
37. carbonyl group	k. organic molecule containing an –OH group
38. carbohydrate	l. molecule containing two alkyl groups on either side of a carbonyl group
39. cycloalkane	m. organic molecule containing an alkyl group and a nitrogen bonded to a carbonyl group
40. enzyme	n. controls carbohydrate metabolism
41. ester	o. carbohydrate that cannot be broken down by acid hydrolysis
42. ether	p. organic molecule that contains a $-C(=O)-O-R$
43. fatty acid	q. containing a benzene ring
44. hydrocarbon	r. hydrocarbon containing only single bonds
45. ketone	s. protein that catalyzes biological reactions
46. lipid	t. molecule that contains only hydrogen and carbon
47. insulin	u. sugar component in RNA
48. monosaccharide	v. hydrocarbon containing at least one triple bond
49. peptide	w. having as many hydrogens as possible, containing only single bonds
50. conjugated system	x. naturally occurring molecule that is soluble in nonpolar solvents
51. saturated	y. sugar component in DNA
52. ribose	z. not having as many hydrogens as possible: containing one or more multiple bonds
53. deoxyribose	aa. delocalized electrons in a π system
54. unsaturated	bb. chain of 2 to 100 amino acids

Problems

55. Identify the functional groups in the following molecules:

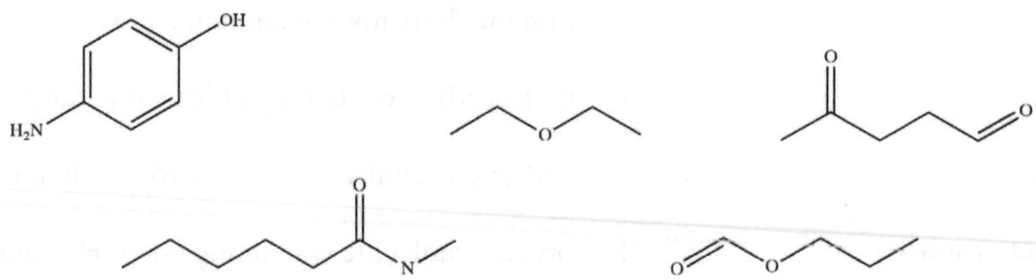

56. Draw the two possible dipeptides that can be formed by alanine and serine.

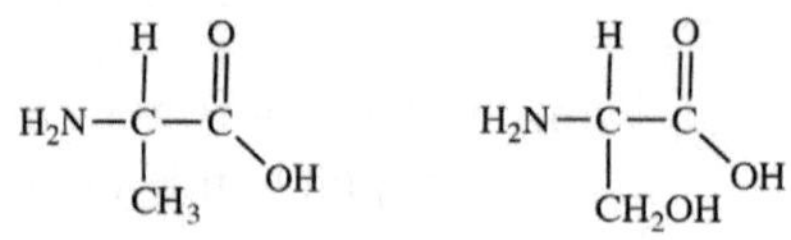

alanine serine

57. List the six possible tripeptides that can be formed by alanine (ala), leucine (leu), and histidine (his).

58. Classify the following monosaccharides:

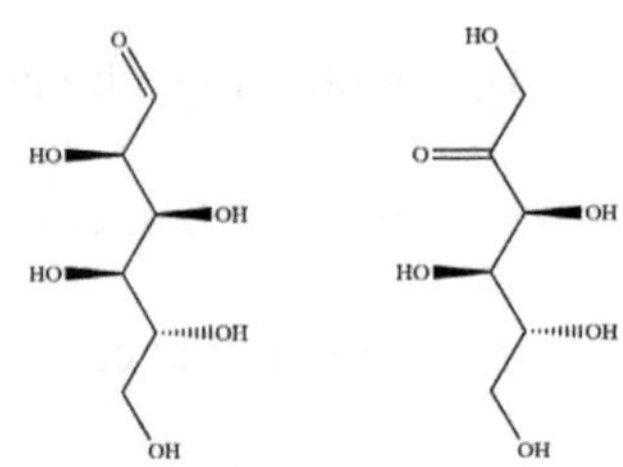

59. Draw the basic structure of a saturated triglyceride.

60. Give the complementary DNA sequence.

TAGCACTGATCAGTAAGC

Challenge Problem

61. Citric acid is an unsaturated, tricarboxylic acid. When 5.00 g of the acid undergoes combustion, 6.87 g of CO_2 and 1.87 g of H_2O are produced. When a 0.312 g sample of citric acid is titrated with 0.100 M sodium hydroxide, 48.8 mL is needed for complete titration. Determine the molecular formula of the acid.

Test Yourself: *Take the Practice Test on pages 1049–1051 of your textbook. Again, use a testing environment using only those items that your instructor will allow you to have access to during the test and mark each question with a star, question mark, or x to denote your confidence level. Once you have finished you can check off the LO at the beginning of this chapter based on which questions you starred and got correct. Go back and spend more time with the topics from the other learning objectives.*

APPENDIX

Solutions to Study Guide Problems

Chapter One—Chemical Tools: Experimentation and Measurement

Workbook Problems

Workbook Problem 1.1

Step 1: Use the formula for the conversion of °F to °C. (Remember, for this conversion, you do a zero-point correction followed by a size correction.)

$$°\mathrm{C} = \frac{5}{9} \times (102 - 32) = 38.9\ °\mathrm{C}$$

Step 2: Use the formula for the conversion of °C to K.

$$\mathrm{K} = 38.9\ °\mathrm{C} + 273.15\ \mathrm{K}/°\mathrm{C} = 312\ \mathrm{K}$$

Workbook Problem 1.2

Step 1: Substitute the information given in the problem into the mathematical equation you set up.

$$\mathrm{Density}\left(\frac{\mathrm{g}}{\mathrm{mL}}\right) = \frac{\text{Mass of sample (g)}}{\text{Volume of sample (mL)}}$$

$$11.34\ \frac{\mathrm{g}}{\mathrm{mL}} = \frac{\text{mass of lead (g)}}{53.43\ \mathrm{mL}}$$

Step 2: Solve for the mass of lead.

$$\left(11.34\ \mathrm{g}/\mathrm{mL}\right)(53.43\ \mathrm{mL}) = 605.9\ \mathrm{g\ Pb}$$

Workbook Problem 1.3

Step 1: Use the equation for density to find the volume of the gas.

$$\mathrm{Density}\left(\frac{\mathrm{g}}{\mathrm{mL}}\right) = \frac{\text{Mass of sample (g)}}{\text{Volume of sample (mL)}}$$

$$1.976\,\frac{\text{g}}{\text{L}} = \frac{4.84365\ \text{g}}{\text{volume of sample (L)}}$$

$$\text{volume of sample (mL)} = \frac{4.84365\ \text{g}}{1.976\,\frac{\text{g}}{\text{L}}} = 2.451\,239\,879\ \text{L}$$

Step 2: Apply the rules for multiplication and division to determine the number of significant numbers in your answer.

When carrying out either multiplication or division, your answer cannot have more significant figures than either of the original numbers. 1.976 has four significant figures; 4.84365 has six significant figures. Your answer can't have more than four significant figures.

Step 3: If necessary, use the rules for to round off numbers.

We must round off 2.451 239 879 L to four significant figures. The first digit we must remove is a 2; therefore the number can simply be truncated at four digits The final answer is 2.451 L.

Workbook Problem 1.4

Step 1: Calculate the time it takes her to run 5 miles.

$$27\ \text{min} \times \frac{60\ \text{s}}{1\ \text{min}} = 1620\ \text{s}$$

$$1620\ \text{s} + 46\ \text{s} = 1666\ \text{s}$$

Step 2: Use conversions to calculate the amount of time it will take for the athlete to run 10 km.

$$10\ \text{km} \times \frac{1\ \text{mi}}{1.6093\ \text{km}} \times \frac{1666\ \text{s}}{5\ \text{mi}} = 2070\ \text{s}$$

$$2070\ \text{s} \times \frac{1\ \text{min}}{60\ \text{s}} = 34.5\ \text{mins}$$

Self-Test

True–False

1. F A theory is formed after a hypothesis has been tested numerous times and is shown to be correct.
2. T
3. F Mass is the amount of matter in an object, while weight is the gravitation pull on that object. Mass and weight are often used interchangeably because it is rare to be making measurements in any gravitational field but Earth's, but it is important to understand the distinction.

4. F The units most commonly used in the laboratory to measure volume are the liter (dm^3) and the milliliter (cm^3).
5. F The prefix μ means 10^{-6}. 1×10^6 g is equal to 1 megagram.
6. T
7. T
8. F The SI unit for temperature is degree Celsius.
9. T
10. F Density describes an object's mass and its volume.

Multiple Choice

11. c	12. d	13. a	14. d
15. c	16. a		

Matching

17. d	18. f	19. a	20. c
21. b	22. e		

Fill in the Blanks

23. theory
24. kilogram, kelvin, meter
25. kelvin, Celsius, Fahrenheit
26. density
27. accurate, precise
28. scientific notation
29. round, significant figures
30. conversion factor
31. the number with the fewest digits to the right of the decimal
32. giga, G
33. mass
34. volume

Problems

35. $246\text{ cm} \times \dfrac{10\text{ mm}}{1\text{ cm}} = 2460\text{ mm}$ $\quad 2460\text{ mm} \times \dfrac{1\text{ m}}{1000\text{ mm}} = 2.46\text{ m}$

$$2.46\text{ m} \times \frac{1\text{ km}}{1000\text{ m}} = 2.46 \times 10^{-3}\text{ km}$$

36. a. $\dfrac{8.3}{2.793} = 3.0$ (The least amount of significant figures, two, is found in the numerator.)

b. $23.945\,62 \times 8.3421 = 199.76$ (The least amount of significant figures, five, is found in 8.3421.)

c. $3.2 + 5.3875 = 8.6$ (There is only one digit to the right of the decimal point in the number 3.2, so the answer must be rounded off to one digit after the decimal point.)

d. $9434.21 - 4.199 = 9430.01$ (There are only two digits to the right of the decimal point in 9434.21.)

37. $°F = \left(\frac{9}{5} \times 56\right) + 32 = 130\ °F$

38. The conversion factor needed to solve this problem is 1 mile = 1.6093 km.

$$290{,}587\ mi^2 \times \left(\frac{1.6093\ km}{1\ mi}\right)^2 = 752{,}576\ km^2$$

39. The conversion factor needed to solve this problem is 1×10^9 nm = 1 m.

$$6.73 \times 10^{-7}\ m \times \frac{1 \times 10^9\ nm}{1\ m} = 673\ nm$$

40. 78.5 °C + 273.15 = 351.6 K $\qquad °F = \left(\frac{9}{5} \times 78.5\right) + 32 = 173°F$

41. The conversion factors needed to solve these problems are 1 mile = 1.6093 km and 1 hour = 3600 s

Sprinter: $$\frac{100\ m}{9.58\ s} \times \frac{3600\ s}{1\ hr} \times \frac{1\ km}{1000\ m} \times \frac{1\ mi}{1.6093\ km} = 23.4\ mi/hr$$

Swimmer: $$\frac{100\ m}{47.85\ s} \times \frac{3600\ s}{1\ hr} \times \frac{1\ km}{1000\ m} \times \frac{1\ mi}{1.6093\ km} = 4.68\ mi/hr$$

42. density = mass/volume

$$\text{density} = \frac{9.634\ g}{(16.2\ mL - 11.5\ mL)} = 2.049787\ g/mL = 2.05\ g/mL$$

43. It is important to realize that the speed of light, 3.00×10^8 m/s, can also be used as a conversion factor. 3.00×10^8 m = 1 s

$$19.1 \times 10^9\ km \times \frac{10^3\ m}{1\ km} \times \frac{1\ s}{3.00 \times 10^8\ m} \times \frac{1\ h}{3600\ s} = 17.7\ h$$

44. The first problem is figuring out the volume of the pycnometer based on the mass of the water.

37.82 g − 23.75 g = 14.07 g of water. The mass of this water is 14.07g/0.9982g/mL = 14.095 mL of water. This is the volume of the pycnometer.

When the beryllium is put in, we need to solve for the mass of the remaining water:
Total mass − mass of pycnometer − mass of beryllium = mass of remaining water
40.72 g − 23.75 g − 6.345 g = 10.625 g water.

The volume of this water will be 10.625 g/0.9982 g/mL = 10.644 mL

The difference between this volume and the total volume of the apparatus is the volume of the metal.

Beryllium volume = 14.095 mL − 10.644 mL = 3.451 mL beryllium.

Density = mass/volume = 6.345 g/3.451 mL = 1.84 g/mL (At this point, the answer is rounded off to the correct number of significant figures.)

45. Volume = mass/density. 10.67g/4.55 g/mL = 2.35 mL

46. Use the equation density = mass/volume. We know that m = 7.13 g. To find volume, subtract the final volume from the initial volume, 22.7 mL − 20.0 mL = 2.7 mL. Substitute the values into the equation for density, d = 7.13 g/2.7 mL = 2.6 g/mL.

47. 0 °C is the freezing point of water, while 0 K is the coldest possible temperature, also known as absolute zero.

48. The numbers in a calculation are not rounded or truncated until the final step. When a number is rounded because of significant figures, it can be changed slightly. If each number along the way is rounded, error will be introduced at each step. For example, if a box is measured as being 2.57 cm long, 3.634 cm wide, and 1 cm tall, the volume would properly be calculated 2.57 cm × 3.634 cm × 1 cm = 9.339 cm^3, which then is rounded to 9 cm^3. If an overzealous student were instead to round all the numbers to one significant figure before beginning the calculation, the answer would be 12 cm^3, a significant error.

Challenge Problem

49. On a mercury thermometer, the liquid range of gallium is 2373.22 °C (the difference in temperature between the melting and boiling points). On the gallium thermometer, the liquid range of gallium is 1000 °C. The size adjustments for these two scales are:

$$1\ {}^\circ\text{C} \times \frac{1000\ {}^\circ\text{Ga}}{2373.22\ {}^\circ\text{C}} = 0.421\ {}^\circ\text{Ga} \qquad 1\ {}^\circ\text{Ga} \times \frac{2373.22\ {}^\circ\text{C}}{1000\ {}^\circ\text{Ga}} = 2.37\ {}^\circ\text{C}$$

Therefore, 1 °C is 0.421 times smaller than 1 °Ga, and 1 °Ga is 2.37 times larger than 1°C. The melting point of gallium is higher by 29.78° on the Celsius scale, so we would first do the size adjustment followed by the zero-point adjustment when converting from °Ga to °C.

$${}^\circ\text{C} = \left(2.37 \times {}^\circ\text{Ga}\right) + 29.78$$

Now convert the melting point of copper from °Ga to °C:

$${}^\circ\text{C} = (2.37 \times 447\ {}^\circ\text{Ga}) + 29.78 = 1089\ {}^\circ\text{C}$$

We can determine the accuracy of the thermometer by comparing the difference between the converted and reported value to the reported value. The temperature is identical.

Chapter Two—Atoms, Molecules, and Ions

Workbook Problems

Workbook Problem 2.1

Step 1: Determine the carbon-to-hydrogen ratio for the first compound.

$$\text{First compound: C:H mass ratio} = \frac{3.43\text{ g C}}{0.857\text{ g H}} = 4.00$$

Step 2: Determine the carbon-to-hydrogen ratio for the second compound.

$$\text{Second compound: C:H mass ratio} = \frac{4.80\text{ g C}}{0.400\text{ g H}} = 12.0$$

Step 3: Divide the ratio found for the first compound by the ratio for the second compound.

$$\frac{\text{C:H mass ratio in 1st compound}}{\text{C:H mass ratio in 2nd compound}} = \frac{4.0}{12.0} = 0.33$$

Step 4: Can your answer be converted to a small whole-number ratio?

Yes. $0.33 = \frac{33}{100} = \frac{1}{3}$

Workbook Problem 2.2

Step 1: Find the atomic number for oxygen.

From the periodic table, we find that the atomic number for O is 8.

Step 2: Calculate the mass number for each isotope.

The mass number can be determined from the following equation.

A (mass number) =– Z (atomic number) + number of neutrons.

$A = 8 + 8 = 16$ $A = 8 + 9 = 17$ $A = 8 + 10 = 18$

Step 3: Write the isotope symbol with the mass number as a superscript and the atomic number as a subscript, both to the left of the symbol.

The isotope symbol is written with the mass number as a superscript and the atomic number as a subscript, both to the left of the symbol.

$^{16}_{8}O$ $^{17}_{8}O$ $^{18}_{8}O$

Workbook Problem 2.3

Step 1: Determine the atomic number from the information given.

We are told that element *X* has 51 protons; therefore, the atomic number of the element is 51.

Step 2: Knowing the atomic number, identify the metal by using the periodic table.

From the periodic table, we find that the element with $Z = 51$ is Sb (antimony).

Step 3: Write the isotope symbol with the mass number as a superscript and the atomic number as a subscript, both to the left of the symbol.

The mass number is determined by adding the number of protons and the number of neutrons.

$A = 51 + 70 = 121$. The isotope symbol is ${}^{121}_{51}\text{Sb}$.

Workbook Problem 2.4

Step 1: Use the information given to determine the abundance of ${}^{204}\text{Pb}$.

The abundance of ${}^{204}\text{Pb}$ is not given, but the abundances of the other isotopes are, making it a straightforward process to find the missing abundance:

Abundance ${}^{204}\text{Pb} + 52.4\% + 22.1\% + 24.1\% = 100\%$
Abundance ${}^{204}\text{Pb} = 1.4\%$

Step 2: Use the information given to determine the average atomic mass.

$$(0.524 \times 207.977\text{ u}) + (0.221 \times 206.976\text{ u}) + (0.241 \times 205.974\text{ u}) + (0.014 \times 203.973\text{ u}) = 207.217\text{ u}$$

Workbook Problem 2.5

Step 1: Use the molar mass of aluminum to convert the grams to moles.

$$14.9\text{ g} \times \frac{1\text{ mol Al}}{26.981\,538\text{ g}} = 0.552\,229\,454\text{ mol Al} = 0.552\text{ mol Al}$$

Step 2: Convert moles to atoms using Avogadro's number.

$$0.552\,229\,454\text{ mol Al} \times \frac{6.022 \times 10^{23}\text{ atoms}}{1\text{ mol Al}} = 3.33 \times 10^{23}\text{ Al atoms}$$

Workbook Problem 2.6

Step 1: When naming compounds, determine if the metal is a main-group metal or a transition metal. If the metal is a transition metal, you must indicate the charge on the metal when naming the compound. The charge on the metal is determined from the number and charge on the anion. (Remember the charges must all add up to zero.)

MgF_2; main-group metal; magnesium fluoride
CaO; main-group metal; calcium oxide
FeF_3; transition metal; iron(III) fluoride
$CuSe$; transition metal; copper(II) sulfide

Step 2: When writing formulas, use the periodic table or the name of the compound to determine the charge on the metal. Use the periodic table to determine the charge on the anion. (Remember the charges must all add up to zero.) Make sure the formula contains the smallest whole-number ratio of cation to anion.

sodium chloride; $NaCl$
chromium(III) nitride; CrN
vanadium(V) oxide; V_2O_5
aluminum sulfide; Al_2S_3

Workbook Problem 2.7

Step 1: When naming a compound, identify the cation and the anion. Write the cation first.

$Fe(NO_3)_3$; iron(III) nitrate
$Al_2(SO_4)_3$; aluminum sulfate

Step 2: Determine the charge on the cation and write the formula to make sure all charges are neutralized.

potassium chromate; K_2CrO_4
magnesium phosphate; $Mg_3(PO_4)_2$
ammonium nitrate; NH_4NO_3

Workbook Problem 2.8

Step 1: When naming the molecules, remember that the element that is more anion-like uses the *-ide* suffix. Also, remember to use numerical prefixes to indicate the number of each atom.

PF_6 dinitrogen tetroxide selenium dioxide NBr_3 $HF(g)$

PF_6 phosphorus hexafluoride

	NBr_3	nitrogen tribromide
	$HF(g)$	The formula was written to indicate that this binary hydrogen compound is a gas. Therefore, it is named as hydrogen fluoride gas.
Step 2:	Add numerical prefixes.	
	dinitrogen tetroxide	N_2O_4
	selenium dioxide	SeO_2

Self-Test

True–False

1. F *Ions* are the same element carrying a different charge; *isotopes* are the same element with a different number of neutrons.
2. T
3. T
4. F Nonmetals have a strong tendency to form multiple compounds—think of carbon monoxide and carbon dioxide—and so must be named using prefixes.
5. F Protons and neutrons are found in the nucleus, electrons in a cloud around the nucleus.
6. T
7. F Isotopes of the same element have different masses.
8. F Numbers are used after the metal if it is a transition element.
9. F Transition metals can carry a variety of charges. Main-group metals have invariant charges.

Multiple Choice

10. c	11. d	12. a	13. b
14. d	15. c	16. b	17. a
18. c	19. b		

Fill in the Blanks

20. atoms, protons, neutron, and electrons
21. protons, atomic number
22. neutrons, isotope, electrons, ion
23. compounds, mixtures
24. covalent, shared
25. metal, nonmetal, ionic, opposite charges
26. polyatomic ion, nitrate, ammonium, sulfate
27. di-, tri-, penta-
28. reactants, law of conservation of mass

Matching

29. f	30. i	31. b	32. m
33. a	34. q	35. j	36. c
37. l	38. o	39. r	40. k
41. d	42. p	43. n	44. e
45. h	46. g		

Problems

47. mass of oxygen $3.2\ \text{L oxygen} \times \dfrac{1.43\ \text{g}}{\text{L}} = 4.58\ \text{g oxygen}$

hydrogen + 4.58 g oxygen → 5.15 g water

mass of hydrogen = 5.15 g − 4.58 g = 0.57 g hydrogen.

$$\text{volume of hydrogen} = \frac{0.57\text{ g}}{0.0893\text{ g/L}} = 6.4\text{ L}$$

48. Compound 1: $\text{O:P} = \frac{11.4\text{ g O}}{4.3\text{ g P}} = 2.65$ Compound 2: $\text{O:P} = \frac{5.33\text{ g O}}{3.34\text{ g P}} = 1.59$

$$\frac{\text{O:P ratio in compound 2}}{\text{O:P ratio in compound 1}} = \frac{1.59}{2.65} = 0.6 = \frac{6}{10} = \frac{3}{5}$$

49. a) p = 42, n = 55, e = 42 b) p = 92, n = 143, e = 86
c) p = 93, n = 142, e = 93 d) p = 80, n = 122, e = 76

50. Abundances must all add up to 100%:
Abundance of Neon-22 = 100% − 90.48% − 0.27% = 9.25%
Solve for mass of Neon-22
$= (19.992\text{ u} \times 0.9048) + (20.993\text{ u} \times 0.0027) + (x\text{ u} \times 0.0925) = 20.18\text{ u}$
x u = (20.18 u − 18.0887616 u − 0.0566811 u)/0.0925 = 21.9952 u
x = 22.00 u

51. Average atomic mass
$= (0.7899 \times 23.985\text{ u}) + (0.1000 \times 24.986\text{ u}) + (0.1101 \times 25.983\text{ u}) = 24.305\text{ u}$

52. Naphthalene = $C_{10}H_8$

53. a) LiCl–lithium chloride b) Na_2S–sodium sulfide
c) CaO–calcium oxide d) Mg_3N_2–magnesium nitride

54. a) carbon monoxide–covalent b) strontium chloride–ionic
c) carbon tetrachloride–covalent d) platinum (IV) oxide–ionic

55. Use Avogadro's number in calculating the mass.

$$9500\text{ atoms} \times \frac{1\text{ mole}}{6.022 \times 10^{23}\text{ atoms}} \times \frac{196.97\text{ g}}{1\text{ mole}} = 3.107 \times 10^{-18}\text{ g}$$

56. In dichloromethane, there are two chlorine atoms, two hydrogen atoms, and one carbon atom. From the periodic table, the masses of these atoms add up to:

(2 × 35.45) + (2 × 1.00) + 12.01 = 84.91 u

The percentage of chlorine will be (2 ×35.45)/84.91× 100 = 83.50% Cl
The percentage of hydrogen will be (2 × 1.00)/84.91 ×100 = 2.36% H
The percentage of carbon will be 12.01/84.91 × 100 = 14.14% C

57. a. CaF_2 b. $Ca_3(PO_4)_2$ c. $Ca(NO_3)_2$ d. $CaCO_3$

58. a. tin (IV) chloride
b. sodium phosphate
c. magnesium acetate
d. hydrogen chloride
e. rubidium sulfide
f. carbon dioxide
g. chromium (III) phosphate
h. difluorine monoxide
i. phosphorous pentoxide
j. nickel (II) chloride
k. iron (III) hydroxide
l. hydrogen chloride
m. lead (II) iodide
n. sodium nitrate
o. barium bromide
p. aluminum nitrate

59. a. $Ca(NO_2)_2$
b. $NaCl$
c. Cl_2O
d. H_2O
e. NiH_2
f. BBr_3
g. $CuCl_2$
h. XeF_4
i. NO_2
j. Ba_3N_2
k. CBr_4
l. PCl_3
m. $LiMnO_4$
n. Na_3PO_3
o. Na_2HPO_4

Challenge Problem

60. To calculate the volume of the sphere, it is necessary first to find the radius. It is also a good idea to convert the radius into centimeters, as the final desired unit is cm^3.

Diameter = 0.2 mm
Radius = 0.2 mm/2 × 1cm/10mm = 0.01cm
Volume of a sphere = $4/3\pi r^3 = 4.19 \times 10^{-6}$ cm^3

With the volume and the densities of the two different metals, it is possible to calculate the masses of the spheres, then convert the masses to atomic masses and the atomic masses to atoms:

$$\text{Osmium} = 22.61\text{ g/cm}^3 \times 4.19\times10^{-6}\text{ cm}^3 \times \frac{1\text{ u}}{1.6605\times10^{-24}\text{ g}} \times \frac{1\text{ atom}}{190.23\text{ u}} = 2.999\times10^{17}\text{ atoms}$$

$$\text{Aluminum} = 2.70\text{ g/cm}^3 \times 4.19\times10^{-6}\text{ cm}^3 \times \frac{1\text{ u}}{1.6605\times10^{-24}\text{ g}} \times \frac{1\text{ atom}}{26.98\text{ u}} = 2.525\times10^{17}\text{ atoms}$$

If these atoms were the size of sand grains, 1 mm square, what volume would they take up?

Volume per "atom" = (0.1 mm × 1 cm/10 mm)^3 = (0.01 cm)^3 = 1 ×10^{-6} cm^3
Osmium: 2.999 ×10^{17} atoms ×1 × 10^{-6} cm^3 = 2.999 × 10^{11} cm^3 = 299,900,000 liters
Aluminum: 2.525 ×10^{17} atoms ×1 ×10^{-6} cm^3 = 2.525 × 10^{11} cm^3 = 299,900,000 liters

To put this in perspective, an Olympic-sized swimming pool holds about 2,500,000 liters, so if the atoms in this miniscule sphere were expanded to the size of small grains of sand, they would be equivalent in volume to around 120 Olympic-sized swimming pools. Atoms are really, really tiny.

Chapter Three—Mass Relationships in Chemical Reactions

Workbook Problems

Workbook Problem 3.1

Step 1: Write the unbalanced chemical equation.

Remember, the term *combustion* is used to indicate reaction with oxygen. When hydrocarbons (compounds containing primarily C and H) undergo a combustion reaction, carbon dioxide and water are the products.

$$C_6H_{14} + O_2 \rightarrow CO_2 + H_2O$$

Step 2: Use coefficients to balance the equation. (Remember, it helps to save oxygen for last.)

Begin with carbon. There are 6 carbons on the reactant side, but only 1 carbon on the product side. Place 6 in front of CO_2.

$$C_6H_{14} + O_2 \rightarrow 6\ CO_2 + H_2O$$

There are 14 hydrogens on the reactant side, but only 2 hydrogens on the product side. Place a 7 in front of H_2O.

$$C_6H_{14} + O_2 \rightarrow 6\ CO_2 + 7\ H_2O$$

Now balance the oxygens. There are 2 oxygens on the reactant side and a total of 19 on the product side. This odd number of oxygens will require a doubling of everything we have done so far to make it even:

$$2\ C_6H_{14} + O_2 \rightarrow 12\ CO_2 + 14\ H_2O$$

Now there are 2 oxygens on the reactant side, and 38 oxygens on the product side. Place a 19 in front of the oxygen on the reactant side:

$$2\ C_6H_{14} + 19\ O_2 \rightarrow 12\ CO_2 + 14\ H_2O$$

Step 3: Reduce the coefficients to their smallest whole-number ratio.

The ratio is 2:19:12:14. This cannot be reduced, as 19 is a prime number.

Step 4: Check your answer.

Reactant side	*Product side*
12 C	12 C
28 H	28 H
38 O	38 O

Workbook Problem 3.2

Step 1: Write an unbalanced chemical equation from the information given in the problem.

$Fe + O_2 \rightarrow Fe_2O_3$ (This provides a little nomenclature review!)

Step 2: Balance the equation.

$$4Fe + 3O_2 \rightarrow 2Fe_2O_3$$

Step 3: From the balanced chemical equation, determine the mole-to-mole ratio of iron-to-iron(III) oxide.

4 mol Fe: 2 mol Fe_2O_3 can be reduced (or not) to 2 mol Fe: 1 mol Fe_2O_3

Step 4: Find the atomic mass of iron.

From the periodic table, Fe = 55.845 g/mol

Step 5: Determine the molecular mass for iron (III) oxide.

$$(55.845\text{ g/mol} \times 2) + (15.999 \times 3) = 159.687\text{ g/mol}$$

Step 6: From the information in Steps 3, 4, and 5, create conversion factors. Using these conversion factors, set up a dimensional-analysis problem to convert from grams of iron to moles of iron, to moles of iron oxide, to grams of iron oxide.

$$14\text{ g Fe} \times \frac{1\text{ mol Fe}}{55.845\text{ g}} \times \frac{2\text{ mol Fe}_2\text{O}_3}{4\text{ mol Fe}} \times \frac{159.687\text{ g Fe}_2\text{O}_3}{1\text{ mol Fe}_2\text{O}_3} = 20.02\text{ g}$$

Step 7: Using the law of mass conservation, how many grams of oxygen were added to the iron?

There are two ways to solve this: the easy way, as indicated in the question, and the hard way. The easy way to solve this is to see that on the process of rusting, the iron has gained 6.0 grams, and that must have come from the only other reagent, oxygen.

The hard way is to do another stoichiometry problem like the one above:

$$14\text{ g Fe} \times \frac{1\text{ mol Fe}}{55.845\text{ g}} \times \frac{3\text{ mol O}_2}{4\text{ mol Fe}} \times \frac{31.988\text{ g O}_2}{1\text{ mol O}_2} = 6.014\text{ g}$$

As you continue through chemistry, you will often find situations like this, where there are numerous ways to get to the same answer. Always look for the easiest: $20 - 14 = 6$ beats the alternative any day.

Workbook Problem 3.3

Step 1: Balance the equation.

$$3H_2 + N_2 \rightarrow 2NH_3$$

Step 2: Calculate the theoretical yield using dimensional analysis.

$$18.0 \text{ g } N_2 \times \frac{1 \text{ mol } N_2}{28.02 \text{ g}} \times \frac{2 \text{ mol } NH_3}{1 \text{ mol } N_2} \times \frac{17.01 \text{ g } NH_3}{\text{mol } NH_3} = 21.85 \text{ g}$$

Step 3: Calculate the percent yield using the actual yield stated in the problem and the theoretical yield just calculated.

$$\%\text{yield} = \frac{\text{actual yield}}{\text{theoretical yield}} \times 100 = \frac{14.6 \text{ g}}{21.85 \text{ g}} \times 100 = 66.8\%$$

Workbook Problem 3.4

Step 1: Using the information given, write an unbalanced chemical equation.

$$Al + HCl \rightarrow AlCl_3 + H_2$$

Step 2: Balance the chemical equation.

$$2Al + 6HCl \rightarrow 2AlCl_3 + 3H_2$$

Step 3: Determine the number of moles of each reactant present.

$$5.3 \text{ g Al} \times \frac{1 \text{ mol Al}}{27.0 \text{ g Al}} = 0.196 \text{ mol Al}$$

$$9.2 \text{ g HCl} \times \frac{1 \text{ mol HCl}}{36.5 \text{ g HCl}} = 0.252 \text{ mol HCl}$$

Step 4: Determine which reactant gives the smallest number of moles of product. This gives the limiting reactant and the theoretical yield.

$$0.196 \text{ mol Al} \times \frac{2 \text{ mol } AlCl_3}{2 \text{ mol Al}} = 0.196 \text{ mol } AlCl_3$$

$$0.252 \text{ mol HCl} \times \frac{2 \text{ mol } AlCl_3}{6 \text{ mol HCl}} = 0.084 \text{ mol } AlCl_3$$

The limiting reactant is HCl, and the theoretical yield is 0.084 mol $AlCl_3$.

Step 5: Determine how many moles of the excess reactant were used.

$$0.252 \text{ mol HCl} \times \frac{2 \text{ mol Al}}{6 \text{ mol HCl}} = 0.084 \text{ mol Al used}$$

Step 6: Determine how many moles of the excess reactant remain.

$$0.196 \text{ mol Al} - 0.084 \text{ mol Al} = 0.112 \text{ mol Al remaining}$$

Step 7: Determine how many grams of the excess reactant remain.

$$0.112 \text{ mol Al} \times \frac{27.0 \text{ g Al}}{\text{mol Al}} = 3.02 \text{ g Al left over}$$

Step 8: Determine the percent yield of the reaction.

First, convert the theoretical yield to grams. Then calculate percent yield using the equation.

$$0.084 \text{ mol AlCl}_3 \times \frac{133.34 \text{ g AlCl}_3}{1 \text{ mol AlCl}_3} = 11.2 \text{ g AlCl}_3$$

$$\%\text{yield} = \frac{8.3 \text{ g AlCl}_3}{11.2 \text{ g AlCl}_3} \times 100 = 74.1\%$$

Note: the same answer could be obtained by converting the actual yield to moles:

$$8.3 \text{ g AlCl}_3 \times \frac{1 \text{ mol AlCl}_3}{133.34 \text{ g AlCl}_3} = 0.622 \text{ mol AlCl}_3$$

$$\%\text{yield} \times \frac{0.0622 \text{ mol AlCl}_3}{0.084 \text{ mol AlCl}_3} = 74.0\%$$

Workbook Problem 3.5

Step 1: Convert the grams of each element into moles of each element.

$$18.9\% \text{ Li} = 18.9 \text{ g Li} \times \frac{1 \text{ mol Li}}{6.94 \text{ g Li}} = 2.72 \text{ mol Li}$$

$$16.2\% \text{ C} = 16.2 \text{ g C} \times \frac{1 \text{ mol C}}{12.01 \text{ g C}} = 1.35 \text{ mol C}$$

$$64.9\% \text{ O} = 64.9 \text{ g O} \times \frac{1 \text{ mol O}}{16.0 \text{ g O}} = 4.06 \text{ mol O}$$

Step 2: Find the mole ratios.

There are no obvious ratios here—nothing is 1:1—so it makes sense to determine how many there are of each element relative to carbon, the element of which there is the least. Assume there is one carbon, and go from there:

$$\frac{2.72 \text{ mol Li}}{1.35 \text{ mol C}} = 2\frac{\text{Li}}{\text{C}}$$

$$\frac{4.06 \text{ mol O}}{1.35 \text{ mol C}} = 3\frac{\text{O}}{\text{C}}$$

Step 3: If necessary, convert the ratios to whole numbers, and write the empirical formula.

Li_2CO_3—this is lithium carbonate.

Workbook Problem 3.6

Step 1: Determine the formula of the compound.

NaOCl—this is common household bleach.

Step 2: Determine the molecular weight and the weights of each component.

Na	=	22.989 769
O	=	15.999 4
Cl	=	35.453
Molecular weight		74.442 169

Step 3: Determine the percent composition by dividing the mass of each component.

$$\% \text{ Na} = \frac{22.989\,769}{74.442\,169} \times 100\% = 30.88\%$$

$$\% \text{ O} = \frac{15.999\,4}{74.442\,169} \times 100\% = 21.49\%$$

$$\% \text{ Cl} = \frac{35.453}{74.442\,169} \times 100\% = 47.62\%$$

Workbook Problem 3.7

Step 1: Find the molar amounts of C and H in CO_2 and H_2O.

$$5.70 \text{ g } CO_2 \times \frac{1 \text{ mol } CO_2}{44 \text{ g } CO_2} = 0.129 \text{ mol } CO_2 \times \frac{1 \text{ mol C}}{\text{mol } CO_2} = 0.129 \text{ mol C}$$

$$2.31 \text{ g } H_2O \times \frac{1 \text{ mol } H_2O}{18 \text{ g } H_2O} = 0.128 \text{ mol } H_2O \times \frac{2 \text{ mol H}}{\text{mol } H_2O} = 0.257 \text{ mol H}$$

Step 2: Carry out mole-to-gram conversions to find the number of grams of C and H in the original sample.

$$0.129 \text{ mol C} \times \frac{12 \text{ g C}}{\text{mol C}} = 1.55 \text{ g C}$$

$$0.257 \text{ mol H} \times \frac{1.008 \text{ g C}}{\text{mol H}} = 0.259 \text{ g H}$$

Step 3: Subtract the masses of C and H from the mass of the starting sample to determine the mass of S.

$$2.85 \text{ g} - 1.55 \text{ g} - 0.259 \text{ g} = 1.04 \text{ g}$$

Step 4: Convert the mass of S to moles of S.

$$1.04 \text{ g S} \times \frac{1 \text{ mol S}}{32 \text{ g S}} = 0.0325 \text{ mol S}$$

Step 5: Find the mole ratios.

$$\frac{0.129 \text{ mol C}}{0.0325 \text{ mol S}} = 4\frac{\text{C}}{\text{S}}$$

$$\frac{0.257 \text{ mol H}}{0.0325 \text{ mol S}} = 8\frac{\text{H}}{\text{S}}$$

Step 6: Use the ratios above to write the empirical formula of the compound.

C_4H_8S

Step 7: Compare the empirical formula weight to the molecular formula weight to see if a conversion is needed.

The empirical formula weight is 88 u—no conversion is needed.

Step 8:	Write the molecular formula.
	The molecular formula is C_4H_8S. Phew!

Self-Test

True–False

1. F The number of atoms of each element must be the same, but the number of molecules is often different.
2. T
3. T
4. F The coefficients in a balanced chemical equation show the *mole* ratios of the components.
5. F Theoretical yields of chemical reactions are determined through mathematical calculation, not experiment.
6. T
7. F Combustion analysis can only be used to determine the empirical formula of a compound.
8. T

Matching

9. g	10. c	11. i	12. d
13. j	14. a	15. b	16. f
17. e	18. h		

Fill in the Blanks

19. balanced
20. coefficients, subscripts
21. molecular mass or formula mass
22. limiting reactant
23. excess reactant
24. theoretical yield, actual yield, percent yield
25. empirical formula, molecular formula, molecular mass
26. carbon dioxide, water
27. Avogadro's number, 6.022×10^{23}

Problems

28. a. $2N_2 + 5O_2 \rightarrow 2N_2O_5$
    b. $C_3H_8 + 5O_2 \rightarrow 3CO_2 + 4H_2O$
    c. $H_3PO_4 + 3NaOH \rightarrow Na_3PO_4 + 3H_2O$
    d. $Ba(NO_3)_2 + 2NaOH \rightarrow Ba(OH)_2 + 2NaNO_3$

29. a. formula is $CaCl_2$, molecular mass $= 40.078 + (2 \times 35.453) = 110.984$ g/mol; $45/110.984 = 0.41$ mol $CaCl_2$
    b. molecular mass $= (12 \times 12.010\,7) + (22 \times 1.007\,94) + (11 \times 15.999\,4) = 342.29648$ g/mol; $23/342.296\,48 = 0.067$ mol sucrose
    c. formula is PF_3, molecular mass $= 30.973\,762 + (3 \times 18.998\,403) = 87.968\,971$ g/mol; $67/87.968\,971 = 0.76$ mol PF_3
    d. formula is $(NH_4)_2CO_3$, molecular mass is $(2 \times 14.006\,7) + (8 \times 1.007\,94) + 12.010\,7 + (3 \times 15.999\,4) = 96.085\,82$ g/mol; $5/96.085 = 0.05$ mol $(NH_4)_2CO_3$

30. To determine whether there is a limiting reactant, run the reaction mathematically using both reagents.

$$2N_2 + 5O_2 \rightarrow 2N_2O_5$$

$$4 \text{ mol } N_2 \times \frac{2 \text{ mol } N_2O_5}{2 \text{ mol } N_2} = 4 \text{ mol } N_2O_5$$

$$10 \text{ mol } O_2 \times \frac{2 \text{ mol } N_2O_5}{5 \text{ mol } O_2} = 4 \text{ mol } N_2O_5$$

The reactants are present in stoichiometric ratios, so the theoretical yield is 4 mol of N_2O_5. The molecular mass of dinitrogen pentoxide is 108 g/mol, so the theoretical yield in grams is:

$$4 \text{ mol } N_2O_5 \times 108\frac{\text{g}}{\text{mol}} = 432 \text{ g}$$

If the actual yield is 376 grams, the percent yield is calculated:

$$\frac{376 \text{ g}}{432 \text{ g}} \times 100 = 87\%$$

31. Balance the equation.

$$2NaHCO_3 \rightarrow Na_2CO_3 + CO_2 + H_2O$$

Convert the 20.0 g of sodium hydrogen carbonate to moles: the formula weight is 84 g/mol.

$$20.0 \text{ g } NaHCO_3 \times \frac{1 \text{ mol } NaHCO_3}{84.007 \text{ g } NaHCO_3} = 0.238 \text{ mol } NaHCO_3$$

Set up a stoichiometric ratio to determine the theoretical yield of water:

$$0.238 \text{ mol } NaHCO_3 \times \frac{1 \text{ mol } H_2O}{2 \text{ mol } NaHCO_3} = 0.119 \text{ mol } H_2O$$

$$0.119 \text{ mol } H_2O \times \frac{18 \text{ g } H_2O}{\text{mol } H_2O} = 2.14 \text{ g } H_2O = 2.1 \text{ g } H_2O$$

If the reaction has a typical percent yield of 87%, then only 87% of the theoretical yield can be expected.

$$2.14 \text{ g } H_2O \times 0.87 = 1.87 \text{ g } H_2O$$

32. Write and balance the equation.

$$CH_4 + 2O_2 \rightarrow CO_2 + 2H_2O$$

Convert the quantities given to moles:

$$6.7\ g\ CH_4 \times \frac{1\ mol\ CH_4}{16.042\ g\ CH_4} = 0.418\ mol\ CH_4$$

$$20.0\ g\ O_2 \times \frac{1\ mol\ O_2}{31.9988\ g\ O_2} = 0.625\ mol\ O_2$$

Run the reaction with each quantity to determine the limiting reagent and the theoretical yields.

$$0.418\ mol\ CH_4 \times \frac{1\ mol\ CO_2}{1\ mol\ CH_4} = 0.418\ mol\ CO_2 \times \frac{44.009\,5\ g\ CO_2}{1\ mol\ CO_2} = 18.38\ g\ CO_2$$

$$0.418\ mol\ CH_4 \times \frac{2\ mol\ H_2O}{1\ mol\ CH_4} = 0.835\ mol\ H_2O \times \frac{18.015\ g\ H_2O}{1\ mol\ CO_2} = 15.05\ g\ H_2O$$

$$0.625\ mol\ O_2 \times \frac{1\ mol\ CO_2}{2\ mol\ O_2} = 0.313\ mol\ CO_2 \times \frac{44.009\,5\ g\ CO_2}{1\ mol\ CO_2} = 13.75\ g\ CO_2$$

$$0.625\ mol\ O_2 \times \frac{2\ mol\ H_2O}{2\ mol\ O_2} = 0.625\ mol\ H_2O \times \frac{18.015\ g\ H_2O}{1\ mol\ H_2O} = 11.25\ g\ H_2O$$

The lower theoretical yields come from the 20.0 g of oxygen, so this is the limiting reagent, and the theoretical yields are 13.8 g of carbon dioxide and 11.3 g of water.

To determine how many grams of methane will be left over, set up another stoichiometry problem to determine how many moles are consumed.

$$0.625\ mol\ O_2 \times \frac{1\ mol\ CH_4}{2\ mol\ O_2} = 0.313\ mol\ CH_4\ \text{consumed}$$

0.418 moles of methane were available, 0.313 were consumed, leaving 0.105 moles.

$$0.105\ mol\ CH_4 \times \frac{16.042\ g\ CH_4}{1\ mol\ CH_4} = 1.7\ g\ CH_4\ \text{left over}$$

33. Carbon dioxide
CO_2 molecular weight = 44.009 5 u

$$\%C = \frac{12.0107\ u\ \text{of}\ C}{44.009\,5\ u\ CO_2} \times 100\% = 27.29\ \%C$$

$$\%O = \frac{31.9988\ u\ \text{of}\ O}{44.009\,5\ u\ CO_2} \times 100\% = 7.27\ \%O$$

Sodium phosphate

Na_3PO_4 molecular weight = 164.940 669 u

$$\%Na = \frac{68.969\,37 \text{ u of Na}}{163.940\,669 \text{ u } Na_3PO_4} \times 100\% = 42.07\ \%Na$$

$$\%P = \frac{30.973\,762 \text{ u of P}}{163.940\,669 \text{ u } Na_3PO_4} \times 100\% = 18.89\ \%P$$

$$\%O = \frac{63.997\,6 \text{ u of O}}{163.940669 \text{ u } Na_3PO_4} \times 100\% = 39.04\ \%O$$

CH_3COOH
Acetic acid molecular weight = 60.337 8 u

$$\%C = \frac{24.021\,4 \text{ u of C}}{60.337\,8 \text{ u } CH_3COOH} \times 100\% = 39.81\ \%C$$

$$\%H = \frac{4.031\,76 \text{ u of H}}{60.337\,8 \text{ u } CH_3COOH} \times 100\% = 6.682\ \%H$$

$$\%O = \frac{31.9988 \text{ u of O}}{60.337\,8 \text{ u } CH_3COOH} \times 100\% = 53.03\ \%O$$

Adenosine triphosphate: $C_{10}H_{16}N_5O_{13}P_3$
Molecular weight = 507.181 026 u

$$\%C = \frac{120.107 \text{ u of C}}{507.181\,026 \text{ u } C_{10}H_{16}N_5O_{13}P_3} \times 100\% = 23.68\ \%C$$

$$\%H = \frac{16.127\,04 \text{ u of H}}{507.181\,026 \text{ u } C_{10}H_{16}N_5O_{13}P_3} \times 100\% = 3.180\ \%H$$

$$\%N = \frac{70.006\,7 \text{ u of N}}{507.181\,026 \text{ u } C_{10}H_{16}N_5O_{13}P_3} \times 100\% = 13.81\ \%N$$

$$\%O = \frac{207.9922 \text{ u of O}}{507.181\,026 \text{ u } C_{10}H_{16}N_5O_{13}P_3} \times 100\% = 41.01\ \%O$$

$$\%P = \frac{92.921286 \text{ u of P}}{507.181\,026 \text{ u } C_{10}H_{16}N_5O_{13}P_3} \times 100\% = 18.32\ \%P$$

34. Convert percentages to mole ratios assuming a 100 g sample of the unknown compound.

$$40\%\ \text{C} = 40\ \text{g C} \times \frac{1\ \text{mol C}}{12.0107\ \text{g C}} = 3.33\ \text{mol C}$$

$$6.7\%\ \text{H} = 6.7\ \text{g H} \times \frac{1\ \text{mol H}}{1.00794\ \text{g H}} = 6.65\ \text{mol H}$$

$$53.3\%\ \text{O} = 53.3\ \text{g O} \times \frac{1\ \text{mol O}}{15.9994\ \text{g O}} = 3.33\ \text{mol O}$$

Carbon and oxygen have a 1:1 ratio, and carbon and hydrogen are 1:2, making the empirical formula CH_2O (this empirical formula is the origin of the term "carbohydrate"—you know from this that you are looking at a sugar molecule). The empirical formula weight is 30, so there are six empirical formula units in the molecular formula, which must be $C_6H_{12}O_6$. For the record, this is glucose.

35. Convert grams of water and carbon dioxide to moles, then grams of carbon and hydrogen.

$$3.71\ \text{g CO}_2 \times \frac{1\ \text{mol CO}_2}{44.009\,5\ \text{g CO}_2} \times \frac{1\ \text{mol C}}{1\ \text{mol CO}_2} = 0.084\,29\ \text{mol C} \times \frac{12.0107\ \text{g C}}{1\ \text{mol C}} = 1.01\ \text{g C}$$

$$1.51\ \text{g H}_2\text{O} \times \frac{1\ \text{mol H}_2\text{O}}{18.015\ \text{g H}_2\text{O}} \times \frac{2\ \text{mol H}}{1\ \text{mol H}_2\text{O}} = 0.167\,63\ \text{mol H} \times \frac{1.007\,94\ \text{g H}}{1\ \text{mol H}} = 0.169\ \text{g H}$$

From this, it is possible to solve for the grams of oxygen in the sample.

$$1.57\ \text{g} - 1.01\ \text{g} - 0.169\ \text{g} = 0.391\ \text{g O}$$

$$0.391\ \text{g O} \times \frac{1\ \text{mol O}}{15.9994\ \text{g O}} = 0.024\ \text{mol O}$$

It is now possible to develop mole ratios for the formula.

$$\frac{0.084\ \text{mol C}}{0.024\ \text{mol O}} = 3.5\frac{\text{C}}{\text{O}} \qquad \frac{0.167\ \text{mol H}}{0.024\ \text{mol O}} = 7\frac{\text{H}}{\text{O}}$$

From this, an initial formula can be developed: $C_{3.5}H_7O$. This can now be refined to the proper formula: $C_7H_{14}O_2$.

Chapter Four—Reactions in Aqueous Solution

Workbook Problems

Workbook Problem 4.1

Step 1: First, determine the number of moles found in 500 mL of a 0.30 M solution.

$$\text{moles} = \text{M} \times \text{L} = 0.3\frac{\text{moles}}{\text{L}} \times 0.0500\text{L} = 0.150 \text{ moles}$$

Step 2: Determine the number of grams of sodium chloride from the number of moles.

$$0.150 \text{ mole NaCl} \times \frac{58.5 \text{ g}}{\text{mol}} = 8.8 \text{ g NaCl}$$

Workbook Problem 4.2

Step 1: Rearrange the equation $M_i \times V_i = M_f \times V_f$ to solve for the final molarity.

$$M_f = \frac{M_i \times V_i}{V_f} = \frac{50 \text{ mL} \times 10 \text{ M}}{1000 \text{ mL}} = 0.5 \text{ M}$$

Note that here, the conversion of volumes *was* needed, so that they would match and cancel out.

Workbook Problem 4.3

Step 1: Determine the total number of moles of ions formed when $AlCl_3$ completely dissociates in water.

$$AlCl_3 \rightarrow Al^{3+} + 3Cl^-$$

Step 2: Create a conversion factor that relates the total number of moles of ions in solution to 1 mol of $AlCl_3$.

Four ions are formed in the dissociation of aluminum chloride, so the conversion factor is as given below:

$$\frac{4 \text{ mol ions}}{\text{mol } AlCl_3}$$

Step 3: Use the conversion factor to calculate the molar concentration of ions in solution.

$$\frac{0.75 \text{ mol } AlCl_3}{\text{L}} \times \frac{4 \text{ mol ions}}{\text{mol } AlCl_3} = 3.0 \text{ M ions}$$

Workbook Problem 4.4

Step 1: Write and balance the molecular equation.

$$HNO_3(aq) + KOH(aq) \rightarrow H_2O(l) + KNO_3(aq)$$

Step 2: Write the complete ionic equation (from Step 1).
Ionic Equation:

$$H^+(aq) + NO_3^-(aq) + K^+(aq) + OH^-(aq) \rightarrow H_2O(l) + K^+(aq) + NO_3^-(aq)$$

Step 3: List the spectator ions in the complete ionic equation.

spectator ions: K^+ and NO_3^-

Step 4: Write the net ionic equation.

Net ionic equation: $H^+(aq) + OH^-(aq) \rightarrow H_2O(l)$

This net ionic equation is characteristic of acid–base neutralization reactions.

Workbook Problem 4.5

Step 1: Write the balanced molecular equation for each reaction.

$$2LiOH + H_2SO_4 \rightarrow Li_2SO_4 + 2H_2O$$
$$NH_4Cl + AgNO_3 \rightarrow NH_4NO_3 + AgCl$$

Step 2: Determine the presence of any strong electrolytes. Write the ionic equation, showing the strong electrolytes in terms of their free ions.

$$2Li^+ + 2OH^- + 2H^+ + SO_4^{2-} \rightarrow 2Li^+ + SO_4^{2-} + 2H_2O$$
$$NH_4^+ + Cl^- + Ag^+ + NO_3^- \rightarrow NH_4^+ + NO_3^- + AgCl$$

Step 3: Write the net ionic equation by removing spectator ions.

$2OH^- + 2H^+ \rightarrow 2H_2O$ This is an acid–base reaction.
$Cl^- + Ag^+ \rightarrow AgCl$ This is a precipitation reaction.

Workbook Problem 4.6

Step 1: Write a balanced chemical equation.

$$Zn(s) + 2\ HCl(aq) \rightarrow ZnCl_2(aq) + H_2(g)$$

Step 2: Determine the number of moles of zinc present.

$$\text{mol Zn} = 4.75\ \text{g Zn} \times \frac{1\ \text{mol Zn}}{65.4\ \text{g}} = 0.073\ \text{mol Zn}$$

Step 3: Determine the number of moles of HCl needed.

$$0.073 \text{ mol Zn} \times \frac{2 \text{ mol HCl}}{\text{mol Zn}} = 0.145 \text{ mol HCl needed}$$

Step 4: Calculate the volume of 5.0 M HCl needed for this reaction.

$$V = \frac{0.145 \text{ mol}}{5.0\frac{\text{mol}}{\text{L}}} \times \frac{1000 \text{ mL}}{\text{L}} = 29 \text{ mL HCl}$$

Step 5: To calculate the volume of H_2 gas produced, we first need to calculate the moles, then grams of H_2 gas produced.

$$0.073 \text{ mol Zn} \times \frac{1 \text{ mol } H_2}{\text{mol Zn}} \times \frac{2.016 \text{ g } H_2}{\text{mol } H_2} = 0.147 \text{ g } H_2$$

Step 6: Using the density of H_2 gas, determine the volume produced.

$$V = \frac{0.147 \text{ g } H_2}{0.0899\frac{\text{g}}{\text{L}}} = 1.64 \text{ L } H_2$$

Workbook Problem 4.7

Step 1: Write a balanced chemical equation for the reaction.

$$2NaOH + H_2SO_4 \rightarrow Na_2SO_4 + 2H_2O$$

Step 2: Calculate the number of moles of NaOH present.

$$0.153 \text{ g NaOH} \times \frac{1 \text{ mol NaOH}}{40.0 \text{ g NaOH}} = 0.00368 \text{ mol}$$

Step 3: Determine the number of moles of sulfuric acid needed to react with that quantity of NaOH.

$$\text{moles } H_2SO_4 = 0.00383 \text{ mol NaOH} \times \frac{1 \text{ mol } H_2SO_4}{2 \text{ mol NaOH}} = 0.00191 \text{ mol } H_2SO_4$$

Step 4: Calculate the molarity of the acid solution.

$$\text{Molarity} = \frac{0.00191 \text{ mol } H_2SO_4}{0.01637 \text{ L}} = 0.117 \text{ M } H_2SO_4$$

Workbook Problem 4.8

Step 1: Identify the elements that have a fixed oxidation state.

VO_4^{3-}—oxygen carries an oxidation state of −2.

$NaVO_3$—oxygen carries an oxidation state of −2, Na carries an oxidation state of +1.

Step 2: Determine the oxidation number of the remaining elements, keeping in mind that the sum of all oxidation numbers must be equal to the ionic charge or to zero for a neutral molecule.

VO_4^{3-}—oxygen carries an oxidation state of −2.
The overall charge is −3, so $(4 \times -2) + V = -3$, vanadium has an oxidation state of +5.

$NaVO_3$—oxygen carries an oxidation state of −2, Na carries an oxidation state of +1. This is a neutral molecule, so $+1 + (3 \times -2) + V = 0$, vanadium is again +5.

Workbook Problem 4.9

Step1: Identify the species that have a change in oxidation number.

$$\overset{0}{Cu}(s) + 2\overset{+1}{Ag^{+}}(aq) + 2\overset{-1}{NO_3^{-}}(aq) \rightarrow 2\overset{0}{Ag}(s) + \overset{+2}{Cu^{2+}}(aq) + 2\overset{-1}{NO_3^{-}}(aq)$$

Copper goes from an oxidation number of 0 to an oxidation number of 2+.
Silver goes from an oxidation number of 1+ to and oxidation number of 0.

Step 2: Based on the change in oxidation number, identify the species oxidized, the species reduced, and the oxidizing and reducing agent.
Copper loses electrons—it is oxidized and is the reducing agent in this reaction.
Silver gains electrons—it is reduced and is the oxidizing agent in this reaction.

Workbook Problem 4.10

Step 1: Predict the outcome of these reactions.

a. $Cu(s) + HCl(aq) \rightarrow$ no reaction: copper is below H^+ in the activity series. Most metals will dissolve in acid, but mercury, platinum, and what are sometimes called the coinage metals—silver, copper, and gold—will not.
b. $Mn(s) + ZnCl_2(aq) \rightarrow MnCl_2(aq) + Zn(s)$
c. $Mg(s) + Al(NO_3)_3(aq) \rightarrow Mg(NO_3)_2(aq) + Al(s)$

Step 2: Balance the equation.

a. the equation is balanced
b. the equation is balanced
c. $3\ Mg(s) + 2\ Al(NO_3)_3(aq) \rightarrow 3\ Mg(NO_3)_2(aq) + 2\ Al(s)$

Workbook Problem 4.11

Step 1: Use mole ratios to determine the amount of iodine present in the reaction mixture.

$$\text{moles } S_2O_3^{2-} \text{ ion} = 0.15\frac{\text{mol}}{\text{L}} \times 0.0157 \text{ L} = 2.36 \times 10^{-3} \text{ mol or } 2.36 \text{ mmol}$$

$$2.36 \times 10^{-3} \text{ mol } S_2O_3^{2-} \text{ ion} \times \frac{1 \text{ mol } I_2}{2 \text{ mol } S_2O_3^{2-} \text{ ion}} = 1.18 \times 10^{-3} \text{ mol } I_2$$

Step 2: Determine the original concentration of I_2 in solution.

$$\frac{1.18 \times 10^{-3} \text{ mol } I_2}{0.050 \text{ L solution}} = 0.024 \text{ M } I_2$$

Self-Test

True–False
1. F An acid and a base react to form water and a salt.
2. F Both sodium and nitrate indicate solubility.
3. T
4. T
5. T
6. F Acids release hydrogen ions or protons into solution.
7. T
8. F Oxidation numbers can be equal to ionic charge, but are not always equal.
9. F An activity series makes it possible to predict the results of *redox* reactions.
10. T

Matching

11. h	12. f	13. l	14. a
15. b	16. o	17. j	18. n
19. d	20. m	21. e	22. k
23. i	24. c	25. g	26. r
27. p	28. q		

Fill in the Blanks
29. soluble, insoluble, precipitation
30. acid, base, water, salt
31. redox or oxidation/reduction
32. strong electrolytes, weak electrolytes
33. molecular, ionic, net ionic
34. hydrogen or hydronium, hydroxide
35. loses, gains

Problems

36. a. $HNO_3(aq) + KOH(aq) \rightarrow KNO_3(aq) + H_2O(l)$.

Neutralization: net ionic equation is $H^+ + OH^- \rightarrow H_2O$.

b. $Cu(NO_3)_2(aq) + 2NaOH(aq) \rightarrow Cu(OH)_2(s) + 2NaNO_3(aq)$

Precipitation: two soluble components generate and insoluble product.

c. $Cu(NO_3)_2(aq) + Zn(s) \rightarrow Zn(NO_3)_2(aq) + Cu(s)$

Redox: Cu and Zn are transferring electrons—Zn is oxidized, Cu is reduced.

37. Convert grams to moles.

$$15\text{ g } Na_3PO_4 \times \frac{1\text{ mol}}{164\text{ g}} = 0.0915\text{ mol}$$

Convert to molar concentration of ions.

$$\frac{0.0915\text{ mol } Na_3PO_4}{0.250\text{ L}} \times \frac{4\text{ ions}}{Na_3PO_4} = 1.46\text{ M}$$

38. a. $Pb(NO_3)_2(aq) + 2NaCl(aq) \rightarrow PbCl_2(s) + 2NaNO_3(aq)$
$Pb^{2+}(aq) + 2NO_3^-(aq) + 2Na^+(aq) + 2Cl^-(aq) \rightarrow PbCl_2(s) + 2Na^+(aq) + 2NO_3^-(aq)$
$Pb^{2+}(aq) + 2Cl(aq) \rightarrow PbCl_2(s)$

b. $H_2SO_4(aq) + 2LiOH(aq) \rightarrow Li_2SO_4(aq) + 2H_2O(l)$
$2H^+(aq) + SO_4^{2-}(aq) + 2Li^+(aq) + 2OH^-(aq) \rightarrow 2Li^+(aq) + SO_4^{2-}(aq) + 2H_2O(l)$
$H^+(aq) + OH^-(aq) \rightarrow H_2O(l)$

c. $KCl(aq) + AgCH_3CO_2(aq) \rightarrow KCH_3CO_2(aq) + AgCl(s)$
$K^+(aq) + Cl^-(aq) + Ag^+(aq) + CH_3CO_2^-(aq) \rightarrow K^+(aq) + CH_3CO_2^-(aq) + AgCl(s)$
$Ag^+(aq) + Cl^-(aq) \rightarrow AgCl(s)$

39. Write a balanced net ionic equation.

$$Ca^{2+}(aq) + 2OH^-(aq) \rightarrow Ca(OH)_2(s)$$

Determine the number of moles of calcium hydroxide required:

$$1.8\text{ g } Ca(OH)_2 \times \frac{1\text{ mol}}{74.1\text{ g}} = 0.0243\text{ mol}$$

From the balanced equation above, 0.0243 mol of Ca ions are required, and 0.0486 mol of hydroxide ions are required.

From the definition of *molarity*, solve for the volumes needed.

$$M = \frac{\text{mol}}{L} \therefore V = \frac{\text{mol}}{M}$$

$$V_{OH} = \frac{0.0486\text{ mol}}{1\text{ M}} = 0.0486\text{ L or } 48.6\text{ mL}$$

$$V_{Ca} = \frac{0.0243\text{ mol}}{1\text{ M}} = 0.0243\text{ L or } 24.3\text{ mL}$$

48.6 mL 1 M $Ca(NO_3)_2$ solution would need to be combined with 24.3 mL 1 M KOH to produce 1.8 g $Ca(OH)_2$.

40. a. No reaction. Possible products are lead acetate (soluble) and copper (II) nitrate (soluble).
    b. No reaction. Possible products are sodium sulfate (soluble) and ammonium nitrate (soluble).
    c. $Ag^+(aq) + Cl^-(aq) \rightarrow AgCl(s)$

41. a. Cl_2 Elemental forms, even in molecules, always have oxidation numbers of 0.
    b. NaCl Na = +1, Cl = −1
    c. CO_3^{2-} O = −2, C = +4 $4 + (3 \times -2) = -2$, which is the charge on the ion.
    d. Cr_2O_3 O = −2, Cr = +3 $(3 \times -2) + (2 \times +3) = 0$
    e. ClO_3^- O = −2, Cl = +5 $(3 \times -2) + 5 = -1$

42. a. $C_2H_6 + O_2 \rightarrow CO_2 + H_2O$ (Combustion is a form of redox reaction).
    Oxidation number of C goes from −3 to +4 oxidized, reducing agent
    Oxidation number of O goes from 0 to −2 reduced, oxidizing agent
    b. $Na + Cl_2 \rightarrow NaCl$
    Oxidation number of Na goes from 0 to +1 oxidized, reducing agent
    Oxidation number of Cl goes from 0 to −1 reduced, oxidizing agent
    c. $Mg + Fe^{2+} \rightarrow Mg^{2+} + Fe$
    Oxidation number of Mg goes from 0 to +2 oxidized, reducing agent
    Oxidation number of Fe goes from +2 to 0 reduced, oxidizing agent
    d. $Ca + H^+ \rightarrow Ca^{2+} + H_2$
    Oxidation number of Ca goes from 0 to +2 oxidized, reducing agent
    Oxidation number of H goes from +1 to 0 reduced, oxidizing agent

43. a. chromium and copper (II) nitrate
    $2Cr(s) + 3Cu(NO_3)_2(aq) \rightarrow 2Cr(NO_3)_3(aq) + 3Cu(s)$
    b. tin and silver (I) nitrate
    $Sn(s) + 2AgNO_3(aq) \rightarrow Sn(NO_3)_2(aq) + 2Ag(s)$
    c. copper (II) nitrate and cobalt
    $Co(s) + Cu(NO_3)_2(aq) \rightarrow Co(NO_3)_2(aq) + Cu(s)$
    d. silver and HCl—no reaction.

44. Determine the moles of potassium chromate.

$$0.0237 \text{ L} \times 0.124 \text{ M} = 0.00294 \text{ mole } K_2Cr_2O_7$$

Determine moles of Na_2SO_3 using the balanced equation.

$$0.00294 \text{ mol } K_2Cr_2O_7 \times \frac{3 \text{ mol } Na_2SO_3}{1 \text{ mol } K_2Cr_2O_7} = 0.00882 \text{ mol } Na_2SO_3$$

$$M = \frac{0.00882 \text{ mol } Na_2SO_3}{0.015 \text{ L}} = 0.588 \text{ M}$$

45. a. $K_2SO_4(aq) + Pb(NO_3)_2(aq) \rightarrow 2KNO_3(aq) + PbSO_4(s)$
$2K^+(aq) + SO_4^{2-}(aq) + Pb^{2+}(aq) + 2NO_3^-(aq) \rightarrow 2K^+(aq) + 2NO_3^-(aq) + PbSO_4(s)$
$Pb^{2+}(aq) + SO_4^{2-}(aq) \rightarrow PbSO_4(s)$

b. Determine the number of moles of each reactant.

$$5.3\text{ g } K_2SO_4 \times \frac{1\text{ mol}}{174\text{ g } K_2SO_4} = 0.030\,460\text{ mol } K_2SO_4$$

$$4.9\text{ g } Pb(NO_3)_2 \times \frac{1\text{ mol}}{331\text{ g } Pb(NO_3)_2} = 0.014\,803\text{ mol } Pb(NO_3)_2$$

Reactants react in a 1:1 ratio: $Pb(NO_3)_2$ is the limiting reagent.

$$0.014\,803\text{ mol } Pb(NO_3)_2 \times \frac{1\text{ mol } PbSO_4}{1\text{ mol } Pb(NO_3)_2} \times \frac{303\text{ g}}{1\text{ mol } PbSO_4} = 4.49\text{ g } PbSO_4$$

c. Subtract the moles of limiting reagent used from the moles of excess reagent:

$$0.0304\,597\text{ mol} - 0.0148\,036\text{ mol} = 0.015\,656\text{ mol } K_2SO_4$$

$$0.015\,656\text{ mol} \times \frac{174\text{ g } K_2SO_4}{1\text{ mol}} = 2.72\text{ g } K_2SO_4$$

d. $\text{percent yield} = \frac{4.37\text{ g}}{4.49\text{ g}} \times 100 = 97\%$

e. For this, the number of moles of K_2SO_4 in excess and the moles of KNO_3 need to be converted to ions and diluted.

$$0.015\,656\text{ mol } K_2SO_4 \times \frac{3\text{ mol ions}}{1\text{ mol } K_2SO_4} = 0.046\,968\text{ mol ions}$$

$$0.0148\,036\text{ mol } Pb(NO_3)_2 \times \frac{2\text{ mol } KNO_3}{1\text{ mol } Pb(NO_3)_2} = 0.029\,607\text{ mol } KNO_3$$

$$0.029\,607\text{ mol } KNO_3 \times \frac{2\text{ mol ions}}{1\text{ mol } KNO_3} = 0.059\,214\text{ mol ions}$$

$$M_{ions} = \frac{0.059\,214 + 0.046\,968}{0.200\text{ L}} = 0.530\text{ M}$$

Challenge Problem

46. To approach this problem, first solve for the mole quantity of hydroxide generated by the reaction.

$$HCl + MOH \rightarrow H_2O + MCl$$

$$0.0597\text{ L HCl} \times 7.5\frac{\text{mol}}{\text{L}} \times \frac{1\text{ mol OH}}{\text{mol HCl}} = 0.448\text{ mol OH}$$

This is also equivalent to the number of moles of the alkali metal. There is now enough information to set up a system of equations.

$$\text{Li} + \text{Na} = 4.00\text{ g}$$

$$\frac{\text{Li}}{6.9\text{ g/mol}} + \frac{\text{Na}}{23\text{ g/mol}} = 0.448\text{ mol}$$

This second equation is equivalent to 0.145Li + 0.0435Na = 0.448. To solve by substitution,

$$\text{Li} = 4 - \text{Na, so}$$

$$0.145(4 - \text{Na}) + 0.0435\text{Na} = 0.448$$

$$0.58 - 0.145\text{Na} + 0.0435\text{Na} = 0.448$$

$$-0.145\text{Na} + 0.0435\text{Na} = 0.448 - 0.58$$

$$-0.102\text{Na} = -0.132$$

$$\text{Na} = 1.3\text{ g}$$

$$\text{Na} + \text{Li} = 4$$

$$\text{Li} = 2.7\text{ g}$$

Chapter Five—Periodicity and the Electronic Structure of Atoms

Workbook Problems

Workbook Problem 5.1

Step 1: Choose the appropriate form of the equation and solve:

$$\lambda = \frac{c}{\upsilon} = \frac{3.00 \times 10^8\text{ m/s}}{88.5 \times 10^6\text{ /s}} = 3.39\text{ m}$$

As this demonstrates, wavelengths can be quite long. Although visible light is in the nanometer range, it would not make sense to convert this wavelength to nanometers.

Workbook Problem 5.2

Step 1: Determine the energy of one photon, then convert to one mole of photons.

$$E = h\nu = 6.626\times10^{-34}\ J\cdot s \times 4.4\times10^{14}\ /s = 2.91\times10^{-9}\ J$$

$$2.91\times10^{-19}\ J \times 6.022\times10^{23}\ /mol = 176{,}000\ J/mol = 176\ kJ/mol$$

Step 2: Convert the frequency to wavelength.

$$c = \lambda\nu \therefore \lambda = \frac{c}{\nu} = \frac{3.00\times10^{8}\ m/s}{4.4\times10^{14}\ /s} = 6.82\times10^{-7}\ m = 682\ nm$$

Step 3: Consult your textbook for the color of this wavelength of light.

Light of 682 nm will be reddish-orange.

Workbook Problem 5.3

Step 1: Determine the values of *n* that will make λ the shortest and longest.

The longest wavelength line will be at the smallest possible value of *n*. If *m* = 5, and *n* must be greater than 5, the longest wavelength will occur when *n* = 6. The shortest wavelength will occur with the largest possible value of n. If *n* = ∞, 1/*n* = 0.

Step 2: Use the Balmer–Rydberg equation with *m* = 5 and solve for the shortest wavelength.

$$\frac{1}{\lambda} = R\left[\frac{1}{m^2} - \frac{1}{n^2}\right] = 1.097\times10^{-2}\ nm^{-1}\left[\frac{1}{25} - \frac{1}{\infty}\right] = 4.36\times10^{-4}\ m$$

$$\frac{1}{\lambda} = \frac{1}{4.39\times10^{-4}\ m} = 2280\ nm$$

Step 3: Solve for λ using the Balmer–Rydberg equation and the values of *m* that make λ the longest.

$$\frac{1}{\lambda} = R\left[\frac{1}{m^2} - \frac{1}{n^2}\right] = 1.097\times10^{-2}\ nm^{-1}\left[\frac{1}{25} - \frac{1}{36}\right] = 1.34\times10^{-4}\ nm^{-1}$$

$$\frac{1}{\lambda} = \frac{1}{1.34\times10^{-4}\ nm^{-1}} = 7460\ nm$$

Workbook Problem 5.4

Step 1: Identify the subshell with the value of n and the letter designation for $l = 0$.

When $l = 0$, the subshell is s. This is a 3s subshell.

Step 2: Determine the number of orbitals in this subshell.

The number of orbitals in a subshell is determined by the magnetic quantum number, m_l. For $l = 0$, only one value of m_l is available, so there is only one orbital.

Workbook Problem 5.5

Step 1: Determine the number of electrons in calcium.

Calcium has atomic number 20, so it has 20 electrons.

Step 2: Use the Aufbau principle to determine the ground-state electronic configuration.

Begin building up the electron configuration from $n = 1$

Ca: $1s^2\ 2s^2 2p^6 3s^2 3p^6 4s^2$

Step 3: Determine the noble gas in the previous row, and remove the electrons associated with it to a shorthand notation. Specifically name the electrons in unfilled subshells.

The noble gas with fewer electrons than calcium is argon. With the argon atoms delineated using shorthand, the electron configuration becomes:

Ca: $[Ar]4s^2$

Step 4: Draw the orbital-filling diagram:

↑↓	↑↓	↑↓ ↑↓ ↑↓	↑↓	↑↓ ↑↓ ↑↓	↑↓
1s	2s	2p	3s	3p	4s

Self-Test

True–False

1. F Amplitude is the height of a wave from the midline.
2. T
3. T
4. F Above the atomic level, de Broglie wavelengths are irrelevant.
5. F The Bohr model only holds for hydrogen.
6. T
7. F The angular momentum quantum number must be less than the principal quantum number.
8. F f orbitals are only found in the lanthanide and actinide metals.
9. T
10. F Electrons that share an orbital must have opposite spins.

Multiple Choice

11. d	12. b	13. c	14. a
15. b	16. c	17. b	18. a
19. d	20. c		

Matching

21. b	22. g	23. e	24. d
25. c	26. i	27. a	28. f
29. h			

Fill in the Blanks

30. wavelength, frequency
31. line spectra
32. photons
33. emission
34. quantum numbers
35. *s* orbitals
34. *p* orbitals, two, six
37. degenerate, parallel spins
38. valence electrons
39. atomic radius

Problems

40. First, convert from frequency to wavelength. Remember that Hz = 1/s.

$$c = \lambda v, \text{ so } \lambda = \frac{c}{v} = \frac{3.00 \times 10^8 \text{ m/s}}{835.6 \times 10^6 \text{ /s}} = 0.359 \text{ m or } 359 \text{ nm}$$

Then calculate the energy of 1 mol of these photons.

$$E = hv = 6.626 \times 10^{-34} \text{ J} \cdot \text{s} \times 835.6 \times 10^6/\text{s} = 5.54 \times 10^{-25} \text{ J/photon}$$

$$5.54 \times 10^{-25} \text{ J/photon} \times 6.022 \times 10^{23} \text{ photons/mol} = 0.333 \text{ J/mol}$$

41. First, convert from wavelength to frequency, taking care with the units of wavelength.

$$c = \lambda v, \text{ so } v = \frac{c}{\lambda} = \frac{3.00 \times 10^8 \text{ m/s}}{2.15 \times 10^{-14} \text{ /m}} = 1.40 \times 10^{22} \text{ /s}$$

Then calculate the energy of 1 mol of these photons.

$$E = hv = 6.626 \times 10^{-34} \text{ J} \cdot \text{s} \times 1.40 \times 10^{22}/\text{s} = 9.28 \times 10^{-12} \text{ J/photon}$$

$$9.28 \times 10^{-12} \text{ J/photon} \times 6.022 \times 10^{23} \text{ photons/mol} = 5.59 \times 10^{12} \text{ J/mol}$$

42. First, determine the energy of one photon at this wavelength.

$$E = \frac{hc}{\lambda} = \frac{6.626 \times 10^{-34} \text{ J} \cdot \text{s} \times 3.00 \times 10^8 \text{ m/s}}{432 \times 10^{-9} \text{ m}} \; 4.60 \times 10^{-19} \text{ J}$$

From this, determine the number of photons in 2.5 mJ of energy.

$$\frac{2.5\times10^{-3}\ \text{J}}{4.60\times10^{-19}\ \text{J / photon}}=5.43\times10^{15}\ \text{photons}$$

43. First, calculate the energy needed to dislodge one electron. This will be the same energy found in a photon capable of dislodging an electron.

$$495.8\frac{\text{kJ}}{\text{mol}}\times\frac{1\ \text{mol}}{6.022\times10^{23}\ \text{photons}}\times\frac{1000\ \text{J}}{\text{kJ}}=8.233\times10^{-19}\frac{\text{J}}{\text{photon}}$$

A photon with this much or more energy can dislodge an electron from sodium. From this, the threshold frequency and wavelength can be calculated.

$$\text{E}=\text{h}\nu,\ \nu=\frac{\text{E}}{\text{h}}=\frac{8.233\times10^{-19}\ \text{J}}{6.626\times10^{-34}\ \text{J}\cdot\text{s}}=1.243\times10^{15}\ /\text{s}$$

$$\lambda=\frac{\text{c}}{\nu}=\frac{3.00\times10^{8}\ \text{m / s}}{1.243\times10^{15}\ /\text{s}}=2.414\times10^{-7}\ \text{m}=241.4\ \text{nm}$$

If the threshold frequency corresponds to a wavelength of 241 nm, any wavelength that gives a higher-energy photon—that is, any wavelength lower than 241 nm—will eject electrons. 540 nm is below the threshold frequency and will not eject electrons. 195 nm is above the threshold frequency and will eject photons.

44.

$$(\Delta\text{x})(\Delta\text{mv})\geq\frac{\text{h}}{4\pi}$$

$$\Delta\text{v}\geq\frac{\text{h}}{4\pi\text{m}\Delta\text{x}}\geq\frac{6.626\times10^{-34}\ \frac{\text{kg}\cdot\text{m}^2}{\text{s}}}{4\times3.14\times9.11\times10^{-34}\ \text{kg}\times327\times10^{-12}\ \text{m}}\geq1.77\times10^{8}\ \text{m / s}$$

45.

$$\frac{1}{\lambda}=R\left[\frac{1}{1^2}-\frac{1}{6^2}\right]=1.097\times\frac{10^{-2}}{\text{nm}\left[1-\frac{1}{36}\right]}=0.010\,665/\text{nm}$$

$$\lambda=93.762\ \text{nm}$$

$$\text{E}=\frac{\text{hc}}{\lambda}=\frac{6.626\times10^{-34}\ \text{J}\cdot\text{s}\times3.00\times10^{8}\ \text{m/s}}{93.762\times10^{-9}\ \text{m}}=2.12\times10^{-18}\ \text{J}$$

46. Y = $1s^2 2s^2 2p^6 3s^2 3p^6 4s^2 3d^{10} 4p^6 5s^2 4d^1$

47. Neon has electrons in two energy levels, $n = 1$ and $n = 2$. It has both s and p electrons and a total of 10 electrons. They will have the following quantum numbers:

1	0	0	+½	$1s^1$
1	0	0	−½	$1s^2$
2	0	0	+½	$2s^1$
2	0	0	−½	$2s^2$
2	1	−1	+½	$2p^1$
2	1	−1	−½	$2p^2$
2	1	0	+½	$2p^3$
2	1	0	−½	$2p^4$
2	1	+1	+½	$2p^5$
2	1	+1	−½	$2p^6$

48. The highest energy electron in bromine will be a $4p$ electron. The quantum numbers associated with a $4p$ electron are $n = 4$, and $l = 1$.

49.

	$1s$	$2s$	$2p$	$3s$	$3p$
Si =	↑↓	↑↓	↑↓ ↑↓ ↑↓	↑↓	↑ ↑ _
P =	↑↓	↑↓	↑↓ ↑↓ ↑↓	↑↓	↑ ↑ ↑
Cl =	↑↓	↑↓	↑↓ ↑↓ ↑↓	↑↓	↑↓ ↑↓ ↑

Phosphorus has three unpaired electrons, making it the most paramagnetic. Chlorine has one unpaired electron and is therefore the least paramagnetic element.

50. Solve for the wavelength of the photon, then solve the Balmer–Rydberg equation for n.

$$E = \frac{hc}{\lambda}, \lambda = \frac{hc}{E} = \frac{6.626\times10^{-34}\ J\cdot s\times3.00\times10^{8}\ m/s}{4.58\times10^{-19}\ J} = 434\ nm$$

$$\frac{1}{434\ nm} = 1.097\times10^{-2}\ nm^{-1}\left[\frac{1}{2^2} - \frac{1}{n^2}\right]$$

$$\frac{1}{434\ nm\times1.097\times10^{-2}\ nm^{-1}} = \left[\frac{1}{4} - \frac{1}{n^2}\right]$$

$$0.210 = 0.25 - \frac{1}{n^2}$$

$$0.04 = \frac{1}{n^2}$$

$$n^2 = 25$$

$$n = 5$$

51. $\lambda = \dfrac{h}{mv} = \dfrac{6.626 \times 10^{-34} \ \dfrac{kg \cdot m^2}{s}}{9.109 \times 10^{-31} \ kg \times 2.57 \times 10^6 \ m/s} = 2.83 \times 10^{-10} \ m = 0.283 \ nm$

52. $[Ar]4s^13d^{10}$ = 29 electrons – Cu
$[Kr]\ 4d^{10}$ = 46 electrons – Pd
$[Xe]6s^24f^{14}5d^{10}6p^5$ = 85 electrons – At
$[Ar]4s^13d^5$ = 24 electrons – Cr

53. These elements are in different periods, so they need to be ranked from highest to lowest in the periodic table.

$$Fe < Ru < Os < Hs$$

54. These elements are in different groups, so they need to be ranked from right to left in the periodic table.

$$Cl < S < Mg < Na$$

Challenge Problems

55. First, solve for the energy of one photon at 550 nm.

$$E = \frac{hc}{\lambda} = \frac{6.626 \times 10^{-34} \ J \cdot s \times 3.00 \times 10^8 \ m/s}{550 \times 10^{-9} \ m} = 3.61 \times 10^{-19} \ J/photon$$

The light bulb is emitting 40 J/s, and is allowed to do so for 4 hours. The energy emitted is as given below:

$$40 \frac{J}{s} \times 4 \ hours \times \frac{3600 \ s}{h} = 576{,}000 \ J$$

Solve for photons emitted.

$$\frac{576{,}000 \ J}{3.61 \times 10^{-19} \ J/photon} = 1.60 \times 10^{24} \ photons$$

For the second part of the problem, solve for the energy needed to heat the water.

$$energy = 200 \ g \times 4.184 \frac{J}{g \cdot °C} \times (95 \ °C - 20 \ °C) = 62{,}760 \ J$$

If the light bulb is emitting 576,000J of energy in 4 hours, and 95% of this is heat, the bulb is emitting 547,000 J of heat in 4 hours, or 137,000 J/hr.

If all of this heat went into the 200 g of water, solve for the time.

$$\text{time} = \frac{62{,}760\text{ J}}{137{,}000\text{ J / hr}} = 0.459\text{ hours} \times \frac{60\text{ min}}{\text{hr}} = 27.5\text{ min}$$

56. 15 W = 15 J/s.
If the laser were 100% efficient, 2.73×10^{27} photons would equal 15 J. The energy of each photon would therefore be:

$$\frac{15\text{ J}}{2.73 \times 10^{19}\text{ photons}} = 5.49 \times 10^{-19}\text{ J / photon}$$

$$E = \frac{hc}{\lambda}, \lambda = \frac{hc}{E} = \frac{6.626 \times 10^{-34}\text{ J}\cdot\text{s} \times 3.00 \times 10^{8}\text{ m / s}}{5.49 \times 10^{-19}\text{ J}} = 3.62 \times 10^{-7}\text{ m or 362 nm}$$

If the laser were 80% efficient, the wavelength would be:

$$\lambda = \frac{hc}{E} = \frac{6.626 \times 10^{-34}\text{ J}\cdot\text{s} \times 3.00 \times 10^{8}\text{ m / s}}{5.49 \times 10^{-9}\text{ J} \times 0.8} = 4.52 \times 10^{-7}\text{ m or 452 nm}$$

If 20%

$$\lambda = \frac{hc}{E} = \frac{6.626 \times 10^{-34}\text{ J}\cdot\text{s} \times 3.00 \times 10^{8}\text{ m / s}}{5.49 \times 10^{-19}\text{ J} \times 0.2} = 1.81 \times 10^{-6}\text{ m or 1810 nm}$$

To solve for the efficiency if its wavelength is 1542, rearrange the equation to solve for the percentage.

$$x\% = \frac{hc}{E\lambda} = \frac{6.626 \times 10^{-34}\text{ J}\cdot\text{s} \times 3.00 \times 10^{8}\text{ m / s}}{5.49 \times 10^{-19}\text{ J} \times 1.542 \times 10^{-6}\text{ m}} = 0.235$$

The laser is 23.5% efficient.

Chapter Six—Ionic Compounds: Periodic Trends and Bonding Theory

Workbook Problems

Workbook Problem 6.1

Step 1: The largest jump in ionization energies is between the fourth and fifth ionization energies. This shows that there are five valence electrons, so this must be a Group 5A element. In fact, it is phosphorus. ($P \rightarrow P^{+} \rightarrow P^{2+} \rightarrow P^{3+} \rightarrow P^{4+} \rightarrow P^{5+}$)

Workbook Problem 6.2

Step 1: Identify the group number of each element, and identify them as metals or nonmetals.

Rb:	Rubidium	Group 1A metal
Sc:	Scandium	Group 3B transition metal
Se:	Selenium	Group 6A nonmetal
As:	Arsenic	Group 5A nonmetal

Step 2: Find the nearest noble gas.

Rb:	Krypton (36 electrons)
Sc:	Argon (18 electrons)
Se:	Krypton (36 electrons)
As:	Krypton (36 electrons)

Step 3: Identify the number of electrons that must be gained/lost to achieve the electron configuration of that noble gas.

Rb:	37 electrons $\rightarrow$ Krypton (36 electrons)	One electron must be lost.
Sc:	21 electrons $\rightarrow$ Argon (18 electrons)	Three electrons must be lost.
Se:	34 electrons $\rightarrow$ Krypton (36 electrons)	Two electrons must be gained.
As:	33 electrons $\rightarrow$ Krypton (36 electrons)	Three electrons must be gained.

Step 4: Write the ionized form of the element.

Rb^+ Sc^{3+} Se^{2-} As^{3-}

Workbook Problem 6.3

Step 1: Write each step of the reaction with its energy required or produced per mole.

Sublimation of sodium	$Na(s) \rightarrow Na(g)$	+107.3 kJ/mol
Vaporization of bromine	$\frac{1}{2}Br_2(l) \rightarrow \frac{1}{2}Br_2(g)$	+ 15.4 kJ/mol
Splitting of bromine molecules	$\frac{1}{2}Br_2(g) \rightarrow Br(g)$	+112 kJ/mol
Ionization of sodium	$Na(g) \rightarrow Na^+(g) + e^-$	+495.8 kJ/mol
Ionization of bromine	$Br(g) + e^- \rightarrow Br^-(g)$	−325 kJ/mol
Formation of ionic solid	$Na^+(g) + Br^-(g) \rightarrow NaBr(s)$	−747 kJ/mol

Step 2: Sum the energy terms to find the overall energy change of the reaction.

Total = −341.5 kJ/mol

Note: the values given for bond breaking and vaporizing bromine are divided in half. As given, they are for the following reactions: $Br_2(l) \rightarrow Br_2(g)$ and $Br_2(g) \rightarrow 2Br(g)$.

Workbook Problem 6.4

Step 1: Identify the relative charges in the two sets of ions, and how they change from one set to the other.

NaCl, KCl	same charges
AlF_3, $MgCl_2$	Al – 3+, Mg – 2+, aluminum will provide stronger lattice energy.
$CuBr_2$, CuCl	Cu (II) – 2+, Cu (I) – 1+, copper (II) will provide stronger lattice energy.

Step 2: If necessary, identify the relative ion sizes in the two sets of ions, and how they change from one to the other.

NaCl, KCl have same charges. Sodium is smaller than potassium, so will provide stronger lattice energy.

Step 3: Determine the relative lattice energies of the ion pairs.

NaCl > KBr $AlF_3 > MgCl_2$ $CuBr_2 > CuCl$

Self-Test

True–False

1. T
2. F Nonmetals gain electrons to obtain noble gas electron structures.
3. F Transition metals will sometimes form more than one ion.
4. T
5. T
6. F Ionization energies decrease down the periodic table.
7. F Nonmetals tend to have large negative electron affinities.
8. T
9. F The Born–Haber cycle provides information about ionic bonding.
10. F Lattice energies are related to atomic size and charge.

Multiple Choice

11. d	12. b	13. c	14. b
15. a	16. c	17. d	18. a

Matching

19. e	20. d	21. a	22. b
23. h	24. g	25. f	26. c

Fill in the Blanks

27. argon	28. decreases	29. core, easy
30. increase, valence	31. added, gaseous	32. charge, Coulomb's
33. eight, octet rule		

Problems

34. P: $1s^22s^22p^63s^23p^3$
P^{3-}: $1s^22s^22p^63s^23p^6$ Electron structure is that of argon.

Ti: $1s^22s^22p^63s^23p^64s^23d^2$
Ti^{2+}: $1s^22s^22p^63s^23p^63d^2$

Mn: $1s^22s^22p^63s^23p^64s^23d^5$
Mn^{2+}: $1s^22s^22p^63s^23p^63d^5$

K: $1s^22s^22p^63s^23p^64s^1$
K^+: $1s^22s^22p^63s^23p^6$ Electron structure is that of argon.

Se: $[Ar]4s^23d^{10}4p^4$
Se^{2-}: $[Ar]4s^23d^{10}4p^6$ Electron structure is that of krypton.

35. Remembering that cations are smaller than neutral atoms, anions are larger than neutral atoms, and that atoms decrease in size from left to right across the periodic table and increase in size down the periodic table:

$$Li^+ < Li < Al^{3+} < Al < Cl < Cl^-$$

36. There is a very large jump (~4x) between the third and fourth ionization energies. The fourth electron removed is a core electron, meaning that there are three valence electrons. In the third period of the periodic table, this is aluminum.

37.

		4s	3d	4p
Ga:	[Ar]	↑↓	↑↓ ↑↓ ↑↓ ↑↓ ↑↓	↑ _ _
Ge:	[Ar]	↑↓	↑↓ ↑↓ ↑↓ ↑↓ ↑↓	↑ ↑ _
Br:	[Ar]	↑↓	↑↓ ↑↓ ↑↓ ↑↓ ↑↓	↑↓ ↑↓ ↑

Based on its electron structure, gallium is far more likely to lose an electron to form a +1 ion than to gain an electron. Looking at the periodic table, this is also confirmed by the position of gallium, which is a main-group metal. Germanium can adopt a half-filled shell, which is also a fairly stable arrangement, but it does not have the stability of a noble gas configuration. Bromine is the closest to having a noble gas configuration, so it will have the highest electron affinity.

38. Use the Born–Haber cycle. However, this time, the net energy change is known, and we need to solve for the heat of sublimation of aluminum.

$Al(s) \rightarrow Al(g)$	? kJ/mol	
$Al(g) \rightarrow Al^+(g) + e^-$	578 kJ/mol	578 kJ/mol
$Al^+(g) \rightarrow Al^{2+}(g) + e^-$	1,817 kJ/mol	1817 kJ/mol
$Al^{3+}(g) \rightarrow Al^{3+}(g) + e^-$	2,745 kJ/mol	2745 kJ/mol
$3/2\ Cl_2(g) \rightarrow 3\ Cl(g)$	3/2(244 kJ/mol)	366 kJ/mol
$3\ Cl(g) \rightarrow 3\ Cl^-(g)$	3(−348.6 kJ/mol)	−1046 kJ/mol
$Al^{3+}(g) + 3\ Cl^-(g) \rightarrow AlCl_3(s)$	−5,492 kJ/mol	−5492 kJ/mol

Overall net energy change −704.2 kJ/mol

−704.2 = ? + 578 kJ/mol + 1817 kJ/mol + 2745 kJ/mol + 366 kJ/mol − 1046 kJ/mol − 5492 kJ/mol

? = 327.8 kJ/mol

39. $Mg(s) + Cl_2(g) \rightarrow MgCl_2(s)$
$2\ Mg(s) + O_2(g) \rightarrow 2\ MgO(s)$
$3\ Mg(s) + 2\ P(s) \rightarrow Mg_3P_2(s)$

$2\ Li(s) + Cl_2(g) \rightarrow 2\ LiCl(s)$
$4\ Li(s) + O_2(g) \rightarrow 2\ Li_2O(s)$
$3\ Li(s) + P(s) \rightarrow LiP_3(s)$

$2\ Al(s) + 3\ Cl_2(g) \rightarrow 2\ AlCl_3(s)$
$4\ Al(s) + 3\ O_2(g) \rightarrow 2\ Al_2O_3(s)$
$Al(s) + P(s) \rightarrow AlP(s)$

40.

$K(s) \rightarrow K(g)$	+89.2 kJ/mol
$K(g) \rightarrow K^+(g) + e^-$	+418.8 kJ/mol
$\frac{1}{2}\ F_2(g) \rightarrow F(g)$	½(+158 kJ/mol)
$F(g) \rightarrow F^-(g)$	−328 kJ/mol
$K^+(g) + F^-(g) \rightarrow KF(s)$	−821 kJ/mol
Total:	−562 kJ/mol

41. $AlCl_3 > MgCl_2 > LiCl > KCl > CsCl$
Using Coulomb's law, an increase in ionic charge on the cation will dominate other factors, so the cation with the highest charge will have the highest lattice energy. With ions that have charges of 1, size becomes a factor: smaller ions can get closer together and will thus have a larger lattice energy than larger ions, so these can be ranked in order of cation size, smallest to largest.

Challenge Problem

42. The chemical equation for the reaction of the alkaline earth chloride with silver nitrate is as follows:

$$MCl_2 + 2\ AgNO_3 \rightarrow M(NO_3)_2 + 2\ AgCl$$

a. The mass of chlorine can be determined from the mass of silver chloride produced.

$$18.8\text{ g AgCl} \times \frac{1\text{ mol AgCl}}{143.3\text{ g}} \times \frac{1\text{ mol Cl}}{1\text{ mol AgCl}} \times \frac{35.5\text{ g Cl}}{\text{mol Cl}} = 4.66\text{ g Cl}$$

$$\text{mass \% Cl in metal chloride} = \frac{4.66\text{ g Cl}}{5.25\text{ g MCl}_2} \times 100 = 88.8\%$$

b. The identity of the metal can be determined by calculating the molar mass of the metal chloride, then subtracting out the mass of the chloride.

$$\text{mol MCl}_2 = 18.8\text{ g AgCl} \times \frac{1\text{ mol AgCl}}{143.3\text{ g AgCl}} \times \frac{1\text{ mol MCl}_2}{2\text{ mol AgCl}} = 0.0656\text{ mol MCl}_2$$

$$\text{molar mass MCl}_2 = \frac{5.25\text{ g}}{0.0656\text{ mol}} = 80\frac{\text{g}}{\text{mol}}$$

Molar mass of M = Molar mass of MCl_2 − Molar mass of Cl_2 = 80 g/mol − 71 g/mol = 9 g/mol

From the periodic table, the metal must be beryllium.

c. For the reaction of beryllium with chlorine, $Be(s) + Cl_2(g) \rightarrow BeCl_2(s)$
For the reaction of $BeCl_2$ with $AgNO_3$, $BeCl_2 + 2\ AgNO_3 \rightarrow Be(NO_3)_2 + 2\ AgCl$

d. To determine the reactant that is the limiting reactant, the moles of beryllium and the moles of chlorine gas must be calculated.

$$1.53\ \text{g Be} \times \frac{1\ \text{mol}}{9.01\ \text{g}} = 0.170\ \text{mol}$$

$$4.75\text{L Cl}_2 \times 3.17\ \frac{\text{g}}{\text{L}} \times \frac{1\ \text{mol}}{71\ \text{g}} = 0.212\ \text{mol}$$

Chlorine was in excess: the liters left unreacted can now be calculated.
0.212 mol Cl_2 present − 0.170 mol Cl_2 reacted = 0.042 mol Cl_2 unreacted.

$$0.042\ \text{mol Cl}_2 \times \frac{71\ \text{g}}{\text{mol}} \times \frac{1\ \text{L}}{3.17\ \text{g}} = 0.94\ \text{L}$$

Chapter Seven—Covalent Bonding and Electron-Dot Structures

Workbook Problems

Workbook Problem 7.1

Step 1: Predict the polarity of the bonds based on their locations, then check with electronegativity values from Figure 7.4 in the textbook:

H – F—nonmetals, far apart: polar covalent
F – Cl—nonmetals, close together: nonpolar
C – O—nonmetals, close together: nonpolar
K – Cl—metal and nonmetal: ionic
Ge – Br—semimetal and nonmetal: polar covalent

Step 2: Find the difference between the numbers to check.

H – F: 4.0 − 2.1 = 1.9, polar covalent
F – Cl: 4.0 − 3.0 = 1.0, polar covalent (estimate is off—fluorine is *really* electronegative!)
C – O: 3.5 − 2.5 = 1.0, polar covalent
K – Cl: 3.0 − 0.8 = 2.2, ionic
Ge – Br: 2.8 − 1.8 = 1.0, polar covalent

Note: what this exercise shows is that the farther apart on the periodic table two elements are, the more polar/ionic their bond. Also, almost all nonmetal—nonmetal bonds are polar.

Workbook Problem 7.2

Strategy: Determine the number of valence electrons in the compounds, combine atoms, then form complete octets.

Step 1: Determine the number of electrons on each atom based on its position on the periodic table:

H—1 valence electron F—7 valence electrons C—4 valence electrons
Br—7 valence electrons Cl—7 valence electrons

Step 2: Determine how the atoms are connected.

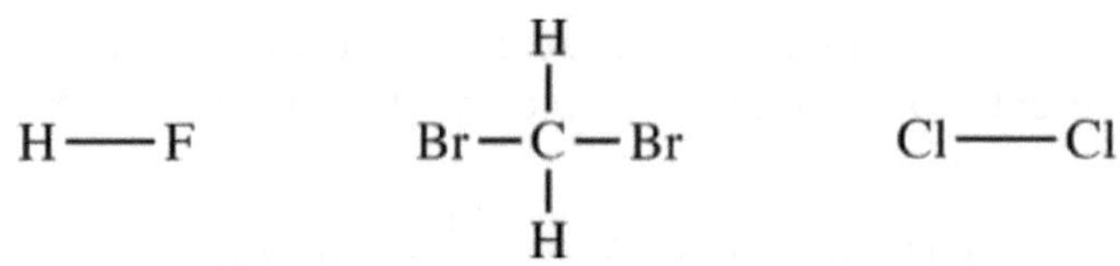

Steps 3 and 4: Subtract the number of valence electrons used for bonding from the total number calculated in Step 1 and assign the rest to complete the octets of all atoms.

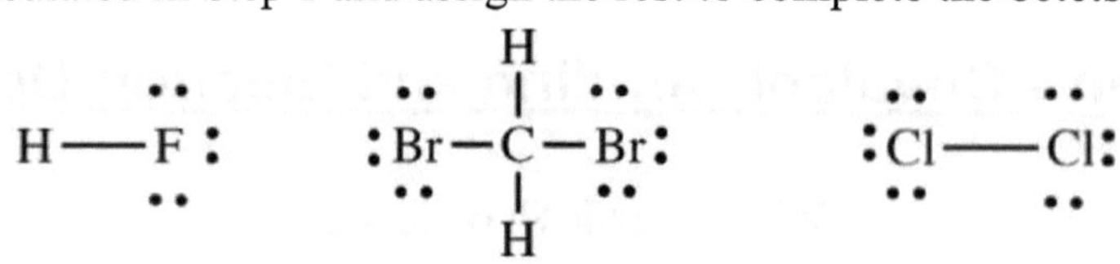

Workbook Problem 7.3

Step 1: Calculate the total number of valence electrons.

$$N - 2 \text{ atoms} \times 5\ e^- = 10\ e^-$$

Step 2: Determine how the atoms are connected.

N–N

Steps 3 and 4: Subtract the number of valence electrons used for bonding from the total number calculated in Step 1 and assign the rest to complete the octets of all atoms.

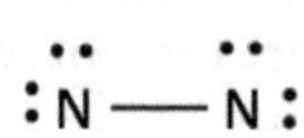

Step 5: Form multiple bonds as needed to complete the octet of the central atom.

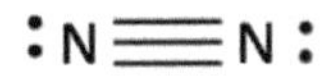

There are not enough electrons to complete the octets on the nitrogen atoms. To complete the octets, the nitrogen atoms share these electrons forming a triple bond.

Workbook Problem 7.4

Step 1: Calculate the total number of valence electrons.

$CH_3CH_2CH_3$

C – 3 atoms × 4 e^- = 12 e^-
H – 8 atoms × 1 e^- = 8 e^-
20 valence electrons

CH_3OH

C – 1 atom × 4 e^- = 4 e^-
H – 4 atoms × 1 e^- = 4 e^-
O – 1 atom × 6 e^- = 6 e^-
14 valence electrons

C_2H_4

C – 2 atoms × 4 e^- = 8 e^-
H – 4 atoms × 1 e^- = 4 e^-
12 valence electrons

Step 2: Determine how the atoms are connected.

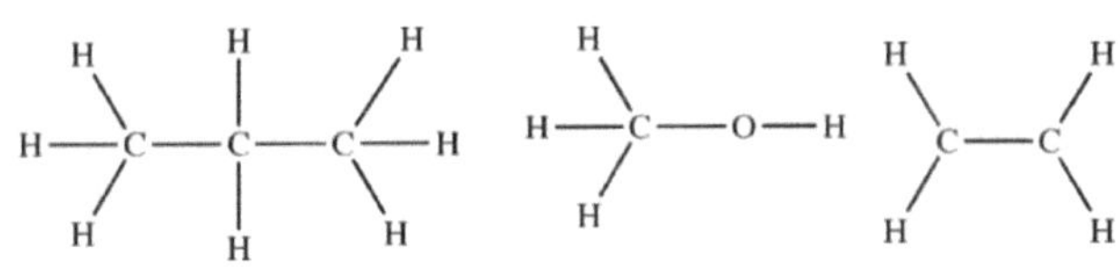

Steps 3 and 4: Subtract the number of valence electrons used for bonding from the total number calculated in Step 1 and assign the rest to complete the octets of all atoms.

CH_3OH and C_2H_4 had electrons to add.

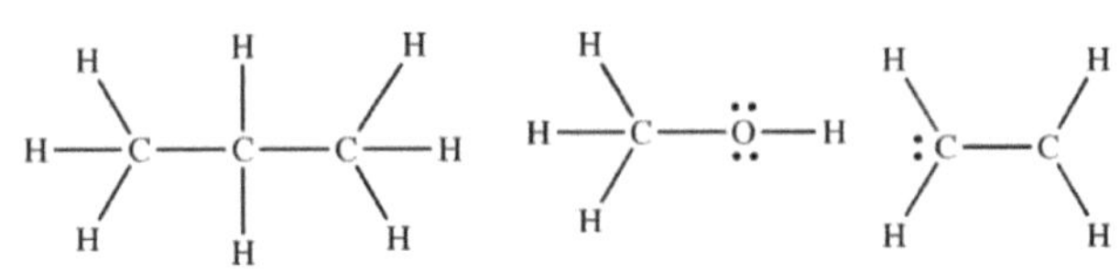

Step 5: Form multiple bonds as needed to complete the octet of the central atom.

A carbon in C_2H_4 did not have an octet so a multiple bond was formed from the lone pair.

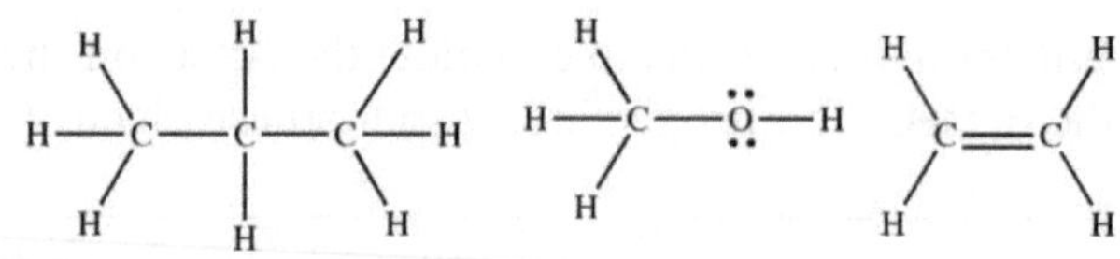

There are not enough electrons to complete the octets on the nitrogen atoms. To complete the octets, the nitrogen atoms share these electrons forming a triple bond.

Workbook Problem 7.5

Step 1: Calculate the total number of valence electrons.

$$
\begin{aligned}
C - 1 \times 4\,e^- &= 4\,e^- \\
O - 3 \times 6\,e^- &= 18\,e^- \\
-2\text{ charge} - 2 \times 1\,e^- &= 2\,e^- \\
&\quad 24 \text{ valence } e^- \text{ total}
\end{aligned}
$$

Step 2: Determine how the atoms are connected.

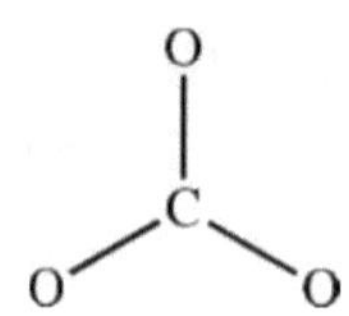

Carbon is the central atom for symmetry's sake, as well as it being the most electronegative atom.

Steps 3 and 4: Subtract the number of valence electrons used for bonding from the total number calculated in Step 1 and assign the rest to complete the octets of all atoms.

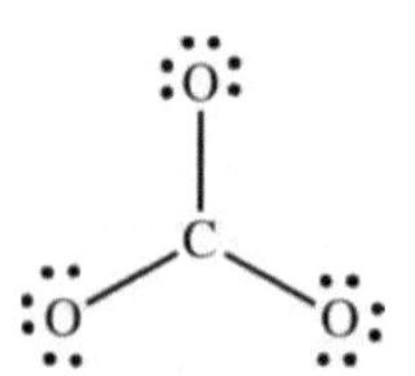

This uses 6 valence electrons, leaving 18.

Step 5: Form multiple bonds as needed to complete the octet of the central atom.

There are no electrons remaining, but carbon has an incomplete octet and will need to share a pair of electrons from one of the oxygens. This leads to a number of resonance structures:

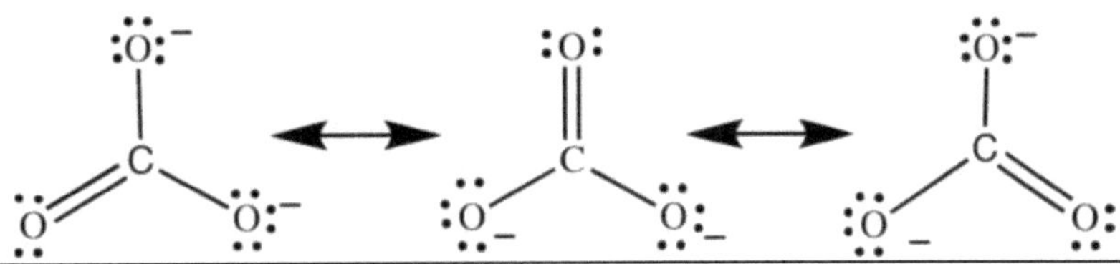

Workbook Problem 7.6

Step 1: Determine the number of valence electrons in the ion.

Total number of valence-shell electrons = 4 (from C) + 6 (from O) + 5 (from N) + 1 (from the 1− charge) = 16

Step 2: Determine three reasonable electron-dot structures with carbon as the central atom and this number of valence electrons.

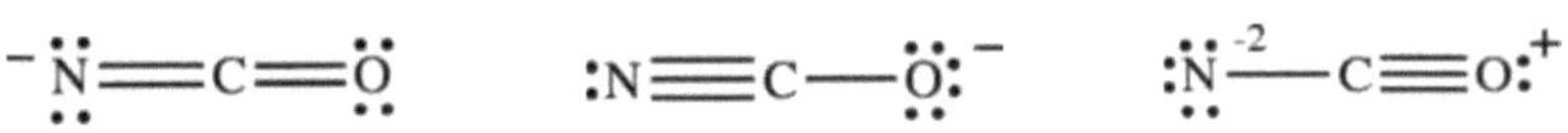

Step 3: Calculate the formal charge on each of the atoms in each of the structures.

Structure 1:	Structure 2:	Structure 3:
N = 5 − 2 − 4 = −1	N = 5 − 3 − 2 = 0	N = 5 − 1 − 6 = −2
C = 4 − 4 − 0 = 0	C = 4 − 4 − 0 = 0	C = 4 − 4 − 0 = 0
O = 6 − 2 − 4 = 0	O = 6 − 1 − 6 = −1	O = 6 − 3 − 2 = +1

Step 4: Choose the most stable structure based on your comparison of the formal charges.
Notice that all three have a net charge of −1, appropriate to this ion. Structure 3 (which is already violating rules about how many bonds oxygen likes to have) has formal charges on two atoms, and is therefore unstable. Structures 1 and 2 both have single negative charges on one atom. To choose between these two, choose the structure in which the negative charge is assigned to the most electronegative atom. This is structure 2.

Self-Test

True–False

1. F Covalent bonds involve sharing of electrons.
2. T
3. T
4. T
5. T
6. F Valences of main group elements can be easily predicted using the periodic table.

7. F Multiatom structures can be predicted using electron-dot models.
8. F The structures of charged species can be predicted by removing or adding electrons as needed.
9. T
10. T

Fill in the Blanks

11. bond lengths, bond, nuclei
12. bond-dissociation energy, positive, released
13. polar covalent, ionic
14. nonbonding, bonding electrons
15. octet rule, expanded octets
16. formal charges

Matching

17. h	18. a	19. i	20. b
21. d	22. g	23. f	24. e
25. c			

Problems

26. LiBr—ionic (metal/nonmetal) HCl—covalent (nonmetal/nonmetal)
MgO—ionic (metal/nonmetal) NF_3—covalent (nonmetal/nonmetal)

27.

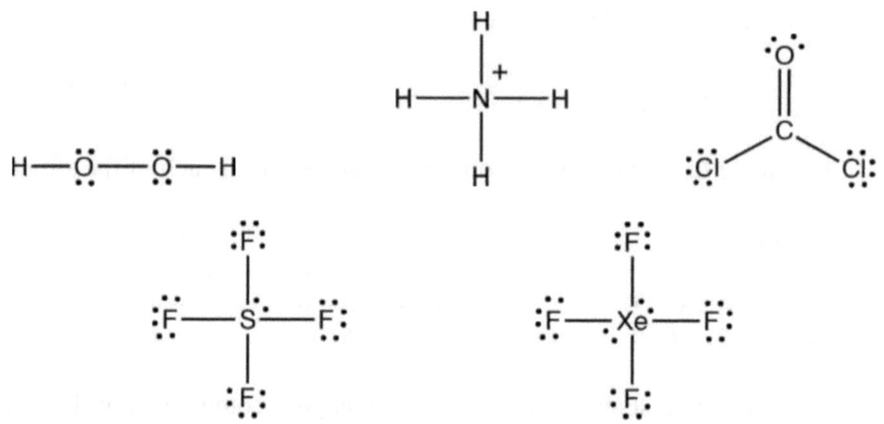

28. The high melting point and conductivity in solution indicate that the first compound is ionic. The second compound, with its low melting point and nonconducting solution, is covalent. The solubilities of the compounds provide no information about bonding.

29.

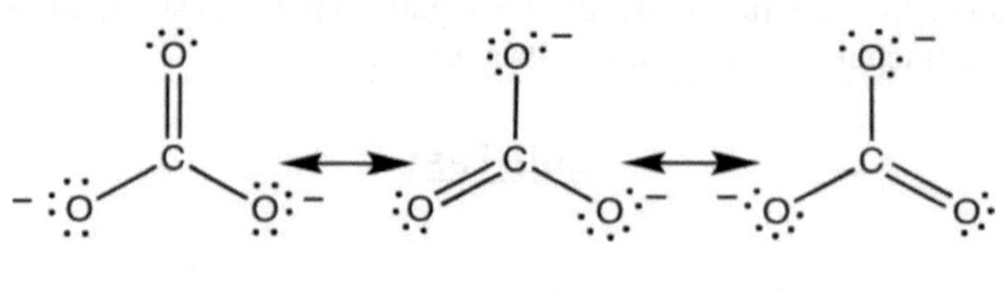

The three resonance structures are all equivalent.

Formal charges:	Central C	$= 4 - 4 - 0 = 0$
	Double-bonded O	$= 6 - 2 - 4 = 0$
	Single-bonded O	$= 6 - 1 - 6 = -1$

The formal charges match the ionic charge.

30. Li–Cl > Al–F > H–Cl > C–H > Cl–Cl. The atoms that are farther apart on the periodic table will have more-polar bonds.

31. Cs < Fe < O < F. Fluorine is the most electronegative element: proximity to fluorine gives a good approximation of electronegativity of an element.

32. The most symmetrical structure is usually the most stable, and formal charges bear this out.

Leftmost structure:
- Leftmost nitrogen = 5 − 1 − 6 = −2
- Central nitrogen = 5 − 4 − 0 = +1
- Oxygen = 6 − 3 − 2 = +1

Center structure:
- Leftmost nitrogen = 5 − 2 − 4 = −1
- Central nitrogen = 5 − 4 − 0 = +1
- Oxygen = 6 − 2 − 4 = 0

Rightmost structure:
- Leftmost nitrogen = 5 − 3 − 2 = 0
- Central nitrogen = 5 − 4 − 0 = +1
- Oxygen = 6 − 1 − 6 = −1

None of these structures is perfectly satisfactory, but the rightmost structure is the only one in which there is a negative formal charge on oxygen, the most electronegative element in the structure.

Chapter Eight—Covalent Compounds: Bonding Theories and Molecular Structure

Workbook Problems

Workbook Problem 8.1

Step 1: Draw the electron-dot structure for the molecules, using the number of valence electrons and the most symmetrical structure.

CS_2 has 4 valence electrons from carbon and 6 from each of the sulfurs, for a total of 16.

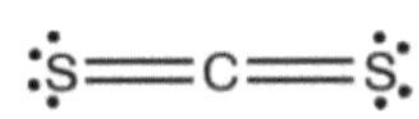

Double bonds were needed to complete the octet around carbon. Although sulfur will form an expanded octet, here it does not need to.

SF_4 has 6 valence electrons from the sulfur, and 28 from the four fluorines, for a total of 34.

Step 2: Determine the number of charge clouds around the central atom. How many are used for bonding and how many are nonbonding?
CS_2 has two electron clouds around the central atom.
SF_4 has five electron clouds around the central atom, four bonding, and one nonbonding pair.

Step 3: Based on the chart, determine the shape.
CS_2 is linear. SF_4 is seesaw shaped.

Workbook Problem 8.2

Given the following structure:

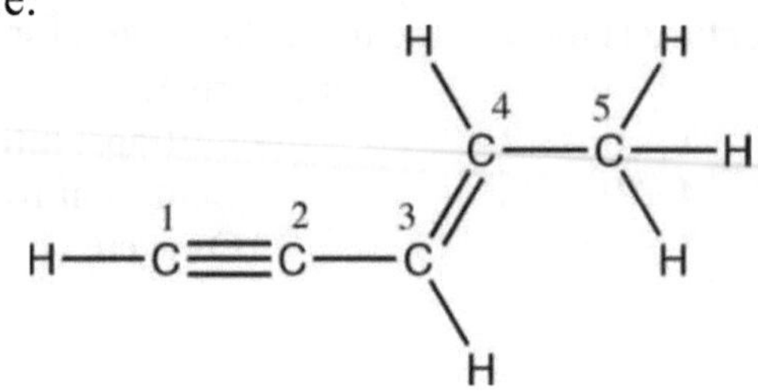

Step 1: Determine the number of electron clouds around each carbon and its hybridization.

Carbon 1: two electron clouds—*sp* hybridization
Carbon 2: two electron clouds—*sp* hybridization
Carbon 3: three electron clouds—sp^2 hybridization
Carbon 4: three electron clouds—sp^2 hybridization
Carbon 5: four electron clouds—sp^3 hybridization

Step 2: Determine the bond types present.

Carbon 1: σ bond to hydrogen, σ and 2π bonds to carbon 2
Carbon 2: σ and 2π bonds to carbon 1, σ bond to carbon 3
Carbon 3: σ bond to carbon 2, σ bond to hydrogen, and σ and π to carbon 4
Carbon 4: σ and π to carbon 3, σ to hydrogen and σ to carbon 5
Carbon 5: σ bonds to carbon 4 and the three hydrogens

Workbook Problem 8.3

Step 1: Draw the electron structure for each molecule.

All of these molecules are asymmetrical, even trimethylamine, which has a lone pair of electrons on the nitrogen atom.

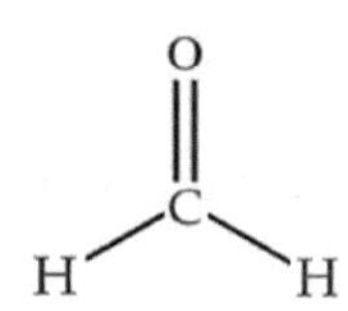

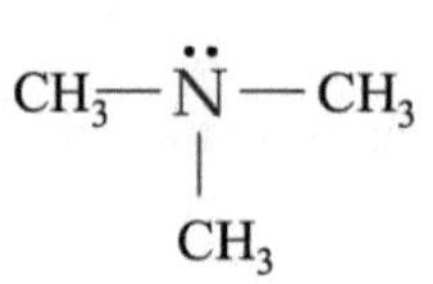

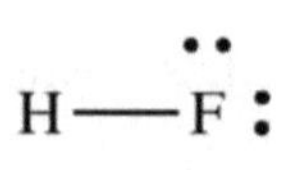

Step 2: Draw arrows that represent the flow of electrons in each bond.

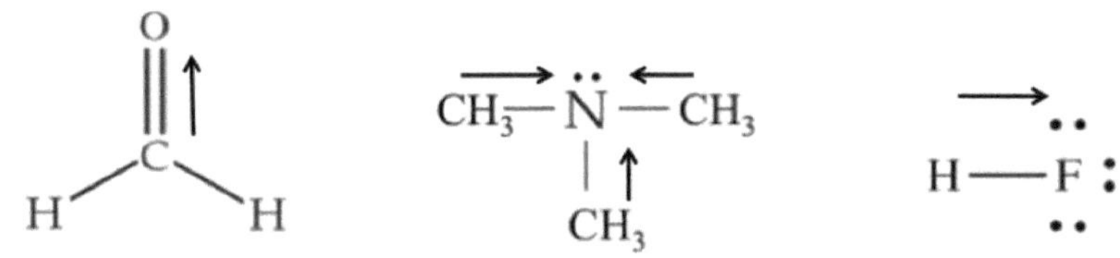

Step 3: Consider the net sum of these bond dipoles to find the molecular dipole.

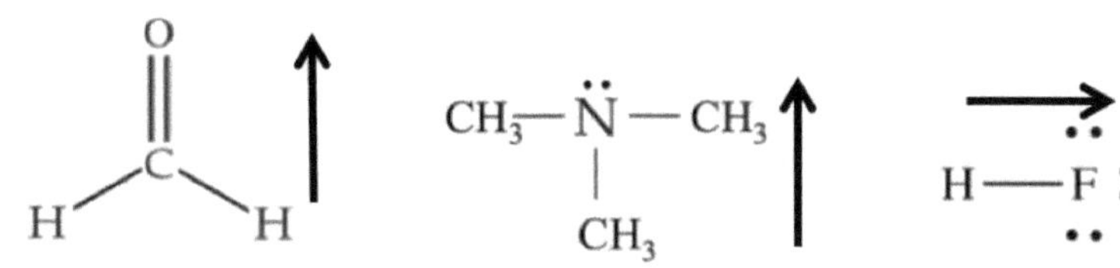

H_2CO–oxygen $N(CH_3)_3$–nitrogen HF–fluorine

All of these molecules will have dipoles.

Workbook Problem 8.4

Step 1: List the interactions experienced by each molecule:

CH_3OH: hydrogen bonding, dipole–dipole, and dispersion forces

Xe: dispersion forces only

CH_3Cl: dipole–dipole and dispersion forces

Step 2: Rank the molecules in order of increasing strength of interactions.

$Xe < CH_3Cl < CH_3OH$

Self-Test

True–False

1. T
2. T
3. T
4. F Methane molecules are tetrahedral.
5. T
6. F When orbitals overlap head-on, σ bonds are formed.
7. T
8. T
9. F The polarity of individual bonds sometimes cancels out, leaving a molecule with polar bonds without a dipole.
10. F Intermolecular forces range from very weak dispersion forces to fairly strong.
11. T

Fill in the Blanks

12. tetrahedral, bent, trigonal pyramidal
13. sp^2
14. σ, π
15. Electronegativity
16. dipole–dipole, London dispersion forces, hydrogen bonds

Matching

17. i	18. h	19. g	20. f
21. e	22. d	23. c	24. b
25. a			

Problems

26. NF_3—four electron clouds provides tetrahedral electron geometry. Three bonds lead to trigonal pyramidal molecular geometry.

    H_2S—four electron clouds again provides tetrahedral electron geometry, with two bonds leading to bent molecular geometry.

    IF_4^-—here, there is an expanded octet to consider with 36 electrons spread over five atoms. There are six electron clouds in this molecule, with two nonbonding pairs on the iodine. This molecule will have octahedral electron geometry and square planar molecular geometry.

    PF_5—here, there is another expanded octet. There are five electron clouds in this molecule, with no nonbonding pairs. Both the electronic and molecular geometries are trigonal bipyramidal.

27. The carbon with four bonds has tetrahedral geometry, which indicates sp^3 hybridization and σ binding. The carbon bonded to the oxygen has three electron groups, which indicates trigonal planar geometry and sp^2 hybridization. This atom has σ bonds to carbon and hydrogen, and one σ and one π bond to oxygen.

28.

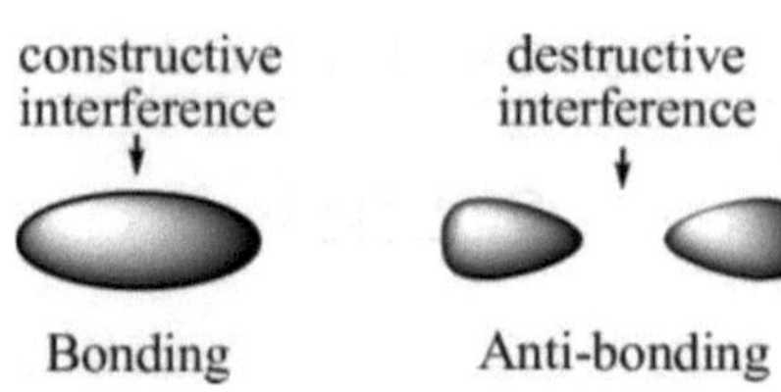

Challenge Problem

29. Determine the empirical formula of the acid.

    26.1 g C/12.01 g/mol = 2.17 mol C
    69.7 g O/15.999 g/mol = 4.36 mol O
    4.35 g H/1.008 g/mol = 4.32 mol H

    Empirical formula = CO_2H_2

    Determine the molecular formula.

The moles of NaOH needed to neutralize the 0.36 grams of acid are:

$$0.07824 \text{ L} \times 0.10 \text{ mol/L} = 0.007824 \text{ mol}$$

The molecular weight of the compound is 0.360 g/0.007824 mol = 46.01.

The empirical formula weight is 46.01, so the empirical formula is the same as the molecular formula.

Lewis structure:

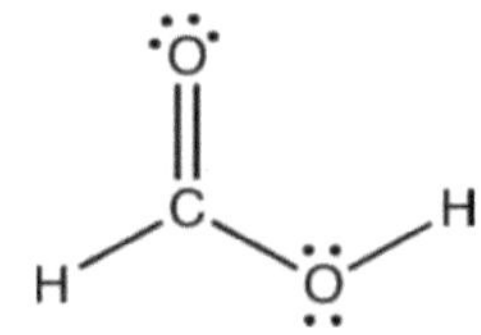

The central carbon atom is sp^2 hybridized, with trigonal planar geometry

The oxygen bonded to the C and H atoms is sp^3 hybridized, with tetrahedral geometry.

C–H = σ   C–O = σ   O–H = σ   C=O = π

Chapter Nine—Thermochemistry: Chemical Energy

Workbook Problems

Workbook Problem 9.1

Step 1: Use $w = -P\Delta V$

$$w = -(13 \text{ atm} \times [7.4 \text{ L} - (7.4 \text{ L} + 3.7 \text{ L})]) = 48.1 \text{ L} \cdot \text{atm}$$

Step 2: Convert from L · atm to joules to kilojoules.

$$48.1 \text{ L} \cdot \text{atm} \times 101 \text{ J/L} \cdot \text{atm} = 4858 \text{ J} = 4.9 \text{ kJ}$$

Step 3: What does the sign tell you? Was work done by or on the system?

The positive value of the work shows that the system gained: work was done on the system. The same thing can be determined by looking at the initial and final volume of the reactions: 11.1 L of reactants provided 7.4 L of products. The overall system contracted, so work was done on it.

Workbook Problem 9.2

Step 1: Use $\Delta E = q - P\Delta V$

$$\Delta E = -79.9 \text{ kJ} - 4.9 \text{ kJ} = -84.8 \text{ kJ}$$

Workbook Problem 9.3

Step 1: Determine the ΔH for the reaction given the amount of carbonic acid that dissociated.

$$0.5 \text{ mol} \times -20.7 \text{ kJ/mol} = -10.4 \text{ kJ}$$

Step 2: Determine $P\Delta V$ and convert from L · atm to kJ.

$$1 \text{ atm} \times 11.2 \text{ L} = 11.2 \text{ L} \cdot \text{atm} \times 101 \text{ J/L} \cdot \text{atm} = 1131 \text{ J} = 1.13 \text{ kJ}$$

Step 3: Determine ΔE from the equation: $\Delta H = \Delta E + P\Delta V$. Rearrange the equation to $\Delta E = \Delta H - P\Delta V$.

$$\Delta E = -10.4 \text{ kJ} - (1.13 \text{ kJ}) = -11.5 \text{ kJ}$$

Think it through to make sure this makes sense. Heat was released from the exothermic equation, leading to a negative ΔH. Work was done by the system when the volume increased, which increases the magnitude of the overall energy change. Therefore, it makes sense.

Workbook Problem 9.4

Step 1: Convert the grams of water to moles.

$$50 \text{ g} \times \frac{1 \text{ mol}}{18 \text{ g}} = 2.8 \text{ mol}$$

Step 2: Multiply by the heat of vaporization.

$$2.8 \text{ mol} \times 40.7 \frac{\text{kJ}}{\text{mol}} = 113 \text{ kJ}$$

Step 3: Make sure the sign of the heat transfer makes sense: the water is the system.

Vaporizing water is the same as boiling it on the stove. This is a way of adding heat, so it is expected that heat will need to be added to the system to vaporize the water. In fact, lots of heat will need to be added because vaporizing water requires breaking all of the hydrogen bonds that hold it together as a liquid. The sign on the energy transfer is positive, so heat is being added to the system, as expected.

Workbook Problem 9.5

Step 1: Determine the number of moles of calcium.

$$3.24\text{ g} \times \frac{1\text{ mol}}{40.08\text{ g}} = 0.0808\text{ mol}$$

Step 2: Using the stoichiometric ratio, determine the heat involved in the reaction.

$$0.0808\text{ mol Ca} \times \frac{2\text{ mol CaO}}{2\text{ mol Ca}} \times -634.9\frac{\text{kJ}}{\text{mol CaO}} = -51.3\text{ kJ}$$

Step 3: Based on the sign of the heat transfer, determine whether the reaction is exothermic or endothermic.

Based on the negative sign for the energy transfer, this is an exothermic reaction.

Workbook Problem 9.6

Step 1: Determine the amount of heat gained by the water using the equation $q = mc\Delta T$.

$$q = 50\text{ g} \times 4.184\text{ J/g °C} \times (30.70\text{ °C} - 22.73\text{ °C}) = 1667\text{ J}$$

Step 2: Determine the amount of heat released by the reaction.

As the dissociation released the energy absorbed by the water, it will have the same magnitude, but be opposite in sign: −1667 J.

Step 3: Determine the number of moles of NaOH involved.

$$1.5\text{ g} \times \frac{1\text{ mol}}{40\text{ g}} = 0.0375\text{ mol}$$

Step 4: Determine the heat of dissociation per mole of NaOH.

$$\frac{-1667\text{ J}}{0.0375\text{ mol}} = 44\,453\frac{\text{J}}{\text{mol}} \times \frac{1\text{ kJ}}{1000\text{ J}} = -44.5\frac{\text{kJ}}{\text{mol}}$$

Workbook Problem 9.7

Step 1: Place reactants on the left, multiplying, if necessary.

$Fe_2O_3(s) \rightarrow 2\ Fe(s) + 3/2\ O_2(g)$ $\Delta H° = +824.2$ kJ (reversed, and multiplied by ½)

$3\ CO(g) + 3/2\ O_2(g) \rightarrow 3\ CO_2(g)$ $\Delta H° = -848.1$ kJ (multiplied by 3/2)

Step 2: Place products on the right, multiplying, if necessary.
Products are already on the right as a result of the manipulations above.

Step 3: Stack up the rewritten reactions, and add them together.

$$Fe_2O_3(s) \rightarrow 2\ Fe(s) + 3/2\ O_2(g) \qquad \Delta H° = +824.2\ kJ$$
$$3\ CO(g) + 3/2 O_2(g) \rightarrow 3\ CO_2(g) \qquad \Delta H° = -848.1\ kJ$$
$$Fe_2O_3(s) + 3\ CO(g) + 3/2 O_2(g) \rightarrow 2\ Fe(s) + 3/2\ O_2(g) + 3\ CO_2(g) \quad \Delta H° = -23.9\ kJ$$

Step 4: Cancel terms that appear on both sides.

$$Fe_2O_3(s) + 3\ CO(g) \rightarrow 2\ Fe(s) + 3\ CO_2(g) \qquad \Delta H° = -23.9\ kJ$$

Workbook Problem 9.8

Step 1: Write and balance the equation.

$$C_6H_{12}O_6(s) + 6\ O_2(g) \rightarrow 6\ H_2O(l) + 6\ CO_2(g)$$

Step 2: Determine the overall heat of reaction using $\Delta H° = \sum H°_{\text{products}} - \sum \Delta H°_{\text{reactants}}$.

$$\Delta H° = [(6 \times -285.8\ kJ/mol) + (6 \times -393.5\ kJ/mol)] - (-1273.3\ kJ/mol)$$
$$= -2802.5\ kJ/mol$$

Step 3: Convert the grams of glucose to moles.

$$10\ g \times \frac{1\ mol}{180\ g} = 0.055556\ mol\ glucose$$

Step 4: Determine the energy released by 10 g of glucose.

$$0.055\ 556\ mol \times -2802.5\ kJ/mol = -155.696\ kJ$$

Step 5: Using $q = mc\Delta T$, determine the temperature change of 500 g of water.

$$155{,}696\ J = 500\ g \times 4.184\ J/g \cdot °C \times \Delta T$$

$$\Delta T = \frac{155{,}696\ J}{500\ g \times 4.184\ J/g \cdot °C} = 74.42\ °C$$

Workbook Problem 9.9

Step 1: Write and balance the equation.

$$CH_4(g) + 2\ O_2(g) \rightarrow CO_2(g) + 2\ H_2O(l)$$

Step 2: Determine the energy of the bonds on the reactant side.

$$(4 \times 414\ kJ/mol) + (2 \times 498\ kJ/mol) = 2652\ kJ/mol$$

Step 3: Determine the energy of the bonds on the product side.

$$(2 \times 799 \text{ kJ/mol}) + (4 \times 464 \text{ kJ/mol}) = 3454 \text{ kJ/mol}$$

Step 4: Determine the energy released by the reaction.

$$\text{Reactants} - \text{products} = 2652 \text{ kJ/mol} - 3454 \text{ kJ/mol} = -802 \text{ kJ/mol}$$

Workbook Problem 9.10

Step 1: Write and balance the equation.

$$H_2(g) + Cl_2(g) \rightarrow 2\ HCl(g)$$

Step 2: Determine ΔH.

ΔH_f = is given for 1 mole of HCl. Since 2 moles are produced in the reaction multiply it by 2

$$2 \times -92.3 \text{ kJ/mol} = -184.6 \text{ kJ/mol}$$

Step 3: Write the value for ΔS.

$$20 \text{ J/ K} \cdot \text{mol} = 0.020 \text{ kJ/K} \cdot \text{mol}$$

Step 4: Determine the free energy of the reaction at 25°C, first converting the temperature to K.

$$\Delta G = -184.6 \text{ kJ/mol} - 298\text{K}(0.020 \text{ kJ/K} \cdot \text{mol}) = -190.56 \text{ kJ/mol} = -191 \text{ kJ/mol}$$

Step 5: Determine the spontaneity of the reaction.

The negative sign on the free-energy term indicates that this reaction is spontaneous at 25°C. If the value here is written per mole of HCl formed, the value is −95.3 kJ/mol. This matches the value for the standard free energy of formation of HCl.

Self-Test

True–False

1. T
2. F Changes in energies are always calculated final − initial.
3. F State functions are path independent.
4. T
5. F Standard state is measured at 25 °C, which is room (or lab) temperature.
6. F It is possible for a reaction to both transfer heat and do work.
7. F Melting is an endothermic process—heat must be added.
8. F Specific heat is an intensive property: it can be used to identify materials.
9. T
10. T
11. T
12. F A positive change in entropy—an increase in disorder—is favorable.

13. F Gases have higher entropy than liquids because they are more disordered.
14. T
15. T

Matching

16. e	17. h	18. d	19. m
20. g	21. l	22. q	23. n
24. p	25. i	26. c	27. k
28. j	29. o	30. a	31. b
32. f			

Fill in the Blanks

33. kinetic energy, potential energy
34. temperature, heat
35. negative, exothermic
36. reversible
37. volume, work
38. thermodynamic standard state
39. specific heat
40. surroundings, system, positive
41. bond-dissociation energy, positive
42. entropy, increase
43. Gibbs free energy, equilibrium, less, greater

Problems

44. To easily solve this problem, set the two energies equal to one another to solve for the needed velocity of the marble. Watch your units: they must match for the answer to come out correctly.

$$\frac{1}{2}\times 1.3\text{ kg}\times\left(5\frac{\text{m}}{\text{s}}\right)^2=\frac{1}{2}\times 0.002\text{ kg}\times\left(\text{v}\frac{\text{m}}{\text{s}}\right)^2$$

$$16.25\frac{\text{kg}\cdot\text{m}^2}{\text{s}^2}=0.001\,\text{v}^2\text{ kg}$$

$$16250\frac{\text{m}^2}{\text{s}^2}=\text{v}^2$$

$$\text{v}=127\frac{\text{m}}{\text{s}}$$

45. Work is being done on the system, since the change in volume is negative. Knowing q and w makes it possible to calculate ΔE.

$w = -P\Delta V = -8\text{atm} \times (0.550\text{L} - 0.950\text{L}) = 3.2\text{ L}\cdot\text{atm} \times 101\text{ J/(L}\cdot\text{atm)} = 323\text{ J} = 0.323\text{ kJ}$

$q = 12$ kJ (positive, as it is being absorbed by the system)

For the system, $\Delta E = q + w$; $\Delta E = 12\text{ kJ} + 0.323\text{ kJ} = 12.3\text{ kJ}$.
For the surroundings, $\Delta E = -12.3$ kJ

46. Using $q = m \times c \times \Delta T$:

$$150\ \text{J} = 15\ \text{g} \times 0.140\ \text{J/g} \cdot {}^\circ\text{C} \times \Delta T$$
$$\Delta T = 150\ \text{J}/(15\ \text{g} \times 0.140\ \text{J/g} \cdot {}^\circ\text{C})$$
$$\Delta T = 71.4\ {}^\circ\text{C}$$

47. For this reaction, both w and q need to be calculated to be able to determine the overall energy change of the system.

$$w = -P\Delta V = -1\ \text{atm} \times (2\text{L} - 3\text{L}) = 1\ \text{L} \cdot \text{atm} \times 101\ \text{J/(L} \cdot \text{atm)} = 101\ \text{J} = 0.101\ \text{kJ}$$

To calculate q, it is necessary to know how many moles of chlorine gas were involved in the reaction:

$$3.17\frac{\text{g}}{\text{L}} \times 2.00\ \text{L} = 6.34\ \text{g}$$

$$6.34\ \text{g} \times \frac{1\ \text{mol}}{70.906\ \text{g}} = 0.089\ 414\ 15\ \text{mol}$$

Multiply by the heat involved per mole of chlorine.

$$0.089\ 414\ 15\ \text{mol} \times 80.3\ \text{kJ/mol} = 7.179\ 957\ \text{kJ}$$
$$\Delta E = q + w;\ \Delta E = 7.179\ 957\ \text{kJ} + 0.101\ \text{kJ} = 7.28\ \text{kJ}$$

48. Heat is being absorbed from the water by the salt as it dissolves (indicating an entropy-driven process). The ΔH for the solution is per mole, not per gram, so it is first necessary to convert the grams of Epsom salts to moles.

$$15\ \text{g} \times \frac{1\ \text{mol}}{246.46\ \text{g}} = 0.0609\ \text{mol} \times 16.11\frac{\text{kJ}}{\text{mol}} = 0.980\ \text{kJ} = 980\ \text{J}$$

$$q_{\text{rxn}} = -q_{\text{soln}} = -\ [(\text{sp. heat}) \times (\text{mass of soln}) \times (\Delta T)]$$
$$980\ \text{J} = -4.184\frac{\text{J}}{\text{g} \cdot {}^\circ\text{C}} \times 100\ \text{g} \times \Delta\text{T}$$

$$\Delta T = -2.34\ {}^\circ\text{C}$$

The final temperature will be 24.36 °C + (−2.34 °C) = 22.0 °C

49. A number of calculations are required for this problem: the ice and water must be warmed and two phase changes completed.

$$50\ \text{g}\ H_2O = 2.8\ \text{moles}\ H_2O$$

−10 °C to 0 °C:

$$q = mc\Delta T = 50\ \text{g} \times 2.108\ \text{J/g}{}^\circ\text{C} \times 10\ {}^\circ\text{C} = 1054\ \text{J} = 1.054\ \text{kJ}$$

Ice to water:

$$q = 2.8 \text{ moles} \times 6.01 \text{ kJ/mol} = 16.8 \text{ kJ}$$

0 °C to 100 °C:

$$q = 50 \text{ g} \times 4.184 \text{ J/g°C} \times 100 \text{ °C} = 20{,}900 \text{ J} = 20.9 \text{ kJ}$$

Water to steam:

$$q = 2.8 \text{ moles} \times 40.7 \text{ kJ/mol} = 114 \text{ kJ}$$

$$\text{Total energy} = 153 \text{ kJ} = 150 \text{ kJ}$$

Of this energy, 75% is used in the phase change from liquid to gas. This makes sense because this phase transition requires a breaking of all intermolecular attractions. Warming ice or water is simply adding kinetic energy, and the phase change from solid to liquid loosens, but does not break, the interactions between the molecules.

50. This problem requires a bit of algebra. Let the amount of copper = Cu and the amount of gold = Au. The specific heat of the metal chunk can be determined by seeing how much heat it released to the water.

$$q_{H_2O} = c_{H_2O} \cdot m \cdot \Delta T$$

$$q_{H_2O} = 25.0\,\text{g} \times \frac{4.184\,\text{J}}{\text{g}\cdot°\text{C}} \times (26.65\,°\text{C} - 25.00\,°\text{C}) = 172.59\,\text{J}$$

$$-q_{H_2O} = q_{\text{metal}} = -172.59\,\text{J}$$

$$q_{\text{metal}} = c_{\text{metal}} \cdot m \cdot \Delta T$$

$$-172.59\,\text{J} = 6.73\,\text{g} \times c_{\text{metal}} \times (26.65\,°\text{C} - 100\,°\text{C})$$

$$c_{\text{metal}} = 0.34962 \frac{\text{J}}{\text{g}\cdot°\text{C}}$$

To solve for the amount of each, it is necessary to set up a system of equations.

$$\text{Cu} + \text{Au} = 6.73 \text{ g}; \ \text{Cu} = 6.73 \text{ g} - \text{Au}$$

$$\%\text{Cu} \cdot c_{\text{Cu}} + \%\text{Au} \cdot c_{\text{Au}} = c_{\text{metal}}$$

$$\frac{\text{Cu}}{6.73\,\text{g}} \cdot c_{\text{Cu}} + \frac{\text{Au}}{6.73\,\text{g}} \cdot c_{\text{Au}} = c_{\text{metal}}$$

Substituting:

$$\frac{(6.73\,\text{g} - \text{Au})\left(0.385 \frac{\text{J}}{\text{g}\cdot°\text{C}}\right) + \text{Au}\left(0.129 \frac{\text{J}}{\text{g}\cdot°\text{C}}\right)}{6.73\,\text{g}} = 0.34962 \frac{\text{J}}{\text{g}\cdot°\text{C}}$$

$$2.59105\frac{\text{J}}{^\circ\text{C}} - 0.385\text{Au}\frac{\text{J}}{\text{g}\cdot{}^\circ\text{C}} + 0.129\text{Au}\frac{\text{J}}{\text{g}\cdot{}^\circ\text{C}} = 2.35297\frac{\text{J}}{^\circ\text{C}}$$

$$-0.256\text{Au}\frac{\text{J}}{\text{g}\cdot{}^\circ\text{C}} = -0.23808\frac{\text{J}}{^\circ\text{C}}$$

$$\text{Au} = 0.93001\text{ g}$$

$$\text{Cu} = 6.73 - 0.93001 = 5.80\text{ g}$$

51. $2\ Fe(s) + 3/2\ O_2(g) \rightarrow Fe_2O_3(s)$

Given the following:

$Fe_2O_3(s) + 3\ CO(g) \rightarrow 2\ Fe(s) + 3\ CO_2(g)$ $\quad \Delta H^\circ = -26.7$ kJ

$CO(g) + \frac{1}{2}\ O_2(g) \rightarrow CO_2(g)$ $\quad \Delta H^\circ = -283.0$ kJ

The first reaction needs to be turned so as to give the proper product. Once this is done, the second reaction is fine in the direction it is written, but needs to be multiplied through by 3 to give the proper number of CO_2 molecules.

$2\ Fe(s) + 3\ CO_2(g) \rightarrow Fe_2O_3(s) + 3\ CO(g)$ $\quad \Delta H^\circ = +26.7$ kJ
$3CO(g) + 3/2\ O_2(g) \rightarrow 3CO_2(g)$ $\quad \Delta H^\circ = -849.0$ kJ
$2\ Fe(s) + 3CO_2(g) + 3CO + 3/2\ O_2 \rightarrow Fe_2O_3(s) + 3CO(g) + 3CO_2(g)$ $\quad \Delta H^\circ = -822.3$ kJ

Cancel terms that appear on both sides.

$2\ Fe(s) + 3/2\ O_2 \rightarrow Fe_2O_3(s)$ $\quad \Delta H^\circ = -822.3$ kJ

52. The approximate ΔH is calculated by subtracting the sum of the heats of formation of the reactants from the sum of the heats of formation of the products.

[(−1131 kJ/mol) + (−285.9 kJ/mol) + (−393.5 kJ/mol)] − (2 × −947.7 kJ/mol) = 85.00 kJ/mol

53. Using the bond-dissociation energies given, calculate the enthalpy of the following reaction:

$2\ Cl_2(g) + CH_4(g) \rightarrow 2\ H_2(g) + CCl_4(g)$

Cl–Cl = 243 kJ/mol H–H = 436 kJ/mol
C–H = 410 kJ/mol C–Cl = 330 kJ/mol

Reactants:
2 Cl–Cl bonds: 486 kJ/mol
4 C–H bonds: 1640 kJ/mol
2126 kJ/mol bonds broken

Products:
2 H–H bonds: 872 kJ/mol
4 C–Cl bonds: 1320 kJ/mol
2192 kJ/mol bonds formed

reactants − products = −66 kJ/mol

54. $\Delta G° = \Delta H° - T\Delta S°$, when $\Delta G° = 0$, $T = \Delta H/\Delta S$

$\Delta H = +66$ kJ/mol for the reverse reaction,

$\Delta S = +158.8$ J/mol · K = 0.1588 kJ/mol · K for the reverse reaction

$$T = 66{,}000/158.8 = 415.6 \text{ K} = 143 \text{ °C}$$

55. First, write and balance the equation for the combustion of pentane.

$$C_5H_{12}(l) + 8\ O_2(g) \rightarrow 5\ CO_2(g) + 6\ H_2O(l)$$

[5(−393.5 kJ/mol) + 6(−285.9 kJ/mol)] − [(−146.3 kJ/mol) + 8(0 kJ/mol)] = −3537 kJ/mol

$$-3537\frac{\text{kJ}}{\text{mol}} \times \frac{1\text{ mol}}{72\text{ g}} = -49\frac{\text{kJ}}{\text{g}}$$

$$-49\frac{\text{kJ}}{\text{g}} \times 0.626\frac{\text{g}}{\text{mL}} = -30.8\frac{\text{kJ}}{\text{mL}}$$

Chapter Ten—Gases: Their Properties and Behavior

Workbook Problems

Workbook Problem 10.1

Step 1: Determine the pressure of the gas in mm Hg.

The pressure is higher inside the manometer. Pressure = 760 mm Hg + 73 mm Hg = 833 mm Hg

Step 2: Convert from mm Hg to atmospheres.

$$833\text{ mm Hg} \times \frac{1\text{ atm}}{760\text{ mm Hg}} = 1.10\text{ atm}$$

Workbook Problem 10.2

Step 1: Determine which gas law to use.

This problem involves only volume and temperature: use Charles's law.

Step 2: Convert the temperatures to K.

$$37\text{ °C} + 273.15 = 310\text{ K}$$
$$350\text{ °C} + 273.15 = 623\text{ K}$$

Step 3: Solve for the final volume.

$$\frac{V_1}{T_1} = \frac{V_2}{T_2} \qquad \frac{27\ \text{L}}{310\ \text{K}} = \frac{V_2}{623\ \text{K}}$$

$$V_2 = \frac{27\ \text{L} \times 623\ \text{K}}{310\ \text{K}} = 54.3\ \text{L}$$

Workbook Problem 10.3

Step 1: Rearrange the ideal gas law.

$$PV = nRT \text{ becomes } P = \frac{nRT}{V}$$

Step 2: Convert grams to moles and °C to K.

$$6.7\ \text{g}\ F_2 \times \frac{1\ \text{mol}\ F_2}{38\ \text{g}\ F_2} = 0.176\ \text{mol}\ F_2$$

$$25.0\ °\text{C} + 273.15 = 298\ \text{K}$$

Step 3: Solve for the final pressure.

$$P = \frac{nRT}{V} = \frac{0.176\ \text{mol} \times 0.08206 \frac{\text{L} \cdot \text{atm}}{\text{mol} \cdot \text{K}} \times 298\ \text{K}}{3.0\ \text{L}} = 1.43\ \text{atm}$$

Workbook Problem 10.4

Step 1: Determine the form of gas law needed.

Although this problem again requires solving for one variable, it is a very different type of problem. Whenever a change is indicated, this is an indication that the full ideal gas law is not needed. Instead, a version of the simple gas law will be required.

Here, the temperature, pressure, and volume are all changing, but the number of moles is not.

$$\frac{P_1V_1}{T_1} = \frac{P_2V_2}{T_2}$$

Step 2: Convert grams to moles and °C to K.
There is no need to convert to moles of fluorine.

$$25\ °\text{C} + 273.15 = 298\ \text{K}$$

$$750\ °\text{C} + 273.15 = 1023\ \text{K}$$

Step 3: Solve for the final pressure.

$$\frac{1\ \text{atm} \cdot 3.2\ \text{L}}{298\ \text{K}} = \frac{18\ \text{atm} \cdot \text{V}_2}{1023\ \text{K}}$$

$$\text{V}_2 = \frac{1\ \text{atm} \cdot 3.2\ \text{L} \cdot 1023\ \text{K}}{298\ \text{K} \cdot 18\ \text{atm}} = 0.61\ \text{L}$$

Workbook Problem 10.5

Step 1: Determine which gas law to use.

This is Avogadro's law, or at least a corollary of it. Since the volume of a molar amount of gas is constant, gases can be made to react in equivalent volumes. The volumes will be in stoichiometric ratios.

Step 2: Solve for the volume of oxygen.

$$7.45\ \text{L NO} \times \frac{1\ \text{L O}_2}{2\ \text{L NO}} = 3.73\ \text{L O}_2$$

Step 3: Solve for the volume of nitrogen dioxide.

$$7.45\ \text{L NO} \times \frac{2\ \text{L N}}{2\ \text{L NO}} = 7.45\ \text{L}$$

Workbook Problem 10.6

Step 1: Convert the temperature and pressure.

$$100\ °\text{C} + 273.15 = 373.15\ \text{K}$$

$$300\ \text{mm Hg} \times \frac{1\ \text{atm}}{760\ \text{mm Hg}} = 0.395\ \text{atm}$$

Step 2: Use the ideal gas law to calculate the volume of the gas.

$$\text{V} = \frac{\text{nRT}}{\text{P}} = \frac{1\ \text{mol} \times 0.08206 \frac{\text{L} \cdot \text{atm}}{\text{mol} \cdot \text{K}} \times 373\ \text{K}}{0.395\ \text{atm}} = 77.5\ \text{L}$$

Step 3: Determine the density of the gas.

$$\text{Mass of 1 mol of xenon} = 131.3\ \text{g}$$

$$\text{Density} = \frac{131.3\ \text{g}}{77.5\ \text{L}} = 1.69\frac{\text{g}}{\text{L}}$$

Workbook Problem 10.7

Step 1: Determine the number of moles in 1 L of gas.

$$n = \frac{PV}{RT} = \frac{3.75\ \text{atm} \times 1\ \text{L}}{0.08206 \frac{\text{L} \cdot \text{atm}}{\text{mol} \cdot \text{K}} \times (700\ °\text{C} + 273.15\ \text{K})} = 0.0470\ \text{mol}$$

Step 2: Solve for the mass in 1 L of gas by using the gas's density.

Mass = density × volume = 0.798 g/L × 1 L = 0.798 g

Step 3: Determine the molar mass by dividing g/moles.

MW = 0.798 g/0.0470 mol = 17.0 g/mol.

Ammonia is a gas that fits this molar mass.

Workbook Problem 10.8

Step 1: Determine the number of moles of each gas:

$$18.3\ \text{g}\ CO_2 \times \frac{1\ \text{mol}}{44.0\ \text{g}} = 0.416\ \text{mol}\ CO_2$$

$$23.1\ \text{g Ne} \times \frac{1\ \text{mol}}{20.2\ \text{g}} = 1.14\ \text{mol Ne}$$

$$5.26\ \text{g}\ H_2 \times \frac{1\ \text{mol}}{2.02\ \text{g}} = 2.61\ \text{mol}\ H_2$$

Step 2: Use the total number of moles to calculate the total pressure:

$$P = \frac{nRT}{V} = \frac{(0.416\ \text{mol} + 1.14\ \text{mol} + 2.61\ \text{mol})0.08206 \frac{\text{L} \cdot \text{atm}}{\text{mol} \cdot \text{K}}(25\ °\text{C} + 273.15)\text{K}}{30\ \text{L}}$$

$$= \frac{4.17\ \text{mol} \times 0.08206 \frac{\text{L} \cdot \text{atm}}{\text{mol} \cdot \text{K}} \times 298\ \text{K}}{30\ \text{L}} = 340\ \text{atm}$$

Step 3: Use the mole fraction (X) of each component to determine the partial pressures:

$$X_{CO_2} = \frac{0.416\ \text{mol}}{4.17\ \text{mol}} = 0.100 \qquad X_{Ne} = \frac{1.14\ \text{mol}}{4.17\ \text{mol}} = 0.273$$

$$X_{H_2} = \frac{2.61\ \text{mol}}{4.17\ \text{mol}} = 0.626$$

$$P_{CO_2} = 0.100 \times 3.40\ \text{atm} = 0.340\ \text{atm} \qquad P_{Ne} = 0.273 \times 3.40\ \text{atm} = 0.928\ \text{atm}$$

$$P_{H_2} = 0.626 \times 3.40\ \text{atm} = 2.13\ \text{atm}$$

These problems are particularly easy to check because the mole fractions must add up to 1, and the partial pressures must add up to the total pressure, as they do here.

Workbook Problem 10.9

Step 1: Determine the speed of H_2 at 25 °C.

$$u = \sqrt{\frac{3RT}{M}} = \sqrt{\frac{3 \times 8.314 \frac{\text{kg} \cdot \frac{\text{m}^2}{\text{s}^2}}{\text{K} \cdot \text{mol}} \times (25 + 273)\text{K}}{0.00202 \frac{\text{kg}}{\text{mol}}}} = \sqrt{3{,}680{,}000 \frac{\text{m}^2}{\text{s}^2}} = 1918 \frac{\text{m}}{\text{s}}$$

Step 2: Determine the temperature required for the iodine molecules to achieve the same speed.

$$1918 \frac{\text{m}}{\text{s}} = \sqrt{\frac{3 \times 8.314 \frac{\text{kg} \cdot \frac{\text{m}^2}{\text{s}^2}}{\text{K} \cdot \text{mol}} \times \text{T}}{0.2538 \frac{\text{kg}}{\text{mol}}}} = \sqrt{98.27 \frac{\frac{\text{m}^2}{\text{s}^2}}{\text{K}} \times \text{T}}$$

Square both sides to solve for T:

$$3{,}680{,}000 \frac{\text{m}^2}{\text{s}^2} = 98.27 \frac{\frac{\text{m}^2}{\text{s}^2}}{\text{K}} \times \text{T}$$

$$\text{T} = 37{,}450\ \text{K}$$

Workbook Problem 10.10

Step 1: Solve using Graham's law:

$$\frac{\text{Rate of effusion of } ^{16}\text{O}}{\text{Rate of effusion of } ^{18}\text{O}} = \sqrt{\frac{35.998\ \text{g}\ ^{18}\text{O/mol}}{31.990\ \text{g}\ ^{16}\text{O/mol}}} = 1.061$$

Workbook Problem 10.11

Step 1: Solve using the ideal gas law:

$$P = \frac{nRT}{V} = \frac{1\ \text{mol} \times 0.08206\ \frac{\text{L}\cdot\text{atm}}{\text{mol}\cdot\text{K}} \times 300\ \text{K}}{0.100\ \text{L}} = 246\ \text{atm}$$

Step 2: Solve using the van der Waals equation:

$$P = \frac{nRT}{V - nb} - \frac{an^2}{V^2} = \frac{1\ \text{mol} \times 0.08206\ \frac{\text{L}\cdot\text{atm}}{\text{mol}\cdot\text{K}} \times 300\ \text{K}}{0.100\ \text{L} - (1\ \text{mol} \times 0.0318\ \text{L/mol})} - \frac{(1\ \text{mol})^2 \times 1.38\,(\text{L}^2\cdot\text{atm})/\text{mol}^2}{(0.100\ \text{L})^2}$$

$$= \frac{24.6\ \text{L}\cdot\text{atm}}{0.0682\ \text{L}} - \frac{1.38\ \text{L}^2\cdot\text{atm}}{0.0100\ \text{L}^2} = 361\ \text{atm} - 138\ \text{atm} = 223\ \text{atm}$$

Step 3: Compare the two values:

The ideal and real values are reasonably close, given the large pressures involved. Whether the real value would be acceptable would depend on the precision required in the work.

Self-Test

True–False

1. F The atmosphere is approximately 70% nitrogen.
2. T
3. T
4. T
5. T
6. F The smaller a gas particle is, the faster it travels.
7. F The ratio is inverse.
8. F The ideal gas law begins to fail at high pressures.
9. F We breathe troposphere.
10. T

Matching

11. j	12. a	13. l	14. i
15. h	16. o	17. m	18. b
19. e	20. n	21. c	22. f
23. d	24. p	25. r	26. q
27. g	28. k		

Fill in the Blanks

29. pressure, collisions
30. pressure, volume, temperature, moles
31. mole fraction, partial pressure
32. volume, negligible, pressure
33. low, faster

Problems

34. To determine the mm of silicone oil that the pressure will support, first determine the mm of Hg that it will support by converting from atmospheres to mmHg.

$$0.983 \text{ atm} \times \frac{760 \text{ mm Hg}}{1 \text{ atm}} = 747 \text{ mm Hg}$$

Then convert from mm Hg to mm silicone oil using the relative densities as a conversion factor.

$$747 \text{ mm Hg} \times \frac{13.6 \text{ mm oil}}{0.970 \text{ mm Hg}} = 10{,}500 \text{ mm oil}$$

35. a) 340 mm ethyl alcohol greater than atmospheric pressure

$$\text{Pressure in bulb} = 340 \text{ mm EtOH} \times \frac{0.789 \text{ mm Hg}}{13.6 \text{ mm EtOH}} = 19.7 \text{ mm Hg} \text{ above}$$

atmospheric pressure.

Total pressure in the bulb = 760 mm Hg + 19.7 mm Hg = 780 mmHg

b) 257 mm ethyl alcohol less than atmospheric pressure

$$\text{Pressure in bulb} = 257 \text{ mm EtOH} \times \frac{0.789 \text{ mm Hg}}{13.6 \text{ mm EtOH}} = 14.9 \text{ mm Hg} \text{ below}$$

atmospheric pressure

Total pressure in the bulb = 760 mm Hg − 14.9 mm Hg = 745 mmHg

36. PV = nRT; n = PV/RT

$$n = \frac{5.78 \text{ atm} \times 0.050 \text{ L}}{0.08206 \frac{\text{L} \cdot \text{atm}}{\text{mol} \cdot \text{K}} \times (-20\ ^\circ\text{C} + 273.15)\text{K}} = 0.0139 \text{ mol}$$

If the mass of 0.0139 mol is 0.281 g, the molecular weight is 0.281 g/0.0139 mol = 20.2 g/mol. The gas is neon.

37. Assume that you have 1 L of the gas and determine the number of moles in that liter. Alternatively, you can do the reverse: assume that you have 1 mole of gas, and determine the number of liters. Both will get you the correct answer.

$$n = \frac{4.00 \text{ atm} \times 1.00 \text{ L}}{0.08206 \frac{\text{L} \cdot \text{atm}}{\text{mol} \cdot \text{K}} \times (100\ ^\circ\text{C} + 273.15)\text{K}} = 0.131 \text{ mol}$$

$$\text{density} = \frac{0.131 \text{ mol}}{\text{L}} \times \frac{16.0 \text{ mol}}{\text{mol}} = 2.10 \frac{\text{g}}{\text{L}}$$

38. a. Assume a 100 g sample, and convert 92.3 g C and 7.69 g H to moles.

$$92.3\text{ g C}\times\frac{1\text{ mol C}}{12.0\text{ g C}}=7.69\text{ mol C; }7.69\text{ g H}\times\frac{1\text{ mol H}}{1.0\text{ g H}}=7.69\text{ mol H}$$

This gives rise to a 7.69:7.69 C to H mole ratio.

There is one carbon atom for every hydrogen atom in the molecule, so the empirical formula is CH.

b. To solve for molar mass, we need to determine the moles of gas.

$$n=\frac{\frac{763}{760}\text{ atm}\times 0.250\text{ L}}{0.08206\frac{\text{L}\cdot\text{atm}}{\text{mol}\cdot\text{K}}\times(35+273)\text{ K}}=0.00993\text{ mol}\qquad\frac{0.258\text{ g}}{0.00993\text{ mol}}=26.0\text{ g / mol}$$

c. The empirical formula weight is 13 g/mol, so the molecular formula is two empirical formula units: C_2H_2.

39. Determine the number of moles in 1 L under these conditions.

$$n=\frac{15\text{ atm}\times 1.00\text{ L}}{0.08206\frac{\text{L}\cdot\text{atm}}{\text{mol}\cdot\text{K}}\times 573\text{ K}}=0.319\text{ mol}$$

14.04 g/0.319 mol = 44.0 g/mol

40. Write and balance the equation.

$$2\ C_4H_{10}(g) + 13\ O_2(g) \rightarrow 8\ CO_2(g) + 10\ H_2O(g)$$

Determine the number of moles of butane burned.

$$25\text{ mL}\times 0.6014\frac{\text{g}}{\text{mL}}\times\frac{1\text{ mol}}{58.12\text{ g}}=0.26\text{ mol}$$

Determine the moles of oxygen consumed, then convert to liters.

$$0.26\text{ mol }C_4H_{10}\times\frac{13\text{ mol }O_2}{2\text{ mol }C_4H_{10}}=1.69\text{ mol }O_2$$

$$V=\frac{nRT}{P}=\frac{1.7\text{ mol}\times 0.08206\frac{\text{L}\cdot\text{atm}}{\text{mol}\cdot\text{K}}\times(30+273)\text{ K}}{1\text{ atm}}=42.2\text{ L}$$

Determine the number of moles of carbon dioxide produced, then convert to liters.

$$0.26\ \text{mol C}_4\text{H}_{10} \times \frac{8\ \text{mol CO}_2}{2\ \text{mol C}_4\text{H}_{10}} = 1.04\ \text{mol CO}_2$$

$$V = \frac{nRT}{P} = \frac{1.04\ \text{mol} \times 0.08206\,\frac{\text{L}\cdot\text{atm}}{\text{mol}\cdot\text{K}} \times (30+273)\ \text{K}}{1\ \text{atm}} = 25.9\ \text{L}$$

Determine the number of moles of water vapor produced, then convert to liters.

$$0.26\ \text{mol C}_4\text{H}_{10} \times \frac{10\ \text{mol H}_2\text{O}}{2\ \text{mol C}_4\text{H}_{10}} = 1.30\ \text{mol H}_2\text{O}$$

$$V = \frac{nRT}{P} = \frac{1.30\ \text{mol} \times 0.08206\,\frac{\text{L}\cdot\text{atm}}{\text{mol}\cdot\text{K}} \times (30+273)\ \text{K}}{1\ \text{atm}} = 32.3\ \text{L}$$

Determine the overall energy change of the reaction.

Volume of products − volume of reactants = (32.3 L + 25.9 L) − (42.2 L + 0.025 L) = +16.0 L

Work is done by this reaction.

41. Find the number of moles of each gas.

$$7.03\ \text{g Ne} \times \frac{1\ \text{mol}}{20.2\ \text{g}} = 0.348\ \text{mol} \qquad 11.7\ \text{g N}_2 \times \frac{1\ \text{mol}}{28.0\ \text{g}} = 0.418\ \text{mol} \qquad 15.7\ \text{g He} \times \frac{1\ \text{mol}}{4.00\ \text{g}} = 3.93\ \text{mol}$$

The total moles of gas = 0.348 mol + 0.418 mol + 3.93 mol = 4.70 moles

Find the total pressure.

$$P = \frac{nRT}{V} = \frac{(0.348\ \text{mol} + 0.418\ \text{mol} + 3.93\ \text{mol}) \times 0.08206\,\frac{\text{L}\cdot\text{atm}}{\text{mol}\cdot\text{K}} \times 298\ \text{K}}{5.000\ \text{L}} = 22.97\ \text{atm}$$

Find the mole fraction and partial pressure of each component.

$$\text{Ne}: \frac{0.348\ \text{mol}}{4.70\ \text{mol}} \times 23.0\ \text{atm} = 1.70\ \text{atm} \quad \text{N}_2: \frac{0.418\ \text{mol}}{4.70\ \text{mol}} \times 23.0\ \text{atm} = 2.05\ \text{atm}$$

$$\text{He}: \frac{3.93\ \text{mol}}{4.70\ \text{mol}} \times 23.0\ \text{atm} = 19.2\ \text{atm}$$

42. $$\frac{\text{rate He}}{\text{rate H}_2} = \frac{\sqrt{1.00794 \times 2}}{\sqrt{4.002602}} = \frac{1.4198}{2.00065} = 0.7096 \text{ or } \frac{\text{rate H}_2}{\text{rate He}} = \frac{\sqrt{4.002602}}{\sqrt{1.00794 \times 2}} = \frac{2.00065}{1.4198} = 1.409$$

43. $$\frac{P_1V_1}{T_1} = \frac{P_2V_2}{T_2} \qquad T_2 = \frac{P_2V_2T_1}{P_1V_1} = \frac{1\ \text{atm} \times 0.375\ \text{L} \times 298\ \text{K}}{7.8\ \text{atm} \times 0.050\ \text{L}} = 286.538\ \text{K} = 13.5\ ^\circ\text{C}$$

Chapter Eleven—Liquids and Phase Changes

Workbook Problems

Workbook Problem 11.1

Step 1: Convert temperatures to Kelvin.

$78.4\ ^\circ\text{C} + 273.15 = 351.55\ \text{K}$ $\qquad$ $63.5\ ^\circ\text{C} + 273.15 = 336.65\ \text{K}$

Step 2: Solve for ΔH.

$$\ln\left(\frac{400\ \text{mm Hg}}{760\ \text{mm Hg}}\right) = \frac{\Delta H_{vap}}{8.3145\ ^{J}\!/_{mol\cdot K}}\left(\frac{1}{78.4+273.15} - \frac{1}{63.5+273.15}\right)$$

$$\ln\left(\frac{400\ \text{mm Hg}}{760\ \text{mm Hg}}\right) = \frac{\Delta H_{vap}}{8.3145\ ^{J}\!/_{mol\cdot K}}\left(\frac{1}{351.55\ \text{K}} - \frac{1}{336.65\ \text{K}}\right)$$

$$-0.641853886 = \frac{\Delta H_{vap}}{8.3145\ ^{J}\!/_{mol\cdot K}}\left(\frac{0.0028445}{\text{K}} - \frac{0.00297044}{\text{K}}\right)$$

$$-0.641853886 = \frac{\Delta H_{vap}}{8.3145\ ^{J}\!/_{mol\cdot K}}\left(-\frac{0.000125898}{\text{K}}\right)$$

$$\frac{-0.641853886}{0.000125898} = \frac{\Delta H_{vap}}{8.3145\ ^{J}\!/_{mol}}$$

$$5098.1854 = \frac{\Delta H_{vap}}{8.3145\ ^{J}\!/_{mol}}$$

$$\Delta H_{vap} = (5098.1854)\left(8.3145\ ^{J}\!/_{mol}\right) = 42{,}388.86278\ ^{J}\!/_{mol} = 42.4\ ^{kJ}\!/_{mol}$$

Workbook Problem 11.2

Step 1: Use the Clausius–Clapeyron equation to solve for the vapor pressure.

$$\ln\left(\frac{P_1}{760\ \text{mm Hg}}\right) = \frac{42{,}388.86278\ ^{J}\!/_{mol}}{8.3145\ ^{J}\!/_{mol\cdot K}}\left(\frac{1}{78.4+273.15} - \frac{1}{25+273.15}\right)$$

$$\ln\left(\frac{P_1}{760\ \text{mm Hg}}\right) = \frac{39355.3941\ ^{J}\!/_{mol}}{8.3145\ ^{J}\!/_{mol\cdot K}}\left(\frac{1}{351.55\ \text{K}} - \frac{1}{298.15\ \text{K}}\right)$$

$$\ln\left(\frac{P_1}{760\text{ mm Hg}}\right) = 5098.1854\text{ K}\left(\frac{0.0028445}{\text{K}} - \frac{0.00335402}{\text{K}}\right)$$

$$\ln\left(\frac{P_1}{760\text{ mm Hg}}\right) = 5098.1854\text{ K}\left(\frac{-0.00050947}{\text{K}}\right)$$

$$\ln\left(\frac{P_1}{760\text{ mm Hg}}\right) = -2.59738$$

$$e^{\ln\left(\frac{P_1}{760\text{ mm Hg}}\right)} = e^{-2.59738}$$

$$\frac{P_1}{760\text{ mm Hg}} = 0.7446866$$

$$P_1 = 56.59618\text{ mm Hg} = 56.6\text{ mm Hg}$$

Workbook Problem 11.3

Strategy: Rearrange the equation $\Delta G = \Delta H - T\Delta S$ to solve for the entropy change.

Step 1: Rearrange to solve for ΔS.

$$0 = \Delta H - T\Delta S$$

$$\Delta S = \frac{\Delta H}{T}$$

Step 2: Solve for ΔS, being careful to maintain the correct sign.

$$\Delta S = \frac{\Delta H}{T} = \frac{-5970\,\frac{\text{J}}{\text{mol}}}{(-107 + 273.15)\text{ K}} = -35.9\,\frac{\text{J}}{\text{mol}\cdot\text{K}}$$

It is important to remember that in freezing, heat is released and order increases. This means using $-\Delta H_{\text{fus}}$. It will be immediately apparent if this was not done correctly, because an increase in entropy will result, while freezing decreases entropy.

Self-Test

True–False

1. T
2. F Vapor pressure is temperature dependent. The closer the liquid is to its boiling point, the higher the vapor pressure will be.
3. F Anisotropic properties depend on the orientation of the molecule.
4. F Supercritical fluids exist when both the temperature and pressure are high.
5. T
6. T
7. T
8. T
9. F Bugs can walk on water because of surface tension.

Matching

10. b	11. g	12. e	13. f
14. d	15. a	16. h	17. c

Multiple Choice

18. d	19. c	20. c	21. b
22. b	23. a		

Fill in the Blanks

24. enthalpy
25. phase changes
26. sublimation
27. decreases, increase

Problems

28. $CH_3CH_2CH_2OH$ — this molecule is capable of hydrogen bonding. Because hydrogen bonding is the strongest intermolecular force, this intermolecular force will predominate.

    $CH_3(CH_2)_4CH_3$ — this molecule is nonpolar, so it will be unable to interact except with London dispersion forces.

    PCl_3 — this is a polar molecule with a dipole. The most important interactions will be dipole–dipole.

29. a. $\mathbf{Br_2}$ or Cl_2—Bromine is larger, so will have more dispersion forces and a higher boiling point.

    b. CH_3OH or $\mathbf{CH_3CH_2OH}$—both will hydrogen bond, but the second will also have higher dispersion forces.

    c. CH_4 or $\mathbf{CH_3OH}$—the second molecule can hydrogen bond, while the first cannot: it will have the highest boiling point.

30. $T = \dfrac{\Delta H_{vap}}{\Delta S_{vap}}$; $\quad T = \dfrac{565{,}300 \text{ J / mol}}{150.8 \text{ J / K} \cdot \text{mol}} = 3749 \text{ K} = 3476\ °\text{C}$

31. $q = n \times \Delta H_{vap}$; ΔH_{vap} is negative because the water is condensing.

$$0.237 \text{ g} \times \frac{1 \text{ mol}}{18.0 \text{ g}} \times -40.7 \frac{\text{kJ}}{\text{mol}} = -0.536 \text{ kJ}$$

Since this is a negative number, heat is released. Heat is released from the water to the iron and q = −q.

$$q = m \times C_{iron} \times \Delta T$$

$$536\ \text{J} = 50.0\ \text{g} \times 0.449 \frac{\text{J}}{\text{g} \cdot {}^\circ\text{C}} \times \Delta T$$

$$\Delta T = 23.9\ {}^\circ\text{C}$$

The final temperature is 25.0 °C + 23.9 °C = 48.9 °C

32. To heat the ice cube from −20 °C to 0.0 °C:

$$q = 45\ \text{g} \times 1\ \text{mol}/18\ \text{g} \times 36.57\ \text{J/mol} \cdot {}^\circ\text{C} \times 20\ {}^\circ\text{C} = 1830\ \text{J} = 1.83\ \text{kJ}$$

To melt the ice cube at 0.0 °C:

$$45\ \text{g} \times \frac{1\ \text{mol}}{18\ \text{g}} \times 6.01 \frac{\text{kJ}}{\text{mol}} = 15\ \text{kJ}$$

To warm the water to body temperature:

$$q = 45\ \text{g} \times 4.184\ \text{J/g} \cdot {}^\circ\text{C} \times 37\ {}^\circ\text{C} = 7.0\ \text{kJ}$$

Total heat absorbed by water/lost by victim = 1.83 kJ + 15 kJ + 7.0 kJ = 24 kJ.

33. $\ln P_2 = \ln P_1 + \frac{\Delta H_{vap}}{R}\left(\frac{1}{T_1} - \frac{1}{T_2}\right)$

$$\ln P_2 = \ln(760\ \text{mm Hg}) + \frac{30{,}800 \frac{\text{J}}{\text{mol}}}{\left(8.3145 \frac{\text{J}}{\text{mol} \cdot \text{K}}\right)}\left(\frac{1}{(80.1 + 273.15)\ \text{K}} - \frac{1}{(75.0 + 273.15)\ \text{K}}\right) = 6.48;$$

$$P_2 = e^{6.48} = 652\ \text{mm Hg}$$

34. normal freezing point = 197.5 K
critical point = 304.25 K
sublimation = line A
triple point = 216.6 K
boiling = line C
melting = line B

The forward slope of the central line (the freezing/melting line) shows that the solid form of the substance is more dense than the liquid form. As pressure increases, it is more likely to be a solid.

Challenge Problem

35. To solve this problem, first determine the number of joules available.

$$320\text{ g} \times \frac{1\text{ mol}}{44.0\text{ g}} \times 2220\,\frac{\text{K}}{\text{mol}} = 16{,}145\text{ kJ}$$

Now set up an algebraic equation to solve for the mass of water through the phase changes.

x = mass of water

$x/18$ = moles of water

$$16145\text{ kJ} = \left(0.00203\,\frac{\text{kJ}}{\text{g}\cdot{}^\circ\text{C}}\right)20\,{}^\circ\text{Cx} + \left(6.01\,\frac{\text{kJ}}{\text{mol}}\right)\frac{\text{x}}{18\,\frac{\text{g}}{\text{mol}}} + \left(0.004184\,\frac{\text{kJ}}{\text{g}\cdot{}^\circ\text{C}}\right)100\,{}^\circ\text{Cx}$$

$$+\left(40.67\,\frac{\text{kJ}}{\text{mol}}\right)\frac{\text{x}}{18\,\frac{\text{g}}{\text{mol}}} + \left(0.00208\,\frac{\text{kJ}}{\text{g}\cdot{}^\circ\text{C}}\right)15\,{}^\circ\text{Cx} = 16145\text{ kJ}$$

$$= 0.0406\text{ kJ/g}\cdot\text{x} + 0.334\text{ kJ/g}\cdot\text{x} + 0.4184\text{ kJ/g}\cdot\text{x} + 2.259\text{ kJ/g}\cdot\text{x} + 0.0312\text{ kJ/g}\cdot\text{x}$$

$$16145\text{ kJ} = 3.083\text{ kJ/g}\cdot\text{x}$$

$$\text{x} = 5237\text{ g}$$

Chapter Twelve—Solids and Solid-State Materials

Workbook Problems

Workbook Problem 12.1

Step 1: Determine the mass of a unit cell.

A simple cubic structure has a primitive cubic unit cell, which contains one atom.

$$1\text{ atom} \times \frac{1\text{ mol}}{6.022\times10^{23}\text{ atoms}} \times \frac{209\text{ g}}{\text{mol}} = 3.470\times10^{-22}\text{ g}$$

Step 2: Solve for the volume of a unit cell.

$$\frac{3.470\times10^{-22}\text{ g}}{9.3\,\frac{\text{g}}{\text{cm}^3}} = 3.73\times10^{-23}\text{ cm}^3$$

Step 3: Determine the edge length of a unit cell.

$$\text{Edge length} = \sqrt[3]{3.731\times10^{-23}\text{ cm}^3} = 3.34\times10^{-8}\text{ cm}\times\frac{1\text{ m}}{100\text{ cm}}\times\frac{1\times10^{12}\text{ pm}}{\text{m}} = 334\text{ pm}$$

Step 4: Determine the atomic radius.

For a primitive cubic unit cell, the radius of an atom is half the edge length.

Atomic radius = 0.5 × 334 pm = 167 pm.

Workbook Problem 12.2

Step 1: Determine the numbers of electrons of each type:

Mo: $4d^45s^2$—6 electrons: bonding orbitals are filled and antibonding orbitals are empty.

Cd: $4d^{10}5s^2$—12 electrons: bonding and antibonding orbitals are filled.

Y: $4d^15s^2$—3 electrons: bonding orbitals half-filled, antibonding orbitals are empty.

Step 2: Rank in order of expected bonding strength:

$$\text{Cd} < \text{Y} < \text{Mo}$$

The published melting points are 321 °C for Cd, 1523 °C for Y, and 2623 °C for Mo, which means this order is correct.

Workbook Problem 12.3

Step 1: Determine the numbers of valence electrons in each element:

Si: Group 4A—four valence electrons	Ge: Group 4A—four valence electrons
Ga: Group 3A—three valence electrons	Sb: Group 5A—five valence electrons
As: Group 5A—five valence electrons	P: Group 5A—five valence electrons

Step 2: Determine whether the dopant has more or fewer valence electrons than the semiconductor:

a. Si > Ga b. Ge < Sb c. Si < As d. Ge < P

Step 3: Identify the doped semiconductors as *n*- or *p*-type:

a. *p*-type b. *n*-type c. *n*-type d. *n*-type

Workbook Problem 12.4

Step 1: Determine the energy of an individual photon:

$$178\,\frac{\text{kJ}}{\text{mol}} \times \frac{1000\text{ J}}{\text{kJ}} \times \frac{1\text{ mol}}{6.022 \times 10^{23}\text{ photons}} = 2.96 \times 10^{-19}\text{ J/photon}$$

Step 2:	Convert the photon energy to wavelength:
	$E = \frac{hc}{\lambda}; \lambda = \frac{hc}{E} = \frac{6.626 \times 10^{-34}\ J \cdot s \times 3.00 \times 10^{8}\ m/s}{2.96 \times 10^{-19}\ J} = 673\ nm$
	This is red light.

Self-Test

True–False

1. T
2. F X-ray crystallography is used to examine the structure of crystalline solids.
3. F The coordination number is the number of other atoms each atom is touching.
4. F Fullerene and graphite are both allotropes of carbon.
5. F There is no good theory for superconductivity at the present time.
6. T
7. T
8. F Heating semiconductors increases conductivity. Heating metals decreases conductivity.
9. T
10. F The primary drawback of ceramics is that they are brittle.

Matching

11. e	12. a	13. h	14. f
15. k	16. c	17. d	18. j
19. m	20. b	21. o	22. g
23. i	24. q	25. l	26. n
27. p			

Multiple Choice

28. a	29. b	30. b	31. d
32. b	33. a	34. c	35. c
36. b			

Fill in the Blanks

37. ionic solid, molecular solid, covalent network solid
38. malleability and ductility, localized, brittle
39. close, band, valence, conduction, larger
40. below, resistance, nitrogen
41. metals, density, insulators, corrosion, malleable
42. advantageous, strengthened

Problems

43. Face-centered cubic unit cells contain four total atoms.

Mass of unit cell:

$$4\ \text{atoms} \times \frac{1\ \text{mol}}{6.022 \times 10^{23}\ \text{atoms}} \times \frac{196.97\ g}{\text{mol}} = 1.308 \times 10^{-21}\ g$$

Volume of unit cell:

$$\frac{1.308\times10^{-21}\ \text{g}}{19.3\ \frac{\text{g}}{\text{cm}^3}}=6.779\times10^{-23}\ \text{cm}^3$$

Edge length of unit cell =

$$\sqrt[3]{6.779\times10^{-23}\ \text{cm}^3}=4.077\times10^{-8}\ \text{cm}=407.7\ \text{pm}$$

A face-centered cubic cell has an edge diagonal equal to 4r. Using the Pythagorean theorem:

$$(4r)^2 = 2(407.7\ \text{pm})^2$$
$$r = 144\ \text{pm}$$

44. A face-centered cubic unit cell has four total atoms, and the diagonal of a side is equal to 2x the atomic diameter.

$$\text{Mass of unit cell} = 4\ \text{atoms}\times\frac{1\ \text{mol}}{6.022\times10^{23}\ \text{atoms}}\times\frac{106.42\ \text{g}}{\text{mol}}=7.069\times10^{-22}\ \text{g}$$

Volume of unit cell:

Diagonal = 137 pm × 4 = 548 pm = (548×10^{-10} cm)

Edge length:

$$(2d)^2 = 2(548\ \text{pm})^2$$
$$4d^2 = 600{,}608\ \text{pm}^2$$
$$d^2 = 150{,}152\ \text{pm}^2$$
$$d = 3.875\times10^{-8}\ \text{cm}$$
$$\text{Volume} = (3.875\times10^{-8}\ \text{cm})^3 = 5.818\times10^{-23}\ \text{cm}^3$$

$$\text{Density of metal} = \frac{7.069\times10^{-22}\ \text{g}}{5.81\times10^{-23}\ \text{cm}^3}=12.15\frac{\text{g}}{\text{cm}^3}$$

45. As fourth-period elements, these all have access to 4*s* orbitals and 3*d* orbitals. Twelve electrons can be accommodated, 6 in bonding orbitals, and 6 in nonbonding orbitals. The more electrons there are in bonding orbitals, and the fewer electrons in the nonbonding orbitals, the harder the element and the higher the melting point.

Co: [Ar] $4s^23d^7$—six electrons in bonding orbitals, three electrons in antibonding orbitals

Cr: [Ar] $4s^23d^4$—six electrons in bonding orbitals, no electrons in antibonding orbitals

Cu: [Ar] $4s^23d^9$—six electrons in bonding orbitals, five electrons in antibonding orbitals

Highest to lowest melting point: Cr > Co > Cu

The observed melting points are Cr: 1857 °C Co: 1495 °C Cu: 1083 °C

This confirms the determination made using molecular orbital theory.

46. Determine the number of valence electrons in each of the materials:

Si—Group 4A—four valence electrons
P—roup 5A—five valence electrons
B–Group 3A—three valence electrons
Ge—Group 4A—four valence electrons
As—Group 5A—five valence electrons

Identify the combinations and their types:

SiP—4A5A—*n*-type
SiB—4A3A—*p*-type
SiAs—4A5A—*n*-type
GeP—4A5A—*n*-type
GeB—4A3A—*p*-type
GeAs—4A5A—*n*-type

Chapter Thirteen—Solutions and Their Properties

Workbook Problems

Workbook Problem 13.1

Strategy: Determine which is the major component and which is the minor component.

In a 3% solution of hydrogen peroxide, hydrogen peroxide is the solute and the water, which makes up the remaining 97%, is the solvent.

Workbook Problem 13.2

Strategy: Charge and molecular size determine the hydration energy.

Smaller, more highly charged ions have the greatest hydration energy:

$$Cl^- < Li^+ < Mg^{2+}$$

Workbook Problem 13.3

Step 1: Determine mass percent.

$$\text{mass \%} = \frac{15.43\text{ g}}{315.43\text{ g}} \times 100 = 4.89\%\text{ glucose}$$

Step 2: Calculate the moles of solute and the molality of the solution.

$$15.43\text{ g} \times \frac{1\text{ mole}}{180\text{ g}} = 0.0857\text{ mol}$$

$$\text{m} = \frac{0.0857\text{ mol}}{0.300\text{ kg}} = 0.286\text{ m}$$

Step 3: Calculate the volume of the solution and the molarity of the solution.

$$\frac{315.43\text{ g}}{1.09\frac{\text{g}}{\text{mL}}} = 289\text{ mL}$$

$$\text{M} = \frac{0.0857\text{ mol}}{0.289\text{ L}} = 0.296\text{ M}$$

Workbook Problem 13.4

Step 1: Calculate the moles of dissolved CO_2 under pressure.

$$\text{Solubility} = 0.0032\frac{\text{mol}}{\text{L}\cdot\text{atm}} \times 2.3\text{ atm} = 0.00736\frac{\text{mol}}{\text{L}}$$

$$\text{Dissolved } CO_2 = 0.00736\text{ mol/L} \times 0.500\text{ L} = 0.00368\text{ mol}$$

Step 2: Calculate the moles of dissolved CO_2 after pressure is released and equilibrium with the atmosphere reached:

$$\text{Solubility} = 0.0032\frac{\text{mol}}{\text{L}\cdot\text{atm}} \times 0.30\text{ mm Hg} \times \frac{1\text{ atm}}{760\text{ mm Hg}} = 1.26\times10^{-6}\frac{\text{mol}}{\text{L}}$$

$$1.26\times10^{-6}\frac{\text{mol}}{\text{L}} \times 0.5\text{ L} = 6.32\times10^{-7}\text{ mol}$$

Step 3: Calculate the change in dissolved CO_2—this is the CO_2 that has been released.

$$0.00368\text{ mol} - 0.000000632\text{ mol} = 0.00368\text{ mol.}$$

For all practical purposes, all of the CO_2 has bubbled away.

Step 4: Determine the volume of gas at 25°C and 1 atm pressure.

$$V = \frac{nRT}{P} = \frac{0.00368\text{ mol} \times \left(0.08206\frac{\text{L}\cdot\text{atm}}{\text{K}\cdot\text{mol}}\right) \times (25 + 273.15)\text{ K}}{1\text{ atm}} = 0.0900\text{ L} = 90.0\text{ mL}$$

Workbook Problem 13.5

Step 1: Calculate the mole fraction of water in a solution with this vapor pressure lowering:

$$230.7\text{ mm Hg} = 233.7\text{ mm Hg} \times X_{H_2O}$$
$$0.987 = X_{H_2O}$$

Step 2: Calculate the moles of dissolved ions that will lead to this mole fraction:

$$500\text{ g}/18.0\text{ g/mol} = 27.78\text{ mol } H_2O$$

$$X_{H_2O} = \frac{27.78\text{ mol}}{(27.78\text{ mol} + \text{x mol})} = 0.987$$

$$27.78\text{ mol} = 0.987x(27.78\text{ mol} + x\text{ mol})$$
$$27.78 - 27.42 = 0.987x$$
$$x = 0.3659\text{ mol ions}$$

Step 3: Convert to moles, then grams of $CaCl_2$:

$$0.3659\text{ mol ions} \times \frac{1\text{ CaCl}_2}{3\text{ ions}} = 0.1220\text{ mol CaCl}_2$$

$$0.1220\text{ mol CaCl}_2 \times 111\text{ g/mol} = 13.5\text{ g CaCl}_2$$

Step 4: Convert to moles, then grams of LiCl:

$$0.3659\text{ mol ions} \times \frac{1\text{ LiCl}}{2\text{ ions}} = 0.1830\text{ mol LiCl}$$

$$0.1830\text{ mol LiCl} \times 42.5\text{ g/mol} = 7.78\text{ g LiCl}$$

Workbook Problem 13.6

Step 1: Solve for the mass and moles of each component:

$$0.13 \times 150\text{ g} = 19.5\text{ g ethanol}/46\text{ g/mol} = 0.424\text{ mol ethanol}$$

$$0.87 \times 150\text{ g} = 130.5\text{ g water}/18.0\text{ g/mol} = 7.25\text{ mol water}$$

Step 2: Determine the mole fraction of each component:

$$X_{H_2O} = \frac{7.25\text{ mol}}{7.25\text{ mol} + 0.424\text{ mol}} = 0.945$$

$$X_{EtOH} = \frac{0.424\text{ mol}}{7.25\text{ mol} + 0.424\text{ mol}} = 0.0552$$

Step 3: Determine the partial pressure of each component and the total pressure:

$$P_{H_2O} = 0.945 \times 23.8 \text{ mm Hg} = 22.5 \text{ mm Hg}$$

$$P_{EtOH} = 0.0552 \times 61.2 \text{ mm Hg} = 3.38 \text{ mm Hg}$$

$$P_{total} = 22.5 \text{ mm Hg} + 3.38 \text{ mm Hg} = 25.9 \text{ mm Hg}$$

Workbook Problem 13.7

Step 1: Determine the molality of ions needed to generate this freezing point depression.

$$15.0\ °C = 1.86\ °C/m \times X\ m$$

$$X = 8.06\ m$$

Step 2: Determine the moles of $(NH_4)_2SO_4$ needed to generate this quantity of ions.

$$8.06\ m = \frac{x \text{ moles}}{0.300 \text{ kg}}$$

$$8.06 \frac{\text{moles}}{\text{kg}} \times 0.300 \text{ kg} = 2.42 \text{ mol ions}$$

$$2.42 \text{ mol ions} \times \frac{1 \text{ mol salt}}{3 \text{ mol ions}} = 0.80645 \text{ mol salt}$$

Step 3: Convert to grams of ammonium sulfate needed.

$$0.80645 \text{ mol} \times \frac{132.14 \text{ g}}{\text{mol}} = 106.56 \text{ g}$$

Workbook Problem 13.8

Strategy: Using the equation for osmotic pressure; solve for the molarity of the solution:

$$7.34 \text{ atm} = M \times 0.08206 \frac{\text{L} \cdot \text{atm}}{\text{mol} \cdot \text{K}} \times 298\ K$$

$$M = 0.300\ M$$

Workbook Problem 13.9

Step 1: Calculate the molarity of the solution:

$$\Pi = \text{M} \times \text{R} \times \text{T}$$

$$8.428 \text{ mm Hg} \times \frac{1 \text{ atm}}{760 \text{ mm Hg}} = \text{M} \times \frac{0.08206 \text{ L} \cdot \text{atm}}{\text{mol} \cdot \text{K}} \times 298 \text{ K}$$

$$\text{M} = 0.000453 \text{ M}$$

Step 2: Calculate the number of moles in the solution:

$$\frac{0.00453\text{ mol}}{\text{L}} \times 0.100\text{ L} = 0.0000453 = 45\ \mu\text{mol}$$

Step 3: Determine the molecular weight:

$$23 \times 10^{-3}\text{ g} = 4.53 \times 10^{-5}\text{ mol} \times \text{MW}$$

$$\text{MW} = 507.73\text{ g/mol}$$

Self-Test

True–False

1. T
2. T
3. F ppm and ppb are the most convenient units for environmental work.
4. F Potassium nitrate will dissolve readily in water (polar solvent).
5. T
6. F Supersaturated solutions are always unstable.
7. F Increasing pressure causes gases to become more soluble.
8. T
9. F Concentrated solutions have a lower freezing point.
10. T

Multiple Choice

11. a	12. d	13. c	14. b
15. c			

Matching

16. k	17. f	18. a	19. m
20. d	21. l	22. b	23. g
24. c	25. n	26. h	27. i
28. e	29. j		

Fill in the Blanks

30. solvent–solvent, solute–solute	31. solute, solution, solute, solvent
32. density	33. more
34. drop, increase, decrease, osmotic pressure	35. molecular mass
36. solvated, hydrated	37. exothermic, endothermic

Problems

38. The rankings are from most polar (least soluble) to most nonpolar (most soluble). Iodine is completely nonpolar, so will be readily soluble in benzene. Butanol $CH_3(CH_2)_2CH_2OH$ is primarily nonpolar, but has a polar oxygen group that reduces its solubility. NaCl is ionic, and so will be essentially insoluble in benzene.

$$NaCl < CH_3(CH_2)_2CH_2OH < I_2$$

39. a. $\text{mass \%} = \dfrac{28.4\text{ g}}{(28.4\text{ g} + 350\text{ g})} \times 100 = 7.50\%$

b. mole fraction:

moles sucrose = 28.4 g/342.3 g/mol = 0.0830 mol

moles water = 350 g/ 18.0 g/mol = 19.4 mol

$$\text{mol fraction} = \frac{0.0830\text{ mol}}{(19.4\text{ mol} + 0.0830\text{ mol})} = 0.00425$$

c. $\text{molality} = \dfrac{0.0830\text{ mol}}{0.350\text{ kg}} = 0.237\text{ m}$

d. $\text{molarity} = \dfrac{0.0830\text{ mol}}{0.374\text{ L}} = 0.222\text{ M}$

40. 100 g of this solution will contain 3 g of hydrogen peroxide and have a volume of 100 g/1.01 g/mL = 99 mL

$$\text{molarity} = \frac{3\text{ g} \times \dfrac{1\text{ mol}}{34\text{ g}}}{0.099\text{ L}} = 0.891\text{ M}$$

41. $15\text{ ppb} = \dfrac{0.015\text{ mg}}{\text{L}} \times \dfrac{1\text{ g}}{1000\text{ mg}} \times \dfrac{1\text{ mol As}}{74.9\text{ g}} = 2.00 \times 10^{-7}\text{ M}$

$15\text{ ppb} = \dfrac{0.015\text{ mg}}{\text{L}} \times \dfrac{1\text{ g}}{1000\text{ mg}} \times 100{,}000\text{ L} = 1.5\text{ g}$

42. $\text{solubility} = 1.32 \times 10^{-3}\,\dfrac{\text{mol}}{\text{L}\cdot\text{atm}} \times 0.21\text{ atm} = 2.77 \times 10^{-4}\text{ M}$

43. $\text{solubility} = 6.25 \times 10^{-4}\,\dfrac{\text{mol}}{\text{L}\cdot\text{atm}} \times 976\text{ mm Hg} \times \dfrac{1\text{ atm}}{760\text{ mm Hg}} = 8.03 \times 10^{-4}\text{ M}$

44. moles of glucose = 4.5 g/180 g/mol = 0.025 mol
moles of water = 55 g/18 g/mol = 3.06 mol

$$P_{\text{soln}} = 42.175\text{ mm Hg} \times \frac{3.06\text{ mol}}{3.06\text{ mol} + 0.025\text{ mol}} = 41.09\text{ mm Hg}$$

45. moles of $FeCl_3$ = 5.8 g /162.2 g/mol = 0.03576 mol $FeCl_3$
moles of ions = 0.03576 mol × 3.4 = 0.1216 mol ions
moles of water = 72.1 g /18 g/mol = 4.01 mol

$$P_{\text{soln}} = 23.8\text{ mm Hg} \times \frac{4.01\text{ mol}}{4.01\text{ mol} + 0.122\text{ mol}} = 23.1\text{ mm Hg}$$

46. moles of benzene: 10 g/78.0 g/mol = 0.128 mol
moles of toluene: 10 g/92.0 g/mol = 0.109 mol

$$\text{mole fraction benzene} = \frac{0.128 \text{ mol}}{(0.128 \text{ mol} + 0.109 \text{ mol})} = 0.540$$

$$\text{mole fraction toluene} = \frac{0.109 \text{ mol}}{(0.128 \text{ mol} + 0.109 \text{ mol})} = 0.460$$

$$P_{total} = (0.540 \times 93.4 \text{ mm Hg}) + (0.460 \times 26.9 \text{ mm Hg}) = 62.8 \text{ mm Hg}$$

47.

$$\Delta T_b = 80.0\ °C - 76.7\ °C = 3.3\ °C$$

$$\Delta T_b = K_b m$$

$$3.3\ °C = 5.03 \frac{°C}{m} \times X$$

$$X = 0.656\ m$$

$$0.656 \text{ m} = \frac{\text{x mol naphthalene}}{1.0 \text{ kg}} = 0.656 \text{ mol naphthalene} \times \frac{128 \text{ g}}{\text{mol}} = 84.0 \text{ g}$$

48. $$\text{m} = \frac{10.0 \text{ g NaCl} \times \frac{1 \text{ mol}}{58.5 \text{ g}}}{0.090 \text{ kg}} = 1.90 \text{ m} \times \frac{2 \text{ ions}}{\text{mol}} = 3.80 \text{ m ions}$$

$$\Delta T_b = 0.51 \frac{°C}{m} \times 3.80 \text{ m} = 1.94\ °C$$

$$\Delta T_f = 1.86 \frac{°C}{m} \times 3.80 \text{ m} = 7.07\ °C$$

Boiling point = 102 °C Freezing point = −7.07 °C

49. $\Delta T_f = |3.5 - 5.5| = 2.0\ °C$

$$2.0\ °C = 5.12 \frac{°C}{m} \times X$$

$$X = 0.390 \text{ m}$$

0.390 m = x mol/0.075 kg x = 0.0292 mol

MW = 2.50 g/0.0292 mol = 85.62 g/mol

50. moles glucose = 18.0g/180 g/mol = 0.10 mol

molarity = 0.10 mol/1.2 L= 0.0833 M

$$\Pi = 0.0833\ \text{M} \times 0.08206\ \frac{\text{L}\cdot\text{atm}}{\text{K}\cdot\text{mol}} \times (2.73.15 + 15)\ \text{K} = 1.97\ \text{atm}$$

51. $$16.4\ \text{mm Hg} \times \frac{1\ \text{atm}}{760\ \text{mm Hg}} = \text{X M} \times 0.08206\ \frac{\text{L}\cdot\text{atm}}{\text{K}\cdot\text{mol}} \times (2.73.15 + 15)\ \text{K}$$

$$\text{X} = 9.13 \times 10^{-4}\ \text{M} \times 1.00\ \text{L} = 9.13 \times 10^{-4}\ \text{mol}$$

$$\frac{0.250\ \text{g}}{9.13 \times 10^{-4}\ \text{mol}} = 274\ \frac{\text{g}}{\text{mol}}$$

Challenge Problem

52. First, determine the mole fraction of pentane in the vapor.

100 g vapor will be 43.2 g pentane, 56.8 g hexane

43.2 g pentane/72.1453 g/mol = 0.60 mol 56.8 g hexane /86.1963 g/mol = 0.66 mol

$$X_{\text{pentane}} = \frac{0.60\ \text{mol}}{0.60\ \text{mol} + 0.66\ \text{mol}} = 0.476$$

$$X_{\text{hexane}} = \frac{0.66\ \text{mol}}{0.60\ \text{mol} + 0.66\ \text{mol}} = 0.524$$

Now set up a two-variable system to solve for the mole fractions in the solution.

$0.476P_{\text{total}} = 425$ x, where x = mole fraction of pentane in solution

$0.524P_{\text{total}} = 151\ (1 - x)$, where $1 - x$ = mole fraction of hexane in solution

$$\begin{array}{rl} & P = 425\ x/0.476 = 893\ x \\ - & \underline{P = [151\ (1 - x)]/0.524 = 288 - 288x} \\ & 0 = -288 + 1181\ x \end{array}$$

$x = 0.244$ = mole fraction of pentane $1 - x = 0.756$ = mole fraction of hexane

P_{total} = (151 mmHg × 0.756) + (425 mmHg × 0.244) = 218 mmHg

Chapter Fourteen—Chemical Kinetics

Workbook Problems

Workbook Problem 14.1

Step 1: Balance the equation:

$$2\ HBr(g) \rightarrow H_2(g) + Br_2(g)$$

Step 2: Write the relative reaction rates:

$$\text{General rate} = -\frac{\Delta HBr}{2\Delta t} = \frac{\Delta H_2}{\Delta t} = \frac{\Delta Br_2}{\Delta t}$$

Step 3: Determine the rate of formation of bromine from the rate of disappearance of HBr.

$$-\frac{-3.26 \times 10^{-3}\ M/s}{2} = 0.00163\ M/s$$

This is also the rate of formation of hydrogen.

Workbook Problem 14.2

Step 1: Zeroth-order reaction:

Rate = k – units will be M/s.

Step 2: First-order reaction:

Rate = k[A], so M/s = $k \cdot$ M, $k = s^{-1}$

Step 3: Second-order reaction:

Rate = $k[A]^2$, so M/s = $k \cdot M^2$, $k = M^{-1}s^{-1}$

Workbook Problem 14.3

Step 1: Determine the order of the reaction with respect to NO_2.

In reactions 1 and 2, the concentration of NO_2 is doubled, while the concentration of F_2 is held constant. Doubling [NO_2] causes the rate to double, so the reaction is first order in NO_2.

Step 2: Determine the order of the reaction with respect to F_2.

In reactions 2 and 3, the concentration of NO_2 is held constant while the concentration of F_2 is varied. Doubling [F_2] causes the rate of the reaction to double, so the reaction is first order in F_2.

Step 3: Write the equation for the rate law.

$$\text{Rate} = k[NO_2][F_2]$$

Step 4: Calculate the value of *k*.

We can calculate the value of *k* using the data from any one of the experiments. Using the data in experiment 1 gives:

$$0.026\ \text{M/s} = \text{k}(0.10\ \text{M})(0.10\ \text{M})$$

$$\text{k} = 2.6\ \text{M}^{-1}\text{s}^{-1}$$

Workbook Problem 14.4

Step 1: Solve for *k*.

$$\text{Rate} = \text{k}[\text{A}]^{\circ} = \text{k} = 0.026\ \text{M/s}$$

Step 2: Substitute into the integrated rate law and solve for [A] after 20 seconds.

$$[\text{A}]_t = -(0.026\ \text{M/s})\ (20\ \text{s}) + 1.00\ \text{M} = 0.48\ \text{M}$$

Workbook Problem 14.5

Step 1: Solve for *k*:

$$\ln\frac{0.03\ \text{M}}{0.10\ \text{M}} = -k(73\ \text{min})$$

$$k = 0.0165\ \text{min}^{-1}$$

Step 2: Determine the concentration of the reactant at 80% completion:

$$0.10\ \text{M} \times 0.80 = 0.08\ \text{M}$$
$$0.10\ \text{M} - 0.08\ \text{M} = 0.02\ \text{M}$$

Step 3: Solve for *t* at 80% completion:

$$\ln\frac{0.020\ M}{0.10\ M} = -0.0165\ \text{min}^{-1} \times t$$

$$t = 98\ \text{min}$$

Workbook Problem 14.6

Strategy: Substitute into the half-life equation.

$$t_{1/2} = \frac{0.693}{0.0165\ \text{min}^{-1}} = 42\ \text{min}$$

Workbook Problem 14.7

Step 1: Solve for *k*.

$$\frac{1}{[A]_t} = kt + \frac{1}{[A]_0}$$

$$\frac{1}{0.0495\ \text{M}} = k \times 400\ \text{s} + \frac{1}{0.1000\ \text{M}}$$

$$k = 0.0255\ \text{M}^{-1}s^{-1}$$

Step 2: Use the half-life equation to determine the time required for 50% completion.

$$t_{1/2} = \frac{1}{0.0255\ \text{M}^{-1}s^{-1} \times 0.100\ \text{M}} = 392\ \text{s}$$

Step 3: Determine the concentration of NO_2 at 95% completion (5% of NO_2 remaining).

$$0.100\ \text{M} \times 0.05 = 0.00500\ \text{M}$$

Step 4: Substitute the concentrations into the integrated rate law, and solve for *t*.

$$\frac{1}{0.00500\ \text{M}} = 0.0255\ \text{M}^{-1}s^{-1} \times t + \frac{1}{0.100\ \text{M}}$$

$$t = 7450\ \text{s} \times \frac{1\ \text{hour}}{3600\ \text{s}} = 2.07\ \text{hours, or 2 hours 4 minutes}$$

Workbook Problem 14.8

If a reaction has a rate constant of 0.0123 s^{-1} at 700 K and 0.00342 s^{-1} at 250 K, what is the activation energy of the reaction, and what will the rate constant be at 500 K?

Strategy: Solve for E_a using the two-point Arrhenius equation, then use the same equation to solve for the rate constant at 500 K.

Step 1: Solve for E_a using the two-point Arrhenius equation:

$$\ln\left(\frac{0.00342\ \text{s}^{-1}}{0.0123\ \text{s}^{-1}}\right) = \left(\frac{-E_a}{8.314\ \text{J/mol}\cdot\text{K}}\right)\left(\frac{1}{250\ \text{K}} - \frac{1}{700\ \text{K}}\right)$$

$$E_a = 4138\ \text{J/mol}$$

Step 2: Solve for *k* at 500 K:

$$\ln\left(\frac{k_2}{0.0123\ \text{s}^{-1}}\right)=\left(\frac{-4138\ \text{J/mol}}{8.314\ \text{J/mol}\cdot\text{K}}\right)\left(\frac{1}{500\ \text{K}}-\frac{1}{700\ \text{K}}\right)$$

$$\ln\left(\frac{k_2}{0.0123\ \text{s}^{-1}}\right)=-0.284$$

$$\left(\frac{k_2}{0.0123\ \text{s}^{-1}}\right)=e^{-0.284}$$

$$k_2=0.00926\ \text{s}^{-1}$$

Workbook Problem 14.9

Step 1: Write the expected rate law:

$$\text{Rate}=k[\text{A}]^2$$

Step 2: Sum for the overall reaction:

$$2\ \text{A}+\text{A}_2+\text{B}\rightarrow \text{A}_2+\text{A}_2\text{B},\ \text{cancels to}$$

$$2\ \text{A}+\text{B}\rightarrow \text{A}_2\text{B}$$

Self-Test

True–False

1. T
2. F The exponents must be experimentally determined.
3. F The order of a reaction is determined by adding the exponents in the rate law.
4. T
5. T
6. F The half-life is the time required for the concentration to drop to half its initial level.
7. T
8. F Zeroth-order reactions are quite uncommon.
9. F Reaction steps can occur at all different rates.
10. T
11. F Activation energy is always positive, although reactions can be net exothermic.
12. T
13. F Biological reactions almost invariably involve catalysts.
14. T

Multiple Choice

15. b	16. c	17. a	18. a
19. c	20. d	21. d	22. c
23. b	24. d		

Matching

25. k	26. d	27. n	28. b
29. f	30. a	31. m	32. c
33. o	34. i	35. p	36. g
37. h	38. j	39. l	40. e

Fill in the Blanks

41. reaction rate, decrease, increase, product
42. rate constant, k
43. 2, 4, 8, 1
44. reverse reaction
45. first order
46. elementary reactions, rate-determining step, slowest
47. activation energy, transition state
48. lower energy
49. double
50. adsorb, release/desorb

Problems

51. $\frac{\Delta[C]}{\Delta t}$ can be calculated arithmetically.

$$\text{initial rate} = \frac{\Delta[C]}{\Delta t} = \frac{0.005 - 0.000}{15 - 0} = 3.3 \times 10^{-4}\ \text{M/s}$$

$$\text{rate from 30 to 75s} = \frac{\Delta[C]}{\Delta t} = \frac{0.019 - 0.010}{75 - 30} = 2.0 \times 10^{-4}\ \text{M/s}$$

52. Rate of appearance of

$$H_2 = \left[1.67 \times 10^{-2}\ \text{mol PH}_3/(\text{L}\cdot\text{s})\right] \times \frac{6\ \text{mol H}_2}{4\ \text{mol PH}_3} = 2.50 \times 10^{-2}\ \text{mol H}_2/(\text{L}\cdot\text{s})$$

53. When the value of [NO] is doubled, the reaction rate increases by a factor of four. The reaction is second order in NO. When the concentration of Cl_2 is doubled, the reaction rate doubles. The reaction is first order in Cl_2. The overall order is third order.

The rate law is as given below:

$$\text{Rate} = k[\text{NO}]^2[\text{Cl}_2]$$

Using the data in the second experiment:

$$63.6\ \text{mol/}(\text{L}\cdot\text{s}) = \text{k}(0.050\ \text{mol/L})^2(0.025\ \text{mol/L})$$

$$\text{k} = \frac{63.6\ \text{mol/}(\text{L}\cdot\text{s})}{6.25 \times 10^{-5}\ \text{mol}^3/\text{L}^3} = 1.02 \times 10^6\ \text{L}^2/(\text{mol}^2\cdot\text{s})$$

54. When the concentration of A is doubled, the reaction rate doubled, so the reaction is first order in A. When the concentration of B is doubled, the rate is unaffected: the reaction is zero order in B.

The reaction is first order overall.

The rate law is

$$\text{Rate} = k[\text{A}].$$

Using the data in second experiment:

$$0.134 \text{ mol/(L} \cdot \text{s)} = \text{k}(0.300 \text{ mol/L})$$

$$\text{k} = 0.447 \text{ 1/s}$$

$$\text{rate} = (0.447 \text{ 1/s})\ (0.542 \text{ mol/L}) = 0.242 \text{ mol/(L} \cdot \text{s)}$$

55. $\dfrac{1}{[\text{O}_3]} = \left[1.40\times10^{-2}\ \text{L}/(\text{mol}\cdot\text{s})\right]\left(12\ \text{h}\times\dfrac{3600\ \text{s}}{1\ \text{h}}\right)+\dfrac{1}{2.37\ \text{M}} = 605\ \text{L/mol}$

$[\text{O}_3] = 1.65\times10^{-3}\ \text{M}$

56. The reaction order can be determined graphically by plotting the data as [HI] versus time, ln [HI] versus time, and 1/[HI] versus time. The graph that provides a straight line provides the order of the reaction.

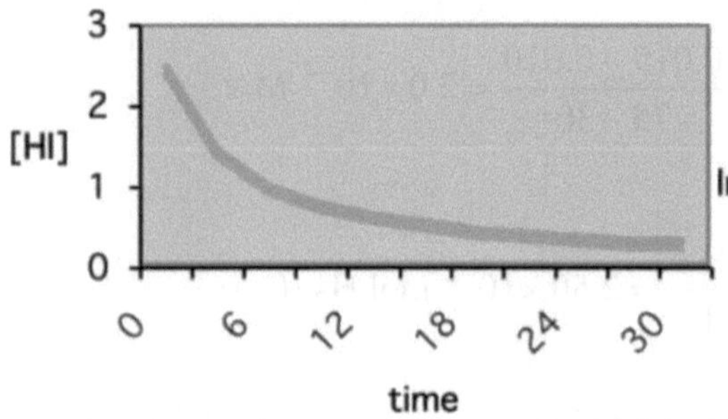

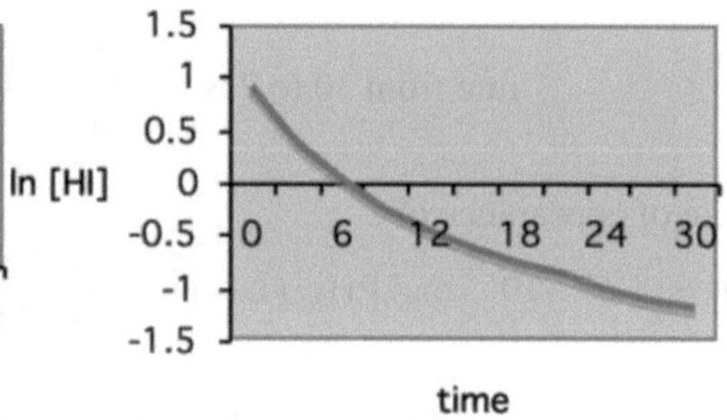

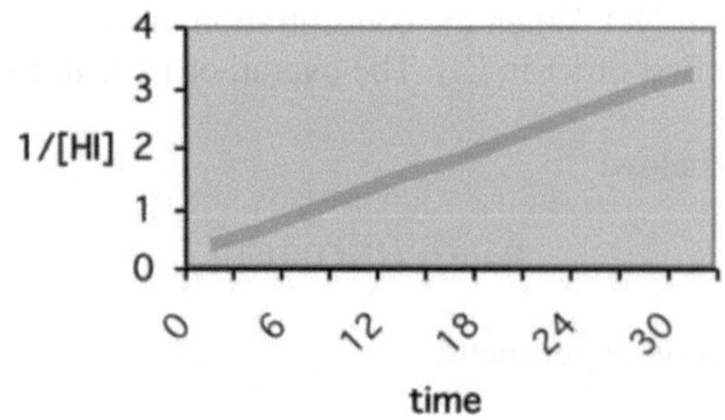

The plots show that the reaction is second order. Using the integrated rate law and the initial rate,

$$\frac{1}{1.45\ \text{M}} = \text{k}(3\ \text{min}) + \frac{1}{2.50\ \text{M}}$$

$$k = \frac{\dfrac{1}{1.45\ \text{M}} - \dfrac{1}{2.50\ \text{M}}}{3\ \text{min}} = 9.66\times10^{-2}\ \text{L}/(\text{mol}\cdot\text{min})$$

$$t_{1/2} = \frac{1}{\left[9.66\times10^{-2}\ \text{L/(mol}\cdot\text{min)}\right]\times 2.50\ \text{mol/L}} = 4.14\ \text{min}$$

57. The overall reaction is

$$2\ NO_2(g) + Cl_2(g) \rightarrow 2\ NO_2Cl(g).$$

Both steps are bimolecular, and Cl(*g*) is a reaction intermediate: it does not appear in the final equation.

The expected rate is

$$\text{Rate} = k[NO_2][Cl_2].$$

This is a second-order reaction.

58. $k = Ae^{\frac{-E_a}{RT}}$

$$A = \frac{k}{e^{\frac{-E_a}{RT}}} = \frac{3.79\times10^{-5}\ \text{L/mol}\cdot\text{s}}{e^{\frac{-1.18\times10^4\ \text{J/mol}}{8.314\frac{\text{J}}{\text{mol}\cdot\text{K}}\times 381\ \text{K}}}} = \frac{3.79\times10^{-5}\ \text{L/mol}\cdot\text{s}}{e^{-3.73}} = 1.57\times10^{-3}$$

59. $\ln\left(\frac{3.2\times10^{-3}}{7.2\times10^{-5}}\right) = \left(\frac{-9.862\times10^{4}\ \text{J/mol}}{8.314\ \text{J}/(\text{mol}\cdot\text{K})}\right)\left(\frac{1}{T_2} - \frac{1}{340\ \text{K}}\right)$

$$3.79 = -1.186\times10^{4}\ \text{K}\left(\frac{1}{T_2} - \frac{1}{340\ \text{K}}\right)$$

$$-3.195\times10^{-4}\ \text{K}^{-1} = \frac{1}{T_2} - \frac{1}{340\ \text{K}}$$

$$\frac{1}{T_2} = 3.26\times10^{-3}\ \text{K}^{-1} \qquad T_2 = 307\ \text{K}$$

Challenge Problem

60. To calculate the initial rate requires a rate law, rate constant, and the initial concentrations of C_2H_5I and/or OH^-. From the information given, we know that the rate = $k[C_2H_5I][OH^-]$ because the doubling of the concentration of each reactant while the other is held constant results in a doubling of the rate: it is first order in each reactant.

The rate constant, *k*, is calculated using the Arrhenius equation, $k = Ae^{-E_a/RT}$.

$$k = 1.09\times10^{9}\ \text{M}^{-1}\text{s}^{-1}e^{-(7.97\times10^4\ \text{J/mol})/(8.314\ \text{J/mol}\cdot\text{K})(318.15\ \text{K})} = 8.956\times10^{-5}\ \text{M}^{-1}\text{s}^{-1}$$

Remembering that the two 250 mL solutions are going to be combined, and assuming that they sum to 500 mL, the molar concentrations of the reactants are

$$\frac{0.475 \text{ g KOH}}{0.500 \text{ L}} \times \frac{1 \text{ mol}}{56.1 \text{ g}} = 1.692 \times 10^{-2} \text{ M OH}$$

$$\frac{1.378 \text{ g } C_2H_5I}{0.500 \text{ L}} \times \frac{1 \text{ mol}}{156 \text{ g}} = 1.768 \times 10^{-2} \text{ M } C_2H_5I$$

Rate = $(8.956 \times 10^{-5}$ 1/Ms$)(1.692 \times 10^{-2}$ M$)(1.768 \times 10^{-2}$ M$) = 2.68 \times 10^{-8}$ M/s.

Chapter Fifteen—Chemical Equilibrium

Workbook Problems

Workbook Problem 15.1

Step 1: Write the equation for the equilibrium constant:

$$K_c = \frac{[H][A]}{[HA]}$$

Step 2: Determine the concentration of each species:

If the solution is 1.73% ionized, the concentrations of H and A will be 0.0173×0.250 M $= 4.325 \times 10^{-3}$ M. The concentration of HA will be 0.245675 M.

Step 3: Solve for K_c:

$$K_c = \frac{(4.325 \times 10^{-3})(4.325 \times 10^{-3})}{(0.245675)} = 7.61 \times 10^{-5}$$

Workbook Problem 15.2

$$K_c = \frac{[PCl_3][Cl_2]}{[PCl_5]}$$

Step 1: Solve for K_c, using the molar concentrations given.

$$K_c = \frac{(1.56)(1.56)}{(2.35)} = 1.04$$

Step 2: Determine Δn.

$$\Delta n = 2 - 1 = 1$$

Step 3: Solve for K_p, using the equation that expresses the relationship between K_c and K_p.

$$K_p = (1.04)\left[\left(\frac{0.0821\ \text{L}\cdot\text{atm}}{\text{mol}\cdot\text{K}}\right)(250\ \text{K})\right]^{1} = 21.3$$

Workbook Problem 15.3

Step 1: Write the balanced equation for the reaction.

$$C(s) + H_2O(g) \leftrightarrows CO(g) + H_2(g)$$

Step 2: Make a table listing the initial concentration, the change in concentration, and the equilibrium concentration. (Let x = the amount of substance that reacts.)

Principal Reaction	$C(s)$	+	$H_2O(g)$	$\leftrightarrows$	$CO(g)$	+	$H_2(g)$
Initial Concentration (M)			11.75 M		0		0
Change (M)			$-x$		$+x$		$+x$
Equilibrium Conc. (M)			$11.75 - x$		$+x$		$+x$

Step 3: From the balanced equation, write the equilibrium equation. Substitute the algebraic expressions for the equilibrium concentrations into the equilibrium equation, and solve for x. Use the quadratic equation, if necessary.

$$K_c = 3.0\times10^{-2} = \frac{[CO][H_2]}{[H_2O]} = \frac{(x)(x)}{(11.75 - x)}$$

$$3.0\times10^{-2}(11.75 - x) = (x)^2$$

$$x^2 + \left(3.0\times10^{-2}\right)x - 0.353 = 0$$

$$x = \frac{-\left(3.0\times10^{-2}\right)^2 \pm \sqrt{\left(3.0\times10^{-2}\right) - 4(-0.353)}}{2}$$

$$x = \frac{-\left(3.0\times10^{-2}\right) \pm \sqrt{\left(3.0\times10^{-2}\right)^2 - 4(1)(-0.353)}}{2(1)}$$

$x = 0.593$ or $x = -0.596$. The meaningful solution is $x = 0.593$. Substituting the value of x back into the equilibrium concentrations, we have:

$[CO] = [H_2] = 0.593$ M $\qquad [H_2O] = 11.75 - 0.593 = 11.16$ M

Workbook Problem 15.4

Step 1: Determine the direction of reaction needed to minimize each stress.

BrNO added: reaction will shift away from BrNO to the right, toward products.

BrNO removed: reaction will shift toward BrNO to the left, toward reactants.

NO added: reaction will shift away from NO, to the left, toward reactants.

Br_2 removed: reaction will shift toward Br_2, to the right, toward products.

Workbook Problem 15.5

Step 1: Determine the direction of reaction needed to minimize each stress.

pressure increased: reaction will shift toward N_2O_4; to the left, toward reactants.

pressure decreased: reaction will shift away from N_2O_4; to the right, toward products.

Ne added: equilibrium will be unaffected.

Workbook Problem 15.6

Step 1: Determine the direction of reaction needed to minimize each stress and its effect on $[CO_2]$.

Increased pressure: increased $[CO_2]$ (3 mol of gas on left, 2 mol on right)

Decreased pressure: decreased $[CO_2]$

Removed catalyst: decreased $[CO_2]$

Lowered temperature: increased $[CO_2]$

Added O_2: increased $[CO_2]$

Self-Test

True–False

1. T
2. F Reaction needs to be run at different concentrations, not temperatures.
3. F Equilibrium constants are unitless.
4. T
5. F There will be more products than reactants.
6. F The reaction will go in the direction of fewer moles of gas.
7. T
8. T
9. T
10. T

Multiple Choice

11. d	12. b	13. c	14. a
15. d	16. a	17. d	18. c
19. a	20. b		

Matching

21. g 22. d 23. a 24. j
25. b 26. h 27. c 28. i
29. f 30. e

Fill in the Blanks

31. dynamic, concentration, constant
32. heterogeneous, liquids, solids
33. minimize, products/right, reactants/left, temperature, equilibrium, catalyst, increases
34. direction, equilibrium, products, reactants, reactants, products
35. lowering

Problems

36. a. $K_p = \dfrac{[O_2][ClO]}{[O_3][Cl]}$ b. $K_p = \dfrac{[NOBr]^2}{[NO]^2[Br_2]}$

c. $K_p = \dfrac{1}{[O_2]}$ d. $K_p = \dfrac{[N_2]^2[O_2]}{[N_2O]^2}$

37. $K_p = \dfrac{(p_{NO})^2}{(p_{N_2})(p_{O_2})}$ $K_p = \dfrac{(0.045)^2}{(0.27)(0.187)} = 4.0 \times 10^{-2}$

38. $K_c = \dfrac{K_p}{(RT)^{\Delta n}}$ $\Delta n = 2 - 2 = 0$; $K_c = K_p = 4.0 \times 10^{-2}$

39. a. mainly products
 b. slightly more products, reactants also present
 c. mainly reactants

40. To determine if a system is at equilibrium, you must first solve for the value of Q_c and then compare it to the value of K_c.

$$Q_c = \frac{[PCl_3][Cl_2]}{[PCl_5]} \qquad Q_c = \frac{(0.83)(0.83)}{(2.43 \times 10^{-3})} = 2.83 \times 10^2$$

Since $Q_c > K_c$, the system is not at equilibrium. The reaction will shift from right to left to achieve equilibrium.

41. To solve this problem, make a table of initial, change, and final concentrations. The equilibrium concentrations of N_2O and O_2 can be determined by allowing x to be the concentration of N_2O that reacts on going to the equilibrium state. Since the mole ratios are 2:3:4, if $2x$ mol/L of N_2O reacts, then $3x$ mol/L of O_2 reacts and $4x$ mol/L of NO_2 are produced.

This can be summarized in the following table.

Principal Reaction	$2\ N_2O(g)$	+	$3\ O_2(g)$	→	$4\ NO_2(g)$
Initial Concentration (M)	0.0342		0.0415		0
Change (M)	$-2\,x$		$-3\,x$		$+4\,x$
Eq Concentration (M)	$0.0342 - 2x$		$0.0415 - 3x$		$+4\,x$

We know that the equilibrium concentration of NO_2 is 0.0298 M, and we can, therefore, solve immediately for the value of x. $4\,x = 0.0298;\ x = 7.45 \times 10^{-3}$

The value of x can be substituted back into the expressions for the equilibrium concentrations of N_2O and O_2.

$$[N_2O] = 0.0342\text{ M} - (2 \times 7.45 \times 10^{-3}\text{ M}) = 1.93 \times 10^{-2}\text{ M};$$

$$[O_2] = 0.0415\text{ M} - (3 \times 7.45 \times 10^{-3}\text{ M}) = 1.92 \times 10^{-2}\text{ M}$$

Now calculate K_c by substituting the equilibrium concentrations of all species into the equilibrium expression.

$$K_c = \frac{[NO_2]^4}{[N_2O]^2[O_2]^3} = \frac{\left(7.45\times10^{-3}\right)^4}{\left(1.93\times10^{-2}\right)^2\left(1.91\times10^{-2}\right)^3} = 1.19$$

42. The initial concentrations of SO_2 and NO_2 are 0.100 M. Let x be the concentration of SO_2 that reacts on going to the equilibrium state. Since the mole ratios are 1:1:1:1, if x mol/L of SO_2 reacts, then x mol/L of NO_2 also reacts and x mol/L of SO_3 and NO are produced. This can be summarized in the following table:

Principal Reaction	$SO_2(g)$	$NO_2(g)$	⇆	$SO_3(g)$ + $NO(g)$
Initial Concentration (M)	0.100	0.100		0
Change	$-x$	$-x$		$+x$
Eq Concentration (M)	$0.100 - x$	$0.100 - x$		$+x$

The algebraic expressions for the equilibrium concentrations from the table can be substituted into this expression, and we can solve for x.

$$K_c = \frac{[NO][SO_3]}{[SO_2][NO_2]}$$

$$85.0 = \frac{(x)(x)}{(0.100 - x)(0.100 - x)} = \left(\frac{x}{0.100 - x}\right)^2$$

Taking the square root of both sides gives:

$$\pm 9.22 = \frac{x}{0.100 - x}$$

Solving for x, there are two possible solutions. The equation with the positive square root of 85 gives:

$+9.22(0.100 - x) = x$ $\quad$ $+0.922 = x + 9.22x$ $\quad$ $x = \frac{0.922}{10.22} = 9.02 \times 10^{-2}$

The equation with the negative square root of 85 gives:

$-9.22(0.100 - x) = x;$ $\quad$ $-0.922 = x - 9.22x$ $\quad$ $x = \frac{-0.922}{-8.22} = 0.112$

Because the initial concentrations of SO_2 and NO_2 are 0.100, x cannot exceed 0.100. Therefore, $x = 0.112$ can be discarded as unreasonable and $x = 9.02 \times 10^{-2}$ is the solution. The equilibrium concentrations are:

$[SO_2] = [NO_2] = 0.100 - (9.02 \times 10^{-2}) = 0.0098$ M; $\quad$ $[SO_3] = [NO] = 0.0902$ M

43. a. reaction shifts to the left
    b. shifts left (reaction that absorbs heat, reverse reaction generates heat)
    c. shifts to the left (greater number of moles of gas)
    d. no effect
    e. shifts to the right

44. $K_c = \frac{k_f}{k_r}$ $\quad$ $K_c = \frac{6.8 \times 10^{-5}}{7.2 \times 10^{-7}} = 94$

45. The initial concentration of $HCONH_2$ is 0.125 M. Let x be the concentration of $HCONH_2$ that reacts on going to the equilibrium state. Since the mole ratios are 1:1:1, if x mol/L of $HCONH_2$ reacts, then x mol/L of NH_3 and CO are produced. This can be summarized in the following table:

Principal Reaction	$HCONH_2(g)$	$\leftrightarrows$	$NH_3(g)$	+	$CO(g)$
Initial Concentration (M)	0.125		0		0
Change	$-x$		$+x$		$+x$
Eq Concentration (M)	$0.125 - x$		$+x$		$+x$

The equilibrium expression for the reaction is $K_c = \frac{[NH_3][CO]}{[HCONH_2]}$. The equilibrium concentrations from the table can be substituted into this expression.

$$4.84 = \frac{(x)(x)}{(0.125 - x)} = \frac{x^2}{0.125 - x}$$

To solve for x, we need to rearrange this expression and use the quadratic equation.

$$4.84(0.125 - x) = x^2$$

$$x^2 + 4.84x - 0.605 = 0;$$

$$x = \frac{-4.84 \pm \sqrt{(4.84)^2 - 4(-0.605)}}{2} \qquad x = -4.96 \text{ or } 0.122$$

The mathematical solution that makes chemical sense is 0.122. This can be used to calculate the equilibrium concentrations.

$$[HCONH_2] = 0.125 - 0.122 = 0.003\text{ M} \qquad [NH_3] = [CO] = 0.122\text{ M}$$

46. The position of equilibrium is to the left, given the very small value of K_c. We can calculate the value of Q_c from the concentrations given.

$$Q_c = \frac{[H_2]^4[CS_2]}{[H_2S]^2[CH_4]} = \frac{(0.450)^4(0.085)}{(2.00)^2(1.75)} = 4.98 \times 10^{-4}$$

$Q_c > K_c$; therefore, the reaction mixture is not at equilibrium. The reaction will proceed from right to left to reach equilibrium.

Challenge Problem

47. The equilibrium expression is

$$K_c = \frac{[N_2O][O_2]}{[NO_2][NO]}.$$

The initial concentrations (mol/L) for each species are

$[NO_2] = 0.100$ M $[NO] = 0.050$ M $[N_2O] = 0.0125$ M $[O_2] = 0.00125$ M.

The mole ratio of NO_2 to NO is 1:1. Therefore, the amount of NO_2 that reacts equals the amount of NO that reacts. Define this amount as x. We can summarize the initial and equilibrium concentrations in the following table:

Principal Reaction	$NO_2(g)$ +	$NO(g)$ ⇆	$N_2O(g)$ +	$O_2(g)$
Initial Concentration (M)	0.100	0.050	0.0125	0.00125
Change (M)	$-x$	$-x$	$+x$	$+x$
Eq. Concentration (M)	$0.100 - x$	$0.050 - x$	$0.0125 + x$	$0.00125 + x$

The algebraic equations for the equilibrium concentrations can be substituted back into the equation for the equilibrium constant:

$$0.914 = \frac{(0.0125 + x)(0.00125 + x)}{(0.100 - x)(0.050 - x)}$$

$$0.914 = \frac{\left(1.56\times10^{-5} + 0.0138x + x^2\right)}{\left(0.0050 - 0.150x + x^2\right)}$$

$$0.00457 - 0.137x + 0.914x^2 = 1.56\times10^{-5} + 0.0138x + x^2$$

$$0 = 0.086x^2 + 0.151x - 0.00455$$
$$x^2 + 1.76x - 0.0529 = 0$$

$$x = \frac{-1.76 \pm \sqrt{1.76^2 - 4(-0.0529)}}{2}$$

$$x = 0.0296, -1.80$$

Substituting the value of x into the equilibrium concentrations leads to:

$[NO_2]_e = 0.100 - 0.0296 = 0.0704$ M
$[N_2O]_e = 0.0125 + 0.0296 = 0.0421$ M
$[NO]_e = 0.050 - 0.0296 = 0.0204$ M
$[O_2]_e = 0.00125 + 0.0296 = 0.0308$ M

Chapter Sixteen—Aqueous Equilibria: Acids and Bases

Workbook Problems

Workbook Problem 16.1

Step 1: Write the reactions in the forward direction.

$HNO_2(aq) + H_2O(l) \leftrightarrows NO_2^-(aq) + H_3O^+(aq)$

$SO_4^{-2}(aq) + H_2O(l) \leftrightarrows HSO_4^-(aq) + OH^-(aq)$

Step 2: Identify the conjugate acid and conjugate base.

Reaction 1: acid = HNO_2, conjugate base = NO_2^-

Reaction 2: base = SO_4^{-2}, conjugate acid = HSO_4^-

Workbook Problem 16.2

Step 1: Write the reactions, transferring protons from the stronger acid to the stronger base:

$$HCl(aq) + CO_3^{2-}(aq) \leftrightarrows Cl^-(aq) + HCO_3^-(aq)$$

$$CN^-(aq) + H_3PO_4(aq) \leftrightarrows HCN(aq) + H_2PO_4^{2-}(aq)$$

Workbook Problem 16.3

Step 1: Use K_w to calculate the ion concentrations:

$$1 \times 10^{-14} = [OH^-](2.3 \times 10^{-5})$$

$$\left[OH^-\right] = \frac{1\times10^{-14}}{2.3\times10^{-5}} = 4.35\times10^{-10}$$

This solution is acidic: the concentration of hydronium ions is much larger than the concentration of hydroxide ions.

$$1 \times 10^{-14} = [H_3O^+](5.73 \times 10^{-9})$$

$$\left[H_3O^+\right] = \frac{1\times10^{-14}}{5.73\times10^{-9}} = 1.75\times10^{-6}$$

This solution is also acidic.

Workbook Problem 16.4

Step 1: Calculate the $[H_3O^+]$, using the expression for the dissociation of water.

$$1\times10^{-14} = \left[H_3O^+\right]\left(2.35\times10^{-6}\right)$$

$$\left[H_3O^+\right] = \frac{1\times10^{-14}}{2.35\times10^{-6}} = 4.25\times10^{-9}$$

Step 2: Calculate the pH.

$$pH = -\log(4.25 \times 10^{-9}) = 8.37$$

Step 3: Determine if the solution is acidic, basic, or neutral.

The pH is higher than 7, so the solution is basic.

Workbook Problem 16.5

Step 1: Determine the concentration of $[OH^-]$.

KOH is a strong base and exists in aqueous solution as K^+ and OH^-. Therefore, $[OH^-]$ = initial concentration of KOH.

$[OH^-] = 7.2 \times 10^{-3}$ M

Step 2: Calculate the $[H_3O^+]$.

The H_3O^+ concentration is calculated from the OH^- concentration.

$$\left[H_3O^+\right] = \frac{K_w}{\left[OH^-\right]} = \frac{1.0 \times 10^{-14}}{7.2 \times 10^{-3}} = 1.4 \times 10^{-12}\ M$$

Step 3: Calculate the pH.

$$pH = -\log(1.4 \times 10^{-12}) = 11.85$$

Workbook Problem 16.6

Step 1: Determine the species present initially.

$HC_3H_5O_3$ and H_2O

Step 2: Write the proton-transfer reaction.

$HC_3H_5O_3(aq) + H_2O(l) \leftrightarrows H_3O^+(aq) + C_3H_5O_3^-(aq)$ $K_a = 1.4 \times 10^{-4}$

Step 3: Make a table showing the principal reaction and the initial and equilibrium concentrations.

Principal Reaction	$HC_3H_5O_3(aq)$	$\rightleftharpoons$	$H_3O^+(aq)$	+	$C_3H_5O_3^-(aq)$
Initial Concentration	1.63		0		0
Change	$-x$		$+x$		$+x$
Eq. Concentration	$1.63 - x$		$+x$		$+x$

Step 4: Substitute the equilibrium concentrations into the equilibrium equation, and solve for x. (Since $K_a < 1.0 \times 10^{-3}$, assume x is negligible and that $1.63 - x \cong 1.63$).

$$K_a = \frac{\left[H_3O^+\right]\left[C_3H_5O_3^-\right]}{\left[HC_3H_5O_3\right]} = \frac{(x)(x)}{(1.63 - x)}$$

$$1.4 \times 10^{-4} = \frac{x^2}{1.63} \qquad x = 1.51 \times 10^{-2}$$

Step 5: Calculate the equilibrium concentrations.

$$[H_3O^+]_e = [C_3H_5O_3^-]_e = 1.51 \times 10^{-2}\ M$$

$$[HC_3H_5O_3]_e = 1.63 - (1.51 \times 10^{-2}) = 1.61\ M$$

Step 6: Calculate the pH.

$$pH = -\log(1.51 \times 10^{-2}) = 1.82$$

Workbook Problem 16.7

Step 1: Determine the species present initially, and write the proton-transfer reaction.

The species present initially are C_6H_5COOH and H_2O (acid or base).

The proton-transfer reaction is as follows:

$C_6H_5COOH(aq) + H_2O(l) \leftrightarrows H_3O^+(aq) + C_6H_5COO^-(aq)$ $\quad K_a = 6.5 \times 10^{-5}$

Step 2: Make a table showing the principal reaction, initial concentration, change in concentration, and the equilibrium concentration.

Principal Reaction	$C_6H_5COOH(aq)$	$\leftrightarrows$	H_3O^+	+	$C_6H_5COO^-$
Initial Concentration	0.79		0		0
Change	$-x$		$+x$		$+x$
Equilibrium Concentration	$0.79 - x$		$+x$		$+x$

Step 3: Substitute the equilibrium concentrations into the equilibrium expression, and solve for x.

$$K_a = 6.5 \times 10^{-5} = \frac{[H_3O^+][C_6H_5COO^-]}{[C_6H_5COOH]} = \frac{(x)(x)}{(0.79 - x)}$$

Given the value of K_a, assume that x is negligible compared with the initial concentration of the acid; therefore, $0.79 - x \approx 0.79$. Using this value in the denominator, solve for x.

$$x^2 \approx (6.5 \times 10^{-5})(0.79)$$
$$x \approx 7.2 \times 10^{-3}$$

Step 4: Calculate the equilibrium concentrations of the major species present.

The equilibrium concentrations of H_3O^+ and $C_6H_5COO^-$ are $[H_3O^+] = [C_6H_5COO^-] = 0.0072$ M. This concentration also represents the amount of C_6H_5COOH that dissociated.

Step 5: Calculate the percent dissociation.

The percent dissociation can now be calculated.

$$\frac{0.0072}{0.79} \times 100 = 0.91\%$$

Workbook Problem 16.8

Step 1: Write the stepwise dissociation for arsenic acid.

The stepwise dissociation for arsenic acid is as given below:

$$H_3AsO_4(aq) + H_2O(l) \leftrightarrows H_3O^+(aq) + H_2AsO_4^-(aq)$$
$$H_2AsO_4^-(aq) + H_2O(l) \leftrightarrows H_3O^+(aq) + HAsO_4^{2-}(aq)$$
$$HAsO_4^{2-}(aq) + H_2O(l) \leftrightarrows H_3O^+(aq) + AsO_4^{3-}(aq)$$

Step 2: Determine the principal reaction.

Since $K_{a1} > K_{a2}$, K_{a3}, and K_w, we know that the principal reaction is the first dissociation step and that all of the H_3O^+ present is produced from this step. Therefore, we only need to consider the first dissociation step when calculating the pH of the solution.

Step 3: Make a table showing the principal reaction, initial concentration, change in concentration, and equilibrium concentration of the reactant and products.

Principal Reaction	$H_3AsO_4(aq)$	$\leftrightarrows$	H_3O^+	+	$H_2AsO_4^-$
Initial Concentration	1.69×10^{-3}		0		0
Change	$-x$		$+x$		$+x$
Equilibrium Concentration	$(1.69 \times 10^{-3}) - x$		$+x$		$+x$

Step 4: Substitute the equilibrium concentrations into the equilibrium expression.

Because K_a is so close to the concentration of the acid, it is unlikely that x will be negligible.

If it is assumed to be negligible, the result is as follows:

$$5.62 \times 10^{-3} = \frac{x^2}{1.69 \times 10^{-3}} \qquad x = 3.08 \times 10^{-3}$$

The quadratic equation must be used to solve for the equilibrium concentration of H_3O^+. Using $(1.69 \times 10^{-4}) - x$ and rearranging the equilibrium expression gives:

$$5.62 \times 10^{-3}[(1.69 \times 10^{-3}) - x] = x^2 \qquad x^2 + (5.62 \times 10^{-3})x - (9.50 \times 10^{-6})$$

$$x = \frac{-\left(5.62 \times 10^{-3}\right) \pm \sqrt{\left(5.62 \times 10^{-3}\right)^2 - 4\left(-9.50 \times 10^{-6}\right)}}{2}$$

$$x = 1.36 \times 10^{-3}$$

Step 5: Calculate the pH of the solution.

$$\text{pH} = -\log(1.36 \times 10^{-3}) = 2.87$$

Workbook Problem 16.9

Step 1: Write the principal reaction.

$K_b > K_w$; therefore, the principal reaction is

$$(CH_3)_3N(aq) + H_2O(l) \leftrightarrows (CH_3)_3NH^+(aq) + OH^-(aq).$$

Step 2: Construct a table with concentrations of the reactant and products.

Principal Reaction	$(CH_3)_3N(aq)$	$\leftrightarrows$	$(CH_3)_3N^+$	+	OH^-
Initial Concentration	0.975		0		0
Change	$-x$		$+x$		$+x$
Equilibrium Concentration	$0.975 - x$		$+x$		$+x$

Step 3: Substitute the equilibrium concentrations into the equilibrium expression, and solve for x.

$$K_a = 6.5 \times 10^{-5} = \frac{\left[(CH_3)_3NH^+\right]\left[OH^-\right]}{\left[(CH_3)_3N\right]} = \frac{(x)(x)}{0.975 - x}$$

Assume that x is negligible compared with the initial concentration of the base; therefore, $0.975 - x \approx x$. Using this value in the denominator, we can now solve for x. Remember: x represents the equilibrium concentration of the OH^- ion.

$$6.5 \times 10^{-5} = \frac{x^2}{0.975} \qquad x = 0.0080$$

Step 4: Determine the equilibrium concentrations of the species present.

The equilibrium concentrations are $[(CH_3)_3N] = 0.975 - 0.0080 = 0.967$ M; $[OH^-] = [(CH_3)_3NH^+] = 0.0080$ M.

Step 5: Calculate the pH of the solution.

Knowing the equilibrium concentration of OH^-, we can calculate the H_3O^+ concentration from the K_w expression.

$$\left[H_3O^+\right] = \frac{1.0 \times 10^{-14}}{0.0080} = 1.3 \times 10^{-12}$$

$$pH = -\log(1.3 \times 10^{-12}) = 11.90$$

Workbook Problem 16.10

Step 1: Write the hydrolysis reaction for this salt.

$$OCl^-(aq) + H_2O(l) \leftrightarrows HOCl(aq) + OH^-(aq)$$

Step 2: Write an equilibrium expression for this reaction.

$$K_b = \frac{[HOCl][OH^-]}{[OCl^-]} \qquad K_b = \frac{1.0\times10^{-14}}{K_a(HOCl)} = \frac{1.0\times10^{-14}}{3.5\times10^{-8}} = 2.9\times10^{-7}$$

Step 3: Construct a table.

Principal Reaction	$OCl^-(aq)$	$\leftrightarrows$	$HOCl(aq)$	+ OH^-
Initial Concentration	0.137		0	0
Change	$-x$		$+x$	$+x$
Equilibrium Concentration	$0.137 - x$		$+x$	$+x$

Step 4: Substitute the equilibrium concentrations into the equilibrium expression.

Assume that x is negligible; therefore, $0.137 - x \approx 0.137$.

$$2.9\times10^{-7} = \frac{(x)(x)}{0.137} \qquad x = 2.0\times10^{-4}$$

Step 5: Calculate the pH of the solution.

The equilibrium concentration of OH^- is 2.0×10^{-4}. Determine the H_3O^+ concentration to calculate pH.

$$[H_3O^+] = \frac{1.0\times10^{-14}}{2.0\times10^{-4}} = 5.0\times10^{-11}$$

$$pH = -\log(5.0\times10^{-11}) = 10.30$$

Self-Test

True–False

1. F The Lewis theory deals with electron transfers, but an electron acceptor is an acid, and an electron donor is a base.
2. F A strong acid fully dissociates. A concentrated solution of a weak acid can be far more harmful than a dilute solution of a strong acid.
3. T
4. T
5. F The strength of an acid and the strength of its conjugate base are inversely related (the stronger the acid, the weaker the conjugate base).
6. T

7. F The principal reaction for a diprotic acid is the dissociation of H_2A.
8. T
9. T
10. F $Mg(OH)_2$ is a strong base.

Multiple Choice

11. c	12. d	13. a	14. b
15. c	16. d	17. b	18. a

Matching

19. f	20. a	21. j	22. d
23. g	24. b	25. k	26. c
27. m	28. o	29. e	30. n
31. h	32. i	33. l	34. q
35. p			

Fill in the Blanks

36. weak, proton
37. hydrogen or hydronium, hydroxide, equal
38. increases, decreases
39. stepwise, smaller
40. electrons, hydrogen

Problems

41. We first need to solve for the concentration of the H_3O^+ ion, then solve for $[OH^-]$.

$$[H_3O^+] = 10^{(-11.78)} = 1.66 \times 10^{-12}\text{ M} \quad \left[OH^-\right] = \frac{1.0\times10^{-14}}{1.66\times10^{-12}} = 6.0\times10^{-3}\text{ M}$$

42. $Ba(OH)_2$ is a strong base and completely dissociates in water according to the reaction:

$$Ba(OH)_2(aq) \rightarrow Ba^{2+}(aq) + 2\ OH^-(aq)$$

Determine the molar concentration of OH^-.

$$1.83\text{ g Ba}(OH)_2 \times \frac{1\text{ mol Ba}(OH)_2}{171.3\text{ g Ba}(OH)_2} \times \frac{2\text{ mol }OH^-}{\text{mol Ba}(OH)_2} = 2.14\times10^{-2}\text{ mol }OH^-;$$

$$\frac{2.14\times10^{-2}\text{ mol }OH^-}{0.150\text{ L}} = 0.142\text{ M}$$

$$\left[H_3O^+\right] = \frac{1.0\times10^{-14}}{0.142} = 7.02\times10^{-14};\ \text{pH} = -\log(7.02\times10^{-14}) = 13.15$$

43. $0.51\% = \dfrac{[\text{acid}]\text{ dissociated}}{7.35\times10^{-3}}\times 100$ [acid] dissociated = 3.75×10^{-5}

The dissociated acid represents the $[H_3O^+]$ and $[A^-]$ at equilibrium.

$$pH = -\log(3.75 \times 10^{-5}) = 4.426$$

The equilibrium concentration of acid is $7.35 \times 10^{-3} - 3.75 \times 10^{-5} = 7.31 \times 10^{-3}$ M:

$$K_a = \frac{(3.75\times10^{-5})(3.75\times10^{-5})}{(7.31\times10^{-3})} = 1.92\times10^{-7}$$

44. Calculate $[H_3O^+]$ at equilibrium from the pH.

$$[H_3O^+] = 10^{(-6.03)} = 9.33 \times 10^{-7}.$$

This concentration also represents the amount of histidine that dissociated, which can be used to calculate the equilibrium concentration of histidine.

$$(1.30 \times 10^{-3}) - (9.33 \times 10^{-7}) = 1.30 \times 10^{-3}$$

$[H_3O^+] = [A^-]$ at equilibrium. Calculate K_a.

$$K_a = \frac{(9.33\times10^{-7})(9.33\times10^{-7})}{(1.30\times10^{-3})} = 6.70\times10^{-10}$$

45. This is an equilibrium problem.

Principal Reaction	$C_6H_5N(aq) + H_2O(l) \leftrightarrows$	$C_6H_5NH^+(aq) +$	$OH^-(aq)$
Initial Concentration	1.24	0	0
Change	$-x$	$+x$	$+x$
Equilibrium Concentration	$1.24 - x$	$+x$	$+x$

Substituting these values into the equilibrium expression gives:

$$K_b = 1.8\times10^{-9} = \frac{[C_6H_5NH^+][OH^-]}{[C_6H_5N]} = \frac{(x)(x)}{(1.24-x)}$$

Assume that $1.24 - x \approx 1.24$. The above expression then simplifies to:

$$x^2 = (1.8\times10^{-9})(1.24);\ x = 4.72\times10^{-5}\text{ M}$$

$$[OH^-] = x = 4.72\times10^{-5}$$

Solve for $[H_3O^+]$ from $[OH^-]$.

$$\left[H_3O^+\right]=\frac{1.0\times10^{-14}}{4.72\times10^{-5}}=2.12\times10^{-10}\text{ M}$$

Calculate pH.

$$pH = -\log(2.12\times 10^{-10}) = 9.67$$

46. Oxalic acid is a diprotic acid and has two possible dissociations:

Principal Reaction	$H_2C_2O_4(aq) + H_2O(l) \leftrightarrows$	$H_3O^+(aq) +$	$HC_2O_4^-(aq)$
Initial Concentration	0.125	0	0
Change	$-x$	$+x$	$+x$
Equilibrium Concentration	$0.125 - x$	$+x$	$+x$

Substituting these values into the equilibrium expression gives:

$$K_{a1}=5.9\times10^{-2}=\frac{\left[H_3O^+\right]\left[HC_2O_4^-\right]}{\left[H_2C_2O_4\right]}=\frac{(x)(x)}{(0.125-x)}$$

Due to the value of K_{a1}, we cannot assume that the value of x is negligible. Therefore, we must use the quadratic equation to solve for x. Rearranging the above expression gives:

$$\left(7.375\times10^{-3}\right)-\left(5.9\times10^{-2}\right)x=x^2$$
$$x^2+\left(5.9\times10^{-2}\right)x-\left(7.375\times10^{-3}\right)=0$$

$$x=\frac{-\left(5.9\times10^{-2}\right)\pm\sqrt{\left(5.9\times10^{-2}\right)^2-4\left(-7.375\times10^{-3}\right)}}{2}$$

$$x=6.13\times10^{-2}$$

$$[H_2C_2O_4] = 0.125 - 0.0613 = 0.064\text{ M};\quad [H_3O^+] = [HC_2O_4^-] = x = 0.0613$$

To calculate the $[C_2O_4^{2-}]$, use the second dissociation step of $H_2C_2O_4$, remembering that the dissociation takes place in the presence of 0.0613 M H_3O^+ from the first dissociation.

$$HC_2O_4^-(aq) + H_2O(l) \leftrightarrows H_3O^+(aq) + C_2O_4^{2-}(aq)$$

$$K_{a2}=6.4\times10^{-5}=\frac{\left[H_3O^+\right]\left[C_2O_4^{2-}\right]}{\left[HC_2O_4^-\right]}$$

Use the concentration of $HC_2O_4^-$ calculated above in the calculation of $[C_2O_4^-]$. The quantities of $C_2O_4^-$ generated will be negligible compared to the ions already present.

$$6.5\times10^{-5}=\frac{(0.061+x)x}{0.061-x}=\frac{0.061x}{0.061}$$

$$x=6.5\times10^{-5}$$

$$[C_2O_4^{2-}]=x=6.5\times10^{-5}\text{ M}$$

$$\text{pH}=-\log(0.0613)=1.21$$

47. Let the formula B represent the formula for benzylamine.

Principal Reaction	$B(aq)$ +	$H_2O(l)$ ⇆	$BH^+(aq)$ +	$OH^-(aq)$
Initial Concentration	0.350		0	0
Change	−x		+x	+x
Equilibrium Concentration	0.350 − x		+x	+x

Substituting these values into the equilibrium expression gives:

$$K_b=2.14\times10^{-5}=\frac{[BH^+][OH^-]}{[B]}=\frac{(x)(x)}{(0.350-x)}$$

Assume that $0.350-x\approx0.350$. The above expression then simplifies to:

$x^2=(2.14\times10^{-5})(0.350)$ $\qquad$ $x=2.74\times10^{-3}$

$$[OH^-]=2.74\times10^{-3}$$

$$[H_3O^+]=\frac{1.0\times10^{-14}}{2.74\times10^{-3}}=3.65\times10^{-12}$$

$$\text{pH}=-\log(3.65\times10^{-12})=11.44$$

48. a. Ascorbic acid: $K_b=\dfrac{1.0\times10^{-14}}{8.0\times10^{-5}}=1.2\times10^{-10}$

b. Hydrazine: $K_a=\dfrac{1.0\times10^{-14}}{8.9\times10^{-7}}=1.1\times10^{-8}$

49. KCN is a salt derived from a weak acid (HCN) and a strong base (KOH). This solution will be basic. The hydrolysis reaction is given below:

$$CN^-(aq)+H_2O(l)\leftrightarrows HCN(aq)+OH^-(aq)$$

The K_b for this reaction can be calculated from K_w and K_a for HCN:

$$K_b = \frac{1.0 \times 10^{-14}}{4.9 \times 10^{-10}} = 2.04 \times 10^{-5}$$

To determine the HCN equilibrium concentration and the pH of the solution, determine the initial concentration of CN^-:

$$1.75 \text{ g KCN} \times \frac{1 \text{ mol KCN}}{65.1 \text{ g KCN}} \times \frac{1 \text{ mol } CN^-}{1 \text{ mol KCN}} = 0.0269 \text{ mol } CN^-$$

$$\frac{0.0269 \text{ mol } CN^-}{0.150 \text{ L}} = 0.179 \text{ M}$$

Principal Reaction	$CN^-(aq) + H_2O(l) \leftrightarrows$	$HCN(aq) +$	$OH^-(aq)$
Initial Concentration	0.179	0	0
Change	$-x$	$+x$	$+x$
Equilibrium Concentration	$0.179 - x$	$+x$	$+x$

Substituting these values into the equilibrium expression gives:

$$K_b = 2.04 \times 10^{-5} = \frac{[HCN][OH^-]}{[CN^-]} = \frac{(x)(x)}{(0.179 - x)}$$

Assume that $0.179 - x \approx 0.179$. The above expression then simplifies to:

$$x^2 = (2.04 \times 10^{-5})(0.179) \qquad x = 1.91 \times 10^{-3}$$

$$x = [HCN] = [OH^-] = 1.91 \times 10^{-3}$$

$$[H_3O^+] = \frac{1.0 \times 10^{-14}}{1.91 \times 10^{-3}} = 5.23 \times 10^{-12}$$

$$pH = -\log(5.23 \times 10^{-12}) = 11.281$$

50. a. Stronger acid: HI—For binary acids, acid strength increases down a group due to the decrease in bond strength because of the increase in the size of the atom.

b. Stronger acid: H_3PO_4—For oxoacids, acid strength increases with increasing number of oxygen atoms.

Challenge Problem

51. Assume 100 g of solution, which is 25 g of H_3PO_4. The volume of the solution can be calculated from the mass and density of the solution.

$$\text{mL of solution} = \frac{100\text{ g}}{1.1667\text{ g/mL}} = 85.7\text{ mL}$$

$$\text{mol } H_3PO_4 = 25\text{ g} \times \frac{1\text{ mol } H_3PO_4}{98\text{ g}} = 0.255\text{ mol}$$

$$M = \frac{0.255\text{ mol } H_3PO_4}{0.0857\text{ L}} = 2.98\text{ M}$$

Write the stepwise dissociation for phosphoric acid.

$H_3PO_4(aq) + H_2O(l) \leftrightarrows H_3O^+(aq) + H_2PO_4^-(aq)$ $\quad K_{a1} = 7.5 \times 10^{-3}$

$H_2PO_4^-(aq) + H_2O(l) \leftrightarrows H_3O^+(aq) + HPO_4^{2-}(aq)$ $\quad K_{a2} = 6.2 \times 10^{-8}$

$HPO_4^{2-}(aq) + H_2O(l) \leftrightarrows H_3O^+(aq) + PO_4^{3-}(aq)$ $\quad K_{a3} = 4.8 \times 10^{-13}$

Since $K_{a1} > K_{a2}$, K_{a3}, we know that the principal reaction is the first dissociation step and that essentially all of the H_3O^+ present is produced from this step. Therefore, we need only to consider the first dissociation step when calculating the pH of the solution.

Make a table showing the principal reaction, initial concentration, change in concentration, and equilibrium concentration of the reactant and products.

Principal Reaction	$H_3PO_4(aq) \leftrightarrows$	H_3O^+	$+\ H_2PO_4^-$
Initial Concentration	2.98	0	0
Change	−x	+x	+x
Eq. Concentration	2.98 − x	+x	+x

Substitute the equilibrium concentrations into the equilibrium expression. Since $K_{a1} > 1.0 \times 10^{-3}$, you cannot assume that x is negligible.

$$7.5 \times 10^{-3} = \frac{x^2}{2.98 - x}$$

$$x^2 + \left(7.5 \times 10^{-3}\right)x - \left(2.2 \times 10^{-2}\right) = 0$$

$$x = \frac{-\left(7.5 \times 10^{-3}\right) \pm \sqrt{\left(7.5 \times 10^{-3}\right)^2 + 4\left(2.2 \times 10^{-2}\right)}}{2}$$

$x = 0.145$ M. This value represents the equilibrium concentration of H_3O^+ and $H_2PO_4^-$.

Calculate the pH from this value.

$$pH = -\log(0.145) = 0.839$$

The equilibrium concentrations are as given below:

$$[H_3PO_4]_e = 2.98\ M - 0.145\ M = 2.83\ M$$
$$[H_3O^+]_e = [H_2PO_4^-]_e = 0.145\ M$$

We now repeat the process, using the chemical equation for $H_2PO_4^-$ and the equilibrium concentration calculated above.

Principal Reaction	$H_2PO_4^-(aq) \leftrightarrows$	$H_3O^+(aq)$	$+ HPO_4^{2-}(aq)$
Initial Concentration	0.145	0.145	0
Change	−x	+x	+x
Eq Concentration	0.145 − x	0.145 + x	+x

The value of K_{a2} is small enough that x is negligible; therefore, $0.145 - x \approx 0.145$ and $0.145 + x \approx 0.145$.

$$6.2 \times 10^{-8} = \frac{(0.145)x}{0.145}$$

$$x = 6.2 \times 10^{-8}$$

This value represents the equilibrium concentration of $[H_2PO_4^-]$.

$$[HPO_4^{2-}]_e = 6.2 \times 10^{-8}\ M$$

Repeat the process to calculate the equilibrium concentration of PO_4^{2-}.

Principal Reaction	$HPO_4^{2-}(aq)$	$\leftrightarrows$	$H_3O^+(aq)$ +	$PO_4^{3-}(aq)$
Initial Concentration	6.2×10^{-8}		0.145	0
Change	−x		+x	+x
Eq Concentration	$(6.2 \times 10^{-8}) - x$		0.145 + x	+x

The value of K_{a3} is small enough that x is negligible; therefore, $(6.2 \times 10^{-8}) - x \approx 6.2 \times 10^{-8}$.

$$4.8 \times 10^{-13} = \frac{(0.145)x}{6.2 \times 10^{-8}}$$

$$x = 2.05 \times 10^{-19}$$

This value of x represents the equilibrium concentration of PO_4^{3-}.

$$[PO_4^{3-}]_e = 2.05 \times 10^{-19}$$

Chapter Seventeen—Applications of Aqueous Equilibria

Workbook Problems

Workbook Problem 17.1

Step 1: Write the net neutralization reaction by writing individual reactions.

$HNO_2(aq) + H_2O(l) \leftrightarrows NO_2^-(aq) + H_3O^+(aq)$ $\quad K_a = 4.5 \times 10^{-4}$

$H_3O^+(aq) + OH^-(aq) \rightarrow H_2O(aq)$ $\quad 1/K_w = 1.0 \times 10^{14}$

$HNO_2(aq) + OH^-(aq) \rightarrow NO_2^-(aq) + H_2O(aq)$

Step 2: Calculate K_n based on the equilibrium constants for the individual reactions in Step 1.

$K_n = K_a \times 1/K_w = (4.5 \times 10^{-4}) \times (1.0 \times 10^{14}) = 4.5 \times 10^{10}$

Step 3: Determine the position of equilibrium from the value of K_n.

Based on the value of K_n, the equilibrium will lie far to the right. The reaction will go to completion, leaving the conjugate base of the weak acid in solution.

Step 4: Predict the pH.

The solution will be basic: the pH will be greater than 7.

Workbook Problem 17.2

Step 1: Write the overall reaction:

$H_3PO_3(aq) + H_2O(l) \leftrightarrows H_2PO_3^-(aq) + H_3O^+(aq)$

Step 2: Make a chart of initial and equilibrium concentrations:

Principal Reaction	$H_3PO_3(aq) \leftrightarrows$	$H_3O^+(aq)$	$H_2PO_3^-(aq)$
Initial Conc.	0.250	0	0.175
Change	$-x$	$+x$	$+x$
Equilibrium Conc.	$0.250 - x$	$+x$	$0.175 + x$

Step 3: Solve for the equilibrium concentrations of H_3O^+, $H_2PO_2^-$:

$$1.0 \times 10^{-2} = \frac{x(0.175 + x)}{0.250 - x}$$

$$x^2 + (0.175x + 0.010x) - 0.0025 = 0$$

$$x^2 + 0.185x - 0.0025 = 0$$

$$x = \frac{-0.185 \pm \sqrt{(0.185)^2 + 4(0.0025)}}{2}$$

$$x = 0.0126$$

$[H_3O^+] = 0.0126$ M

$[H_2PO_3^-] = 0.0126 + 0.175 = 0.188$ M

Step 4: Determine the pH of the solution:

$-\log(0.0126) = 1.90$

Step 5: Determine the percent dissociation of the acid:

$$\%\ \text{dissociation} = \frac{0.0126}{0.250} \times 100 = 5.04\%$$

Workbook Problem 17.3

Step 1: Determine the principal reaction and equilibrium concentrations.

The principal reaction and equilibrium concentration for this solution are shown in the table:

Principal Reaction	$HCOOH(aq) + H_2O(l) \leftrightarrows$	$H_3O^+(aq)$ +	$HCO_2^-(aq)$
Initial Conc.	0.35	0	0.25
Change	$-x$	$+x$	$+x$
Equilibrium Conc.	$0.35 - x$	$+x$	$0.25 + x$

Step 2: Using the equilibrium equation, solve for $[H_3O^+]$, assuming that the change (x) is negligible.

$$K_a = \frac{[H_3O^+][HCO_2^-]}{[HCOOH]}$$

$$[H_3O^+] = K_a \frac{[HCOOH]}{[HCO_2^-]}$$

K_a for formic acid equals 1.8×10^{-4}. Substituting the given information into the equilibrium expression gives:

$$[H_3O^+] = (1.8 \times 10^{-4})\frac{(0.35)}{(0.25)} = 2.5 \times 10^{-4}$$

Step 3: Solve for the pH.

$$pH = -\log (2.5 \times 10^{-4}) = 3.60$$

Step 4: Write the neutralization reaction that occurs on addition of HCl, and set up a table showing the number of moles present before and after the addition of HCl.

Neutralization Reaction	$HCO_2^-(aq)$ +	$H_3O^+(aq)$ ⇆	$HCOOH(aq)$ +	$H_2O(l)$
Before Reaction (mol)	0.25	0.10	0.35	
Change (mol)	−0.10	−0.10	+0.10	
After Reaction (mol)	0.15	0	0.45	

Step 5: Determine the concentrations of the buffer components after neutralization occurs.

Assuming that the volume of the solution does not change and is still 1 L, the concentrations of the buffer components after neutralization are as given below:

$$[HCO_2^-] = 0.15\ M$$

$$[HCOOH] = 0.45\ M$$

Step 6: Substitute the concentrations of the buffer components into the equilibrium expression and calculate the pH.

$$K_a = \frac{[H_3O^+][HCO_2^-]}{[HCOOH]}$$

$$[H_3O^+] = K_a \times \frac{[HCOOH]}{[HCO_2^-]}$$

$$[H_3O^+] = (1.8 \times 10^{-4}) \times \frac{0.45}{0.15} = 5.4 \times 10^{-4}$$

$$pH = -\log (5.4 \times 10^{-4}) = 3.27$$

Workbook Problem 17.4

Step 1: Calculate the pK_a of acetic acid.

$$-\log (1.8 \times 10^{-5}) = 4.74$$

Step 2: Use the Henderson–Hasselbalch equation to calculate the [base]/[acid] ratio.

$$\log\frac{[\text{base}]}{[\text{acid}]} = \text{pH} - \text{pK}_\text{a}$$

$$\log\frac{[base]}{[acid]} = 5.75 - 4.74 = 1.01$$

$$\frac{[base]}{[acid]} = 10^{1.01} = 10.2$$

Step 3: Determine the mass of sodium acetate needed.

$$10.2 \times 0.01 \text{ mol} = 0.102 \text{ mol sodium acetate}$$

$$0.102 \text{ mol} \times \frac{82.03 \text{ g}}{\text{mol}} = 8.37 \text{ g}$$

Workbook Problem 17.5

Step 1: Use the Henderson–Hasselbalch equation.

$$\log\frac{25}{75} = pH - 8.8$$

$$-0.477 = pH - 8.8$$

$$pH = 8.3$$

Workbook Problem 17.6

Step 1: Calculate the moles of OH^- and the moles of H_3O^+.

$$\text{Moles } OH^- = 0.05 \text{ L} \times 0.1 \text{ mol/L} = 0.005 \text{ mol } OH^-$$

$$\text{Moles } H_3O^+ = 0.015 \times 1 \text{ mol/L} = 0.015 \text{ mol } H_3O^+$$

Step 2: Determine the excess quantity and the resultant $[H_3O^+]$.

$$H_3O^+ \text{is in excess: } 0.015 - 0.005 = 0.010 \text{ mol } H_3O^+$$

$$[H_3O^+] = 0.010 \text{ mol}/0.065 \text{ L} = 0.154 \text{ M HCl}$$

Step 3: Determine the pH of the solution.

$$\text{pH} = -\log(0.154) = 0.813$$

Workbook Problem 17.7

Step 1: Calculate the moles of phenol present and thus the moles of OH^- needed:

$$0.100 \text{ L} \times 0.050 \text{ mol/L} = 0.0050 \text{ moles}$$

Step 2: Determine the volume of sodium hydroxide solution required:

$$0.0050 \text{ moles} / 0.1 \text{ mol/L} = 0.050 \text{ L} = 50 \text{ mL}$$

Step 3: Determine the total volume of the resultant solution:

$$100 \text{ mL original volume} + 50 \text{ mL NaOH solution} = 150 \text{ mL}$$

Step 4: Solve for the concentration of salt:

$$0.0050 \text{ mol}/0.150 \text{ L} = 0.033 \text{ M}$$

Step 5: Set up an equilibrium table.

Principal Reaction	$C_6H_5O^-(aq) + H_2O(l) \leftrightarrows$	$C_6H_5OH(aq) +$	$OH^-(aq)$
Initial Conc (mol)	0.033	0	~0
Change (mol)	$-x$	$+x$	$+x$
Equil Conc (mol)	$0.033 - x$	x	x

Step 6: Solve for K_b and [OH]:

$$K_b = \frac{K_w}{K_a} = \frac{1.0 \times 10^{-14}}{1.3 \times 10^{-10}} = 7.7 \times 10^{-5}$$

$$7.7 \times 10^{-5} = \frac{x^2}{(0.033 - x)} \quad \text{assume } 0.033 - x \approx 0.033$$

x = 0.00159 M (assumption was correct)

Step 7: Solve for $[H_3O^+]$ and pH:

$$\left[H_3O^+\right] = \frac{1 \times 10^{-14}}{0.00159} = 6.3 \times 10^{-12}$$

$$pH = -\log 6.3 \times 10^{-12} = 11.20$$

Workbook Problem 17.8

a) milliliters of acid required to reach the equivalence point:

Step 1: Write the neutralization reaction that occurs, and determine the number of mmoles of $(CH_3)_3N$ in the initial solution.

$$(CH_3)_3N(aq) + HCl(aq) \leftrightarrows (CH_3)_3NH^+(aq) + Cl^-(aq)$$

$$\text{mol }(CH_3)_3N = 0.040\text{ L} \times 0.375\frac{\text{mol }(CH_3)_3N}{\text{L }(CH_3)_3N} = 0.015\text{ mol }(CH_3)_3N$$

Step 2: Determine the moles of HCl needed to react with the $(CH_3)_3N$ and the volume of 0.100 M HCl needed.

From the neutralization reaction, we know that the mole ratio of $(CH_3)_3N$:HCl is 1:1. Therefore,

$$\text{Moles HCl} = 0.015$$

$$0.015\text{ mol HCl} \times \frac{\text{L}}{0.100\text{ mol HCl}} = 0.150\text{ L} = 150\text{ mL}$$

b) Initial pH of the $(CH_3)_3N$ solution:

Step 1: Determine the initial pH of the solution:

Principal Reaction	$(CH_3)_3N(aq)$	+	$H_2O(l)$	$\leftrightarrows$	$(CH_3)_3NH^+(aq)$	+	$OH^-(aq)$
Initial Conc.	0.375 M				0		0
Change	$-x$				$+x$		$+x$
Equilibrium Conc.	$0.375\text{ M} - x$				$+x$		$+x$

$$K_b = \frac{\left[(CH_3)_3NH^+\right]\left[OH^-\right]}{\left[(CH_3)_3N\right]}$$

$$6.5\times10^{-5} = \frac{(x)(x)}{(0.375-x)} \cong \frac{x^2}{0.375} \qquad x = [OH^-] = 4.93\times10^{-3}$$

$$\left[H_3O^+\right] = \frac{K_w}{\left[OH^-\right]} = \frac{1.0\times10^{-14}}{4.93\times10^{-3}} = 2.02\times10^{-12}; \qquad pH = -\log(2.02\times10^{-12}) = 11.7$$

c) pH of the solution after addition of 10.00 mL of acid:

Step 1: Calculate the concentration of $(CH_3)_3N$ and its conjugate acid after the addition of 10.00 mL of HCl.

$$\text{mol } (CH_3)_3N = 0.015 \text{ mol} - (0.100 \text{ M})(0.010 \text{ L}) = 0.014 \text{ mmol}$$

The total volume present = 40.00 mL $(CH_3)_3N$ + 10.00 mL HCl added.

$$\left[(CH_3)_3N\right] = \frac{0.014 \text{ mol } (CH_3)_3N}{0.050 \text{ L soln}} = 0.280 \text{ M}$$

The conjugate acid of $(CH_3)_3N$ is $(CH_3)_3NH^+$.

$$\text{mol } (CH_3)_3NH^+ = \text{mol HCl used} = (0.100 \text{ M})(0.010 \text{ L}) = 0.0010 \text{ mol}$$

$$\left[(CH_3)_3NH^+\right] = \frac{0.0010 \text{ mol } (CH_3)_3NH^+}{0.050 \text{ L}} = 0.020 \text{ M}$$

Step 2: Determine the K_a and pK_a of the conjugate acid of $(CH_3)_3N$.

$$K_a\,(CH_3)_3NH^+ = \frac{K_w}{K_b\,(CH_3)_3N} = \frac{1.0\times10^{-14}}{6.5\times10^{-5}} = 1.5\times10^{-10}$$

$$pK_a = -\log(1.5\times10^{-10}) = 9.82$$

Step 3: Using the Henderson–Hasselbalch equation, calculate the pH of the solution.

$$pH = pK_a + \log\frac{[\text{base}]}{[\text{acid}]} = 9.81 + \log\frac{0.280}{0.020} = 11.0$$

d) pH halfway to the equivalence point:

Step 1: Use the Henderson–Hasselbalch equation to calculate the pH. Halfway to the equivalence point, [base] = [acid].

$$pH = pK_a = 9.81.$$

e) pH at the equivalence point:

Step 1: Calculate the concentration of the conjugate acid, using the initial moles of base and the total volume of solution:

At the equivalence point, moles of $(CH_3)_3NH^+$ = moles of $(CH_3)_3N$ reacted (= amount of $(CH_3)_3N$ present initially).

$$\text{Moles } (CH_3)_3NH^+ = 0.015$$

$$\text{mL of HCl} = 0.015 \text{ moles}/0.100 \text{ M} = 0.150 \text{ L} = 150 \text{ mL}$$

$$\text{Total volume} = 40.00 \text{ mL } (CH_3)_3N + 150 \text{ mL HCl to reach eq point} = 190 \text{ mL}$$

$$\left[(CH_3)_3NH^+\right] = \frac{0.015 \text{ mol}}{0.190 \text{ L}} = 0.079 \text{ M}$$

Step 2: Determine the pH of the solution:

Principal Reaction	$(CH_3)_3NH^+(aq)$ +	$H_2O(l)$ ⇆	$(CH_3)_3N(aq)$ +	$H_3O^+(aq)$
Initial Conc.	0.079		0	0
Change	$-x$		$+x$	$+x$
Equilibrium Conc.	$0.079 - x$		$+x$	$+x$

$$K_a\ (CH_3)_3NH^+ = 1.5 \times 10^{-10}$$

$$K_a = \frac{\left[H_3O^+\right]\left[(CH_3)_3N\right]}{\left[(CH_3)_3NH^+\right]} \qquad 1.5 \times 10^{-5} = \frac{(x)(x)}{0.079 - x} \cong \frac{x^2}{0.079}$$

$$x = [H_3O^+] = 1.09 \times 10^{-3} \qquad \text{pH} = -\log(1.09 \times 10^{-3}) = 2.96$$

f) pH after the addition of 200.00 mL of 0.100 M HCl:

Step 1: Determine the number of mmoles of excess HCl.

The first 150 mL titrated the base, the remaining 50 mL are excess.

Moles HCl in excess = 0.050 L × 0.100 mol/L = 0.0050 mol

Step 2: Determine the concentration of excess HCl, and calculate the pH.

$$[HCl]_{xs} = \frac{0.005 \text{ mol}}{0.240 \text{ L}} = 0.021 \text{ M} \qquad \text{pH} = -\log(0.021) = 1.68$$

Workbook Problem 17.9

Step 1: Determine how many moles of proline and how many moles of sodium hydroxide are present:

Moles proline = 0.100 L × 0.050 mol/L = 0.0050 mol proline

$$\text{Moles NaOH} = \frac{0.025 \text{ g}}{40 \text{ g/mol}} = 0.000625 \text{ mol}$$

Step 2: Use the Henderson–Hasselbalch equation to determine the pH:

$$pH = pK_a + \log\frac{[\text{base}]}{[\text{acid}]} = 1.952 + \log\frac{0.000625\ \text{mol} / 0.100\ \text{L}}{(0.005 - 0.000625\ \text{mol}) / 0.100\ \text{L}} = 1.1$$

Step 3: Calculate the isoelectric point:

$$\text{isoelectric point} = \frac{\text{pK}_{\text{a1}} + \text{pK}_{\text{a2}}}{2} = \frac{(1.952 + 10.64)}{2} = 6.30$$

Workbook Problem 17.10

Step 1: Write the solubility equilibrium expression for $CdCl_2$.

The solubility equilibrium for $CdCl_2$ is as follows:

$$CdCl_2(s) \leftrightarrows Cd^{2+}(aq) + 2\ Cl^-(aq)$$

and the K_{sp} expression for this reaction is as given below:

$$K_{sp} = \left[Cd^{2+}\right]\left[Cl^-\right]^2$$

Step 2: Determine the concentration of Cd^{2+} and Cl^- and calculate the K_{sp}.

If $[Cd^{2+}] = 1.10 \times 10^{-5}$ then the $[Cl^-]$ is two times that, 2.20×10^{-5}. Substituting these values into the equilibrium expression gives:

$$K_{sp} = \left(1.10 \times 10^{-5}\right)\left(2.20 \times 10^{-5}\right)^2 = 5.32 \times 10^{-15}$$

Workbook Problem 17.11

Step 1: Write the equilibrium for $Cu(OH)_2$.

The solubility equilibrium for $Cu(OH)_2$ is as given below:

$$Cu(OH)_2(s) \leftrightarrows Cu^{2+}(aq) + 2\ OH^-(aq)$$

Step 2: Write the equilibrium expression.

The equilibrium expression is as given below:

$$K_{sp} = [Cu^{2+}][OH^-]^2$$

Step 3: Let x be the number of mol/L of $Cu(OH)_2$ that dissolves. The saturated solution then contains x mol/L of Cu^{2+} and $2x$ mol/L of OH^-. Solve for K_{sp}.

Substitute this into the equilibrium expression:

$$K_{sp} = 2.18 \times 10^{-20} = \left[Cu^{2+}\right]\left[OH^-\right]^2 = (x)(2x)^2$$

Solve for x:

$$2.18 \times 10^{-20} = 4x^3; \qquad x = 1.75 \times 10^{-7}\ \text{M}$$

Workbook Problem 17.12

Step 1: Write the solubility equation and solubility expression.

$$Sn(OH)_2(s) \leftrightarrows Sn^{2+}(aq) + 2\ OH^-(aq)$$

The equilibrium expression is as follows:

$$K_{sp} = [Sn^{2+}][OH^-]^2$$

Step 2: Using the pH, determine the $[OH^-]$ concentration.

$$[H_3O^+] = 10^{(-9.45)} = 3.55 \times 10^{-10}$$

$$\left[OH^-\right] = \frac{1 \times 10^{-14}}{3.55 \times 10^{-10}} = 2.82 \times 10^{-5}$$

Step 3: Let x be the number of mol/L of $Sn(OH)_2$ that dissolves. Construct a table showing the equilibrium concentrations.

Solubility Equilibrium	$Sn(OH)_2(s)$	$\leftrightarrows$	$Sn^{2+}(aq)$	+	$2\ OH^-(aq)$
Initial Concentration			0		2.82×10^{-5}
Equilibrium Concentration			$+x$		$2.82 \times 10^{-5} + 2x$

Step 4: Calculate the solubility of $Sn(OH)_2$.

Given the very small value of K_{sp}, we can assume that x is negligible. Substituting the equilibrium concentrations into the K_{sp} expression gives:

$$5.4 \times 10^{-27} = (x)(2.82 \times 10^{-5})^2 \quad x = 6.79 \times 10^{-18}$$

Workbook Problem 17.13

$$Zn(NO_3)_2(aq) + Na_2CO_3(aq) \rightarrow 2\ NaNO_3(aq) + ZnCO_3(s)$$

Step 1: Write the solubility equation and IP expression for the precipitate.

$$ZnCO_3(s) \rightarrow Zn^{2+}(aq) + CO_3^{2-}(aq) \qquad K_{sp} = 1.2 \times 10^{-10}$$

The ion product expression is as follows:

$$IP = [Zn^{2+}][CO_3^{2-}]$$

Step 2: Calculate the IP and compare its value to the K_{sp}. Will a precipitate form?

Calculate the IP from the concentrations of Zn^{2+} and CO_3^{2-}, remembering that two solutions were mixed for a total volume of 350 mL:

$$\left[Zn^{2+}\right] = \frac{(0.75\text{ M})(100\text{ mL})}{350\text{ mL}} = 0.21\text{ M} \qquad \left[CO_3^{2-}\right] = \frac{(1.50\text{ M})(250\text{ mL})}{350\text{ mL}} = 1.07\text{ M}$$

$$\text{IP} = (0.21)(1.07) = 0.23$$

IP > K_{sp}; therefore, a precipitate will form.

Self-Test

True–False

1. T
2. F When a weak acid reacts with a strong base the pH of the solution is greater than 7.00.
3. T
4. F The equilibrium shifts to more dissociated acid to neutralize the base.
5. F Half of the acid is dissociated.
6. T
7. F The equivalence point will be lower than 7 because there is conjugate acid left in solution.
8. F Molar solubility and K_{sp} can be mathematically interconverted.
9. T
10. F Na and K ions cannot be precipitated out, but can be identified by flame tests.

Matching

11. d
12. i
13. b
14. g
15. a
16. c
17. h
18. e
19. j
20. f

Fill in the Blanks

21. strong, completion, protons
22. weak, conjugate, pH, equilibrium, buffers
23. metal, coordinate, molecules

Problems

24. a. HNO_3 and NaOH: This is a reaction between a strong acid and a strong base. The net ionic equation is $H_3O^+(aq) + OH^-(aq) \rightarrow 2\ H_2O(l)$; pH = 7.

    b. HCl and NH_2NH_2: NH_2NH_2 is a weak base. The net ionic equation is $H_3O^+(aq) + NH_2NH_2 \rightarrow H_2O(l) + NH_2NH_3^+(aq)$; pH < 7.00.

    c. Acetic acid (CH_3COOH) and NaOH: Acetic acid is a weak acid. The net ionic equation is $CH_3COOH + OH^-(aq) \rightarrow CH_3COO^-(aq) + H_2O(l)$; pH > 7.00.

25. The salt NH_4Cl is 100% dissociated, so the species present initially are NH_3, NH_4^+, Cl^-, and H_2O. NH_3 is the strongest base present ($K_b >> K_w$). This gives the following table:

Principal Reaction	$NH_3(aq) + H_2O(l) \leftrightarrows$	$NH_4^+(aq) +$	$OH^-(aq)$
Initial Concentration (M)	0.10	0.25	0
Change (M)	$-x$	$+x$	$+x$
Eq Concentration (M)	$0.10 - x$	$0.25 + x$	$+x$

The common ion in this problem is NH_4^+. The equilibrium equation for the principal reaction is as follows:

$$K_b = 1.8 \times 10^{-5} = \frac{[NH_4^+][OH^-]}{[NH_3]} = \frac{(0.25 + x)(x)}{(0.10 - x)} \approx \frac{(0.25)(x)}{(0.10)}$$

x is assumed to be negligible.

$$x = [OH^-] = \frac{(1.8 \times 10^{-5})(0.10)}{(0.25)} = 7.2 \times 10^{-6}$$

The assumption concerning the size of *x* is justified. Since we now know the $[OH^-]$, we can calculate the $[H_3O^+]$.

$$[H_3O^+] = \frac{1.0 \times 10^{-14}}{7.2 \times 10^{-6}} = 1.4 \times 10^{-9}$$

$$pH = -\log(1.4 \times 10^{-9}) = 8.85$$

26. The principal reaction and equilibrium concentrations for this solution are as given below:

Principal Reaction	$HClO(aq) + H_2O(l) \leftrightarrows$	$H_3O^+(aq) +$	$ClO^-(aq)$
Initial Concentration (M)	0.45	0	0.25
Change (M)	$-x$	$+x$	$+x$
Equilibrium Concentration (M)	$0.45 - x$	$+x$	$0.25 + x$

If we solve the equilibrium equation for H_3O^+, we obtain:

$$K_a = \frac{[H_3O^+][ClO^-]}{[HClO]} \qquad [H_3O^+] = K_a \frac{[HClO]}{[ClO^-]}$$

K_a for HClO is 3.5×10^{-8}. Substituting gives:

$$[H_3O^+] = (3.5 \times 10^{-8}) \frac{(0.45)}{(0.25)} = 6.3 \times 10^{-8}$$

$$pH = -\log(6.3 \times 10^{-8}) = 7.20$$

After addition of NaOH:

Neutralization Reaction	$HClO(aq)$ +	$OH^-(aq) \leftrightarrows$	$H_2O(l) +$	$ClO^-(aq)$
Before Reaction (mol)	0.45	0.10		0.25
Change (mol)	−0.10	−0.10		+ 0.10
After Reaction (mol)	0.35	0		0.35

Substituting these values into the expression for $[H_3O^+]$, we can then calculate the pH.

$$\left[H_3O^+\right] = \left(3.5\times10^{-8}\right)\frac{(0.35)}{(0.35)} = 3.5\times10^{-8}$$

$$pH = -\log(3.5 \times 10^{-8}) = 7.46$$

To determine the pH of the solution after addition of HCl, we must take into account the neutralization reaction that takes place.

Neutralization Reaction	$ClO^-(aq)$ +	$H_3O^+(aq)$ ⇆	$H_2O(l)$ +	$HClO(aq)$
Before reaction (mol)	0.25	0.10		0.45
Change (mol)	−0.10	−0.10		+0.10
After Reaction (mol)	0.15	0		0.55

Substituting these values into the expression for $[H_3O^+]$, we can then calculate the pH.

$$\left[H_3O^+\right] = \left(3.5\times10^{-8}\right)\frac{(0.55)}{(0.15)} = 1.3\times10^{-7}$$

$$pH = -\log(1.3 \times 10^{-7}) = 6.89$$

27. Rearrange the Henderson–Hasselbalch equation:

$$\log\frac{[\text{base}]}{[\text{acid}]} = pH - pK_a$$

The pK_a for lactic acid is $-\log(1.4 \times 10^{-4}) = 3.85$.

Substituting the pH and pK_a values gives:

$$\log\frac{[\text{lactate}]}{[\text{lactic acid}]} = 4.75 - 3.85 = 0.90$$

Taking the antilog of both sides gives:

$$\frac{[\text{lactate}]}{[\text{lactic acid}]} = 7.94$$

This gives the ratio of lactate ion to lactic acid. To determine the ratio of lactic acid to the lactate ion, take the reciprocal:

$$\text{lactic acid} = 0.126 \times \text{lactate}$$

$$\frac{[\text{lactic acid}]}{[\text{lactate}]} = 0.126$$

If the solution is to be 0.750 M overall, this can be rearranged to solve for the amount of each:

$$\text{lactic acid} + \text{lactate} = 0.750 \text{ mol}$$

$$0.126 \text{ lactate} + \text{lactate} = 0.750 \text{ mol}$$

$$1.126 \text{ lactate} = 0.750 \text{ mol}$$

$$\text{lactate} = 0.666 \text{ mol}$$

$$\text{lactic acid} = 0.750 \text{ mole} - 0.666 \text{ mol} = 0.084$$

0.084 mol lactic acid and 0.666 mol lactate are required. To check, 0.084/0.666 = 0.126.

28. Calculate the pH of the buffer solution using the Henderson–Hasselbalch equation:

$$\text{pH} = \text{p}K_a + \log\frac{[\text{HCO}_2^-]}{[\text{HCOOH}]} \qquad \text{pH} = 3.74 + \log\frac{0.25}{0.50} = 3.44$$

Account for the reaction on addition of HCl:

Neutralization Reaction	$HCO_2^-(aq)$ +	$H_3O^+(aq)$ →	$HCOOH(aq) + H_2O(l)$
Before Reaction (mol)	0.250	0.300	0.50
Change (mol)	−0.250	−0.250	0.250
After Reaction (mol)	0	0.050	0.75

The limiting reactant in this problem is the HCO_2^- ion, which means that this amount of HCl exceeds the capacity of the buffer. The pH will be determined by the excess HCl present.

$$\text{pH} = -\log(0.050) = 1.30$$

29. a. 0.0 mL of NaOH: $[H_3O^+] = 0.0400$ M; pH = 1.398

b. 10.0 mL of NaOH = 0.010 L × 0.100 M = 1 mmol

75 mL of HCl = 0.075 L × 0.040 M = 3 mmol

2 mmol HCl will remain after the reaction.

$$[\text{H}_3\text{O}^+] \text{ after neutralization} = \frac{(2.00 \text{ mmol})}{85.0 \text{ mL}} = 0.0235 \qquad \text{pH} = -\log(0.0235) = 1.629$$

c. 30.0 mL of NaOH = 3mmol. This is the equivalence point: the HCl will be completely neutralized, and there is no NaOH in excess. The pH will be 7.0.

d. 50.0 mL of NaOH = 5 mmol of NaOH.

2 mmol of NaOH will remain after the reaction.

$$\left[OH^-\right] \text{after neutralization} = \frac{(2.00 \text{ mmol})}{125 \text{ mL}} = 0.016 \text{ M}$$

$$\left[H_3O^+\right] = \frac{1.0\times10^{-14}}{0.016} = 6.25\times10^{-13}$$ $\text{pH} = -\log(6.25 \times 10^{-13}) = 12.20$

30. a. 0.0 mL NaOH: pH is calculated in the same manner as a weak acid.

Principal Reaction	$CH_3COOH(aq) + H_2O(l) \leftrightarrows$	$H_3O^+(aq) +$	$CH_3COO^-(aq)$
Initial concentration (M)	0.0500	0	0
Change (M)	$-x$	$+x$	$+x$
Equilibrium Concentration (M)	$0.0500 - x$	$+x$	$+x$

$$K_a = 1.8\times10^{-5} = \frac{\left[H_3O^+\right]\left[CH_3COO^-\right]}{[CH_3COOH]} = \frac{(x)(x)}{(0.0500 - x)} \approx \frac{(x)^2}{(0.0500)}$$

$$x = 9.48 \times 10^{-4}$$

$$\text{pH} = -\log(9.48 \times 10^{-4}) = 3.02$$

b. 10.0 mL of NaOH = 0.010 L × 0.100 M = 1 mmol

50 mL of acetic acid = 0.050 L × 0.050 M = 2.5 mmol

1.5 mmol acetic acid will remain after the reaction, and 1 mmol of sodium acetate will have been formed.

Using the Henderson–Hasselbalch equation:
$\text{pH} = \text{p}K_a + \log 1.0/1.5 = 4.56$

c. 25.0 mL NaOH = 2.5 mmol. This is the equivalence point of the titration. All the acetic acid is neutralized, and what remains is a solution containing 2.5 mmol sodium acetate.

2.5 mmol/75 mL = 0.033 M acetate

$$K_b = \frac{K_w}{K_a} = \frac{1\times10^{-14}}{1.8\times10^{-5}} = 5.56\times10^{-10}$$

The pH is determined from the equilibrium expression for sodium acetate:

$$K_b = 5.56\times10^{-10} = \frac{(x)^2}{3.33\times10^{-2}}$$ $x = [OH^-] = 4.30 \times 10^{-6}$

$$[H_3O^+] = K_w/4.30 \times 10^{-6} = 2.34 \times 10^{-9}.$$

$$\text{pH} = -\log(2.34 \times 10^{-9}) = 8.63$$

d. 30.0 mL of NaOH = 0.030 L × 0.100 M = 3 mmol

50 mL of acetic acid = 0.050 L × 0.050 M = 2.5 mmol

0.5 mmol NaOH will remain after the reaction. This will determine the pH of the solution.

$$\left[OH^-\right] = \frac{0.50\text{ mmol}}{80.0\text{ mL}} = 6.3\times10^{-3}$$

$$\left[H_3O^+\right] = \frac{1\times10^{-14}}{6.3\times10^{-3}} = 1.59\times10^{-12} \qquad \text{pH} = -\log(1.59\times10^{-12}) = 11.80$$

31. The solubility equilibrium for $CaCO_3$ is $CaCO_3(s) \leftrightarrows Ca^{2+}(aq) + CO_3^{2-}(aq)$

If 7.07×10^{-5} mol is the amount of $CaCO_3$ that dissolves in 1.0 L of solution then the $[Ca^{2+}] = 7.07\times10^{-5}$ M and the $[CO_3^{2-}] = 7.07\times10^{-5}$ M. Substituting these values into the equilibrium expression gives:

$$K_{sp} = \left[Ca^{2+}\right]\left[CO_3^{2-}\right] = \left(7.07\times10^{-5}\right)^2 = 5.0\times10^{-9}$$

32.

Solubility Equilibrium	$Ca_3(PO_4)_2(s) \leftrightarrows$	$3\,Ca^{2+}(aq)$	$2\,PO_4^{3-}(aq)$
Equilibrium Concentration (M)		$+3\,x$	$+2\,x$

$$K_{sp} = 2.1\times10^{-33} = \left[Ca^{2+}\right]^3\left[PO_4^{3-}\right]^2 = (3x)^3(2x)^2 = 108x^5$$

$$x = 1.14\times10^{-7}$$

33.

Solubility Equilibrium	$CaSO_4(s) \leftrightarrows$	$Ca^{2+}(aq) +$	$SO_4^{2-}(aq)$
Equilibrium Concentration (M)		$+\,x$	$0.50 + x$

$$K_{sp} = 7.1\times10^{-5} = \left[Ca^{2+}\right]\left[SO_4^{2-}\right] = (x)(0.50+x) \approx (x)(0.50)$$

$$x = 1.42\times10^{-4}$$

34. a. AgBr: Not more soluble. Br^- is the conjugate base of a very strong acid and is very unreactive.

b. Na_2S: More soluble. S^{2-} is the conjugate base of a weak acid.

c. LiCN: More soluble. CN^- is the conjugate base of a very weak acid.

d. CaF_2: More soluble. F^- is the conjugate base of a weak acid.

35. Because K_f for $Zn(NH_3)_4^{2+}$ is large, nearly all the Zn^{2+} from $Zn(NO_3)_2$ will be converted to $Zn(NH_3)_4^{2+}$.

$$Zn^{2+}(aq) + 4\ NH_3(aq) \leftrightarrows Zn(NH_3)_4^{2+}$$

Conversion of 0.50 mol/L of Zn^{2+} to $Zn(NH_3)_4^{2+}$ consumes 2.00 mol/L of NH_3 (due to a 1:4 mole ratio of Zn^{2+} to NH_3). Assuming 100% conversion to $Zn(NH_3)_4^{2+}$ gives the following concentrations:

$[Zn^{2+}] = 0\ M$ $\quad [Zn(NH_3)_4^{2+}] = 0.50\ M$ $\quad [NH_3] = 4.0 - 2.00 = 2.0\ M$

After conversion of Zn^{2+} to $Zn(NH_3)_4^{2+}$, assume that a small amount of the reverse reaction occurs, producing Zn^{2+}.

$$Zn(NH_3)_4^{2+} \leftrightarrows Zn^{2+}(aq) + 4\ NH_3(aq)$$

Dissociation of x mol/L of $Zn(NH_3)_4^{2+}$ produces x mol/L of Zn^{2+} and $4x$ mol/L of NH_3. The equilibrium concentrations are as given below:

$[Zn(NH_3)_4^{2+}] = 0.50 - x$ $\quad [Zn^{2+}] = x$ $\quad [NH_3] = 2.0 + 4x$

The following table summarizes this reasoning under the balanced equation

Principal Reaction	$Zn^{2+}(aq)$ +	$4\ NH_3(aq) \leftrightarrows$	$Zn(NH_3)_4^{2+}(aq)$
Initial Concentration (M)	0.50	4.0	0
After 100% Reaction (M)	0	2.0	0.50
Equilibrium Concentration (M)	x	$2.0 + 4x$	$0.50 - x$

Substitute the equilibrium concentrations into the expression for K_f and make the approximation that x is negligible compared with 0.50 M:

$$K_f = 7.8 \times 10^8 = \frac{\left[Zn(NH_3)_4^{2+}\right]}{\left[Zn^{2+}\right][NH_3]^4} = \frac{(0.50 - x)}{(x)(2.0 + 4x)^4} \approx \frac{(0.50)}{(x)(2.4)^4}$$

$$\left[Zn^{2+}\right] = x = \frac{(0.50)}{\left(7.8 \times 10^8\right)(2.0)^4} = 4.01 \times 10^{-11}\ M$$

$$[Zn(NH_3)_4] = 0.50 - 4.01 \times 10^{-11}\ M = 0.50\ M$$

36. The equation for this reaction is $3\ CaCl_2(aq) + 2\ Na_3PO_4(aq) \rightarrow Ca_3(PO_4)_2(s) + 6\ NaCl(aq)$

After the two solutions are mixed, the total volume is 550 mL, making the concentrations of the relevant ions:

$$[Ca^{2+}] = \frac{350\ mL \times 0.75}{550\ mL} = 0.477\ M \qquad [PO_4^{3-}] = \frac{200\ mL \times 1.50}{550\ mL} = 0.545\ M$$

To determine if a precipitate will form, calculate the ion product and compare its value to K_{sp}.

$IP = [Ca^{2+}]^3[PO_4^{3-}]^2$ $\qquad$ $IP = (0.477)^3(0.545)^2 = 0.0322$

K_{sp} for $Ca_3(PO_4)_2 = 2.1 \times 10^{-33}$; IP >> K_{sp}; therefore, a precipitate will form.

37. K_{spa} (CuS) = 6×10^{-16}; K_{spa} (FeS) = 6×10^2

$$Q_c = \frac{[Cu^{2+}][H_2S]}{[H_3O^+]^2} = \frac{(0.007)(0.10)}{(0.20)^2} = 1.75 \times 10^{-2}$$

Since K_{spa} (CuS) < Q_c < K_{spa} (FeS), Cu^{2+} will selectively precipitate out of solution.

Challenge Problem

38. For CuS

$CuS(s) \leftrightarrows Cu^{2+}(aq) + S^{2-}(aq)$ $\qquad$ $K_{spa} = 6.0 \times 10^{-16}$

The acidic buffer will cause S^{2-} ions to be removed from the solution to form HS^- and $H_2S(aq)$:

$$H^+(aq) + S^{2-}(aq) \leftrightarrows HS^-(aq)$$

$$H^+(aq) + HS^-(aq) \leftrightarrows H_2S(aq)$$

Adding the three chemical equations together, we have:

$$CuS(s) + 2\ H^+(aq) \leftrightarrows Zn^{2+}(aq) + H_2S(aq)$$

This equation is the definition of K_{spa}.

$$K_{spa} = \frac{[Cu^{2+}][H_2S]}{[H_3O^+]^2}$$

To determine the molar solubility of CuS, we need to determine the concentration of Cu^{2+} at equilibrium. The concentration of H_3O^+ can be calculated from the information given about the buffer:

$$[HCHO_2] = 0.45\ M \text{ and } [CHO_2^-] = 0.25\ M;\ K_a = 1.8 \times 10^{-4}$$

$$pK_a = -\log(1.8 \times 10^{-4}) = 3.74$$

$$pH = pK_a + \log\frac{[CH_2O^-]}{[HCH_2O]} = 3.74 + \log\frac{0.25}{0.45} = 3.48$$

$$[H_3O^+] = 10^{-3.48} = 3.31 \times 10^{-4}$$

Principal Reaction	$CuS(s)$ +	$2\,H^+(aq) \leftrightarrows$	$Cu^{2+}(aq)$ +	$H_2S(aq)$
Initial concentration		3.31×10^{-4}	0	0
Change		$-2x$	$+x$	$+x$
Equilibrium Conc.	$(3.31 \times 10^{-4}) - 2x$		$+x$	$+x$

Substituting the equilibrium concentrations into the equilibrium expression, we have:

$$6.0\times10^{-16} = \frac{(x)(x)}{\left[\left(3.31\times10^{-4}\right)-2x\right]^2}$$

The value of x can be considered negligible.

$$(6.0 \times 10^{-16})\,(3.31 \times 10^{-4})^2 = x^2$$

$$x = 8.11 \times 10^{-12}$$

$x = [Cu^{2+}] = 8.11 \times 10^{-12}$. $[Cu^{2+}]$ represents the molar solubility of CuS.

Chapter Eighteen—Thermodynamics: Entropy, Free Energy, and Spontaneity

Workbook Problems

Workbook Problem 18.1

Step 1: Calculate the moles of hydrogen.

$$3\text{ g } H_2 \times \frac{1\text{ mol}}{2.0\text{ g}} = 1.5\text{ mol}$$

Step 2: Calculate ΔS:

$$\Delta S = nR\ln\frac{V_{\text{final}}}{V_{\text{initial}}} = 1.5\text{ mol} \times 8.314\text{ J/K}\ln\frac{20}{5} = 17\ J/K$$

Workbook Problem 18.2

Step 1: Write the formula:

$$2H_2(g) + O_2(g) \rightarrow 2H_2O(g)$$

Step 2: Calculate ΔS:

$$2(188.7) - [2(130.6) + 205.0] = -88.8\text{ J/mol}\cdot\text{K}$$

Workbook Problem 18.3

Step 1: Write the net ionic equation for this reaction.

$$Ag^{+}(aq) + Br^{-}(aq) \rightarrow AgBr(s)$$

Step 2: Calculate ΔS_{system}.

$$\Delta S_{system} = [S°(AgBr] - [S°(Ag^{+}) + S°(Br^{-})]$$

$$\Delta S_{system} = [107.1] - [72.7 + 82.4] = -48.0\ J/K$$

Step 3: Calculate ΔH_{rxn}.

$$\Delta H°_{rxn} = [\Delta H°_{f}(AgBr)] - [\Delta H°_{f}(Ag^{+}) + \Delta H°_{f}(Br^{-})]$$

$$\Delta H°_{rxn} = [-100.4] - [105.6 + (-121.5)] = -84.5\ kJ$$

Step 4: Calculate ΔS_{total}.

$$\Delta S_{total} = -48.0\ J/K - \frac{-84{,}500\ J}{298\ K} = 236\ J/K$$

The reaction is spontaneous.

Workbook Problem 18.4

Step 1: Calculate ΔS:

$$\Delta S = 239.8 - 127 = 112.8\ J/mol{\cdot}K$$

Step 2: Calculate ΔH:

$$\Delta H = -201.0 - (-239.2) = 38.2\ kJ/mol$$

Step 3: Calculate T:

$$T = \frac{\Delta H}{\Delta S} = \frac{38{,}200\ J/mol}{112.8\ J/mol \cdot K} = 338.65\ K = 339\ K \text{ or } 66\ °C$$

Workbook Problem 18.5

Step 1: Determine $\Delta H°_{rxn}$.

$$\Delta H°_{rxn} = [(-127.1) + (-447.5)] - [(-407.3) + (-101.8)] = -65.5\ kJ/mol$$

Step 2: Determine $\Delta S°_{rxn}$.

$$\Delta S°_{rxn} = (96.2 + 205.4) - (219.1 + 115.5) = -33.0\ J/mol \cdot K$$

Step 3: Calculate ΔG°_{rxn}.

$$\Delta G^\circ = \Delta H^\circ - T\Delta S^\circ = -65{,}500 \text{ J/mol} - 298.15\text{K}(-33.0 \text{ J/mol·K}) = -55.7 \text{ kJ/mol}$$

Workbook Problem 18.6

Step 1: Calculate ΔG°_{rxn} and determine if the reaction is spontaneous.

$$\Delta G^\circ_{rxn} = [(-65.3 + (2 \times 0)] - [(-50.8) + (2 \times 0)] = -14.5 \text{ kJ/mol}$$

Reaction is spontaneous.

Step 2: Determine the temperature at which the reaction will become spontaneous.

Calculate ΔH° and ΔS° for this reaction:

$$\Delta H^\circ = \left[(-135.4) + (2\times 0)\right] - \left[(-74.8) + (2\times 0)\right] = -60.6 \text{ kJ/mol}$$

$$\Delta S^\circ = \left[(216.4) + (2\times 130.6)\right] - \left[(186.2) + (2\times 223)\right] = -154.6 \text{ J/mol}\cdot\text{K}$$

$$T = \frac{\Delta H^\circ}{\Delta S^\circ} = \frac{-60{,}600 \text{ J/mol}}{-154.6 \text{ J/mol}\cdot\text{K}} = 392\ K = 119\ ^\circ\text{C}$$

Step 3: Determine the spontaneity at the temperatures listed.

In Step 1, it was determined that the reaction was spontaneous at 25 °C. In Step 2, it was determined that the reaction became spontaneous at 119 °C. Therefore, the reaction is spontaneous below 119 °C, and nonspontaneous above 119 °C.

400 °C = nonspontaneous   45 °C = spontaneous   –20 °C = spontaneous

Workbook Problem 18.7

Step 1: Calculate ΔG°_{rxn} and, determine if the reaction is spontaneous.

$$\Delta G^\circ_{rxn} = 2(51.3) - 2(86.6) = -70.6\ kJ/mol$$

Step 2: Calculate ΔG.

$$\Delta G = \Delta G^\circ + RT \ln Q_p$$

$$\Delta G = -70{,}600\frac{\text{J}}{\text{mol}} + 8.314\frac{\text{J}}{\text{mol}\cdot\text{K}}(298\text{ K})\ln\frac{(0.001)^2}{(0.150)^2(0.250)} = -91{,}993\frac{\text{J}}{\text{mol}} = -92.0\frac{\text{kJ}}{\text{mol}}$$

The reaction is more spontaneous under the nonstandard conditions.

Workbook Problem 18.8

Step 1: Solve for the free energy:

$$\Delta G^\circ = -\text{RT}\ln\text{K} = -8.314\frac{\text{J}}{\text{mol}\cdot\text{K}} \times 298\text{ K} \times \ln 81.9 = -10{,}900 \text{ J/mol} = -10.9 \text{ kJ/mol}$$

Self-Test

True–False

1. F A reaction can be spontaneous and very slow.
2. T
3. T
4. F Standard entropies of reaction can be calculated from standard entropies of formation.
5. T
6. T
7. T
8. F Standard enthalpy of formation for a pure element is always zero.
9. F ΔG determines the spontaneity of a reaction.
10. T

Matching

11. c	12. a	13. f	14. e
15. d	16. h	17. b	18. g

Fill in the Blanks

19. spontaneous, increases
20. increases, decreases
21. motion, entropy
22. complex, less complex, gases, solids
23. entropy/enthalpy, entropy/enthalpy, free energy, temperature
24. positive, unstable
25. reverse/leftward

Problems

26. a. The toy box has the higher entropy. A dishwasher can be arranged in a finite number of ways that could be considered orderly, but toys can be tumbled into a toybox in a multitude of ways.
    b. The sugar dissolved in tea has the higher entropy. The crystals in the bowl are relatively pure and very constrained in their arrangement.
    c. The 1 mol of O_2 gas in 35.5 L has the higher entropy. The 32 g of oxygen gas will have a volume of 22.4 L because it is at STP, and more volume means less order.

27. a. ΔS is positive; an increase in the volume of a gas at constant temperature leads to a decrease in the pressure, which leads to an increase in the entropy.
    b. ΔS is negative; solids have lower entropy than aqueous ions.
    c. ΔS is positive; production of a gas increases disorder.
    d. ΔS is negative; the reactant side of the reaction has more moles of gas.
    e. ΔS is negative; separating the isotopes decreases the randomness of the system.

28. a. $2\ NH_4NO_3(s) \rightarrow 2\ N_2(g) + O_2(g) + 4\ H_2O(g)$

$$\Delta S^\circ_{rxn} = \left[2 \times S^\circ(N_2) + S^\circ(O_2) + 4 \times S^\circ(H_2O)\right] - \left[2 \times S^\circ(NH_4NO_3)\right]$$

$$\Delta S^\circ_{rxn} = \left[(2 \times 191.5) + 205.0 + (4 \times 188.7)\right] - (2 \times 151.1)$$

$$\Delta S^\circ_{rxn} = 1040.6\ \text{J/K}$$

b. $2\ S(s) + 3\ O_2(g) \rightarrow 2\ SO_3(g)$

$$\Delta S°_{rxn} = \left[2 \times S°(SO_3)\right] - \left[2 \times S°(S) + 3 \times S°(O_2)\right]$$
$$\Delta S°_{rxn} = (2 \times 256.6) - \left[(2 \times 31.8) + (3 \times 205.0)\right]$$
$$\Delta S°_{rxn} = -165.4\ \text{J/K}$$

29. Use the equation $\Delta S°_{surr} = -\dfrac{\Delta H°_{rxn}}{T}$.

Solve for $\Delta S°_{surr}$:

$$\Delta S°_{surr} = \Delta S°_{total} - \Delta S°_{rxn} = 1957 - 515.7 = 1441\ \text{J/K}$$

$$\Delta H°_{rxn} = -T\Delta S_{surr} = -298 \times 1441 \frac{\text{J}}{\text{K}} = -4.295 \times 10^5 \frac{\text{J}}{\text{mol}} = -429.5 \frac{\text{kJ}}{\text{mol}}$$

30. Use the equation $\Delta S = R \ln \dfrac{V_{final}}{V_{initial}}$, we can calculate the value of ΔS.

$$\Delta S = 8.314 \frac{\text{J}}{\text{mol} \cdot \text{K}} \ln \frac{45.3\ \text{L}}{22.4\ \text{L}} = 5.85 \frac{\text{J}}{\text{mol} \cdot \text{K}} \times 1\ \text{mol} = 5.85 \frac{\text{J}}{\text{K}}$$

31. The temperature at which ethanol spontaneously goes from liquid to gas (and back again) can be calculated using the formula:

$$T = \frac{\Delta H°}{\Delta S°} = \frac{-235.1\ \text{kJ/mol} - (-277.7\ \text{kJ/mol})}{282.6\ \text{J/mol} \cdot \text{K} - 161\ \text{J/mol} \cdot \text{K}} = \frac{42.6\ \text{kJ/mol}}{121.6\ \text{J/mol} \cdot \text{K}}$$

Take care with the units!

$$T = \frac{42.6\ \text{kJ/mol}}{121.6\ \text{J/mol} \cdot \text{K}} = \frac{42{,}600\ \text{J/mol}}{121.6\ \text{J/mol} \cdot \text{K}} = 350\ \text{K}$$

32. ΔS_{sys} is the same as ΔS_{rxn}, and can be calculated using the standard enthalpies given:

$$\Delta S°_{rxn} = \left[2 \times S°(HCl)\right] - \left[S°(H_2) + S°(Cl_2)\right]$$

$$\Delta S°_{rxn} = [2 \times 186.8] - [130.6 + 223.0] = 20.0\ \text{J}/(\text{mol} \cdot \text{K})$$

The entropy of the surroundings can be calculated using the formula:

$$\Delta S°_{surr} = \frac{-\Delta°_{rxn}}{T} = \frac{-\left(2\ \text{mol}\left(-92.3 \frac{\text{kJ}}{\text{mol}}\right)\right)}{298\ \text{K}} = 619 \frac{\text{J}}{\text{mol} \cdot \text{K}}$$

Once again, when going from enthalpy values to entropy values, be careful with units.

$$\Delta S°_{total} = \Delta S°_{surr} + \Delta S°_{rxn} = 639 \frac{\text{J}}{\text{mol} \cdot \text{K}}$$

33. $\Delta G^\circ_{rxn} = \left[2\times\Delta G^\circ_f(CO_2)+2\times\Delta G^\circ_f(H_2O)\right]-\left[\Delta G^\circ_f(C_2H_4)+3\times\Delta G^\circ_f(O_2)\right]$

$\Delta G^\circ_{rxn} = \left[(2\times-394.4)+(2\times-237.2)\right]-\left[68.1+0\right] = -1331\text{ kJ / mol}$

34. $\Delta G^\circ = \left[2\times\Delta G^\circ_f(CO_2)\right]-\left[2\times\Delta G^\circ_f(CO)+\Delta G^\circ_f(O_2)\right]$

$\Delta G^\circ = (2\times-394.4)-\left[(2\times-137.2)+0\right] = -514.4\text{ kJ / mol}$

$\Delta G = \Delta G^\circ + RT\ln Q$

$$\Delta G = -514{,}400\frac{\text{J}}{\text{mol}}+8.314\frac{J}{\text{K}\cdot\text{mol}}\times 298\,K\times ln\frac{(15)^2}{(0.5)^2(0.1)} = -4918\times10^5\frac{\text{J}}{\text{mol}}$$

$$\Delta G = -491.8\frac{\text{kJ}}{\text{mol}}$$

35. $\Delta G^\circ = -\text{RT}\,ln\,K$

$\Delta G^\circ_{rxn} = \left[\Delta G^\circ_f(CO_2)+\Delta G^\circ_f(H_2O)\right]-\left[\Delta G^\circ_f(H_2CO_3)\right]$

$\Delta G^\circ_{rxn} = \left[(-386.0)+(-237.2)\right]-(-623) = -0.2\text{ kJ / mol}$

$$\ln K = -\frac{\Delta G^\circ}{RT} = -\frac{-200\text{ J / mol}}{8.314\text{ J / K}\times 310\text{ K}} = 7.76\times10^{-2}$$

$$K = e^{7.76\times10^{-2}} = 1.08$$

36. To solve this problem, you need to use the data found in Appendix B in your textbook.

$$\Delta G^\circ_{rxn} = [0+0]-[(2\times-384.2)] = 768.4\text{ kJ/mol}$$

To calculate the temperature at which the reaction becomes spontaneous, we need to first calculate the ΔS°_{rxn} and ΔH°_{rxn}.

$$\Delta S^\circ_{rxn} = [223.0+(2\times51.2)]-[(2\times72.1)] = 181.2\text{ J/mol·K}$$

$$\Delta H^\circ_{rxn} = [0+0]-[\,(2\times-411.2)] = 822.4\text{ kJ/mol}$$

Solve for the temperature (*watch your units*).

$$T = \frac{\Delta H}{\Delta S} = \frac{822{,}400\frac{\text{J}}{\text{mol}}}{181.2\frac{\text{J}}{\text{mol}\cdot\text{K}}} = 4539\,K$$

$$\ln K = \frac{\Delta G^\circ_{rxn}}{-RT} = \frac{768{,}400\ \frac{\text{J}}{\text{mol}}}{-\left(\left(8.314\frac{\text{J}}{\text{mol}\cdot\text{K}}\right)(4539\ K)\right)} = -20.36$$

$$K = e^{-20.36} = 1.44\times10^{-9}$$

Challenge Problem

37. a) $\Delta G^\circ = [(2\times 86.6\ \text{kJ/mol})] = 173.2\ \text{kJ}$

$\Delta G^\circ = -RT\ \ln K;$

$$\ln K = \frac{1.732\times10^5\ \text{J}}{-\left(8.314\frac{\text{J}}{\text{mol}\cdot\text{K}}\right)(298\ \text{K})} = -69.9$$

$$K = 4.39\times10^{-31}$$

b) $\Delta H^\circ = [2\ \text{mol NO}\times 90.2\ \text{kJ/mol}] = 180.4\ \text{kJ}$

$\Delta S^\circ = [2\ \text{mol NO}\times 210.7\ \text{J/mol·K}] - [(1\ \text{mol } N_2\times 191.5\ \text{J/mol·K}) + (1\ \text{mol } O_2\times 205.0\ \text{J/mol}\cdot\text{K})]$

$\Delta S^\circ = 24.9\ \text{J/mol}$

$$\Delta G^\circ = \Delta H - T\Delta S = 1.804\times10^5\ \text{J} - (773\ K)\left(24.9\frac{\text{J}}{\text{K}}\right) = 1.61\times10^5\ \text{J}$$

$$\Delta G^\circ = -RT\ \ln K$$

$$\ln K = \frac{-1.21\times10^4\ \text{J}}{-\left(8.314\frac{\text{J}}{\text{mol}\cdot\text{K}}\right)(773\ \text{K})} = -25.05$$

$$K = e^{-25.05} = 1.32\times10^{-11}$$

c) To determine the partial pressure of N_2, O_2, and NO at equilibrium, we need to first calculate the initial pressures of N_2 and O_2 using the ideal gas law.

$$P_{N_2} = P_{O_2} = \frac{(3.00\ \text{mol})\left(\frac{0.08206\ \text{L}\cdot\text{atm}}{\text{mol}\cdot\text{K}}\right)(773\ \text{K})}{15.0\ \text{L}} = 12.7\ \text{atm}$$

Principal Reaction	$N_2(g)$	+	$O_2(g)$	$\leftrightarrows$	2 NO(g)
Initial Pressure	12.7		12.7		0
Change	−x		−x		+2x
Equilibrium Pressure	$12.7 - x$		$12.7 - x$		$+2x$

Using the equilibrium constant calculated above, we can now solve for the equilibrium pressures.

$$1.32\times10^{-11}=\frac{P_{NO}^2}{P_{N_2}P_{O_2}}=\frac{(2x)^2}{(12.7-x)(12.7-x)}=\frac{(2x)^2}{(12.7-x)^2}$$

Taking the square root of both sides gives:

$$3.63\times10^{-6}=\frac{2x}{12.7-x}$$

$$x=2.31\times10^{-5}$$

The equilibrium partial pressures are $P_{N_2}=P_{O_2}=12.7$ atm and $P_{NO}=2.16\times10^{-5}$ atm.

Chapter Nineteen—Electrochemistry

Workbook Problems

Workbook Problem 19.1

Step 1: Write two unbalanced half-reactions.

Oxidation: $Na_2SO_3 \rightarrow Na_2SO_4$ (S goes from +4 to +6)

Reduction: $KMnO_4 \rightarrow MnO_2$ (Mn goes from +7 to +4)

Step 2: Balance each half-reaction for atoms other than H and O.

$$Na_2SO_3 \rightarrow Na_2SO_4$$

$$KMnO_4 \rightarrow MnO_2 + K^+$$

Step 3: Add H_2O to balance oxygen, and H^+ to balance hydrogen.

$$H_2O + Na_2SO_3 \rightarrow Na_2SO_4 + 2\ H^+$$

$$4\ H^+ + KMnO_4 \rightarrow MnO_2 + K^+ + 2\ H_2O$$

Step 4: Balance each reaction for charge.

$$H_2O + Na_2SO_3 \rightarrow Na_2SO_4 + 2\ H^+ + 2\ e^-$$

$$3\ e^- + 4\ H^+ + KMnO_4 \rightarrow MnO_2 + K^+ + 2\ H_2O$$

Step 5: Make the electron count the same in both reactions.

$$(H_2O + Na_2SO_3 \rightarrow Na_2SO_4 + 2\ H^+ + 2\ e^-) \times 3$$

$$(3\ e^- + 4\ H^+ + KMnO_4 \rightarrow MnO_2 + K^+ + 2\ H_2O) \times 2$$

$$3\ H_2O + 3\ Na_2SO_3 \rightarrow 3\ Na_2SO_4 + 6\ H^+ + 6\ e^-$$

$$6\ e^- + 8\ H^+ + 2\ KMnO_4 \rightarrow 2\ MnO_2 + 2\ K^+ + 4\ H_2O$$

Step 6: Add the two half-reactions together, canceling anything that appears on both sides of the equation.

$$3\ H_2O + 3\ Na_2SO_3 + 6\ e^- + 8\ H^+ + 2\ KMnO_4 \rightarrow 3\ Na_2SO_4 + 6\ H^+ + 6\ e^- + 2\ MnO_2 + 2\ K^+ + 4\ H_2O$$

$$3\ Na_2SO_3 + 2\ H^+ + 2\ KMnO_4 \rightarrow 3\ Na_2SO_4 + 2\ MnO_2 + 2\ K^+ + H_2O$$

Step 7: Make the solution basic by adding 1 OH^- to each side for every H^+.

$$2\ OH^- + 3\ Na_2SO_3 + 2\ H^+ + 2\ KMnO_4 \rightarrow 3\ Na_2SO_4 + 2\ MnO_2 + 2\ K^+ + H_2O + 2\ OH^-$$

$$2\ H_2O + 3\ Na_2SO_3 + 2\ KMnO_4 \rightarrow 3\ Na_2SO_4 + 2\ MnO_2 + 2\ K^+ + H_2O + 2\ OH^-$$

Step 8: Cancel water molecules that appear on both sides of the equation.

$$H_2O + 3\ Na_2SO_3 + 2\ KMnO_4 \rightarrow 3\ Na_2SO_4 + 2\ MnO_2 + 2\ K^+ + 2\ OH^-$$

Step 9: Check your answer to make sure both atoms and charges are balanced.

Reactant side:	2H	6Na	3S	18O	2K	2Mn	net charge: 0
Product Side:	2H	6Na	3S	18O	2K	2Mn	net charge: 0

Workbook Problem 19.2

Step 1: Write the half-reactions, identifying the oxidation and reduction reactions:

$Pb^{2+}(aq) + 2\ e^- \rightarrow Pb(s)$ reduction – cathode

$Cd(s) \rightarrow Cd^{2+}(aq) + 2\ e^-$ oxidation – anode

Pb^{2+} is the oxidizing agent, and Cd is the reducing agent.

Step 2: Identify solutions for each half-cell and a suitable inert electrolyte for the salt bridge:

Cathode solution: $Pb(NO_3)_2$ Anode solution: $Cd(NO_3)_2$

Salt bridge electrolyte: $NaNO_3$

Any soluble salt containing the metal ions is suitable for the half-cell solutions, and a common ion in the salt bridge will prevent any side reactions.

Workbook Problem 19.3

Step 1: Write the half-reaction occurring at the anode. Remember, the reactant in each half-cell is written first, followed by the product.

Oxidation (loss of electrons) occurs at the anode. The half-reaction is as follows:

$$Mg(s) \rightarrow Mg^{2+}(aq) + 2\ e^-$$

Step 2: Write the half-reaction occurring at the cathode.

Reduction (gain of electrons) occurs at the cathode. The half-reaction is as follows:

$$Cr^{3+}(aq) + 3\ e^- \rightarrow Cr(s)$$

Step 3: Combine the two half-reactions to obtain the overall cell reaction (recall that the equation must be balanced in electrons).

For the number of electrons lost to equal the number of electrons gained, we must multiply the anode half-reaction by 3 and the cathode half-reaction by 2. The two half-reactions are then:

$$3\ Mg(s) \rightarrow 3\ Mg^{2+}(aq) + 6\ e^-$$

$$2\ Cr^{3+}(aq) + 6\ e^- \rightarrow 2\ Cr(s)$$

Adding the two half-reactions together gives the overall equation:

$$3\ Mg(s) + 2\ Cr^{3+}(aq) \rightarrow 3\ Mg^{2+}(aq) + 2\ Cr(s)$$

Step 4: Describe the cell indicated in the shorthand notation.

$$Mg(s)\ |\ Mg^{2+}(aq)\ ||\ Cr^{3+}(aq)\ |\ Cr(s)$$

This cell would consist of a strip of magnesium, as the anode, dipping into an aqueous solution containing the Mg^{2+} ion, such as $Mg(NO_3)_2$. The cathode would consist of a strip of chromium dipping into an aqueous solution of Cr^{3+}, such as $Cr(NO_3)_3$.

Workbook Problem 19.4

Step 1: Write the balanced equation for the reaction:

$$Mg(s) + Zn^{2+}(aq) \rightarrow Mg^{2+}(aq) + Zn(s)$$

Step 2: Identify n and use $\Delta G° = -nFE°$ to calculate $\Delta G°$.

In this reaction, two moles of electrons are transferred per mole of reactants.

$$\Delta G° = -2 \times 96{,}500\ \frac{\text{C}}{\text{mol}} \times 1.61\ \text{V} \times 1\frac{\text{J}}{\text{C} \cdot \text{V}} = -311\ kJ/mol$$

Workbook Problem 19.5

Step 1: Determine the position of the reducing agent relative to the position of the oxidizing agent.

In the first reaction, the reducing agent (the species undergoing oxidation) is Cu. Cu lies below the oxidizing agent in the table of standard reduction potentials. In the second reaction, the reducing agent is F^-, which lies above the oxidizing agent, MnO_4^-.

Step 2: Determine the spontaneity of the reactions.

In a spontaneous reaction, the reducing agent must lie below the oxidizing agent in the table of standard reduction potentials. For the first reaction, the reducing agent, Cu, lies below the oxidizing agent, Ag^+. Therefore, the first reaction is spontaneous. For the second reaction, the reducing agent, F^-, lies above the oxidizing agent, MnO_4^-. Therefore, the second reaction is nonspontaneous.

Workbook Problem 19.6

Step 1: Calculate $E°_{cell}$:

$$E°_{cell} = E°_{Sn^{2+} \rightarrow Sn^{4+}} + E°_{Cu^{2+} \rightarrow Cu} = -0.15 \text{ V} + 0.34 \text{ V} = 0.19 \text{ V}$$

Step 2: Using $E°_{cell}$ and E, solve for $[Sn^{4+}]$:

$$E_{cell} = E° - \frac{0.0592}{n} \log \frac{[Sn^{4+}]}{[Sn^{2+}][Cu^{2+}]}$$

$$0.17 = 0.19 - \frac{0.0592}{2} \log \frac{[Sn^{4+}]}{[0.63][0.72]}$$

$$0.02 \times \frac{2}{0.0592} = \log \frac{[Sn^{4+}]}{[0.45]}$$

$$10^{0.676} = \frac{[Sn^{4+}]}{[0.45]}$$

$$[Sn^{4+}] = 2.13\ M$$

Workbook Problem 19.7

Step 1: Determine E_{ref}:

$$E_{cell} = E_{H_2 \rightarrow 2H^+} + E_{Cu^{2+} \rightarrow Cu}$$

$$0.62 \text{ V} = E_{H_2 \rightarrow 2H^+} + 0.34 \text{ V}$$

$$E_{H_2 \rightarrow 2H^+} = 0.28 \text{ V} = E_{cell} - E_{ref}$$

Step 2: Calculate the pH of the anode solution:

$$\text{pH} = \frac{0.28}{0.0592} = 4.73$$

Workbook Problem 19.8

Step 1: Determine $E°$:

$H_2O_2(aq) + 2\,H^+(aq) + 2\,e^- \rightarrow 2\,H_2O(l)$ $E° = 1.78$ V

$2\,Br^-(aq) \rightarrow Br_2(l) + 2\,e^-$ $E° = -1.09$ V

$$E° = -1.09\text{ V} + 1.78\text{ V} = 0.69\text{ V}$$

Step 2: Calculate K from $E°$:

$$\log K = E°\frac{n}{0.0592} = 0.69\frac{2}{0.0592} = 23.3$$

$$K = 10^{23.3} = 1.99 \times 10^{23}$$

Workbook Problem 19.9

Step 1: Determine the possible reactions at the cathode and their standard voltages:

$Li^+(aq) + e^- \rightarrow Li(s)$ $E° = -3.04$ V *or*

$2\,H_2O(l) + 2\,e^- \rightarrow H_2(g) + 2\,OH^-(aq)$ $E° = -0.83$ V

Step 2: Determine the possible reactions at the anode and their standard voltages:

$2\,SO_4^{2-}(aq) \rightarrow S_2O_8^{2-}(aq) + 2e^-$ $E° = -2.01$ V *or*

$2\,H_2O \rightarrow O_2(g) + 4\,H^+(aq) + 4\,e^-$ $E° = -1.23$ V

Step 3: Predict which reactions will occur and calculate the overall reaction:

In both cases, the hydrolysis of water requires far less energy than the reduction or oxidation of the other species, so the overall reaction is as follows:

$$2\,H_2O(l) \rightarrow O_2(g) + 2\,H_2(g)$$

Workbook Problem 19.10

Step 1: Treat the electrons as reactants in the chemical equation, and solve for the time required to produce 12 kg. of magnesium. Remember, the conversion factor of 96,500 C (or A·s) per mole of electrons.

$$Mg^{2+} + 2e^- \rightarrow Mg(s)$$

$$2\,Cl^- \rightarrow Cl_2(g) + 2e^-$$

$$12\text{ kg Mg} \times \frac{1000\text{ g}}{1\text{ kg}} \times \frac{1\text{ mol Mg}}{24.3\text{ g Mg}} \times \frac{2\text{ mol e}^-}{1\text{ mol Mg}} \times \frac{96{,}500\text{ A}\cdot\text{s}}{1\text{ mol e}^-} \times \frac{1}{15\text{ A}} \times \frac{1\text{ h}}{3600\text{ s}} \times \frac{1\text{ day}}{24\text{ h}} = 73.54\text{ days}$$

Step 2: Use the information calculated above to determine the amount of Cl_2 gas.

$$15\ \frac{C}{s} \times 73.54\ \text{days} \times \frac{24\ \textit{hours}}{\textit{day}} \times \frac{60\ \text{min}}{\text{hour}} \times \frac{60\ \text{s}}{1\ \text{min}} \times \frac{1\ \text{mol e}^-}{96{,}500\ \text{C}} \times \frac{1\ \text{mol Cl}_2}{2\ \text{mol e}^-} \times \frac{22.4\ \text{L Cl}_2}{\text{mol}} = 11{,}062\ \text{L Cl}_2$$

Self-Test

True–False

1. T
2. F A wire allows the transfer of electrons; the salt bridge maintains electrical neutrality.
3. F Positive E corresponds to negative free-energy change. Both indicate spontaneity.
4. T
5. F The Nernst equation allows calculation of E under nonstandard conditions.
6. T
7. T
8. T
9. F Corrosion can be prevented by contact with a metal *below* the metal of interest in the table of standard reduction potentials.
10. F Experimentation shows that more voltage than what standard reduction potentials predict is always needed for electrolysis.

Multiple Choice

11. b	12. a	13. a	14. c	15. b
16. c	17. a	18. d	19. a	20. b

Fill in the Blanks

21. lead storage, alkaline dry cells
22. corrosion, galvanization, sacrificial electrode.
23. sodium or aluminum, aluminum or sodium, electricity, hydroelectric plants.
24. reduction, decreasing, increasing, standard hydrogen electrode
25. equilibrium, cell potentials, small or large, small or large, completion

Matching

26. f	27. c	28. h	29. a	30. b
31. i	32. d	33. e	34. g	

Problems

35. Half-reactions:

Anode: $Mg(s) \rightarrow Mg^{2+}(aq) + 2\ e^-$

Cathode: $Ni^{2+}(aq) + 2\ e^- \rightarrow Ni(s)$

Shorthand notation:

$Mg(s)\ |\ Mg^{2+}(aq)\ ||\ Ni^{2+}(aq)\ |\ Ni(s)$

36. a. $2\ Al(s) \rightarrow 2\ Al^{3+}(aq) + 6\ e^-$ $E° = 1.66$ V
$3\ Pb^{2+}(aq) + 6\ e^- \rightarrow 3\ Pb(s)$ $E° = -0.13$ V
overall $E° = 1.53$ V – spontaneous as written

b. $Br_2(aq) + 2\ e^- \rightarrow 2\ Br^-(aq)$ $E° = 1.09$ V
$2\ Cl^-(aq) \rightarrow Cl_2(g) + 2\ e^-$ $E° = -1.36$ V
overall $E° = -0.27$ V – spontaneous in reverse

c. $Ni(s) \rightarrow Ni^{2+}(aq) + 2\ e^-$ $\quad E° = 0.26$ V
$Pb^{2+}(aq) + 2\ e^- \rightarrow Pb(s)$ $\quad E° = -0.13$ V
overall $E° = 0.13$ V – spontaneous as written

d. $Fe(s) \rightarrow Fe^{2+}(aq) + 2\ e^-$ $\quad E° = 0.45$ V
$2\ H^+ + 2\ e^- \rightarrow H_2(g)$ $\quad E° = 0.00$ V
overall $E° = 0.45$ V – spontaneous as written

37. The two half-reactions (balanced in e^-) and their potentials for this cell are as follows:

Anode: $3\ Mg(s) \rightarrow 3\ Mg^{2+}(aq) + 6\ e^-$ $\quad E° = 2.37$ V

Cathode: $2\ Fe^{3+}(aq) + 6\ e^- \rightarrow 2\ Fe(s)$ $\quad E° = 0.77 - 0.45 = 0.32$ V

a. At standard state, $E° = 2.69$ V

b. $[Mg^{2+}] = 0.1$ M, $[Fe^{3+}] = 2.0$ M

$$E_{cell} = E°_{cell} - \frac{0.0592}{6}\log\frac{\left[Mg^{2+}\right]^3}{\left[Fe^{3+}\right]^2} = 2.69 - \frac{0.0592}{6}\log\frac{[0.1]^3}{[2.0]^2} = 2.72\text{ V}$$

c. $[Mg^{2+}] = 2.0$ M, $[Fe^{3+}] = 0.1$ M

$$E_{cell} = E°_{cell} - \frac{0.0592}{6}\log\frac{\left[Mg^{2+}\right]^3}{\left[Fe^{3+}\right]^2} = 2.69 - \frac{0.0592}{6}\log\frac{[2.0]^3}{[0.1]^2} = 2.66\text{ V}$$

38. $$\Delta G° = -nFE = -2 \times 96{,}500\frac{C}{mol\ e^-} \times 0.27\text{ V} \times \frac{1\text{ J}}{C \cdot V} = -52.1\frac{kJ}{mol}$$

$$E° = \frac{0.0592V}{n}\log K$$

$$\log K = \frac{2 \times 0.27\text{ V}}{0.0592\text{ V}} = 9.12$$

$$K = 10^{9.12} = 1.32 \times 10^9$$

39. The half-reactions for the cell are as follows:

$Zn^{2+}(aq) + 2\ e^- \rightarrow Zn(s)$ $\quad E° = -0.76$ V

$E°_{cell} = E°_{anode} + E°_{cathode}$ $\quad +1.61\text{ V} = E°_{anode} + (-0.76\text{ V})$ $\quad E°_{anode} = 2.37$ V

From the table of standard reduction potentials, we find that Mg(*s*) has a reduction potential of −2.37 V. The half-reaction at the anode is as follows:

$$Mg(s) \rightarrow Mg^{2+}(aq) + 2\ e^-$$

40. $E_{ref} = -0.26$ V for $Ni^{2+}(aq) + 2\ e^- \rightarrow Ni(s)$

$$pH = \frac{E_{cell} - E_{ref}}{0.0592\ V} = \frac{0.14-(-0.26)}{0.0592\ V} = 6.75$$

41. For a metal to be able to protect iron from corrosion, it must have a reduction potential below that of Fe^{2+}. Any metal below that reaction will able to protect iron because any iron atoms that are oxidized will be reduced again by the metal.

$$Fe^{2+}(aq) + 2\ e^- \rightarrow Fe(s) \qquad E° = -0.45$$

a. $Pb^{2+}(aq) + 2\ e^- \rightarrow Pb(s)$ — $E° = -0.13$ V. This is *above* the reduction potential of iron, so would speed corrosion.

b. $Mg^{2+}(aq) + 2\ e^- \rightarrow Mg(s)$ — $E° = -2.37$ V. This is *below* the reduction potential of iron, so would protect against corrosion.

c. $Cu^{2+}(aq) + 2\ e^- \rightarrow Cu(s)$ — $E° = 0.34$ V. This is *above* the reduction potential of iron, so would speed corrosion.

42. Using the table of standard reduction potentials, we can break this overall reaction into the following two half-reactions:

$$3\ I_2(s) + 6\ e^- \rightarrow 6\ I^-(aq) \qquad E° = 0.54\ V$$

$$2\ Al(s) \rightarrow +\ 2\ Al^{3+}(aq) + 6\ e^- \qquad E° = 1.66\ V$$

$$E_{cell} = 0.54 + 1.66 = 2.20\ V$$

$$2.20\ V = \frac{0.0592}{6}\log K$$

$$222.97 = \log K$$

$$K = 10^{223}$$

43. $Na^+(l) + e^- \rightarrow Na(s)$

$2\ Br^-(l) \rightarrow Br_2(g) + 2\ e^-$

To determine the current required to produce 3 g of sodium per hour, set up a unit analysis to convert 3 g Na/hr to C/s:

$$\frac{3\ g\ Na}{hr} \times \frac{1\ mol\ Na}{23.0\ g} \times \frac{96{,}500\ C}{mol\ e^-} \times \frac{1\ mol\ e^-}{mol\ Na} \times \frac{1\ hr}{3600\ s} = 3.5\frac{C}{s} = 3.5\ A$$

44. $15 \text{ g Ag} \times \frac{1 \text{ mol Ag}}{108 \text{ g}} \times \frac{96{,}500 \text{ C}}{\text{mol e}^-} \times \frac{1 \text{ mol e}^-}{\text{mol Ag}} \times \frac{1 \text{ s}}{3 \text{ C}} \times \frac{1 \text{ min}}{60 \text{ s}} = 74 \text{ min}$

45. To begin this problem, you will need to write and balance the equation. To make it easier, take the half-reaction for the reduction of MnO_4^- straight from a table to determine the number of electrons involved:

$$5 \text{ Zn}(s) \rightarrow 5 \text{ Zn}^{2+}(aq) + 10 \text{ e}^-$$

$$2 \text{ MnO}_4^- + 16 \text{ H}^+(aq) + 10 \text{ e}^- \rightarrow 2 \text{ Mn}^{2+}(aq) + 8 \text{ H}_2\text{O}(l)$$

The overall reaction is $5 \text{ Zn}(s) + 2 \text{ MnO}_4^- + 16 \text{ H}^+(aq) \rightarrow 5 \text{ Zn}^{2+}(aq) + 2 \text{ Mn}^{2+}(aq) + 8 \text{ H}_2\text{O}(l)$

The problem can now be set up as a unit-analysis problem:

$$3.27 \text{ g Zn} \times \frac{1 \text{ mol Zn}}{65.4 \text{ g Zn}} \times \frac{2 \text{ mol MnO}_4^-}{5 \text{ mol Zn}} \times \frac{142 \text{ g NaMnO}_4}{\text{mol NaMnO}_4} = 2.84 \text{ g}$$

This is the minimum amount needed for the reaction.

46. To calculate the volume of O_2, first calculate the moles of O_2 produced and then use the ideal gas law. The half-reaction for the electrolysis of water that is relevant here is as follows:

$$2 \text{ H}_2\text{O}(l) \rightarrow \text{O}_2(g) + 4 \text{ H}^+(aq) + 4 \text{ e}^-$$

$$2.97 \times 10^3 \text{ C} \times \frac{1 \text{ mol e}^-}{96{,}500 \text{ C}} \times \frac{1 \text{ mol O}_2}{4 \text{ mol e}^-} = 7.69 \times 10^{-3} \text{ mol O}_2$$

$$V = \frac{nRT}{P} = \frac{(7.69 \times 10^{-3})(0.0821)(298)}{\left(356 \text{ mm Hg} \times \frac{1 \text{ atm}}{760 \text{ mm Hg}}\right)} = 0.402 \text{ L}$$

Challenge Problem

47. Separate the overall reaction into half-reactions and calculate K for the reduction of lead. $K_{red} \times K_{sp} = K_{overall}$. From this, calculate $E°$ for the reaction.

$\text{PbSO}_4(s) \rightarrow \text{Pb}^{2+}(aq) + \text{SO}_4^{2-}(aq)$	K_{sp}
$\text{Pb}^{2+}(aq) + 2 \text{ e}^- \rightarrow \text{Pb}(s)$	K_{red} (calculated from $E°$)
$\text{PbSO}_4(s) + 2 \text{ e}^- \rightarrow \text{Pb}(s) + \text{SO}_4^{2-}(aq)$	$K_{overall} = K_{sp} \times K_{red}$

Calculate $K_{overall}$, then $E_{overall}$.

$$E°_{red} = \frac{0.0592}{n} \log K_{red}$$

$n = 2$ for this reaction.

$$\log K_{red} = \frac{2(-0.126)}{0.0592} = -4.26$$

$$K_{red} = 5.50 \times 10^{-5}$$

Calculate K_{net} for the overall reaction.

$$K_{net} = (1.8 \times 10^{-8}) \times (5.5 \times 10^{-5}) = 9.9 \times 10^{-13}$$

Now calculate $E°_{overall}$:

$$E°_{net} = \frac{0.0592}{n} \log K_{net}$$

$$E°_{net} = \frac{0.0592}{2} \log\left(9.9 \times 10^{-13}\right) = -0.355 \text{ V}$$

Chapter Twenty—Nuclear Chemistry

Workbook Problems

Workbook Problem 20.1

Step 1: First, find the chemical symbol for oxygen.

The chemical symbol for oxygen is O.

Step 2: Use the periodic table to determine the atomic number for oxygen.

From the periodic table, we find that the atomic number for O is 8.

Step 3: Determine the number of neutrons from the definition for mass number.

A (mass number) = *Z* (atomic number) + number of neutrons.
number of neutrons = $A - Z$

For oxygen-16: number of neutrons = $16 - 8 = 8$
For oxygen-17: number of neutrons = $17 - 8 = 9$
For oxygen-18: number of neutrons = $18 - 8 = 10$

Step 4: The standard symbol is written with the mass number as a superscript and the atomic number as a subscript, both to the left of the symbol.

$${}^{16}_{8}O \qquad {}^{17}_{8}O \qquad {}^{18}_{8}O$$

Workbook Problem 20.2

Strategy: Determine the atomic number from the information given.

We are told that element X has 51 protons; therefore, the atomic number of the element is 51.

Step 1: From the atomic number, identify the element by using the periodic table.

From the periodic table, we find that the element with $Z = 51$ is Sb (antimony).

Step 2: The standard symbol is written with the mass number as a superscript and the atomic number as a subscript, both to the left of the symbol.

The mass number is determined by adding the number of protons and the number of neutrons.

$A = 51 + 70 = 121.$ The standard symbol is ${}^{121}_{51}\text{Sb}$.

Workbook Problem 20.3

Step 1: Write the nuclide symbols for polonium-214 and the alpha particle.

polonium-214: ${}^{214}_{84}\text{Po}$ alpha particle: ${}^{4}_{2}\text{He}$

Step 2: Write an incomplete nuclear equation representing the alpha decay of polonium-214.

$${}^{214}_{84}\text{Po} \rightarrow {}^{4}_{2}\text{He} + \_\_\_$$

Step 3: Determine the missing mass number.

$$214 - 4 = 210$$

Step 4: Determine the missing atomic number and the symbol of the new nucleus.

$84 - 2 = 82$ The atomic number is lead (Pb).

Step 5: Write the balanced, complete nuclear equation.

$${}^{214}_{84}\text{Po} \rightarrow {}^{4}_{2}\text{He} + {}^{210}_{82}\text{Pb}$$

Workbook Problem 20.4

Strategy: Use the periodic table to write the nuclide symbols given in the problem. Using this information, write an incomplete nuclear equation and solve for the missing particle.

Step 1: Write the incomplete nuclear equation.

$$^{58}_{28}\text{Ni} + ^{1}_{1}\text{p} \rightarrow ^{4}_{2}\text{He} + \underline{\quad\quad}$$

Step 2: Determine the missing atomic mass.

$$(58 + 1) - 4 = 55$$

Step 3: Determine the missing atomic number and the identity of the new nucleus.

$(28 + 1) - 2 = 27$ This is Cobalt (Co)

Step 4: Write the complete balanced nuclear equation.

$$^{58}_{28}\text{Ni} + ^{1}_{1}\text{p} \rightarrow ^{4}_{2}\text{He} + ^{55}_{27}\text{Co}$$

Workbook Problem 20.5

Step 1: Solve for k:

$$5730 \text{ years} = \frac{\ln 2}{k}$$

$$k = 1.21 \times 10^{-4} \text{ year}^{-1}$$

Step 2: Substitute into the integrated rate law and solve for t after 23% decay:

$$\ln (1.0 - 0.23) = -1.21 \times 10^{-4}/\text{year} \times t$$

$$t = 2160 \text{ years old}$$

Workbook Problem 20.6

Step 1: Determine the number of nucleons and their mass.

First, calculate the total mass of the nucleons and the mass defect:

Xe-130 has 54 protons, and 76 neutrons

Mass of 54 protons = (54)(1.007 28 u) = 54.393 12 u
Mass of 76 neutrons = (76)(1.008 67 u) = 76.658 92 u
Mass of 54 protons and 76 neutrons =131.052 04 u

Mass of xenon-130 =129.903 u
Mass of electrons = (54)(5.486×10^{-4} u) = 0.030 u
Mass of xenon-130 nucleus =129.873 u

Mass defect = 131.052 04 u − 129.873 u = 1.179

Step 2: Convert the mass defect to grams.

$$1.179\text{ u} \times \frac{1.6605 \times 10^{-24}\text{ g}}{\text{u}} = 1.958 \times 10^{-24}\text{ g}$$

Step 3: Determine ΔE and convert to MeV/nucleon.

$$\Delta E = 1.958 \times 10^{-24} \times \frac{1\text{ kg}}{1000\text{ g}} \times \left(3.00 \times 10^{8}\ \frac{\text{m}}{\text{s}}\right)^2 = 1.762 \times 10^{-10}\text{ J}$$

Convert to MeV/nucleon:

$$1.761 \times 10^{-10}\ \frac{\text{J}}{\text{nucleus}} \times \frac{1\text{ MeV}}{1.60 \times 10^{-13}\text{ J}} \times \frac{1\text{ nucleus}}{130\text{ nucleons}} = 8.47\ \frac{\text{MeV}}{\text{nucleon}}$$

Workbook Problem 20.7

Step 1: Write and balance the equation.

$$^{235}\text{U} + {}^{1}\text{n} \rightarrow {}^{136}\text{Xe} + {}^{99}\text{Ru} + {}^{1}\text{n}$$

Step 2: Calculate the mass defect in u and convert to kilograms.

235.043 92 u − (135.907 2195 u + 98.905 939 u) = 0.230 76 u

$$0.230\ 76\text{ amu} \times \frac{1.6605 \times 10^{-27}\text{ kg}}{\text{u}} = 3.8318 \times 10^{-28}\text{ kg}$$

Step 3: Determine ΔE and convert to J/mol.

$$\Delta E = 3.8318 \times 10^{-28}\text{ kg} \times \left(3.00 \times 10^{8}\ \text{m}/\text{s}\right)^2 = 3.44861 \times 10^{-11}\text{ J}$$

$$\frac{3.44861 \times 10^{-11}\text{ J}}{\text{reaction}} \times \frac{6.022 \times 10^{23}\text{ reactions}}{\text{mol}} = 2.08 \times 10^{13}\ \text{J}/\text{mol}$$

Workbook Problem 20.8

Step 1: Write and balance the equation.

$$^{2}\text{H} + {}^{3}\text{H} \rightarrow {}^{4}\text{He} + {}^{1}\text{n}$$

Step 2: Calculate the mass defect in u and convert to kilograms.

(2.014 10 u + 3.016 05 u) − (4.002 60 u +1.008 67 u) = 0.018 88 u

$$0.01888\text{ u} \times \frac{1.6605 \times 10^{-27}\text{ kg}}{\text{u}} = 3.1350 \times 10^{-29}\text{ kg}$$

Step 3:	Determine ΔE and convert to J/mol.
	$\Delta E = 3.1350 \times 10^{-29}\ \text{kg} \times (3.00 \times 10^{8} \text{m/s})^2 = 2.82 \times 10^{-12}\ \text{J}$
	$2.82 \times 10^{-12}\ \frac{\text{J}}{\text{reaction}} \times 6.022 \times 10^{23}\ \frac{\text{reactions}}{\text{mol}} = 1.70 \times 10^{12}\ \text{J / mol}$

Workbook Problem 20.9

Step 1: Determine the number of protons in the new element.

The total number of protons in the new element is 98; 92 from uranium plus 6 from carbon.

Step 2: Determine the mass number of the new element.

The mass of the new element will be the mass numbers of the uranium and the carbon, minus the mass of the neutrons: 238 + 12 − 6 = 244.

Step 3: Determine the identity of the new element using the number of protons.

The element with 98 protons is californium, or Cf.

Step 4: Write the balanced nuclear equation.

The final equation is: ${}^{238}_{92}\text{U} + {}^{12}_{6}\text{C} \rightarrow {}^{244}_{98}\text{Cf} + 6\ {}^{1}_{0}n$.

Self-Test

True–False

1. F Radioactive decay is a first-order process.
2. F Light elements have a 1:1 ratio of protons to neutrons, but heavier elements require more "glue" and thus more neutrons.
3. F A small amount of mass is lost during its conversion to energy.
4. T
5. T
6. T
7. T
8. T
9. T
10. T

Multiple Choice

11. b	12. c	13. b	14. a
15. b	16. c	17. a	18. d
19. c	20. b		

Fill in the Blanks

21. glue, proton, proton, nucleus
22. mass defect, $\Delta E = \Delta mc^2$
23. maximum, fusion, fission, decay
24. transmutation, particles, nucleus, element
25. alpha, beta, internal, gamma, X, external, stopped, lead
26. in vivo, diseased, internal, external, imaging

Matching

27. e 28. g 29. d 30. a
31. c 32. b 33. f

Problems

34. Europium-130

Mass of 63 protons = (63)(1.007 28 u)	= 63.458 64 u
Mass of 67 neutrons = (67)(1.008 67 u)	= 67.580 89 u
Mass of 63 protons and 67 neutrons	= 131.039 53 u
Mass of europium-130 atom	= 129.963 57 u
Mass of electrons = (63) (5.486×10^{-4})	= 0.034 56 u
Mass of europium-130 nucleus	= 129.929 01 u

Mass defect: 131.039 53 u − 129.929 01 u = 1.110 52 u

Europium-153

Mass of 63 protons = (63)(1.007 28 u)	= 63.458 64 u
Mass of 90 neutrons = (90)(1.008 67 u)	= 90.780 30 u
Mass of 63 protons and 90 neutrons	= 154.238 94 u
Mass of europium-153 atom	= 152.921 23 u
Mass of electrons = (63) (5.486×10^{-4})	= 0.034 56 u
Mass of europium-153 nucleus	= 152.886 67 u

Mass defect: 154.238 94 u − 152.886 67 u = 1.352 27 u

Europium-167

Mass of 63 protons = (63)(1.007 28 u)	= 63.458 64 u
Mass of 104 neutrons = (104)(1.008 67 u)	= 104.901 68 u
Mass of 63 protons and 104 neutrons	= 168.360 32 u
Mass of europium-167 atom	= 166.953 21 u
Mass of electrons = (63) (5.486×10^{-4})	= 0.034 56 u
Mass of europium-167 nucleus	= 166.918 65 u

Mass defect: 168.360 32 u − 166.918 65 u = 1.442 27 u

Convert this mass to kilograms:

$$\text{Europium-130}: 1.110\,52\ \text{u} \times \frac{1.6605 \times 10^{-27}\ \text{kg}}{\text{u}} = 1.84402 \times 10^{-27}\ \text{kg}$$

$$\text{Europium-153}: 1.352\,27\ \text{u} \times \frac{1.6605 \times 10^{-27}\ \text{kg}}{\text{u}} = 2.24544 \times 10^{-27}\ \text{kg}$$

$$\text{Europium-167}: 1.442\,27\ \text{u} \times \frac{1.6605 \times 10^{-27}\ \text{kg}}{\text{u}} = 2.39489 \times 10^{-27}\ \text{kg}$$

Use the Einstein equation to convert the mass defect to binding energy:

Europium-130: $\Delta E = 1.84402 \times 10^{-27}\ \text{kg} \times (3.00 \times 10^{8}\ \text{m/s})^2 = 1.65962 \times 10^{-10}\ \text{J}$

Europium-153: $\Delta E = 2.24544 \times 10^{-27}\ \text{kg} \times (3.00 \times 10^{8}\ \text{m/s})^2 = 2.02090 \times 10^{-10}\ \text{J}$

Europium-167: $\Delta E = 2.39489 \times 10^{-27}\ \text{kg} \times (3.00 \times 10^{8}\ \text{m/s})^2 = 2.15540 \times 10^{-10}\ \text{J}$

Convert to MeV/nucleon:

$$\text{Europium-130}: \frac{1.65962\times10^{-10}\ \text{J}}{\text{nucleus}}\times\frac{1\ \text{MeV}}{1.60\times10^{-13}\ \text{J}}\times\frac{1\ \text{nucleus}}{130\ \text{nucleons}}=7.98\frac{\text{MeV}}{\text{nucleon}}$$

$$\text{Europium-153}: \frac{2.02090\times10^{-10}\ \text{J}}{\text{nucleus}}\times\frac{1\ \text{MeV}}{1.60\times10^{-13}\ \text{J}}\times\frac{1\ \text{nucleus}}{153\ \text{nucleons}}=8.26\frac{\text{MeV}}{\text{nucleon}}$$

$$\text{Europium-167}: \frac{2.15540\times10^{-10}\ \text{J}}{\text{nucleus}}\times\frac{1\ \text{MeV}}{1.60\times10^{-13}\ \text{J}}\times\frac{1\ \text{nucleus}}{167\ \text{nucleons}}=8.07\frac{\text{MeV}}{\text{nucleon}}$$

According to this data, from most to least stable these are as given:

E-153 > E-167 > E-130

The observed half-lives of these isotopes are as given:

E-130: 1.1 ms E-153: stable E-167: 200 ms

This bears out the calculated results.

35. Rearrange the equation $E = mc^2$ and solve:

$$\Delta m=\frac{\Delta E}{c^2}=\frac{-498{,}400\ \text{J/mol}}{\left(3.00\times10^{8}\ \text{m/s}\right)^2}=-5.54\times10^{-12}\ \text{kg/mol}=-5.54\times10^{-9}=\text{g/mol}$$

36. mass of products = 91.9262 u + 140.9144 u + 3(1.008 67 u) = 235.8667 u

mass of reactants = 235.0439 u + 1.008 67 u = 236.0526 u

Δm = 236.053 u − 235.867 u = 0.1859 u

$$\text{Convert mass to kg}: 0.1859\ \text{u}\times\frac{1.6605\times10^{-27}\ \text{kg}}{\text{u}}=3.0869\times10^{-28}\ \text{kg}$$

Convert mass to energy: $E = 3.0869\times10^{-28}\ \text{kg}\times(3.00\times10^{8}\ \text{m/s})^2 = 2.78\times10^{-11}$ J/reaction

$$\text{Convert to }\frac{\text{J}}{\text{mole}}: 2.78\times10^{-11}\ \frac{\text{J}}{\text{reaction}}\times\frac{6.022\times10^{23}\ \text{reactions}}{\text{mole}}=1.67\times10^{13}\ \frac{\text{J}}{\text{mole}}$$

37. mass of products = 4.002 60 u + 2(1.008 67 u) = 6.019 94 u
mass of reactants = 2 (3.016 049 u) = 6.032 10 u

Δm = 6.032 10 u − 6.019 94 u = 0.012 16 u

Convert mass to kg:

$$0.01216\,\text{u} \times \frac{1.6605 \times 10^{-27}\ \text{kg}}{\text{u}} = 2.019168 \times 10^{-29}\ \text{kg}$$

Convert mass to energy:

$$E = 2.019168 \times 10^{-29}\ \text{kg} \times \left(3.00 \times 10^{8}\ \text{m/s}\right)^2 = 1.81725 \times 10^{-12}\ \frac{\text{J}}{\text{reaction}}$$

Convert to J/mol:

$$1.81725 \times 10^{-12}\ \frac{\text{J}}{\text{reaction}} \times \frac{6.022 \times 10^{23}\ \text{reactions}}{\text{mol}} = 1.09435 \times 10^{12}\ \text{J/mol}$$

38. a. $^{9}_{4}\text{Be} + {}^{1}_{1}\text{H} \rightarrow {}^{6}_{3}\text{Li} + {}^{4}_{2}\text{He}$

b. $^{239}_{94}\text{Pu} + {}^{4}_{2}\text{He} \rightarrow {}^{242}_{96}\text{X} + {}^{1}_{0}\text{n}$. With an atomic number of 96, this is Curium, Cm-96.

c. $^{208}_{82}\text{Pb} + {}^{58}_{26}\text{X} \rightarrow {}^{265}_{108}\text{Hs} + {}^{1}_{0}\text{n}$. With an atomic number of 26, this is Fe-58.

39. The decay constant, *k*, can be calculated from the half-life:

$$t_{1/2} = \frac{0.693}{k};\ k = \frac{0.693}{t_{1/2}} = \frac{0.693}{6.3 \times 10^{26}\ \text{s}} = 1.1 \times 10^{-27}/\text{s}$$

40. Use the integrated first-order rate law:

$$k = \frac{0.693}{t_{1/2}}$$

$$\ln \frac{N_t}{N_o} = -\left(\frac{0.693}{t_{1/2}}\right) t$$

Substituting:

$$\ln \frac{3.5}{15.3} = -0.693\left(\frac{t}{5715\ \text{years}}\right) \qquad \frac{\ln 0.229 \times 5715\ \text{years}}{-0.693} = 12{,}165\ \text{years}$$

The bone fragment is approximately 12,000 years old.

41. Find the mole ratio of lead to uranium:

56 g Pb-207 = 0.271 mol 27 g U-235 = 0.115 mol

The total number of moles of uranium initially will be the moles of uranium remaining plus the moles of lead found in the sample.

Substitute into the rate law:

$$\ln\frac{0.115}{(0.271+0.115)}=-0.693\left(\frac{t}{7.04\times10^{8}\text{ years}}\right)$$

$$\frac{\ln 0.298\times7.04\times10^{8}\text{ years}}{-0.693}=1.23\times10^{9}\text{ years}$$

Challenge Problem

42. The information given in the problem can be summarized by the partial equation:

$$^{10}_{5}\text{B}+{}^{1}_{0}\text{n}\rightarrow{}^{?}_{3}\text{Li}+?$$

No protons are being added to the system, but two are being lost by boron, so the radiation produced must contain two protons. Only α particles meet this requirement.

$$^{10}_{5}\text{B}+{}^{1}_{0}\text{n}\rightarrow{}^{?}_{3}\text{Li}+{}^{4}_{2}\text{He}$$

Balancing the masses leaves Li-7.

Radiation therapies are constantly being refined so that damage to surrounding tissues can be minimized. This therapy allows a nonradioactive material to be injected, removing the need for surgery. The neutron bombardment can be tightly focused to avoid the areas around the tumor so that the α-particles produced can destroy the tumor with minimum collateral damage.

It may not be surprising that this therapy is being investigated primarily for inoperable brain cancers. Unfortunately, as elegant as it is, so far it does not appear to improve outcomes.

Chapter Twenty-One—Transition Elements and Coordination Chemistry

Workbook Problems

Workbook Problem 21.1

Step 1: Write the electron configuration of the neutral atom:

Ti: $[\text{Ar}]3d^{2}4s^{2}$ Mn: $[\text{Ar}]3d^{5}4s^{2}$ Ni: $[\text{Ar}]3d^{8}4s^{2}$

Step 2: Determine the oxidation state of the metal:

Ti: 0—no electrons need be removed, the answer is complete.
Mn: each oxygen is −2, one negative charge remains, so oxidation state is +7.
Ni: 2+

Step 3: Remove the necessary electrons:

Ti: $[\text{Ar}]3d^{2}4s^{2}$ remove 0 electrons, $[\text{Ar}]3d^{2}4s^{2}$
Mn: $[\text{Ar}]3d^{5}4s^{2}$ remove 7 electrons, [Ar]
Ni: $[\text{Ar}]3d^{8}4s^{2}$ remove 2 electrons, $[\text{Ar}]3d^{8}$

Workbook Problem 21.2

Step 1: Determine the oxidation state of the metals:

$Ag + 2(-1) = -1$ $Ag = +1$	$Mn + 4(-1) = -2;$ $Mn = +2$	$Fe + 6(-1) = -4;$ $Fe = +2$

$Co + 4(0) + 2(-1) = +1$ (the complex has a +1 charge, balanced by a Br^- ion);
$Co = +3$

Step 2: Determine the coordination numbers:

$AgCl_2^-$	Coordination number is 2.
$[MnCl_4]^{-2}$	Coordination number is 4.
$[Fe(CN)_6]^{4-}$	Coordination number is 6.
$[Co(NH_3)_4Br_2]Br$	Coordination number is 6.

Workbook Problem 21.3

Step 1: Determine the energy of the photons absorbed.

$$E = 6.6261 \times 10^{-34}\ J \cdot s \times \frac{3.00 \times 10^{8}\ m/s}{550 \times 10^{-9}\ m} = 3.61 \times 10^{-19}\ J$$

Step 2: Convert to kJ/mol.

$$\frac{3.61 \times 10^{-19}\ J}{photon} \times \frac{1\ kJ}{1000\ J} \times \frac{6.022 \times 10^{23}\ photons}{1\ mole} = 218\ kJ / mole$$

Workbook Problem 21.4

Step 1: Determine the electron configuration of the metal alone:

The free Co^{3+} ion has the valence $[Ar]3d^6$.

Co^{3+} : ↑↓ ↑ ↑ ↑ ↑ (3*d*)   __ (4*s*)   __ __ __ (4*p*) $3d^6$

The d^2sp^3 configuration has fewer unpaired electrons and is also lower energy overall. Both support the choice of this configuration.

Step 2: Identify the possible hybridization schemes:

This is an octahedral complex, so will use either d^2sp^3 or sp^3d^2 hybrid orbitals. There are four valence electrons from the cobalt ion and two from each chloride ion.

Step 3:	Draw electron diagrams with both hybridizations:
	$[Co(Cl)_6]^{3-}$: ↑↓ ↑↓ ↑↓ ↑↓ ↑↓ (3*d*) ↑↓ (4*s*) ↑↓ ↑↓ ↑↓ (4*p*) d^2sp^3
	or
	$[Co(Cl)_6]^{3-}$: ↑↓ ↑ ↑ ↑ ↑ (3*d*) ↑↓ (4*s*) ↑↓ ↑↓ ↑↓ (4*p*) ↑↓ ↑↓ (4*d*) sp^3d^2
Step 4:	Choose the hybridization that maximizes electron spin:
	sp^3d^2 has the greater number of unpaired electrons.

Self-Test

True–False

1. T
2. F Mercury (Hg) is a liquid.
3. F Densities have a maximum in the center of the row.
4. F Transition metals are commonly found in multiple oxidation states.
5. T
6. T
7. F The oxidation number is always given as part of the complex name.
8. T
9. T
10. F Crystal-field theory assumes that all interactions are electrostatic.

Multiple Choice

11. b	12. c	13. a	14. d	15. c
16. c	17. a	18. d	19. b	20. b

Fill in the Blanks

21. oxidation, oxidizing, reducing
22. Coordination, ligands, coordinate covalent
23. isomers, geometric, stereoisomers
24. wavelength or frequency, energy, color
25. Valence bond theory, hybrid, crystal-field theory, electrostatic

Matching

26. h	27. m	28. a	29. f	30. b
31. g	32. c	33. k	34. n	35. e
36. q	37. i	38. p	39. j	40. l
41. d	42. o			

Questions

43. Mn^{5+}: [Ar] $4s^2$ Mo: [Kr] $5s^1\ 4d^5$
    Cr^{3+}: [Ar] $3d^3$ Fe^{2+} : [Ar] $4s^1 3d^5$

44. Mn, Rh, La

45. Zn > Cu > Co > Mn. These metals are all better reducing agents than oxidizing agents.

46. $MnO_4^- > CrO_3^- > Fe_2O_3 > CoO$

47. 3, 8, 5, 2

48. a. Ligands: H_2O (donor atom—O); NH_3 (donor atom—N).
Coordination number is 6; oxidation state of metal is +3; has the general formula MA_2B_4; therefore can be geometrical isomers.
b. Ligands: NH_3 (donor atom—N); Br (donor atom—Br); H_2O (donor atom—O).
Coordination number is 6; oxidation state of metal is +2; has the general formula MA_4BC; therefore can be geometrical isomers. This compound can also have ionization isomers were the Br and NO_3 to switch places.
c. Ligands: NH_3 (donor atom—N), SCN^- (donor atom either N or S).
Coordination number is 6; oxidation state of metal is +2. Because SCN^- can bond through either the N or S, this compound can display linkage isomerism.
d. Ligands: ox (donor atom—O)
Coordination number is 6 (ox is a bidentate ligand); oxidation state of metal is +3; charge on complex ion is −3. Because the ligands can spiral either to the right or to the left, this compound can exist as enantiomers.
e. Ligands: en (donor atoms—N)
Coordination number is 6 (en is a bidentate ligand); oxidation state of metal is +3. The en ligand can spiral either to the right or to the left, indicating the compound can exist as enantiomers.

49. a. diamminetetraaquacobalt(III) chloride
b. tetraammineaquabromocobalt(II) nitrate
c. tetraamminedithiocyanatochromate(II) or tetraamminediisothiocyanatochromate(II) depending on the linkage isomer.
d. ammonium trioxaloferrate(III)
e. tris(ethylenediamine)cobalt(III) iodide

50. a. $[Cr(NH_3)_4(H_2O)_2]Cl_3$
b. $[Cu(en)(SCN)_2]$
c. $[Mn(CO)_5Cl]$
d. $K_4[Fe(CN)_4Cl_2] \cdot 2\ H_2O$
e. $[Cu(H_2O)_4]SO_4$

51 a. Since $[Zn(NH_3)_4]^{2+}$ is tetrahedral, the hybridization on the metal ion is sp^3. The Zn^{2+} ion has the electron configuration [Ar] $4s^2 3d^8$. The orbital diagram for this ion is as follows:

[Ar] ↑↓ ↑↓ ↑↓ ↑ ↑ (3d) ↑↓ (4s) ___ ___ ___ (4p)

To share the four electron pairs provided by the ammonia molecules, Zn^{2+} must use the empty 4*s* and 4*p* orbitals to form hybrid orbitals. The orbital diagram for the complex is

[Ar] ↑↓ ↑↓ ↑↓ ↑↓ ↑↓ (3d) ↑↓ (4s) ↑↓ ↑↓ ↑↓ (4p)

This complex is low spin, since it has no unpaired electrons.
b. $[Ni(CN)_4]^{2-}$ is square planar; therefore, the hybrid orbitals being used by the Ni^{2+} ion are dsp^2. The Ni^{2+} ion has the electron configuration [Ar] $3d^8$. The orbital diagram for this ion is as follows:

[Ar] ↑↓ ↑↓ ↑↓ ↑ ↑ (3d) ↑↓ (4s) ___ ___ ___ (4p)

For Ni^{2+} to use *dsp*^{2} hybrid orbitals, the two unpaired electrons must first pair up. Ni^{2+} can then accept a share in the four electron pairs provided by the CN^- ion. The orbital diagram for the complex is as follows:

$$[Ar]\ \underset{}{\underline{\uparrow\downarrow}}\ \underline{\uparrow\downarrow}\ \underline{\uparrow\downarrow}\ \underline{\uparrow\downarrow}\quad \underline{\uparrow\downarrow}\ \underline{\uparrow\downarrow}\ \underline{\uparrow\downarrow}\ \underline{\uparrow\downarrow}\quad \underline{}$$

3*d* — *dsp*^{2} — 4*p*

The complex is low spin, since there are no unpaired electrons.

52. The crystal-field splitting in tetrahedral complexes is half of the crystal-field splitting in octahedral complexes because none of the d orbitals points directly at the ligands and there are only four ligands, as opposed to six in an octahedral complex.

<u>*Challenge Problem*</u>

53. Chromium forms octahedral complexes, and there are nine possible ligands. Therefore, the compound can be a hydrate or a salt: not all nine ligands will be complexed to the chromium. The experimental data must be used to determine what is bonded and what is hydrate water or balancing ions.

In each case, the moles of compound dehydrated are the same. The molecular weight of all three is 266.45 g/mol.

$$0.500\text{ g} \times \frac{1\text{ mol}}{266.45\text{ g}} = 1.88 \times 10^{-3}\text{ mol}$$

The moles of compound in each solution are also the same:

$$0.125\text{ mol/L} \times 0.050\text{ L} = 6.25 \times 10^{-3}\text{ mol}$$

If the mass of the compound does not change when it is dehydrated, all six waters are complexed to the chromium. This is seen with compound C. Therefore, it is likely that this is $[Co(H_2O)_6]Cl_3$. This can be confirmed by determining whether 3 mol of AgCl are produced per mol of compound:

$$0.809\text{ g AgCl} \times \frac{1\text{ mol}}{143.4\text{ g}} = 5.64 \times 10^{-3}\text{ mol AgCl}$$

$$\frac{5.64 \times 10^{-3}\text{ mol AgCl}}{1.88 \times 10^{-3}\text{ mol C}} = 3.00$$

This confirms that the chlorides are balancing ions and will separate in solution. Compound C is $[Co(H_2O)_6]Cl_3$.

If there is a mass change on dehydration, the number of hydrate waters can be determined. The mass change for compound A is 0.500 g − 0.432 g = 0.068 g water. The number of hydrate waters can be determined by comparing the moles of water to the moles of compound.

$$0.068\text{ g } H_2O \times \frac{1\text{ mol}}{18.01\text{ g}} = 3.76 \times 10^{-3}\text{ mol } H_2O$$

$$\frac{3.76 \times 10^{-3}\text{ mol } H_2O}{1.88 \times 10^{-3}\text{ mol A}} = 2.00$$

If there are two hydrate waters, the compound will likely be $[Co(H_2O)_4Cl_2]Cl \cdot 2H_2O$. This can be confirmed by the number of moles of AgCl produced per mole of compound.

$$0.269\text{ g AgCl} \times \frac{1\text{ mol}}{143.4\text{ g}} = 1.88 \times 10^{-3}\text{ mol AgCl}$$

$$\frac{1.88 \times 10^{-3}\text{ mol AgCl}}{1.88 \times 10^{-3}\text{ mol A}} = 1.00$$

This confirms one balancing chloride ion. This is $[Co(H_2O)_4Cl_2]Cl \cdot 2H_2O$.

The mass change for compound B is 0.500 g – 0.466 g = 0.034 g water. The number of hydrate waters can be determined by comparing the moles of water to the moles of compound.

$$0.034\text{ g } H_2O \times \frac{1\text{ mol}}{18.01\text{ g}} = 1.88 \times 10^{-3}\text{ mol } H_2O$$

$$\frac{1.88 \times 10^{-3}\text{ mol } H_2O}{1.88 \times 10^{-3}\text{ mol B}} = 1.00$$

If there is one hydrate water, the compound is likely to be $[Co(H_2O)_5Cl]Cl_2 \cdot H_2O$. This can be confirmed by the number of moles of AgCl produced per mole of compound.

$$0.539\text{ g AgCl} \times \frac{1\text{ mol}}{143.4\text{ g}} = 3.76 \times 10^{-3}\text{ mol AgCl}$$

$$\frac{3.76 \times 10^{-3}\text{ mol AgCl}}{1.88 \times 10^{-3}\text{ mol B}} = 2.00$$

This confirms two balancing chloride ions. This is $[Co(H_2O)_5Cl]Cl_2 \cdot H_2O$.

Chapter Twenty-Two—The Main-Group Elements

Workbook Problems

Workbook Problem 22.1

Step 1: Predict the traits of the elements based on their position on the periodic table:

N: ionization energy increases from left to right and from bottom to top

N: electronegativity increases from left to right and from bottom to top

Si: radius increases from right to left and top to bottom

Si: metallic character increases from right to left and top to bottom

Workbook Problem 22.2

Step 1: Determine what reaction, if any, will occur.

$Sr(s) + O_2(g) \rightarrow$ strontium oxide will form spontaneously.

$Ca(s) + H_2O(l) \rightarrow$ calcium hydroxide will form and hydrogen gas will be generated.

$Be(s) + H_2O(l) \rightarrow$ no reaction expected. Beryllium is the least reactive of the alkaline earth metals.

Step 2: Write balanced equations for the reactions that occur.

$$2Sr(s) + O_2(g) \rightarrow 2SrO(s)$$

$$Ca(s) + 2H_2O(l) \rightarrow Ca(OH)_2(aq) + H_2(g)$$

Workbook Problem 22.3

Step 1: Determine the oxidation state of carbon.

CO Oxidation state for O is −2, neutral molecule, so the oxidation state for C is +2.

CO_2 Oxidation state for O is −2, neutral molecule, so the oxidation state for C is +4.

C_2O_3 Oxidation state for O is −2, neutral molecule, so the oxidation state for C is +3.

CO_3^{2-} Oxidation state for O is −2, charge on molecule is −2, so the oxidation state for C is +4.

CaC_2 Oxidation state for Ca is +2, neutral molecule, so the oxidation state for C is −1.

Workbook Problem 22.4

Step 1: Determine the silicate structure:

As the SiO_4^{4-} subunits join together, they share oxygens:

$SiO_4^{4-} + SiO_4^{4-} = Si_2O_7^{6-}$ $Si_2O_7^{6-} + SiO_4^{4-} = Si_3O_{10}^{8-}$

Step 2: Determine how many of each cation is needed to balance the charge on the silicate unit:

To balance a charge of −8 with cations having charges of +3 and +2 will require two of the +3, and one of the +2. There is no other way to use both and get to +8

Step 3: Write the formula unit:

$Ca_3Al_2(SiO_4)_3$

Workbook Problem 22.5

Step 1: Write the balanced chemical equation.

$$NH_4NO_3(s) \rightarrow N_2O(g) + 2\ H_2O(l)$$

Step 2: Calculate the pressure of N_2O, using the total pressure and the vapor pressure of water.

$$P_T = P_{H_2O} + P_{N_2O}$$

$$P_{N_2O} = 705 \text{ mm Hg} - 24 \text{ mm Hg} = 681 \text{ mm Hg}$$

Step 3: Calculate the number of moles of N_2O produced from 3.5 g NH_4NO_3.

$$3.5 \text{ g } NH_4NO_3 \times \frac{1 \text{ mol } NH_4NO_3}{80.0 \text{ g } NH_4NO_3} \times \frac{1 \text{ mol } N_2O}{1 \text{ mol } NH_4NO_3} = 4.38 \times 10^{-2} \text{ mol } N_2O$$

Step 4: Calculate the volume of N_2O, using the ideal gas law.

$$V = \frac{nRT}{P} = \frac{\left(4.38 \times 10^{-2}\right)\left(0.0821 \frac{L \cdot atm}{mol \cdot K}\right)(295.15 \text{ K})}{681 \text{ mm Hg} \times \frac{1 \text{ atm}}{760 \text{ mm Hg}}} = 1.18 \text{ L}$$

Self-Test

True–False

1. F Semimetals are on a zigzag line from boron to astatine, not in the center of the periodic table.
2. F Metallic character increases down a column of the periodic table.
3. T
4. F Due to their size, it is uncommon to find double bonds in elements outside of the second row.
5. T
6. T
7. F Silicon is too large to form double bonds: the chains consist of single bonds.
8. T
9. T
10. F Liquid sulfur decreases in viscosity until the S_8 rings begin to break and reform into long chains. The viscosity increases at this point and decreases again when the chains begin to break as well.

Multiple Choice

11. b	12. c	13. a	14. d	15. d
16. c	17. a	18. b	19. b	20. d

Fill in the Blanks

21. increases, increases, decreases, decreases
22. diamond, graphite, fullerene
23. white phosphorus, bent, red phosphorus
24. SiO_4
25. solid, liquid, S_8, rings, chains, viscosity

Matching

26. c	27. d	28. g	29. f
30. a	31. e	32. b	

Questions

33. a) Rb b) In c) Cs

34. a) K b) In c) Sb

35. Ionization energy—decreases down a group; atomic radius—increases down a group; electronegativity—decreases down a group; basicity of oxides—increases down a group.

36. Each boron in diborane uses an sp^3 hybrid orbital to overlap with a terminal hydrogen 1*s* orbital. Each bridging H atom is joined to both boron atoms through a three-center, two-electron bond in which the two electrons in the B–H–B bridge are spread out over three atoms.

37. Diamond: tetrahedral bonding with sp^3 hybridization. It is nonconducting, the hardest known substance, and has the highest melting point of any element.

Graphite: trigonal planar bonding with sp^2 hybridization, with delocalized π bonds to adjacent molecules. It conducts electricity and forms sheets that can slide past one another, making it useful as a dry lubricant.

Graphene: prepared by the "Scotch tape method," graphene is a two-dimensional array of carbon molecules with a hexagonal arrangement. It is an excellent conductor of electricity.

Fullerene: this covers a number of structures from carbon nanotubes to spheres, which share a common bonding. They are molecular structures and dissolve in nonpolar solvents, as well as conducting electricity.

38. $-3 = NH_3$ $-2 = N_2H_4$ $-1 = NH_2OH$ $+1 = N_2O$
$+2 = NO$ $+3 = HNO_2$ $+4 = NO_2$ $+5 = HNO_3$

39. $HClO_4$—perchloric acid; $HClO_3$—chloric acid; $HClO_2$—chlorous acid;
$HClO$—hypochlorous acid
$HBrO_4$—perbromic acid; $HBrO_3$—bromic acid; $HBrO_2$ bromous acid;
$HBrO$—hypobromous acid
HIO_4—periodic acid; HIO_3—iodic acid; HIO_2—iodous acid; HIO—hypoiodous acid

40. $Cu(s) + 4\ HNO_3(aq) \rightarrow 2\ NO_2(g) + 2\ H_2O(l) + Cu(NO_3)_2(aq)$

Challenge Problem

41. $CaC_2(s) + H_2O(l) \rightarrow C_2H_2(g) + CaO(s)$

Determine the moles present of each reactant:

$$15.0\ \text{g}\ CaC_2 \times \frac{1\ \text{mol}\ CaC_2}{64.10\ \text{g}} = 0.234\ \text{mol}\ CaC_2$$

$$7.45\ \text{g}\ H_2O \times \frac{1\ \text{mol}\ H_2O}{18.0\ \text{g}} = 0.413\ \text{mol}\ H_2O$$

Calcium carbide is the limiting reactant, as the reactants react in a 1:1 ratio.

$$0.234\ \text{mol}\ C_2H_2 \times \frac{26.04\ \text{g}\ C_2H_2}{\text{mol}} = 6.09\ \text{g}\ C_2H_2$$

Write the balanced reaction for the burning of acetylene:

$$C_2H_2(g) + 5/2\ O_2(g) \rightarrow 2\ CO_2(g) + H_2O(g)$$

Determine overall ΔH_{rxn} for acetylene:

$$[(-241.8\ \text{kJ/mol}) + 2\ (-393.5\ \text{kJ/mol})\ ] - (227.4\ \text{kJ/mol}) = -1256\ \text{kJ/mol}$$

Determine the heat generated from this reaction:

$$0.234\ \text{mol}\ C_2H_2 \times -1256\ \text{kJ/mol} = -294\ \text{kJ}$$

Chapter Twenty-Three—Organic and Biological Chemistry

Workbook Problems

Workbook Problem 23.1

Step 1: Name the functional groups:

a. An oxygen double-bonded to a carbon that is also bonded to an alcohol group is a *carboxylic acid.*
b. A carbon double-bonded to an oxygen and single-bonded to carbons on each side is a *ketone*.
c. An oxygen that is double-bonded to a carbon that is also bonded to a hydrogen is an *aldehyde*.
d. An oxygen that is single-bonded to two carbons is an *ether.*

Workbook Problem 23.2

Step 1: Determine the complementary bases.

A and T are usually complementary, G and C are complementary.

TCGAACTGTGCACAGTGATCT

Self-Test

True–False

1. T
2. F Double and triple bonds are considered functional groups.
3. F Cis–trans isomerization has tremendous practical consequences.
4. T
5. F Carboxylic acids are weak acids.
6. T
7. T
8. T
9. F Cellulose is an indigestible long-chain polysaccharide.
10. F Nucleic acids carry genetic information and provide instructions for protein production.

Multiple Choice

11. b	12. c	13. a	14. c	15. d
16. b	17. d	18. c	19. b	20. d

Fill in the Blanks

21. Bonding, functional groups
22. Hydrocarbons, carbon, alkanes, alkenes, alkynes, triple
23. double, benzene (or aromatic)
24. Carbohydrates, energy
25. amino acids, enzymes
26. adenine, guanine, cytosine, thymine, uracil

Matching

27. k	28. d	29. r	30. i
31. v	32. a	33. m	34. h
35. b	36. q	37. f	38. c
39. e	40. s	41. p	42. g
43. j	44. t	45. l	46. x
47. n	48. o	49. bb	50. aa
51. w	52. u	53. y	54. z

Questions

55.

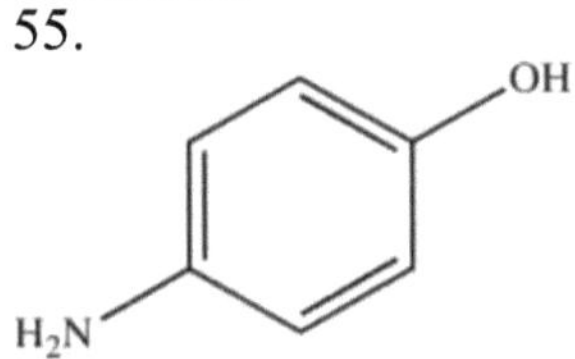

amine, alcohol

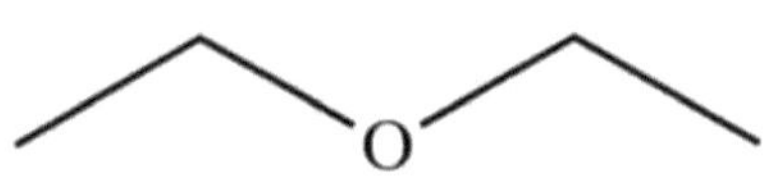

ether

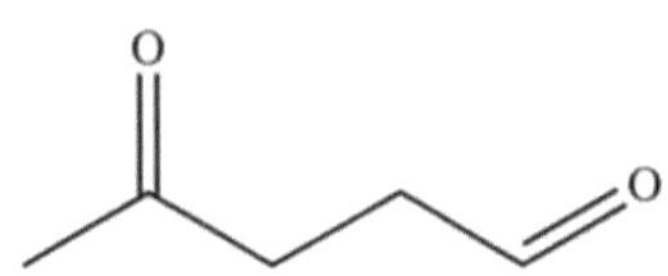

ketone, aldehyde

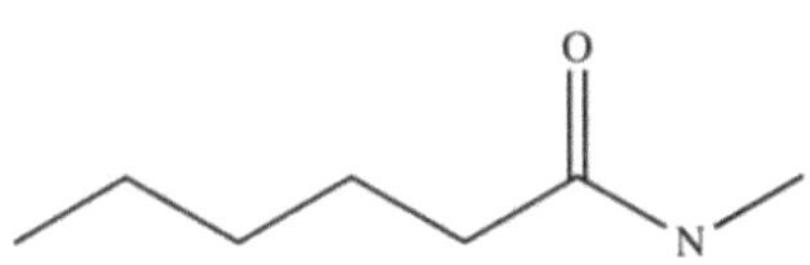

amide

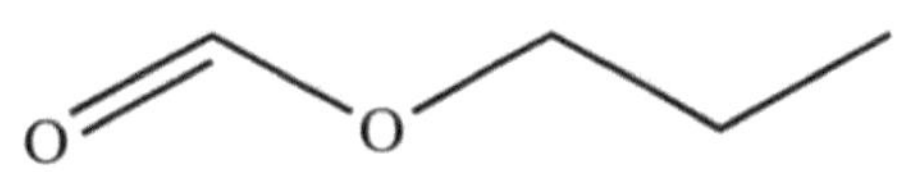

ester

56.

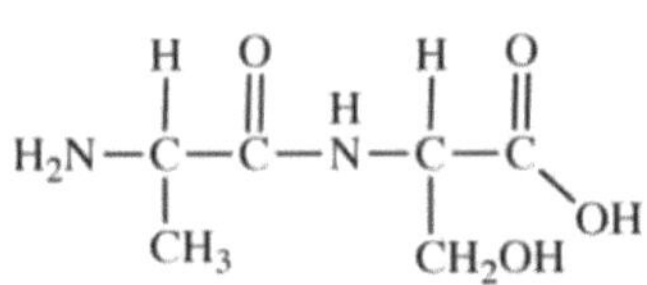

alanine serine

ala-ser

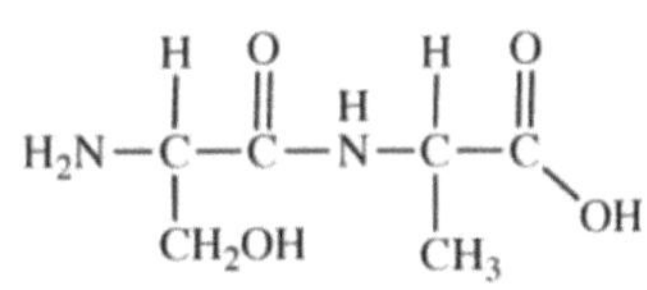

serine alanine

ser-ala

57. Ala-Leu-His, Ala-His-Leu, His-Leu-Ala, His-Ala-Leu, Leu-Ala-His, Leu-His-Ala

58. Aldohexose; ketohexose

59.

60. DNA sequence = ATCGTGACTAGTCATTCG

Challenge Problem

61.

$$6.87\text{ g }CO_2 \times \frac{12\text{ g C}}{44.01\text{ g }CO_2} = 1.87\text{ g C}$$

$$1.87\text{ g }H_2O \times \frac{2\text{ g H}}{18.015\text{ g }H_2O} = 0.208\text{ g H}$$

5.00 grams of original compound − 0.208g − 1.87 g = 2.92 g left for oxygen.

2.92 g oxygen × 1 mol/16.0 g = 0.182 mol O

$$\text{moles C} = \frac{0.156}{0.156} = 1 = \frac{3}{3} = \frac{6}{6}$$

$$\text{moles H} = \frac{0.208}{0.156} = 1.3 = \frac{4}{3} = \frac{8}{6}$$

$$\text{moles O} = \frac{0.182}{0.156} = 1.17 = \frac{7}{6}$$

Empirical formula = $C_6H_8O_7$

The molecular molar mass is determined from the moles of NaOH used in the titration and the fact that citric acid is a triprotic acid. Three moles of NaOH will react for each mole of citric acid.

$$\text{mol citric acid} = 0.0488\text{ L} \times \frac{0.100\text{ mol NaOH}}{1\text{ L}} \times \frac{1\text{ mol citric acid}}{3\text{ mol NaOH}} = 1.63 \times 10^{-3}\text{ mol citric acid}$$

$$\text{molar mass} = \frac{0.312\text{ g citric acid}}{1.63 \times 10^{-3}\text{ mol maleic acid}} = 192\text{ g/mol}$$

The molecular weight corresponds to the empirical formula. Therefore, the empirical and molecular formulas are the same.

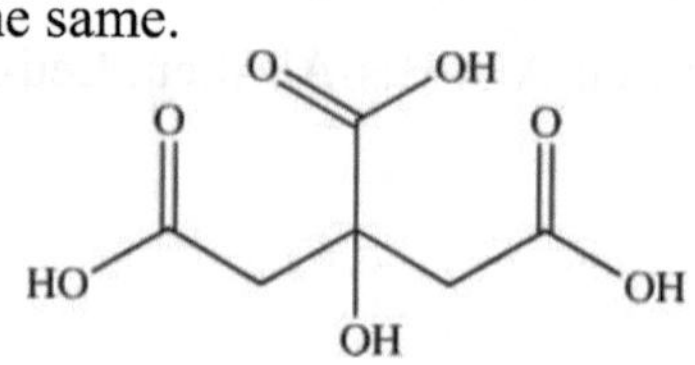